Ecology in Action

Taking a fresh approach to integrating key concepts and research processes, this undergraduate textbook encourages students to develop an understanding of how ecologists raise and answer real-world questions. Four unique chapters describe the development and evolution of different research programs in each of ecology's core areas, showing students that research is undertaken by real people who are profoundly influenced by their social and political environments.

Beginning with a case study to capture student interest, each chapter emphasizes the linkage between observations, ideas, questions, hypotheses, predictions, results, and conclusions. Discussion questions, integrated within the text, encourage active participation and a range of end-of-chapter questions reinforce knowledge and encourage application of analytical and critical thinking skills to real ecological questions. Students are asked to analyse and interpret real data, with support from online tutorials that show them how to use the R programming language for statistical analysis.

FRED D. SINGER is Professor Emeritus of Biology at Radford University, where he began teaching in 1989. A committed teacher, he developed research programs on the behavioral and community ecology of spiders, dragonflies, and zebrafish, while also collaborating with several colleagues to promote active learning as part of an ongoing research program on new approaches to teaching. In 2000, in recognition of his dual research programs, he was awarded the Radford Foundation Award for Creative Scholarship. He has taught approximately 20 different courses, including general ecology, field ecology, and climate change ecology, using the philosophy that the best learning occurs when students deal with real experiments and real data.

"In contrast to most other science textbooks, Singer's book... emphasizes the roles of curiosity and careful observation in discovering hypotheses that are worth testing. This will be an ideal ecology text for anyone who would like to help students appreciate the excitement of scientific creativity."

Robert Askins, Connecticut College, USA

"... an excellent resource for students and young professionals... Singer provides a detailed overview of the foundations of ecology while he seamlessly incorporates an illuminating insight into how ecology is done. This is a welcome perspective in an ecological textbook since many contemporary titles heavily focus on abstract ecological concepts."

Joris van der Ham, George Mason University, USA

"... written in a very student-friendly manner... Case studies incorporated into the text provide a much needed basis for the comprehension of difficult ecological concepts."

Troy Ladine, East Texas Baptist University, USA

"It is very refreshing to read a new textbook, rather than a new edition of an old textbook... The case studies bring the chapters to life, which contributes to making this a very interesting read."

Judith Lock, University of Southampton, UK

"... I'm excited by this book. It is refreshing and interesting with unique examples and clean artwork... The inclusion of an assortment of end of chapter questions... that run the gamut from application to data analysis will give students practice and insight into both critical thinking and quantitative skills that most books do not."

Lynn Mahaffy, University of Delaware, USA

"This book is a breath of fresh air. Singer has provided a clear and compelling text that will engage students at every level of knowledge."

Holly Porter-Morgan, City University of New York, USA

Ecology in Action

Fred D. Singer
Radford University, Virginia

CAMBRIDGE
UNIVERSITY PRESS

CAMBRIDGE
UNIVERSITY PRESS

University Printing House, Cambridge CB2 8BS, United Kingdom

Cambridge University Press is part of the University of Cambridge.

It furthers the University's mission by disseminating knowledge in the
pursuit of education, learning, and research at the highest international levels
of excellence.

www.cambridge.org
Information on this title: www.cambridge.org/9781107115378

First published 2016

Printed in the United Kingdom by Bell and Bain Ltd

A catalog record for this publication is available from the British Library

Library of Congress Cataloging in Publication data
Singer, Fred, 1952–
Ecology in Action / Fred D. Singer, Radford University, Virginia.
   pages cm
Includes bibliographical references and index.
ISBN 978-1-107-11537-8 (Hardback : alk. paper)   1. Ecology–Research.   2. Ecology–Case
studies.   I. Title.
QH541.2.S59 2015
577–dc23   2015020028

ISBN 978-1-107-11537-8 Hardback

Additional resources for this publication at www.cambridge.org/ecologyinaction

To Cindy for true love and grace, to Jed and Alison who kept the questions coming
when they were young and have metamorphosed into awesome adults, and to my canine companions,
Maia, Sheik, and Cheyanne, who somehow knew when my face needed a good licking.

CONTENTS

PREFACE

I taught at a university where teaching is a professor's major action. Coming from a research background, I was somewhat surprised at this emphasis, but after immersing myself in teaching for 25 years, I decided that it was time to share my ideas. Over those years, I learned that students (and people in general) learn best when they are actively doing things. One of my goals in writing this text is to get students actively involved in what ecologists do, to learn where knowledge comes from, to learn how to ask questions, how to approach answering these questions, and to expect that the answer will be somewhat unsatisfactory and will generate a new series of stimulating questions. *Ecology in Action* is the title of this book but it is also a philosophy of learning, and a statement of hope for future times when scientifically sound ecological action will be essential.

Humans are natural ecologists. We are great observers of phenomena of all types, we develop expectations of what will happen under a set of conditions, and we recognize patterns. We also love to do experiments. Unfortunately, some of the experiments we are conducting today are very risky. We are increasing atmospheric carbon dioxide levels and collecting data on the effects of climate change. We have doubled the amount of fixed nitrogen in global systems and are collecting data on ecosystem response. We are breaking up large continuous habitat into small fragments and observing how populations of endangered species respond to their diminished habitat. The good news is that we are collecting at least some of the relevant data. My hope for the future is that everyone who reads this book will be able to interpret these data effectively, so they can take appropriate action as scientists or as private citizens.

MY GOALS FOR THE READER

I wrote this book for biology majors who have a year of introductory biology and at least one semester of general chemistry. Ecology is a young and growing science, and the informational content is increasing explosively.

After reading this book, students should be able to:
- understand the key concepts in ecology.
- learn how to learn, how to identify what is important, and how to go about finding answers to questions.
- understand how everything we know about ecology started as an idea and metamorphosed into knowledge; in other words, each reader will learn about ecological process. The approach throughout the book is to integrate content and process in a manner that promotes a much deeper understanding of core areas of ecology.
- appreciate how research can be an extension or reflection of an ecologist's personality, values, or social situation. The history

of an idea is presented, so that students can follow the logical development of an ecological concept through time. Students learn how ecologists think of their questions, and students then become empowered to think of their own questions.
- recognize that the natural world is awesome and fascinating. Examples are carefully chosen to engage students, and introduce the reader to novel species from parts of the world that they may have never heard of. At the same time students learn that humans are influencing natural systems in many profound ways. Numerous explicit connections between ecological concepts and ecological applications are made, enabling students to use their knowledge to treat the world in a sustainable manner.

ORGANIZATION AND STRUCTURE OF THE BOOK

Ecology in Action has 24 chapters, beginning with an introductory part that presents major ecological questions and a discussion of the physical environment. Tony Sinclair's research in the Serengeti is highlighted in the first chapter, as the questions that he asks span the entire biological hierarchy, from molecules to global ecology. The remainder of the text is divided into four primary parts: organismal and evolutionary ecology, population ecology, community ecology, and ecosystem and global ecology. Each part has two major components. The bulk of each part (four or five major concepts chapters) presents students with the major concepts of each topic, and the supporting experimentation/ observations related to each major concept. The final chapter of each part immerses the students in a research program of one individual or group of individuals that made important contributions to the field.

Major concepts chapters

Each major concepts chapter begins with an introduction that describes one element of a case study that will be discussed in some detail in the chapter. In addition, the introduction provides a roadmap of the major concepts that will be explored and concludes with a list of 3–5 key questions that will be addressed as the chapter's major focus. I gave preference to case studies that highlighted fascinating natural history, that taught an important ecological concept, that demonstrated important aspects of ecological process, and that showed how ideas changed over time. The main body of each chapter answers the initial 3–5 questions in the context of research studies, which were chosen based on three criteria: (1) Do they clearly demonstrate an important concept? (2) Are they classics that show where ideas or new methods originated? (3) Are they recent or ongoing? Recent research is emphasized to convey the message that ecological knowledge is always evolving, that though we have just made a new discovery, there is now a new pressing question that needs to be addressed. At the end of each chapter is a *Revisit* section, which returns to the initial case study and address some of the issues it raises in light of what students have learned in the chapter.

All chapters have a strong emphasis on data analysis and interpretation. In my experience, the greatest challenge for many students is drawing reasonable conclusions from an experiment or observation. All chapters contain questions that prompt students to analyse data. These questions require certain analytic skills that students usually learn in an introductory statistics course, or in their science laboratories. To make sure all students can solve these problems, the text provides several opportunities to learn or review all of the necessary analytic approaches.

Research program chapters

One distinguishing feature of this text is the chapters that introduce students to research programs. Because we tend to do labs in 3 hour blocks of time, students gain a mistaken impression about how science is actually done. My goal with these research program chapters is to allow students to gain a deeper understanding of answers to the following questions:

1. How does a research program differ from a research project or question?
2. How do ecologists think of their questions?
3. How do ecologists and the scientific community decide what is important, and what is trivial?
4. How are research findings evaluated?
5. What factors determine a successful research program?
6. How is ecological research an extension or reflection of a scientist's personality?
7. How do ecologists work at different levels of the ecological hierarchy, and how do they collaborate with other ecologists and other scientists from different fields?

I present the research programs of four different ecologists or groups of ecologists who are major contributors to the discipline. Diversity of approach is important, so the following researchers are featured, in part because they used philosophically different approaches to ask and answer questions: organismal and evolutionary ecology – Bernd Heinrich; population ecology – Jane Goodall/Anne Pusey; community ecology – Dan Janzen/Winnie Hallwachs; ecosystem and global ecology – Jane Lubchenco. In addition to reading many of their papers and books, I conducted lengthy in-person interviews with each researcher, and also traveled with Anne Pusey to Tanzania to meet the famous chimpanzees of Gombe. On a personal level, it was awesome to meet these people (and chimpanzees) and to gain an understanding of why they are so successful.

PEDAGOGICAL FEATURES

Emphasis on ecological processes

Having read all of the books on the market (numerous times), I am quite confident that none explores ecological process as much as *Ecology in Action*. The four research program chapters have a heavy emphasis on ecological process, and this is carried on throughout the major concepts chapters to show students where ideas come from and how they develop over time. Through repeated exposure, students understand the relationship between a hypothesis and a prediction, how to analyse data, and how to draw logical conclusions.

Strong evolutionary foundation

Evolution is introduced early – initially in Chapter 1, and extensively in Chapter 3. Students learn skills that are rarely taught in ecology texts, for example, how to make and interpret an evolutionary tree and how to use molecular data to make evolutionary inferences. Having established a strong evolutionary foundation, students can then understand more advanced topics that are presented in almost every other chapter.

Historical perspective

People are natural storytellers and story listeners. The historical perspective used in the book immerses students in the history of an idea, so that they understand not only where an idea originated, but also how it changed over time and how it is related to other ideas that exist today. In the process, they also learn how to ask scientific questions and get considerable experience with how ecologists come up with their answers. Students learn that many answers are tentative, or only apply under restricted conditions.

Case studies

Engaging case studies begin each major concepts chapter and are revisited at the end of the chapter in light of the issues raised in the main text.

Key questions

Key questions are listed at the beginning of each chapter and provide a roadmap to help students prepare for and structure their learning. Summaries at the end of each chapter recap the key points.

Artwork

Illustrations are carefully designed to allow students to easily take away the key information they need and photographs help bring the subject to life.

Key terms

Key terms are highlighted in bold so that students can easily identify them. Definitions are provided in a comprehensive glossary at the end of the book.

Thinking ecologically questions

Each chapter has about three or four *Thinking ecologically* questions embedded in the text, allowing students to think about a problem while reading about it, rather than coming back to it at a later time. When I teach, I often require students to work on a *Thinking ecologically* question in class as soon as I finish the topic as a way of reinforcing the concept. These questions are usually open-ended; they develop a student's critical thinking skills and they can be easily read around without disrupting the flow of the narrative, which allows instructors to choose which questions to assign.

End-of-chapter questions

Review questions

About five or six review questions require students to review and summarize the major concepts, helping them retain what they have learned.

Synthesis and application questions

About 5–10 questions require students to use higher order thinking skills to apply what they've learned in the chapter. Students may be asked to design an experiment or a series of systematic observations to test a particular hypothesis, and to articulate predictions generated by the hypothesis. Alternatively, students may be asked to develop an analogy between what they just learned and a concept that they learned in a previous chapter; or students may be asked to apply a concept that they learned in the chapter to a different context.

Analyse the data

These end-of-chapter questions present students with real data sets (or occasionally graphs) and ask them to answer a question based on these data. In some cases the data were already presented in the text in a different context, while in other cases students are provided with new data sets to work from. Students are usually asked to interpret their findings in the context of a concept that was introduced in the chapter.

Dealing with data

To help students deal with data analysis a special embedded box feature, *Dealing with data*, is included primarily in the early chapters. *Dealing with data* boxes present analytic concepts such as extrapolation, error reduction, and data transformation, and applied statistical techniques beginning with calculating means and standard errors, branching into t-tests, one-way ANOVA, simple correlation and regression analysis, chi-square, and several other commonly used statistical approaches. There are also discipline-specific *Dealing with data* boxes that teach students how particular methodologies, such as molecular biology databases, geographic information systems, and stable isotope analyses, are used by ecologists to answer important questions.

Further reading

Each chapter concludes with 4–6 suggested readings that are a suitable basis for further discussion. These readings are mostly recent publications that present novel findings, but a few classics are also included to give students a sense of ecological history.

ONLINE RESOURCES

Supporting online resources for the book can be found at www.cambridge.org/ecologyinaction

For students and instructors

- All of the figures in jpeg (and PowerPoint) format
- Online glossary
- Statistics tutorial: an introduction to R.

One feature that distinguishes this text from some others is the *Analyse the data* feature that asks students to analyse real data sets and draw conclusions based on their analysis. To do so requires using elementary statistical analyses, so the website has a tutorial that, in conjunction with the *Dealing with data* feature, uses data sets in the text to teach the basics of statistics using the R programming language and software environment. R is free, and its use is expanding tremendously within the ecological community. Being able to use R will be a major boon to any ecology student moving forward. Much thanks to Dr Edd Hammill for developing a superb tutorial that integrates seamlessly with *Ecology in Action*.

For instructors only

- Solutions to questions.

The website provides answers to the *Analyse the data* questions.

Thinking ecologically and *Synthesis and application* questions are, for the most part, open-ended. As such, it is impossible to give an answer key, but some suggested approaches to dealing with each question are provided. I am also hoping to hear interesting ideas from students and instructors, for suggestions on the book and additional online resources that might be useful for teaching and learning.

ACKNOWLEDGMENTS

I'd like to thank ecologists around the world for doing the research that I've discussed, for doing so much more research that I could not discuss, and for persistently pushing the frontiers of our understanding of the natural world. Jane Goodall, Winnie Hallwachs, Bernd Heinrich, Dan Janzen, Jane Lubchenco, Anne Pusey and Tony Sinclair gave me large chunks of their time, so that I could hear their story and share it with you. I am very grateful to them.

My wife, Cindy Miller, carefully read the first draft and identified numerous mistakes, contradictions, omissions, and general cases of fuzziness before I passed the first draft to my development editor, Mary Catherine Hager. Mary Catherine further refined the first and second drafts, helping to me to write with a more consistent voice and coercing me to use a more linear approach in my writing. The folks at Cambridge University Press were awesome at every stage of the process. Dom Lewis adopted the project and has steadfastly encouraged me at every step. Claire Eudall coordinated my writing and research, and mostly kept me on schedule. Claire and Natasha Lewis sourced images and permissions, Charlie Howell co-ordinated the illustrations, text design and cover, and Jess Murphy and Rachel Cox managed the book through production with support from Megan Waddington.

Approaches to teaching don't develop within a vacuum; instead they are nurtured by experiences with students, colleagues, and even administrators. Edd Hamill of Utah State University wrote a spectacular tutorial that guides students through the R programming language and software environment. At Radford University, Joel Hagen, Chuck Kugler, and later Donna Boyd, Rich Murphy, Bob Sheehy, and about 5000 students taught me many things about teaching. One of those 5000 students, Daniel Metz, wrote out answers to many of the *Thinking ecologically* and *End-of-chapter questions*. Bud Bennett chased down hundreds of obscure articles and books through interlibrary loans. Matt Lee provided iconic help with some design issues. Lastly, Sam Minner, Orion Rogers, and Joe Scartelli provided financial support during sabbaticals and some summers.

When I was concerned about a particular chapter, I turned it over to other readers for help. These informal reviewers were David Blockstein, Judy Guinan, Joel Hagen, Chuck Kugler, and Mac Post. When I was puzzled about a study, I asked the study's lead author for help, and these researchers were almost universally delighted to provide clarification. Particularly helpful were Stanton Braude, Tim Brodribb, Jim Clark, William Cooper, Charlie Crisafulli, Jim Elser, Marc Fourrier, Jim Galloway, Philip Groom, Elizabeth Howard, Scott Kraus, William Laurance, Kim Mouritsem, Vojtech Novotny, Karen Oberhauser, Susan Riechert, Bill Ripple, W. H. Schlessinger, Rich Kliman, Paul Sherman, and Martin Wikelski.

Finally, I would like to thank the following formal reviewers who read multiple chapters and suggested many important changes:

Professor Daniel R. Ardia, Franklin and Marshall College
Professor Robert Askins, Connecticut College
Associate Professor Kimberley Bolyard, Bridgewater College
Professor Robert S. Boyd, Auburn University
Associate Professor Judith Bramble, DePaul University
Associate Professor Joseph Coelho, Quincy University
Associate Professor Clay Green, Texas State University
Professor John Jaenike, University of Rochester
Associate Professor Jacob Kerby, University of South Dakota
Adjunct Professor Ned Knight, Lichfield College
Dr Christopher H. Kodani, Clayton State University
Assistant Professor Brenda Koerner, Emporia State University
Dr Troy Ladine, Texas Baptist College
Assistant Professor Lynn Mahaffy, University of Delaware
Assistant Professor David Pindel, Corning Community College
Assistant Professor Holly Porter-Morgan, CUNY LaGuardia Community College
Professor Erik Scully, Towson University
Dr Daniela Shebitz, Kean University
Assistant Professor Jose Rodolfo Valdez-Barillas, Texas A&M University
Dr Joris Van der Ham, George Mason University.

PART I
Introduction and the physical environment

Chapter 1
What is ecology in action?

INTRODUCTION

The Serengeti-Mara ecosystem in northern Tanzania and southern Kenya is a showcase of vertebrate life, home to over 650 species of birds and 79 species of large mammals. Equally impressive, many of the species are present in enormous quantities. In the early 1960s, there were approximately 250 000 wildebeest (*Connochaetes taurinus*), 200 000 zebras (*Equus burchelli*), 30 000 buffalo (*Syncerus caffer*), and 750 000 Thomson's gazelles (*Gazella thomsoni*) roaming through the ecosystem. Every year, the zebras, wildebeest, and gazelles spend the wet season in the southeast portion of the Serengeti, eating the grasses and herbaceous plants that grew during the wet season, which extends from November through May. These grasses have very high levels of nutrients, which are particularly important to pregnant females who give birth in the middle of the rainy season and begin the nutrient-demanding process of lactation. As rainfall diminishes in June, many of the zebras, wildebeest, and gazelles migrate to the northwest portion of the ecosystem, munching on taller but less nutritious grasses that grow in the much wetter northern and western plains. When the wet season returns in November, so do the large mammals that have survived the dry season, to begin the cycle anew.

Many researchers have converged on the Serengeti to answer important ecological questions. We will follow in their footsteps, using the Serengeti ecosystem to gain an understanding of what types of questions ecologists ask. We will then see how ecologists use the entire scientific toolbox to answer these questions, but that ecological questions, because they tend to be so broad, may require a researcher willing to tackle numerous levels of the biological hierarchy. Let's go back to the beginnings of ecosystem studies in the Serengeti.

KEY QUESTIONS

1.1. What are ecological questions?

1.2. How do ecologists test hypotheses about ecological processes?

1.3. How do ecologists use observation, modeling, and experimentation?

1.4. How do ecologists ask questions that link different levels of the biological hierarchy?

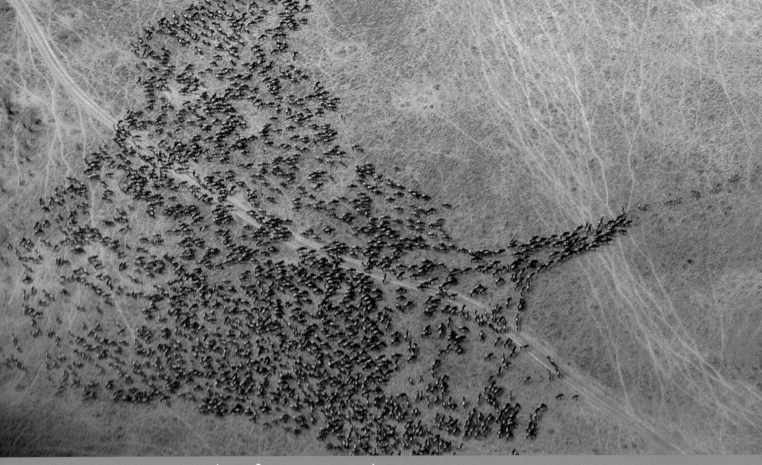

CASE STUDY: Birth of a research program

The western world was introduced to the Serengeti by a German father-and-son team, Bernhard and Michael Grzimek, who together created a movie, *Serengeti Shall not Die*, which was wildly popular around Europe. Tragically, in 1959, as Michael filmed from a low-flying airplane (Figure 1.1a), a griffon vulture flew into the plane, causing it to crash and killing him. Michael was buried atop Ngorongoro Crater, east of the Serengeti; his epitaph reads "He gave all he possessed including his life for the wild animals of Africa." His father joined him at the same site almost 30 years later (Figure 1.1b).

Though Michael gave his life for the Serengeti, he and his father gave birth to a growing global awareness of the beauty, drama, and potential plight of this vast ecosystem, and in particular they uncovered many features of the large mammal migrations. Bernhard, as director of the Frankfurt Zoological Garden, used proceeds from the movie to help fund Serengeti research. In addition, John Owen, then director of Tanzania National Park, used his position to establish the Serengeti Research Institute in 1961, which funded three scientists to continue working on the wildebeest migrations. Within a few years, several other researchers joined the Institute and began investigating other parts of the ecosystem.

Meanwhile, in England, Tony Sinclair was having some issues about what he was going to do for the rest of his life. His father, a New Zealander employed as a judge for the British government, was stationed primarily in Tanzania during the first 10 years of Sinclair's life. There, Tony spent much of his time outdoors hanging out with his friends, learning about the culture, becoming fluent in Swahili, and discovering the large diversity of non-human animals that lived in this ecoregion. At age 10, Tony was shipped off to England to continue his education. To him a career in biology meant medicine, and given his disdain for illnesses of all kinds, he entered a program in engineering, math, and physics. But these fields did not captivate him, and one day, at age 16, he just said, "What am I doing? I can do biology that's not medicine – I can do zoology."

Having switched his career path, he knew he needed to get back to Africa, so when he entered the University of Oxford he immediately sought out Professor Arthur Cain, asking him, "How do I get to Africa?" Cain informed him of his plans to go there the following summer to study bird

Figure 1.1 A. Michael Grzimek's airplane. B. Gravesite of Michael and Bernhard Grzimek atop Ngorongoro Crater.

migration (which Sinclair already knew something about), and Sinclair made enough of a nuisance of himself that he was invited to come along as a research assistant for the Serengeti Research Institute. That did it for Sinclair – he was hooked.

Recall that Sinclair was more experienced in surviving the Serengeti experience than most Oxford expatriates. He was fluent in Swahili, so if he needed help, he could explain his needs to the local people. He was also familiar with the wildlife and the environment, so that he knew where to find animals for observation, and also how to avoid getting eaten or gored by his subjects. After observing Sinclair's abilities to survive in the bush while collecting data under challenging circumstances, the researchers at the Institute presented him with a problem that was to be the focus of his PhD dissertation. They thought that the number of buffalo and wildebeest in the Serengeti-Mara ecosystem had been increasing in the past few years, but they did not know why. Could he come back and figure that out? He responded, "I can do anything you want – just get me there." And so was born a research program that has continued for over 40 years. We will use Sinclair's research in the Serengeti to help us understand what ecologists do, and how they do it.

1.1   WHAT ARE ECOLOGICAL QUESTIONS?

To get started on his project, Sinclair needed to answer two of the most basic questions addressed by population ecologists. The first question is: What is the abundance of a population? Or, how many individuals are there in the population? The second question is: What is the distribution of the population? Or, where are the individuals actually located in space and time? Later in this chapter, we will explore how Sinclair addressed a third important ecological question: How do interactions with the environment influence the distribution and abundance of organisms?

The buffalo challenge

The question of why the buffalo and wildebeest populations were increasing proved very challenging for several reasons. First, the buffalo were petrified of humans, because they had recently been targets of very heavy poaching. So anytime Sinclair got remotely close to a herd, they took off in terror – and motivated buffalo can run very quickly. Second, before discovering why buffalo abundance was increasing, Sinclair first needed to show that abundance was indeed higher than in previous years. Even if he could ultimately figure out how to count the current buffalo population, it would be challenging to go back in time to measure their abundance in previous years. Last, if he could show that buffalo were increasing, there were

several potential causes of the increase. Each cause would be difficult to explore with any degree of rigor. We'll look at how Sinclair solved these problems sequentially, keeping in mind that he was actually working on all three problems at the same time.

Though his PhD research was on buffalo, Sinclair was also keeping track of the migrating wildebeest, in cooperation with other researchers at the Institute. Upon completing his dissertation research in 1970, Sinclair began working as a postdoctoral associate for the Institute, focusing on the wildebeest populations. Thus in the remainder of the chapter we will discuss research on buffalo, wildebeest, and other large Serengeti mammals.

Estimating buffalo and wildebeest abundance

Because buffalo were so afraid of humans, Sinclair's only option was to survey them from the air. Trained pilots and observers photographed the buffalo while flying about 200 m above the surface. Buffalo range almost exclusively in open woodlands, so there was no need for Sinclair to census areas that were exclusively grassland. As a result, all suitable habitats were screened in just a few days. Fortunately, buffalo are huge, but even from 200 m it was challenging, yet possible, to distinguish individual animals.

Wildebeest congregate in much larger herds than buffalo, and are much more tolerant of humans and their flying machines. Because wildebeest congregate so close together and are smaller than buffalo, the researchers needed a higher magnification on their telephoto lenses to see them, and thus only a small fraction of the herd could be photographed in one picture frame. One complete survey required 2770 frames and took 9 months to count.

 Thinking ecologically 1.1

What types of problems might a researcher encounter when conducting aerial surveys? How might a researcher deal with these problems?

Estimating historical buffalo and wildebeest abundance

To estimate sizes of past populations, Sinclair did historical research on populations of both buffalo and wildebeest. Some of his information came from interviews with people who had lived in the Serengeti, but most came from books, articles, travelogues, and especially photographs taken in previous decades. As one example, Sinclair knew that Martin and Osa Johnson had written a travelogue called *Safari* back in the 1920s, which described their experiences in Africa. He knew

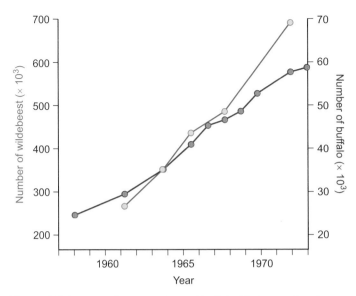

Figure 1.2 Increases in populations of buffalo (dark green) and wildebeest (light green) in 1958–1972.

1.2 HOW DO ECOLOGISTS TEST HYPOTHESES ABOUT ECOLOGICAL PROCESSES?

To answer questions about biological processes, ecologists test hypotheses that are provisional explanations for their observations. To be worthy of study, hypotheses must be plausible and generate testable predictions. Predictions are logical outcomes that are likely to be true if the hypothesis is true. Philosophers of science have extensive and sometimes passionate discussions about the relationship between hypotheses and predictions, which we will not delve into. Rather, we will simply point out that if a prediction is shown to be true, we feel somewhat more inclined to accept the hypothesis, but if we are good scientists, we will continue testing the hypothesis by exploring other predictions it generates. If the prediction is shown to be false, the hypothesis, as stated, is very unlikely to be true, unless there is something wrong in the methods we used to test the prediction.

This relationship between hypotheses and predictions will become clearer as we consider three hypotheses for why buffalo and wildebeest populations increased so sharply in the 1960s and 1970s. Sinclair's research also brings home the point that, ideally, researchers will consider all hypotheses when attempting to understand a biological process.

The food availability hypothesis

Perhaps herbivore populations were increasing because their food was becoming better in quality, or more abundant. If this hypothesis was correct, Sinclair predicted that food quality and abundance would have increased sharply in the early 1960s and remained high during that entire decade.

Sinclair and his colleagues suspected that the amount of rainfall would profoundly influence grass production – the amount of grass that was available to the grazers. To test this hypothesis, the researchers established a series of fenced areas, or exclosures, that prevented grazers from accessing the vegetation within the fences. They periodically harvested the vegetation, and measured the amount of grass that had grown in relation to the amount of rainfall that had recently fallen in the area. They predicted that there would be a positive correlation between rainfall and grass production (see Dealing with data 1.1 for a discussion of correlation). The results were striking – as rainfall increased, grass production also increased (Figure 1.3). The researchers concluded that rainfall has a significant positive effect on grass availability (Sinclair 1975).

This strong correlation between rainfall and food availability allowed Sinclair to use rainfall as an index of food availability. Sinclair did not begin his buffalo research until 1966, so he relied on measures of rainfall as his indication of grass availability during the early and middle 1960s. If increased food availability was causing the increase in buffalo and wildebeest abundance, he

that for every book there are thousands of journal writings, pictures, and stories that don't make it into the book, and in 1981 Sinclair discovered the Martin and Osa Johnson Museum in Chanute, Kansas, Osa's hometown. It was a repository for much of the Johnsons' travel documents and pictures, and Sinclair was able to convince the curator to grant him access to all of their materials. In return, he organized the collection, so the curator and future patrons would know which materials were from the Serengeti.

Because he knew the Serengeti so well, Sinclair could identify exactly where many of the photographs were from. So he now had historical records of the vegetation from 1926, 1928, and 1933. One set of aerial photographs included a series of contiguous pictures that encompassed the entire wildebeest migration, which allowed Sinclair to estimate an abundance of approximately 90 000 animals. He was not so fortunate with historical research on buffalo, so we have no good early estimates of buffalo abundance.

More recent estimates of buffalo and wildebeest populations, beginning in the late 1950s, are more accurate, though they have a margin of error as well. Based on estimates by previous researchers, and Sinclair's more refined techniques in the late 1960s, the buffalo and wildebeest populations more than doubled in the 1960s, and continued to increase through the mid-1970s (Figure 1.2).

Having established that his colleagues at the Serengeti Research Institute were correct about a sudden increase in the buffalo and wildebeest herds, Sinclair's next task was to figure out why this was happening. He considered several hypotheses to explain why population sizes were increasing. We will use this question to explore how ecologists use hypotheses and predictions to answer questions.

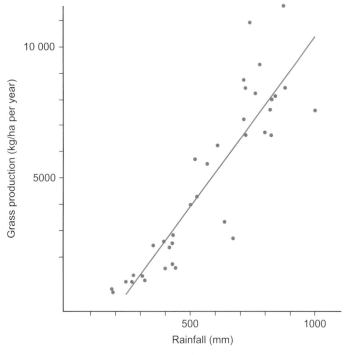

Figure 1.3 Relationship between rainfall and grass production.

 Thinking ecologically 1.2

Draw a scatterplot showing how rainfall in the Serengeti varied over time. Year should be the x-axis label, and mean monthly rainfall (mm) should be the y-axis label. Is this a strong or weak correlation? Is this a positive or negative correlation?

 Dealing with data 1.1 Correlation analysis

When both variables are numeric or continuous (have values that can be counted or measured), we can use a correlation analysis to evaluate the relationship between the two variables. In the exclosure example described previously, Sinclair and his colleagues predicted a positive correlation between the amount of rainfall and grass production. The researchers found

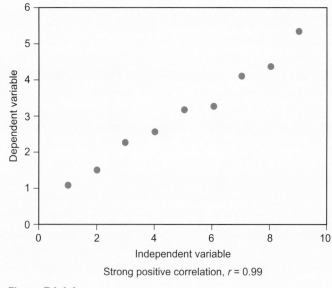

Strong positive correlation, $r = 0.99$

Figure D1.1.1

a strong positive correlation, which means that as rainfall increased, so did grass production. Figures D1.1.1 and D1.1.2 are examples of strong and weaker positive correlations. Statisticians use the **correlation coefficient (r)** to describe the strength of the correlation. If r is equal to 1.0, then all the data points will line up perfectly together to make a straight line. If r is less than 1.0, there will be a general upward trend. The closer r is to 1.0, the less scatter there is in the data (the closer the data points are to forming a line). The closer r is to 0, the more scatter there is in the data. If $r = 0$, there is no correlation between the two variables. A graph showing this type of relationship is often called a scatterplot or scattergram. For Figure 1.3, $r = 0.96$, which indicates a strong positive relationship between rainfall and grass production.

Many relationships between numeric variables have negative correlations. These are sometimes called inverse correlations. In this case, as one variable increases, the other variable decreases. As one example, Figure 1.10 illustrates a negative correlation between wildebeest abundance and the percentage of burned area. The exact same rules apply as for a positive correlation: As r approaches -1.0, the negative correlation grows stronger, and there is less scatter among the points. Figures D1.1.3 and D1.1.4 show strong and weak negative correlations. We will not discuss the mathematical formulas that calculate r, but you can refer to Gotelli and Ellison (2004) to learn how this is done. You will be expected to use simple statistical software packages for all statistical analyses in this text. These packages will do the work for you – in this case they will generate an r-value.

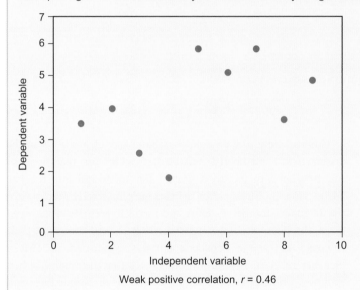

Weak positive correlation, $r = 0.46$

Figure D1.1.2

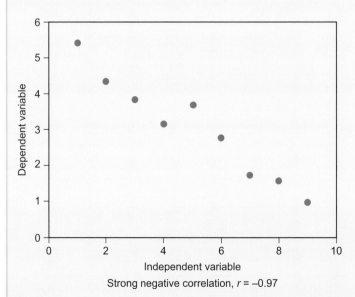

Strong negative correlation, $r = -0.97$

Figure D1.1.3

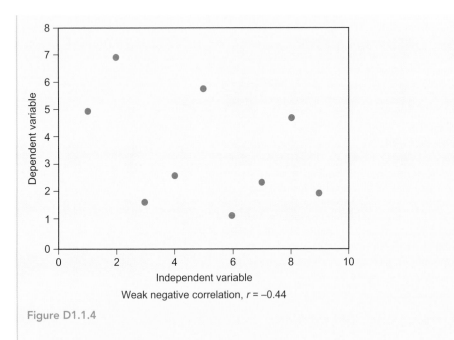

Figure D1.1.4

Cautionary note: A strong correlation does not necessarily mean that one variable caused a change in the other variable. Correlation analyses simply measure the strength of association between two variables.

predicted that there would have been higher than average rainfall during that decade.

Historically, dry-season rainfall in the central and northern Serengeti woodlands has averaged 37.5 mm per month. Table 1.1 gives rainfall data for 1962–1969. Restricting ourselves to the 1960s, you can see that average monthly dry-season rainfall during the 1960s – 36.96 mm – was very close to the historical average. Sinclair concluded that the abundance of buffalo and wildebeest was increasing for reasons other than increased food availability.

Given the lack of support for the food availability hypothesis, Sinclair considered an alternative hypothesis.

The predator release hypothesis

Perhaps herbivore populations increased during the 1960s because there was a reduction or release from high levels of predation. If this hypothesis was correct, Sinclair predicted that the abundance of predators capable of killing buffalo and wildebeest would have declined during the 1960s.

Unfortunately, the data on the abundance of predators in the Serengeti is a bit spotty, but all indications are that predator numbers actually increased during the 1960s and into the late 1970s. Lions and hyenas are the two most important predators in the Serengeti. Jeannette Hanby and David Bygott (1979) surveyed all of the lions in the Serengeti in 1974–7, and compared their numbers to George Schaller's surveys conducted in 1966–8 (Schaller 1972). Hanby and Bygott showed that the number of lion groups or prides increased from 18 to 24 between the mid-1960s and the mid-1970s. In addition, the mean number of lions per pride increased from about 15 to 19. Hanby and Bygott also surveyed the number of hyenas in the mid-1970s, and estimated 3391 hyenas in 1977 in comparison to Hans Kruuk's estimate of 2117 for the same area in 1964–8 (Kruuk 1972). In contrast to the prediction of the reduction in predation pressure hypothesis, the number of lions and hyenas was actually increasing during the same time period that wildebeest and buffalo populations were increasing rapidly.

A final hypothesis focused on the complicated effects of a disease – rinderpest – on buffalo and wildebeest populations.

Table 1.1 Estimated abundance of wildebeest and mean monthly dry-season rainfall for the central and northern Serengeti woodland in 1962–1969.

	1962	1963	1964	1965	1966	1967	1968	1969
Wildebeest abundance	309 743	*	397 624	439 124	461 208	483 292	535 663	588 034
Mean monthly rainfall (mm)	38.75	*	54.25	32.50	38.00	29.25	33.50	32.50

*There were no measurements in 1963.

The rinderpest release hypothesis

Having rejected two important hypotheses as explanations for the increase in buffalo and wildebeest populations during the 1960s and early 1970s, Sinclair was left with one other option to consider. Rinderpest is a measles-like virus that attacks and kills cattle and other *ruminants*. Ruminants are mammals, such as buffalo and wildebeest, that digest plant-based food by initially softening it within their rumen, where it ferments with the help of microorganisms that live there. They then regurgitate the semi-digested mass, chew it, and swallow it again. Sinclair knew that the Great Rinderpest Plague of 1890 originated in Europe and killed about 95% of the cattle in southern and eastern Africa. He also knew that several other waves of rinderpest had caused serious damage to the African ruminant populations in the early and mid-twentieth century. He reasoned that perhaps the increase in wildebeest and buffalo populations in the 1960s resulted from a reduction or release from high levels of rinderpest infection. Perhaps rinderpest infection had been keeping the populations of buffalo and wildebeest unnaturally low during the 1950s, and that somehow, the animals were no longer being infected by rinderpest in the early 1960s. Sinclair's task was to test the predictions of the rinderpest release hypothesis.

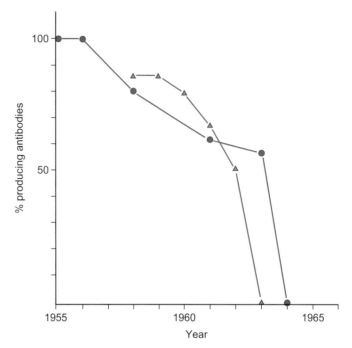

Figure 1.4 Percentage of buffalo (circles) and wildebeest (triangles) producing antibodies to rinderpest from 1955–1964.

Prediction 1: A negative correlation between rinderpest infection and ruminant abundance

One prediction of the rinderpest release hypothesis is that rinderpest infection should have declined substantially in association with an increase in the abundance of buffalo and wildebeest. If this was true, then blood from animals born in the early and mid-1960s, when the populations began increasing, should have fewer antibodies to the rinderpest virus than blood from animals born in the 1950s, when the populations were more stable.

Veterinarians were very interested in rinderpest because it killed cattle owned by local tribesmen and devastated the local economy. Walter Plowright, a veterinarian working for the East African Veterinary Research Organization, helped develop a vaccine against rinderpest, and began inoculating cattle in East Africa in 1956. He knew that wildebeest also contracted the disease, so he carried out a wildebeest-monitoring program in the early 1960s. He discovered that juvenile wildebeest received passive immunity from their mother's milk, but by 7 months of age were highly susceptible to infection. He discovered that wildebeest from Tanzania showed no evidence of rinderpest antibodies in 1962, in contrast to an infection rate of about 70% in 1959–61 (Plowright and McCulloch 1967).

Though Plowright moved back to England in 1964, his paper indicates that there was no evidence of major rinderpest infection in wildebeest from 1963 to 1967. But Plowright was a veterinarian interested in eradicating rinderpest, and he was not working on the question of why the wildebeest and buffalo populations were increasing. He was delighted that rinderpest was no longer present in the population but warned that it was likely to return, as it had several times in the past.

In contrast, Tony Sinclair was profoundly interested in the correlation between rinderpest release in wildebeest and population growth. But he also knew that a simple correlation between rinderpest release and population growth did not mean that wildebeests were increasing because they were no longer being infected by rinderpest. He needed more confirmation of the rinderpest release hypothesis.

One of Sinclair's first actions was to work with other veterinarians to measure rinderpest levels in the buffalo. The rinderpest release hypothesis predicts that rinderpest should also have disappeared in buffalo in the early 1960s. Fortunately, veterinarians had supplies of buffalo blood in the freezer, and they also knew (much to Sinclair's delight) the age of each animal that had provided the sample. The results of their analysis showed that rinderpest had completely disappeared from the buffalo population by 1964 (Figure 1.4).

Prediction 2: No correlation between rinderpest infection and non-ruminant abundance

Encouraged, Sinclair proceeded to test other predictions of the rinderpest hypothesis. He argued that Serengeti mammals that were not susceptible to rinderpest (animals that were not ruminants) would not show a trend of population growth over the 1960s, because rinderpest release would, of course, not affect them in any significant way. Zebra were the only large non-ruminant for which there were good survey data and, as predicted, there was no trend for zebra populations to increase over the 1960s (Figure 1.5).

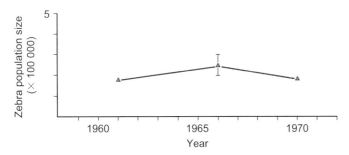

Figure 1.5 Estimates of zebra abundance in the 1960s.

Prediction 3: Increased survival rate in juvenile ruminants

Sinclair also knew that rinderpest had historically killed juveniles after the passive immunity from their mother's milk wore off. If rinderpest release was responsible for the population increase, he predicted an increase in the survival rate of juveniles. One way of measuring juvenile survival is using aerial surveys to measure what percentage of the population is made up of 1-year-old juveniles. An early survey of wildebeest indicated that juveniles made up 8% of the population (Talbot and Talbot 1963). After 1963, that percentage was much higher, usually between 14 and 17% (Sinclair 1977b).

Recall that Sinclair considered two other hypotheses – the food availability and the predator release hypotheses – to explain the increase in buffalo and wildebeest abundance, but the data did not support the predictions of these two alternative hypotheses. When Sinclair came up with the rinderpest release hypothesis, he systematically tested each prediction generated by the hypothesis and was able to support each prediction with data based on observations, models, and experimentation. As each prediction was confirmed, Sinclair's confidence in the rinderpest release hypothesis increased. As Sinclair's experiences reflect, ecology is no different than any other science, in that observation, modeling, and experimentation lie at its heart.

1.3 HOW DO ECOLOGISTS USE OBSERVATION, MODELING, AND EXPERIMENTATION?

In some ways, ecology is a very complex science because it happens in the real world, where controlling variables is difficult, and where replication may be impossible. Sinclair's question of why the buffalo and wildebeest were increasing was especially challenging, because it was an event that happened only once and sample sizes of one are very difficult to test with any degree of certainty. Working under this handicap, a successful ecologist must be a keen observer, able to pick up on small nuances of patterns, and able to extract small amounts of information from a large amount of background noise.

Observations

Scientists use three types of observations. First, they observe actual processes with their senses, or with devices that are extensions of their senses. To estimate abundance and proportions of juveniles, Sinclair used airplanes and cameras. Second, scientists observe and learn from the published literature, which was essential for Sinclair's knowledge of abundance levels prior to his study, in the 1950s and early 1960s. Last, they observe from what other people are doing or saying. Sinclair's ability to speak Swahili helped with his historical research into wildebeest abundance, and his social skills enabled him to establish a rapport with the veterinarians and collaborate with them on the buffalo antibody analyses. Perhaps most importantly, Sinclair got to hang out with a dozen or so senior researchers at the Serengeti Research Institute, bounce ideas off of them, and benefit from their many years of accumulated knowledge. Sinclair states that one of his golden rules is: "If you want to conserve and manage an ecosystem, you need to know all there is to know about it." Much of this knowledge comes from these three types of observations described above. These observations can also be used to construct scientific models.

Scientific models

There are many types of models with very different goals, and you will learn – and hopefully master – some of them as you work through this text. Models seek to describe a system, or to predict what the system will do in the future. All models are simplifications of reality, but ideally each model contains the essential attributes of what it seeks to describe or predict. For example, a map has some of the essential attributes of a landscape, while a global climate model has some of the essential attributes of the world's climatic conditions. But both are simplifications of reality. And both aspire to have enough of the essential attributes to accomplish their goal. In the case of the map, the goal is primarily descriptive, so that the user will be able to make good decisions when on an unfamiliar route. In the case of a global climate model, the goal is primarily predictive, so that citizens can understand the repercussions of their actions and make informed decisions.

Sinclair's research used many models of population growth and of ecosystem function. Sinclair hypothesized that with rinderpest release, buffalo and wildebeest abundance would continue to increase until the populations were limited by grass availability during the dry season. At that point, the populations would begin to level off. One of the problems of making this prediction is that rainfall is highly variable; for example, dry-season rainfall increased sharply in the early and mid-1970s from a mean of about 150 mm to about 250 mm. Based on the correlation between rainfall and food availability (Figure 1.3), and making certain assumptions about predation rates, Ray Hillborn and Sinclair (1979) created a simple mathematical model that predicted wildebeest abundance in relation to dry-season rainfall (Figure 1.6A).

Based on this model, wildebeest abundance could exceed 4 million if dry-season rainfall remained above 250 mm. However, Hillborn and Sinclair issue three warnings in association with this model. First, it would take several decades for the population to reach equilibrium. Second, rainfall levels were

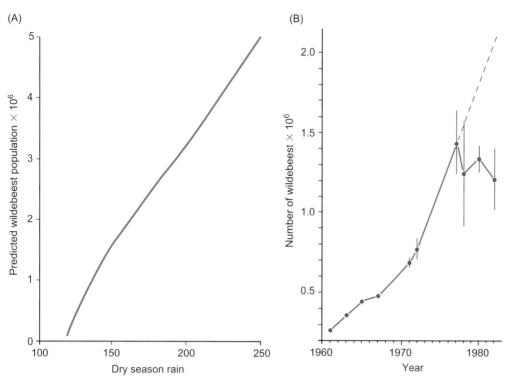

Figure 1.6 A. Graph generated by Hilborn and Sinclair's mathematical model predicting the number of wildebeest in relation to dry-season rainfall. B. Wildebeest abundance through the early 1980s. Dotted line shows predicted abundance if rainfall stayed at 250 mm.

unusually high in the early 1970s, and would likely be lower in the future. Third, this model was an oversimplification – in their paper the authors explicitly state, "We will be wrong in some places, but we invite criticism because this is the most rapid means of improving our understanding of Serengeti dynamics."

How well has this model withstood the test of time? As it turns out, dry-season rainfall in 1977–82 dropped to a mean of 149 mm – almost exactly its historical average during the 1960s. According to the model, the population, under these rainfall conditions, should stabilize at about 1.3 to 1.4 million. Figure 1.6B shows that the population increase of the early and mid-1970s did level out abruptly in association with the decrease in dry-season rainfall (Sinclair *et al*. 1985).

Many observational studies designed to test models will run into problems of getting a large enough sample. Also, just because the prediction of a model is validated does not mean that the model will be correct the next time it is tested. Scientific experiments are designed to get around this problem by controlling as many variables as possible. However, you should recognize that even well-designed experiments are subject to numerous sources of error.

Designing and conducting experiments

Scientists do not control all the variables in a controlled experiment. The simplest experiments attempt to keep all of the variables at similar levels, excepting the one variable or factor the experiment is designed to test. More complex experiments can evaluate the effects of more than one factor at a time, but these are more difficult to design and interpret.

Ecologists make liberal use of carefully designed experiments, both in the laboratory and in the field. As one example, Sinclair wanted to know whether buffalo had preferences for certain

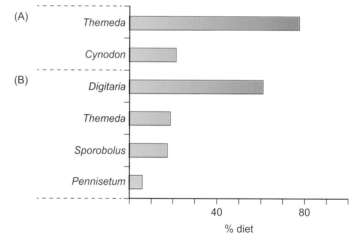

Figure 1.7 Percentage of food eaten when a choice of grass species was presented to tame buffalo: when buffalo were given a choice of two grass species (A); when buffalo were given a choice of four grass species (B).

types of grasses, reasoning that such preferences, if they existed, would be important for buffalo survival, and could help with conservation efforts. Presumably, buffalo would tend to prefer the most nutritious foods, and evaluation of habitat quality by conservation ecologists should be based on how much nutritious food was available to the buffalo.

Sinclair maintained a herd of 10 tame buffalo to test for food preferences. In one series of experiments, he put individual buffalo in a 10-m-diameter enclosure, and gave them a choice of two different species of grasses that were available to them in their natural habitat, *Themeda triandra* and *Cynodon plectostachyus*. The buffalo ate much more *Themeda* than *Cynodon* (Figure 1.7A). In a second series of experiments, the buffalo were given a choice of four grasses, including *Themeda* and three other species. In this case, *Digitaria*, a species that Sinclair knew from circumstantial evidence was heavily grazed by buffalo, was the clear choice (Figure 1.7B). A third series of experiments had to be prematurely terminated because lions broke into the enclosure and seriously damaged the subjects.

The results of this experiment were important, because they showed that food quality is important to the buffalo, and suggested that buffalo survival may depend on their having the correct type of food in the environment. Sinclair's research has shown that dry-season food quality can drop to such a low level that the food-digesting bacteria within the rumen die, at which point the buffalo begin to starve, even while surrounded by an ocean of grass (Sinclair 1977a).

All of the sciences use observation, modeling, and experimentation. All biologists, including ecologists, borrow liberally from the tool bags developed by the other sciences, including chemistry, geology, physics, and mathematics. But ecologists differ from many other biologists, in that their research addresses issues that span and tie together many different levels of the biological hierarchy.

1.4   HOW DO ECOLOGISTS ASK QUESTIONS THAT LINK DIFFERENT LEVELS OF THE BIOLOGICAL HIERARCHY

Biologists study life on all size scales, from subatomic interactions that influence the function of the nervous system, to climatic changes occurring on a global scale. Because ecologists focus their attention on interactions, ecologists must be able to understand processes from all levels of the biological hierarchy. However, ecologists generally devote most of their attention to processes occurring on the level of the organism and above, up to the level of the biosphere. We can use the Serengeti ecosystem to understand the types of questions that can be asked at different levels of the biological hierarchy.

Processes occurring at or above the level of the organism

At the level of the organism, we can ask how an individual buffalo forages in a way that meets its daily needs for energy and nutrients. As we have already seen, buffalo have food preferences, and these preferences are tied to underlying physiological requirements. At a higher level, we can explore questions about the distribution and abundance of populations of a species at a particular location. How many individuals are there, what is their distribution, and how are both abundance and distribution changing over time? Questions about the vast migrations of wildebeest, zebra, and Thompson's gazelles are all addressed at the level of the populations.

But of course these populations interact with each other, and form biological communities. Samuel McNaughton (1976) showed

that wildebeest grazing actually benefited Thomson's gazelles by stimulating the regrowth of succulent grasses and herbaceous plants, which were then eaten by gazelles who followed in wildebeest footsteps. This interaction among wildebeest, gazelles, grasses, and herbaceous plants is an example of a community-level interaction. Linkages extend, of course, beyond the living organisms, and include the nonliving, or abiotic elements of the environment. This brings us into a discussion of the Serengeti ecosystem, which includes both the community of organisms and the abiotic (nonliving) features that interact with the community. Thus the nutrients in the soil determine the grasses that grow there, which influence the distribution of the mammalian community. Conversely, the mammals return nutrients, which are broken down by the burying beetles, to the soil and thus the nutrients can be reused by the next year's population of grasses. These are examples of linkages that occur at the ecosystem level (Figure 1.8A). We will explore some of these linkages in more detail.

At a still higher level, ecologists explore landscapes, groups of interacting, and usually spatially connected, ecosystems. As one example, just to the west of the Serengeti is an ecosystem dominated by agriculture. The human population there is increasing at a minimum of 3% per year (about three times the global growth rate), with the result that humans supplement their resources by hunting elephants, wildebeest, and other ungulates that move in during the dry season (Sinclair *et al.* 2008). Landscape ecologists want to understand how different elements of the landscape, in this case savanna in the Serengeti, and cropland to its west, influence interactions across ecosystems (Figure 1.8B). However, landscapes are parts of larger geographical regions that experience a common set of environmental and evolutionary influences. Thus regional ecologists might ask why large mammal abundance and species richness (the number of species) are so high in the East African region, in comparison to similar regions in North America or Australia (Figure 1.8C).

The highest level is the biosphere itself – the part of the world that supports life. Studies of microorganisms are revealing that the biosphere is almost as large as the globe itself, and extends many kilometers below and above the surface (see Chapter 2). One recurring issue is how global climate change – a biosphere-level process – will influence the functioning of ecosystems such as the Serengeti. Given all the linkages that occur among biotic and abiotic components of the Serengeti ecosystem, there is profound concern that disrupting these linkages could negatively affect its conservation. Let's explore some of those linkages in more detail.

Serengeti abiotic linkages

The three primary abiotic factors that have been studied in the Serengeti are nutrients, fire, and rain, which, as you might imagine, are tightly linked to each other. Nutrient levels are highest in the southeastern plain, as nearby volcanoes have recently erupted and blanketed the region with inorganic minerals. However, rainfall levels are lowest there.

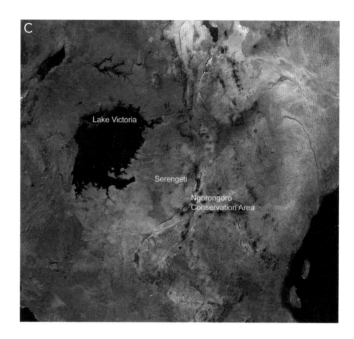

Figure 1.8 A. Serengeti ecosystem. B. Diversity of Serengeti landscape. C. East African region – satellite view.

Nutrient levels are lowest in the northwest, but rainfall levels are highest there. As a result of these differences, migratory wildebeest, zebras, and gazelles tend to spend the wettest months in the southeast, then migrate to the northwest, where there is still some forage available, even during the dry season. Thus the distribution of rainfall and nutrients influences the distribution and abundance of the organisms (Figure 1.9).

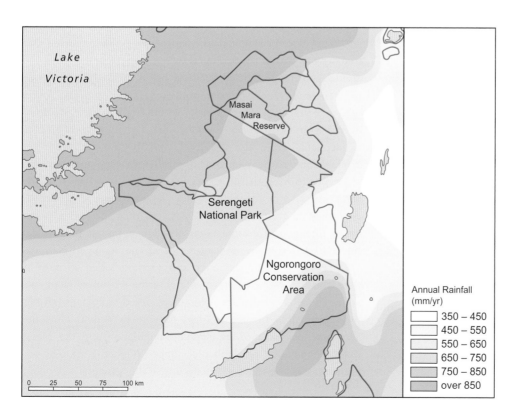

Figure 1.9 Mean annual rainfall (mm) in Serengeti and surrounding ecosystems.

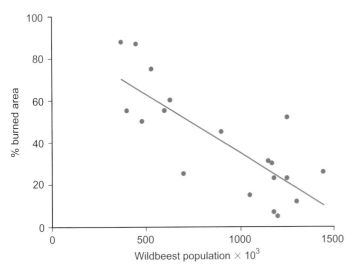

Figure 1.10 Relationship between wildebeest abundance and percentage of the Serengeti that is burned annually. ($r = -0.793$ $P = 0.00005$).

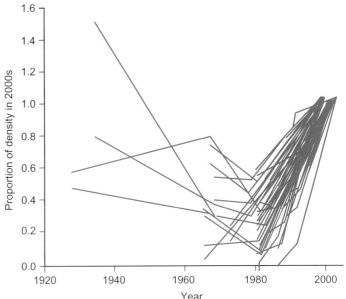

Figure 1.11 Changes in tree density over time as a proportion of tree density values measured in 2000–2003. Each line represents one study site.

The story with fire is a bit more complex. Almost all of the Serengeti fires are started by humans. Some fires are set to improve grazing pastures, others are set by Serengeti National Park employees under park management plans, and others are started by honey hunters smoking out bees from their hives. Fires only occur during the dry season, and require fuel (dried grass) to burn substantial areas. Michael Norton-Griffiths (1979) showed that fires were actually more common in wetter areas of the park, because wet areas grow more grass and hence provide more dry-season fuel. As wildebeest populations have increased since 1960, there has been a sharp reduction in the percentage of the Serengeti that burns annually (Figure 1.10), because the wildebeest have removed much of the fuel. This is an unusual example of a biotic process (grazing by wildebeest) profoundly influencing an abiotic process. This type of relationship can be analysed through correlation or regression analysis (see Dealing with data 1.2).

Sinclair emphasizes that many of the Serengeti linkages are between different biotic components of the ecosystem. We will explore some of these and consider the evidence supporting them.

Serengeti biotic linkages

We've discussed the differences in grass quality in the Serengeti, but we have not mentioned that much of the western and northern portion of the ecosystem is open woodland, primarily *Acacia* species, with canopy cover – the percentage of the ground shaded by the tree canopy – varying between 2% and 30%. Because the woodland is so open, abundant grass grows between the trees, and the shade and cover harbor large numbers of herbivores and carnivores.

Sinclair used historical data from photographs to assess the density of trees in the Serengeti over time. The data are sparse for early years (until about 1962), but there was a trend of high tree density in the 1920s and 1930s, with declining density in the 1960s and 1970s, followed by a sharp increase in the 1980s until 2003 (Figure 1.11). Sinclair and his fellow Serengeti researchers suspected that fire played a role in changing tree density.

Fire is lethal for young trees, but once trees become more mature, they can survive a grass fire. Michael Norton-Griffiths (1979) developed a simple mathematical model that predicted that tree density within the Serengeti would stop declining when the annual percentage burned declined below 30%. This level was reached in 1972. His model predicted that if burning levels remained below 30%, tree density would begin to increase. However, Norton-Griffiths also cautioned that high numbers of giraffes or elephants could also kill trees, and that no data were yet available on whether elephants or giraffes ate very small trees.

Despite Norton-Griffiths' predictions, the woodlands in the Serengeti, and in the Masai Mara National Reserve in Kenya, continued to decline into the 1980s, but then the two adjacent parks diverged. Tanzania went through a period of political turmoil, the economy and authority of the Serengeti National Park Service were undermined, and poachers killed off most of the elephants in the Serengeti. Many of the remaining elephants migrated across the border (which was closed to humans but not to elephants) to the Masai Mara Reserve in Kenya. Thus the Serengeti lost most of its elephant population and the Masai Mara's population of elephants climbed (Figure 1.12). The co-occurrence in the 1980s of the decline in the Serengeti's elephant

Dealing with data 1.2　Variables, regression analyses, and *P*-values

For this text, I will adopt a suggestion by Gotelli and Ellison (2004) that correlation analysis and regression analysis are functionally equivalent. These analyses look for associations between two continuous variables, and construct a *regression line* – a line of best fit – that highlights that relationship.

In the example above, Sinclair argues that high wildebeest abundance reduces the frequency of fire. In this case, wildebeest abundance is the **independent variable** because Sinclair proposes that wildebeest abundance influences the value of the percentage of burned area. Conversely, percentage of burned area is the **dependent variable**, because Sinclair proposes that its value is influenced by wildebeest abundance (the independent variable). Usually the dependent variable is graphed on the *y*-axis and the independent variable is graphed on the *x*-axis. The data appear to show an impressively negative correlation between wildebeest abundance and fire frequency.

Qualitatively, the regression line is drawn through the data points in a way that minimizes the total distance between the points and the line. The regression line, like all lines, has a formula that can be described as $Y = \beta_0 + \beta_1 X$, where β_0 is the *y*-intercept, and β_1 is the slope of the regression line. The regression line is determined mathematically, but the mathematical description is beyond the scope of this book. Fortunately, spreadsheet graphics and statistical programs will draw regression lines for you, if you provide the data in the appropriate format. For Figure 1.10, the regression formula is $y = 91.1 - 0.0561x$. This formula tells us that with each increase of one on the *x*-axis (and remember that a value of 1 represents 1000 wildebeest), the percentage of burned area decreases by 0.0561%. Usually, researchers also cite the coefficient of determination (R^2), which measures the proportion of the variance of the data explained by the line of best fit in a regression analysis. R^2 ranges between 0 and 1 – higher R^2 values indicate closer fit between the observed data and the regression.

The hypothesis we wish to test is called the research hypothesis (H_R). In this case H_R states that there truly is a negative correlation between wildebeest abundance and fire frequency. To establish some credibility for the research hypothesis, we first define a null hypothesis (H_0) that directly opposes H_R. In this case H_0 states that if we collected more data, we would find that there is no relationship between the two variables.

To test correlations, we test the H_R that the slope of the regression line is significantly different from 0 (Figure D1.2.1). This opposes H_0 that the slope of the regression line = 0. The *P*-value in this case is 0.00005. *P* stands for probability. In essence, the **P-value** is telling us what the probability is that H_0 is actually true. Based on these data, there is very low probability (0.00005 or 5 in 100 000) that, if we sampled more years or more wildebeest densities (in other words, if we had a larger data set), β_1 would ultimately equal zero.

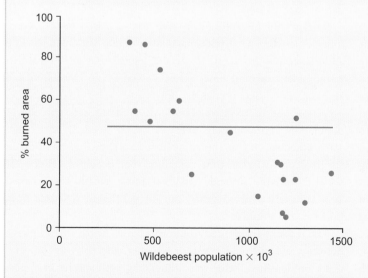

Figure D1.2.1 Regression line with $\beta_1 = 0$. Notice how some of the data points are very far from the regression line, which makes this hypothesis (H_0) very unlikely.

P-values are used in many of the statistical analyses in this text, and in ecology and most other sciences in general. Very low (small) P-values indicate that there is a very small probability that H_0 is correct. The question then arises, how confident do we need to be that H_0 is false, before we have support for H_R? It is up to the researcher to decide how unlikely H_0 must be before it can be rejected. This is known as the critical value (α). The most common α is 0.05. But I emphasize that you need to be very careful about interpreting P-values, there is nothing magical about $P = 0.05$. So if you find $P = 0.07$, you should suspect that there may be a correlation between two variables, but your confidence level for this relationship would be low. And, in contrast, if $P = 0.0001$, then you can be much more certain that there is a correlation between the two variables.

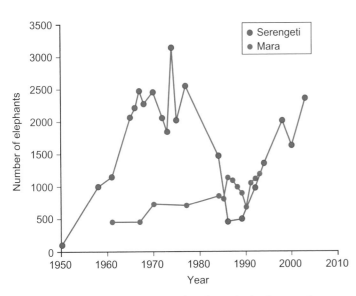

Figure 1.12 Elephant population abundance in the Serengeti woodlands and in the Masai Mara Reserve. Remember that the Masai Mara is much smaller than the Serengeti.

Recent Serengeti studies and unexpected biotic and abiotic linkages

Sinclair and his colleagues (2013) have recently turned up a new series of biotic and abiotic linkages. During his early years in the Serengeti, Sinclair noticed that the abundance of the Natal multimammate rat, *Mastomys natalensis*, oscillated over time. Sometimes populations were very low, and other times they were much higher. As he was not initially studying rats, Sinclair's first observations were not systematic; he simply wrote down in his field notes that he came across some rats on a particular date.

However, Sinclair was somewhat more systematic about noting the presence of an important rat predator – the black-winged kite, *Elanus caeruleus*. These predators move into areas of high rat density, and can reproduce very quickly when rat abundance climbs to high levels. Recall that Sinclair's first visit to the Serengeti involved studying bird migrations, which led him into the habit of keeping records of all the common birds he observed. Thus he has a record of black-winged kite density dating back to 1965.

Sinclair and his colleagues were not surprised to discover a linkage between natal multimammate rat and black-winged kite abundance (Figure 1.13). It seemed logical that an increase in rat abundance would be closely followed by an increase in the abundance of its predator. But there were two puzzling aspects about this system. First, why did the rats explode in population approximately every 5 years? Second, why were some other species in the ecosystem also showing variation in survival on a 5-year time schedule? For example, wildebeest showed variable juvenile survival rates, which also peaked approximately every 5 years, but on a different schedule from peaks in rat abundance.

The researchers wanted to know whether there was a linkage between rats and wildebeest, or whether, alternatively, a previously unidentified factor was influencing both species in opposing directions. Fortunately, researchers at the Serengeti have been keeping monthly rainfall data since 1937. By combining rainfall and temperature data with the surveys of rats, kites, and wildebeest, the researchers were able to show that rat populations exploded following unusually heavy rains in November

population and the increase in the Masai Mara Reserve elephant population can be seen more clearly if we transform the data (Dealing with data 1.3).

Holly Dublin (1995) reported that the woodlands in the Mara preserve continued to decline during the 1980s and 1990s. In contrast, woodland abundance in the Serengeti began to rebound sharply, so that, at present, woodlands are back to their estimated levels from the 1930s. By carefully watching elephant feeding behavior, Dublin was able to demonstrate that elephants at the Mara ate large numbers of small trees. Dublin and her colleagues (1990) argued that ecosystems such as the Serengeti have two stable states. In the case of the Serengeti–Mara ecosystem, there can be either grassland or open woodland. To change a woodland to a grassland, there must be both high fire and abundant elephants. To change a grassland to a woodland, there must be both low fire and sparse elephants. Thus once an ecosystem is in one state, it remains there unless it is disturbed by a profound change in both fire conditions and browsing by elephants.

Dealing with data 1.3 Data transformations

When comparing two variables – for example, elephant abundance in the Serengeti over time versus elephant abundance in the Mara over time – it is sometimes helpful to transform the data so they can be compared more meaningfully. In the case of Figure 1.12, it might make more sense to redraw the graph with the y-axis being the relative number of elephants in each ecosystem. Eyeballing the graph, you can see that the maximum number of elephants in the Serengeti was about 3200 in 1973. Based on that estimate, we can redraw the Serengeti curve so that 1973 has a relative number of elephants of 1.0. In 1950, the relative number would be about 100/3200 = 0.03, while in 1958, the relative number would have climbed to 1000/3200 = 0.31.

We can then repeat the process for the Mara ecosystem, noting that the maximum abundance is about 1200 elephants in 1993. Thus the value for 1993 has a relative number of 1.0. In 1961, the first Mara elephant census has a relative number of 450/1200 = 0.38. Figure D1.3.1 shows the completed transformation of the data for both curves. What relationships are easier to visualize when the data are transformed in this way?

There are many types of data transformations you will encounter is this book. Some transformations make the data more visually understandable, while others prepare the data for a particular type of analytical test.

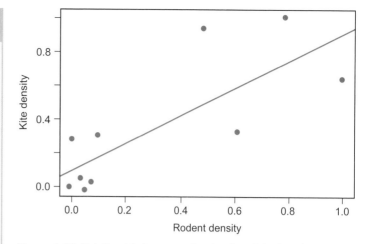

Figure 1.13 Relationship between the density of the Natal multimammate rat and the black-shouldered kite. Data are normalized so that a value of 1.0 represents the greatest value of kite and rodent density observed in the years of the surveys.

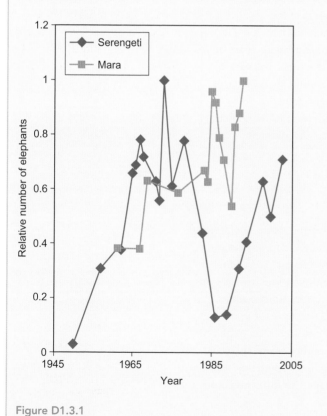

Figure D1.3.1

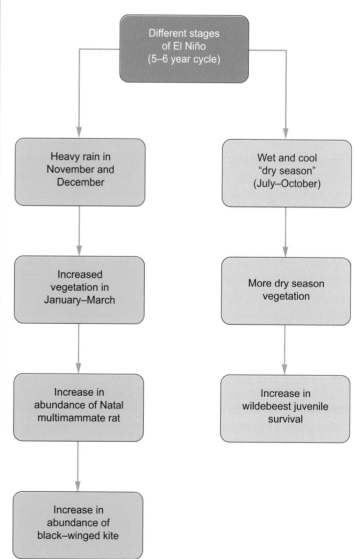

Figure 1.14 Abiotic and biotic linkages in the Serengeti.

and December of the previous year, presumably because this provided more vegetation for rats to eat during a time of year when food availability usually limited their survival and reproduction. In contrast, wildebeest juvenile survival was greatest when the dry season (July through October) was unusually wet and cool, because grass production was greater under those conditions.

In the Serengeti, fluctuations in weather are influenced by the El Niño Southern Oscillation (ENSO), a large-scale atmospheric system that affects climate worldwide. Over the course of about 5 years, the air pressure and sea surface temperatures across the south Pacific and Indian oceans oscillate in a somewhat predictable fashion. During one phase of ENSO, the sea surface temperature increases and atmospheric temperature decreases near the west coast of South America, spawning heavy rains across much of the Americas and Western Europe, and drought in Africa and Australia. In the opposing phase (La Niña), the sea surface temperature decreases and atmospheric temperature increases near the west coast of South America, bringing drought to much of the Americas and Western Europe, and heavy rains to Africa and Australia. The upshot is that about every 5 years there are heavy rains in the Serengeti during November and December, which is, in part, responsible for an explosion in the population of *M. natalensis*. At a different stage of ENSO, there is unusually heavy dry-season rainfall, which increases grass production, and leads to very high juvenile wildebeest survival (Figure 1.14)

REVISIT: The Serengeti narrative

The Serengeti is home to a diverse community of species that respond differently to the wide variety of factors that may influence survival and reproductive success. By studying these processes over a span of several decades, researchers are beginning to uncover some of these interactions. Though Sinclair has recently retired, many other researchers in the Serengeti continue to uncover and deepen our understanding of the ecological linkages.

We began this discussion of Sinclair's research by considering his basic work that estimated the distribution and abundance of two species – wildebeest and buffalo – within the Serengeti ecosystem. But Sinclair wanted to know how these abundances were changing over time, and what factors were causing them to change. This search required Sinclair to consider interactions among these two species and both biotic and abiotic elements of the environment. His most recent work shows that periodic variation in an abiotic factor (rainfall) causes changes in an abiotic factor (grass abundance), which has a cascading influence on a variety of other species. Like a great detective or mystery story, uncovering these linkages provides clues to general patterns of past and present species distribution and abundance, as viewed from different levels of the biological hierarchy. Perhaps most important, understanding these linkages allows ecologists to make predictions about what will happen to species in the future, which is particularly important in a rapidly changing environment.

SUMMARY

Over the years, the Serengeti has been a model ecosystem for answering basic ecological questions about the distribution and abundance of organisms, populations, and species, and about how different species interact with each other and with their environment. Tony Sinclair and many other researchers have addressed some of these questions, and continue to work on understanding important biotic and abiotic linkages that influence ecosystem functioning.

In common with all types of scientific inquiry, ecologists use predictions to test hypotheses about ecological processes; this approach is highlighted by Sinclair's research that explored why buffalo and wildebeest populations were rapidly expanding. Like other scientists, ecologists use observation, modeling, and experimentation to generate and test hypotheses. However, in contrast with much biological inquiry, ecologists ask questions that link numerous levels of the biological hierarchy, from molecular to global ecology.

FURTHER READING

Goulson, D. 2014. Pesticides linked to bird declines. *Nature* 511: 295–296.

A short understandable summary on the effect of a commonly used class of pesticides on ecosystems. A great opportunity to gain familiarity with different biotic and abiotic linkages. The actual paper is:

Hallmann, C. A., Foppen, R. P., van Turnhout, C. A., de Kroon, H., and Jongejans, E. 2014. Declines in insectivorous birds are associated with high neonicotinoid concentrations. *Nature* 511: 341–344.

Sinclair, A. R. E. 1977b. The eruption of the ruminants. In *Serengeti: Dynamics of an Ecosystem*, ed. A. R. E. Sinclair and M. Norton-Griffiths. Chicago, IL: University of Chicago Press, pp. 82–103.

A very readable summary of Sinclair's early work exploring why some large mammal species increased in abundance while others did not.

Sinclair, A. R. E and 17 others. 2013. Asynchronous food-web pathways could buffer the response of Serengeti predators to El Niño Southern Oscillation. *Ecology* 94: 1123–1130.

A challenging but fascinating study of how El Niño may be influencing the dynamics of the Serengeti ecosystem, creating a 5-year periodicity to shifts in abundances of different species.

White, K. S., Barten, N. L., Crouse, S., and Crouse, J. 2014. Benefits of migration in relation to nutritional condition and predation risk in a partially migratory moose population. *Ecology* 95: 225–237.

Good introduction to hypothesis testing in this very accessible study that evaluates the costs and benefits of migration in a marked moose population. Nice discussion of the limitations of the findings in the context of a rapidly changing environment.

END-OF-CHAPTER QUESTIONS

Review questions

1. Outline the wildebeest migration pattern and tie that together with rainfall patterns, the availability of nutritious grass, and the timing of when wildebeest give birth to their offspring.
2. What were Sinclair's three hypotheses for increases in buffalo and wildebeest abundance in the 1960s and 1970s. How did he test the predictions of each hypothesis, and what did he conclude based on each test?
3. Give two examples of community-level biotic interactions discussed in this chapter. In what ways may abiotic factors be linked to each example?
4. What are the levels of the biological hierarchy most often considered by ecologists? Give an example of a question that an ecologist might ask that pertains to each level of the hierarchy that you cited.
5. What are three general questions that ecologist ask?

Synthesis and application questions

1. In Sinclair's (1975) exclosure experiments (Figure 1.3), grasshoppers and rodents were able to access the exclosures, and eat some of the grass. This might decrease grass growth within the exclosures. Is this a problem for Sinclair's use of rainfall as a measure of food availability? Why or why not?
2. Dennis Normile (2008) wrote an article entitled "Driven to extinction," in which he argues that rinderpest may be the second disease that has been driven to extinction by an aggressive

vaccination campaign (smallpox being the first). What types of evidence would you need before you felt confident that an infectious virus had indeed been driven to extinction? How would you collect those data?

3. What do you think would happen to lion cub survival following an unusually wet and cool dry season in the Serengeti? Explain your reasoning.

4. Think of a model that influences your everyday life. Why is this model useful? In what way is this model a simplification of reality? Is the model primarily descriptive, or predictive, or both?

5. If you were given 3 years to do research in the Serengeti, what question might you ask? How would you go about answering it?

Chapter 2
The physical environment

INTRODUCTION

Sunlight filtering down through the forest is intercepted by branches, so that by the time it reaches the surface of the ground, its intensity is literally only a shadow of its former self. The branches are arranged in multiple layers, with the top layers accessing the most sunlight, and lower layers receiving greatly diminished intensity. The branches host many types of plants and animals, including photosynthetic organisms that harvest the sunlight to power their cellular activities. Brilliant butterflies flit through the specks of light, foraging, mating, and creating new generations, as they have done in similar environments for millions of years. Interacting with some species, but unaffected by others, these animals and plants – and a host of organisms less conspicuous to our eyes – are part of a biome found in tropical latitudes.

A biome is a large geographical area with characteristic groups of organisms adapted to that particular environment. Terrestrial biomes are most influenced by temperature, moisture, and soils, while aquatic biomes are most influenced by temperature, chemical composition of the water, and water current. The boundaries and species composition of biomes change constantly, through both natural causes and human impact. Our opening case study looks at similarities and differences in how some terrestrial and aquatic biomes are structured and in how they are threatened by human activities. We will then consider the role climate plays in structuring biomes. We will explore the diversity of terrestrial and aquatic biomes, and discover that even though biomes are large geographical areas, researchers continue to find new biomes in unexpected places. In a world that has been so extensively changed by human actions, some ecologists question whether the traditional biome concept is still useful, or should be replaced by an approach that accounts for the effects of human actions.

KEY QUESTIONS

2.1. How do physical principles influence climatic variation across the globe?

2.2. What are terrestrial biomes?

2.3. How do biomes change over time?

2.4. What are aquatic biomes?

CASE STUDY: Biomes compared

Biomes that appear very different may have striking similarities in form and function. Reading this introduction, you may have surmised that the first paragraph described a tropical rainforest. You are correct! But this same paragraph also describes a coral reef, where the branches are coral and the photosynthetic organisms are zooxanthellae, algae that have a symbiotic – living together – relationship with the coral reef (Figure 2.1A). Ecologists describe this relationship as a mutualism, because both species benefit (see Chapter 15). Corals provide zooxanthellae with a secure place to live and a steady supply of nitrogenous waste to fuel their metabolism. In return, zooxanthellae share with the corals some of the organic compounds they construct from photosynthesis. Similar to coral reefs, tropical rainforests also harbor photosynthetic organisms on their branches. These epiphytes use tree branches for support, allowing them to access the Sun's rays for photosynthesis. Some epiphytes don't have any impact on the success of their tree hosts, while others actually provide the trees with nutrients, in a mutualism resembling the coral/zooxanthellae relationship. To complete the comparison of biomes, the brilliant butterflies in the opening paragraph are butterflyfish in the ocean – highly conspicuous animals in many coral reef ecosystems (Figure 2.1B).

Tropical forests and coral reefs share many other similarities. Both biomes have an extraordinarily high number of species, or high species richness. Both biomes also have a very high level of primary production – a term ecologists use to describe the chemical energy generated by the autotrophs within an ecosystem. Both trees and zooxanthellae are autotrophs, because they convert energy from sunlight to chemical bond energy.

Marine and aquatic biomes also face similar challenges from human activity. As one example, since the beginning of the Industrial Revolution, human activity has caused atmospheric CO_2 levels to increase about 100 times faster than they have in the past 650 000 years. About 30% of these emissions have been absorbed by the oceans, and transported by ocean currents, shifting to the right the equilibrium favoring the following reaction:

$$CO_2 + H_2O \rightleftharpoons H_2CO_3 \text{(carbonic acid)}$$

Figure 2.1 A. Coral with zooxanthellae. B. Butterflyfish in coral reefs.

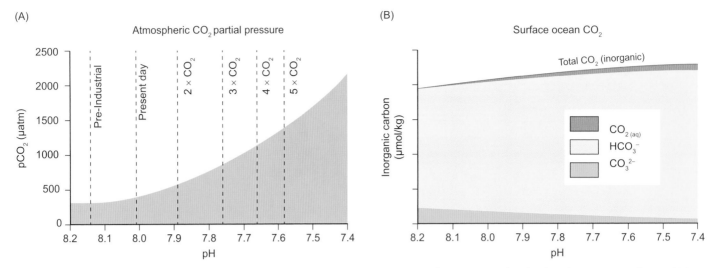

Figure 2.2 A. Increasing atmospheric CO_2 causes an increase in the partial pressure of CO_2 (pCO_2) dissolved in the ocean's surface, which increases the concentration of carbonic acid and reduces ocean pH. B. Reduced ocean pH increases the total amount of dissolved inorganic carbon, by increasing the concentration of dissolved carbon dioxide and bicarbonate. Note, however, that even though total dissolved inorganic carbon increases, the concentration of carbonate declines sharply with acidification.

This carbonic acid increases hydrogen ion concentration through the following reaction:

$$H_2CO_3 \rightleftharpoons HCO_3^- (\text{bicarbonate}) + H^+$$

This increases the acidity, or more properly, decreases the alkalinity of the ocean. Since the onset of industrialization, the mean ocean pH has declined about 22%, from 8.16 to 8.05.

Coral and many other marine organisms use calcium carbonate to build their skeletons with the following reaction:

$$Ca^{2+} + CO_3^{2-} (\text{carbonate}) \rightleftharpoons CaCO_3 (\text{calcium carbonate})$$

But as the ocean becomes more acidic, the higher concentration of dissolved hydrogen ions shifts the equilibrium of the following reaction to the right:

$$CO_3^{2-} + H^+ \rightleftharpoons HCO_3^-$$

This robs coral of the carbonate needed for making skeletons. Figure 2.2 shows the relationship between pH and levels of CO_2, HCO_3^-, and CO_3^{2-} in the ocean.

How bad is this trend for the future of the ocean's most productive biome? Joan Kleypas and her colleagues (2006) summarized the results of 12 laboratory studies conducted on 11 species of scleractinian corals, the most important components of coral reefs. Each study found that a doubling of CO_2 levels from preindustrial times, which is projected to occur within the next 50 years, led to a reduction in calcification rates; the average reduction was about 30%. As they also experience coral bleaching brought on by rising ocean temperatures, and damage from fishing trawlers, coral reefs are one of the most endangered biomes.

Terrestrial ecosystems are also suffering from human-induced acidification. In many industrial temperate zones, coal-fueled power plants emit large quantities of sulfur dioxide and nitrogen oxides. In the atmosphere, these substances react to form sulfuric acid (H_2SO_4) and nitric acid (HNO_3), strong acids that descend in rain, fog, or dust over large areas downwind from these sources. Acid deposition reduces soil pH, and can kill soil microorganisms that have mutualistic interactions with plants. Acid deposition can also release toxic metals that are bound to rocks, and can leach away vital soil nutrients. In many areas, numerous trees are dead and overall species richness has declined dramatically from acid deposition.

Acid deposition is a global problem (Figure 2.3). Though primarily affecting temperate biomes, the problem is now spreading as countries in tropical regions become more industrialized and dependent on fossil fuel. One recent study indicates that in the near future, tropical regions will lose biological diversity at a much higher rate from acid deposition than will any other terrestrial biome (Azevedo *et al.* 2013).

As we have just seen, prevailing currents and the chemical environment have a major influence on species richness and overall ecosystem functioning. Other physical conditions, such as air and water temperature, also influence which organisms live where. Let's explore some of the basic principles underlying global variation in these defining factors.

2.1 HOW DO PHYSICAL PRINCIPLES INFLUENCE CLIMATIC VARIATION ACROSS THE GLOBE?

In almost every respect, the Sun is the primary driver of life on Earth. *Photons* – particles or waves of solar radiation – stream into space in all directions. Though only a miniscule fraction of these photons are intercepted by Earth's atmosphere, enough radiation penetrates to Earth's surface to support most

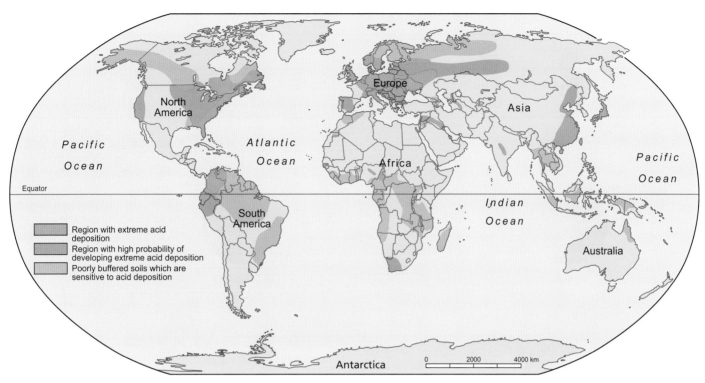

Figure 2.3 Terrestrial regions of the world with problematic levels of acid deposition.

Region with extreme acid deposition

Region with high probability of developing extreme acid deposition

Poorly buffered soils which are sensitive to acid deposition

ecosystems. Though the Sun is an unbiased emitter, showing no favor in the direction it sends its photons, Earth's curvature makes the planet a very biased receiver, favoring some latitudes over others.

Influence of latitude and season on surface temperature

It should come as no surprise that it is hotter near the equator than at the poles and that, outside of the tropics, it is hotter in the summer than the winter. Figure 2.4 demonstrates two reasons for this relationship. The arrows represent photons as they strike Earth's atmosphere. On average, photons will be an equal distance apart as they travel from the Sun. However, because of Earth's curvature, photons headed toward the equator will be more closely packed than those destined for polar regions. Thus, on average, more photons strike a unit of area of the equatorial atmosphere than the polar atmosphere. A second influence of Earth's curvature is that there is more atmosphere for a photon to travel through near the poles than near the equator (Figure 2.4). Overall, almost 50% of the photons never reach Earth's surface, either being absorbed or reflected back into space by molecules or atmospheric particles. Over the course of a year, about twice as many photons reach Earth's surface at the equator in comparison to the amount at 60° latitude (Buffo *et al*. 1972). This contrast is illustrated by global differences in direct radiation, a measure of the amount of energy striking a defined area over time (Table 2.1). Such data can be extrapolated to estimate direct radiation at other

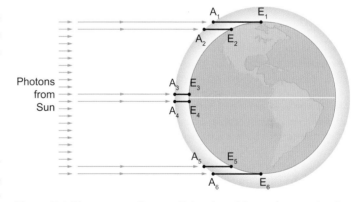

Figure 2.4 Photons traveling parallel and equidistant from each other are further apart when striking the atmosphere because of Earth's curvature. Compare the distance between A3 and A4 to A1 and A2 or A5 and A6. Then notice that photons passing through the atmosphere near the poles have much more atmosphere to go through (A1 to E1, or A6 to E6) before reaching the Earth's surface than do photons near the equator (e.g. A3 to E3).

locations (see Dealing with data 2.1 for a discussion of extrapolation).

Because Earth's rotational axis is approximately 23.5°, the position of the greatest direct solar radiation varies seasonally. On December 21, the most direct rays strike at 23.5° S (the Tropic of Capricorn). As the year progresses, the position of the most direct solar radiation gradually moves northward, reaching the equator on March 21, and 23.5° N (the Tropic of Cancer) on June 21. After June 21, the position of the most direct solar radiation returns southward, revisiting the equator on September

Table 2.1 Annual direct radiation striking Earth at different latitudes.

Latitude (°)	0	10	20	30	40	50	60
Direct radiation (Cal./cm^2/day)	266 271	261 965	249 369	228 998	201 947	169 828	135 941

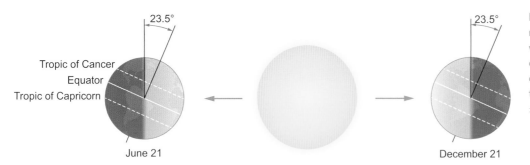

Figure 2.5 The position of Earth in relation to the Sun determines which latitude receives the most direct solar radiation over the course of one solar revolution. All dates in the figure and text above are ±2 days.

Dealing with data 2.1 Extrapolation

Extrapolation estimates new data points that lie outside the range of existing data points. For example, if you had a series of data points 2, 3, 4, 5, and 6, you might extrapolate, with a fair degree of confidence, that the next data point will have the value of 7. The confidence of your extrapolation depends on how well you understand the mechanism that is creating the data points. For example, if you had the data points 1, 2, and 4, you could extrapolate, but with much lower confidence, that the next value is 8, reasoning that each point was the previous point multiplied by two. However, it is equally likely that the next data point would be 7, with each point being the previous point plus the position the previous point occupies in the sequence.

In the case of Buffo's data, we can see quite clearly from Figure 2.4 that the number of photons striking the Earth should continue to decrease more sharply with increasing latitude, because the influence of the curvature of the Earth becomes more significant as we approach the North Pole. Without knowing all the mathematical computations, we could not come up with a precise extrapolation/projection, but we could probably come pretty close to the actual number. Buffo and his colleagues only computed to 60° north, because the computers they had available in 1972 were overwhelmed by the task of dealing with the absence of light during some winter days in the polar regions.

The confidence in your extrapolation also decreases as you move further outside the range of existing data points. If you were to estimate direct radiation values for 70° N and for the North Pole, you would expect your estimate for 70° N to be closer to the actual value.

Thinking ecologically 2.1

John Buffo and his colleagues (1972) wanted to know how much direct radiation struck the Earth at different latitudes. Can you extrapolate from Table 2.1 an estimate of the direct radiation striking the North Pole over a 1-year time period (See Dealing with data 2.1)?

21, and completing the cycle on December 21. Thus December 21 heralds the beginning of summer for the southern hemisphere and winter for the northern hemisphere, while the reverse is true for June 21 (Figure 2.5).

Only a small portion of direct solar radiation actually heats Earth's surface. Much of the solar radiation striking Earth is radiated back into the atmosphere in the form of longwave radiation. Much of this energy is absorbed by atmospheric molecules such as water vapor, carbon dioxide, and ozone, and reradiated back to the Earth, helping to warm Earth's surface even more. This greenhouse effect (Chapter 23) is responsible for the approximately linear decrease in temperature with increasing distance away from Earth and into the troposphere (Figure 2.6).

As we will discuss in later chapters, many other factors influence global temperature patterns as well. Now we'll investigate some of the primary factors that influence global precipitation patterns.

Influence of latitude and season on precipitation

The bases of global precipitation patterns have confounded scientists for many centuries. Some aspects are very well understood, while others continue to be mysterious, even to professional meteorologists.

The relationship between precipitation and atmospheric pressure

Earth's atmosphere is made up of molecules of gas, primarily nitrogen (78%) and oxygen (21%), with many other gases present in much smaller quantities. The atmospheric pressure per m^2 exerted by the weight of the air above Earth's surface averages 1013.25 millibars (mb) or 101.3 kilopascals at sea level. There are

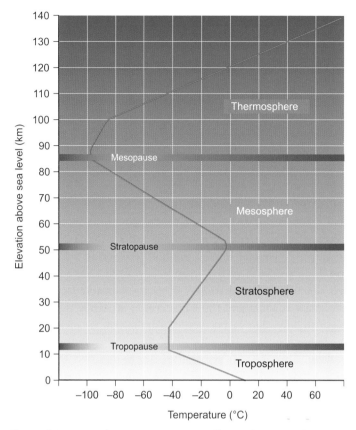

Figure 2.6 Atmospheric temperature profile in relation to elevation above sea level. Note the nearly linear temperature decrease in relation to elevation in the troposphere and the abrupt transition to different relationships in the other atmospheric layers.

many more gas molecules near the surface, in part because of Earth's gravitational attraction, so the atmospheric pressure decreases sharply with elevation above sea level (Table 2.2).

But air masses vary in pressure. Areas of low pressure form from warm air that rises, because warm molecules have higher kinetic energy, and are spaced farther apart than cooler molecules. As the air cools at higher altitudes in the troposphere, the water vapor condenses and forms clouds and rain. Some parts of Earth get a lot of rain, because these low-pressure systems consistently form over them.

Variation in precipitation from the equator to the poles

One important factor influencing precipitation patterns is differential heating of Earth's surface. For simplicity, let's consider Earth at one of the equinoxes, with direct solar radiation striking the equator. The resulting hot air mass is forced upward, cooling as it rises. Because cool air holds less water than hot air, the water vapor molecules condense, forming clouds and ultimately producing the heavy rains characteristic of equatorial regions. This zone of upwelling, warm moist air is the Intertropical Convergence Zone (ITCZ). Because hot air is being generated below it, this initial air mass, now high in the troposphere, is forced away from the equator, both north and south.

At around 30° latitude, north and south, this air mass, now much cooler, begins its descent toward the surface from the upper troposphere. As it descends, it warms, and because it has already lost much of its moisture on its ascent near the equator, it tends to create a very dry environment at this latitude. Many of the world's

Table 2.2 Relationship between mean atmospheric pressure (kPa = kilo Pascals) and elevation above sea level for selected geographical landmarks (many are discussed in this chapter).

Geographic landmark	Elevation (m)	Atmospheric pressure (kPa)
Dead Sea	-423	107
Sea level	0	101
Lake Yoa	378	97
Ulaanbaatar	1350	86
Mount Marcy (Adirondack State Park)	1629	83
San Pedro de Atacama (Atacama desert)	2407	75
Mauna Loa	4169	60
Mount Kilimanjaro	5895	48
Mount Everest	8850	31

(A)

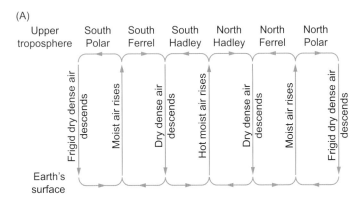

(B)

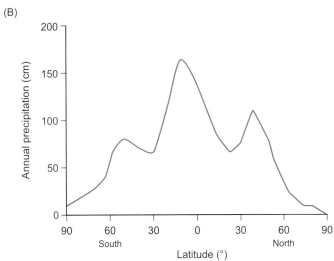

Figure 2.7 A. Formation of the Hadley, Ferrel, and polar cells. These cells are constantly shifting, so the latitudes are only approximate. B. Annual precipitation in relation to latitude. The peaks in the graph result from rising air masses cooling as they rise, causing condensation and precipitation. The low points in the graph result from descending air masses warming as they descend, and becoming even drier.

great deserts are found at about 30° latitude. Now at the surface, some of this dry air mass flows toward the equator, completing a cell of circulating air known as the Hadley cell, and the rest flows away from the equator, forming the surface component of a second circulatory cell, the Ferrel cell (Figure 2.7A). We will leave the Ferrel cell for a brief digression to the polar regions.

The key to understanding the formation of a polar cell is to recognize that the poles receive the least solar radiation, and have the coldest temperatures. As the cold air masses over the poles sink toward the surface, they absorb heat and become warmer, which increases their ability to hold moisture. Thus polar regions receive very little rain (or snow). As this cold air reaches the surface, it is forced away from the poles by the cold air descending from above. As it moves away from the poles, it tends to warm somewhat. At about 60° latitude, this polar air mass collides with the warmer air from the Ferrel cell. As we've already described for air at the equator, warm air tends to rise. As it rises to higher elevations it cools down, forming a second region of heavier precipitation at 60° latitude (Figure 2.7B).

As happened at the equator, some of the air mass, now high in the troposphere, where it has become much cooler and has lost much of its moisture, is forced toward 30° latitude, completing the circulation of the Ferrel cell. The remainder of the air mass is forced toward the poles, completing the circulation of the polar cell. Thus both the northern and southern hemispheres have three major circulatory cells, which result in a pattern of high precipitation near the equator and 60° latitude, and lower precipitation at 30° latitude, and especially at each pole (Figure 2.7B).

Seasonal shifts in precipitation peaks

Our previous discussion of air cells neglects the periodicity of direct solar radiation. As you might expect, the ITCZ migrates north of the equator during the northern spring and summer, and migrates south of the equator during the southern spring and summer. This migration is in part responsible for the great monsoons in India and the Sahel region of Africa that usually begin in July, and in northern Australia that usually begin in December.

Because water has a very high specific heat (4.1855 J/g or 1 cal/cm^3), it takes much more solar radiation to heat water than land. In addition, water absorbs about 2260 J/g as it vaporizes, which slows down its rate of heating during the summer. As a result, landmasses that receive the Sun's direct rays tend to heat up much more rapidly than nearby oceans. These large masses of warm air develop into strong low-pressure systems, which when coupled with moisture-laden prevailing winds can cause tremendous amounts of rain to fall within relatively short time periods. As the ITCZ moves away from these regions, the disproportionate heating subsides, and monsoon rains abate for another year.

Meteorology would be a very simple science if the air cells and precipitation shifts described thus far made up the entire picture. However, many other factors influence the movement of air masses, which in turn also influences the movement of water in the oceans. One important factor is the Coriolis effect, which tends to deflect air masses away from a linear path.

Deflection of air masses by the Coriolis effect

If we were to view the Earth from a stationary reference point in space, an object on the equator would speed from west to east at a velocity of 1674 km/h. It is moving so quickly, because it will do one complete rotation along the equator (c. 40 176 km) every 24 h. But this same object at 45° latitude has only 28 290 km to cover in the same time period, so it will travel a somewhat more mundane 1179 km/h. Meanwhile, an object at the poles will appear to not travel at all, merely making one complete rotation every 24 h while fixed in space.

This has important implications for understanding wind currents, and the formation and movement of storms. For example, if we consider the north Hadley cell, the air mass from 30° N is moving near the surface toward the equator, and encountering

Figure 2.8 Coriolis effect deflects air currents to the right in the northern hemisphere and to the left in the southern hemisphere.

an Earth that is spinning from west to east faster than it is. So it tends to lag behind, or curve to the right as it moves south. Conversely, if we consider the south Hadley cell, the air mass from 30° S will also lag behind as it moves toward the equator, curving to the left as it moves north. Wind currents are named by the direction from which they come. Thus the northeast trade winds in the northern Hadley cell originate in the northeast and flow to the southwest (Figure 2.8).

Overall, air masses in the northern hemisphere tend to deflect toward the right, while air masses in the southern hemisphere tend to deflect toward the left. If you are having trouble visualizing this, you should do the following experiment. Grab a ball and a marker pen. Circumscribe the ball at the approximate locations of the equator, 30° N, and 30° S, to mark the outlines of the two Hadley cells. Rotate the ball from west to east along the equatorial plane. If you view it from above, you will notice it is rotating counterclockwise, but if you view it from below, you will note it is rotating in a clockwise direction. Thus the direction of rotation is a matter of perspective. While rotating your ball, draw a line due south from 30° N to the equator. You will note that your "line" is actually a curve that is deflected to the right, analogous to the northeast trade winds in the north Hadley cell. Then repeat this experiment to simulate the southeast trade winds of the south Hadley cell.

The Coriolis effect also influences the ocean's currents. The ocean's gyres, or large-scale surface circulations, tend to be deflected to the right in the northern hemisphere, and toward the left in the southern hemisphere.

We've only scratched the surface in describing the relation between solar radiation, precipitation, and currents. We have focused our discussion on temperature and moisture, because these two factors are viewed by most ecologists as responsible for the formation of the world's terrestrial biomes.

2.2   WHAT ARE TERRESTRIAL BIOMES?

A terrestrial biome is a large land area with characteristic groups of organisms adapted to its environment. In his classic book, *Communities and Ecosystems*, Robert Whittaker (1975) recognizes 26 types of terrestrial biomes. You will learn the major characteristics of only eight types of terrestrial biomes, but should appreciate that condensing this list necessitates that there will be substantial variation within biomes regarding climate and types of organisms (Figure 2.9).

Defining and distinguishing biomes: an exercise in ambiguity

Few ecologists agree on how many terrestrial biomes there actually are, for several reasons. First, subjectivity enters into deciding whether two geographical areas have sufficiently different characteristic organisms to be classified as defining two distinct biomes. For example, Whittaker (1975) described one biome as a tropical seasonal forest, and a second with somewhat less precipitation as a woodland. Is this distinction biologically meaningful? Second, the boundaries between biomes are not distinct, and one biome will usually grade into another. This again makes it difficult to evaluate whether there is one highly variable biome in a region, or two adjacent biomes that differ from each other only marginally.

Finally, new biomes are still being discovered. For example, Li-Hung Lin and his colleagues (2006) described a subterranean biome dominated by one group of Firmicutes bacteria living at a depth of 3–4 km below the surface. These organisms reduce SO_4^{2-} (sulfate) to H_2S (hydrogen sulfide) in a many-step process that produces enough ATP for their metabolic needs. The sulfate is leached from volcanic rocks. The researchers estimate that this particular site has been functioning for at least 3 million years. There is no reliance on photosynthesis as an energy source, in contrast to the plant-dominated biomes, so some ecologists argue that this ecosystem should be classified as a distinct biome.

How climate influences Earth's terrestrial biomes

Rainfall and temperature are the two most important factors influencing the development of terrestrial biomes. Whittaker (1975) summarized how these two variables interact to create the eight biomes that we discuss below (Figure 2.10). But several other factors, such as soil characteristics and the frequency and type of disturbances, such as fires or monsoon floods, can also influence the type of biome characteristic of a region.

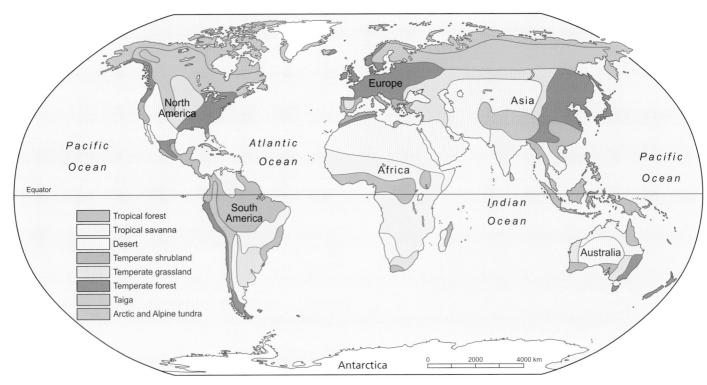

Figure 2.9 Terrestrial biomes of the world.

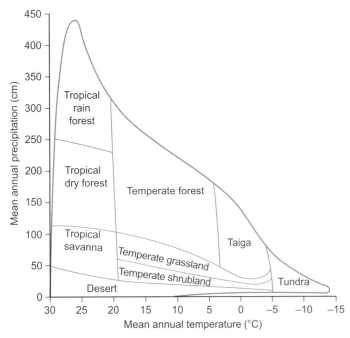

Figure 2.10 Pattern of terrestrial biomes in relation to mean annual temperature and precipitation.

Whittaker's summaries were a breakthrough in understanding the relationship between climate and biomes. However, you will note that his axes are mean annual temperature and mean annual precipitation. As Whittaker acknowledges, it is not only the annual data that are important, but also how the temperature and precipitation are distributed over the course of a year that influence the establishment of a particular biome in a given region. For each biome, we will provide a representative graph that illustrates this variation over the course of a year by providing mean values for each month (see Dealing with data 2.2).

Tundra: frigid and dry

If you like extremes, you'll love tundra. This is the dominant biome of the northern rims of Asia, North America, and, increasingly, sections of Antarctica. Most tundra is quite flat and very dry, with extremely cold winter temperatures and cool summers. Most tundra has a permanently frozen soil layer called *permafrost*, and a thin, nutrient-poor upper soil layer that freezes each fall and thaws each spring.

Species diversity is low, but abundances of species adapted to the tundra may be extraordinarily high. Because of the permafrost, plant root systems are shallow, and cold temperatures and low levels of solar radiation create a short growing season. Thus plants tend to be small, with dwarf willows, birches, and alder the commonest trees. Ground cover is provided by lichens, *Sphagnum*, grasses, and forbs – nonwoody plants other than grasses. Most insects go through periods of *diapause* – metabolic and developmental dormancy – during the winter, and then metamorphose into huge swarms in the spring. Some bird species undergo vast migrations to reap the benefits of this entomological bounty, in some cases producing several clutches of babies over the brief breeding season. Small mammals are often abundant but may go through regular

Dealing with data 2.2 Means, sampling, and why we do statistics

The sample mean is probably the most commonly used statistic in all of science, in part because it is so easy to calculate, and in part because it is intuitively meaningful. We use the sample mean to estimate the population mean, which is often unmeasurable (usually because you can't find or measure the entire population). For example, if we want to estimate how much it rains in July in the tundra, we might find rainfall data from four different tundra locations in a database, add those values up and divide by four (our sample size), and that result would be our mean. These four different sites on the tundra are our samples, and we use these samples to make inferences about our population, which is theoretically made up of all the different locations on the tundra.

More formally:

$$\bar{X} = \frac{\sum X}{N}$$

where $\bar{X}$ = the sample mean,
$\sum X$ = the sum of all the samples, and
N = the number of values in the sample.

In this case my research found the following July rainfall values (in mm) from our four locations: 118, 67, 44, 41.

Then the mean ($\bar{X}$) = (118 + 67 + 44 + 31)/4 = 65

There are three major problems with the preceding procedure.

1. I am actually interested in a representative number for the tundra biome as a whole. The tundra is a population of many (theoretically infinitely) different sites, each of which has its own characteristic rainfall totals for July. I should definitely use a much larger sample size.
2. To estimate the population mean, I must choose my samples randomly. This means that every location on the tundra should have an equal probability of being chosen for my calculation. From a practical standpoint, this is impossible. But I could have been more random by using a very large database of tundra locations, and then randomly selecting within that database the values for my calculation. This procedure is still not completely random, because sites that were easy to measure were more likely to be in my database than were less accessible sites.
3. There is a great deal of variation among samples. The mean gives no indication of this variation. See Dealing with data 2.3 page 36 for a discussion of variation around a mean.

population cycles, while large mammals such as caribou (*Rangifer tarandus*) and musk ox (*Ovibos moschatus*) form herds that eat lichens, grasses, and new woody growth (Figure 2.11). Major mammalian predators are the wolf (*Canis lupus*) and Arctic fox (*Alopex lagopus*).

Alpine tundra occurs in high mountain ranges, just above the treeline. Climatic conditions for alpine tundra are also very cold, but moisture levels range from dry to quite wet. Alpine tundra generally lacks permafrost.

Taiga: cold, moist, and covered with forest

The taiga, or boreal forest, is a forester's dream. Spruce, fir, and pine trees dominate the landscape that occupies a wide swath just south of the tundra throughout most of North America, Europe, and Asia. The climate can be as extreme as the tundra directly to its north, with long frigid winters, and short but often warm summers. Precipitation levels are moderate, and moisture is generally abundant, because low temperatures and thick forest cover reduce evaporation rates. Soils are generally nutrient poor because of low decomposition rates in the cold climate. Though taiga plant diversity is very low, the animal communities there are somewhat more diverse than in the tundra, with many spectacular additions to the faunal landscape, including moose (*Alces alces*) and lynx (*Lynx canadensis*) (Figure 2.12).

Temperate forest: moderate climate, fertile soils, and diverse forests

The name of this biome implies moderation, and indeed temperate forests grow under relatively moderate conditions: warm summers, cool winters, and significant rainfall – generally 600 to 2000 mm per year. Soils are usually fertile, and the growing season is much longer than that of the taiga. Deciduous trees dominate most temperate forests, with the exception of southeast Australia, where evergreen eucalyptus are abundant,

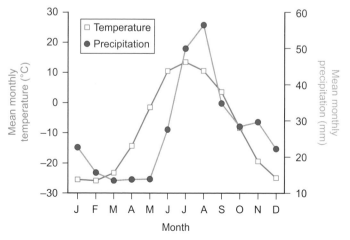

Figure 2.11 Reindeer grazing on Siberian tundra (top). Climate diagram of Markovo, Russia (bottom).

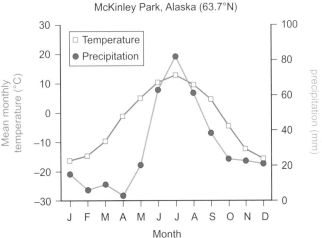

McKinley Park, Alaska (63.7°N)

Figure 2.12 Bull moose bugling his mating call in an opening in the Alaskan taiga (top). Climate diagram of McKinley Park, Alaska (bottom).

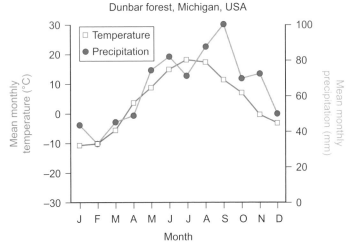

Dunbar forest, Michigan, USA

Figure 2.13 Temperate forest in Michigan, USA, highlighted by maple trees showing off their fall foliage (top). Climate diagram of Dunbar forest (bottom).

and the northwest United States, which is primarily dominated by conifers. Most temperate forests are stratified, with a canopy layer above one or several layers of understory trees, and a layer of shrubs and herbaceous plants just above the surface. With warmer temperatures and high moisture levels, there is substantial decomposition and generally fertile soils. All of this structure and relatively benign climate support a moderately diverse animal community (Figure 2.13).

Tropical forest: hot, humid, and species rich

The area between the two tropics is dominated by forests of two types: tropical rainforest (Figure 2.14A) and tropical dry forest (Figure 2.14B). Due to constant proximity to the ITCZ, the region within 10° of the equator is constantly warm and wet, supporting primarily rainforest, and stratified with luxurious deciduous evergreen trees looming over several layers of subcanopy trees, and shrubs, herbaceous plants, and ferns on the forest floor. Competition for light is intense within the rainforest; common growth forms adapted to this biome include lianas, vines that use the trees for support as they climb toward

the canopy, and epiphytes. Tree species richness is extraordinarily high – a single hectare may support over 500 species. Stratification and high plant diversity create tremendous habitat diversity for other organisms. For example, 90% of primate species are native to tropical rainforests, and estimates of insect diversity range into the tens of millions of species.

Terry Erwin (1982) made an early attempt to estimate species richness in a tropical forest. Working in Panama, he fumigated the canopy of 19 individual linden trees, *Luehea seemanni*, and collected all the dead insects that drifted down onto sheets he laid next to the trees (Figure 2.15). He found over 1200 species of beetles, and estimated that 163 species were unique to that species of tree. His research indicated that approximately two-thirds of the species live in the canopy and one-third live in the bark and roots, so he extrapolated that the linden tree probably had about 245 unique beetle species. Zoologists estimate that approximately 40% of the forest's arthropods are beetles, so Erwin estimated that the linden tree probably had about 612 unique arthropods. Finally, he knew that tropical forests have perhaps 50 000 different tree species. If *L. seemanni*

(A)

Manokwari, Indonesia (0.9°S)

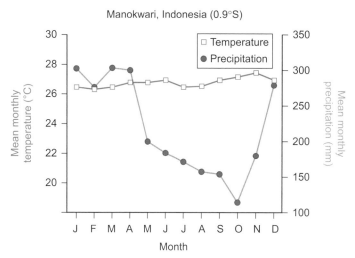

(B)

Guanacaste, Costa Rica (10.5 °N)

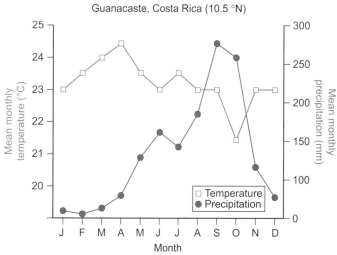

Figure 2.14 A. Understory of a tropical rainforest in Gunung Meja Nature Reserve, Irian Jaya, Indonesia (top, left). Climate diagram of Manokwari, Indonesia (bottom, left). B. Tropical dry forest during the dry season (top, right) and the rainy season (middle, right). Climate diagram of Guanacaste, Costa Rica (bottom, right).

Figure 2.15 Terry Erwin fumigates the forest canopy, and captures the dead insects on sheets placed below a linden tree.

supports average species richness, then the tropical forest should be home to over 30 million (612 × 50 000) arthropod species!

Panama, where Erwin did his research, is at about 9° N latitude. Because the ITCZ migrates seasonally, tropical regions somewhat distant from the equator tend to have substantial seasonal variation in rainfall, often with a pronounced dry season. Consequently, tropical rainforests tend to grade into tropical dry forests at about 10 to 15° latitude, north or south. Most tree species in tropical dry forests shed their leaves during the dry season as an adaptation to drought. Trees tend to be shorter, the forest is less stratified, and habitat diversity is lower. As a result, overall species richness is lower in dry forests than rainforests, but is still higher than in other terrestrial biomes. Surprisingly, soils in tropical rainforests and dry forests tend to be nutrient poor; ecologists have demonstrated that most of the available nutrients are quickly taken up by the abundant vegetation. Thus these forests are slow to recover from deforestation.

 Thinking ecologically 2.2

What assumptions does Erwin's extrapolation make? How might you test whether these assumptions are valid?

Tropical savanna: prolonged dry season, abundant grasses, and occasional trees

Tropical dry forests grade into **tropical** savannas at about 15 to 20° latitude, north or south. Temperatures are warm all year, but there is a prolonged dry season, again partly in response to the movement of the ITCZ. Because tropical savanna is drier than most tropical dry forests, periodic fire has historically prevented the intrusion of woody growth there. However, savannas often have trees that are adapted to both fire and the relatively impermeable clay subsoil that prevents water from draining during the rainy season. This lack of drainage forms temporary lakes, which help irrigate the vast herds of herbivores and the surrounding vegetation. Grasses that escape a herbivore's attention create large numbers of seeds, most of which are harvested by an abundant and diverse assemblage of ants. There are many species of insectivorous birds that live in the grasses. The large mammals of the African savanna are legendary (and discussed in Chapter 1). It is puzzling that the savannas of South America and Australia have a much lower diversity and abundance of large mammals (Figure 2.16).

Desert: hot and dry with a wide range of species diversity

Many of Earth's great deserts, are found near 30° latitude, north and south, where moisture-starved air masses complete the Hadley cell. In the most extreme cases, some deserts, such as the Atacama Desert along the West Chilean coast, receive almost no rain. Under these conditions, the primary organisms are *hypolithic cyanobacteria*–photosynthetic microorganisms able to subsist on the moisture absorbed by stones (Warren-Rhodes *et al.* 2006), reflected sunlight, and tiny quantities of nutrients. But even these organisms are very limited by moisture availability (Figure 2.17).

However, most deserts are much rainier – some average up to 300 mm of rain annually. Desert soils are generally poor because the sparse vegetation does not contribute much organic matter to the soil when it decomposes. Sparse assemblages of salt-tolerant shrubs are common in the saline soils found in some deserts. Many plants imbibe and store water during the brief rainy season, while others lose their leaves during the dry season and are dormant during periods of water shortage. Many animals avoid temperature extremes by seeking out microhabitats that are protected from the Sun. But if you really want to experience the diversity of the desert fauna, you should spend a night there during the spring or early summer. As the Sun goes down, the

Dodoma, Tanzania (6.2°S)

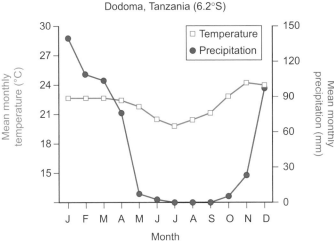

Gila Bend, Arizona, USA (33.0°N)

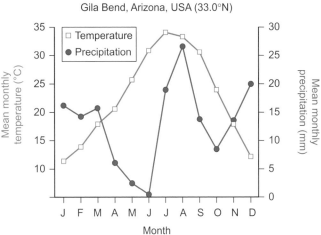

Figure 2.18 A coyote surrounded by desert vegetation in Joshua Tree National Park (top). Climate diagram of Gila Bend, Arizona (bottom).

Figure 2.16 An impala joins a group of wildebeest under the shade of a tree in a savanna in Mikumi National Park, Tanzania (top). Climate diagram of Dodoma, Tanzania (bottom).

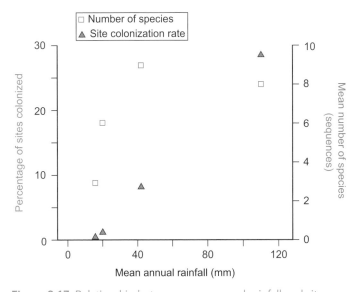

Figure 2.17 Relationship between mean annual rainfall and site colonization rate (triangles) and number of species (squares) of hypolithic cyanobacteria per site. Number of species is estimated based on the number of distinct DNA sequences (see Chapter 3).

desert transforms to a frenzied arena of animal activity. You may hear immense groups of insects singing, frogs calling out their mating choruses, or kangaroo rats thumping the ground with their huge hind feet. Reptiles and mammals that had been inactive during the day are now present in astonishingly large numbers (Figure 2.18).

Temperate grassland: cold winters, warm summers, and lots of grass

Temperate grasslands dominate the interior of Eurasia, North America, and to a lesser extent, South America. Far removed from the moderating effect of ocean proximity, this biome has cold winters and warm summers, but there is considerable variation when comparing the temperature profile of different grasslands (Dealing with data 2.3). Precipitation is moderate – generally 300 to 1000 mm annually. As the biome name implies, grasses and forbs are the dominant vegetation. Grasslands are similar to savannas in that fire helps to prevent woody vegetation from taking over. Temperate grassland is perhaps the most endangered terrestrial biome, because it has deep fertile soils, is easy to clear

Alma-Ata, Kazakhstan (43.2°N)

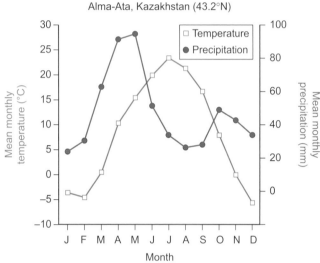

Figure 2.19 A Kazakh yurt stands in the Assy plateau grasslands in Almaty, Kazakhstan (top). Climate diagram of Alma-ata, Kazakhstan (bottom).

for agriculture, and has a relatively benign climate. Though the landscape is dotted with some grassland/prairie preserves, most people have never experienced the spring bouquet of flowers presented by an undamaged grassland. Also missing are the vast herds of large herbivores and the packs of wolves which either directly or indirectly derived their sustenance from the abundant vegetation (Figure 2.19).

Temperate shrubland: mild wet winters, hot dry summers, and occasional fires

In October, 2007, southwestern California, USA, and Baja California, Mexico, were ablaze, courtesy of very strong winds, a prolonged dry season, and various incendiary devices, including lightning, power lines knocked down by the winds, and matchsticks. All told, over 2000 km^2 of temperate shrubland burned over approximately 2 weeks. Over 1500 homes were destroyed, and many others were damaged. Though these fires were somewhat unusual in their coverage, fires are very common in temperate shrubland biomes.

Only nine people died in the fires, a relatively small number considering the devastation. People survived in part because they had been through this ordeal many times before (Figure 2.20), and many recently built residences were well adapted to fires, with well-insulated and fire-resistant windows and concrete roof tiles. The plants in this biome are fire adapted as well; they produce well-insulated seeds that germinate rapidly after fire, and shrubs that re-sprout after fire from a resilient root system.

Temperate shrublands are characteristic of the southwest sides of large landmasses at latitudes of 30 to 40°, north and

Dealing with data 2.3 Measures of variation

Each temperate grassland site will have a different climate, soil type, and chemical environment. But how different are the grassland sites from each other? Scientists use several related measures of variation to describe and analyse the amount of variation in their samples. Table D2.3.1 shows the mean monthly temperatures of five temperate grassland sites from around the world.

Whenever you see a new data set, your first step should be to look for trends. Obviously, June, July, and August are the warmest months, while December, January, and February are coldest. But which months are more variable from site to site? One approach is to consider the range – the difference between the largest and smallest value. It looks like May has the smallest range, 8.0°C, which is equal to Topeka (17.8°C) minus Ulaanbaatar (9.8°C). In this data set, the warmest months generally have a smaller range than the coldest months.

However, the range is not particularly useful, as it only considers the extreme values in a sample. Much more useful are the variance, standard deviation, and standard error. The formula for the sample variance is:

$$\text{Sample variance} = \frac{\sum (X - \overline{X})^2}{N - 1}$$

Σ indicates that we will sum up the values
X = the value of each observation,
$\overline{X}$ = the mean value for the sample,
and N = the sample size (5 in this case).

Table D2.3.1 Mean monthly temperatures (°C) from five different cities in temperate grassland biomes.

City	Country	Jan	Feb	Mar	Apr	May	Jun	Jul	Aug	Sep	Oct	Nov	Dec
Volgograd	Russia	−11.0	−8.0	−2.5	8.5	15.0	20.0	23.0	20.0	15.0	6.5	0.5	−2.5
Ulaanbaatar	Mongolia	−21.1	−17.8	−8.9	1.4	9.8	15.0	16.9	15.6	8.9	0.3	−10.3	−18.9
Debrecen	Hungary	−2.5	0.6	5.6	11.1	15.8	18.9	20.1	20.0	16.4	11.1	4.7	0
Edmonton	Canada	−14.4	−10.8	−5.6	3.6	10.3	14.2	15.8	15.0	10.0	4.4	−5.8	−12.2
Topeka	USA	−2.8	0.8	6.7	12.5	17.8	23.3	25.8	24.7	20.0	14.2	5.8	−0.3
	Mean for each month	−10.4	−7.0	−0.9	7.4	13.7	18.3	20.3	19.1	14.1	7.3	−1.0	−6.8

Data from "Weather Underground" website (various dates).

Let's use this formula to calculate the variance for May.
1. We will calculate $(X - \overline{X})$ for each observation.
2. We will square each value of $(X - \overline{X})$.
3. We will add up (sum) the squares.

$$
\begin{array}{ll}
(X - \overline{X}) & (X - \overline{X})^2 \\
15 - 13.7 = (1.3) & (1.3)^2 = 1.69 \\
9.8 - 13.7 = (-3.9) & (-3.9)^2 = 15.21 \\
15.8 - 13.7 = (2.1) & (2.1)^2 = 4.41 \\
10.3 - 13.7 = (-3.4) & (-3.4)^2 = 11.56 \\
17.8 - 13.7 = (4.1) & (4.1)^2 = 16.81 \\
& \sum (X - \overline{X})^2 = 49.68
\end{array}
$$

$$\text{The sample variance} = \frac{\sum (X - \overline{X})^2}{N - 1} = \frac{49.68}{5 - 1} = 12.42$$

There are two other commonly used measures of variation.

$$\text{The standard deviation (SD)} = \sqrt{\text{Variance}} = \sqrt{12.42} = 3.52$$

$$\text{The standard error of the mean (SE)} = \frac{SD}{\sqrt{N}} = \frac{3.52}{\sqrt{5}} = 1.57$$

To make sure you understand how this works, try calculating the three measures of variation for January. Looking at the data before you do the analysis, you should expect January to show considerably more variation than May.

south. This same biome goes by many different names in different parts of the world, such as Mediterranean shrubland, chaparral, maquis, matorral, and fynbos (Figure 2.21). The winters are wet and mild, while the summers are hot and very dry, leading to a short growing season. Consequently, production is usually low and decomposition rates are moderate in the cool winter, and moisture limited in the summer drought. Low production and low decomposition rates tend to create soils that are nutrient poor. The plants in this biome have small, thick, and waxy leaves, which help conserve moisture during the regular summer droughts. Their stomata, organs that regulate water and CO_2 balance, tend to be sunken and covered with

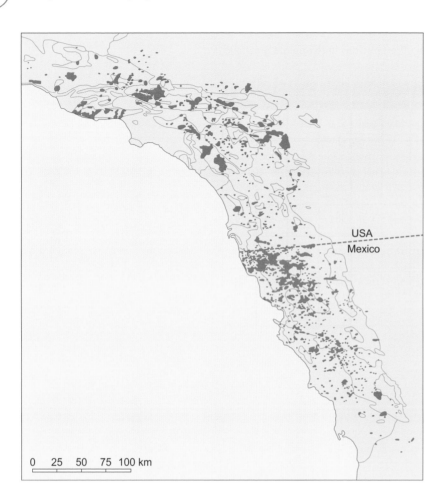

Figure 2.20 Fire boundaries (red) in 1972 to 1980 in southwestern California, USA, where fires had been suppressed historically, and Baja California, Mexico, where fires were allowed to burn historically.

Marseille, France (43.5°N)

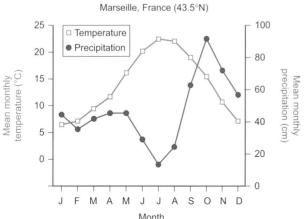

Figure 2.21 Temperate shrubland (maquis) near the southern coast of France (top). Climate diagram of Marseille, France (bottom).

trichomes or hairs that reduce water loss. Temperate shrublands host a high diversity and abundance of animals, including many mammals that browse leaves and branches from the shrubs and forbs.

Biomes, their borders, and their species composition are constantly changing. We will use an unlikely lake in the middle of the arid Sahara Desert as a portal to see how these changes occur.

2.3   HOW DO BIOMES CHANGE OVER TIME?

The Sahara is the world's largest desert, at about 9 million km². Recent evidence indicates that beginning about 15 000 years ago, the Sahara was primarily tropical savanna, laced with numerous lakes and wetlands, and even some tropical vegetation (Holmes 2008). What happened to convert the "Green Sahara" into the massive desert we know today? How can we be sure that the events I just described actually happened, and when did this conversion from savanna to desert actually take place? These are simple questions with complex answers, and we will explore a few of them here.

Figure 2.22 A. Researchers collecting core sample from the bottom of Lake Yoa. B. Part of 9 m long core sample, with 5 cm of sample in inset.

Inferring ecological history from sediment cores

Lakes have geological histories, which we can infer from several sources. Perhaps the most useful source of information is the layer of sediment that forms on the bottom of a lake, because it contains a record of conditions in the lake, and – equally important for our questions about the Sahara – of conditions in the surrounding terrestrial environment.

Paleoecologists reconstruct the history of the distribution and abundance of species. When working on lakes, researchers can

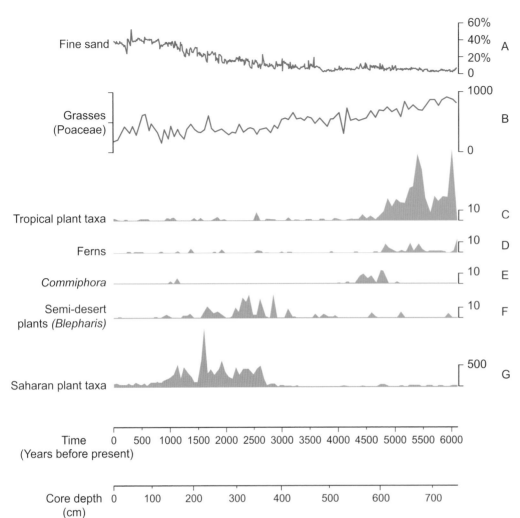

Figure 2.23 Changes to the terrestrial ecosystem surrounding Lake Yoa in the past 6000 years as inferred from sediment cores. All pollen measures are scaled as number of grains of pollen present per cm² of the core sample over a year time span. A. Steady increase in fine sand (measured as percent of the sample). B. Decrease in pollen from grasses (genus *Poa*). C. Decrease in tropical plant taxa. D. Decrease in ferns. E. Near extinction of flowering plants in genus *Commiphora* beginning 4200 years before the present (YBP). F. Increase in *Blepharis* beginning 2700 YBP. G. Increase in Sahara-type desert plants beginning 2500 YBP.

drive a core sampler into the sediment that has collected over thousands of years at the bottom of a lake, to extract a core sample that contains pollen grains and animal microfossils from different depths. The hollow tube of the core sampler is designed to collect sediment and its contents with a minimum of disturbance to the sample (Figure 2.22). From this core sample, researchers estimate the abundance of each species, and the type of soil prevalent in the surrounding terrestrial environment. They then use either relative or absolute dating techniques to estimate the age of material collected at varying levels of the core sample. Pollen grains and microfossils buried deep in the lake sediment represent more ancient species, while materials near the surface represent more recent species.

Lake Yoa in northeastern Chad has an area of about 3 km² and a maximum depth of about 25 m. Despite a tremendous evaporation rate of about 6 m per year, it has not dried up because it receives a continuous influx of water from an *aquifer* – an underground layer of water-bearing rock or soil – that was formed during the Green Sahara age. So long as the aquifer continues to supply Lake Yoa with water, it will continue to exist. But when the aquifer dries up, the lake will evaporate very quickly.

Inferring Sahara transitions from Lake Yoa sediment cores

Stephan Kröpelin and his colleagues (2008) used the information buried in the Lake Yoa sediment to infer both the history of the terrestrial environment and the history of the lake itself over the past 6000 years. Their cores showed a steady increase in fine sand particulate coupled with a fairly steady decrease in grass pollen, indicating a gradual transition from the Green Sahara to more arid conditions (Figure 2.23A and B). Tropical trees and ferns were well represented in their fossil record beginning at 6000 years ago, until dropping rather abruptly about 4500 years ago (Figure 2.23C and D). Shrubs (genus

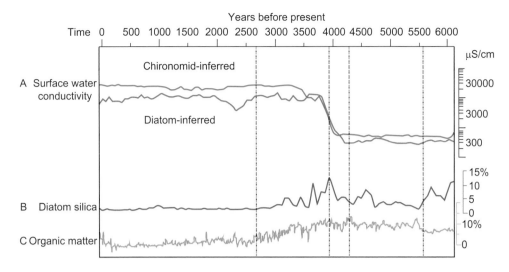

Figure 2.24 Chemical and biological changes to Lake Yoa in the past 6000 years inferred from sediment cores. A. Surface-water conductivity (µS/cm), a measure of salinity, as inferred from the record of diatoms and chironomids. B. Abundance of diatoms as measured by quantity of silica in the sample (diatom cell walls are made from silica). C. Decline in the amount of organic matter in the sample beginning about 4000 YBP.

Thinking ecologically 2.3

There were many different types of diatoms and chironomids at different depths of the Lake Yoa core. Modern African lakes vary in salinity and in the species composition of diatoms and chironomids, with lakes of similar salinity levels tending to have similar species composition. The researchers inferred salinity levels in the past 6000 years by comparing the species composition of the core with salinity levels and species composition of modern lakes. What assumptions does this inference make, and does it seem valid to you?

Commiphora) characteristic of the current Sahel region (a dry tropical savanna biome about 500 km south) increased about 4700 years ago but abruptly passed out of the record about 4200 years ago (Figure 2.23E). Lastly, semi-desert plants (genus *Blepharis*) became more common about 3000 years ago, while plants in the current desert biome became common about 2700 years ago (Figure 2.23F and G). So we see a gradual transition from tropical savanna, to shrubland, to semi-desert, and lastly to the arid desert of today.

Kröpelin and his colleagues used the same cores to reconstruct the chemical and biological history of the lake. The fossil evidence indicates that about 4000 years ago Lake Yoa transitioned from a stable freshwater lake to a salt lake over a period of about 200–300 years. Over that time span, salt-tolerant fly larvae (from the family Chironomidae) and salt-tolerant diatoms became abundant (Figure 2.24A). About 3300 years ago, as salinity levels continued to rise, even the most salt-tolerant freshwater organisms disappeared. Diatoms are scarce in the fossil record beginning 2700 years ago

(Figure 2.24B). Because diatoms are highly productive phytoplankton of lake ecosystems, a decline in diatoms usually accompanies a decline in production. In addition, the decline in overall production is implied by the decrease in organic matter (Figure 2.24C). By 2700 years ago, diatoms completely disappeared and the salt-loving chironomids dominated the fossil record.

This study highlights that both terrestrial and aquatic biomes can undergo relatively rapid transitions. Let's turn our attention to the aquatic realm, and investigate the diversity of biomes we find there.

2.4 WHAT ARE AQUATIC BIOMES?

As was the case with terrestrial biomes, the number of aquatic biomes is highly subjective (Figure 2.25). Analogous to terrestrial biomes, temperature and sunlight are critical factors influencing the development of aquatic biomes. Nutrient levels are particularly important, because they are so variable even within the same body of water. However, new variables enter into the equation, most importantly salinity, water current, and water depth.

Water's unique properties

Oceans and lakes have a much narrower temperature range over the course of a year than does a landmass at the same latitude. Water has a very high specific heat – it takes about 4.2 J of energy to heat 1 cc (= 1 ml) of water 1°C, while an

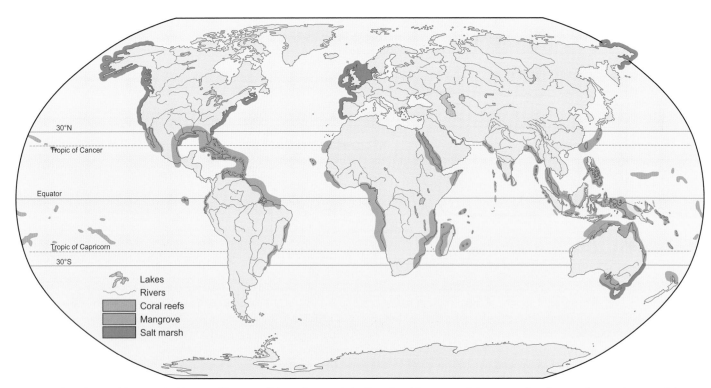

Figure 2.25 The world's aquatic biomes. The term "biome" was originally coined for terrestrial systems, but for consistency I am applying it to aquatic systems as well. Other authors use the following terms instead: realm, zone, ecosystem, and environment.

equivalent piece of land, depending on its composition, would generally require only 10–30% as much solar energy. Most large bodies of water do not heat up much above 30°C, nor cool below −2°C.

Water has another important property that influences life in some aquatic biomes. As water cools down to 4°C, its density increases. Other fluids show the same property. But below 4°C, water's density begins to decrease, and when it forms ice at 0°C, its density declines abruptly. Because of its low density, ice floats above the surface, helping to insulate organisms below from temperature extremes in polar climates. In more temperate or tropical climates, the water that is warmed by the Sun's rays remains at the surface, while cooler, denser water sinks, leading to thermal stratification – the formation of distinct layers along a temperature profile.

Lastly, light intensity and quality vary with depth. In oceans, less than 1% of the available light penetrates to 200 m under ideal conditions (with no interfering organisms or sediment). Thus almost all of the ocean's production occurs in this relatively narrow photic or epipelagic zone.

Variation in nutrient availability

Aquatic autotrophs have similar nutrient needs as terrestrial autotrophs. Nutrients are in short supply near the surface in much of the open ocean and in many deep lakes, because dead organisms tend to sink to the bottom – the benthic zone – where they decompose, and their nutrients are released. However, the problem is that the benthic zone is often very cold, while the photic zone, where production takes place, is much warmer. Cold water is denser than warm water, so there may be very little mixing between the nutrient-rich benthic zone and the potentially productive photic zone.

There are two important cases when mixing does occur and nutrients become widespread in the photic zone. The first case, turnover, occurs during fall or early winter in temperate and polar regions, when water-surface temperatures become cooler than deep-water temperatures. At that temperature, surface waters, now denser than the deeper waters, will sink, forcing the nutrient-rich deeper waters to the surface. The second case, upwelling, occurs in oceans near the western shores of landmasses, where the prevailing winds tend to move surface waters away from the shore, causing deeper, nutrient-rich waters to move in, where they form a highly productive region. Many of the world's great fisheries depend on upwelling to provide nutrients for the phytoplankton, which are the foundation of the food web.

Variation in salinity

Lakes and rivers are primarily freshwater, but as we've already discussed, some lakes develop salinity levels that are even higher than are found in the ocean. Rivers too, where they enter the ocean, may develop high salinity levels, which vary with the

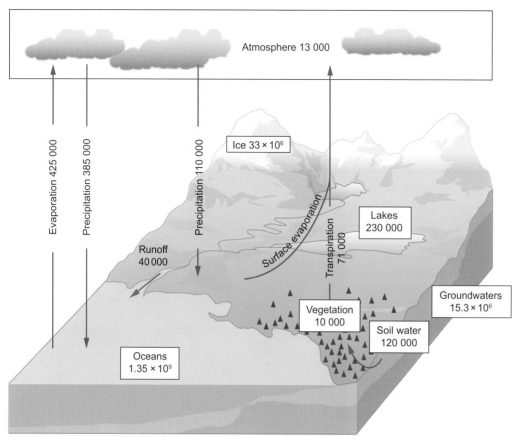

Figure 2.26 The hydrological cycle showing fluxes (km³/year) and reservoirs (km³). Atmospheric turnover time = atmospheric reservoir size divided by total flux (from precipitation) = 13 000 km³/year divided by 496 000 km³ (combined precipitation over the ocean and lands) = 0.0262 year or almost 10 days.

influence of the ocean tides. The ocean is saline because solutes are constantly washing in from streams and runoff, and of course the ocean has no outflow. In the open ocean, salinity averages about 35 g of salt per kg of water (parts per thousand or ‰). But there is some variation: areas with high precipitation (such as the tropics) have somewhat lower salinity, while areas with low precipitation and high evaporation rates (such as at 30° latitude) have somewhat higher salinity. Salinity in smaller seas covers a much wider range; the salinity level of the Dead Sea is about 330‰. Some of the nearshore aquatic biomes we will discuss are highly variable in their salinity levels, because they have considerable freshwater input.

Let's next consider how water moves through the world, and then we will turn our attention to the specific aquatic biomes.

The hydrological cycle

The vast majority of the world's water is stored in liquid form in the ocean, while the second greatest reservoir or storage is in frozen form, primarily on the island of Greenland and the continent of Antarctica. The hydrological cycle describes the basic patterns of water's journey through the atmosphere, between the atmosphere and the surface (in both directions), and between different surface reservoirs (Figure 2.26).

The primary flux or flow of water from the atmosphere to the Earth's surface is in the form of precipitation, while water returns to the atmosphere from the Earth's surface by evaporation and transpiration. We can measure turnover time – the average amount of time a water molecule remains in its reservoir – as the size of the reservoir (in km³) divided by the flux (in km³/year). Though the atmospheric reservoir seems pitifully small (13 000 km³), it has a very quick average turnover time of about 10 days, which is why the atmosphere has such an important influence on global climate (Oki and Kanae 2006).

The short turnover time of atmospheric water helps charge the world's lakes and streams. We begin our description of aquatic biomes by following the fate of freshwater as it flows over land.

Freshwater biomes

Ecologists have identified up to 14 different freshwater biomes. For simplicity, we will discuss only two of them, which are distinguished from each other based on whether they contain relatively still or flowing water.

Rivers and streams

Streams and creeks are fed by many sources – headwaters may originate from melting glaciers, as springs or seeps from underlying rock or soil, or as outflows from lakes. The rest is simple: gravity rules. Most streams get larger as they follow the path dictated by gravity, as they are joined by other streams and water inputs. There is no rule about how large a stream or creek must be before it transitions into a river. On a small scale, many streams alternate between *riffles*, which are areas of high flow rate, and *pools*, where deep water moves at a very low flow rate. Riffles are highly oxygenated and very productive, while decomposition of sediment that settles in pools makes CO_2 available for photosynthesis.

A stream's character changes over its course, as the stream typically widens while its current slows near the end of its course. Mountain streams are usually rich in oxygen and contain large numbers of leaves that feed a diverse community of invertebrates. *Shredders*, such as caddisflies, stoneflies, and amphipods, consume leaves and large particulate matter, and *collectors*, such as mayflies and water beetles, feed on the smaller organic matter. Moving downstream, the waters become more productive from nutrient inflow and higher temperatures. A diverse group of *grazers*, including many species of snails and caddisflies, feed off the algae and other abundant aquatic vegetation. Of course, invertebrate and vertebrate predators are present throughout the stream – they too are adapted to differences in flow rate, oxygen availability, and prey species (Figure 2.27).

The stream or river ends its journey at a body of relatively still water, either a lake or an ocean, where it dumps all of its contents, including its load of sediment. We will first turn our attention to lakes.

Lakes

If a stream runs into an inland basin or depression, it (and any accompanying streams) will fill the depression up to the level of its outflow – either a stream or groundwater – and form a lake. Many lakes in temperate biomes and in the taiga formed when retreating glaciers left depressions; some are still fed by glacier melt. Lakes also can be formed by biotic processes; beavers damming up rivers form small ponds, while humans damming up rivers can form much larger lakes. Lakes formed from dams can have serious impacts on ecosystems; for example, four dams along the Snake River in Idaho, USA, have destroyed about one-third of the world's salmon runs, and led to the extinction of several salmon subspecies (Duncan 2001).

Some of the world's largest and deepest lakes were formed by impressive geological processes. Crater Lake, in Oregon, USA, formed from a massive eruption of Mt. Mazama about 7700 years ago. The resulting crater filled in with water, forming a lake with a maximum depth of almost 600 m. More impressive is Lake Baikal, which fills an ancient rift valley in

Figure 2.27 Stream organisms. A. Stonefly (shredder) from the Duerna River that flows through León, Spain. B. A caddisfly collecting small prey items from its net.

southern Siberia, Russia. This rift valley formed when a tectonic plate began pulling apart; the rift is still pulling apart at the rate of 2 cm per year. At 1637 m, Baikal is the deepest freshwater lake in the world, and contains about 20% of the

Figure 2.28 A. The Holy Nose Peninsula juts into Lake Baikal. B. The Lake Baikal seal.

world's supply of liquid freshwater (Figure 2.28). Researchers estimate that over 80% of the species in Lake Baikal are endemic to the lake – found there and nowhere else. Over 90% of the zooplankton in the lake are one endemic species, *Epischura baikalensis*. On a larger scale, the endemic Baikal seal (*Phoca sibirica*) is one of only three freshwater species of seal in the world.

Lakes are classified according to their nutrient status and consequent level of production. An oligotrophic lake has low levels of nutrients and is relatively unproductive. A eutrophic lake has high levels of nutrients and is much more productive, courtesy of large populations of photosynthetic phytoplankton. Both climate and lake structure influence a lake's trophic status. Deep mountain lakes in unpolluted areas are often oligotrophic for several reasons. First, mountain environments tend to be cool, with short growing seasons, limiting production by phytoplankton. Second, mountain lakes are often situated in a relatively unpolluted environment, so fewer nutrients enter from the surrounding environment. In addition, mountain lakes are often relatively deep in relation to

their surface area, which results in the dilution of nutrients that enter the lake in the relatively large volume of water. In contrast, shallow lakes in warm environments tend to be more eutrophic, as they have longer growing seasons and a greater nutrient concentration to support a highly productive phytoplankton community.

David Schindler and Everett Fee (1974) wanted to identify which nutrient components of the sediment were most important in the process of eutrophication. In one experiment, the researchers added 0.48 g/m^2 of phosphorus, and 6.29 g/m^2 of nitrogen over a period of 5 years to an Experimental Lake 227, located in a research area in the Canadian Shield region of Ontario. Figure 2.29A shows the dramatic increase in production in response to the treatment, in comparison to untreated lakes. To separate out the effects of nitrogen and phosphorus, the researchers used a curtain to divide a second lake, Experimental Lake 226, into two sections. One part of the lake was fertilized with carbon, nitrogen, and phosphorus, while the second part was fertilized with only carbon and nitrogen. Production increased dramatically in the part that had all three nutrients added (Figure 2.29B). Based on these experiments, Schindler and Fee concluded that phosphorus is the limiting nutrient for production in these Canadian lakes and may play an important role in eutrophication of freshwater lakes. Furthermore, carbon and nitrogen do not appear to limit production in these freshwater lakes.

Some lakes, both natural and artificial, tend to become shallower as nutrient-rich sediments drain into the lake and accumulate on the lake bottom. Over time, these lakes may transition from oligotrophic to eutrophic, particularly if the sediments have high levels of phosphorus. Desert lakes, as we saw earlier, can transition from freshwater to saline (Figure 2.22). All water has at least some dissolved salts, so continuing evaporation of freshwater should increase salinity. Most lakes in other biomes have both inflows and outflows (usually streams or rivers), so the salts are carried away before they can build up to high levels. But many desert lakes lack significant outflows, so salinity increases to very high levels. Kröpelin and his colleagues (2008) suggest that Lake Yoa's outflows dried up about 4300 years ago, resulting in the freshwater to saltwater transition.

Not all lakes have obvious inflows and outflows. There is a group of about 400 identified freshwater lakes beneath the Antarctic ice sheet, that are fed by ice sheet meltwater created by ambient heat deep within Earth. Recently, Brent Christner and his colleagues (2014) were able to bore through the West Antarctic Ice Sheet to collect 30 l of water samples from Lake Whillans, which lies about 800 m below the surface. The researchers discovered a community of almost 4000 species of bacteria and archaeans, which form a closed ecosystem that operates in total darkness. Energy is produced by chemosynthesis – the conversion of energy from chemical compounds in the environment into chemical bond energy stored in carbohydrates. In Lake Whillans, microorganisms

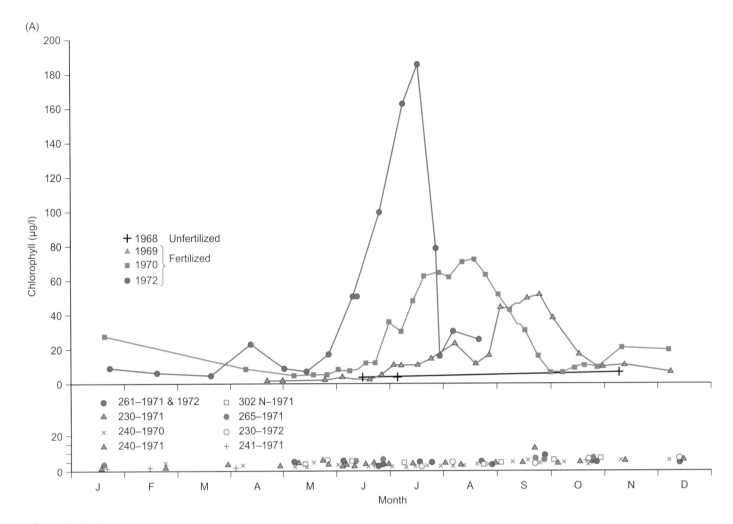

(A)

B

Figure 2.29 A. Production as measured by the chlorophyll level in Experimental Lake 227 (top graph) in comparison to numerous readings in 1970–2 from seven nearby control lakes (bottom graph). Only 3 years are included in the graph to reduce clutter, but chlorophyll levels rose sharply in all 5 years of the study. Note that chlorophyll levels never increased above 15 µg/l in any of the control lakes, nor in Experimental Lake 227 during the year before fertilization. B. Experimental Lake 226 is divided by a curtain to separate unfertilized section (foreground) from fertilized section (background).

Figure 2.30 Organisms from open ocean ecosystems. A. Bloom of *Trichodesmium* cyanobacteria. B. Chemosynthetic bacteria form mats of tangled filaments on seafloor lava. C. Ocean vent communities feature giant tubeworms (*Riftia pachyptila*) that may be up to 1 m long. D. Glass sponge (*Heterochone calyx*) from Fraser Ridge.

oxidize ammonium to generate the energy to create carbohy-drates. The researchers speculate that most of the ammonium originated from old marine sediments that accumulated in much warmer time when Antarctica was covered by a shallow sea. Given how large Antarctica is, and that 55% of its surface is estimated to harbor lakes or rivers beneath the ice shield, there is vast potential for discovering new and important ecosystems operating in this unexplored environment.

Let's turn our attention now to marine systems, which can be divided into several different biomes.

Marine biomes

The ocean contains over 97% of the world's water. We will divide the ocean into several different biomes, each of which has different physical and chemical processes and tends to have a different assemblage of characteristic organisms.

The open ocean

The open ocean is by far the world's vastest biome – vast in that it covers about 70% of the Earth's surface, and vast in its three-dimensionality. The average ocean depth is about 4000 m, while the deepest trench is over 11 000 m deep. Conventionally, the open ocean, or oceanic zone, begins at the end of the continental shelf, a shallow, highly productive region of the ocean that surrounds the continents and reaches a maximum depth of about 200 m before giving way to the open ocean.

Most of the open ocean is relatively unproductive because most of the nutrients are near the ocean floor, the primary site of decomposition. However, there are sufficient nutrients near the surface to support phytoplankton of various types. The most prevalent phytoplankton are cyanobacteria, but many other forms are also present in huge numbers (Figure 2.30A). Many different species of herbivorous zooplankton eat the phytoplank-ton, and many different species of carnivorous zooplankton eat both carnivorous and herbivorous zooplankton. An accurate description of who eats whom gets very messy very quickly, as zooplankton are eaten by tiny larval fish and massive whales. The picture is even more complex than it first appears, because bacteria are actually the most abundant organisms in the ocean, and they are eaten by a diverse array of predators of different shapes and sizes. Most ecologists have abandoned the linear concept of a food chain, replacing it with the more descriptive concept of a food web – a summary of the feeding relationships within an ecological community. (See Chapter 16 for a more thorough exploration of food webs.)

Until recently, ecologists believed that the open ocean's benthic zone was relatively unproductive, with a low abundance of deep-sea organisms being nourished by the organic matter raining down from above. All this changed when Peter Lonsdale (1977) discovered a benthic ecosystem in some of the ocean's deepest trenches, where tectonic plates are separating and molten rock heats the water to over 400°C. Hot water passing through the underlying rocks releases many nutrients, including vast quantities of hydrogen sulfide. Chemosynthetic sulfur-oxidizing bacteria are adapted to the high pressures in the ocean depths and to the glut of a dependable sulfur source, oxidizing the hydrogen sulfide and using the energy released from this process to synthesize carbohydrates. A diverse array of organisms either feed on these bacteria or have symbiotic relationships with them (Figure 2.30B and C).

The deep-sea trench, however, was not the final ocean benthic zone ecosystem to be discovered. In 1984, a seafloor mapping expedition using sonar imaging revealed large mounds on the seafloor off the coast of British Columbia, Canada, at depths ranging from 150 to 250 m. In 1987, Kim Conway and Vaughan Barrie captured these mounds on film, which revealed the mounds to be massive reefs of glass sponges . Some of these reefs are at least 6000 years old, 18 m high, and 700 km² in surface area (Figure 2.30D). Sponges use silica to construct their glass skeletons, so these reefs are only found in water with high levels of dissolved silica. We know very little about their ecology, as they have not been studied extensively. But like coral reefs, these sponges provide a foundation that attracts many other species of animals. More recently, in 2007, Paul Johnson discovered a glass sponge reef about 50 km off the Washington coast (Dybas 2008). This reef is immediately adjacent to a huge methane gas seep, and Johnson proposed that these sponges get their energy from bacteria that use methane gas as their energy source. These sponges grow very slowly, only about 1–2 cm per year. One concern is that four of seven reefs surveyed in the Georgia Basin off the British Columbia coast had been damaged by fishing trawlers (Cook *et al.* 2008).

Ecologists eagerly anticipate discovering other benthic oases of species diversity. Yet much of the ocean's species diversity occurs in shallow waters, where high production can support a complex food web. Coral reefs are the most species-rich aquatic biome.

Figure 2.31 Numerous fish live in a rich coral reef community near Sulawesi, Indonesia.

Coral reefs

As we discussed in the opening case study, most coral reefs occupy warm shallow waters, usually growing within 50 m of the surface, where temperatures range between 20° and 30°C. In addition to corals themselves, coral reefs are home to a dazzling assemblage of invertebrate and vertebrate animals, including some of the world's great fisheries. Because coral reefs are declining worldwide, many of them have been closed to commercial fishing (Figure 2.31).

Many corals reproduce sexually by broadcasting immense quantities of eggs and sperm into the ocean. Successfully fertilized eggs develop into larvae, which if they are amazingly fortunate, will settle into the sediment and develop into adult polyps. Only a tiny fraction of eggs are fertilized, and a much smaller fraction of larvae make it to the polyp stage. Polyps can reproduce asexually, then develop into the characteristic adult colony.

Like their terrestrial high-diversity counterparts – tropical rainforests – coral reefs are associated with low nutrient levels. One hypothesis for the characteristic low nutrient level is that the high production in shallow and warm waters ties up most available nutrients in living organisms. Christian Wild and his colleagues (2008) tested this hypothesis by studying a dramatic event on Australia's Great Barrier Reef by Heron Island – the annual coral mass spawning that occurs a few days after November's full moon. Most of the gametes released into the ocean never achieve fertilization, but they do fertilize the ocean by adding nutrients, including organic carbon and nitrogen compounds, into the water column, ultimately getting incorporated into other organisms or into the sediment as they pass through a complex food web. The researchers predicted that nitrogen levels would increase immediately after the mass spawning, but that high production would return nutrient levels to their low-nutrient pre-spawning conditions very quickly.

In general, the data support these predictions. Immediately after spawning, nitrogen levels in the water column shot up

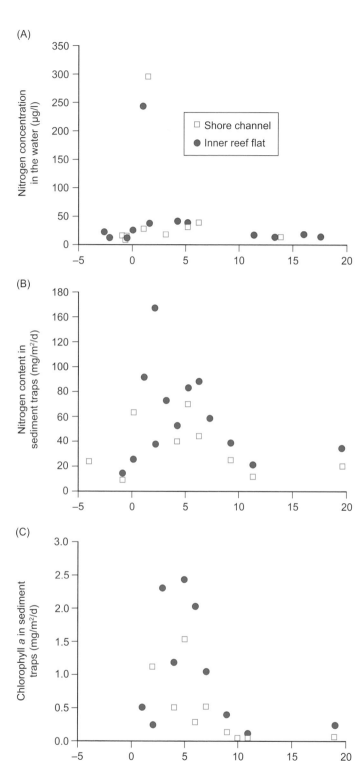

(A)

(B)

(C)

Days after first coral spawning night

Figure 2.32 Effects of coral mass spawnings on nutrient levels in and near a coral reef. A. Nitrogen levels in water in relation to time since spawning. B. Nitrogen levels in sediments in relation to time since spawning. C. Chlorophyll a levels in sediment in relation to time since spawning.

Figure 2.33 Kelp forest in Monterey Bay, California, USA.

water column to pre-spawning levels could result from three factors. First, the tide might have washed the nitrogen away. Second, the nitrogen may have sunk down into the sediment. Third, the nitrogen may have been consumed by highly productive reef organisms, particularly phytoplankton, which have the capacity to reproduce very rapidly. It is very likely that all three factors are important. The slow return of nitrogen levels to pre-spawning values in the sediment indicates that nitrogen compounds did settle in the sediment. Moreover, the increase in chlorophyll *a* levels beginning about 2 days post spawning indicates a much higher phytoplankton abundance, mostly of the dinoflagellate *Prorocentrum* (Figure 2.32C).

Despite this nitrogen limitation, coral reefs have very high species richness. However, global warming is causing some ocean temperatures to rise rather rapidly. High ocean temperatures are associated with corals ejecting their symbiotic zooxanthellae, which leads to the death of the coral and the appearance of bleaching. We will investigate this phenomenon more closely in Chapter 15.

Because most reef-forming corals do poorly in temperatures below 20°C, their distribution is primarily restricted to the tropics. As we move into cooler regions, corals are replaced by kelp forests.

Kelp forests

Kelp are large brown algae that live in shallow nutrient-rich waters in temperate latitudes, in water temperatures generally between 5° and 20°C. Kelp are anchored to the seafloor with holdfasts – structures that superficially resemble roots. Leaf-like blades extend from the stem or stipes, and, in some species, gas-filled pneumatocysts at the base of the blades provide buoyancy. Some kelp can reach heights of greater than 50 m, with growth rates of up to 0.5 m per day, forming dense forests (Figure 2.33).

Analogous to coral reefs, kelp forests have very high levels of production, and kelp are a foundation for complex communities. Many herbivores, including sea urchins and many species of

dramatically, but returned to baseline levels after only 2 days (Figure 2.32A). Nitrogen levels in the sediment increased more slowly and took about 10 days to return to pre-spawning levels (Figure 2.32B). The rapid decrease of nitrogen compounds in the

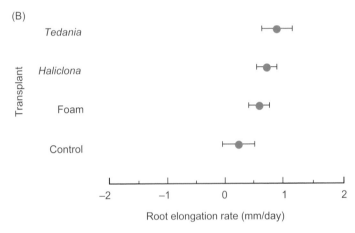

Figure 2.34 A. Red mangrove forest in south Florida, USA. B. Mean root elongation rate (mm/day) in relation to type of transplant.

herbivorous fish, graze on the kelp. Of course carnivorous fish graze on the herbivores, so depletion of carnivorous fish by fishermen can lead to overgrazing by herbivores. Because kelp grow so quickly, they can usually recover from overgrazing, and are considered less threatened than coral reefs.

Both kelp forests and coral reefs occur in shallow waters near the edge of continents or islands. Our final three aquatic biomes – mangrove forests, salt marshes, and estuaries – occupy the dynamic zone where land and sea meet.

Mangrove forests

If you're wandering along a tropical shoreline, particularly one with soft sediment and calm wave action, you may be surprised to discover yourself not on a sandy beach, but struggling through what appears to be a tropical forest. The trees in this mangrove forest are very specialized for life in challenging environmental conditions, which include low oxygen levels and limited nutrient availability. Mangroves produce aerial roots that exchange gas above the waterlogged environment (Figure 2.34A). Though relatively few plant species live in mangrove forests, they do harbor a very rich community of animals.

Aaron Ellison and his colleagues (1996) were particularly interested in the interaction between red mangrove (*Rhizophora mangle*) and two species of giant sponge (*Tedania ignis* and *Haliclona implexiformis*). The sponges attach themselves to the mangrove roots, covering an average of 30% of the root surface. For food, they filter small particles of organic matter from the ocean. Mangroves benefit from this symbiosis. In a previous study, when Ellison and Farnsworth (1990) removed sponges from mangrove roots, there was a 55% decrease in root growth as a result of the bare roots being attacked by isopods.

To more directly assess how this symbiosis actually works, Ellison and his colleagues subjected newly formed mangrove roots to one of four treatments: (1) *Tedania* transplant, (2) *Haliclona* transplant, (3) artificial sponge transplant made from polyurethane foam insulation, and (4) unmanipulated control. Roots with sponge or foam insulation all showed significantly higher elongation rates than did the unmanipulated controls (Figure 2.34B). This was not surprising, because the sponges were protecting the roots against isopod attack. But the researchers wanted to know whether there were additional benefits to the mangroves from this association.

One piece of evidence supporting the additional benefits hypothesis was that the majority of the mangroves with live sponges produced *adventitious rootlets* – fine rootlets projecting laterally from the main cable root – that permeated the body of the sponge. The researchers suspected that these adventitious rootlets functioned in nutrient exchange. To test this hypothesis, they conducted stable isotope analysis on the carbon and nitrogen present in cable roots. Stable isotope analysis (described in detail in Chapter 3) allows researchers to determine the source of elements such as carbon or nitrogen that are present in the body of an organism. This analysis revealed that sponges are transferring substantial quantities of nitrogen to the adventitious mangrove rootlets, while the rootlets are leaking substantial quantities of carbon compounds that are absorbed by the sponges.

Overall, sponges benefit from this association through increased carbon uptake, and by having a solid substrate to which they can adhere. Mangroves benefit directly from increased nitrogen uptake, and indirectly by being protected against predacious isopods. Ellison and his colleagues predict that many more mutualisms remain to be discovered in mangrove forests. We will now continue our journey along the coastline to temperate climates, where salt marshes replace mangrove forests as the dominant biome.

Salt marshes

In the introduction to their classic book, *Life and Death of the Salt Marsh*, John and Mildred Teal (1969) describe a seductive and elusive biome.

The ribbon of green marshes, part solid land, part mobile water, has a definite but elusive border, now hidden, now

Figure 2.35 Numerous shorebirds live in this salt marsh along the Texas, USA coast.

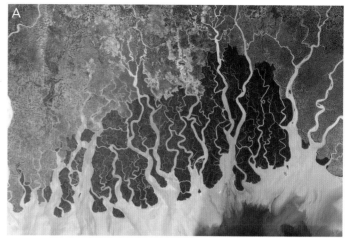

Figure 2.36 A. A portion of the Ganges river estuary and delta, as viewed from the space shuttle. B. Satellite image of northern Gulf of Mexico showing the dead zone (brown, green, and light blue water) formed from runoff, primarily from the Mississippi River delta.

exposed, as the tides of the Atlantic fluctuate ... At low tide, the wind blowing across *Spartina* grass sounds like wind on the prairie. When the tide is in, the gentle music of moving water is added to the prairie rustle. There are sounds of birds living in the marshes ...You can hear the tiny, high pitched rustling thunder of the herds of crabs moving through the grass.

This evocative description of the salt marsh and its thundering crabs captures the ever-changing spirit of this biome. Salt marsh plants must deal with the dual challenges of fluctuating salinity and water levels. While the highly productive marsh plants trap sediment and lay down a layer of decaying organic matter, the tides and wave action erode away this developing soil layer. Plant species richness is very low, dominated by only a few very abundant plants that tend to grow in somewhat distinct zones parallel to the shoreline. These highly productive marsh plants – primarily grass species – form the foundation of the salt marsh ecosystem. Grazers, such as snails and amphipods, ingest the leaves while fungi and bacteria begin the decomposition process. Much of the production is decomposed by bacteria and fungi, rather than eaten by herbivores, but there is nonetheless a significant army of herbivores that feed on the leaves, seeds, and fruits, including mammals, birds, arthropods, and crustacea (Figure 2.35).

It's important to understand that not all of the shoreline biomes occur where land meets water. Our last biome exists where water meets water.

Estuaries

Estuaries are hydrologically and chemically very dynamic. The amount of freshwater entering the ocean from a river may vary tremendously over the course of the year, and also from year to year. For example, the flow rate of the Ganges River as it enters the Bay of Bengal is usually about 10 000 times higher in August

than it is in April (Mondal and Wasimi 2006). The amount of salt in an estuary can fluctuate dramatically in relation to the tides. High tide may bring an influx of salty ocean water well upstream, which then recedes with low tide. Saline ocean water is denser than riverine freshwater, so the two flows may stratify, with ocean water moving upriver along the bottom of a river's course, and freshwater moving downstream over the top of the ocean water (Figure 2.36A).

Concurrently, rivers may carry with them very high levels of materials that they have accumulated from runoff, including sediment, nutrients, and toxic pollutants such as heavy metals. High nutrient levels lead to high phytoplankton productivity, supporting abundant populations of invertebrates and fish. However, species richness is usually very low, because most species are not adapted to the highly variable chemical environment. Primary production in estuaries is often limited by nitrogen availability, so periodic increases in nutrients can significantly increase primary production in the form of algal growth. Massive algal death leads to rapid decomposition by microorganisms, which may consume most of the available oxygen for their own cellular respiration. This process has created vast dead zones in some of the world's largest estuaries (Figure 2.36B). In dead zones, the dissolved oxygen levels drop below 2.0 ml/l for an extended period of time, causing the death of most aerobic benthic organisms.

Humans use fertilizers for agriculture, which run off into rivers, and ultimately into estuaries. As a result, the number of reported dead zones on Earth has approximately doubled every decade since the 1960s (Diaz and Rosenberg 2008). Of course, estuaries are not the only biome that has been degraded by human activities. But biomes that are both readily accessible to people and are in relatively benign climates are most threatened by human disturbance. Some biomes have become so disrupted that many of the species adapted to them have gone extinct or are severely threatened.

REVISIT: Biomes reconsidered

Some ecologists propose modifying our portrait of the world's biomes by acknowledging the profound human impact on ecosystems operating within these biomes. They ask, is it meaningful to talk about the primary productivity of grassland biomes, when almost all native grasslands have been converted to croplands, many of which are supplemented with irrigation and artificial fertilizers? Erle Ellis and Navin Ramankutty (2008) recommend incorporating this huge human impact into a new taxonomy of biomes.

In their model, Ellis and Ramankutty considered three factors as most important in defining their biomes: human population (urban and non-urban), land use (percent in pasture, crops, irrigation, rice, and urban) and land cover (percent vegetated). They then used a statistical analysis to identify the distinct groupings that emerged based on these factors. Their analysis identified six major groupings of biomes: dense settlements, villages, croplands, rangelands (for livestock), inhabited (by humans) forests, and uninhabited wildlands. The five inhabited groupings yielded 18 anthropogenic biomes – biomes whose nature and character were determined by human impact. The uninhabited wildlands were classified into three biomes: wild forests, sparse trees, and barren biomes (Figure 2.37).

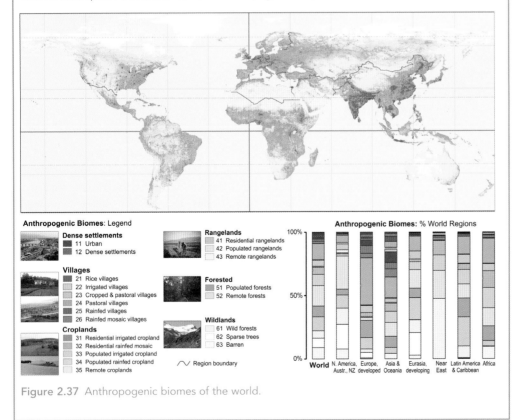

Figure 2.37 Anthropogenic biomes of the world.

Table 2.3 Mean population density, land use, land cover, and net primary production (NPP[a]) for each village biome.

Village-biome	Description of major components of biome	Population density (per km²)	Pasture	Crops	Cover (%)				NPP[a] (g/m²)
					Irrigated	Rice	Trees	Bare ground	
Rice	Paddy rice	774	1.9	71.9	40.4	62.3	6.8	2.1	550
Irrigated	Irrigated crops	500	7.0	67.6	60.1	9.1	4.4	7.6	380
Crop and pastoral	Mix of crops and pasture	300	29.8	42.3	15.9	1.0	1.2	43.0	180
Pastoral	Rangeland	256	68.8	26.4	8.3	2.1	11.7	7.7	500
Rainfed	Rainfed agriculture	243	8.1	62.7	8.4	10.3	8.4	6.6	440
Rainfed mosaic	Mix of trees and crops	230	8.3	18.9	3.6	4.3	27.8	1.1	750

[a]NPP measures the amount of new biomass produced by autotrophs (primarily plants) per year.

Each biome grouping comprises two to six biomes. As one example, villages, which are defined as densely populated agricultural populations, are broken down into six biomes. Together the village biomes have about 50% of the world's non-urban population. They are most common in Asia, where they cover more than 25% of the land mass. Table 2.3 summarizes the data used to break the village grouping down into six distinct biomes.

Are Ellis and Ramankutty intentionally making life more difficult for ecologists by imposing a new taxonomy of biomes? The two researchers argue that their taxonomy allows ecologists to make more realistic models of what occurs at the level of the ecosystem, region, and biosphere, by acknowledging the huge impact humans have on ecosystem-level processes. For example, if we want to predict an ecosystem's productivity over an extended period of time, we need to consider not only the patterns of precipitation, temperature, and natural nutrient levels, but also human influences operating over short and long timescales. Humans may increase nutrient levels by adding fertilizers, or decrease nutrient levels with land-use practices that lead to soil erosion. Similarly, humans may increase available moisture by irrigating, or decrease available moisture by cutting down vegetation that would otherwise reduce rates of evaporation and runoff. The researchers argue that, given the growing human population, anthropogenic biomes will become even more relevant in the future. For example, increasing populations may use more fossil fuels, leading to higher atmospheric CO_2 levels, and contributing to global warming. As we discussed in the opening case study, burning fossil fuels and higher atmospheric CO_2 levels also cause acidification of terrestrial and aquatic biomes, threatening their continued existence. As we discuss in Chapter 23, global climate change has already led to significant shifts in the geographical distributions of many different species.

SUMMARY

A biome is a large geographical region with characteristic groups of organisms adapted to its environment. Terrestrial biomes are influenced primarily by precipitation and temperature, but also by fire, soils, and nutrients. Aquatic biomes are also influenced by temperature and nutrients, but additionally by salinity, water depth, and current. Consequently, in order to understand why a certain biome exists in a particular region, it is essential to understand factors that influence that region's climate, and the processes that are responsible for the distribution of nutrients.

Though tropical rainforests are terrestrial and coral reefs are aquatic, both biomes share many important features. They are hubs of species richness, they have numerous similarities in their structure and functioning, and both are threatened by a variety of human activities, including acidification. All biomes are dynamic, with shifting borders and changing species composition. Several new biomes have been discovered in recent years, bringing home the message of how much we still don't know about how our planet functions. Humans have changed Earth's properties in so many ways that some researchers argue that ecologists should shift their focus to exploring how human activities influence the distribution of species.

FURTHER READING

Bednaršek, N., Feely, R. A., Reum, J. C. P., *et al.* 2014. *Limacina helicina* shell dissolution as an indicator of declining habitat suitability due to ocean acidification in the California Current Ecosystem. *Proceedings of the Royal Society B* 251 (1785) 20140123.

Challenging and informative paper on changes that are occurring to pteropods (sea snails) as a result of ocean acidification. Also provides predictions of future changes based on assumed reduction of ocean pH by 2050.

Fox, D. 2014. Antarctica's secret garden. *Nature* 512: 244–246.

This feature news summary gives a clear and fascinating behind-the-scenes account of the challenges facing the group of researchers that ultimately collected water samples from Lake Whillans beneath the Antarctic ice sheet. This could be read in conjunction with the actual research article:

Christner, B. C., and 37 others. 2014. A microbial ecosystem beneath the West Antarctic ice sheet. *Nature* 512: 310–313.

Kröpelin, S. and 14 others. 2008. Climate-driven ecosystem succession in the Sahara: the past 6000 years. *Science* 320: 765–768.

This is a great study for highlighting how ecologists can make inferences about the historical development of biomes in a particular region.

Lawler, A. 2014. In search of green Arabia. *Science* 345: 994–997.

In analogy to the Lake Yao study, Lawler describes how a large research team headed by Michael Petraglia is compiling evidence that the Arabian Peninsula has historically harbored a much more hospitable habitat, and may have served as an early migration route for humans leaving Africa, and that early humans, attracted to the benign conditions in the Arabian Peninsula, may have settled there for extended periods of time.

END-OF-CHAPTER QUESTIONS

Review questions

1. Which terrestrial and aquatic biomes have greatest species richness? How are these two biomes similar to each other, and how do they differ?
2. How does climate influence the distribution and structure of Earth's terrestrial biomes?
3. What factors are most important in determining the distribution and structure of each of Earth's aquatic biomes?

4. Describe five different ways in which nutrients influence the structure and functioning of terrestrial or aquatic biomes.

5. What factors do Ellis and Ramankutty identify as defining anthropogenic biomes? Do you believe their definition of biomes is useful? Why or why not?

Synthesis and application questions

1. The angle of the Earth's axis fluctuates between 22.5° and 24° over a regular 41 000-year cycle. How do you think this periodic fluctuation would affect global temperature patterns?

2. There is a dispute among conservation biologists regarding whether fire suppression leads to larger fires. As it turns out, fire suppression has been practiced extensively in the USA, but only to a very small extent in Baja California, Mexico. Based on Figure 2.20, can you conclude that fire suppression has led to larger fires? Why or why not?

3. Use Figure 2.26 to calculate the turnover time for ocean water. Why is this value so much larger than the value for atmospheric water?

4. Reconsider Schindler and Fee's (1974) experimental design for Experimental Lakes 226 and 227. Would the results be more convincing if the researchers had tried adding only carbon (C) to one lake, phosphorus (P) to a second lake, nitrogen (N) to a third lake, C and P to a fourth, C and N to a fifth, P and N to a sixth, and C, P, and N to a seventh lake? What would be the advantages to such a design, and in what way(s) would it be inferior to Schindler and Fee's experiments?

5. Do you think that mass spawning of coral will lead to an increase or a decrease in oxygen levels in the reef? Explain your rationale.

6. The mangrove/sponge interaction (Ellison *et al.* 1996) is an example of a mutualism. Do you think that mutualisms should be more common in stressed biomes like mangroves, or in less-stressed biomes, like coral reefs? Explain your reasoning and suggest how you might go about answering this question.

7. Based on Table 2.3, which of the village biomes seems most distinct from the other five? Which two village biomes seem most similar? Explain your answers.

Analyse the data 1

Reconsider the data in Table 2.1. This time do a linear regression analysis of the data. If you entered the numbers properly, you will get a regression equation that looks something like this: direct solar radiation = 283 048 − 2224 (latitude). If you plug in the value for the North Pole, your regression equation will estimate the direct solar radiation there. How does this estimate compare to your extrapolation estimate? Why is your extrapolation estimate likely to be more accurate than the value predicted by the linear regression equation?

Analyse the data 2

Refer back to Table D2.3.1. Graph the relationship between elevation and atmospheric pressure. What atmospheric pressure would you predict at 15 km? Did you use extrapolation or linear regression? Explain your choice.

PART II

Evolutionary and organismal ecology

Chapter 3
Evolution and adaptation

INTRODUCTION

Retreating glaciers at the end of the last ice age formed numerous freshwater lakes and streams, which were quickly colonized by the threespine stickleback (*Gasterosteus aculeatus*), a small fish migrating from the oceans. As the fish colonized freshwater habitats, new morphs or distinct forms of the stickleback evolved. One new morph ("low-plated") had many fewer dermal plates lining both sides of its body than did the marine stickleback. This same morphological change evolved independently in numerous lakes and streams. Why did this low-plated morph evolve in freshwater systems from the ancestral ("complete-plated") morph prevalent in the ocean, and how was this evolutionary event tied to ecological processes experienced by the stickleback?

As we describe in the opening case study of the stickleback fish, ecologists use diverse approaches to address evolutionary questions. While the processes of gene flow, genetic drift, mutation, and natural selection work together to bring about evolutionary change, only natural selection can lead to adaptation by selecting for traits that benefit individuals within a population. Studies of rapid evolution in species as different as sticklebacks, disease-causing bacteria, and Darwin's finches, illustrate how traits beneficial in one environment may be detrimental in a second environment.

KEY QUESTIONS

3.1. How do ecologists use genetic and molecular approaches to study evolution?

3.2. What four processes interact to bring about evolutionary change?

3.3. Does a changing environment influence the costs and benefits of adaptation?

3.4. How do natural selection and sexual selection influence an individual's fitness?

3.5. How might natural selection cause new species to evolve?

3.6. How do evolutionary ecologists unravel evolutionary relationships?

CASE STUDY: Stickleback plate evolution

We begin this chapter by considering how ecologists answer questions about evolution as they apply to sticklebacks in a novel environment. Many research teams in the scientific community are exploring the question of how and why the low-plated morph has evolved time and time again after a population of complete-plated sticklebacks colonized freshwater systems (Figure 3.1). But there are several distinct components of this problem; consequently researchers can be asking any of the following questions:

1. What genes code for the production of dermal plates, and how are they different in the two morphs?
2. What proteins help construct the dermal plates during development, and how are they different in the two morphs?
3. How do proteins interact with other factors in the stickleback's environment to construct the dermal plates? How does this process differ in the two morphs?

Or they could be asking:

4. What events occurred over the course of evolutionary history that led to the presence of two morphs in this species?
5. Is there something about the freshwater environment that favors the survival and/or reproductive success of the low-plated morph? If so, what is it?

Note that the first three questions address *how* the two morphs construct their plates. These questions look for genetic and physiological factors, operating over the course of an individual stickleback's lifetime, that led to the formation of two different morphs. The last two questions address *why* the low-plated morphs have evolved, and investigate events and ecological processes that occur over generations. As we will illustrate, investigating both "how" and "why" questions provides a more complete understanding of evolution. Later in this chapter we will explore why the low-plated morph has evolved repeatedly in threespine stickleback populations that migrated to freshwater systems.

For now we will turn to the question of how differences in genetic makeup can influence developmental differences among individuals in a population. Researchers who study the evolution of developmental changes use a combination of classical genetic crosses and modern

Figure 3.1 A. Complete-plated stickleback and B. low-plated stickleback. Specimens are stained to highlight the bone.

molecular biological techniques to address questions like how the low-plated morph evolved from the ancestral complete-plated morph.

3.1 HOW DO ECOLOGISTS USE GENETIC AND MOLECULAR APPROACHES TO STUDY EVOLUTION?

Pamela Colosimo and her colleagues (2005) have conducted research addressing the first four questions listed above. First, they needed to know how the genes influencing the expression of dermal plates were transmitted from generation to generation. When they crossed complete-plated sticklebacks from the ocean off the coast of Japan with low-plated sticklebacks from Paxton Lake in British Columbia, all of the offspring were complete-plated. One simple hypothesis for this finding is that marine complete-plated individuals are homozygous for a dominant allele (genotype AA), and freshwater low-plated individuals are homozygous for a recessive allele (genotype aa). Recall that alleles are variant forms of a gene, so according to this hypothesis there are two alleles within the stickleback population, with the A allele coding for complete plates, and the a allele coding for low plates. Presumably the cross between AA and aa parents generated first-generation offspring that were all heterozygous (genotype Aa), and expressed the dominant (complete-plated) morph (Figure 3.2).

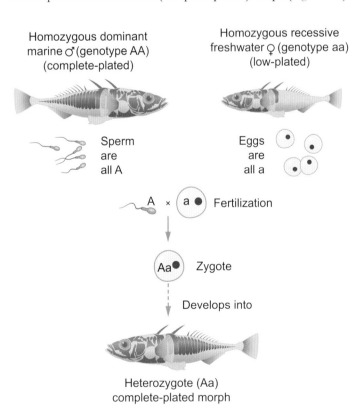

Figure 3.2 Cross of marine fish (genotype AA) with freshwater fish (genotype aa) should yield heterozygous offspring (Aa), all of which are expected to be complete-plated.

These heterozygous offspring were raised to adulthood, and crossed with each other. If our simple hypothesis for the transmission of dermal plates is correct, we would predict that 75% of these second-generation offspring would be complete-plated and 25% of the offspring would be low-plated (Table 3.1).

Table 3.1 Genotype frequencies of offspring produced when two heterozygotes are crossed. Fifty percent (0.50) of the eggs produced by a heterozygous (Aa) female carry the A allele, while 50% carry the a allele. Fifty percent of the sperm produced by a heterozygous (Aa) male carry the A allele, while 50% carry the a allele. If two heterozygotes are crossed with each other, 25% of the offspring are expected to be AA, 50% are expected to be Aa, and 25% are expected to be aa.

Sperm \ Eggs	A = 0.50	a = 0.50
A = 0.50	AA = 0.25 (25%)	aA = 0.25 (25%)
a = 0.50	Aa = 0.25 (25%)	aa = 0.25 (25%)

Table 3.2 Expected phenotype frequencies from crosses of two heterozygotes (Aa) based on the simple model of plate expression influenced by one gene, with complete plates dominant over low plates.

| Genotype | Expected phenotype | |
	Low	Complete
AA	0	25%
Aa	0	50%
aa	25%	0

Because the complete-plated allele is dominant, all of the individuals with at least one A allele are predicted to be complete-plated, as shown in Table 3.2.

The actual results of the experimental cross between two heterozygotes show that the prediction of our simple hypothesis is false and require that we modify our original hypothesis (Table 3.3). Colosimo and her colleagues discovered a third morph in almost 20% of the offspring – the partial-plated morph – which has fewer plates than the complete-plated morph, but more plates than the low-plated morph.

When a population has two or more forms of a particular trait, it is said to be **polymorphic** (poly = many, morph = form) for that trait. This laboratory population had three morphs of the dermal plate trait. The researchers concluded that the pattern of inheritance was more complex than predicted by our simple one-gene, two-allele hypothesis.

Table 3.3 Observed number of offspring of each morph from the cross of two heterozygotes. Frequencies are in parentheses.

Genotype	Observed phenotype		
	Low	Partial	Complete
AA	0	3 (0.8)	79 (22.6)
Aa	6 (1.7)	66 (18.9)	104 (29.7)
aa	90 (25.7)	2 (0.6)	0

Dealing with data 3.1 Biological databases and the molecular biology community

Nucleic acids and proteins are very long polymers – molecules composed of repeating subunits. The sequence of nitrogenous bases identifies the DNA and RNA, while the sequence of amino acids identifies a protein. These sequences are extraordinarily long; for example, the human genome is approximately 3 billion base pairs long (some salamander genomes may be approximately 30 billion base pairs long!). Base pair information is essential for biomedical researchers, as well as researchers in ecology and evolutionary ecology. Several organizations have formed to help deal with this complexity. These groups share all of the sequence information, and make it readily available to researchers around the world, including students such as you. For more information, you can check out the following websites:

PubMed: http://www.ncbi.nlm.nih.gov/entrez/query.fcgi?
   DB=pubmed
GenomeNet: http://www.genome.ad.jp/
ENA: http://www.ebi.ac.uk/embl/

Research by Colosimo and colleagues (2004) indicates that at least three other genes influence the expression of plate number in these fish. But these genes (called modifier genes) primarily influence plate number in heterozygous (Aa) fish. Heterozygotes that had primarily freshwater modifier alleles tended to be partial-plated, while heterozygotes with primarily marine modifier alleles tended to be complete-plated.

Colosimo and her colleagues then used molecular techniques to isolate a region of chromosome 4 that contained the major gene influencing the number of plates. Using computer databases of DNA sequences from many study organisms, they searched and compared the region they isolated on chromosome 4 to known sequences on these other organisms (see Dealing with data 3.1). When the researchers entered their base sequence information, they discovered a close match of a portion of their stickleback DNA sequence to the DNA that codes for ectodysplasin. In mammals, this protein is required for the proper development of ectodermal tissue such as teeth, hair, and some bones. In another species of fish, *Oryzias latipes* (the medaka), a mutation in this gene causes loss of scales. Given the similarity in the base sequence, and its homologous function in other species, the *Eda* gene – the gene that codes for the ectodysplasin protein – is a convincing candidate for influencing the expression of plate formation in sticklebacks. One important lesson in this study is that the same gene may have a similar function in even distantly related species.

Researchers now agree that *Eda* is the plate gene in fish, with Eda^C/Eda^C coding for complete-plated fish, Eda^L/Eda^L coding for low-plated fish, and Eda^C/Eda^L coding for either partial or complete-plated fish, depending on the action of modifier alleles. They now have a good understanding of the linkage between a stickleback's genetic makeup and how these plates develop. Having observed the same process, the loss of dermal plates, occurring repeatedly as marine sticklebacks invade freshwater lakes, evolutionary ecologists reason that something about the freshwater environment must be favoring a rapid increase in the Eda^L allele in stickleback populations that move into lakes. This system provides an unusual opportunity to examine the process of evolution from the perspective of genetics, development, and ecology. Being able to observe evolution in real time allows researchers to understand evolutionary processes more clearly.

3.2 WHAT FOUR PROCESSES INTERACT TO BRING ABOUT EVOLUTIONARY CHANGE?

Two important components of evolutionary theory are (1) that species change over time, and (2) that species are related to other species. Darwin did not know about genes, thus his vision of species change and species relatedness was different from modern evolutionary biology's genetic perspective. We define genetic evolution as a change in the frequency of alleles of a gene over time. There are four processes that can cause allele frequencies to change over time: gene flow, genetic drift, mutation, and natural selection.

Gene flow

Migration of individuals out of their population (emigration) and into a new population (immigration) causes gene flow, the successful movement of alleles from one population to another. If there were substantial gene flow between two populations, we would expect the allele frequencies of the two populations to be similar. For example, a high level of stickleback migration from marine to freshwater populations in Paxton Lake in British Columbia, Canada, would increase the frequency of the Eda^C allele in Paxton Lake, perhaps approaching the frequency of Eda^C found in marine systems. We don't see this, thus we can infer that gene flow is limited between Paxton Lake and the nearby marine systems.

Gene flow is particularly important for understanding evolution, because it can bring new genes or new combinations of genes into a population. Genetic drift has the opposite effect, reducing the genetic variation within a population.

Genetic drift

Genetic drift is a change in allele frequency within a population from chance events. Genetic drift is a much more important mechanism of evolutionary change in small populations than it is in large populations. We will consider one hypothetical example. Let's imagine that a retreating glacier forms a new lake off the Canadian coast, which is colonized by 10 marine stickle-backs – nine homozygous for the Eda^C allele (genotype Eda^C/Eda^C), and one heterozygous (genotype Eda^C/Eda^L). The frequency of the Eda^C allele is 0.95 (19 of 20 alleles in the population are Eda^C), while the frequency of the Eda^L allele is 0.05. This would not be a surprising distribution, as Eda^C is the most common allele in marine populations. What will be the frequency of the Eda^C allele in the next generation?

There is no way to answer this question with any degree of certainty, because chance will play an important role in such a small population. For example, the heterozygous individual might die before it has the opportunity to breed, eliminating the Eda^L allele from the population by a chance event. Alternatively, the heterozygote may, by chance, be an unusually successful breeder, and the Eda^L allele frequency may increase substantially in the next generation. Over time, random fluctuations in allele frequency from genetic drift will tend to reduce genetic variation in small populations, because some alleles will, by chance fluctuation, go extinct. Our discussion of conservation biology in Chapter 11 will consider this problem more thoroughly.

Loss of genetic variation from genetic drift may be countered by the introduction of genetic variation from mutations.

Mutation

Occasionally, a mutation – a random change in the base sequence of the genetic material – will arise within an organism. In the case of asexual single-celled organisms, such as bacteria and many protists, mutations are transmitted to offspring when the organism undergoes binary fission to produce two daughter cells. In multicellular organisms, mutations are only transmitted to offspring if they occur in the cells that ultimately produce offspring. Because they are relatively rare events, mutations have a negligible direct effect on allele frequencies. However, mutations are essential to evolution because they introduce new combinations of genetic material into the population.

Let's consider the three processes of evolution we've discussed thus far from the perspective of genetic variation within a population. Gene flow may reduce genetic variation if individuals are emigrating from a population, but can increase genetic variation if individuals with novel alleles are immigrating into a population. Genetic drift will reduce genetic variation, because some alleles will, by chance, drift to extinction over generations. Mutations, in contrast, will almost always increase genetic variation within a population. Because individual survival and reproductive success can vary with differences in genetic makeup, a population with genetic variation has the potential to experience evolutionary change through natural selection.

Natural selection and adaptation

HERE IS A TRUE STORY In 2007, life was looking pretty good for Andrew Speaker, a personal-injury lawyer from Atlanta, Georgia, USA. He had a job, and was in love with Sarah. Unfortunately he took a fall, bruised a rib, and his doctor took a chest X-ray to assess the damage. Seeing some lesions in Speaker's lungs, the doctor used a bronchoscope to get a better look, confirmed a diagnosis of tuberculosis, and began Speaker on an antibiotic regimen that included rifampicin and isoniazid (Berlin 2008). Shortly thereafter, Speaker, whose rib had healed nicely, was off to Greece to wed Sarah, and then to Rome for his honeymoon. While in Europe, tests revealed that Speaker was carrying a rare form of *extensively drug-resistant tuberculosis*, which resists most antibiotics typically used to treat the disease. Like all forms of tuberculosis, extensively drug-resistant tuberculosis is transmitted through the air, and thus is highly contagious. The Centers for Disease Control became very concerned at this point, and claims to have instructed Speaker not to travel – a claim that Speaker denies. The story gets muddled here, but Speaker then flew on a commercial jet from Rome to Prague, Czech Republic, then to Montreal, Canada, and lastly rented a car to cross back into the USA, slipping past a US Customs official. In the process of his travels, Speaker came into contact with numerous people who could have contracted this disease from him. He was sued by several airplane passengers, and even by a relative of one passenger who claimed to be traumatized by the event. Later he apologized to all passengers, explaining that he did not intend to endanger them. Later still, he underwent surgery that removed a small piece of infected lung, and more recently he sued the Centers for Disease Control for publicizing his case.

So what is all the fuss about tuberculosis? And how does this case tie in to our discussion of evolution?

Some of you reading this chapter are infected with *Mycobacterium tuberculosis* – the bacterium that causes tuberculosis. The World Health Organization (2012) estimates a global infection rate of one in three people. Approximately 10% of people carrying the bacterium will actually become symptomatic over the course of their lifetimes, presenting symptoms that include fever, a bad cough (sometimes with blood), night sweats, and weight loss over the course of a few weeks. Tuberculosis is most prevalent in Africa and Asia (Figure 3.3), and is most likely to be expressed in people with immune systems that are already weakened by diseases such as HIV infection, and by poor nutrition. Over 1.5 million people die from tuberculosis each year. Thankfully, the death rate is dropping as improved treatments become more available in developing countries. However,

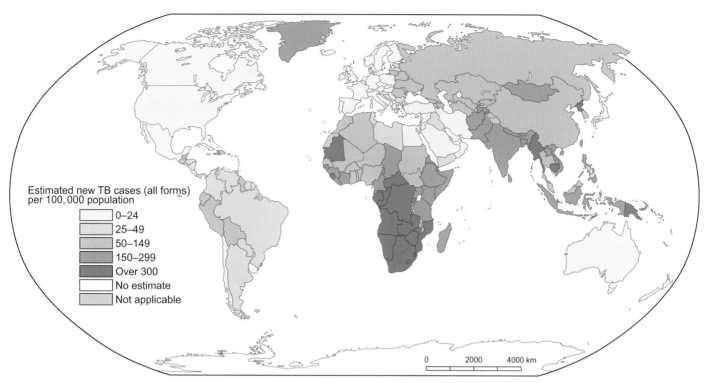

Figure 3.3 New cases of tuberculosis in 2011 as estimated by the World Health Organization.

antibiotic treatments must be administered with great care, particularly as antibiotic resistance can rapidly evolve within bacterial populations, courtesy of two evolutionary processes, mutation and natural selection.

There is one prerequisite for natural selection to occur within a population, and two sequential logical outcomes.

Prerequisite for and outcomes of natural selection

For natural selection to occur, there must be heritable variation within a population in a trait that influences an individual's ability to survive or reproduce in a given environment. The heritability of a trait is the proportion of the variation in that trait that is due to genetic variation. When we say that a trait has high heritability, we are simply saying that the variation we see among individuals with that trait is primarily a result of genetic differences among individuals, rather than of environmental differences experienced by those individuals.

Once the prerequisite is met two outcomes are likely.

1. Individuals with a trait that improves their ability to survive or reproduce in their environment will leave more offspring than individuals lacking this trait. A trait that increases an individual's reproductive success in a particular environment in comparison to individuals without that trait is an adaptation.

2. Because variation in these beneficial traits is heritable, the individual's offspring and their offspring's offspring will tend to have these adaptations. Consequently, alleles that code for

adaptive traits will tend to increase in frequency over generations of natural selection.

We will discuss two cases of rapid evolution, which we can see happening in real time. We will then discuss how ecological factors influence the process of natural selection.

CASE 1. TUBERCULOSIS AND THE EVOLUTION OF ANTIBIOTIC-RESISTANT BACTERIA. Let's return to our discussion of tuberculosis, and explore the evolution of antibiotic resistance in a population of *M. tuberculosis*. The prerequisite for natural selection is satisfied, because the population of bacteria has heritable variation in how well individual bacteria survive in an environment with antibiotics. Most infected people have lungs that are colonized primarily by *M. tuberculosis* that are sensitive to antibiotics. But, in a large bacterial population, there may be a few bacteria that are somewhat resistant, and others that are highly resistant to specific antibiotics. Rifampicin, one of the two most important antibiotics used to treat tuberculosis, diffuses through the *M. tuberculosis* cell membrane, binds to RNA polymerase molecules, and effectively blocks transcription of RNA, killing the bacterium. Most of the genes for rifampicin resistance are on the *rpoB* gene, which codes for the β-subunit of RNA polymerase. These mutations operate by changing the shape of the β-subunit of the RNA polymerase molecule so that rifampicin can no longer bind to it, and thus it no longer inhibits transcription of bacterial RNA (Da Silva and Palamino 2011).

Bacteria with these mutant alleles for rifampicin resistance produce offspring who also have these alleles for resistance,

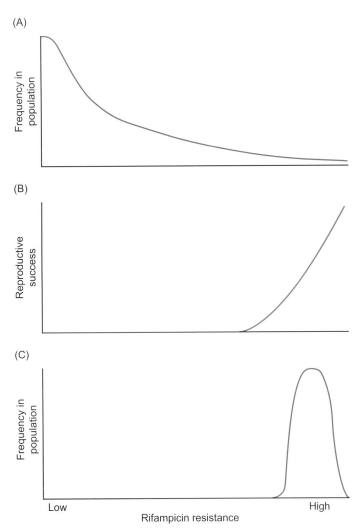

(A)

Frequency in population

(B)

Reproductive success

(C)

Frequency in population

Low   High

Rifampicin resistance

Figure 3.4 A. Starting population with most of the *M. tuberculosis* bacteria showing low levels of resistance to rifampicin. B. After rifampicin is added to the patient's system, resistant *M. tuberculosis* bacteria have much higher survival and reproductive success. C. In subsequent generations, the frequency of resistant *M. tuberculosis* bacteria has increased by natural selection.

so there is heritable variation in a trait essential for survival and reproductive success within the environment. If the starting population contained primarily rifampicin-sensitive bacteria, but also a few somewhat resistant bacteria and a few highly resistant bacteria, the stage would be set for natural selection to increase the frequency of an allele that codes for resistance.

Once rifampicin is used for treatment, the rifampicin-resistant bacteria are at a selective advantage, because they have the adaptation that prevents the rifampicin from binding their RNA polymerase. A much higher proportion of rifampicin-resistant bacteria survive, while the majority of rifampicin-sensitive bacteria die. Because the rifampicin-resistant trait is heritable, it is passed on to the offspring of the resistant parent. Consequently the allele coding for rifampicin resistance increases in frequency within the population (Figure 3.4).

CASE 2. THE RAPID EVOLUTION OF BEAK MORPHOLOGY IN DARWIN'S FINCHES. When Charles Darwin made his 5-year journey around the world, he spent around 5 weeks on various islands in the Galápagos archipelago, located about 1000 km off the west coast of South America. He noticed a group of finches (now called Darwin's finches), which ultimately were shown to comprise 14 different species. This group has been the focus of intense research over the past 70 years, with the most detailed data collection courtesy of Peter and Rosemary Grant and their many graduate students and postdoctoral associates. For now, we will confine our discussion to the tiny island of Daphne Major and to one species – the medium ground finch, *Geospiza fortis*.

As evolutionary ecologists, the Grants knew it was critical for them to investigate whether there was heritable variation within the population in traits that could influence a finch's ability to survive or reproduce on Daphne Major. The Grants (2003) measured all of the characters they believed might be important, and found that the depth of a finch's beak is highly variable, approximately fitting a normal distribution (Figure 3.5A and B). This variation could be important to ground finches, which get their food from cracking and eating seeds. The Grants reasoned that deep beaks provide more power for cracking large, hard seeds, while shallower beaks provide more dexterity for processing smaller, softer seeds.

The Grants hypothesized that this variation in beak depth was heritable. If their hypothesis was correct, they predicted that measurements of the beak depth of parents and their offspring should be positively correlated. The Grants did in fact observe this strong positive correlation (Figure 3.5C), supporting the hypothesis that variation in beak depth is heritable.

On Daphne Major, environmental change can be very abrupt, primarily due to variation in the ENSO (Chapter 1). There was a terrible drought on Daphne Major in 1977, which killed over 80% of the finches on the island. During the course of the drought, seed quantity and quality declined dramatically, so that by the end of 1977 the primary remaining seeds were from the caltrop plant (*Tribulus cistoides*). These seeds are nutritious, but they are nestled inside very large, hard, and thorny fruits, requiring so much energy to extract that finches usually rejected them in most years (Figure 3.6A). But as the year (and the drought) progressed, most of the more accessible seeds were eaten, so that most of the remaining seeds were caltrop.

It seems likely that finches with deeper beaks would be better able to open the caltrop's fruit and access the seeds than finches with shallower beaks. If so, we would predict that birds with deeper beaks would have higher survival rates than birds with shallower beaks. Figure 3.6B shows that mean beak depth was about 6% greater among survivors of the drought in comparison to the pre-drought population. Thus deep beaks were adaptive during this drought.

(A)

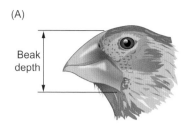

Beak depth

Figure 3.5 A. Measuring a finch's beak depth. B. Variation in finch beak depth in 1976. Arrow indicates mean beak depth. The curve fits a normal distribution, but the variation in beak depth is greater than that found in most other species of birds. C. Mean offspring beak depth in relation to the mean beak depth of the parents in 1978.

(B)

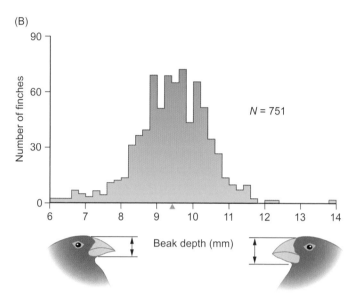

$N = 751$

(C)

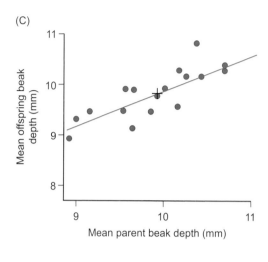

(B)

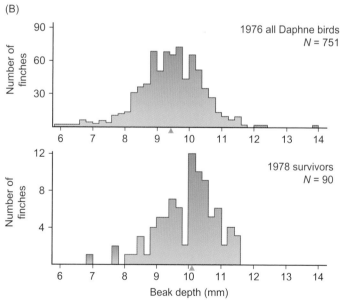

Figure 3.6 A. Caltrop (*Tribulus cistoides*). B. Frequency distribution of *G. fortis* beak depths in 1976 before the drought, and in drought survivors in 1978 after the drought. (Arrow indicates mean beak depth.)

Because of the drought, finches produced no offspring in 1977, but they returned to reproductive activity in 1978. A comparison of juvenile beak depth in 1976 (the year before the drought) with juvenile beak depth in 1978 shows that mean beak depth increased after the drought (Figure 3.7). Thus the adaptation of deep beaks increased in frequency over a relatively short time span as a result of natural selection.

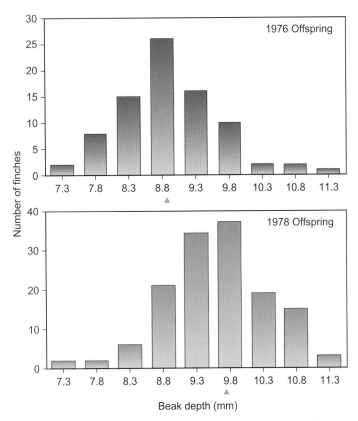

Figure 3.7 Frequency distribution of beak depth in 8-week-old juveniles in 1976, the year before, and 1978, the year after the drought. A finch's beak stops growing at 8 weeks. (Arrow indicates mean beak depth.)

 Thinking ecologically 3.1

How does the mean beak depth of juveniles from 1978 in Figure 3.7 (bottom) compare to the mean beak depth of drought survivors (Figure 3.6B)? Given that drought survivors were the parents of the juveniles in 1978, would you expect the mean beak depth to be identical? Why or why not?

These two cases highlight several important features of how natural selection brings about evolutionary change within populations.

Directional selection

Both cases show **directional selection**, in which natural selection favors the evolution of one extreme phenotype, such that the mean phenotype shifts in one direction along a range of phenotypes. In the case of *M. tuberculosis*, the bacteria most resistant to rifampicin had highest survival and reproductive success. In the case of Darwin's finches, the birds with deepest beaks had highest survival rates under drought conditions, and thus were able to produce offspring when better environmental conditions returned. Directional selection is the most common form of natural selection leading to adaptive evolutionary change.

EVOLUTION WITHIN POPULATIONS. In both of these examples of rapid evolution, antibiotic resistance in bacteria and beak depth in ground finches, there was no genetic change within individuals. *M. tuberculosis* individuals did not become more resistant to rifampicin when put into a high-rifampicin environment. Similarly, finches did not grow deeper beaks so they could crack larger seeds in a drought environment. Rather, individuals either survived to a developmental stage at which they could reproduce, or else they died before they could reproduce. In both cases, natural selection brought about genetic changes in populations by influencing the survival, and ultimately the reproductive success, of individuals.

DIRECTIONAL SELECTION AND VARIATION. There are two important responses of a trait to directional selection: a shift in the mean value of the trait within the population, and a reduction in the total amount of variation within the population. Both of these responses will be greater for traits with high heritability.

The mean value for a trait will shift more (and more quickly) in response to selection for traits with high (compared to low) heritability. If we consider a mixed population of *M. tuberculosis* bacteria, with both rifampicin-sensitive and rifampicin-resistant individuals present, we know that with the exception of rare mutations, the offspring of rifampicin-sensitive individuals will be rifampicin sensitive, and the offspring of rifampicin-resistant individuals will be rifampicin resistant. Thus rifampicin resistance is highly heritable. When placed in an environment in which only the most resistant bacteria can survive, there will be a rapid shift in resistance in future generations (see Figure 3.4C). In contrast, we can readily imagine that beak size is influenced by genetic and environmental factors. For example, well-fed finch chicks may develop deeper beaks than poorly fed chicks. Thus we might expect to observe a weaker response to selection in the finches, with a somewhat slower and smaller shift in mean beak depth in drought years, than would be expected if variation in beak size depth resulted only from variation in genes. Research by Lukas Keller and his colleagues (2001) indicates that about 65% of the variation in beak depth is heritable.

Directional selection will tend to reduce genetic variation within a population, particularly in traits with high heritability. Again consider a mixed population of *M. tuberculosis* bacteria, with rifampicin-sensitive, somewhat resistant, and highly resistant individuals present (as in Figure 3.4A). Once the human host of these bacteria started a course of rifampicin, two changes would occur. First, there would be a shift in the mean rifampicin resistance in the population of *M. tuberculosis* bacteria over a very brief time span, as described above. Additionally, almost all of the rifampicin-sensitive and somewhat resistant individuals would be eliminated from the population. The existing variation in resistance will be greatly reduced (see Figure 3.4C). Thus while directional selection shifts the mean value of a trait (in this case rifampicin resistance), it also reduces the variation existing within a population. As we will discuss in Chapter 11, this reduction in variation can have important consequences for conservation ecology.

If the genetic variation in a population is reduced by the action of directional selection, we should question why so much genetic variation exists within natural populations. There are several answers to this question. Recall that earlier in this chapter, we discussed that both mutation and gene flow can introduce new alleles into a population, which can restore some of the genetic variation. But a second answer to this question is that an organism's environment can change, and the direction and intensity of selection may fluctuate in space and time.

Selection and a fluctuating environment

While natural selection can favor certain adaptations in a given environment, the environment itself is highly dynamic. *M. tuberculosis* in a person's lungs may initially experience a relatively benign environment, which may become much harsher if the patient begins a regimen of four different types of antibiotics. The more natural selective environment on Daphne Major also fluctuates drastically in relation to global and local weather patterns. Different weather patterns change food availability in ways that can make deep beaks advantageous one year and disadvantageous the next year.

For example, rainfall on the Galápagos in 1983 was in dramatic contrast to the drought of 1977. Rain began falling much earlier in the year than normal and tremendous thunderstorms regularly lashed the islands, transforming the desert island of Daphne Major into a lushly vegetated landscape. Birds bred early and often; many records of finches' reproductive success were established in that year. The culprit underlying this phenomenon was the ENSO, which can influence rainfall and temperatures across tropical and temperate biomes.

The rains of 1983 caused an increase in many types of plants that had not been observed on Daphne Major for years immediately prior to that time. Most of these plants had small, soft seeds that could be cracked by a ground finch of any size. Many of the larger cacti – producers of large seeds – were swept away by the floods. Thus the selection environment that had favored deep-beaked finches in 1977 shifted. The survival advantage no longer went to the large, deep-beaked finches (Figure 3.8).

On the surface, you might think that mean beak depth should have leveled off in the 1980s as a result of the returning prevalence of small seeds. In reality, mean beak depth and overall beak size actually decreased during this time of plenty. This reinforces an important principle: a trait that is advantageous in one environment may actually be disadvantageous in a different environment.

3.3 DOES A CHANGING ENVIRONMENT INFLUENCE THE COSTS AND BENEFITS OF ADAPTATION?

Adaptations cannot be optimal solutions to all selective environments. Rather, they are tradeoffs, or compromises between different

Figure 3.8 Daphne Major during a dry season (top) and during an El Niño rainy season (bottom).

functions that may be more or less effective in different environments. Let's explore how evolutionary tradeoffs are resolved by returning to our three case studies: the evolution of beak depth in finches, the evolution of rifampicin resistance in *M. tuberculosis*, and the evolution of the low-plated phenotype in sticklebacks.

Beak size in Darwin's finches and the selection environment

Deep beaks are costly to medium ground finches in two ways. First, while a deep beak allows access to large, hard fruits, it actually makes handling small delicate seeds more difficult. As an analogy, try picking up a coin or playing a guitar while wearing gloves. In the post-drought selection environment on Daphne Major, with small seeds the primary food item, shallow-beaked birds could process food more efficiently. Second, beak depth and body size are positively correlated in these ground finches, and deep-beaked finches need more food to support their large bodies. The combination of inefficient processing and high food

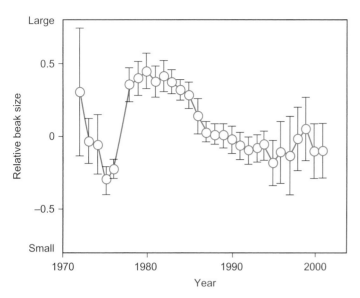

Figure 3.9 Changes in beak size over a 30-year period. The dependent variable on the y-axis is relative beak size, which is a function that combines beak depth, width, and length.

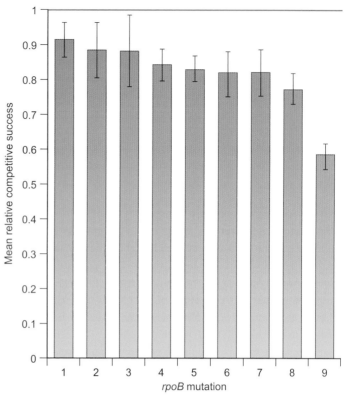

Figure 3.10 Mean relative competitive success of rifampicin-resistant mutants in comparison to wild-type rifampicin-sensitive *M. tuberculosis* cultured in medium without any rifampicin. A value of 1 indicates that the mutant and wild type proliferated equally well, while a value of <1 indicates that the wild type proliferated more quickly in the rifampicin-free medium. All nine mutations affected the *rpoB* gene, which codes for the β subunit of bacterial RNA polymerase.

needs made large deep-beaked finches poorly equipped to meet their metabolic needs.

The deluges of 1983 favored the evolution of smaller finches with smaller and shallower beaks. These finches required less food, which they could easily process with their small beaks that were well adapted to opening soft, small seeds. Because the deluge killed many of the plants that produced large seeds, large finches with large beaks were at a selective disadvantage. Thus we can observe a decrease in beak size in the medium ground finch since 1984 (Figure 3.9). A similar decrease also occurred in body size (Grant and Grant 2002).

Beak morphology is an example of an evolutionary tradeoff. A beak that is well adapted to cracking open hard, large seeds will be poorly adapted to opening soft, small seeds, and vice versa. Next we consider costs and benefits associated with the evolution of antibiotic resistance in *M. tuberculosis*.

Antibiotic resistance and the selection environment

Antibiotic resistance in bacteria is another example of an evolutionary tradeoff, with different populations and different species of bacteria experiencing varying costs of resistance. Sebastien Gagneux and his colleagues (2006) isolated nine mutant strains of rifampicin-resistant *M. tuberculosis* in the laboratory. When the researchers grew the rifampicin-resistant and rifampicin-sensitive *M. tuberculosis* together in a culture without rifampicin, in each case the resistant bacteria grew more slowly (Figure 3.10). They concluded that there was a competitive cost to rifampicin resistance in *M. tuberculosis*.

There were substantial differences in competitive success when the researchers compared different mutant phenotypes. The error bars in Figure 3.10 represent the 95% confidence intervals for the mean competitive success of rifampicin-resistant mutants in a culture without rifampicin (See Dealing with data 3.1). Because these confidence intervals are all below 1, the researchers concluded that all nine rifampicin-resistant mutants were experiencing some type of cost that reduced their growth rate.

Gagneux and his colleagues were interested in what happened once these rifampicin-resistant strains became established in a population. Did natural selection favor adaptations in these resistant strains that reduced the costs associated with rifampicin resistance? Mutations that compensated for the costs of rifampicin resistance – *compensatory mutations* – would be expected to spread throughout the population, because bacteria with compensatory mutations would reproduce more rapidly than bacteria without compensatory mutations.

To test this hypothesis, Gagneux and his colleagues used the same protocol as before, excepting that they measured competitive success of rifampicin-resistant bacteria isolated from 10 patients who acquired resistance during the course of their treatment. Because the resistant bacteria had reproduced for many generations before they were isolated, there was sufficient time for compensatory mutations to arise and be favored by

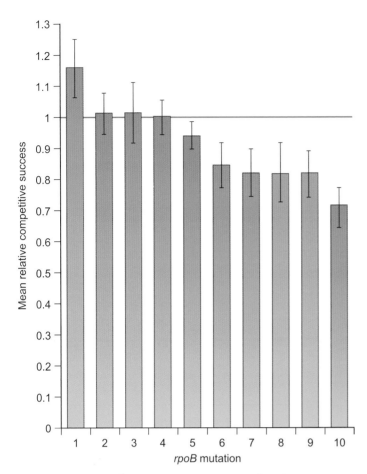

Figure 3.11 Mean relative competitive success of *rpoB* rifampicin-resistant mutants taken from patients who developed rifampicin resistance during the course of treatment, in comparison to wild-type rifampicin-sensitive *M. tuberculosis* cultured in medium without any rifampicin.

Dealing with data 3.2 Confidence intervals

As we've already discussed, scientists use data collected from a sample to make inferences about a population. For example, researchers may be interested in comparing the mean competitive success of rifampicin-resistant *Mycobacterium tuberculosis* bacteria to rifampicin-sensitive *M. tuberculosis* bacteria using the same approach as Gagneux *et al.* (2006). But there is no way to evaluate the entire population of bacteria, so they must be content with collecting a reasonable number of samples. Time, budget, and enthusiasm all influence what constitutes reasonable.

Let's say you discover a new rifampicin-resistant mutant and measure the following values for the competitive success of the rifampicin-resistant bacteria: 0.79, 0.96, 0.88, 1.04, 0.84, 0.92, and 0.87. The question you want to ask is whether you can conclude that the mean competitive success of the resistant strain is less than 1.0 (which is the number you would get if the resistant and successful strains were equal competitors). That question is easy to evaluate for the sample: the mean is equal to 0.90.

Of course, we are not interested in the mean for the sample, but rather we want the true value for the population! Recall in Dealing with data 2.3 we discussed three related measures of variation around the mean: the variance, standard deviation (SD), and standard error (SE). As it turns out, if your sample has a normal distribution, you can use these measures of variation to calculate a **confidence interval** (CI) – a measure of our confidence that an interval that we calculate (based on the data that we collected) includes the true population mean. For example, you can calculate a 95% confidence interval of 0.90 ± 0.051, which indicates that based on your sample there is a 95% probability that the true population mean is in the interval of 0.90–0.051 and 0.90 + 0.051, or (0.849 → 0.951). If you have a normal distribution (and there are ways of testing for that that we will not go into), the 95% confidence interval is calculated as follows:

$$95\% \, CI = mean \pm (1.96 \times SE)$$

If you want higher confidence that your interval includes the true population mean, you will calculate a broader interval using a larger coefficient. For example, the 99% confidence interval is calculated as follows:

$$99\% \, CI = mean \pm (2.58 \times SE)$$

Gagneux and his colleagues used the 95% confidence interval to test the hypothesis that rifampicin-sensitive bacteria have a lower competitive success than rifampicin-resistant bacteria. Because the 95% confidence intervals fell below 1 for all resistant strains in Figure 3.10 they could conclude that there was a cost to resistance that was making them grow more slowly than the rifampicin-sensitive strains. However, if the 95% CI had extended above 1.0, as in four strains in Figure 3.11, such a conclusion would not have been warranted.

natural selection. In contrast to the laboratory-isolated strains in Figure 3.10, the researchers discovered that 4 of 10 rifampicin-resistant strains had growth rates equal to or greater than equivalent rifampicin-sensitive bacteria (Figure 3.11). The researchers concluded that compensatory mutations may have arisen and been selected for over many generations in these patients.

Compensatory mutations are an important public health problem because they reduce or eliminate the costs of resistance incurred by antibiotic-resistant bacteria, allowing these bacteria to proliferate. Evolutionary biologists and public health officials had hoped that a careful program of reduced antibiotic use would create a selection environment that favored the success of sensitive strains, rather than resistant strains. Under these conditions, antibiotics would regain their effectiveness against pathogens such as *M. tuberculosis*. At present we still don't know how common compensatory mutations are, but one study identified 54 compensatory mutations in rifampicin-resistant *M. tuberculosis* (Comas *et al.* 2012).

Our discussion of adaptation in bacteria and in Darwin's finches emphasizes that traits may be adaptive in one environment and maladaptive in another as a result of evolutionary tradeoffs. Let's explore similar tradeoffs in sticklebacks.

Lateral plates in sticklebacks and the selection environment

Why do populations of complete-plated sticklebacks evolve over relatively few generations into populations of low-plated sticklebacks? We see this process happening over and over again.

Table 3.4 Frequencies of the three morphs of sticklebacks in Loberg Lake, Alaska.

Year	N	Morph frequency		
		Low	Partial	Complete
1990	59	0.00	0.04	0.96
1991	100	0.16	0.08	0.76
1992	621	0.15	0.07	0.78
1993	334	0.36	0.25	0.39
1994	3240	0.47	0.13	0.40
1995	2287	0.51	0.12	0.37
1996	1948	0.51	0.12	0.37
1997	1113	0.60	0.13	0.28
1998	770	0.60	0.14	0.26
1999	6582	0.67	0.13	0.20
2000	3348	0.71	0.12	0.17
2001	3304	0.77	0.12	0.11

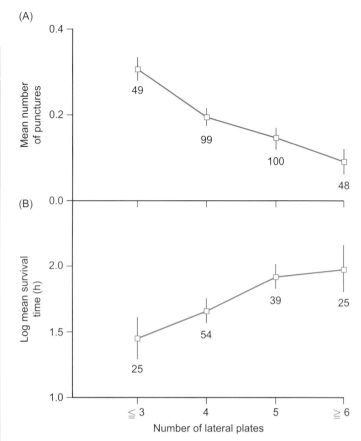

Figure 3.12 A. Mean number of puncture wounds in relation to number of lateral plates on one side. B. Log number of hours survived in relation to number of lateral plates on one side. Numbers below data points indicate sample size.

For example, Michael Bell and his colleagues (2004) documented the change in phenotype frequencies in a recently established stickleback population in Loberg Lake, Alaska (Table 3.4). Loberg Lake was treated with rotenone (a pesticide) in 1982 to kill all of its fish, so that the lake could subsequently be restocked with fish that would appeal to fishermen. Trout and salmon were introduced annually through 1990, and somehow a population of threespine sticklebacks became established – presumably via a temporary outlet that allowed some marine sticklebacks to colonize. Following establishment, the frequency of low-plated sticklebacks increased very rapidly.

We have seen this same process in many lakes, so we can exclude random processes, such as genetic drift as being responsible for this change. Instead, the most reasonable proposal is that the selection environment in the oceans favors complete-plated fish, while the selection environment in freshwater systems favors low-plated fish.

Adaptive significance of complete plates

Why are dermal plates adaptive in the ocean, but not in freshwater lakes? Several researchers had argued that dermal plates function as body armor to protect individuals against puncture

wounds. To test this hypothesis, Thomas Reimchen (1992) captured sticklebacks from a lake in western Canada to examine whether variation in number of plates could influence how well sticklebacks survived attacks by cutthroat trout, the major stickleback predator in that lake. His experimental protocol was to place a stickleback in a fish tank with six cutthroat trout. The body armor hypothesis predicted that sticklebacks with more plates would have higher survival rates than sticklebacks with fewer plates. Reimchen discovered that uneaten fish with more dermal plates survived longer and suffered fewer puncture wounds than uneaten fish with fewer dermal plates (Figure 3.12)

In a second experiment, Reimchen (2000) compared survival rates against predators among low-plated and complete-plated sticklebacks in relation to their relative size. When sticklebacks are relatively large, complete-plated sticklebacks have a significantly greater probability of escaping than do low-plated sticklebacks, but the protection provided by the plates is somewhat reduced for relatively small sticklebacks. Reimchen proposed that the posterior plates (lacking in the low-plated morph) actually interfere with the swallowing action of the predator's jaw.

These studies suggest that plates are advantageous in the presence of predators. But there are predators in freshwater as well in the ocean. So why has there been a tendency for the evolution of fewer protective plates in freshwater?

Adaptive significance of low plates

Carolyn Bergstrom (2002) and Reimchen (1992) argue that we need to consider the types of predators threatening sticklebacks. In some freshwater systems, sticklebacks are subject to very sudden attacks by swooping and diving birds. Presumably marine sticklebacks are much less subject to predation by birds because they tend to inhabit deeper waters. Thus, in freshwater systems, there is a premium on fast-start performance, an ability to react instantaneously to a sudden unexpected disruption. Bergstrom (2002) conducted a study to test the hypothesis that low-plated fish have better fast-start performance than complete-plated fish.

To determine whether the number of plates influenced fast-start performance, Bergstrom startled sticklebacks varying in number of dermal plates with the handle of a net, and recorded the response on a video recorder. Among the variables she considered were maximum escape velocity – the maximum velocity attained by the fish in the first 0.1 s after startling, and displacement – the total distance traveled in the 0.1 s after startling. Analysis of each frame allowed her to measure these variables related to fast-start performance. She found that fish with fewer plates had greater maximum escape velocity ($r = -0.267$, $P = 0.049$) and greater displacement ($r = -0.290$, $P = 0.015$).

Based on these results, Bergstrom argues that dermal plates influence the escape response in threespine sticklebacks. She suggests that the dermal plates stick out sufficiently to impose drag on an escaping fish. Thus in nature there exists a tradeoff between numerous plates, which make the fish more puncture resistant and more difficult to swallow, and fewer plates, which make the fish more difficult for certain types of predators to catch.

Rowan Barrett and his colleagues (2009) suggest a second advantage to low plates in freshwater. They note that dissolved nutrients such as calcium and phosphorus that are needed to build plates may be limiting in freshwater but readily available in saltwater. If these nutrients are limiting, then complete-plated sticklebacks in freshwater systems may grow more slowly than their low-plated counterparts.

To test this hypothesis, the researchers crossed heterozygous fish collected from a wild population of sticklebacks, resulting in a population of offspring containing individuals that were approximately 25% Eda^C/Eda^C (complete-plated), 50% Eda^C/Eda^L (mostly partial-plated) and 25% Eda^L/Eda^L (mostly low-plated). They then raised the fish in aquaria filled with either freshwater or saltwater, and measured individual growth rates. If nutrients are limiting in freshwater systems, the researchers predicted that complete-plated fish would grow more slowly in freshwater than partial-plated fish, which would grow more slowly than low-plated fish. In saltwater tanks, with a higher concentration on dissolved nutrients, they expected to see no substantial differences in growth rates among the three morphs.

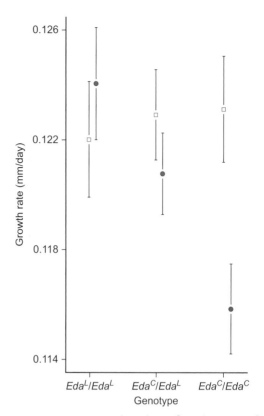

Figure 3.13 Growth rate of Eda^L/Eda^L, Eda^C/Eda^L, and Eda^C/Eda^C sticklebacks in saltwater (open squares) and freshwater (closed circles) tanks.

The data support these predictions. In freshwater, complete-plated fish grew much more slowly than partial-plated fish, which grew somewhat more slowly than low-plated fish. There were no differences in growth rates among morphs in the saltwater tanks (Figure 3.13). The researchers argue that in freshwater, sticklebacks must divert limiting nutrients to plate development, and thus they grow more slowly than they would in a nutrient-rich environment.

We know that natural selection favors the low-plated allele in natural populations of sticklebacks that move into freshwater systems. We now have two hypotheses to explain the benefits of the low-plated trait in freshwater systems: a more flexible body allows for quicker escape from birds, and fewer plates allow the sticklebacks to grow more rapidly. To understand how natural selection leads to adaptation, we need to understand how a trait such as flexibility or a high growth rate influences an individual's reproductive success.

3.4 HOW DO NATURAL SELECTION AND SEXUAL SELECTION INFLUENCE AN INDIVIDUAL'S FITNESS?

Darwin (1859) describes individuals with traits that enhance survival and reproduction as having higher fitness than individuals without (or with reduced quantities of) those traits. Our modern concept of fitness is a measure of the number of genes an individual contributes to the next generation. Evolutionary biologists may estimate fitness by measuring an individual's reproductive success – the number of genetic offspring an individual produces that survive until reproductive age. In practice this may be difficult to measure in natural populations.

Superficially, you might assume that a trait that decreases the probability of surviving will always reduce an individual's fitness. But natural selection may favor a trait that decreases the probability of survival if that same trait increases the probability of achieving higher reproductive success.

Reproductive success and fitness

Let's consider the redback spider, *Latrodectus hasselti*, in which males weigh an average of 4.4 mg, while females weigh an average of 256.0 mg. During sperm transfer, the male actually somersaults his body onto the fangs of his mate, effectively committing suicide approximately 65% of the time (Figure 3.14). In cannibalistic matings, females begin to eat the males within a

Figure 3.14 Male (left) and female (right) redback spiders showing extreme dimorphism.

few seconds of beginning copulation – nonetheless, copulation may continue for more than a half hour. Maydianne Andrade (1996) investigated this phenomenon to see whether male self-sacrifice actually increased reproductive success, at the cost of the male's future survival

Andrade tested two hypotheses for how self-sacrifice might enhance a male's overall reproductive success. One hypothesis, the *paternal effort hypothesis*, was that male self-sacrifice increases male fitness by providing additional nutrients to the female, who can subsequently lay higher quality or more numerous eggs. The second hypothesis, the *mating effort hypothesis*, was that male self-sacrifice enhances male fertilization success. In nature, females can mate with more than one male, and on average females are visited by a median number of two suitors per day. Thus there is strong competition between males for females, and self-sacrifice would be favored by natural selection if it substantially enhanced fertilization success.

These two hypotheses lead to different predictions. The paternal effort hypothesis predicts that the females who eat their mates will lay more eggs, or higher quality eggs, than females who don't eat their mates. The mating effort hypothesis predicts that females who eat their mates will have a higher proportion of their eggs fertilized by the copulating males than will females who don't eat their mates.

Andrade used laboratory studies to test the predictions generated by both hypotheses. She found no significant difference in egg sac mass or number of eggs when comparing the results of cannibalistic and non-cannibalistic matings (Table 3.5). Thus the results did not support the predictions of the paternal effort hypothesis. But Andrade did find that eaten males mated for a much longer period of time than uneaten males, and that long-duration copulations had higher fertilization success than short-duration copulations. In addition, Andrade found that females seldom mated with a second male after eating their first mate, while they almost always mated with a second male when they didn't eat their first mate (Table 3.5).

Andrade concluded that male self-sacrifice is adaptive in redback spiders, because it increases the number of eggs that a male

Table 3.5 Results of laboratory matings testing the predictions of the paternal effort hypothesis and the mating effort hypothesis.

	Cannibalistic matings		Non-cannibalistic matings		*P*-value
	N	Value	N	Value	
Egg sac mass (median mg)	9	198	13	179	0.082
Number of eggs (median)	9	256	13	249	0.526
Duration of matings (median minutes)	5	25	12	11	0.035
Probability female re-mates after first mating	9	0.33	23	0.96	0.001

ultimately fertilizes. Selection favoring suicidal behavior would be stronger if, in nature, males are unlikely to survive the journey to find a second mate. To measure male mortality in the field, Andrade captured males, marked them with unique paint markings, and released them back onto their webs. Shortly thereafter, males left their webs in search of receptive females. By recording every mating event within her study site, Andrade was able to show that about 85% of the males died before finding a receptive female (Andrade 2003). Given this high mortality rate, primarily from predators, it is unlikely that males would survive the journey to find a second mate. In this case, natural selection operating on the mating success of self-sacrificial males has favored the evolution of males somersaulting into the fangs of females during copulation.

Andrade's study demonstrates that understanding the evolution of adaptive traits through the process of natural selection requires a clear picture of how these traits influence an individual's reproductive success. Adaptive traits can enhance survival, as we discussed with antibiotic resistance in bacteria and beak depth in finches. Or adaptive traits can enhance mating success, as we just discussed in redback spiders. A trait that enhances mating success may also enhance survival. But in cases like the redback spider, an adaptation can arise that actually has a negative impact on survival. Charles Darwin used the term sexual selection to describe selection that arises when individuals vary in their ability to achieve mating success in a particular environment.

Sexual selection

Darwin (1859) describes sexual selection as arising when "individual males have had, in successive generations, some slight advantage over other males in their weapons, means of defense, or charms, and have transmitted these advantages to their male offspring." Modern biologists classify differences in "weapons or means of defense" under the category of intrasexual selection, which we can define as differential mating success among individuals of one sex arising from competition among individuals of the same sex. In contrast, Darwin's differences in "charms" may result in intersexual selection, which is differential mating success among individuals of one sex arising from interactions with individuals of the opposite sex. Both forms of sexual selection are simply selection for traits that enhance mating success (Figure 3.15).

Sometimes it is difficult to distinguish between intra- and intersexual selection in field populations. Darwin's finches provide a good example of this problem. Medium ground finches on Daphne Major with very deep beaks tended to survive the 1977 drought because they could process large, hard fruits. But males tend to be larger than females, either because larger males compete more effectively with each other for territories (an example of intrasexual selection), or because large male size is preferred by females (an example of intersexual selection). It is also possible that both types of sexual selection have favored larger males. In any case, the result of this sexual dimorphism – difference in form between females and males of the same species – was that drought survival was six times greater in males than females.

The drought broke the following year, 1978, leaving a population of deep-beaked finches with a highly skewed, male-biased sex ratio on Daphne Major. Body size and beak size are positively correlated in these birds. Because males were six times as common as females, male–male competition was intense and the largest males with the deepest beaks secured the best territories. Females tended to mate with these unusually deep-beaked survivors – thus the influence of sexual selection reinforced natural selection acting on survival.

Thinking ecologically 3.2

Though natural selection favored shallow beaks after the El Niño of 1983, sexual selection might still favor large-beaked males. What types of data could you collect to understand the relative influence of natural selection and sexual selection in this population?

The evolutionary changes we've discussed thus far all have an important influence on the relationship between organisms and their environment. But these changes are relatively minor, for example, in comparison to the difference between a finch and a hummingbird. Medium ground finches on Daphne Major all belong to the same species regardless of beak depth. *M. tuberculosis* is still *M. tuberculosis*, regardless of whether it is sensitive or resistant to a particular type of antibiotic. We will continue our inquiry into the mechanisms of evolutionary change by addressing how natural selection may lead to speciation – the origin of new species.

3.5 HOW MIGHT NATURAL SELECTION CAUSE NEW SPECIES TO EVOLVE?

To discuss speciation, we must first address the question of what constitutes a species. According to the *biological species concept*, the important factor for whether two populations are different species is whether they are potentially reproductively isolated from each other (Mayr 1942). Potentially reproductively isolated populations have sufficiently different genetic constitutions so that if they were brought together, individuals would not interbreed, or else if they did interbreed, they would be unable to produce viable offspring. This concept works moderately well for sexual animal species but is obviously problematic for the millions of species of organisms that are not sexual, and for the millions of species (sexual or not) that have gone extinct.

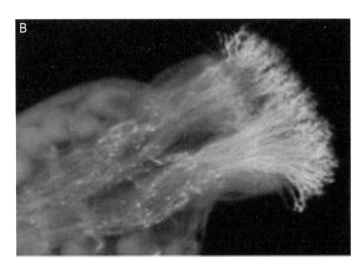

Figure 3.15 Sexual selection. A. *Strategus rhinocerus* beetles, with female on left and male on right. B. Pollen tubes generated by individual pollen compete to reach a female *Arabidopsis thaliana* ovule. C. Two bighorn sheep, *Ovis canadensis*, wage battle. D. Male peacock, *Pavo cristatus*, displaying his train of tail feathers. Females prefer males with the most elaborate feathers. E. The female guppy, *Poecilia reticulata*, chooses a mate with vibrant colors and high display rates. F. Reverse sexual selection in phalaropes, *Phalaropus fulicarius*, in which females are larger and more colorful than males, and compete with each other for access to males.

How geographic separation increases the probability of speciation

Why is there such an emphasis on reproductive isolation in the definition of a species? Let's consider a population of birds from a moist tropical habitat that colonizes a new desert island. In the new environment, natural selection would favor novel adaptations in that population. For example, birds on the desert island may have physiological adaptations associated with dissipating excess heat, and feeding adaptations that allow them to process cactus fruits, seeds, or flowers. Will the desert island population diverge enough genetically from the ancestral tropical habitat population to form a new species? Several factors will influence the answer to this question, but one crucial part of the answer is how reproductively isolated the desert population is from the original population. If reproductive isolation is nearly complete, there will be almost no gene flow – or exchange of alleles – between the two populations. And without gene flow, natural selection can select for the most advantageous alleles for the desert population. Genetic drift may also be important in this process, particularly for small populations. If there is gene flow, the effect of natural selection and drift will be diluted by immigrants from the ancestral population bringing in their alleles for adaptation to a moist tropical habitat; these alleles may prove maladaptive in their new environment.

Thus the most commonly cited model of speciation – allopatric speciation – requires two or more populations to be geographically separated for a long enough period of time for genetic divergence to occur between populations. If the two populations experience different environmental conditions, as may be the case on two different islands, then selection can lead to genetic divergence. If reproductive isolation continues, sufficient divergence may develop so that two or more species are formed. Speciation is considered complete if the two populations come back together, but no longer interbreed (Figure 3.16). Even before he developed his theory of natural selection, Darwin was impressed with similarities between organisms on the Galápagos and their presumed ancestors from mainland South America. He writes in *The Voyage of the Beagle* (1839), "It was most striking to be surrounded by new birds, new reptiles, new shells, new insects, new plants, and yet by innumerable trifling details of

structure . . . to have the temperate plains of Patagonia, or the hot dry deserts of Northern Chile, vividly brought before my eyes." Darwin was beginning to recognize that the species he was discovering in the Galápagos had evolved from ancestral populations native to the South America mainland. Given spatial separation of 1000 km, there was almost no gene flow between the ancestral South American species and their descendant species on the Galápagos Islands. The physical isolation of each island from the other islands, and the wide range of environmental conditions within and between islands, provided an ideal setting for multiple cases of allopatric speciation. Some of these immigrants to the Galápagos Islands, such as Darwin's finches, went through an adaptive radiation – divergence of an ancestral species into numerous descendant species adapted to different habitats available in the new environment.

Speciation in the absence of geographical separation between populations

Research on a small insectivorous bird, the European blackcap (*Sylvia atricapilla*), indicates that speciation may occur in the absence of physical barriers. This research may be capturing the early stages of sympatric speciation – the formation of new species within a single geographical area. Similar to allopatric speciation, sympatric speciation requires genetic divergence, but in this case the genetic divergence is, by definition, within a population. This genetic divergence arises because natural selection favors two or more alternative genotypes within the population. The problem is that genetic divergence can be maintained within a population only under conditions of *assortative mating*, in which individuals of similar genotypes tend to mate with each other. Without assortative mating,

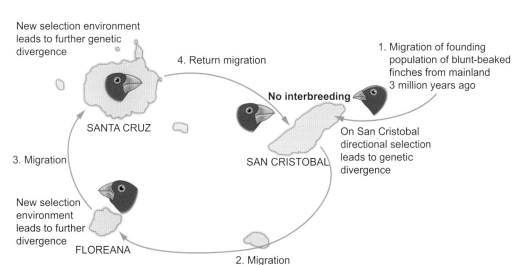

Figure 3.16 General model of allopatric speciation in Darwin's finches. First, a small founding population migrates from South America to one of the newly arisen islands (San Cristobal) 3 mya. Later, there are further migrations to one or more islands, and the migrant populations adapt to a different selection environment on new islands. When a descendant population migrates back to San Cristobal, they do not interbreed with the descendants of the original San Cristobal populations. Speciation has happened in allopatry.

New selection environment leads to further genetic divergence

4. Return migration

1. Migration of founding population of blunt-beaked finches from mainland 3 million years ago

No interbreeding

SANTA CRUZ

On San Cristobal directional selection leads to genetic divergence

SAN CRISTOBAL

3. Migration

New selection environment leads to further divergence

FLOREANA

2. Migration

genetic divergence will be lost when genes from each genotype mix during sexual reproduction. Let's investigate, first, whether there is genetic divergence in the migratory populations of blackcaps, and, second, whether there is evidence for assortative mating within these populations.

European blackcap migration routes

Until 1950, European blackcaps migrated seasonally, spending spring and summer in central Europe during the breeding season, and fall and winter in the Iberian Peninsula and North

A

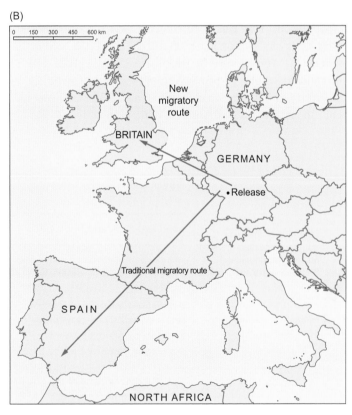

(B)

Figure 3.17 A. European blackcap. B. Alternate migratory paths from Germany northwest into Britain, or southwest into the Iberian Peninsula.

Africa. Since then, a second migratory pattern has arisen, with some blackcaps spending the fall and winter in Britain, before returning to central Europe to breed (Figure 3.17). Peter Berthold and his colleagues (1992) hypothesized that this new migration route was a heritable change in the behavior of a subpopulation of blackcaps. If their hypothesis is correct, offspring from British migrants should have a genetically influenced tendency to fly slightly north of west in the fall migration, while offspring from Iberian migrants should have a genetically influenced tendency to fly southwest in the fall.

To test the predictions of the hypothesis, Berthold and his colleagues captured 20 pairs of wintering blackcaps in Britain and moved them to outdoor aviaries in Germany. These British migrants produced 41 offspring during the study, which were raised with a control group of 49 offspring from parents that had overwintered in Iberia. To test their migratory orientation, the researchers placed each bird in aluminum funnels lined with typewriter correction paper, so that all attempts to fly left scratch marks on the paper liner in the direction of the flight attempt. They conducted their flight studies during the peak of migratory season.

British adults and their offspring tended to migrate slightly north of west, while the nestlings from Iberian parents tended to migrate in a southwesterly direction (Figure 3.18). Berthold and his colleagues concluded that heritable variation has evolved within a breeding population of blackcaps over a very short time period. Consider that there were no geographical

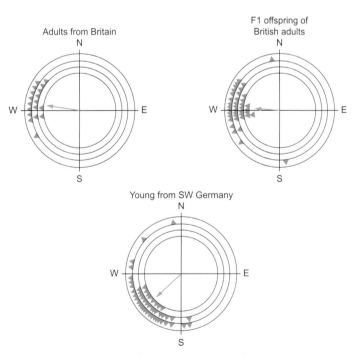

Figure 3.18 Migratory pathways of European blackcap adults from Britain, the mature offspring of adults (hand-raised from the nest) from Britain, and the mature offspring (hand-raised from the nest) from southwest Germany. The arrow indicates the mean direction of all the birds in the group.

barriers separating the British migrants from the Iberian migrants during the breeding season in Germany. So how could the population have diverged genetically in such a short period of time?

 Thinking ecologically 3.3

Propose two potential benefits of the alternative migration route to Britain. What data would you collect to test the predictions of each proposed benefit.

As we've indicated, for genetic divergence to occur within a population, there must be assortative mating. Otherwise there will be no barrier to gene flow, and the differences in genetic constitution will be eliminated by genetic recombination arising from sexual reproduction. British migrants must tend to mate with British migrants, and Iberian migrants must tend to mate with Iberian migrants. Research by Stuart Bearhop *et al.* (2005) used stable isotope analysis to address this question.

Stable isotope analysis can be used to investigate potential differences between British and Iberian migrants, because the relative amount of ^{2}H varies slightly among geographical regions. We can tell where an organism has been, or what it has eaten, by comparing the quantity of ^{2}H in a tissue sample from that organism to the quantity of ^{2}H in a universally accepted standard. The actual calculation, the stable isotope ratio, δ^2H in this example, is a bit more complex (see Dealing with data 3.3).

Let's see how stable isotope analysis can be used to answer questions about European blackcap behavior in the field.

Investigating assortative mating in blackcaps

Bearhop and his colleagues knew from previous research that δ^2H was lower in the rain that falls on Britain than on the rain that falls on the Iberian Peninsula. They also knew that Keith Hobson and his colleagues (2004) had shown a strong correlation between δ^2H in rain and the amount incorporated by birds into their tissues. Bearhop reasoned that if assortative mating was occurring, there should be a strong positive correlation between δ^2H values of paired males and females at their breeding grounds in Germany. The researchers found that the values of δ^2H from the claws of the mated pairs are strongly correlated, indicating that there is a strong tendency for assortative mating in European blackcaps (Figure 3.19A).

The stable isotope ratio evidence for assortative mating raises a host of questions. First, what is the mechanism of assortative mating? Do British migrants have a way of communicating with each other, and choose to mate with each other? Second, what is the adaptive significance of the British migratory route? Is there a fitness advantage to spending the fall and winter in Britain

versus on the Iberian Peninsula? Third, will this "migratory divide" ultimately lead to the formation of two separate species? Finally, what is the genetic basis underlying the differences between individuals that undergo two different migratory routes?

Bearhop and his colleagues have begun to address some of these questions. For example, to determine the mechanism of assortative mating in blackcaps, the researchers have investigated the relationship between δ^2H of males and the arrival date of the males on the territory. In most migratory birds, males arrive first, and those that arrive earliest get the best territories. In this case the British males, which have more negative values of claw δ^2H, arrive before the Iberian males and presumably get the best territories (Figure 3.19B). Similarly, the British females arrive before the Iberian females, and presumably mate with the males possessing the best territories.

If the earliest arriving males are from Britain, and the earliest arriving males get the best territories, it would stand to reason that females paired with these males would enjoy higher fitness. A regression analysis of clutch size and male δ^2H supported this hypothesis: males with low values of δ^2H, presumably British migrants, tended to mate with females who laid larger clutches.

Of course, we still don't know whether European blackcaps will diverge into two species. We have some data to indicate that the new migratory route is adaptive, in that birds taking this route tend to be associated with larger clutches of eggs. We do know that there is a genetic basis for taking these two migratory paths, but we do not know what genes are responsible for the behavioral variation. Unfortunately, there are very few studies that have unraveled the genetic basis of evolutionary change and that have also correlated the genetic differences with the adaptive benefits of the change. Let's consider some of the other tools evolutionary ecologists use to unravel both the processes of speciation, and the results of speciation – the evolutionary relationships among species.

3.6 HOW DO EVOLUTIONARY ECOLOGISTS UNRAVEL EVOLUTIONARY RELATIONSHIPS?

Let's study Darwin's finches in more detail to understand how allopatric speciation led to an adaptive radiation of 14 species of finches over the past 3 million years on the Galápagos Islands. The challenge is that we can't go back in time to watch the process unfold. Before the revolution in molecular biology, ornithologists used morphological and behavioral characters such as beak size and shape, plumage, body size, and foraging and mating behavior to infer evolutionary relationships. These types of characters are still used, but molecular technology is enjoying greater popularity as it has become more available.

Dealing with data 3.3 Stable isotope ratios

In nature, many elements vary in the number of neutrons. For example the air we breathe is about 80% nitrogen (N). All N atoms have seven protons, but approximately 99.63% have seven neutrons (^{14}N), while 0.37% have eight neutrons (and are designated ^{15}N). ^{14}N and ^{15}N are stable isotopes because they don't undergo radioactive decay. Similarly, approximately 99.985% of hydrogen (H) atoms have no neutrons (^{1}H), while 0.015% have one neutron (^{2}H). These percentages are based on globally accepted standards (Table D3.3.1).

Table D3.3.1 Average terrestrial abundances of the stable isotopes of major elements of interest in ecological studies.

Element	Isotope	Abundance (%)
Hydrogen	^{1}H	99.985
	^{2}H	0.015
Carbon	^{12}C	98.89
	^{13}C	1.11
Nitrogen	^{14}N	99.63
	^{15}N	0.37
Oxygen	^{16}O	99.759
	^{17}O	0.037
	^{18}O	0.204
Sulfur	^{32}S	95.00
	^{33}S	0.76
	^{34}S	4.22
	^{36}S	0.014

While we've discussed stable isotope analysis in the context of hydrogen, the same type of analysis can be used for any of the isotopes in the table above.

Two individuals will vary in their stable isotope ratios (for example, values of δ^2H) for one of two reasons. First, they may obtain an element from two different sources. In the example discussed in this chapter, European blackcaps that overwinter in two different locations have different stable isotope ratios because they have assimilated hydrogen from two different geographical sources (either Britain or Iberia). Second, individuals may vary because they have different patterns of assimilation of the isotopes, due to preferential ingestion or excretion of one of the isotopes. We will begin to appreciate the power of stable isotope analysis when we evaluate how food webs are structured and how nutrients cycle through ecosystems.

To calculate δ^2H we need the following information:

R_{sample} = the ratio of ^{2}H to ^{1}H in the sample = ^{2}H/^{1}H. Measuring this ratio usually involves combusting the test material (in this case a section of claw) in a furnace connected to a type of elemental analyser (a stable isotope mass spectrometer).

$R_{standard}$ = the ratio of ^{2}H to ^{1}H in a universally agreed upon standard (called Standard Mean Ocean Water).

$\delta^2 H = [(R_{sample}/R_{standard}) - 1] \times 1000$

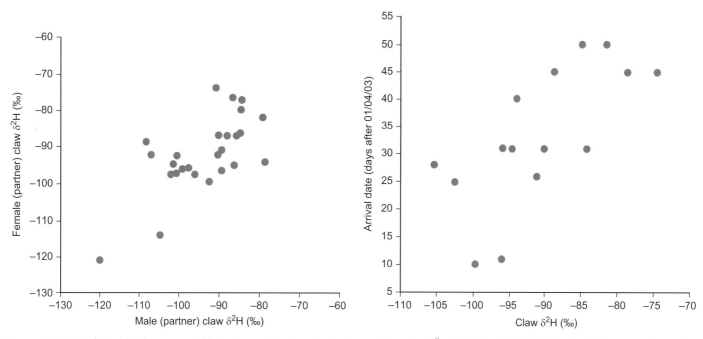

Figure 3.19 A. Relationship between stable isotope ratios in paired males and females ($R^2 = 0.50$, $P = 0.001$). B. Relationship between the stable isotope ratio of migrating males and their arrival date on the breeding grounds ($R^2 = 0.44$, $P = 0.005$).

A **phylogeny** is the evolutionary history of ancestry and descent of a group of related organisms. Hypotheses of evolutionary relationships are often represented with a *phylogenetic tree* – a diagram that shows the phylogeny of a group. Based on a detailed analysis of differences in morphology and behavior, David Lack (1947) proposed a hypothesis about the evolutionary relationship among all 14 species of Darwin's finches, including the four species in the genus *Geospiza*. The phylogenetic tree in Figure 3.20 summarizes Lack's hypothesis for four species of *Geospiza*.

As represented by this phylogenetic tree, Lack hypothesized that the ancestral population diverged at node A, with one lineage giving rise to *Geospiza scandens* (and two other species not pictured here), and the second lineage giving rise to what would ultimately be three other species – *Geospiza fortis* (the medium ground finch from Daphne Major), *Geospiza fuliginosa*, and *Geospiza magnirostris*. *G. scandens* differs from the other three species behaviorally, as it breeds, nests, and feeds on the prickly pear cactus (*Opuntia* spp.). Morphologically, its beak is much shallower and longer in relation to its body size. As we move closer

to the present time, we approach node B, which diverged into one lineage that gave rise to the present day *G. fuliginosa*, and a second lineage that most recently (at node C) diverged into *G. magnirostris* and *G. fortis*. Thus Lack hypothesized that *G. magnirostris* and *G. fortis* shared a common ancestor most recently (at node C). Lack was not certain about this phylogeny, and acknowledged that the relationship among these species was not clear based on differences in morphology and behavior.

Kenneth Petren and his colleagues (1999) used variation in microsatellite length to help resolve the relationships among these species, and in fact among all 14 species of Darwin's finches. Microsatellites are sections of DNA that are made up of repetitive DNA base sequences of variable length surrounded by unique regions of constant length (Figure 3.21A). They can be used to help resolve evolutionary relationships in much the same way as Lack (1947) used morphology and behavior. Because the length of the repetitive region is variable, each population can be polymorphic for each microsatellite. Figure 3.21B shows four morphs of a microsatellite, each of which is a distinct allele of that microsatellite.

Figure 3.20 A. Four species of seed-eating ground finches. Small ground finch, *Geospiza fuliginosa* (upper left), Medium ground finch, *G. fortis* (upper right), Large ground finch, *G. magnirostris* (lower left) and Cactus finch, *G. scandens* (lower right). B. David Lack's (1947) hypothesis for the relationship among four species in the genus *Geospiza*.

(B)

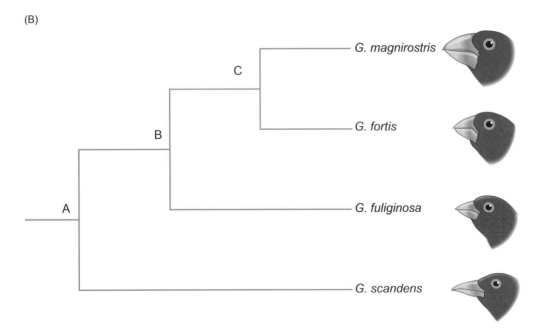

(A) [*AATCCTTCTCGTCCCTCTTGG*]**CACACACACACACACACACACACACACA**[*TTTGAGTGTGCAGCAGTTGG*]

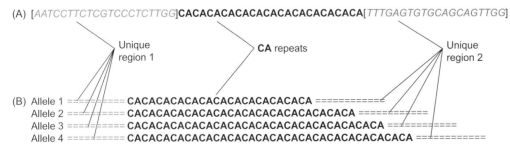

Figure 3.21 A. One microsatellite from the Petren *et al.* (1999) analysis of 16 microsatellites. The repetitive DNA is in **bold** print, and the unique sequences on either side are *italicized*. B. Four alleles of the microsatellite from A that vary in the length of the repetitive DNA sequence, as a result of variation in the number of CA repeats. The symbol ========= is shorthand for the unique sequences.

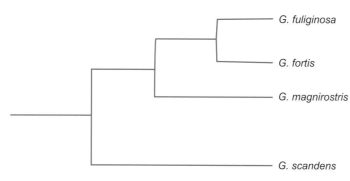

Figure 3.22 Petren *et al.*'s (1999) phylogenetic tree based on microsatellite data.

Petren and his colleagues identified 16 microsatellites from the DNA of Darwin's finches. They hypothesized phylogenetic relationships based on how similar species were at each microsatellite. They treated each microsatellite as an individual character that could be present in several alleles, with each allele corresponding to a particular length of the repetitive region. They then analysed the microsatellite data to construct the phylogenetic tree shown in Figure 3.22. Notice that this phylogenetic tree is different from Lack's (Figure 3.20), as it shows *G. fuliginosa* and *G. fortis* sharing a common ancestor most recently. Our challenge is to evaluate who is correct, Lack or Petren.

Currently, we don't know which hypothesis is correct. As our molecular tools improve, we may become more confident

answering that question. The phylogenies of Lack and Petren *et al.* are alternative hypotheses for how and when Darwin's finches evolved. Lack used the tools of observation and inference, and Petren the more advanced techniques available through molecular biology, which also require inference. Modern evolutionary ecologists rely primarily on phylogenies based on molecular characters, in part because there are so many more characters to use for the analyses.

Charles Darwin, David Lack, Peter and Rosemary Grant, and Kenneth Petren were all impressed by how the Galápagos Islands provided a theater for the adaptive radiation of Darwin's finches. Dolph Schluter, one of the Grants' graduate students, recognized that there were several similarities between the island environment experienced by Darwin's finches and the lake environment experienced by threespine sticklebacks. For example, lakes, like islands, are often geographically isolated from each other, providing the potential for reproductive isolation of different populations. Second, lakes provide different types of food to sticklebacks, depending on whether they forage on the bottom, near the surface of the open lake, or near the shore. Appreciating that finches on islands and sticklebacks in lakes experienced similar ecological variation, Schluter decided to investigate whether there was any evidence of speciation that could be occurring among stickleback populations specializing on different feeding options that were available within the lakes.

REVISIT: Evolution in sticklebacks

The lakes that host threespine sticklebacks along the west coast of North America have formed since the last ice age, so any evolutionary changes in stickleback populations that live in these lakes have occurred relatively recently and quickly. While most of the lakes have only one stickleback morph, some of the lakes have two. The smaller *limnetic* morph inhabits open water (a region known as the limnetic zone), feeding on very small prey by sifting prey-infused water through its numerous long gill rakers. The larger *benthic* morph feeds on much larger invertebrate prey that it captures with its much larger mouth on or near the bottom of the lake. Just as the size of a finch's beak is related to the type of food it eats, so too the size of a stickleback's mouth, and the number and size of its gill rakers, predispose a stickleback to a specific feeding regimen (Figure 3.23A).

Schluter (1993) wanted to know whether these differences in morphology and feeding preferences were biologically significant. In other words, were these two morphs much more successful at feeding in their chosen environment? So he captured individuals of both morphs from Paxton Lake and brought them into the lab, then tested them in tanks that were designed to closely mimic feeding conditions in either a limnetic or benthic environment. Using several different measures of foraging efficiency, he discovered that limnetic fish were much more effective at feeding in the limnetic environment, while benthic fish were much more effective feeders in the benthic environment (Figure 3.23B). Interestingly, hybrids between the two morphs performed relatively poorly in both environments.

Looking at Figure 3.23A, you will note that the limnetic morph has much more robust spines than the benthic morph. One hypothesis for this difference is that limnetic fish live out in open water, which coincides with the habitat used by cutthroat trout (*Oncorhynchus clarki*), the

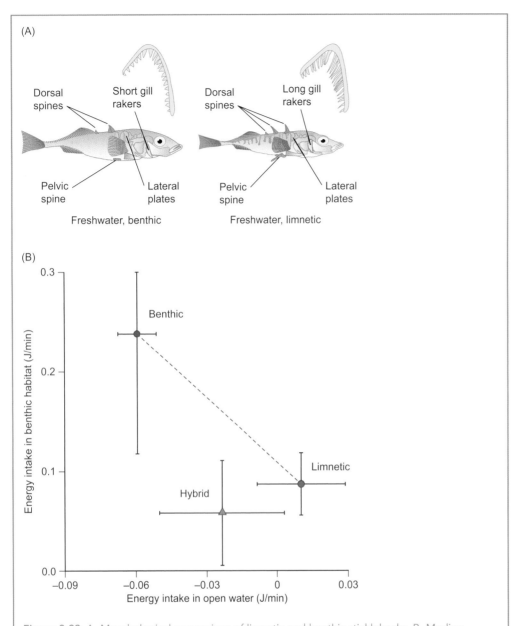

Figure 3.23 A. Morphological comparison of limnetic and benthic sticklebacks. B. Median energetic intake (±1 SE) of benthic, hybrid, and limnetic sticklebacks feeding in open water (x-axis) and on the bottom (benthos) (y-axis).

stickleback's major predator. Steven Vamosi and Schluter (2002) hypothesized that such robust spines would only be present in fish that suffered heavy predation. Consequently they predicted that limnetic sticklebacks are more likely to be preyed upon than benthic sticklebacks. They tested this prediction by establishing populations of sticklebacks in artificial ponds in which they controlled the presence or absence of predaceous cutthroat trout. After 3 months, they found that limnetic sticklebacks survived better than benthic ones when trout were absent, but that benthic sticklebacks survived better than limnetic ones when trout were present. The researchers concluded that trout were feeding preferentially on limnetic stickleback, presumably because limnetic stickleback and trout occupy similar types of habitats, and have much higher encounter rates. Hybrids did relatively poorly in the presence or absence of trout.

Do these two morphs represent two distinct species? Many researchers argue that they do. When the two morphs are raised in the laboratory, they retain most of their differences over numerous generations, indicating that the differences between morphs are genetic rather than environmentally induced. Researchers are beginning to understand some of the genetic bases for morphological differences between the two morphs. For example, Yingguang Chan and his colleagues (2010) have demonstrated that mutation in a regulatory gene is responsible for loss of pelvic spines in benthic morphs from three different lakes. In each case, the mutation inactivated the regulatory gene, and also deleted some additional base pairs alongside the gene. Given that all benthic morphs studied thus far lack these spines, and that loss of spines is a result of independent mutations occurring at the same gene, it is likely that there is strong selection favoring the loss of pelvic spines in benthic morphs.

There is also strong selection operating against hybridization between the two morphs. Todd Hatfield and Schluter (1999) crossed the two morphs to create hybrid sticklebacks, which were released into experimental ponds along with benthic and limnetic individuals. They found that the hybrids grew about 25% more slowly than the limnetic morphs when they foraged in open habitat, and about 25% more slowly than the benthic morphs when they foraged on the bottom near the edge of the pond.

Finally, there are numerous lines of evidence demonstrating that there is a strong tendency toward assortative mating in the field, with limnetic morphs more likely to mate with other limnetic morphs, and benthic morphs more likely to mate with other benthic morphs. Taken together, all these lines of evidence provide evidence that, as Schluter and others predicted, these glacial lakes are an ideal arena for rapid evolution and speciation.

SUMMARY

Threespine sticklebacks, numerous species of disease-causing bacteria, and Darwin's finches have all shown rapid evolutionary change in response to changing environments. Evolutionary ecologists use a variety of genetic and molecular approaches to study evolutionary change in these and other species. Gene flow, genetic drift, mutation, and natural selection can cause evolutionary change within a population, but natural selection is the only evolutionary process that can lead to adaptation. The benefits and costs of adaptations are environment dependent and reflect evolutionary tradeoffs, so a trait may be beneficial in one environmental context and costly in a second. Natural selection may lead to speciation when genetic divergence is maintained either by physical barriers to gene flow, or by assortative mating of similar genotypes within a population. Evolutionary ecologists compare morphological, behavioral, and, most commonly, molecular characters in related groups of organisms, and use similarities in these characters to create phylogenetic trees that reflect evolutionary relationships.

FURTHER READING

Carroll, S. P., and 8 others. 2014. Applying evolutionary biology to address global challenges. *Science* 346: 313 (summary). Full article at http://dx.doi.org/10.1126/ science.1245993.
This exhaustive article identifies and describes how evolutionary approaches can be used to address some of our most pressing environmental issues.

Grant, R. B., and Grant, P. R. 2003. What Darwin's finches can teach us about the evolutionary origin and regulation of biodiversity. *Bioscience*, 53(10): 965–975.
This is an excellent and accessible summary of the research conducted by Rosemary and Peter Grant up to 2003. It explores within-species beak size evolution, and larger processes of speciation and adaptive radiation.

Laxminarayan, R. 2014 Antibiotic effectiveness: Balancing conservation against innovation *Science* 345, 1299–1301.

This article describes ways that we can best use currently available antibiotics by reducing the evolution of resistance, and how we can find and develop new and effective antibiotics.

Sakata, Y., Yamasaki, M., Isagi, Y., and Ohgushi, T. 2014. An exotic herbivorous insect drives the evolution of resistance in the exotic perennial herb. *Ecology* 95: 2569–2578.

This study provides another example of rapid evolution of an adaptive trait – resistance to herbivory – in an invasive plant in Japan.

Weiner, J. 1994. *The Beak of the Finch. A Story of Evolution in Our Own Time*. New York: Alfred A. Knopf, Inc.

This Pulitzer Prize winning book describes in fascinating detail the story of scientists exploring how evolutionary processes operate on Darwin's finches. I once used this book instead of an evolution text in my upper level evolution class and did not regret it (nor did the students).

END-OF-CHAPTER QUESTIONS

Review questions

1. Describe how directional selection has caused rapid evolution in threespine sticklebacks, pathogenic bacteria, and Darwin's finches. For each example, identify the role played by evolutionary tradeoffs in different environments.
2. What are the four major evolutionary processes? Which of these processes can (a) lead to adaptation, (b) increase genetic variation within a population, and (c) reduce the probability that speciation will occur.
3. Use the redback spider example to describe circumstances in which natural selection may tend to decrease the survival of individuals within a population? How does this tie in with the concept of fitness?
4. What is an adaptive radiation and when is it most likely to occur?
5. According to (a) David Lack's and (b) Kenneth Petren *et al.*'s phylogenetic trees, which two species of *Geospiza* are most closely related? According to Lack's tree, which species is/are most closely related to *G. fuliginosa*? According to Petren *et al.*'s tree, which species is/are most closely related to *G. magnirostris*? Explain your reasoning.

Synthesis and application questions

1. Why might some species of bacteria show antibiotic resistance, even if their parent cells never encountered antibiotics?
2. If you were to take eggs from large-beaked finch parents and exchange them with eggs from small-beaked finch parents, what size will the beaks of these offspring be when the birds reach adulthood? What information would you need in order to answer this question?
3. Why is it important to consider antibiotic resistance as a local or global phenomenon? How might the answer to this question influence the way your physician might decide how to treat an infection?
4. What strategies would you recommend for curtailing the spread of antibiotic-resistant strains of bacteria?
5. Finding a positive correlation between parent and offspring beak depth supports the hypothesis that variation in beak depth is heritable. Can you think of alternative (nongenetic) explanations for a positive correlation? What other factors, besides genetics, might influence the size of a body part?
6. As discussed in the text, we have evidence that European blackcaps mate assortatively. We don't know if they are in the early stages of speciation. Imagine you were given the task of evaluating whether these two different migratory routes will ultimately lead to speciation.

What hypotheses would you test, and what data would you collect to help you evaluate the predictions of these hypotheses?

7. Reconsider Andrade's (1996) data testing the predictions of the paternal effort hypothesis and the mating effort hypothesis (Table 3.5). Are the two hypotheses mutually exclusive? How confident are you with rejecting the mating effort hypothesis?

8. Consider Bell *et al.*'s (2004) data (Table 3.4) on the changes in phenotype frequency in Logan Lake. What would be your prediction for the frequency of the low-plated phenotype in 2002?

9. Does the association between overwintering in Britain and larger clutch size mean that British blackcap males are providing more for their females than are Iberian blackcap males? Can you think of alternative explanations for this finding? What might be a better measure of fitness than clutch size?

Analyse the data 1

A study by Phillips *et al.* (2006) suggests that a plague of cane toads in Australia is being supported by natural selection favoring the evolution of long legs. The researchers reason that long legs support long-distance migration in these toads. Phillips *et al.* implanted radio transmitters in 24 toads, measured the length of the toads' legs, and allowed the animals to wander for 3 days. For a second study, Phillips *et al.* tested the prediction that longer-legged toads should be more common among the first wave of arrivals at newly colonized sites. Figure 3.24 shows the results of both studies. Based on these two studies, can we conclude that the spreading of the cane toad plague is being supported by natural selection favoring the evolution of long legs? Why or why not?

Analyse the data 2

Reimchen (2000) was also interested in the relation between handling time and escape rate of the two morphs of sticklebacks. He defined handling time as the interval from first capture to final swallowing or final rejection of the prey by the trout. He reasoned that a long handling time probably indicates that the trout is experiencing some difficulty with the stickleback prey. (a) What factors, besides number of stickleback plates, might influence handling time? (b) Do the data summarized below indicate that handling time is related to escape rate of the two morphs?

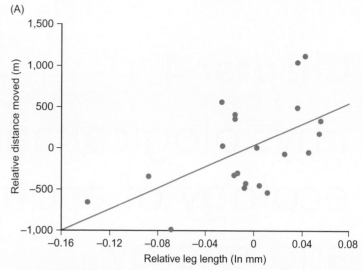

(A)

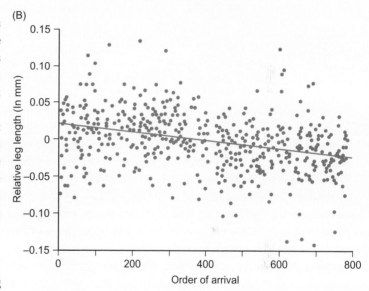

(B)

Figure 3.24 A. Relationship between relative leg length and distance moved in an advancing group of cane toads. B. Relative leg length in relation to order of arrival at sites that are newly colonized by cane toads.

Handling time (sec)	# Escapes by low morph	# Low morph eaten	# Escapes by complete morph	# Complete morph eaten
0–15	4	181	7	166
15–30	2	69	9	64
30+	13	46	24	45

Chapter 4

Physiological and evolutionary ecology of acquiring nutrients and energy

INTRODUCTION

In 1998, the Sea-viewing Wide Field-of-view Sensor (SeaWiFS) project sponsored by the National Aeronautic and Space Association (NASA) launched a satellite into space with a mission of obtaining color data of the ocean's surface. The color-enhanced picture of the Earth sent back by the SeaWiFS satellite showed that the ocean is not a uniform blue color but also has various shades of green, yellow, and orange. These non-blue colors represent high concentrations of phytoplankton, a diverse group of free-floating microorganisms. These diverse groups share one important feature: They all carry out photosynthesis. The ocean is green because each phytoplankton individual uses molecules of the pigment chlorophyll to trap light energy, and chlorophyll molecules reflect green light. NASA scientists use the intensity of the ocean's green color as an assay for how many phytoplankton live at or near the ocean's surface. But what can be gained from knowing phytoplankton abundance?

Our exploration of the ecology of acquiring nutrients and energy begins with a case study that questions the prevailing understanding of which nutrients are often in short supply in oceans. Plants require certain nutrients in order to carry on photosynthesis, a process that is the foundation of most ecosystems, because it converts inorganic carbon to energy-rich carbohydrates, which most organisms use to fuel their metabolic processes. Of course many species don't photosynthesize, and get their energy by eating other organisms or by decomposing other organisms or their products. But prey species, including plants, have evolved a diversity of defenses to avoid being eaten.

KEY QUESTIONS

4.1. How does energy enter an ecosystem?

4.2. What is the impact of nutrient availability on species distribution and abundance?

4.3. What are energy sources for heterotrophs?

4.4. How does the ratio of available nutrients influence ecological processes within a community?

4.5. How can organisms avoid being eaten?

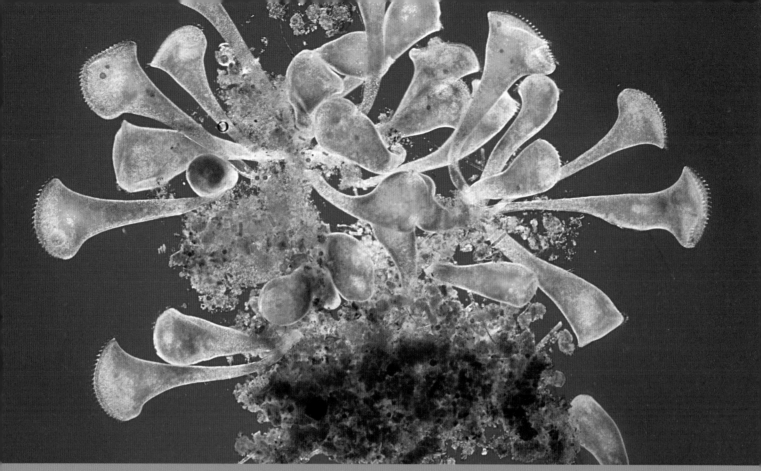

Like all organisms, phytoplankton in the oceans need energy to carry out their metabolic processes, and nutrients – chemical substances that support growth, development, and reproduction. The energy comes from photosynthesis, while the nutrients are taken up directly from the ocean. In many cases, the abundance of a particular species of phytoplankton is limited by the availability of certain nutrients. In those cases, the nutrient in short supply is called a limiting resource. In many marine ecosystems, inorganic nitrogen – in the form of nitrite, nitrate, or ammonium – is thought to be a limiting resource for phytoplankton populations.

Thus marine ecologists were puzzled by large regions of the northern and southern Pacific Ocean and Antarctic Ocean that had high inorganic nitrogen concentration but low chlorophyll concentration. John Martin and Steve Fitzwater (1988) suggested that perhaps nitrogen was not a limiting resource for these phytoplankton populations, as many researchers had previously thought, but that iron might be deficient. Their rationale was that iron would primarily enter marine ecosystems through windblown dust, which would be abundant near shorelines, but much less abundant in waters far removed from landmasses.

To test their hypothesis, Martin and Fitzwater journeyed to the northeast Pacific Ocean and collected seawater, which of course also contained nutrients and numerous species of phytoplankton. On board their ship, they transferred this seawater to 30-l bottles, and established the following experimental groups: one group received an infusion of 1 nM Fe (nanomoles of iron) per kilogram of water (1 Fe treatment); a second group received a 5-nM infusion (5 Fe treatment); and a third group received a 10-nM infusion (10 Fe treatment). A fourth group, the control, received no iron supplementation. Martin and Fitzwater did two replications of each experimental treatment, including the control. They predicted that if iron is a limiting resource for phytoplankton abundance, adding iron would increase phytoplankton abundance, as measured by chlorophyll abundance. They also predicted that greater iron supplementation would lead to greater phytoplankton abundance.

In accordance with Martin and Fitzwater's prediction, chlorophyll abundance increased substantially in the iron-supplemented groups, with the greatest increase occurring after 4 days

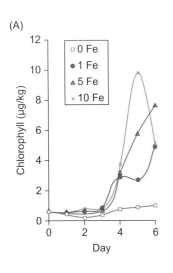 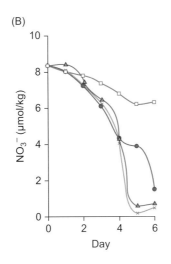 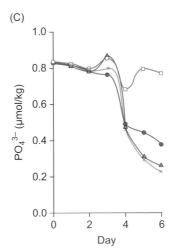

Figure 4.1 A. Change in phytoplankton abundance as measured by chlorophyll levels in Martin and Fitzwater's iron supplementation experiments. B. Change in nitrate levels. C. Change in phosphate levels.

in the 10 Fe treatment (Figure 4.1A). Interestingly, there was a correlated decrease in dissolved nitrate (Figure 4.1B) and phosphate (Figure 4.1C) levels as well. Martin and Fitzwater argued that the growing populations of phytoplankton in the experimental groups were using up these nutrients.

Martin and Fitzwater recognized that conclusions were limited by two weaknesses in their experimental design common to large-scale experiments: very few replications, which reduces confidence in their findings, and the likelihood that the response of organisms to iron supplementation will probably be different in a 30-l bottle than in the open ocean. At the end of this chapter we will revisit experimental iron supplementation with a discussion of follow-up studies to this initial work.

In the process of photosynthesis, phytoplankton use light energy from the Sun to convert CO_2 dissolved in water into complex molecules that build their bodies, and provide energy for their growth, development, and reproduction. Based on the SeaWiFS project color imagery, scientists estimate that these tiny organisms are responsible for approximately 50% of the global photosynthetic activity (Falkowski 2002). Consequently, phytoplankton are critical components of marine ecosystems, because the vast majority of other marine organisms use phytoplankton as their source of energy and nutrients, consuming them either directly or indirectly and leading to a web of connections among marine organisms (Chapter 16).

On land, plants use photosynthesis to create complex molecules that build their bodies and provide them with energy. Like their marine counterparts, plants are the basis of terrestrial ecosystems, providing energy and nutrients to terrestrial consumers. Let's look at the process of photosynthesis in more detail so we can understand how organisms use photosynthesis to meet their energetic needs.

4.1 HOW DOES ENERGY ENTER AN ECOSYSTEM?

Phytoplankton are autotrophs – organisms that synthesize complex carbon molecules from simple inorganic molecules. These complex carbon molecules can then be reassembled to synthesize other complex molecules such as proteins, nucleic acids, or fats. They may also be broken down to generate ATP to fuel metabolic processes, including the synthesis of other complex molecules. Thus autotrophs synthesize molecules that provide energy both to the autotrophs themselves and, indirectly, to nearly all other organisms. All photosynthetic autotrophs use light energy to manufacture complex carbon molecules from CO_2.

Photosynthesis is arguably the most important ecological process on Earth. The vast majority of Earth's ecosystems depend on photosynthesis as their basis for harnessing energy. The basic chemical reaction of photosynthesis converts radiant energy from the Sun into chemical bond energy in sugar molecules:

$$6CO_2 + 12H_2O + \text{radiant energy} \rightarrow C_6H_{12}O_6 (\text{a 6C sugar}) + 6H_2O + 6O_2 (\text{waste product})$$

Realize that the sugar is the product that is useful to the plant, and the oxygen is actually a waste product to the plant.

Photosynthesis is conventionally divided into two stages; the light reactions and the carbon fixation reactions, also known as the Calvin cycle. In the light reactions, the radiant energy from sunlight is converted or transduced into the chemical bond energy of ATP and NADPH, two high-energy molecules. In addition, oxygen is a waste product of the light reactions, and is exported into the surrounding environment. In the carbon fixation reactions, the chemical bond energy in ATP and NADPH is used to power the reactions that create complex carbohydrates from CO_2 (Figure 4.2).

You may recall from your introductory biology course that both reactions are very complex. I will review only enough detail to allow us to address a few questions about how the process of photosynthesis operates under differing environmental conditions.

Converting solar energy into chemical bond energy

Light is a form of electromagnetic radiation, or electromagnetic energy. Within a typical leaf, transduction, the conversion of light energy to chemical bond energy, occurs at one of many photosystems – light-capturing units which comprise a central

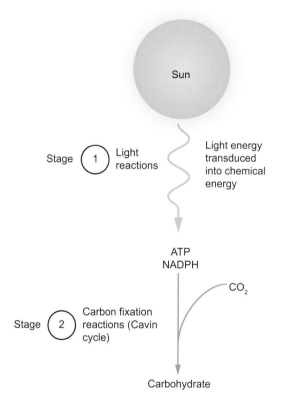

Figure 4.2 Overview of light reactions and carbon fixation reactions (Calvin cycle).

complex with two chlorophyll *a* molecules surrounded by 200–300 *accessory pigment molecules*. The chlorophyll *a* molecules in the central complex are closely associated with a primary electron acceptor molecule. The accessory pigment molecules include other chlorophyll molecules and carotenoids. In contrast to chlorophylls, the carotenoids are effective at absorbing green light. The net effect of numerous spectrally diverse accessory pigment molecules is that the leaf can harvest much more light than it would without these accessory pigment molecules. When light strikes an accessory pigment molecule, the energy is passed to a nearby accessory molecule, ultimately reaching a chlorophyll *a* molecule in the central complex. One electron of the central pair is boosted to a higher energy level, where it is captured by the special electron acceptor molecule; this is the actual act of transduction. The high-energy electron goes through a series of exergonic, or energy releasing reactions, to generate the high-energy end products of the light reactions, ATP and NADPH. Water comes into play in this process because it supplies the replacement electron to the chlorophyll *a* molecule that donated its electron to the primary electron acceptor molecule (Figure 4.3).

Earlier, I described oxygen as a waste product of photosynthesis because it plays no role in synthesizing organic molecules – it is simply a byproduct of splitting water to generate replacement electrons. However, this waste product is essential to all life forms that use aerobic respiration, including most photosynthetic organisms. The oxygen generated by photosynthesis also produces ozone, which, when present in the stratosphere, protects organisms from high-energy radiation that kills cells. We will discuss this process in considerable detail in Chapter 23.

Identifying the details underlying the light-dependent reactions is beyond our present goals, but you should recognize that this process requires the integrated interaction of numerous enzymes and other types of molecules, all of which must be ingested or synthesized by the organism. For example, the light reactions use iron, manganese, and copper to synthesize NADPH. Martin and Fitzwater (1988) were aware that NADPH synthesis

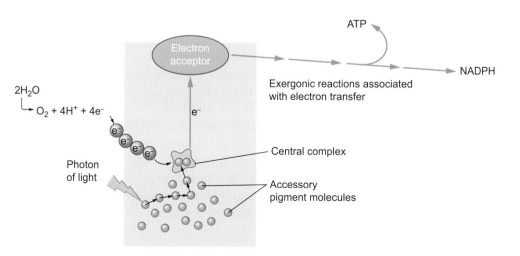

Figure 4.3 Very simplified outline of the light reactions.

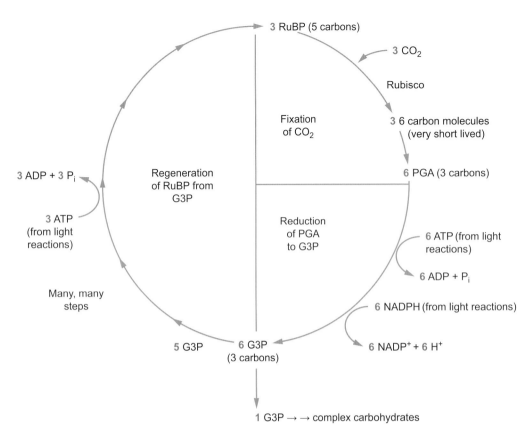

Figure 4.4 The Calvin cycle. P_i refers to inorganic phosphates.

required large quantities of iron when they proposed that iron limited phytoplankton abundance in the open oceans.

In the carbon fixation reactions, CO_2 is converted into sugar with the help of the ATP and NADPH molecules created by the light reactions. Differences in air temperature and moisture availability have selected for plants that use different mechanisms for carbon fixation.

Using chemical bond energy to power the construction of complex carbohydrates

Because they are generally rooted to a specific location, terrestrial plants are, to some extent, captives of rapid environmental change. To survive and reproduce, terrestrial plants must be adapted to the temperature, light, nutrient, and moisture levels in their environment, and must have regulatory mechanisms for dealing with variation in these critical environmental factors. However, regulatory mechanisms that allow a plant to adjust to variation in one environmental factor may interfere with its ability to adjust to variation in other environmental factors.

As one example, plants rely on passive diffusion to import CO_2 into their cells for carbon fixation. Most plants can regulate CO_2 import by opening and closing stomata, which are pores on the leaf's surface through which gas exchange occurs. However, at high temperatures and in dry conditions, open stomata can cause rapid loss of water. But if stomata are shut to reduce water loss,

CO_2 import slows down dramatically and the plant may not produce enough carbohydrate to fuel its metabolic processes. This tradeoff between carbon dioxide import and water loss has selected for alternative mechanisms of carbon fixation in different environments.

C_3 photosynthesis: the most common carbon fixation pathway

Most species of plants use C_3 photosynthesis, employing the Calvin cycle in the mesophyll cells of the leaf, just below the epidermal layer. Open stomata in the epidermal layer allow CO_2 to diffuse into the mesophyll cells. In the *fixation or carboxylation phase*, CO_2 from the atmosphere combines with a 5-carbon ribulose bisphosphate (RuBP) molecule to produce a six-carbon molecule, which is quickly broken down to two 3-carbon molecules of 3-phosphoglycerate (PGA) (Figure 4.4). (These products are the source of the name of this type of photosynthesis.) This reaction is catalysed by the enzyme ribulose-1,5-bisphosphate carboxylase oxygenase, which we will mercifully shorten to rubisco. Fixation is followed by the *reduction phase*, in which PGA is reduced into a three-carbon sugar glyceraldeaphyde-3-phosphate (G3P). This is a highly endergonic (energy-absorbing) reaction, and consumes ATP and NADPH that were generated in the light reaction. Some of this sugar will go through a series of reactions to synthesize other carbohydrates such as glucose, cellulose, or starch that are essential for the plant. But most of the G3P will go through a third phase of *RuBP regeneration*, in which new molecules of RuBP are created from the G3P molecules. These new RuBP molecules can start the cycle anew and fix more CO_2.

Rubisco is a very unusual enzyme for several reasons. First, by enzyme standards it is glacially slow, catalysing only a few reactions per second, which is less than 1% the rate of many other enzymes. Related to its slowness is rubisco's tremendous abundance; it is the most abundant enzyme on Earth, making up about 50% of the leaf protein in many plant species. Presumably a leaf needs a great deal of slow-acting rubisco to catalyse the synthesis of sufficient carbohydrate for maintenance and growth. A final point about rubisco is that it can catalyse two very distinct reactions: it can function both as an enzyme that catalyses carbon dioxide uptake (carboxylase) and also as an enzyme that catalyses oxygen uptake (oxygenase). This oxygenation initiates the process of photorespiration, producing an equal amount of PGA, which is used in the Calvin cycle, and 2-phosphoglycolate, which may be toxic to plants in high concentration, and when broken down results in a loss of CO_2 and ATP. In the process of breaking down 2-phosphoglycolate, 25% is lost as CO_2 and the remaining 75% of the carbon is used to make a molecule of PGA. This highly complex, energy-consuming process is coordinated among several organelles within the cell.

Thus rubisco catalyses two competing reactions. One (carboxylation) is clearly beneficial, while the second (photorespiration) is, at least on the surface, detrimental, in its losses of CO_2 and ATP. Oxygenation and, thus, photorespiration are more likely to occur at high temperatures, for two reasons. First, as we will discuss in Chapter 5, many enzymes work best at high temperatures up to about 40°C. Thus it is not surprising that the rate of the carboxylation reaction and the oxygenation reaction both increase with temperature. Importantly, as temperatures increase to 40°C, the rate of rubisco's catalysis of O_2 uptake increases more rapidly than its catalysis of CO_2 uptake (Jordan and Ogren 1984). A second consideration is that high temperature reduces the solubility of CO_2 in the mesophyll cells more than it does the solubility of oxygen. Thus the CO_2/O_2 ratio in mesophyll cells decreases at high temperatures, which makes it more likely that rubisco will take up oxygen rather than CO_2.

Clearly, hot and dry environments create serious physiological problems for plants that use C_3 photosynthesis. Under these conditions the solubility of CO_2 in mesophyll cells decreases, and rubisco's relative rate of catalysis of CO_2 uptake is reduced, leading to an increase in photorespiration. Increasing CO_2 uptake by opening stomata is not a viable option, because that will cause too much water loss. In a hot and dry environment, plants would be at a selective advantage if they could reduce these negative effects. Two important mechanisms have evolved that do exactly that.

C_4 photosynthesis: moving the Calvin cycle to a better location

One alternative photosynthetic pathway is C_4 photosynthesis. An estimated 5% of all plant species, in at least 19 plant families, use this form of photosynthesis. Because several of these families are only distantly related, C_4 photosynthesis is likely to have evolved independently numerous times (Sage 2004).

C_4 photosynthesis uses the enzyme phosphoenol pyruvate carboxylase, **or** PEPCase, which has a higher affinity for CO_2 than does rubisco, and does not catalyse photorespiration. Under hot and dry conditions, C_4 plants can partially close their stomata, reducing water loss while still importing enough CO_2 into the leaf for carboxylation. In fact, most C_4 plants have very high photosynthetic rates under hot and dry conditions.

Functionally, there are three important processes in C_4 photosynthesis. First, a mesophyll cell takes up CO_2 using PEPCase to convert phosphoenol pyruvate, through a several-step process, into a 4-carbon acid (the source of the name of the process). Second, the 4-carbon acid travels to a nearby bundle sheath cell, where it dumps its CO_2. Lastly, the CO_2 concentrated in the bundle sheath cell goes through the Calvin cycle, catalysed by rubisco (Figure 4.5).

This seems like a great deal of complexity. It is. As we have discussed several times, adaptations usually have tradeoffs associated with them. Increased complexity associated with

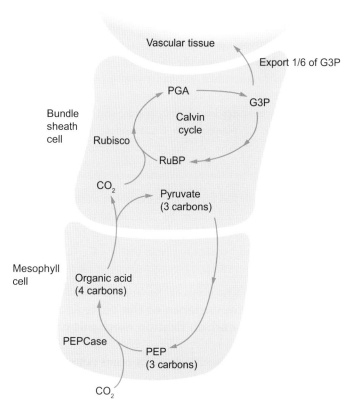

Figure 4.5 C_4 carbon fixing reactions. See Figure 4.4 for more details on the Calvin cycle.

photosynthesis is costly to the plant because C_4 plants require energy to transport the CO_2 to the bundle sheath cell, and to go through different sets of reactions in the mesophyll and bundle sheath cells. But the benefits can be substantial. Carbon dioxide that is transported to the bundle sheath cells can become more than ten times as concentrated as atmospheric CO_2. This improved CO_2/O_2 ratio greatly reduces the frequency of photorespiration, and increases the efficiency of carbon fixation. Concentrating CO_2 is costly, however, because many C_4 plants give up some of their hard-earned carbon to produce a waxy layer in their bundle sheath cells that restricts the diffusion of CO_2 back into the mesophyll cells.

There is one other set of carbon fixation reactions that creates conditions to support growth and reproduction under hot and dry conditions.

CAM photosynthesis: moving the Calvin cycle to a better time

Plant species in the succulent family Crassulaceae, and approximately 25 other families as well, use a third process of carbon fixation known as Crassulacean acid metabolism photosynthesis (or CAM photosynthesis). CAM photosynthesis is extremely efficient in minimizing water loss (Figure 4.6A).

Plants using CAM photosynthesis open their stomata at night when temperatures are cool and relative humidity is high, and close them during the day. They use exactly the same metabolic pathways as C_4 plants, but they store their products – primarily malate – in vacuoles within the mesophyll cells. Then during the day, with their stomata closed, CAM plants break down the malate into pyruvate and CO_2, and go through a similar series of steps as C_4 plants, but the steps are carried out in their mesophyll cells rather than in bundle sheath cells (Figure 4.6B). By keeping their stomata closed during the day, CAM plants have much lower water loss than C_3 plants, and somewhat lower water loss than C_4 plants. Both CAM and C_4 plants conserve water by separating the process of incorporating CO_2 into the cell (by forming organic acids) from the process of producing sugar.

Natural selection has favored the evolution of three different mechanisms of carbon fixation. To understand the adaptive significance of these three mechanisms, we need to explore the respective costs and benefits to plants that use C_3, C_4, and CAM photosynthesis (Table 4.1).

Environmental dependence of the costs and benefits of each carbon fixation mechanism

This diversity of photosynthetic mechanisms raises a series of interesting questions. Why has selection favored three different processes of carbon fixation? Why don't all plants use C_4 photosynthesis, which seems to offer the best of both worlds, in terms of conserving water and avoiding the costs of photorespiration?

Why do most plants use rubisco as the initial enzyme for CO_2 capture, given that rubisco has some significant shortcomings?

To answer these questions, we need to consider the costs and benefits of each process in the context of the environment that each plant experiences. Furthermore, we also need to remember that natural selection does not always lead to adaptations that maximize benefits and minimize costs, but to ones that work best under prevailing environmental constraints. This is particularly true given that environmental conditions on Earth are constantly changing. We will begin with a comparison of energy and water use associated with each mechanism.

Energy and water

C_3 plants require fewer steps and less energy to assimilate one CO_2 molecule than do C_4 or CAM plants. In particular, C_4 and CAM plants must resynthesize PEP after the 4-carbon acids release CO_2, which requires two extra ATP molecules. Consequently, at low temperatures, C_3 photosynthesis has the potential to convert more CO_2 into plant biomass. However, as temperatures rise and photorespiration increases, C_4 plants are, overall, more efficient than C_3 plants at converting CO_2 into biomass, because they lose very little CO_2 and energy to photorespiration. The performance of CAM in relation to C_3 plants also improves at high temperatures; however, growth rates of CAM plants may be limited by their limited storage capacity for the CO_2 they take in during the night.

To survive, plants must regulate their water levels (see Chapter 5). CAM plants only open their stomata at night, so they experience substantially reduced water loss in comparison to C_4 plants, which keep their stomata slightly open during the day. Both CAM and C_4 plants have substantially reduced water loss in comparison to C_3 plants, which need stomata open to maintain high photosynthetic rates (see Table 4.1).

The costs and benefits of the three photosynthetic mechanisms are environment dependent. Consequently, we would expect to see differences in the spatial distribution of C_3, C_4, and CAM plants that reflect these varying costs and benefits.

Ecological constraints on terrestrial plant distribution

Many C_4 plants are grasses. If C_4 photosynthesis is an adaptation to hot environments, that adaptation should be reflected in the global distribution of C_4 plants. We would predict that C_4 plants would tend to be most prevalent near the equator while C_3 plants would be more common in higher latitudes. An analysis by Rowan Sage and Monson (1999) of data collected by numerous researchers across the world generally supports this prediction, with C_4 grasses completely replacing C_3 grasses in many locations within the tropics (Figure 4.7).

Because they use so little water (see Table 4.1), we would expect CAM plants to be most common in very dry environments. This is true up to a point. Almost 2000 species of CAM plants are desert succulents in the family Cactaceae and Agavaceae (see Figure 4.6A). However, more than 20 000 species of

(A)

(B)

Figure 4.6 A. Plants that use CAM photosynthesis. (Upper left) *Sempervivum montanum*, Family Crassulaceae, a succulent plant adapted to a dry, cold habitat. (Upper right) The cardon cactus, *Trichocereus pasacana*, in Argentina. (Lower left) The pineapple, *Ananas comosus*, native to tropical South American forests, and (lower right) an epiphytic orchid from Thailand (genus *Bulbophyllum*). B. CAM photosynthesis occurs within the mesophyll cell.

CAM plants have been identified as rainforest species (Luttge 2004). This seems to fly in the face of the hypothesis that CAM photosynthesis is an adaptation to reduce water loss.

Surprisingly, some plants in rainforests do experience severe water limitation. This is particularly true for epiphytes – plants that grow upon other plants, and depend on their hosts for support (see Figure 4.6A). Most tropical CAM plants are epiphytes and must assimilate moisture from the air or moisture that runs down tree trunks. Particularly if there is a dry season, both moisture sources may be extremely limited, creating selection pressure among epiphytes for the evolution of adaptations, such as CAM photosynthesis, that conserve water.

Table 4.1 Costs and benefits of the three photosynthetic mechanisms.

Measure of cost or benefit	Photosynthetic mechanism		
	C_3	C_4	CAM
# ATP per CO_2 molecule assimilated	3	5	5
# NADPH per CO_2 molecule	2	2	2
Rate of photorespiration under standard atmospheric conditions	c. 25%	very low	very low
CO_2 conversion efficiency at low and moderate temperatures	high	intermediate	low
CO_2 conversion efficiency at high temperatures	relatively low	high	moderate
Moles water lost per mole of CO_2 assimilated	400–500	250–300	50–100

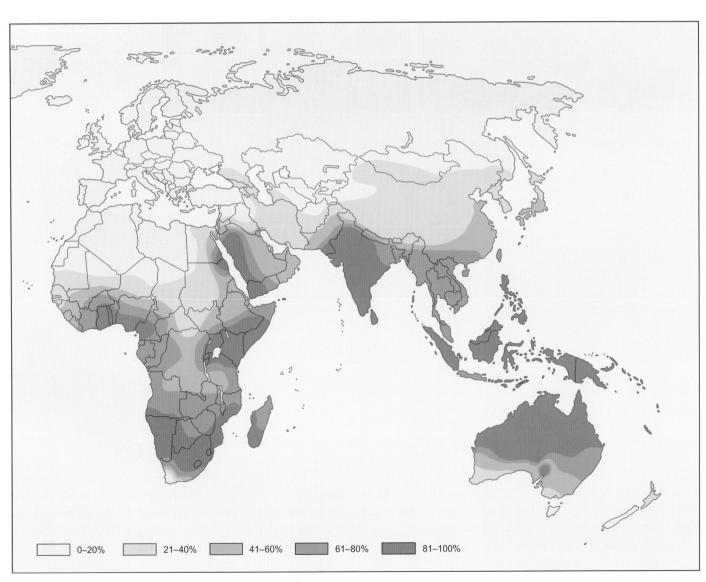

Figure 4.7 Percentage of C_4 grasses in the eastern hemisphere.

Thinking ecologically 4.1

We've discussed that C_4 plants have lower water loss than C_3 plants. Hypothesize how moisture considerations might influence the global distribution of C_4 plants in comparison to C_3 plants. Does Figure 4.7 support your hypothesis?

Terrestrial plants are biochemically well adapted to their environments: C_3 plants, which are most abundant, live in areas with sufficient water and low to moderate temperatures, whereas C_4 and CAM plants are adapted to the demands of drier and hotter environments.

Given the costs of each carbon fixation mechanism, why hasn't a better alternative to rubisco evolved to catalyse the first step of the Calvin cycle? Plants would then have high photosynthetic efficiency in all environments without paying the costs associated with photorespiration.

Evaluating photorespiration

Why do C_3 plants go through photorespiration, in spite of its disadvantages? To answer this question, we should consider an important lesson from Chapter 3, and recognize that "why" can mean many different things in different contexts. First, we could be asking the question of why C_3 plants use such a costly process in which carboxylation and oxygenation are in competition with each other. Second, we might be asking whether rubisco has improved its carboxylation efficiency over time. This question addresses the evolutionary history of the rubisco enzyme, and the genes that code for it. Lastly, we might ask whether there are some important benefits to going through the oxygenation reaction. We will begin by considering the evolution of rubisco, and then discuss the potential benefits of photorespiration.

EVOLUTION OF RUBISCO AND PHOTORESPIRATION. One key part of this puzzle is that photosynthesis evolved in a world with virtually no atmospheric O_2. Ancestral cyanobacteria, which became numerous in the oceans about 2.7 billion years ago, most likely used rubisco to fix CO_2 or bicarbonate from the water, releasing large quantities of O_2 as a waste product of photosynthesis. Atmospheric CO_2 levels were much higher than at present, so the ratio of CO_2/O_2 favored carboxylation. Atmospheric oxygen is first detected from about 2.5 billion years ago, and even 1 billion years ago, oxygen levels were less than 10% of their current value (Lyons *et al.* 2014). Thus for the majority of its history, rubisco has primarily catalysed the carboxylation reaction.

Of course, the world has changed and by about 600 million years ago, oxygen levels were close to present values, and natural selection would have favored a more efficient form of rubisco. In reality, there are at least three very distinct forms of rubisco, and within each form there is substantial variation in the amino acid

sequence and structure. Consequently, among species that use different forms of rubisco, there is a great amount of variation in the *specificity factor*, which measures how efficiently the rubisco enzymes bind CO_2 in preference to O_2, when both molecules are present in equal concentrations.

If rubisco has evolved as a result of natural selection, we would expect that rubisco's specificity factor might be highest in organisms that experience high O_2 levels, and lowest in organisms that experience low O_2 levels. Species such as vascular plants have lived in high-O_2 environments throughout their evolutionary history. Consequently, their rubisco has experienced strong selection pressure to bind CO_2 efficiently – to have high specificity. In contrast, photosynthetic anaerobic bacteria have experienced almost no O_2 throughout their evolutionary history, so their rubisco has experienced virtually no selection pressure to bind CO_2 efficiently and their specificity factors are among the lowest measured. Rubisco from cyanobacteria have intermediate specificity factors, but cyanobacteria also use transport systems to concentrate CO_2 at the fixation site (Buchanan *et al.* 2000). These findings suggest that plants with greatest exposure to O_2 have evolved mechanisms for increasing rubisco's specificity.

BENEFITS OF PHOTORESPIRATION. But perhaps photorespiration does more for plants than rescue a molecule of PGA while wasting considerable energy and CO_2. Several lines of evidence support the hypothesis that photorespiration is actually beneficial to plants. In one experiment, researchers moved tobacco plants from normal atmospheric conditions to a very low (<2%) oxygen atmosphere, to see how changing the CO_2/O_2 ratio would influence photosynthesis and plant growth. The most straightforward hypothesis is that severe reduction in O_2 should increase the rate of carbon fixation because photorespiration is reduced dramatically under these conditions. This is precisely what happened; 1 h after plants were moved to a low-oxygen atmosphere, the rate of carbon fixation increased by 62% in comparison to control plants that remained in a standard oxygen atmosphere. However, by 3 days, the rate of carbon fixation declined to only 77% of control plants. Plants left in this low-oxygen atmosphere for 7 days had leaves with a smaller surface area and fewer starch granules stored in their cells, presumably a result of lower photosynthesis rates (Migge *et al.* 1999).

So why does a reduction of oxygen cause such a serious decline in plant function? One problem for plants is that their leaves are constantly bombarded by high-energy violet and ultraviolet radiation. This can lead to *photoinhibition* – damage to the photosystems used in the light reaction of photosynthesis. Under normal circumstances, this damage is quickly repaired. One hypothesis is that photorespiration actually benefits plants by protecting the photosystems used in the light reactions from being damaged by high-energy light that strikes the leaves. A second hypothesis is that photorespiration does not reduce

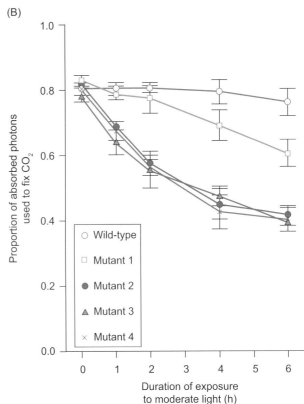

Figure 4.8 A. *Arabidopsis thaliana*. B. Photoinhibition of photosystem II under moderate light (200 µmol photons/m²/s) in wild-type and four mutant strains of *Arabidopsis* that have impaired photorespiration pathways. The y-axis represents the maximum efficiency of photosynthesis, as measured by the proportion of photons absorbed by photosystem II that are actually used to fix CO_2 molecules. The maximum under most conditions is about 0.85 (or 85%).

photosystem damage, but instead, it is essential for the photosystem repair process.

Shunichi Takahashi and his colleagues (2007) knew from previous work that there was a link between high light intensity and photoinhibition in *Arabidopsis thaliana*, a small flowering plant that has become a model organism in developmental biology and genetics (Figure 4.8A). *Arabidopsis* has a very small genome, which simplifies making connections between development and the underlying genetics. Numerous mutants that interfere with photorespiration have been identified in *Arabidopsis*; each mutation may interfere with a different stage of photorespiration. The researchers investigated the performance of these mutant plants under different light conditions to see whether photorespiration helps plants by protecting their photosystems from damage caused by high-energy radiation, or alternatively, whether photorespiration facilitates the repair process after the photosystem is already damaged.

In one experiment the researchers exposed wild-type and mutant leaves to continuous levels of moderate light, and measured the efficiency of photosystem II, one of two photosystems in the light reaction for most plants. As they expected, photosystem performance was much poorer in the mutants, declining sharply in three of the four and moderately in one, while wild-type photosynthetic efficiency remained high (Figure 4.8B). These results reassured the researchers that the loss of

photorespiration in these mutants was associated with substantial photoinhibition.

Next, Takahashi and his colleagues blasted *Arabidopsis* leaves with very high light intensities that they knew would cause photoinhibition even in the wild-type plants. They then allowed the plants to recover in either low or moderate light. The researchers reasoned that if photorespiration mutants were deficient in the repair process, then exposing them to different light environments after severe photoinhibition might reveal differences in the rate of recovery. Assuming that the mutants could do some photosystem repair, they would be most successful when recovering under low light intensities, because there would be little new damage to the photosystem during recovery. However, at moderate light intensities, the mutants might be unable to repair the existing damage at a high enough rate to keep up with the rate of new photosystem damage caused by the moderate-intensity light during recovery. This is precisely what they discovered. At low light intensity, the mutants recovered photosystem efficiency almost as quickly as the wild-type plants. But at moderate light intensity, the mutants showed very little recovery (Figure 4.9). The researchers concluded that photorespiration mutants suffer from photoinhibition because they have reduced rates of photosystem repair. By extension, at least in *Arabidopsis*, photorespiration does not prevent photosystem damage, but rather it facilitates the repair of damage caused by high-energy light waves.

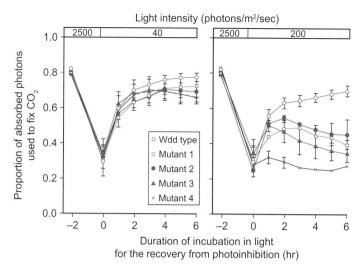

Figure 4.9 Photosynthesis efficiency in wild-type and four mutant strains of *Arabidopsis* following a 2 h exposure to very high intensity light (2500 µmol photons/m²/s). A. Recovery occurs in very low light (40 µmol photons/m²/s). B. Recovery occurs in moderate light (200 µmol photons/m²/s).

Though light energy is the primary source of energy for autotrophs, there are other energy sources that can be exploited for carbon fixation.

Alternatives to light as energy sources for carbon fixation

Autotrophic bacteria and archeans can transform CO_2 into sugar via six different paths. Perhaps the most glamorous and well-known example is the sulfur-oxidizing bacteria of thermal vent communities that generate ATP by oxidizing hydrogen sulfide (H_2S) several kilometers below the ocean surface. These bacteria use their ATP to form complex organic molecules. The giant tube worm, *Riftia pachyptila*, is filled with these bacteria, and uses them as its source of organic molecules. Other chemosynthetic autotrophs oxidize iron, diatomic hydrogen (H_2), carbon monoxide (CO), nitrite (NO_2^-), or ammonium (NH_4^+), to generate energy. Recall from Chapter 2 that an entire ecosystem is based on ammonium oxidation that occurs in Antarctica's subterranean lakes.

Autotrophs support ecosystems by transducing radiant energy or the chemical bond energy of simple inorganic substances into the chemical bond energy of complex organic molecules. But, as we discussed at the beginning of this chapter, all organisms need nutrients, as well as energy, to construct their essential complex molecules and power their biochemical reactions.

4.2 WHAT IS THE IMPACT OF NUTRIENT AVAILABILITY ON SPECIES DISTRIBUTION AND ABUNDANCE?

Martin and Fitzwater's study introduced us to some of the nutrients that are essential to marine phytoplankton. An essential

Table 4.2 Some common essential nutrients, and their major functions in organisms.

Element	Function
Nitrogen (N)	Nucleic acid, protein, and ATP structure
Phosphorus (P)	Backbone for nucleic acids, structure of ATP, phospholipids
Potassium (K)	Cofactor for protein synthesis, solute controlling osmoregulation, ion important in neuronal function
Calcium (Ca)	Structure of cell walls and membranes, regulates cell activity
Magnesium (Mg)	Component of chlorophyll, activates enzymes
Sulfur (S)	Protein structure
Chlorine (Cl)	Regulates many biochemical processes such as photosynthesis, osmoregulation, neuronal function
Iron (Fe)	Chlorophyll synthesis, electron transport, protein structure

nutrient is one required for an organism to carry out normal metabolic processes, and/or to build structures required for growth or reproduction. Martin and Fitzwater demonstrated that nitrogen and phosphorus as well as iron are consumed by growing phytoplankton populations, suggesting that all three nutrients are essential to the phytoplankton in their study. Table 4.2 lists some of the most common essential nutrients for many organisms.

Limiting nutrients for many plant species: phosphorus and nitrogen

Both phosphorus and nitrogen are essential nutrients for all organisms, because they are prevalent in so many types of biologically important molecules (Table 4.2). In terrestrial plants, phosphorus and nitrogen are probably the two most common limiting nutrients. Nitrogen enters soils through the process of nitrogen fixation, a biological process in which bacteria convert nitrogen gas to ammonia. Nitrogen fixation is also a physical process powered by lightning, which converts nitrogen gas to nitrogen oxides, which then form nitrates in rainwater and fall to Earth, but this physical process contributes much less nitrogen to soils than the biological process. Once in soils, nitrogen is recycled through uptake by plants, consumption by hetero-trophs, and decomposition (see Chapter 20).

Phosphorus enters the soils primarily through the weathering of rocks. The presence of rock, such as limestone, that is rich in phosphorus, will lead to high phosphorus soil concentrations. Alternatively, soils that originated from sandstones, or very ancient soils from wet climates that have experienced extensive leaching over the years, tend to be low in phosphorus. Compli-cating the picture, phosphorus in the soil is often bound tightly to clay particles, to organic matter, or to various cations such as Ca, Al, or Fe. Consequently, even if soils are high in phosphorus, very little of the mineral may be available for uptake by plants.

As one example of the relationship between nutrient availabil-ity and plant distributions, Rien Aerts and F. Stuart Chapin (2000) did a meta-analysis of nitrogen and phosphorus concen-trations in the leaves of deciduous species, whose leaves live for less than a year, and evergreen species, whose leaves persist for

considerably more than a year. Meta-analyses are systematic analyses of data collected independently by other researchers. Meta-analyses allow researchers to compile the data from a large number of studies, and to evaluate each species or group of species as one data point (see Dealing with data 4.1).

Aerts and Chapin (2000) hypothesized that habitats with abundant nutrients would favor plant species that grew very quickly so they competed well for light for photosynthesis. In contrast, habitats with limiting availability of essential nutrients would favor plant species that were very efficient in conserving these limiting resources, at the expense of rapid growth. In support of this hypothesis, their meta-analysis shows that the relatively short-lived leaves of deciduous species have, on aver-age, a significantly higher concentration of nitrogen and phos-phorus than do leaves of evergreen species (Figure 4.10). Nutrient-rich habitats are, in effect, selecting for deciduous trees that invest in constructing proteins used in photosynthesis, such

Dealing with data 4.1 Meta-analyses, box plots, and effect sizes

Ecologists use meta-analysis when they want to achieve some consensus about a problem that has been studied by numerous researchers in many different contexts. This usually involves compiling the results of many studies, and establishing some relatively unbiased method for accepting or rejecting studies into the analysis. For example, a researcher may stipulate that they will include all studies published in a certain group of journals in the past 20 years. While such criteria may seem unbiased, they discriminate against studies in other journals, or perhaps more problematically, against studies that were never published.

In some meta-analyses, including the Aerts and Chapin study, researchers present their findings with a box-plot graph, which reports a measure of central tendency for each treatment, such as the mean, median, or geometric mean, and percentiles of interest. Percentiles are the value of a variable below which a certain percentage of observations fall. In the meta-analysis by Aerts and Chapin, the 25th percentile is represented by the left side of each box, indicating that 25% of the studies showed mean concentrations below that value. The 75th percentile is represented by the right side of each box, indicating that 75% of the studies had mean values below that value. The left bar represents the 5th percentile, and the right bar represented the 95th percentile (Figure D4.1.1).

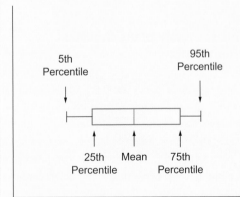

Figure D4.1.1

However, most meta-analyses calculate an effect size for each treatment or category being compared. For example, let's assume that one study considered by Aerts and Chapin compared nitrogen levels in one species of evergreen to those of one species of deciduous tree. Let's assume they found that the deciduous tree species had a mean of 20 mg N/g of biomass, and the evergreen species had a mean of 12 mg/g of biomass, with a combined or pooled standard deviation (s) of 4.0. We can calculate the effect size (Cohen's d) with the following formula:

$$d = \frac{\overline{x_1} - \overline{x_2}}{s},$$

where

x_1 is the mean N concentration for deciduous trees
x_2 is the mean N concentration for the evergreen tree species in that study, and
s is the combined or pooled standard deviation of N concentration of all studies used for the meta-analysis.

In this case,

$$d = (20 - 12)/4 = 2.0$$

The beauty of calculating effect size is that it allows you to add the results of this one study to the results of many other studies. If the mean effect size of the combined studies is positive, we can conclude from the sample that deciduous species in the studies had, on average, greater N content than evergreen species in the study.

However, we want to know if the combined effect is statistically significant. The most common approach we will present in this text is to calculate the mean effect size and in addition, the 95% confidence interval. If the mean is greater than 0, and the 95% confidence interval does not overlap 0, then we have a statistically significant effect (Figure D4.1.2). If the 95% confidence interval does overlap 0, then we have a statistically insignificant effect. In addition, many meta-analyses have more complex formulations that allow researchers to give more weight to studies with larger sample size or more replication.

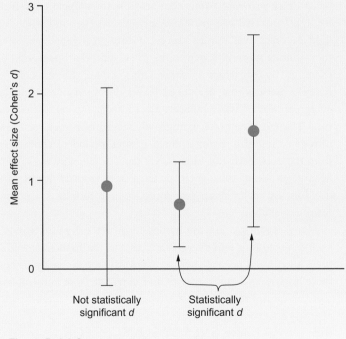

Figure D 4.1.2

as rubisco, and phosphorus-rich molecules, such as various RNAs, that transcribe and translate these proteins. On the other hand, nutrient-poor habitats favor evergreen trees that grow slowly and make efficient use of each nutrient molecule, by retaining leaves for as long as possible. But in order to retain leaves for a long period of time, an evergreen tree must invest a substantial percentage of its nutrients into molecules that help to stabilize its cell walls. Thus the differences in nutrient

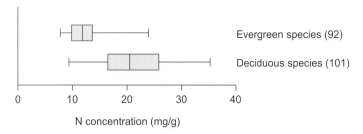

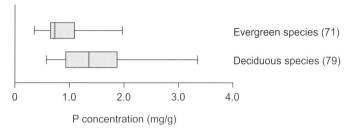

Figure 4.10 Box plots showing the distribution of leaf nitrogen (A) and leaf phosphorus concentrations (B). Values in parentheses are the number of observations or studies. The differences between evergreen and deciduous trees for both nutrients were statistically significant at $P < 0.05$.

concentrations in leaves may be a byproduct of the types of habitats that tend to be occupied by each of the two types of trees.

In reality many factors influence the nutrient content of leaves, including phylogeny, morphology, physiology, and nutrient availability in the local habitat. Thus it is difficult to be certain which factors determine patterns of nutrient utilization. We will discuss one study that addresses this problem.

Tree distributions and a tradeoff in nitrogen partitioning

Teruyuki Takashima and his colleagues (2004) were able to minimize these phylogenetic and habitat effects by comparing two species of deciduous oak, *Quercus crispula* and *Q. cerrata,* with two species of evergreen oak, *Q. acuta* and *Q. glauca*, native to the same cool temperate forest in Japan. First, the researchers tested the hypothesis that photosynthetic rates would be higher in the two deciduous oak species and lower in the two evergreen oak species. Then they tested the hypothesis that the deciduous oak species would allocate more nutrient resources to photosynthetic processes, and less to stabilizing cell walls, than would the evergreen oak species. As a result, the overall efficiency of photosynthesis per unit of nitrogen in the leaf would be higher for the deciduous species than for the evergreen species.

To test the first hypothesis, Takashima and his colleagues grew numerous individuals of each species of oak tree in a greenhouse under both low and high nitrogen conditions, so their trees would experience different levels of nitrogen availability. They measured the amount of CO_2 that was taken up by each tree over time. They predicted that the two deciduous species would have higher CO_2 uptake rates than the two

evergreen species across all nutrient levels. Their results show a clear trend supporting a higher photosynthetic capacity (as measured by CO_2 uptake) in the two deciduous species across the entire range of leaf nitrogen content (Figure 4.11A).

To test the hypothesis that the deciduous oak species would allocate more nutrient resources to photosynthetic processes and less to stabilizing cell walls than would the evergreen oak species, Takashima and his colleagues ground up leaves from each species, and used biochemical methods to extract and isolate the different proteins from the leaves. One finding was that the deciduous and evergreen species allocated similar amounts of nitrogen to protein (Figure 4.11B). This was not surprising, since these trees came from the same habitat and were raised under similar conditions in the greenhouse. The big difference between the two types of trees was that the deciduous species allocated more nitrogen to photosynthetic proteins such as rubisco (Figure 4.11C), while the evergreen species allocated more nitrogen to proteins that stabilized cell walls (Figure 4.11D). The researchers concluded that the evergreen species were less efficient photosynthesizers than the deciduous species because the nitrogen allocated to cell walls is not used for photosynthesis.

Takashima and his colleagues explain this relationship from an evolutionary perspective. Where nutrient availability is high, plants are more limited by access to light than they are by nutrient availability. In these habitats, natural selection will favor quick-growing plants that are able to compete effectively for light, and that can grow new leaves wherever and whenever light becomes available to harvest. Plants grow more quickly if they allocate more nitrogen to the photosynthetic apparatus, for example, by constructing more rubisco. But the tradeoff for quick growth is that deciduous leaves don't survive as long as evergreen leaves, which allocate more nitrogen to cell wall structure. Though they grow more slowly, evergreen trees retain the nitrogen they absorb for a much greater period of time. Thus from the perspective of biomes, we tend to see more evergreens in nutrient-poor soils such as taiga and tropical forest, and deciduous trees in temperate forests where nutrients are more readily available.

Many organisms use the tissues constructed by autotrophs – such as deciduous and evergreen trees – as the basis for meeting their nutritional needs. There is extraordinary diversity in how these organisms acquire energy and nutrients from autotrophs.

4.3 WHAT ARE ENERGY SOURCES FOR HETEROTROPHS?

Heterotrophs consume the organic materials produced by autotrophs. Examples include animals, which may feed in three basic ways. Herbivores eat plant tissues or fluids, and carnivores eat other animals, while omnivores eat both plant and animal material. **Saprotrophs** are heterotrophs that use already dead

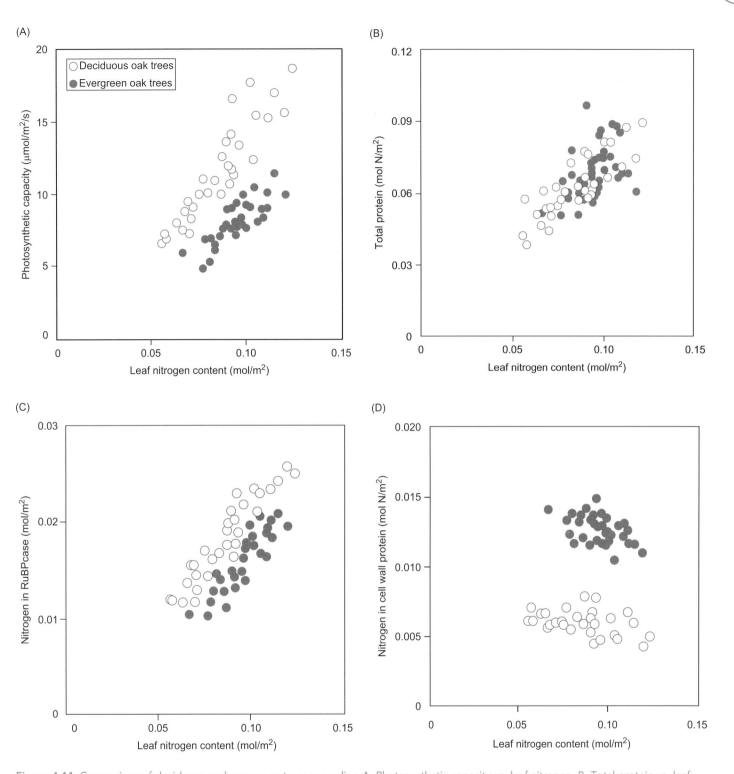

Figure 4.11 Comparison of deciduous and evergreen trees regarding A. Photosynthetic capacity vs. leaf nitrogen. B. Total protein vs. leaf nitrogen. C. Rubisco vs. leaf nitrogen. D. Cell wall protein vs. leaf nitrogen.

organisms or organic matter produced by organisms to meet their nutritional needs. Saprotrophs are further distinguished as either detritivores or decomposers. Detritivores actually ingest the dead matter, **detritus** – primarily the remains of plants, but also of other species. Decomposers are bacteria and fungi that break down organic matter through physical and chemical means into energy and nutrients, which they use for metabolism and nutrition, and CO_2, which is released as waste. Together autotrophs

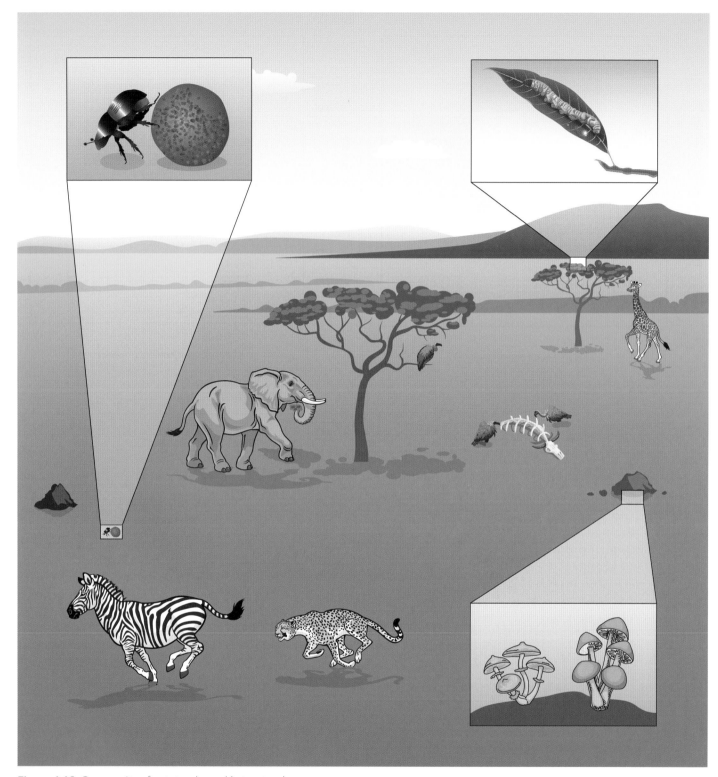

Figure 4.12 Community of autotrophs and heterotrophs.

and the various types of heterotrophs comprise the organisms of a biological community (Figure 4.12). We will discuss in Chapter 16 how energy and nutrients are transferred throughout a biological community.

Herbivores: adaptations and energy

A quick trip to the zoo will allow you to gaze upon an array of large herbivores – heterotrophs that eat plant tissues.

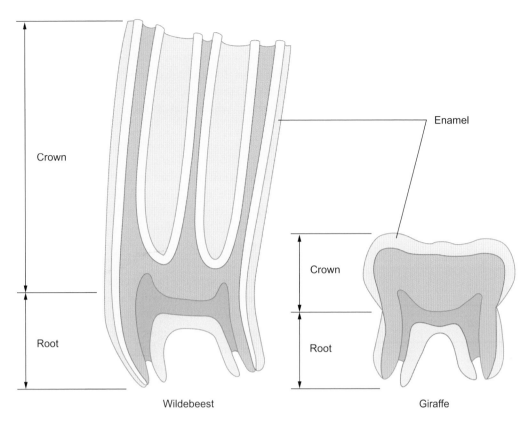

Figure 4.13 A. High-crowned (hypsodont) molar from a wildebeest, a grazer (eats grass) in comparison to low-crowned (brachydont) molar of a giraffe, a browser (eats leaves).

Enamel

Crown

Root

Wildebeest

Crown

Root

Giraffe

Lost in the zoological wonder is the impressive impact these animals can have on an ecosystem. While an individual elephant can consume several hundred kilograms of foliage each day, a migrating herd of tens of thousands of wildebeest can have an even greater impact on the grasses that comprise their primary sustenance in the eastern African ecosystem. But grass is a problematic food source, because it is difficult to digest.

Heterotrophs are challenged to process grass because it is rich in silica and cellulose. High levels of abrasive silica can protect grasses by the wear and tear they cause on a grazer's teeth, discouraging grass consumption. Mammalian grazers such as wildebeests and horses have *hypsodont* teeth – with high crowns and considerable abrasion-resistant enamel extending well above the gum line – which help resist excessive wear (Williams and Kay 2001)(Figure 4.13). To process grasses, many grazers have relatively large stomachs that serve as fermentation vats. Within this stomach live anaerobic bacteria and protozoa that can synthesize *cellulase*, an enzyme that breaks down the cellulose that is locked within the grasses. Large stomachs are associated with overall large body size in ruminants (see Chapter 1), which is why grazers comprise some of the largest terrestrial animals. In East Africa, large body size is strongly correlated with percentage of grass in a ruminant's diet (Table 4.3).

Thinking ecologically 4.2

Use the data from Table 4.3 to determine the strength of the correlation between body weight and the percentage of grass in the diet (see Dealing with data 1.1). What other factors could influence body weight in ruminants?

Very few birds are grazers. One possible explanation is that the large digestive system associated with grazing is (in most cases) incompatible with the aerodynamic demands of flight. Of course, some small animals are able to meet their nutritional needs from foods that are difficult to process or low in nutrients. Examples include aphids that suck on plant vascular fluids, termites that eat wood, and rabbits that eat grass. Each of these cases involves interesting adaptations that make this low-nutrient lifestyle possible. For example, termites have a symbiotic relationship with protistans that are able to digest lignin and cellulose. When termites molt, they shed the lining of their hindgut, which eliminates most of the hindgut intestinal flora. To replenish these lost symbionts, recently molted termites feed on fluids excreted from the hindguts of older termites. Rabbits (and many other mammals) use a very different approach to extract nutrients from grass – *coprophagy* – reingestion of their feces. Microorganisms living within the rabbit gut produce

Table 4.3 Rumen contents and body weights of 14 East African ruminants. From Hoppe *et al.* (1977).

Species	% grass in rumen	Body weight (kg)
Coke's hartebeest	96	120
Topi	96	114
Wildebeest	94	200
Zebu cattle	93	200
Haired sheep	92	25
Thomson's gazelle	79	18
Massai goat	72	24
Impala	64	51
Grant's gazelle	24	49
Bushbuck	10	27
Steenbok	6	10
Suni	6	4
Grey duiker	0	13
Dikdik	0	4

substantial metabolic waste, and several types of B vitamins and lactic acid. Feces-deprived rabbits show depressed weight gain and reduced protein and nitrogen utilization (Schmidt-Nielsen 1997).

Carnivores: adaptations and energy

Carnivores eat herbivores or other carnivores. Because animal tissues are relatively nutrient rich, carnivores tend to have short digestive tracts in comparison to those of herbivores. Hence the most striking adaptations in carnivores are associated with prey capture, rather than with processing the prey. Some of these adaptations will be considered in Chapter 14. Carnivores use two major predatory strategies, active searching and sit-and-wait predation (Figure 4.14). Active searching usually has a relatively low prey-capture rate, in part because natural selection may be stronger on prey to avoid capture than it is on predators to capture prey (Dawkins and Krebs

1979). Physiological adaptations in active searchers may include acute sensory systems for finding prey, and very fast running bursts for short-duration hunts (typical of cats) or high endurance rates for extended hunts (as employed by canids). Sit-and-wait predators, such as spiders, often have low metabolic rates, which makes them more resistant to starvation.

Omnivores: adaptations and energy

Omnivores eat both animal and plant matter. Many omnivores are opportunistic in their approaches to feeding, and will shift their diets in relation to what is available in the environment. As an example, during the summer months, ravens in Maine experience an environment with a high diversity of food items, including scavenged meat, seeds, and fruit. However, during the frigid winters, ravens have one primary food source, the carcasses of dead animals (Heinrich 1989). We will discuss details of the switch from an omnivorous to a detritivorous diet, and many other questions related to raven foraging patterns in Chapter 7.

What factors favor omnivory as a foraging strategy? We can begin to answer this question from both an ecological and evolutionary perspective. As we will see, both perspectives are illustrative.

Ecology of omnivory in fish

Ecologists seek to understand global trends in feeding patterns. Ivan González-Bergonzoni and his colleagues (2012) knew from other studies that omnivory in marine fish is correlated with latitude; species nearer the poles tend to be almost exclusively carnivorous, whereas species nearer the equator have a much higher frequency of omnivory and herbivory. The researchers wondered whether the same pattern was true for freshwater fish, so they did a meta-analysis of the frequency of omnivory in relation to latitude, for both freshwater and marine species. The researchers applied a liberal definition of omnivory – requiring a minimum of 1% plant biomass in the diet. Their results confirmed the increase in omnivory near the tropics in marine systems, and demonstrated a more impressive pattern for freshwater systems, with the mean proportion of omnivorous species approaching 90% in the tropics (Figure 4.15)

One hypothesis to explain this trend is that plant primary production is lower at high latitudes, and particularly in cold months there is not enough plant biomass in the environment to sustain ominvory. A second hypothesis is that fish diversity, overall, is much greater in tropical regions, so competition between different species for prey leads to an increase in eating vegetation as alternative food sources. A third hypothesis is that metabolic rates are greater at elevated temperatures (Chapter 5), so tropical fish need more food and thus rely on plants as acceptable alternatives to eating animals. These hypotheses are not mutually exclusive, and the researchers argue that all of these factors could underlie the relationship between omnivory and latitude.

Figure 4.14 Two predation strategies. A. Frog-eating bat, *Trachops cirrhosus*, swoops down on potential prey. B. Funnel web spider, *Agelenopsis* sp., a sit-and-wait predator.

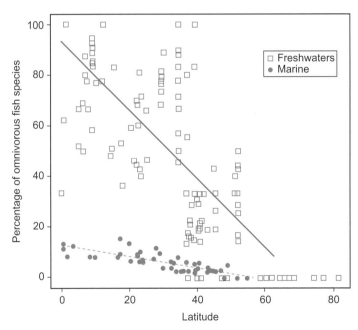

Figure 4.15 Percentage of omnivorous fish species in relation to latitude. $R^2 = 0.51$ for freshwater and 0.67 for marine fish. $P < 0.001$ for both regressions.

Evolution of omnivory in heteropterans

Mickey Ewbanks and his colleagues (2003) wanted to know how physiological and morphological factors influenced the evolution of omnivory in insects in the order Heteroptera. These insects have piercing mouthparts that in many cases are used for sucking liquids from plants. Common examples are aphids and cicadas. However, nature provides many examples of predaceous heteropterans, including giant water bugs and backswimmers, both of which are carnivorous and use their piercing mouthparts to suck out nutrients from their prey. Within the Heteroptera are numerous omnivorous species that can make a living from both plant and animal matter. Not surprisingly, some of the morphological adaptations associated with herbivory and carnivory take intermediate forms in the omnivorous heteropterans (Table 4.4). The researchers wanted to know whether there were ecological factors that favored the evolution of omnivory in some species of Heteroptera.

Ewbanks and his colleagues argued that there are some physiological hurdles that must be overcome by an omnivorous heteropteran. For example, omnivorous species must be capable of synthesizing a much more diverse array of digestive enzymes to process the variety of food they consume. For natural selection to favor the evolution of omnivory, the benefits of this increased physiological complexity must outweigh the cost. The researchers proposed two factors that could predispose heteropterans to evolving an omnivorous foraging strategy. First, they suggested that natural selection would favor the evolution of omnivory in seed and pollen feeders much more than in foliage feeders. Seeds and pollen may contain up to 10% nitrogen, while the plant vascular fluids that are consumed by foliage feeders such as aphids contain much less than 0.1% nitrogen. Seeds and pollen have levels of nitrogen and of other essential nutrients that are similar to the levels found in insect prey. Consequently, heteropterans shifting from seeds and pollen to omnivory would not require as many changes to their digestive machinery as would heteropterans shifting from foliage to omnivory. Second, Ewbanks and colleagues

Table 4.4 Morphological adaptations related to foraging mode in Heteroptera. From Ewbanks *et al.* (2003).

Morphological adaptation	Foraging mode		
	Herbivores	Carnivores	Omnivores
Length of digestive tract	Very long	Relatively short	Intermediate
Mass of salivary gland	Heavy	Light	Intermediate
Primary types of digestive enzymes	Amylases, pectinases	Proteinases, phospholipases	Amylases, pectinases, proteinases, phospholipases
Type of mouthparts	Smooth stylets to penetrate plants	Toothed or curved stylets to hold prey	Piercing sucking intermediate forms

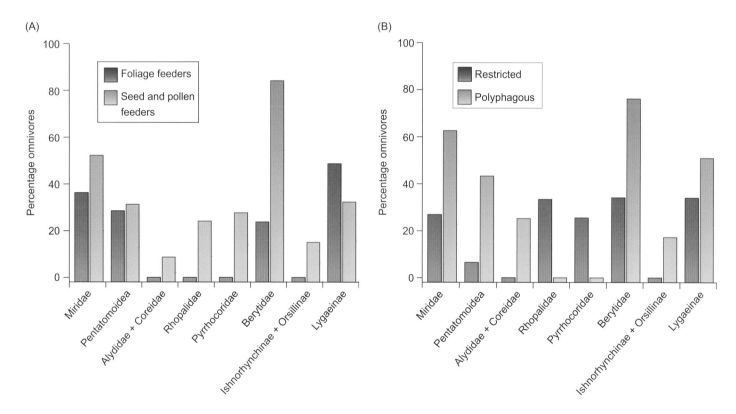

Figure 4.16 Percent of species that are omnivorous in eight heteropteran families, in relation to whether the species are foliage feeders or seed and pollen feeders (A), or polyphagous or have restricted diets (B).

hypothesized that *polyphagous* insects – those that consume a broad range of plant species – may be more predisposed to omnivory. Their rationale was that heteropteran species capable of consuming multiple species of plants may already have diverse digestive enzymes that will enable them to handle animal tissue as well. These polyphagous species will travel more, and be more likely to have a higher encounter rate with potential prey species.

In support of the first hypothesis, heteropteran species that fed on seed and pollen were much more likely than foliage feeders to also feed on prey. This was particularly pronounced in heteropterans from the family Berytidae, in which 85% of the seed and pollen feeders also took animal prey (Figure 4.16A). In support of the second hypothesis, in six of eight families, polyphagous species also showed a greater tendency toward omnivory, with polyphagous Berytidae (and several other families as well) being associated with high levels of omnivory (Figure 4.16B).

To summarize our discussion of foraging, heterotrophs get both energy and nutrients from the organisms they eat. Their

Table 4.5 Molar C:N:P ratios for selected groups of organisms.

Group	Relative molar abundance of			Reference
	C	N	P	
Bacteria (soil)	46	7	1	Cleveland and Liptzin (2007)
Fungi	165	11	1	Cleveland and Liptzin (2007)
Trees (foliage)	1212	28	1	McGroddy *et al.* (2004)
Marine animals	64	12	1	Brey *et al.* (2010)
Marine plants	265	18	1	Brey *et al.* (2010)
Humans	84	6.3	1	Sterner and Elser (2002)

prey varies substantially in the quality and quantity of nutrients it contains. This variation has important implications for physiological processes operating within organisms and for ecological processes operating within communities.

4.4 HOW DOES THE RATIO OF AVAILABLE NUTRIENTS INFLUENCE ECOLOGICAL PROCESSES WITHIN A COMMUNITY?

In your first chemistry course, you probably had the pleasure of learning about *chemical stoichiometry* – the branch of chemistry that studies the quantitative relationship among the constituents of chemical substances. You applied laws of conservation of matter to balance numerous chemical reactions. Ecological stoichiometry studies the balance of the multiple chemical substances that play roles in ecological processes. Ecologists recognize that the ratio of important elements available in the environment (for autotrophs) and in the food they eat (for heterotrophs) has a very important influence on which species can live and reproduce within a particular environment.

Variation in ratios of essential nutrients within organisms

Ecological stoichiometry focuses primarily on the ratio of three elements: carbon (C), nitrogen (N), and phosphorus (P). These three elements are important in many biological processes (see Table 4.2), but are relatively rare in the Earth's crust, or in seawater. Consequently these three elements are often limiting resources for organisms within their environment.

Ecologists often report the relative abundance of C:N:P in the form of a *molar ratio*, which describes the relative number of atoms of each element within an organism. Depending on the question of interest, ecologists might break the ratio down further, discussing the C:N, C:P, or N:P ratio. These ratios vary in relation to nutrients present in the environment, and in relation to other factors such as temperature, moisture, and latitude. Despite this variation, ecologists have estimated the C:N:P ratios for some major groups of organisms (Table 4.5). One important finding is that, compared to other organisms, plants have extraordinarily high C:N ratios.

The high C:N ratio in plants is in part a result of the abundance of two carbohydrates, cellulose and lignin. Cellulose is the most abundant organic molecule in the world, and forms the major structural component of plant cell walls. Lignin is also very abundant, adding mechanical strength to cell walls, and reducing the permeability of vascular tissue, so it can conduct water more effectively. As carbohydrates, both molecules are made up exclusively of carbon, hydrogen, and oxygen. This, of course, provides problems for heterotrophs that eat plants, as their bodies need less carbon, and greater concentrations of other nutrients than are found in the plants they eat. As one example, reconsider the termite that eats wood. The C:N ratio in wood is often greater than 1000 to 1, while the C:N ratio in a termite's body is very similar to values for other invertebrates: approximately 10 to 1. Thus termites show extreme *elemental imbalance* – a dissimilarity in the relative supply of an element between the organism and its resource supply.

 Thinking ecologically 4.3

Some researchers argue that plants don't necessarily need such a carbon-biased stoichiometric ratio, but that natural selection has favored high carbon content, which makes plants less palatable to herbivores. What experiments or observations could you do to investigate this hypothesis?

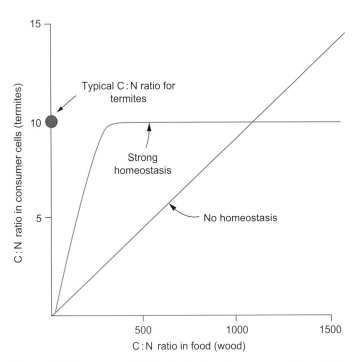

Figure 4.17 Response of C:N ratio in cells of a consumer (a termite) in relation to changes of C:N ratio in its diet.

Two approaches for dealing with elemental imbalance

One simple approach to dealing with elemental imbalance is to assimilate only a fraction of the nutrient that is abundant, and all of the limiting nutrient. The C:N ratio in the wood a termite eats may be roughly 100 times greater than the C:N ratio in its body. Thus a termite could, in theory, obey a processing rule that assimilates 1% of ingested carbon and all of the ingested nitrogen. If the stoichiometric ratio in an organism's food were constant, that would result in a constant stoichiometric ratio within the consumer's body. But if the stoichiometric ratio of an organism's food varies, that processing rule would cause the consumer stoichiometric ratio to vary proportionally (Figure 4.17).

However, many organisms, including termites, are homeostatic. In physiology, homeostasis is the maintenance of a constant internal environment within an organism's cells, tissues, organs, and organ systems. In organisms with stoichiometric homeostasis, the C:N:P ratio in the organism's body is relatively constant, and thus less dependent on the C:N:P ratio in the organism's food (Figure 4.17). These organisms must have mechanisms for regulating their relative assimilation of particular nutrients, so that if too little of a nutrient is ingested, it is assimilated with very high efficiency, and if too much of a particular nutrient is ingested, it can be eliminated from the body.

In reality, termites have a diverse array of adaptations that allow them to regulate their C:N ratio. Many termite species have mutualistic interactions with nitrogen-fixing microorganisms

that reside in their digestive tracts, which provide termites with supplemental nitrogen. Termites conserve nitrogen by eating their molted exoskeletons and the dead colony members. When nitrogen is unusually scarce, some termites become cannibalistic, and when nitrogen is added to their diet, cannibalism levels are reduced. Finally, some termites cultivate fungi on their feces, which metabolize a substantial portion of the carbon from the feces, reducing the fecal C:N ratio. Termites then re-ingest these modified feces, and some of the fungus, both of which have a much lower C:N ratio (summarized in Sterner and Elser 2002).

Many organisms are stoichiometrically homeostatic, but some are not. However, even homeostatic organisms have limits to their ability to maintain a constant internal stoichiometry when the ratio of elements in their diet varies. Thus nutrient distribution can influence species composition – which species are present in a community.

Ecological stoichiometry and the growth rate hypothesis

Stoichiometric nutrient ratios vary among animals, but some ratios vary more than others. Bob Sterner and Jim Elser (2002) compared biomass stoichiometric ratios in two taxonomic groups of organisms: terrestrial herbivorous insects and freshwater zooplankton. Both groups showed very similar patterns, with mean C:N ratios slightly greater than 6.0, mean C:P ratios of approximately 120, and mean N:P ratios of about 25. However, the percentage of phosphorus, in particular, varied greatly among species in both taxonomic groups. For example, some insects and zooplankton species had C:P ratios of less than 50, while others had C:P ratios of greater than 250. Sterner and Elser argue that we can use this variation in phosphorus levels to understand how stoichiometric ratios can influence patterns of growth, development, and reproduction.

Jim Elser and his colleagues (1996) knew that ribosomes were one potential source of high phosphorus levels, because ribosomes are constructed of large quantities of ribosomal RNA (rRNA), a phosphorus-rich molecule. They also knew that there was a very nice correlation between how fast an organisms grows (its growth rate) and RNA composition across a variety of arthropods and microorganisms (Sutcliffe 1970). These considerations led them to propose the *growth rate hypothesis*, that differences in C:N:P ratios among species are caused by differential allocations to RNA necessary to meet the protein synthesis demands of rapid growth and development.

One prediction of this hypothesis is that within taxonomic groups, species with high growth rates should have a higher percent P (and consequently lower N:P ratios) than species with lower growth rates. This hypothesis was tested by Traci Main and her colleagues (1997), who grew five different species of zooplankton from the crustacean order Cladocera in the laboratory, under conditions of high nutrient availability (so that

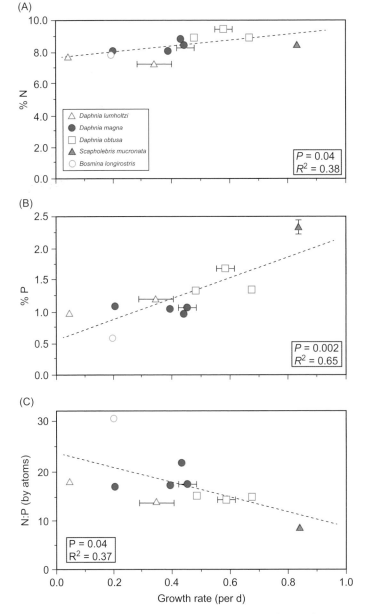

Figure 4.18 Relationship between growth rate (per day) and A. % nitrogen, B. % phosphorus. C. Ratio of N:P in five zooplankton species. Data represent two size classes of *D. lumholtzi*, four size classes of *D. magna*, and three size classes of *D. obtusa*.

One prediction of the growth rate hypothesis is that soils with high phosphorus levels may be able to host a greater abundance of relatively large soil invertebrates in comparison to soils with low phosphorus levels, because large soil invertebrates need to grow rapidly to achieve their large size. Christian Mulder and Elser (2009) tested this prediction by analysing both the abundance and size of all of the invertebrates, fungi, and bacteria in a grassland ecosystem in the Netherlands that varied dramatically in the C:N:P ratio. Their findings support this prediction of the growth rate hypothesis – the relative abundance of larger organisms was greatest at low N:P and C:P ratios, or when phosphorus was most abundant in the soil.

In their book, Sterner and Elser (2002) present ecological stoichiometry as a way of linking molecular ecology to physiological ecology, and ultimately to ecosystem and global ecology by virtue of resource limitation influencing the abundance of species with different growth rates in an ecosystem. But, of course, while resources influence whether an individual can exist in a particular habitat, other factors also play a major role. For example, organisms must avoid getting eaten in order to survive and reproduce.

4.5 HOW CAN ORGANISMS AVOID BEING EATEN?

Prey species don't just allow themselves to be eaten. Natural selection has favored the evolution of an impressive diversity of adaptations that reduce the probability of organisms being preyed upon. Physiological adaptations in animals include high levels of toxicity, which are often associated with *aposematic* or warning coloration (Figure 4.19). For example, the poison-dart frog in Columbia is a brilliant yellow – a color commonly associated with high levels of toxicity. Indigenous hunters are known to rub the tip of a dart in the highly toxic mucus of the frog's skin, creating an effective weapon for killing large prey.

Most animals have an antipredator advantage in comparison to many other types of organisms, in that they can use behavioral means of escape. Plants and fungi are not so fortunate, and must rely on other mechanisms to avoid becoming prey. Many of the antipredator adaptations we see in these two groups involve structural defenses such as spines and thorns, as well as adaptive toxicity.

One of the problems associated with producing substances toxic to herbivores is that, in many cases, these same substances are toxic to the plants as well, and may also be costly to produce. Thus a plant that produces a toxin is usually experiencing some type of metabolic cost. As a consequence, it may be adaptive for a plant to produce a defensive substance only when injured or threatened with injury – an induced defense. Many research studies have shown just how complicated these defense systems can be.

Some of the most common defense compounds produced by plants are tannins, and researchers have known for many years

nutrients were not limiting). The researchers calculated the growth rate for each species by weighing or measuring a sample of individuals at time 0, and another sample 48 h later. Standard laboratory techniques were used to estimate the percentages of N and P for each species. The researchers discovered that rapidly growing species had a slightly higher percentage of N than slowly growing species, but this range in %N was very narrow – 7 to 10% (Figure 4.18A). Most striking was that rapid-growing species had a dramatically greater percentage of P than slow-growing species, with %P ranging from 0.5 to 2.5% (Figure 4.18B). This resulted in a much greater N:P ratio in slow-growing versus fast-growing species (Figure 4.18C).

Figure 4.19 Aposematic coloration. A. The aptly named poison-dart frog, *Phyllobates terribilis*, which is loaded with a potent neurotoxin. B. The burnet moth, *Zygaena fausta*, which carries high levels of hydrogen cyanide.

that oak leaves increase their concentrations of tannins and other defensive compounds in response to wounding (Rossiter *et al.* 1988). Jack Schultz and Heidi Appel wanted to know whether animals that prey on oak leaves were capable of somehow

reducing this induced defense. The researchers began their study with some tantalizing but inconclusive information:

1. A hormone, prostaglandin E$_2$ (PGE$_2$), is produced by midgut cells of the sphinx moth caterpillar (Büyükgüzel *et al.* 2002).
2. PGE$_2$ is produced by blood-feeding species of insects and ticks; it suppresses the host wound responses and promotes vasodilation and consequent blood flow (Stanley 1999).
3. PGE$_2$ is also found in a fluid regurgitated by gypsy moths that eat oak leaves (their own research).

Armed with this knowledge, Schultz and Appel (2004) hypothesized that perhaps the gypsy moth caterpillars regurgitated PGE$_2$ to inhibit the wounding response of oak trees. If so, that would enable the caterpillars to continue feeding on a leaf that was relatively free of tannins. To test this hypothesis, the scientists conducted two experiments. In the first experiment, they treated oak leaves in two different ways. First, they wounded individual leaves by running a pattern wheel – a tool capable of cutting into the leaf – down each side of each leaf's midrib every other day for 7 days. The second treatment was the same as the first, except that Schultz and Appel applied a dilute solution of PGE$_2$ to the wound. The second experiment repeated the wounding treatment but had two additional manipulations, the addition of regurgitated fluids collected from both gypsy moth caterpillars and forest tent caterpillars.

As expected, wounding caused a significant elevation in tannin levels in the leaves. However, this wounding response was completely eliminated by the addition of PGE$_2$ (Figure 4.20A), and it was also eliminated by the addition of regurgitated fluids from the mouths of both species of caterpillars (Figure 4.20B). Schultz and Appel conclude that natural selection has favored the evolution of mechanisms for these herbivores – caterpillars – to override the induced defenses of their prey – oak trees.

 Thinking ecologically 4.4

Shultz and Appel admit in their paper that they possess a very incomplete picture of the interactions between oak trees, predators, and PGE$_2$. What might be a reasonable next step to follow up on their findings?

Obtaining food and avoiding becoming food are two important requirements for success in a particular habitat. But the presence or absence of a species or a group of species may influence the distribution and abundance of other species worldwide. We began this chapter with a discussion of how iron limits the abundance of marine phytoplankton, and how these marine phytoplankton are responsible for about 50% of the world's photosynthesis. These findings suggest that iron limitation may be influencing the distribution and abundance of organisms throughout the biosphere.

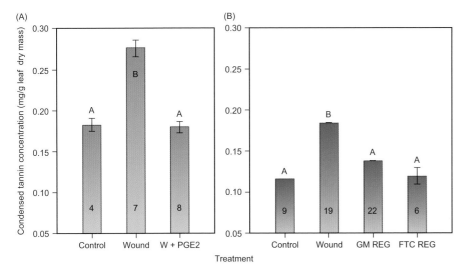

Figure 4.20 Mean (±SE) condensed tannin concentration in red oak leaves. A. Wounded leaves received pattern wheel treatment, and wounded + PGE2 leaves had 1.0 mmol/l PGE_2 added to leaves. B. Wounded leaves received pattern wheel treatment, and GM REG were wounded and treated with 5 μl of gypsy moth regurgitant, while FTC REG were wounded and treated with 5 μl of forest tent caterpillar regurgitant. Numbers in bars represent sample size

REVISIT: Iron limitations on marine phytoplankton

Martin and Fitzwater's initial experiments demonstrated that iron deficiencies limit the population of phytoplankton raised in 30-l glass bottles. But the open ocean has very different dynamics from a 30-l bottle. Ecologists use the laboratory to model processes that they hypothesize are operating within the real world, and to establish parameters or guidelines for real-world experiments. The next stage in this research program was conducting iron-enrichment experiments in the open ocean.

Martin and 43 colleagues (1994) conducted their research on a 64-km^2 site south of the Galápagos Islands. They added enough dissolved iron to raise the concentration of iron in the top 35 m of the ocean from its ambient level of 0.06 nM (nanomolar) to approximately 3.6 nM. Recall that Martin and Fitzwater's 30-l bottle experiments showed that a 1-nM enrichment increased chlorophyll a production, and that a 5-nM enrichment had an even greater effect. Thus they used their laboratory enrichment experiments to determine appropriate levels for the open ocean.

The results of this experiment were a near doubling of chlorophyll a concentrations after 1 day, and continued increases up to about 50 m depth for 3 days (Figure 4.21). After 3 days, ocean currents dragged much of the iron-enriched water to a depth of about 35 m below the surface. In contrast to Martin and Fitzwater's initial laboratory experiments, there was no detectable reduction in nitrate levels, but there was a significant reduction in ammonium levels within the patch.

Kenneth Coale and his colleagues (1996) followed up the open ocean experiment with a slightly different experimental design. Instead of one iron infusion, they added enough dissolved iron to a 72-km^2 area to increase iron concentration from less than 0.02 nM (ambient level) to 2 nM after 1 day, to 3 nM on day 3, and 4 nM on day 7. Measurements taken after day 10 indicated a return to ambient levels of iron concentration.

Figure 4.22 shows a significant increase in chlorophyll a levels and a concurrent decrease in nitrate levels. Much of this increase in chlorophyll a resulted from an 85-fold (8500%) increase in diatom abundance. There was also a reduction in dissolved silicate, which was presumably taken up by diatoms to manufacture their shells. Numbers of zooplankton also increased after 3 days. Thus the entire ecosystem was altered (albeit briefly) by this iron infusion event.

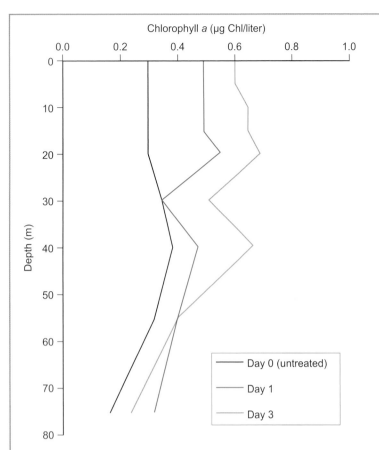

Figure 4.21 Vertical profile of chlorophyll levels (µg/liter) in relation to the depth of the water sample. After 3 days, turbulence caused a subduction event, in which the fertilized column was dragged much deeper below the surface (Figure 3b from Martin *et al.* 1994).

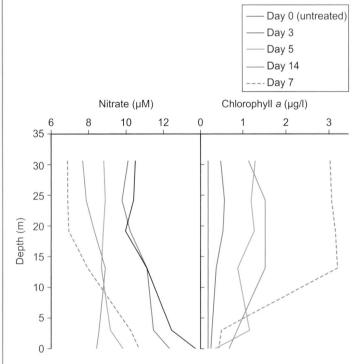

Figure 4.22 Vertical profile of nitrate (A) and chlorophyll levels (B) in relation to the depth of the water sample.

When diatoms and other phytoplankton die, about 85% of the carbon they have fixed from photosynthesis returns very quickly to the atmosphere from microbial decomposition. But about 15% of the carbon sinks deeper in the water column and may remain there for several hundred years. Ultimately, most of this carbon returns to the atmosphere by ocean upwelling. This slow process of CO_2 recirculation is called the **biological pump**. Current levels of atmospheric CO_2 are about 400 parts per million (ppm). Recent estimates are that without the CO_2-holding property of the biological pump, current day atmospheric CO_2 levels would exceed 600 ppm. The biological pump is a little bit leaky, and about 0.5% of the sinking carbon settles to the bottom of the ocean and becomes incorporated into sedimentary rock, with a fraction forming deposits of coal and oil (Falkowski 2002).

This raises the intriguing question of whether we could routinely use iron supplements to increase phytoplankton productivity. Iron supplementation could, in theory, increase the rate that the biological pump withdraws CO_2 from the atmosphere, and either reduce atmospheric CO_2 levels or at least help control the rate of increase. Arguing against this idea, most ecologists maintain that to be effective, iron infusions would be needed at a massive level, and would probably disrupt ecosystems in unforeseen ways.

SUMMARY

The distribution and abundance of species is limited by the availability of nutrients and energy. Adding limiting nutrients to an ecosystem can increase the abundance of some species and may have far-reaching effects on ecosystem functioning. Conversion of light energy into chemical energy by the process of photosynthesis is the primary method by which energy enters an ecosystem. Three different processes of carbon fixation have evolved: C_3, C_4, and CAM. Each process has costs and benefits that are environment dependent, but scientists are still evaluating the relative costs and benefits. C_3 plants are most efficient in moist and cool environments, while C_4 and CAM plants generally do best in warm environments with limited available moisture.

Heterotrophs get energy by consuming autotrophs and other heterotrophs. Herbivores, carnivores, ominvores, detritivores, and decomposers are the major classifications of hetero-trophs and combined with autotrophs make up the organisms within a biological community. There are numerous morphological, physiological, and behavioral adaptations associated with each type of feeding, and, in some taxa, a particular feeding pattern may be favored in specific environments or by a particular evolutionary history. For example, omnivory is more common in tropical fish and in heteropteran insects with an evolutionary history of feeding on seeds and pollen, or of polyphagy.

Species distribution is influenced by how the ratio of nutrients available to species affects their physiological and ecological processes. Species distribution is also influenced by the presence of predators and adaptations of prey species that reduce their probability of being eaten. Some defenses may be induced by the presence of predators.

FURTHER READING

Dawkins, R., and Krebs, J. R. 1979. Arms races between and within species. *Proceedings of the Royal Society B* **205**: 489–511.

A thorough introduction to how the metaphor of an evolutionary arms race is applied to many different ecological contexts, including predator vs. prey, social insect queen vs. worker, and parent vs. offspring.

Fujita, Y. and 11 others. 2014. Low investment in sexual reproduction threatens plants adapted to phosphorus limitation. *Nature* **505**: 82–86.

A challenging but interesting investigation of how N:P stoichiometry influences the frequency of sexual reproduction and the proportion of endangered species within plant communities in Europe and Asia. Not light reading.

Nunes, A. L., Orizaola, G., Laurila, A., and Rebelo, R. 2014. Rapid evolution of constitutive and inducible defenses against an invasive predator. *Ecology* **95**: 1520–1530.

Researchers test the hypothesis that resident prey (tadpoles) evolve constitutive defenses against invasive predators after only 20 years of coevolution with a crayfish predator, and that tadpoles with no crayfish coevolutionary history will use induced defenses when exposed to the crayfish predators.

Schultz, J. C. and Appel, H. M. 2004. Cross-kingdom cross-talk: hormones shared by plants and their insect herbivores. *Ecology* **85**: 70–77.

Nicely laid-out study (understandable experimental design and analysis) that highlights the coevolution of plants and their herbivores.

END-OF-CHAPTER QUESTIONS

Review questions

1. Outline how researchers used different types of experimental approaches to test whether iron was a limiting resource in some ocean communities. What were strengths and weaknesses of each approach?
2. Given its shortcomings, why is rubisco used as a key enzyme in the process of carbon fixation?
3. In what ways are the costs and benefits of C_3, C_4, and CAM environment dependent?
4. What is a meta-analysis? How did Aerts and Chapin (2000) use a meta-analysis to test the hypothesis that nutrient availability influenced the distribution of deciduous and coniferous trees?
5. What are induced and constitutive defense. How does the study by Schultz and Appel (2004) show (a) the existence of induced defenses in oak trees, and (b) that predators on oak trees have evolved adaptations to counter these induced defenses?

Synthesis and application questions

1. When Martin and Fitzwater (1988) looked at their samples under a scanning electron microscope they discovered much greater numbers of diatoms in the experimental samples than in the control samples. Based on what you learned in Figure 4.1 formulate a hypothesis for how dissolved silicate levels might be expected to change over time in each treatment.
2. Assume the unlikely scenario in which the world becomes cooler and drier. What type of carbon fixation process would be favored under these conditions? If your hypothesis is correct, what predictions would you make about the distributions of C_3, C_4, and CAM plants in our present-day biomes?
3. Martin *et al*.'s (1994) iron supplementation experiment in 1994 and Coale *et al*.'s experiment in 1996 had some similarities in their findings, but also some differences. In what ways were their results similar, and in what ways were they different? What factors could possibly account for the differences? (Hint: Compare Figures 4.21 and 4.22. Also refer to the discussion of methods in the text.)
4. In their concluding remarks, Coale *et al*. (1996) state the following "there now exists a preponderance of evidence in support of the 'iron hypothesis' (that iron availability limits phytoplankton growth and biomass in the [. . .] world's oceans). As this working hypothesis has been given such strong support [. . .] it is now time to regard the 'iron hypothesis' as the 'iron theory'." How does a hypothesis differ from a theory? Why is it important for a scientist to make this distinction between hypothesis and theory?

5. Ivan González-Bergonzoni and his colleagues (2012) used the data from 90 papers that considered fish diets within communities across the world (see Figure 4.15). Two issues they discussed were how to decide if a species was truly omnivorous, and whether to consider ecosystems with numerous invasive species as part of their analysis. Why are these problems, and how could they be resolved?

6. Beet armyworm caterpillars feed on corn seedlings. In natural systems, parasitic wasps were observed to lay fertilized eggs in these larvae while the larvae fed on corn seedlings. Ted Turlings and his colleagues (1990) hypothesized that the corn seedlings were emitting volatile compounds to recruit female wasps to attack the caterpillars. To test this hypothesis, they placed female wasps inside two-choice flight tunnels. One choice led to a plant that had been damaged by armyworm larvae the previous day. The second choice led to a plant that had been damaged by researchers the previous day. The researchers replicated this experiment 40 times, using different plants and wasps for each replication. The wasps chose the armyworm-damaged plant 24 times and the researcher-damaged plant 10 times, and made no clear choice six times. Based on these data, can we conclude that wasps are more attracted to armyworm-damaged plants than they are to researcher-damaged plants? Support your answer.

7. Turlings and his colleagues (1990) suspected that a substance in the armyworm saliva was stimulating damaged corn to release a substance that recruited the parasitic wasps. How might you design a choice tunnel experiment to test this hypothesis?

8. Use Table 4.5 to calculate the C:N ratio in tree foliage. Then compare the C:N ratio in tree foliage and wood. What factors might account for these differences?

Analyse the data 1

Martin and Fitzwater (1988) argue that their 30-l bottle experiments show no evidence of a decrease in SiO_3 in the water during the course of the experiment. Based on the data for the 10 nM iron supplementation in Table 4.6, how confident are you about this conclusion? Explain your answer.

Table 4.6 Concentration of SiO_3 (µmol/kg) in 30-l containers of ocean water supplemented with 10 nM dissolved iron.

Day of experiment	1	2	3	4	5	6
SiO_3 concentration in replicate 1	17.01	16.86	18.08	17.40	15.82	15.82
SiO_3 concentration in replicate 2	17.04	17.14	17.96	16.65	15.81	14.43

Modified from Martin and Fitzwater (1988)

Analyse the data 2

Xin-Guang Zhu and colleagues (2008) analysed the maximum efficiency with which photosynthesis can convert solar energy into plant biomass for C_3 and C_4 plants. The results of their analysis are shown in Figure 4.23 for C_3 and C_4 at CO_2 levels of 380 parts per million (ppm). Before the Industrial Revolution, atmospheric CO_2 levels were about 220 ppm, but they have risen, and will continue to rise to an estimated 550–700 ppm by 2100. Based on what you have learned in this chapter, draw in plausible conversion efficiency curves for 220 and 600 ppm for C_3 and C_4 photosynthesis. Explain your reasoning.

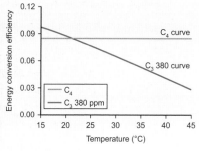

Figure 4.23

Chapter 5

Physiological and evolutionary ecology of temperature and water relations

INTRODUCTION

A central problem in ecology is how environmental factors influence the distribution and abundance of species. While ecologists have been exploring this issue for many decades, the general public has recently become interested in this problem as they become more aware that climate change is happening now. One consequence of climate change is that as climate warms, alpine tundra is being replaced by an advancing line of shrubs and trees. Considering how slowly trees grow, particularly at high altitudes, this process is very difficult to document, and even more difficult to explore in any systematic manner.

Sometimes, spectacular natural events such as tornadoes, hurricanes and typhoons, volcanic eruptions or avalanches can turn our attention to the natural world and how it directly affects our lives. More than 65 years ago, a series of avalanches spawned a unique scientific experiment, that is generating data on how some species survive the challenging environmental conditions near mountaintops. After discussing some results of this experiment, we explore how organisms respond physiologically to changes in their environment in the short term and over evolutionary time. Together, environmental temperature and water availability influence how an organism functions within its environment, and consequently have an important influence on species distribution and abundance.

KEY QUESTIONS

5.1. How do organisms respond to physiological challenges over different time scales?

5.2. How does body temperature influence physiological performance?

5.3. What determines an organism's water balance?

5.4. How are temperature regulation and water regulation functionally linked?

5.5. How do temperature, solute concentration, and water availability affect species distribution and abundance?

5.6. Is Earth's changing thermal and hydrologic environment influencing the distribution and abundance of species?

CASE STUDY: Terror in the European Alps

The "Winter of Terror" in 1950–1 released 649 avalanches in the European Alps, which killed at least 265 people and caused tremendous property damage. This event created impetus to devise methods for controlling avalanches and reducing the damage they cause. One effective approach is *afforestation* – establishment of forests in areas that previously (at least recently) had no forest. In 1975, researchers at the Stillberg long-term afforestation project in the Swiss Alps planted 92 000 conifer trees over a 5-ha site just above the treeline at an elevation that ranged from 2075 to 2230 above sea level (Barbeito *et al.* 2012; Figure 5.1). The researchers wanted to see if they could extend the treeline higher up a mountain to reduce the avalanche threat. They planted 4052 plots with 25 seedlings of Cembran pine (*Pinus cembra*), mountain pine (*Pinus mugo*), or European larch (*Larix decidua*). All three species are common to European Alps treeline communities, and consequently are adapted to some of the extreme conditions that trees – and other organisms – must deal with at those elevations. These conditions include low temperatures, high winds, continuous snow cover for over 7 months of the year, and a very short growing season.

Ignacio Barbeito, who was lead author in the analysis, was not even born when these trees were planted. But a team of researchers followed the survival, growth, and development of the trees for over 30 years, to see how these species responded to the extreme conditions above the treeline. They knew from other research that numerous environmental factors could influence tree survival. For example, some studies indicated that the treeline is limited by low temperatures during the growing season. Other studies showed that the treeline is influenced by wind velocity, solar radiation, or the duration of snow cover. If the soil remains frozen during the early spring, desiccation can damage or kill trees if late-winter water losses are not replaced, because the frozen soil cannot release water to the trees. Snowpacks (accumulated layers of snow) are also associated with a high incidence of fungal diseases, which can kill trees that have been weakened by extreme environmental conditions.

A map comparing tree species distribution in 1975 and 2005 shows some distinct patterns (Barbeito *et al.* 2012). In general, trees survived better at lower elevations. In addition, *Larix decidua* had much higher survival than the two pine species (Figure 5.2). Fortunately, the researchers

Figure 5.1 The Stillberg field site in the European Alps.

(A) 1975

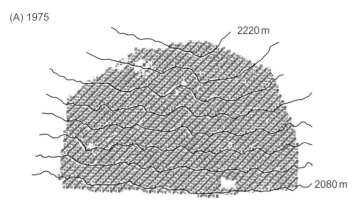

(B) 2005

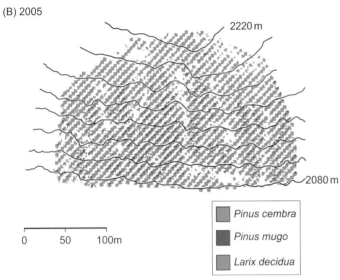

	Pinus cembra
	Pinus mugo
	Larix decidua

0   50   100m

Figure 5.2 Tree species distribution at the Stillberg site in (A) 1975 and (B) 2005. Dark lines indicate 20 m contour intervals.

collected data every year, so they were able to evaluate how a variety of environmental factors interacted to influence these differences in survival. The researchers also measured the height of each tree every 10 years, and could evaluate how environmental factors influenced growth and development. They could then determine whether the environmental factors that influenced survival were different from the environmental factors that influenced growth rates over the course of each species' development.

As we discussed in Chapter 2, temperature and water are the most important influences on the global distribution of terrestrial biomes. But on a much smaller geographic scale, temperature and water availability influence the growth, reproduction, survival, distribution, and abundance of all species. Physiological ecologists attempt to uncover the mechanisms that underlie this influence, and to identify patterns or relationships that are found in natural ecosystems. They often do so by studying organisms that are living in stressful environments, such as near the treeline, where environmental conditions may restrict important physiological processes.

5.1   HOW DO ORGANISMS RESPOND TO PHYSIOLOGICAL CHALLENGES OVER DIFFERENT TIME SCALES?

There are three major responses organisms may show in response to environmental challenges, such as low temperature or insufficient water availability: the stress response, acclimatization (or acclimation), and adaptation. Each occurs over a different time scale.

The stress response: rapid reaction

The stress response is the immediate, usually detrimental, effect of stress on one or more physiological process occurring within an organism. For example, when Hans Selye (1936), an early pioneer of stress research, subjected a group of rats to very low temperatures for 2 days, he discovered that the rats quickly developed a syndrome which included swelling of the adrenal gland, a decrease in the size of the thymus gland, and substantial pitting and ulceration of the stomach lining. These changes were part of a much more complex response to the environmental challenge of cold temperatures. In general, the stress response occurs over a time scale of seconds to days.

The two other responses to environmental challenges, acclimatization (or acclimation) and adaptation, are the basis of our discussion for the rest of this chapter.

Acclimatization: a beneficial physiological response to environmental change

When an individual is subjected to extremes in temperature or water availability, it may change physiologically and/or morphologically in ways that allow it to adjust to the extreme conditions, a process known as acclimatization when it occurs under field conditions, and acclimation when it occurs in the laboratory. Eduardo Sanabria and his colleagues (2012) studied these processes in the toad *Rhinella arenarum*, which lives in the Monte Desert of Argentina, where environmental temperatures and moisture levels fluctuate dramatically over the course of the year. Summers are hot and relatively moist, while winters are dry and cool. The researchers hypothesized that the toads would shift physiologically over the seasons so they could function under the different temperature extremes they experienced in summer and winter.

To test this hypothesis, Sanabria and his colleagues examined the critical minimum (CT_{min}) and critical maximum (CT_{max}) temperatures during the summer and winter months. CT_{min} and CT_{max} are the temperatures at which locomotory behavior becomes disorganized and the animal loses its ability to escape. They predicted that these desert toads acclimatize to seasonal variation in temperature by increasing CT_{max} during the summer, and decreasing CT_{min} during the winter.

The researchers used the righting reflex to estimate the critical temperatures. When placed on its back, a toad will immediately roll over to the feet-down position. At CT_{max} and CT_{min}, toads lose the righting reflex. To estimate CT_{max} in the laboratory, Sanabria and his colleagues gradually increased the ambient (environmental) temperature by $1°C$ per minute until the toad was unable to right itself. In a second experiment, the researchers decreased ambient temperature by $1°C$ per minute to estimate CT_{min}. In accordance with their predictions, mean CT_{max} increased from $35.0°C$ (SE = 0.4) in the winter to $37.8°C$ (SE = 0.3) in the summer, while mean CT_{min} decreased from $4.9°C$ (SE = 0.2) in the summer to $4.1 °C$ (SE = 0.4) in the winter. The *thermal tolerance range*, the difference between CT_{max} and CT_{min}, was $32.9°C$ in the summer and $30.9°C$ in the winter.

Sanabria and his colleagues drew two important conclusions from their study. First, *R. arenarum* acclimatizes to predictable changes in temperature by increasing its tolerance to high temperatures in summer and low temperatures in winter. Tolerance is the ability to survive and function under stressful or extreme environmental conditions. Second, in biomes which have substantial temperature fluctuation, such as deserts, species are expected to have wide thermal tolerance ranges, whereas in biomes with relatively little temperature fluctuation, such as tropical rainforests, species are expected to have much narrower thermal tolerance ranges.

While acclimatization and acclimation to changes in temperature or water availability take place over a relatively short time period, adaptation to environmental conditions occurs over a much grander time scale.

Adaptation: a beneficial evolutionary response that occurs over many generations

As we discussed in Chapter 3, an adaptation is a naturally selected trait that increases an individual's reproductive success in a particular environment in comparison to the reproductive success of an individual without that trait. In the context of physiological ecology, an adaptation is usually viewed as a trait that improves physiological performance within a particular environment. The key to distinguishing adaptation from acclimatization is that adaptation describes a genetic change that occurs within populations over many generations, while acclimatization describes a physiological change that occurs within an individual's lifetime. Neither process changes an individual's genetic constitution.

Different species are adapted to different thermal ranges, and many species can acclimatize to thermal variation over relatively short time periods. To understand the impressive diversity of thermal adaptations and acclimatizations, we will need to first ask the question of why temperature matters, and then consider the mechanisms that underlie these adaptations and acclimations.

5.2 HOW DOES BODY TEMPERATURE INFLUENCE PHYSIOLOGICAL PERFORMANCE?

Michael Angilletta (2009) identified five measures of physiological performance that are often considered by physiological ecologists: survival, locomotion, development, growth, and reproduction. Each of these measures has clear ties to fitness, and has been shown to be sensitive to an organism's body temperature. Let's turn to the question of why body temperature influences physiological performance.

Body temperature and enzyme efficiency

Maintaining a high body temperature generally increases the rate of chemical reactions within an organism. Thus warmer organisms are expected to function more efficiently than cooler organisms. However, natural selection will favor the evolution of biochemical systems that maximize enzyme efficiency over the range of body temperatures most commonly experienced by the organism. Thus we can predict that organisms adapted to cooler climates will have enzymes that operate more efficiently at cooler temperatures than do enzymes of organisms adapted to warmer climates.

Biochemists have several ways of measuring the efficiency of an enzyme. One approach is determining how much substrate it takes to keep an enzyme working at a particular level at a particular temperature. Researchers define the *Michaelis constant* (K_m), which measures how much substrate is necessary to have an enzyme work at 50% its maximum rate. If an enzyme forms a complex with a substrate very quickly, it will have a low value of K_m. Conversely, if an enzyme forms a complex with a substrate relatively slowly, it will have a high value of K_m (Figure 5.3). We

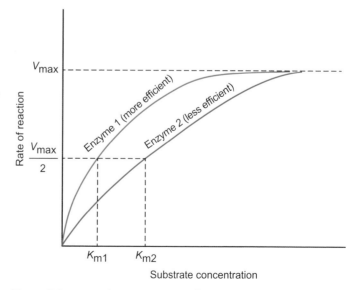

Figure 5.3 V_{max} is the maximum rate for an enzyme-catalysed reaction. K_m (the *Michaelis constant*) is the substrate concentration at which an enzyme is working at 50% of its maximum rate ($V_{max}/2$). Notice that the more efficient enzyme 1 has a lower K_m value than the less efficient enzyme 2.

will discuss two examples that demonstrate how K_m is influenced by environmental temperatures.

Acetylcholinesterase function in relation to normal range of environmental temperature

As our first example, acetylcholine is a neurotransmitter that plays an important role in stimulating muscle contraction in many animals. After binding to receptors on the muscle cell membrane, and possibly stimulating the muscle cell to contract, acetylcholine is released back into the synaptic cleft, and broken down by the enzyme acetylcholinesterase (AChE) into acetate and choline. The acetate is taken up again by the nerve cell, and used to synthesize more acetylcholine. John Baldwin (1971) compared the K_m value of acetylcholine for AChE in several species of fish. He wanted to know whether K_m varies adaptively in response to the environmental temperatures normally experienced by these different species. He hypothesized that each species' normal range of environmental temperature would be associated with a relatively strong tendency of acetylcholine to form a complex with AChE, and therefore have a low value of K_m over that temperature range.

In support of Baldwin's hypothesis, the K_m value is relatively low throughout the environmental temperature range (20–28°C) commonly experienced by mullet, *Mugil cephalus*, and ladyfish, *Elops hawaiensis*. Also notice that the rainbow trout, *Oncorhynchus mykiss,* has two *isozymes*, or forms of the enzyme – one with a lowest value of K_m at approximately 2°C, and a second with a lowest value of K_m at about 20°C (Figure 5.4). To understand the evolution of the dual-enzyme system in rainbow trout, we should compare differences in the thermal environment experienced by rainbow trout and the other two species considered in this research. Rainbow trout live in relatively shallow freshwater streams and rivers, with water temperatures strongly influenced by the air temperature. A rainbow trout's environmental temperature may decline to near 0°C in the winter, and climb above 20°C in the warm summer months. The other two species live in marine environments, which because of the ocean's enormous mass, are buffered against dramatic shifts in environmental temperature. Thus natural selection has favored the evolution of this dual-enzyme system with two very different optimal values of K_m in the rainbow trout – one isozyme for winter and a second for summer.

 Thinking ecologically 5.1

The rainbow trout has two isozymes of AChE while the other species each have one form. (a) What are the biological tradeoffs associated with having two forms of this enzyme? (b) Do the curves in Figure 5.4 support the hypothesis that a dual-enzyme system is costly to the trout?

Lactate dehydrogenase efficiency in relation to mean environmental temperature

A second example of how enzyme efficiency is tied to environmental temperatures involves the enzyme lactate dehydrogenase, which catalyses the conversion of pyruvate to lactate. This is one step in a process that generates energy for muscle cells in the absence of oxygen. If a species' enzymes are adapted to its environment, comparative studies should show that related species experiencing different thermal environments will have enzymes reflecting these environmental differences. To test this hypothesis, John Graves and George Somero (1982) compared the K_m value of lactate dehydrogenase for pyruvate in three East Pacific species of barracuda that experience different mean environmental temperatures as a result of their differences in latitude. In support of this hypothesis they found that K_m at 25°C was lowest in *Sphyraena ensis* – the species whose mean environmental temperature is closest to 25°C, and highest in *S. argentea* – the species whose mean environmental temperature was much lower than that of the other two species and farthest from 25°C. Interestingly, all three species had almost identical values of K_m at their mean environmental temperatures, suggesting that there may be an optimal K_m for this reaction (Table 5.1).

The relationship between temperature and enzyme function is complex, because temperature can influence membrane function indirectly. For example, body temperature influences the fluidity of cell membranes, with low temperatures associated with relatively immobile phospholipids, and high temperatures associated with phospholipids that move around at too fast a rate (Angilletta 2009). Many essential enzymes are embedded in the phospholipid

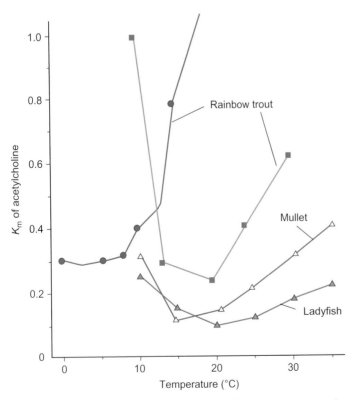

Figure 5.4 K_m of acetylcholine for AChE at different environmental temperatures in three fish species.

Table 5.1 Mean environmental temperatures and K_m value of lactate dehydrogenase for pyruvate in three species of East Pacific barracudas (genus *Sphyraena*). Numbers in parentheses are standard deviations. From Graves and Somero (1982).

	S. argentea	S. lucasana	S. ensis
Mean environmental temperature	18	23	26
K_m at 25 °C	0.34 (0.03)	0.26 (0.02)	0.20 (0.02)
K_m at mean environmental temperature	0.24	0.24	0.23

matrix, and they function best at levels of intermediate membrane fluidity. Organisms can produce membranes suited for warm temperatures by increasing the proportion of saturated fatty acids embedded in the membranes, which tends to reduce the fluidity at high temperatures. Conversely, they can produce membranes suitable for cold temperatures by decreasing the proportion of saturated fatty acids embedded in cell membranes.

Given the relationship between body temperature and cellular and biochemical function, organisms can benefit by regulating their body temperature within an optimal range.

Benefits and costs to temperature regulation

Some organisms maintain a relatively constant temperature by using the heat generated by their metabolic processes (cellular respiration) to elevate body temperature when environmental temperatures are below optimal. Other organisms allow their body temperature to fluctuate with the environment. Organisms that maintain a relatively constant body temperature by conserving metabolic heat are endotherms, while organisms whose body temperature is influenced primarily by the external environment are ectotherms.

Maintaining a constant body temperature has the advantage of allowing an organism to function efficiently over a broad environmental temperature range. However, at suboptimal air temperatures, the organism may need to raise its metabolic rate to regulate its body temperature. At low air temperatures it may need to increase the metabolic heat it generates to keep its body warm. At high air temperatures it may need to increase its metabolic rate so that it can dissipate excess heat through sweating or panting to keep its body temperature below lethal levels. The thermal neutral zone defines an environmental temperature range within which an animal's metabolic rate is constant. Outside of this zone, the metabolic rate will increase substantially (Figure 5.5).

Ectotherms, of course, don't need this complex regulatory machinery. They can, within limits, let their body temperature fluctuate with environmental temperature, so long as their body temperature does not climb above or drop below lethal

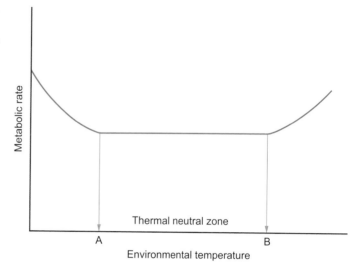

Figure 5.5 The thermal neutral zone ranges between environmental temperatures A and B. Metabolic rates increase at environmental temperatures below A and above B.

temperatures. Because ectotherms do not expend energy for temperature regulation, they generally require less food. On the down side, they may be unable to be active under certain environmental temperatures. As we will discuss shortly, both endotherms and ectotherms have diverse adaptations that allow them to be successful in their thermal environment. We will use a simple conceptual model to explore how different environmental factors influence the ways that organisms deal with their thermal environment.

A general model of heat balance based on heat gain and loss

The total amount of heat stored (H_s) within an organism is influenced by several variables. These are:

H_m metabolic heat generated from cellular respiration

H_{cd} heat gained or lost by conductive heat transfer – transfer of heat energy arising from molecular collisions within or between substances (often solids)

H_{cv} heat gained or lost by convective heat transfer – transfer of heat energy arising from the movement of fluids, such as water and air

H_r heat gained or lost by electromagnetic radiation (such as solar radiation, or longwave radiation from Earth)

H_e evaporative heat loss.

We can put these variables together to create a simple model of heat storage (Schmidt-Nielsen 1997):

$$H_s = H_m \pm H_{cd} \pm H_{cv} \pm H_r - H_e$$

Let's see how this model helps us to understand thermal adaptations in a variety of species. We will begin by considering such adaptations in terrestrial animals.

Thermal adaptations in terrestrial animals

Body size is arguably the most important factor influencing heat balance in all animals. Large animals have lower surface area to volume ratios than small animals, which results in less heat exchange between the animal and its environment, or low values of conductive (H_{cd}) and convective (H_{cv}) heat transfer. We begin by considering a small vertebrate that can gain (and lose) heat remarkably rapidly.

A TERRESTRIAL ECTOTHERMIC ANIMAL IN THE COLD. The iguanid lizard, *Liolaemus multiformis*, lives in a very unusual environment for a lizard, well above 4500 m elevation in the Peruvian Andes. Lizards are uncommon in cold environments, because as ectotherms, it is difficult for them to raise their body temperature to a level of high physiological function at cold environmental temperatures. Oliver Pearson (1954) described a typical summer day in the Andes, when the ground was usually covered with snow in the morning, with bright sunshine melting the snow away by 10:00–11:00 a.m. Clouds gathered in the late morning, and usually by early afternoon hail or snow was falling. And the winter was much worse! Despite this formidable environment, *L. multiformis* manages to thrive. How does it do this?

We can use Pearson's description of the animal's behavior, coupled with our conceptual model of heat balance, to understand this lizard's remarkable abilities. *L. multiformis* spends the night in burrows under rocks, which provide a microclimate that usually stays between 4 and 10°C. When day breaks, the lizard extracts itself from its burrow to begin basking at a very low body temperature – it has a very low critical thermal minimum (CT_{min}). Pearson found that the righting reflex was functional in lizards that were cooled to 1.5°C.

L. multiformis can heat itself very quickly for several reasons. First, solar radiation is abundant at equatorial latitudes, and even more amplified because there is less atmosphere to absorb the radiation at high elevations (see Chapter 2). Thus the lizard enjoys very high heat gain from electromagnetic radiation (H_r), which is further enhanced by its dark color when it is cold. Interestingly, its color becomes much lighter when it warms up. Behaviorally,

the lizard orients itself on an inclined surface that faces the Sun, again maximizing its interception of solar radiation. The lizard also positions itself on a substrate that insulates it from the cold ground, reducing heat loss from conduction (H_{cd}). Once the Sun has warmed the rocks, the lizard switches to basking on these rocks to increase conductive heat gain (Figure 5.6A).

(A)

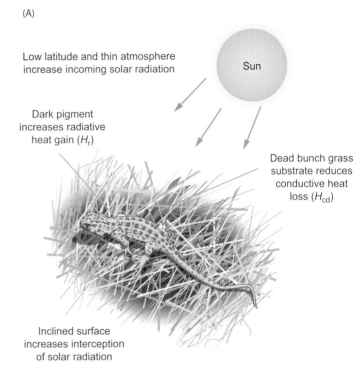

(B)

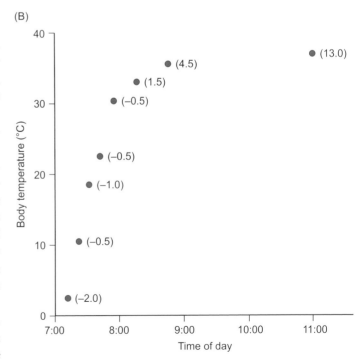

Figure 5.6 A. Adaptations in *Liolaemus multiformis* that allow it to live at high elevations. B. Body temperature of *L. multiformis* in relation to time of day. Values in parentheses are the air temperature at that particular time.

Taken together, these behavioral and morphological adaptations allow *L. multiformis* to heat up surprisingly rapidly, elevating its body temperature well above the environmental temperature (Figure 5.6B). It is then free to forage for food until the afternoon snows send it scurrying back to its burrow. Let's now move to the other extreme and consider heat exchange in some very large animals.

LARGE TERRESTRIAL ENDOTHERMIC ANIMALS IN THE HEAT. Large body size is both a blessing and a curse for the elephant. Large size makes elephants virtually immune to predators, but large bodies retain heat because the low surface area to volume ratio reduces heat exchange from conduction (H_{cd}) and convection (H_{cv}). To help with dumping excess heat, elephants have very large and flattened ears that are highly vascularized. At high temperatures, large quantities of warm blood are shunted to the ears, where heat is dumped to the environment by conductive and convective heat transfer. Elephants can also lessen their heat load by moving into the shade, reducing radiative heat transfer (H_r), and by covering themselves with water or mud, increasing evaporative heat loss (H_e)

Sweating is used to cool a wide variety of large endotherms, for example, humans and other primates, large antelope, and camels (Schmidt-Nielsen 1997). Almost 2500 kJ of heat is dissipated with each kilogram of sweat that is vaporized, so sweating is a highly effective cooling mechanism. Unfortunately, sweating is also associated with considerable salt loss, and can cause dehydration. Many smaller mammals – including carnivores and small antelopes – and many birds use panting to cool themselves. No electrolytes are lost in panting, but a great deal of carbon dioxide is expelled, which can cause a dangerous increase in blood pH. A second problem is that panting increases metabolic rate, which increases endothermic heat production (H_m). In most cases, however, the increase in H_m with panting is relatively small.

Small endothermic animals can dissipate heat very easily, because they have a high surface area to volume ratio, which promotes conductive (H_{cd}) and convective heat transfer (H_{cv}). Their thermoregulatory challenge is retaining heat in cold environments.

SMALL TERRESTRIAL ENDOTHERMIC ANIMALS IN THE COLD. Small endothermic animals must consume large quantities of food to generate enough metabolic heat to thermoregulate. The problem is that in some biomes, low ambient temperature and low food availability occur at the same time (usually the winter). Large animals in cold climates usually have thick fur (mammals) or feathers (birds) that reduce conductive and convective heat loss. However, small body size makes a thick insulating coat unlikely in small endotherms – nonetheless, small endotherms are abundant in cold environments. How do they manage?

One common adaptation in small mammals and birds is periodic torpor – a state of greatly reduced body temperature and greatly reduced metabolic rate. By allowing body temperature to drop, animals need much less energy to fuel their metabolism. The energy savings can be huge – well over 90% in many species that enter hibernation – a state of torpor that lasts for several months throughout the cold season.

Many species of hibernating rodents spend between 80 and 95% of their hibernation time in torpor, which is punctuated by periods of arousal. The current thinking is that arousal is important for dealing with physiological problems that occur during torpor, which include the buildup of body waste, dehydration, and diminished immune function. But arousal is extremely metabolically costly, so we would expect natural selection to favor individuals that minimize these costs.

Behaviorally, a hibernating rodent can conserve heat in three ways to help it survive through the winter. It can curl up into a ball, it can huddle with other individuals, and it can build a nest (Figure 5.7A). All of these behaviors reduce conductive heat loss. Most bats are in a much weaker position to survive hibernation than are rodents, because they tend to be smaller than rodents, and consequently have a greater surface area to volume ratio. In addition, many bat species sleep dangling head-down from the

Figure 5.7 A. Hibernating alpine marmots, *Marmoto marmota*, in Hohe Tauern, Austria. B. Hibernating little brown bats, *Myotis lucifugus*, dangle from a cave in New York, USA.

cave roofs, which prevents them from building nests and from curling up into a heat-saving ball (Figure 5.7B).

There are several ways bats could, in theory, reduce heat loss during hibernation. First, they could spend more time in torpor and have fewer arousals. Second, they could let their body temperatures dip further. Third, rather than arousing to their normal body temperature (about 35°C) they could arouse to a lower body temperature, which presumably would be energetically less costly. Lastly, they could have fewer arousals when their cave is at its thermal minimum, which again would reduce the costs of arousal.

Kristin Jonasson and Craig Willis (2012) tested these hypotheses in the little brown bat, *Myotis lucifugus*. They attached tiny heat sensors, which measured skin temperature – determined to be a good measure of internal body temperature – every 10 min, to 22 torpid bats captured in a cave in Manitoba, Canada. The researchers also recorded the temperature in the cave dome, where the bats were hibernating.

On average, bats spent 99.6% of their hibernation time in torpor, arousing 2–5 times per winter for between 5 min and 4.6 h. This is much greater than the percentage of time spent in torpor by most rodents that have been studied. Jonasson and Willis identified a novel type of arousal period in bats, *heterothermic arousal*, in which skin temperature drops back down partway through the arousal period but then climbs back up. The researchers argue that this temperature drop saves energy by reducing metabolic rate. Lastly, the researchers found that the bats experienced fewer arousal periods when the cave was at its coldest, which they argue was another energy-saving adaptation. Energy use per day in the warmest period (late November–late December) was almost twice as great as in the coldest period (late January–late February). The researchers conclude that this suite of adaptations allows little brown bats to survive the rigors of hibernation, despite their small size and their inability to use some of the behavioral mechanisms that are available to rodents.

Let's turn our attention next to terrestrial plants, which have some surprisingly similar mechanisms for modifying body temperature to those we've already described in terrestrial animals.

Thermal adaptations in terrestrial plants

On the surface, we might expect that terrestrial plants would be out in the cold when it comes to dealing with temperature extremes. Their immobility does limit their options; they can't move into the shade on a hot day, or crawl into a burrow, or huddle with other plants on a cold night. But they do have responses that allow them to deal with these challenges.

For example many plants can acclimate to different temperature regimes, based on the temperatures they experience during development. Wataru Yamori and colleagues (2005) wanted to know if spinach plants, *Spinacia oleracea*, could adjust their photosynthetic rates based on the temperatures they experienced during development. They grew one set of plants at 30°C during the day and 25°C at night (high-temperature plants), and a second

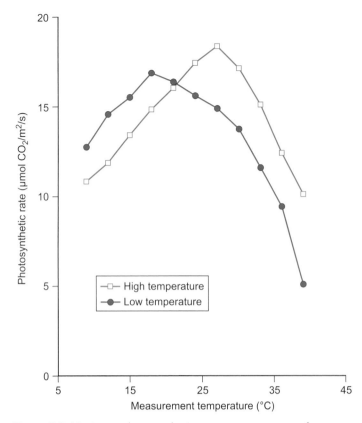

Figure 5.8 Maximum photosynthetic rates across a range of temperatures for spinach plants that were acclimated to high and low temperatures.

set at 15°C during the day and 10°C at night (low-temperature plants). The researchers measured maximum photosynthetic rates across a range of temperatures and found that high-temperature plants had a substantially higher optimum temperature for photosynthesis than did low-temperature plants (Figure 5.8).

In addition to being able to acclimate, plants have diverse adaptations that can permit effective physiological function over a surprisingly wide range of environmental temperatures.

TERRESTRIAL PLANTS IN THE HEAT. Photosynthesis requires plants to absorb light in the form of electromagnetic radiation. At the same time, radiative heat transfer (H_r) can elevate leaf temperature to lethal levels. In tropical biomes this is usually not too much of a problem, because water is passed out from the leaves through stomata, and the subsequent evapotranspiration cools the leaves to acceptable temperatures.

The overheating problem becomes much stickier in the desert, where high transpiration rates are not an option. There, several types of adaptations minimize radiative heat gain (H_r) (Figure 5.9). For example, many desert plants have very small leaves, or even lack leaves entirely, thereby reducing absorption of solar radiation. Some desert plants have reflective surfaces with highly reflective hairs (pubescence) that minimize H_r. Others can fold their leaves up during the heat of the day. Some desert plants have one major growth axis perpendicular to the ground, so that

Figure 5.9 A. Old man cactus, *Cephalocereus senilis*, showing extreme pubescence, loss of leaves, and major growth axis perpendicular to the ground. B. Owl's clover flowers (*Orthocarpus* species) showing open growth form, and reduced leaf size characteristic of desert flowering plants.

very little surface is exposed to solar radiation at the hottest time of the day. Many desert plants have open growth forms, which promote dissipation of excess heat from convection. Most have their leaves elevated sufficiently above the ground so that they absorb very little heat from conduction between the hot ground and their leaves.

Plants that inhabit cold biomes have a complementary set of adaptations that allow them to maximize heat input and minimize heat loss.

TERRESTRIAL PLANTS IN THE COLD. Three major factors prevent tundra plants from achieving high body temperatures. Most obviously, it is very cold in this biome for much of the year, and even during the short growing season air temperature can drop sharply at night. A related problem in high latitudes is that the Sun's oblique angle provides relatively little opportunity

for radiative heat transfer (H_r). Lastly, tundra is very exposed and often windy, so plants tend to lose heat rapidly from convective heat transfer (H_{cv}).

Many cold-adapted plants have dark pigments that increase the absorption of solar radiation (H_r). Some arctic plants orient their leaves perpendicular to the Sun's rays, increasing H_r. Because the ground temperature is much warmer than the air temperature during the day, a growth form that hugs the ground (a cushion growth form) maximizes radiative and conductive heat gain from the surface. In addition, this low growth form minimizes heat loss from convection (H_{cv}), which can be the greatest source of heat loss in tundra (Figure 5.10).

Some plants take a page from the endothermic animal playbook and use their metabolic processes to generate heat and increase body temperature. For example, Eastern skunk cabbage, *Symplocarpus foetidus*, will often bloom under the snow

Figure 5.10 The moss campion, *Silene acaulis*, in the Swiss Alps, has a cushion growth form.

A

Spadix

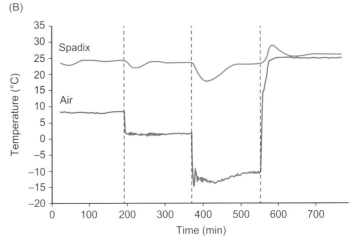

(B)

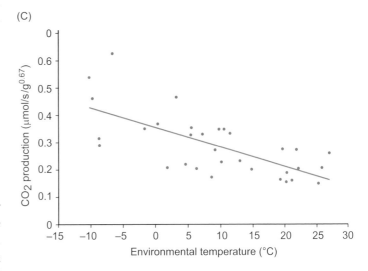

(C)

Figure 5.11 A. The Eastern skunk cabbage, *Symplocarpus foetidus*, emerging through the snow. B. Effect of change in air temperature on spadix temperature. C. Metabolic rate (CO_2 production) in relation to environmental temperature.

cover in very early spring. Roger Seymour (2004) explored the relationship between environmental temperature and the temperature of the skunk cabbage's spadix, the spike-like structure that bears the tiny florets (Figure 5.11A). He wanted to characterize how environmental temperature variation influenced spadix temperature (T_s) under controlled conditions, so he constructed a chamber that allowed him to vary environmental temperature, and to measure T_s and the rate of metabolic CO_2 production by the plant. Spadix temperature regulation was impressive: Decreasing air temperature by $10°C$ caused a mean decrease in spadix temperature of only $0.9°C$ (Figure 5.11B). Of course there was a cost to this endothermic heat generation. Plant respiration rate was about twice as high at an environmental temperature of $-10°C$ than it was at an environmental temperature of $25°C$ (Figure 5.11C).

 Thinking ecologically 5.2

Consider Figure 5.11B. There are some downward dips and upward rises in T_s following the adjustment of environmental temperature. How long do these changes take, and what might be causing them?

Thermal issues in the marine environment

As we described in Chapter 2, the high specific heat of water reduces thermal variation in the marine environment (Figure 5.12A) compared to the terrestrial environment (Figure 5.12B). For example, mean air temperature is much colder than mean sea surface temperature (SST) in northern latitudes. The annual range of SST is much smaller than the annual range of air temperature over land. Finally, the peak in annual range of SST is in temperate waters, while the peak in annual range of air temperature is in polar regions.

(A)

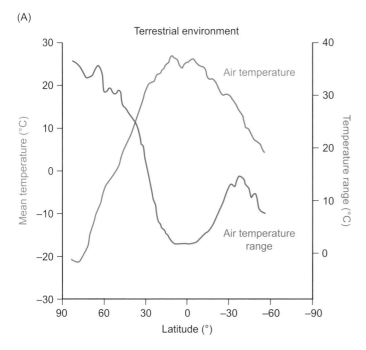

(B)

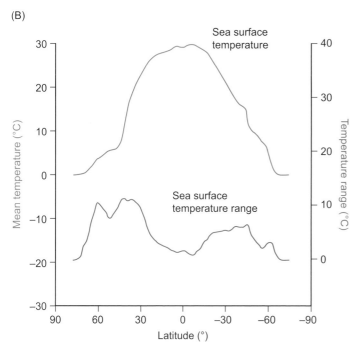

Figure 5.12 A. Mean air temperature and temperature range in terrestrial environment. B. Mean sea surface temperature and sea surface temperature range.

There are some other important thermal considerations as well. Organisms that live in the ocean are unlikely to exchange any substantial heat from radiative or evaporative processes. Consequently, the heat balance model simplifies to:

$$H_s = H_m \pm H_{cd} \pm H_{cv}$$

Despite this apparent simplicity, marine organisms, like their terrestrial counterparts, show extraordinary diversity of thermal adaptations.

Thermal adaptations in marine animals

Many whales live for a significant portion of the year in polar waters. Based on Figure 5.12, you can see that polar waters consistently have temperatures near 0°C. Immersed in water at those temperatures, most endotherms lose body heat at an alarming rate because water has a high heat capacity and high thermal conductance. In addition, rates of convective heat transfer are very high for an immersed animal, so taken together, the potential for heat loss from H_{cd} and H_{cv} is extraordinary.

The whale's secret is that right below its skin is a thick layer of *blubber*, which can comprise up to 50% of its body weight. This remarkable substance is mostly fat and collagen and is surprisingly highly vascularized (in most animals, fat does not contain many blood vessels). This fat is an important energy source, but in the context of our current discussion, it is an important insulator. A 5 cm layer of blubber is generally enough to keep underlying tissue 30°C warmer than the frigid ocean.

These insulative properties could, in theory, lead to a dangerous buildup of heat, which is compounded by the animal's large body size and low surface area to volume ratio. The solution is related to the extensive vascularization that we mentioned earlier. Whales can bypass the insulation afforded by blubber by shunting blood to capillary networks located just under the skin. They can then dump the excess heat with conductive and convective heat exchange. When body temperatures cool down, whales can divert blood flow away from the capillary networks under the skin to a second network that is under the blubber. That way, most of the heat is conserved (Figure 5.13).

It is interesting to contrast fur and blubber as insulators. Fur is located outside the skin surface, so there is no good way to bypass fur in cases of excessive heat buildup. Thus large terrestrial animals have evolved other mechanisms to dissipate excessive heat.

One downside of blubber is that it is not very flexible. Consequently, whale flukes and flippers lack blubber and have the potential to lose a tremendous amount of heat in cold ocean waters. But they don't. The secret is their *countercurrent heat exchanger*.

Each artery in a whale fluke is surrounded by a network of veins. As arterial blood journeys toward the surface, it transfers its heat to the cooler veins that are returning from the surface capillaries. There is a favorable thermal gradient for heat transfer for the entire length of the fluke, so that almost all of the arterial heat is transferred to the veins and almost none is lost to the ocean (Figure 5.14). When an animal exerts itself, the countercurrent heat exchanger can be bypassed so that excess heat can be exchanged with the ocean. Many animals have countercurrent heat exchangers; for example, tuna and sharks use a countercurrent heat exchanger to keep their swimming muscles at high temperature so they can swim quickly and for long periods of time, independent of water temperature.

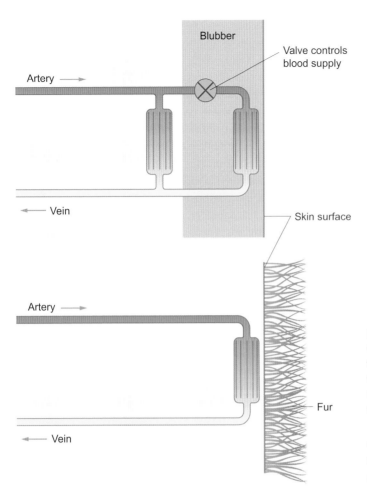

Figure 5.13 Blubber's insulation can be bypassed when the whale needs to dissipate heat by allowing blood to flow to the skin surface. There are no mechanisms to bypass fur's insulation, because it lies outside of the skin's surface.

We have used a few variables to build a conceptual model that allows us to understand a wide range of thermal adaptations in organisms. As we have already seen, thermal adaptations are closely tied to how organisms interact with water.

5.3 WHAT DETERMINES AN ORGANISM'S WATER BALANCE?

In addition to its importance for temperature regulation, water is, of course, essential for all organisms. A typical cell may be 75% water. Water's unique chemical attributes make it the solvent of choice for most biochemical reactions. Because water is such a good solvent, it may contain high concentrations of solutes, and when an organism takes in water, it also takes in solutes. Organisms must balance out their internal solute concentration while balancing out their water concentration.

Animals can absorb water directly through eating or drinking, or they may have specialized organs for taking in water from the environment. Food provides two sources of water: the water

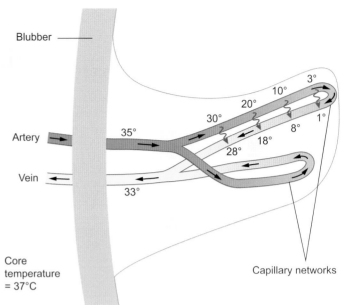

Figure 5.14 Countercurrent heat exchange in a whale flipper. Outgoing arteries are surrounded by a network of veins (for simplicity, the diagram shows only one vein per artery). There is a favorable thermal gradient for heat transfer along the entire length of the network. Rather than losing heat to the environment, the arteries transfer the bulk of their heat to the slightly cooler veins.

content of food and the metabolic water generated by cellular respiration. Most terrestrial animals have the advantage of being able to move to their water source, while aquatic animals are immersed in it. The situation is a bit trickier for terrestrial plants. A plant's access to water is limited by the availability of water in nearby soil. Water enters plants through the roots, which terminate in fine root hairs. These fine hairs have a large surface area to maximize absorption of water from the soil matrix. Terrestrial microorganisms are particularly affected by water availability, as their extremely small size gives them a very high surface area to volume ratio, and a high risk of drying out.

As we did with temperature balance, we will use a simple conceptual model to explore how different environmental factors influence how organisms deal with internal water balance.

A general model of water balance based on water gained and lost

The total amount of water stored by an organism (W_s) will be influenced by several variables. These are:

W_m metabolic water generated from cellular respiration
W_{en} water taken in from the environment
W_{et} water lost from evaporation or transpiration
W_{se} water lost from secretions or excretion.

We can put these variables together to generate a simple model of water regulation:

$$W_s = W_m + W_{en} - W_{et} - W_{se}$$

To understand how these different factors influence water balance, we need to briefly review the relationship between diffusion, osmosis, and energy. We will apply this discussion initially to understanding the movement of water through plants.

Water enters plants primarily through the roots and exits from the stomata during transpiration. Our simple model of water balance reduces to:

$$W_s = W_{en} - W_{et}$$

Water, like all matter, has *potential energy*, energy that exists due to location or structure, and that can be used to do work. In the case of plants, the work is the upward movement of water through the plant in opposition to gravity. Water, like any other matter, flows from conditions of high potential energy to conditions of low potential energy. To understand where water will tend to go in a plant, we can measure the factors that contribute to its potential energy status, or water potential.

Solutes are essential contributors to water potential. Dissolving solutes in water reduces its potential energy; thus pure water has the highest possible *solute potential* (also called osmotic potential), which is set at 0 at standard temperature and pressure. Water with greater solute concentration has increasingly lower (more negative) solute potential. If you have pure water on one side of a selectively permeable membrane, and water with dissolved solute on the other side, water will tend to move from the pure water side to the side with dissolved solutes. The solutes can't move because the membrane is impermeable to them.

Osmosis is the net movement of water across a selectively permeable membrane down its concentration gradient. The term "net movement" implies that water diffuses in both directions, but that more water diffuses down the concentration gradient than up the concentration gradient (Figure 5.15).

Hydrostatic pressure has an important influence on water potential within a plant. When a plant transpires, the exiting water exerts suction tension, or negative pressure on the system, reducing the *pressure potential*, and the overall water potential. A second effect is *matric forces* – in this case the tendency of water to adhere to the walls of xylem vessels, but more generally the tendency for different molecules to adhere to each other as a result of intermolecular forces such as hydrogen bonds. Within the plant, the *matric potential*, the water potential that arises due to matric forces, interacts with the solute potential and the pressure potential to determine the overall water potential. As long as the water potential decreases as water journeys up the plant, water will tend to flow from the soil, move up the plant, and exit through the stomata (Figure 5.16).

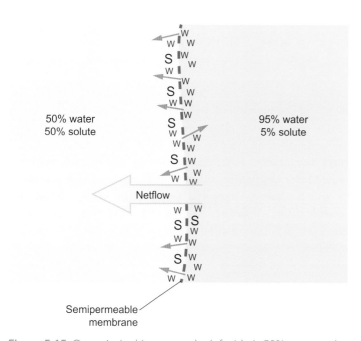

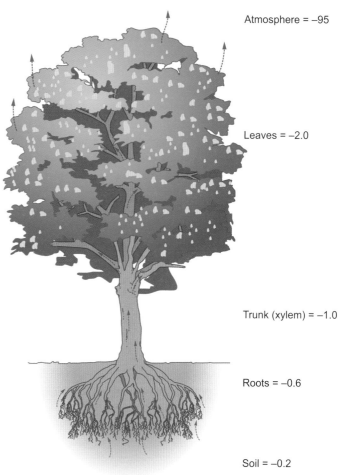

Figure 5.15 Osmosis. In this system, the left side is 50% water and 50% solute, while the right side is 95% water and 5% solute. The two sides are separated by a semipermeable membrane, which allows water, but not solute, to pass through. Overall, each water molecule has an equal chance of crossing the membrane. But there are more water molecules on the right side, so the net flow is from right to left.

Figure 5.16 Water potential gradient between soil, roots, stems, leaves, and atmosphere. Water flows along a potential gradient, from high to low water potential.

Rainwater has very few solutes, so soil water tends to have very high solute potential (just below 0). Soil water potential is also influenced by the nature of the soil itself. Sand particles (>0.05 mm diameter) have large pores between each granule, and exert relatively low matric forces, because most of the water molecules are too distant from the particle surface for adhesion. In contrast, tiny clay particles (< 0.002 mm diameter) have tiny pores between each particle, and exert stronger matric forces, because many more of the water molecules can adhere to the particle surface. So long as the water potential in the soil is greater than the water potential in the roots, water will tend to enter the roots. If the soil water potential drops below the water potential in the root hairs, the flow of water may be stopped and the plant will experience water stress.

Lastly, atmospheric water vapor usually has a much lower water potential than water on stomata, so water readily evaporates from the leaf surface. As we described earlier, this imparts negative pressure on the system, creating a negative pressure potential and drawing more water up through its vascular system. The water molecules are bonded together with hydrogen bonds (cohesion), and usually form an unbroken connected column that extends through the leaves, stem, and roots, and even into the soil. The negative pressure exerted by transpiration even extends a small distance into the soil, helping to bring more water molecules closer to the roots.

For plants to function properly, their cells must be *turgid*, or filled with water. Generally, the water surrounding plant cells will have lower solute concentration (higher water potential) than the water inside of the cells (lower water potential), so water tends to flow into the cells by osmosis until achieving equilibrium with the pressure exerted by the relatively rigid cell walls. A short-term loss of turgor causes the plant to wilt. Most physiological processes such as photosynthesis and protein synthesis are greatly curtailed while the leaves are wilted. If water levels in a particular environment are generally too low for a particular plant species, it will be unable to survive in that environment.

 Thinking ecologically 5.3

When farmers irrigate fields with water from rivers or lakes, salts and other solutes gradually build up in the soil. How might this increase influence crop plants?

Plant species have a diverse array of physiological and morphological adaptations that favor their survival in environments with different moisture levels. Let's consider two studies that investigate adaptations that allow plants to survive in different environments.

Plant adaptations to variable water availability

Small organisms are particularly vulnerable to changes in the environment, as they have a much greater surface area to volume ratio. As a result they are more poorly buffered against environmental changes than larger organisms, which have much less exposed surface area. Seeds are juvenile plants, and may be particularly vulnerable because of their small size.

Tradeoffs in water regulation between desiccation-tolerant and desiccation-sensitive seeds

Though plants are not renowned for their parental care, natural selection should favor the evolution of traits promoting offspring survival in plants to the same extent as in some animal species. In seed plants, the developing embryo is nourished by the rich endosperm, and protected from the elements by a seed coat, which is encased within a fruit. These three morphological adaptations help to provide a thermal and hydrological environment conducive to the survival of the juvenile plant.

John Tweddle and his colleagues (2003) were interested in two challenges faced by plants during the seed stage. First, many seeds are subjected to environments with unpredictable moisture availability. Second, many seeds are shed in the fall, and except for those in the tropics, may experience a harsh thermal regime upon release. Tweddle's research group wanted to know how small, often delicate seeds were able to deal with these challenges.

To begin their research, Tweddle and his colleagues compiled a Seed Information Database, which included information on the seeds of 886 species of woody plants (trees and shrubs) and compared them to the seeds of 517 species of herbaceous (nonwoody) plants (Tweddle *et al.* 2002). This database rated seeds as being either tolerant or sensitive to desiccation. *Desiccation-tolerant* seeds survive drying to a low moisture content (typically below 7%), while *desiccation-sensitive* seeds suffer high mortality if the moisture level is reduced to 20% or below.

From their database, Tweddle and colleagues found that 99.8% of the herbaceous seeds were desiccation tolerant, while 83.4% of the woody plants produced desiccation-tolerant seeds. Confining their analysis to woody plants, the researchers tested three hypotheses. First, they proposed that natural selection would favor desiccation tolerance in plants from highly seasonal or arid habitats as compared to plants in more constant and moister habitats. Second, they hypothesized that even within rainforest habitats, natural selection would favor desiccation tolerance in pioneer species to a greater extent than in non-pioneer species. Pioneer species are adapted to colonizing newly available habitats (such as a newly arisen gap in the rainforest canopy after a large tree has fallen down). Lastly, the researchers hypothesized that desiccation tolerance would be more common in seeds that go through dormancy, an extended period with no growth or development, in comparison to seeds that continue development immediately after their release into the environment.

To test the hypothesis that natural selection has favored desiccation tolerance in plants from highly seasonal or arid habitats

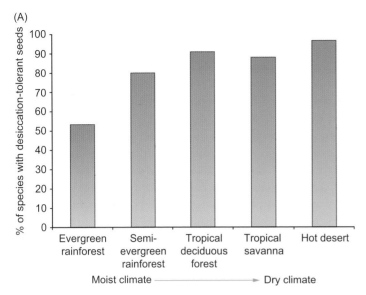

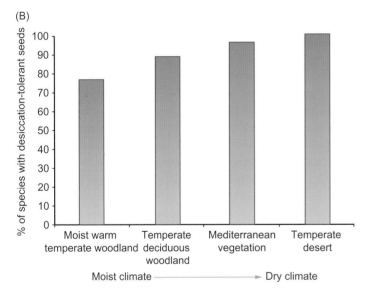

Figure 5.17 Percentage of species with desiccation-tolerant seeds in hot biomes (A), and moderate-temperature biomes (B).

Table 5.2 Number of plant species producing dormant versus non-dormant seeds classified as desiccation tolerant versus desiccation sensitive. From Tweddle *et al.* (2003).

Response to desiccation	Produce dormant seeds	Produce non-dormant seeds
Number (%) tolerant	492 (90.9)	238 (69.0)
Number (%) sensitive	49 (9.1)	107 (31.0)

Lastly, the researchers tested the hypothesis that desiccation tolerance is more common in species that go through a period of dormancy in comparison to seeds that continue development shortly after their release into the environment. As Table 5.2 shows, 541 species were classified as producing seeds that go through a period of dormancy, while 345 species were classified as producing non-dormant seeds. Of the species with dormant seeds, 90.9% were desiccation tolerant, in comparison to 69% desiccation tolerant in species with non-dormant seeds (see Dealing with data 5.1 to learn how these data are analysed).

As with most good studies, Tweddle's results raise many more questions that need to be addressed. One general question deals with the situation that while the overall trends among seeds support Tweddle's three hypotheses, there are many species that don't conform to the predictions generated by the hypotheses. For example, there are many species from moist areas that are desiccation tolerant, and some species from arid areas that are desiccation sensitive. Why does this happen? A second, more basic, question is, why aren't all seeds desiccation tolerant? Are there physiological costs associated with desiccation tolerance, or conversely, are there advantages associated with desiccation sensitivity?

Presumably there are energetic costs associated with producing a thick seed coat, and metabolic costs associated with going into dormancy; both adaptations increase a seed's desiccation tolerance. Given the number of seeds produced by large plants, these costs can be considerable. Tweddle and his colleagues suggest that we can understand the benefits of desiccation sensitivity by considering the greater frequency of desiccation sensitivity in plants that germinate immediately after shedding (plants with non-dormant seeds). They argue that rapid germination may reduce seed mortality by minimizing the time seeds are exposed to predators or pathogens (particularly fungi). In addition, rapid germination may maximize the rate of development at a time of favorable environmental conditions.

Of course, water availability will vary for different species at different stages of development. Next we will consider how more mature plants are adapted to variable conditions of water availability.

as compared to plants in more constant and moist habitats, Tweddle and his colleagues identified the preferred biome of each plant species. In warm and hot biomes, desiccation tolerance was most common in hot desert and decreased to near 50% in evergreen rainforest (which is the wettest and least seasonable of the vegetation zones) (Figure 5.17A). In the temperate biomes, the researchers discovered the same trend, in which desiccation tolerance was more frequent in drier habitats (Figure 5.17B).

Next, Tweddle and his colleagues tested the hypothesis that within rainforest habitats, natural selection has favored desiccation tolerance in pioneer species to a greater extent than in non-pioneer species. All (*N* = 21) of the pioneer species within the evergreen rainforest were desiccation tolerant while less than 50% (75 of 157) of non-pioneer species were desiccation tolerant.

Dealing with data 5.1 Categorical variables, frequencies, and contingency tables

A *categorical variable* (or qualitative variable) describes membership in a defined category or group. Tweddle *et al.*'s (2003) third analysis considers the relationship between two categorical variables. The first variable was seed type (dormant versus non-dormant), while the second variable was response to desiccation (tolerant or resistant). Both variables are categorical, because each species was assigned membership to one of the two groups. Categorical variables can have more than two categories per variable. For example, most taxonomic systems consider three domains of life: Eukarya, Bacteria, and Archaea, and also about 30 orders of insects (which I will not name here).

Tweddle's research group wanted to know whether desiccation tolerance was more common in dormant seeds than in non-dormant seeds. The *frequency* is simply a count of the number of species in each dormancy category that produced desiccation-tolerant versus -sensitive seeds. Often, frequency data are converted to percentages. Before concluding that desiccation tolerance was more common in dormant seeds than in non-dormant seeds, we must recognize that these data consider only a small subset of all the seed plants in the world. Hence we must determine if the difference between 90.9% of the tolerant seeds entering dormancy is statistically significantly greater than the 69% of the tolerant seeds that don't go into dormancy in our sample.

Usually scientists will use the χ^2 (chi-square) contingency test to determine whether frequencies are statistically significantly different from each other. The null hypothesis in this case is that dormant seeds are equally likely to be desiccation tolerant as are non-dormant seeds. Eyeballing the data in Table 5.2, we can see that over 90% of dormant seeds were desiccation tolerant, which is substantially greater than the 69% frequency for non-dormant seeds. The sample size is quite large so intuitively we might think that these differences are statistically significant. Statisticians calculate a χ^2 value to test for the strength of the association (for example, see Whitlock and Schluter 2009). A higher χ^2 value indicates that there is a very small probability that the null hypothesis is correct. Depending on the number of categories in our two variables, a given χ^2 value will generate a *P*-value, which we can interpret as the probability that the null hypothesis is correct. We can reject our null hypothesis, and support the research hypothesis that dormant seeds are more likely to be desiccation tolerant, if $P < 0.05$ (see Dealing with data 1.2 for a review of interpreting *P*-values).

In this case $\chi^2 = 70.01$ and $P < 0.001$. Based on this very low *P*-value we can, with a great deal of confidence, reject our null hypothesis and support the research hypothesis that dormant seeds are more likely than non-dormant seeds to be desiccation tolerant.

Adaptive responses to water availability in mature plants

Plants that live in habitats with periodic droughts may have adaptations that allow them to acclimatize to these conditions. One common response of plants to drought is allocating more resources to root growth during periods of water stress. This creates more absorptive surface area for the root system, and also allows the root system access to water that may be stored deeper in the soil. The root/shoot ratio – a measure of the weight of the root system to that of the shoots – is controlled, in part, by the plant hormone *abscisic acid* (ABA).

Imad Saab and his colleagues (1990) knew that ABA accumulated to high levels in the tissues of plants experiencing water stress. They also knew that when corn, *Zea mays*, experienced water stress, it grew very slowly, allocating most of its growth to roots while the shoots increased very slowly. This increase in root/shoot ratio enabled the corn to take in sufficient water when soil water was scarce. Putting these two findings together, the researchers hypothesized that ABA was responsible for regulating corn's adaptive increase in root/shoot ratio during water stress.

To test this hypothesis, Saab and his colleagues studied the growth of shoots and roots of corn seedlings under either abundant-water or low-water conditions, using either normal seedlings or ABA-deficient mutant seedlings. If ABA plays a major role in controlling corn's adaptive response to water stress, they predicted that the ABA-deficient mutants would fail to decrease shoot growth and fail to increase root growth under conditions of low soil water potential (water stress).

In support of their prediction, the researchers discovered that when water was abundant, shoot growth was greater in seedlings with normal ABA levels than in mutant (ABA-deficient) seedlings, but when water in the environment was limiting, shoot growth was greater in the mutant plants (Figure 5.18A). Thus ABA increased shoot growth in well-watered plants, but depressed shoot growth under drought conditions. In contrast, well-watered roots showed similar growth rates in normal and ABA-deficient plants (Figure 5.18B). However, when water is limiting, normal plants show much higher root growth than ABA-deficient plants. By putting these two results together, we can clearly see that normal corn plants have a significant short-term increase in root/shoot ratio in response to water stress, while ABA-deficient plants do not display this adaptive response to water stress (Figure 5.18C). This adaptive response increases water taken in from the environment (W_{en}) and decreases water loss from evaporation and transpiration (W_{et}).

Some plants are adapted to excess water. Cypress and mangrove trees have *pneumatophores*, extensions of roots that rise above the water table and are hypothesized to provide oxygen to the root systems. Plants may also have short-term responses to flooding. For example, several species within the genus *Rumex* (dockweed and sorrel) grow very quickly in response to long-duration floods, which allows them to keep their photosynthetic leaves above water, and to transport oxygen to their cells (Figure 5.19). Some species within this genus lack this adaptation and consequently are unable to live in areas subject to periodic floods (Voesenek *et al.* 2004).

Our two conceptual models in this chapter present temperature regulation and water regulation as two distinct problems

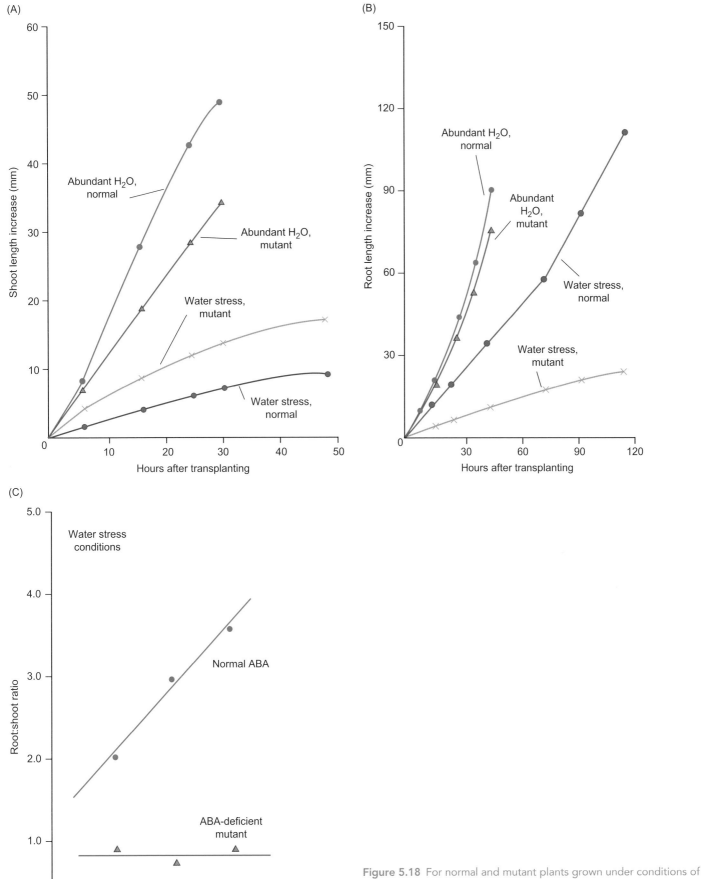

Figure 5.18 For normal and mutant plants grown under conditions of abundant water or water stress, A. Shoot length increase (mm) in relation to time after transplanting. B. Root length increase (mm) in relation to time after transplanting. C. Under water stress conditions, root/shoot ratios for normal and mutant plants in relation to time after transplanting.

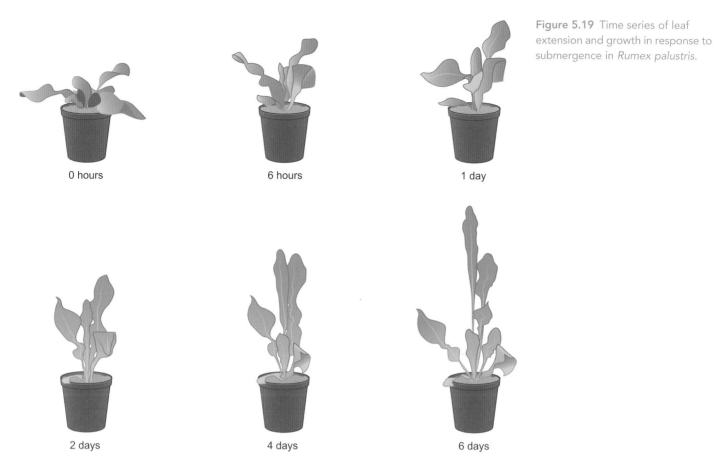

0 hours

6 hours

1 day

2 days

4 days

6 days

facing organisms. In many organisms, however, these two processes are tightly linked.

5.4   HOW ARE TEMPERATURE REGULATION AND WATER REGULATION FUNCTIONALLY LINKED?

Appropriate environmental temperature ranges and sufficient water availability are essential for the survival of all organisms, but appropriate levels are different for different species, and in some cases for distinct populations of the same species. For some species, extreme environmental temperature and low water availability may occur simultaneously. Let's see how three very different species solve the related problems of high environmental temperature and low water availability.

Minimizing heat gain and water loss: the saguaro cactus

The saguaro cactus, *Carnegiea gigantea*, found in the Sonoran Desert of the southwest United States and Mexico, is one of the most distinct desert species. Its large size and unique shape are a contrast to most other desert plants. How can the

saguaro soak up enough water to keep its cells hydrated during the hot summer days, and how can it keep its cells cool enough under heat stress so that they can continue metabolic activities?

Cacti are succulents, which are characterized by widespread but very shallow root systems that can quickly pick up water. With their very large mass and pleated expandable outer skin, saguaros are able to store large quantities of water collected from the occasional rains. The spines are modified leaves, which can provide both shade to reduce radiative heat gain (H_r) and a boundary layer to reduce water lost from evaporation or transpiration (W_{et}) (Figure 5.20). The epidermis has a thick waxy outer layer, and the large superficial roots have a thick corky outer layer, both of which reduce W_{et}. Many cacti also use CAM photosynthesis, which allows them to keep their stomata closed during the day, further reducing W_{et} (see Chapter 4).

Water has a high specific heat, so the saguaro's large mass tends to heat relatively slowly in the desert sun. Being shaped like a column with an orientation perpendicular to the surface exposes relatively little surface area at the hottest time of the day, minimizing H_r and W_{et}. The saguaro has a very high critical thermal maximum (CT_{max}), so it can allow its body temperature to increase above 50°C. At this elevated temperature, the

Figure 5.20 Adaptations to a hot and dry climate in the saguaro cactus. Note dense spines in close-up view.

temperature differential between the cactus's skin and the external environment is usually very small, so there is little potential for any further heat gain from the environment. The cactus then dumps this excess heat during cool desert nights.

Some animals are adapted to hot and dry conditions in ways that are surprisingly similar to the saguaro.

Minimizing heat gain and water loss: the camel

Camels (*Camelus* species) are legendary for their ability to travel through the desert for many days without experiencing water stress. Some calculations and experiments conducted by Knut Schmidt-Nielsen (1997) demonstrate the elegant simplicity of a camel's water conservation mechanisms. First, Schmidt-Nielsen calculated the advantage of large body size for water conservation; a large body has a relatively low surface area to volume ratio (Figure 5.21A), and thus experiences a reduced W_{et}. Then Schmidt-Nielsen engaged in the somewhat more adventuresome activity of taking a camel's rectal temperature during times of water abundance and times

of water deprivation. His results indicated that camels allow their body temperatures to increase markedly during times of water shortage (Figure 5.21B).

Camels benefit in two ways by storing this much heat. First, the camel does not need to dissipate this excess heat by evaporation; it simply waits until the evening, then allows the heat to dissipate naturally through conduction and radiation. Second, when the body is heated to 41°C, the temperature differential between the camel's skin and the air is relatively small, and the consequent rate of heat transfer from the environment into the camel's body is greatly reduced. A fully hydrated camel that maintains a relatively constant body temperature between 36 and 38°C loses about 9.1 l of water during the heat of day. A water-deprived camel that allows its body temperature to rise to 41°C loses about 2.8 l during the same time period. This is analogous to the saguaro cactus, which, as we just discussed, can allow its internal temperature to rise to 50°C. By allowing its body temperature to increase, the camel does not need to lose as much water to cool itself down by sweating. Lastly, Schmidt-Nielsen was able to show that the camel's thick and highly insulative fur

(A)

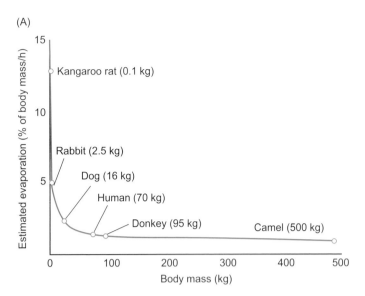

(B)

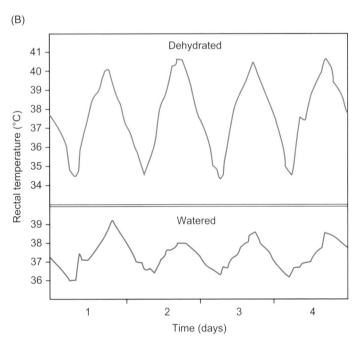

Figure 5.21 A. Theoretical rate of evaporation in relation to body mass, assuming that heat load is proportional to body surface area. B. Body temperature of camels in relation to air temperature for dehydrated camel (top) and well-watered camel (bottom).

confers significant water savings, by reducing radiative heat gain (H_r). When Schmidt-Nielsen sheared a camel's back, its rate of water loss increased approximately 50%.

If you look back at Figure 5.21A you will notice the animal at the other end of the water utilization curve from the camel, the kangaroo rat. Despite its small size, the kangaroo rat survives very well in the desert biome, and with the exception of lactating females, in some desert habitats it manages to survive without taking a drink.

Minimizing heat gain and water loss: the kangaroo rat

One key to the success of the kangaroo rat, *Dipodomys* species, is being active at night when temperatures are cooler and relative humidity is higher. A second piece of the puzzle is the rat's ability to reduce the quantity of water excreted in its urine. The kangaroo rat, along with many other desert mammal species, has an unusually long loop of Henle forming part of each nephron within its kidneys. The nephron creates an osmotic imbalance, so that fluid in the interstitial space just outside the loop has a higher solute concentration than fluid within the loop, and water flows passively from the loop into the interstitial space. Longer loops form a steeper gradient that drives the osmotic flow of water from the loop into the interstitial fluid, creating more concentrated urine. This structural adaptation greatly reduces the water lost to excretion (W_{se}); a kangaroo rat's urine is more than three times as concentrated as a human's and 10 times as concentrated as a beaver's urine.

While saving water is helpful, the kangaroo rat consists of about 66% water, which is very similar to the water content of other animals. Where does this water come from? According to Schmidt-Nielsen's calculations (Table 5.3), the vast majority of its water comes from the oxidation of carbohydrates, proteins, and fats in its diet, which produce a large quantity of metabolic water (W_m).

Randall Tracey and Glenn Walsberg (2002) point out that researchers should recognize the role of geographic variation even within what might usually be considered a single habitat. Schmidt-Nielsen's work with the kangaroo rat was done in a relatively cool and moist region of the Sonoran Desert, so Tracey and Walsberg went to a much hotter and drier region. They hypothesized that hotter and drier conditions might result in

Table 5.3 Overall water metabolism of kangaroo rat per 100 g of barley consumption at air temperature of 25°C. and relative humidity of 20% (Schmidt-Nielsen 1997).

Processes involving water gains	Volume gained (ml)	Processes involving water losses	Volume lost (ml)
Oxidation of food (cellular respiration)	54	Excretion of urine	13.5
Absorbed from food	6	Elimination of feces	2.6
		Evaporation	43.9
Total gain	60	Total loss	60

different environmental pressures, which could lead to either acclimatization or adaptation to these conditions.

In contrast to our previous portrayal of kangaroo rat burrows as cool and relatively moist, Tracey and Walsberg discovered that during most hot summer days, burrow temperature averaged about 35°C. In addition, kangaroo rat activity outside the burrow peaked at 9.00–11.00 p.m., when outside air temperatures were usually above 36°C. Furthermore, burrow relative humidity averaged about 22%. Taken together, there was much greater scope for water loss under these conditions. In fact, the researchers' estimate of water gains and losses shows a very large daily water deficit under these conditions.

Even desert-adapted animals cannot withstand such large water deficits over extended periods. Tracey and Walsberg learned from laboratory studies that water-stressed kangaroo rats consumed crickets, which have a high water content, while kangaroo rats that were not water stressed ignored crickets presented to them. They then collected 20 rats from the field during the heat of summer, and found insect fragments in 11 rat stomachs. They concluded that during the summer, kangaroo rats shift their diets in a manner that helps balance their water budget.

 Thinking ecologically 5.4

How might you determine if the differences between the kangaroo rats studied by Schmidt-Nielsen and by Tracey and Walsberg (2002) are examples of acclimatization or adaptation?

These studies demonstrate that we need to consider temperature and water relations together, so we can understand physiological responses within an ecological context. With this background, we can now consider how environmental temperature, water availability, and solute concentrations influence the distribution and abundance of species.

5.5 HOW DO TEMPERATURE, SOLUTE CONCENTRATION, AND WATER AVAILABILITY AFFECT SPECIES DISTRIBUTION AND ABUNDANCE?

Individuals in a species are constrained and defined in part by their environmental requirements. A species cannot survive outside of the environmental conditions to which it is adapted.

Environmental temperature and the distribution of species

A species' distribution is limited by its adaptations to variations in environmental temperature and water availability. For example, organisms constrained to maintaining a constant body temperature are not expected to live in habitats with environmental temperatures commonly outside of their thermal neutral zone. Interestingly, some of the smallest organisms on Earth are able to tolerate Earth's most extreme conditions. Researchers have identified numerous archaeans and bacteria adapted to conditions that had previously been considered incapable of supporting life.

Let's first consider the *thermophiles* – organisms adapted to very hot conditions. Probably the most famous thermophile is the bacterium *Thermus aquaticus*, known for the DNA polymerase it produces (commonly referred to as taq). This enzyme is active at temperatures up to 95°C, which makes it the enzyme of choice for catalysing DNA replication at high temperatures in the polymerase chain reaction. Some of the most heat-tolerant archaeans, in the genus *Pyrolobus*, can survive temperatures over 120°C. *Pyrolobus fumarii*'s optimal temperature for reproduction is 105°C, and it will cease reproduction if temperatures fall below 90°C.

In contrast to organisms restricted to extremely high temperatures, some species, such as the Antarctic fish *Pagothenia borchgrevinki*, are restricted to water temperatures below or near 0°C. This species has only been collected below the sea ice, where it feeds primarily on zooplankton. All of its physiological processes studied to date function best below or near 0°C and they fail, or show markedly reduced efficiency, at temperatures above 5°C. For example, the rate at which *P. borchgrevinki* breaks down acetylcholine drops precipitously at temperatures above 0°C. Thus this species, and approximately 50 closely related species, are limited in distribution to the Antarctic.

While environmental temperature can influence whether a species can exist in a particular aquatic habitat, many other factors – such as solute concentration, pH, pressure, and the availability of an energy source – affect species distribution as well. Let's consider how solute concentration influences two groups of organisms with very different lifestyles.

Solute concentration and the distribution of species

The osmotic concentration, or osmolarity, of a solution is a measure of the quantity of solutes dissolved in that solution (moles of solute per liter of solution). When an organism is immersed in water, any differences in the osmolarity of the water and the organism will create an osmotic imbalance. If the water (for example ocean water) has more solutes than the organism's tissues, the net movement of water will be out of the organism's cells. In fresh water, which has fewer solutes than the organism's tissues, the net movement of water will be into the cells. Osmoconformers, including most marine invertebrates, have the same solute concentration as their environment. Consequently they don't have problems maintaining water balance, because the net influx and efflux of water are at equilibrium. Though they are

(B)

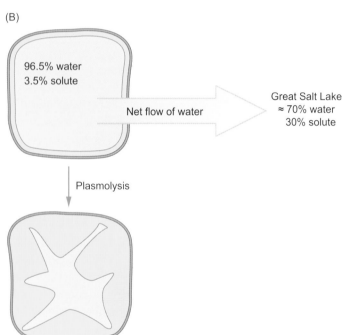

Figure 5.22 A. The Great Salt Lake in Utah, USA, is a hypersaline environment. B. Plasmolysis. In this example, a single-celled organism has a cytoplasm solute concentration equal to that of seawater. If it is placed in the hypersaline Great Salt Lake, there will be a net loss of water, and a shrinkage (plasmolysis) of the cell membrane.

osmoconformers, many invertebrates do need to import specific ions that are in short supply, or to export other ions that may be overabundant within their bodies

Halophilic bacteria and archaeans live in extremely salty environments, such as the Great Salt Lake (Figure 5.22A), or ponds constructed by humans for the purposes of harvesting salt. Many of these environments are also highly alkaline. Most organisms, if placed in these environments, would quickly dehydrate (Figure 5.22B). Halophiles deal with this problem by creating an internal environment that is high in solutes, by either producing solutes internally or importing solutes from their external environment. At high internal solute concentrations, these organisms achieve a balance between water entering their cells and water leaving their cells.

Osmoregulators maintain an internal osmolarity that is different from their external environment. Salmon have two different osmotic worlds to deal with. As juveniles, salmon live in freshwater streams; hence their cells have a higher salt concentration than the surrounding water. As a result, juvenile salmon are constantly losing salt to the environment and gaining water from the environment. They are able to regulate their salt imbalance by ingesting salt from their prey, and by having special chloride cells at the base of their gills that absorb ions from the water. To rid their cells of excess water, they produce copious amounts of very dilute urine. Because they maintain internal salt levels that are greater than those found in their environment, juvenile salmon are called hyperosmotic regulators.

Approximately 90–95% of a salmon's life is spent in the ocean, where it experiences exactly the opposite osmoregulatory challenge. Ocean water is much saltier than the water in a salmon's cells, thus there is a tendency for water to leak out through the cell membrane from osmosis, and for salt to enter the cells down the salt concentration gradient. To deal with the water imbalance, marine salmon drink ocean water, which exacerbates the problem with excess salt. In response, the aforementioned chloride cells secrete excess salt ions directly into the ocean. Because marine salmon maintain internal salt levels that are lower than those found in their environment, they are called hypoosmotic regulators. There is a significant metabolic cost with this (and any) mechanism that works against the concentration gradient. Finally, when salmon return to their natal stream to mate, they rediscover the osmoregulation challenge of their youth, so they must be adapted to be both hypo-osmotic and hyperosmotic regulators.

While the distribution of aquatic and marine organisms is influenced by the composition of solutes in their environment, the distribution of terrestrial organisms is greatly influenced by water availability.

Water availability and the distribution and diversity of species

As we discussed in Chapter 2, temperature and water availability are the two most important factors that determine the types of plant species present in terrestrial biomes. Animals have an advantage over many other groups of organisms, in that they can move about to locate water sources. But they still depend on plants for their food either directly as herbivores, or indirectly as carnivores that eat herbivores. Let's turn to southwest Africa to first explore how water availability influences an unusual group of plants, and then later an understudied group of animals.

Ice plants in the Succulent Karoo

The Succulent Karoo forms a narrow coastal strip in southwestern Namibia and western South Africa. The Aizoaceae are a very large family of approximately 1800 succulent plant species, plants that store water primarily in their leaves and stems, which is an adaptation to the very dry conditions that exist in that region. There are four subfamilies within the Aizoaceae; three of them are relatively species poor, with 39, 100, and 109 species. The fourth subfamily, the Ruschioideae, has been startlingly successful, with 1563 species.

Cornelia Klak and her colleagues (2004) were interested in two related questions. First, how quickly did this large group of related plant species evolve, and second, why was it so successful? The researchers analysed two chloroplast genes and one ribosomal gene as the basis for reconstructing the phylogeny of this species-rich group. A relatively small number of DNA base pair substitutions distinguished each species, which suggests that the adaptive radiation happened very rapidly. Based on their analysis, the researchers estimate that the age of the Ruschioideae is 3.8–8.7 million years. This is a good match to estimates based on the fossil record, which indicates that about 5 million years ago the habitat changed from relatively wet to a much drier desert habitat.

Why was this adaptive radiation so successful? Klak and her colleagues (2004) suggest that three morphological and physiological adaptations are responsible for this successful radiation. First, the Ruschioideae are the only taxonomic group to have *wide-band tracheids* – specialized cells that transport water and have unusually wide secondary cell walls that prevent collapse of the primary cell wall under conditions of extreme water stress (Figure 5.23A). Second, species in the Ruschioideae have

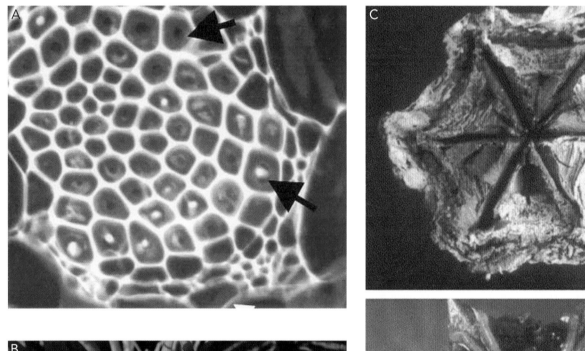

Figure 5.23 A. Transverse section of wide-band tracheids. The arrows point to the wide secondary cell wall that provides support under water stress conditions. B. The narrow-leafed ice plant *Conicosia elongata* with cylindrical leaves (and a mighty nice flower). C. Dried (top) and rain-soaked (bottom) hygrochastic capsules. Seeds are visible in the bottom picture.

cylindrical leaves, which lessen the surface area to volume ratio and reduce water loss under dry conditions (Figure 5.23B). Finally, the Ruschioideae have specialized *hygrochastic capsules*, which are fruits that open during heavy rainfall and are structured so that the force of the raindrops ejects the seeds away from the parent plant. The Ruschioideae are distinct from the other three subfamilies in that their capsules have covering membranes, which release only a few seeds at a time in response to rainfall, and thus increase the opportunity for successful dispersal and germination. As a result, these plants are more likely to have at least a few seeds that are released at an optimal time for successful germination (Figure 5.23C).

About 40% of the plants in the Succulent Karoo are endemic, found there and in no other part of the world. Their suite of adaptations to arid conditions allowed them to thrive and speciate rapidly in this unique region of Africa.

Staying within southwest Africa, let's consider how precipitation influences an unlikely group of animals, the helminth parasites that live inside the gut of the striped mouse, *Rhabdomys pumilio*.

Effect of annual precipitation on parasite distribution and abundance

Precipitation levels can vary considerably over a relatively small geographical region in southwest Africa. Götz Froeschke and his colleagues (2010) wanted to know how these pronounced differences in precipitation would influence the richness and abundance of helminths, parasitic worms within the striped mouse's digestive tract. They knew that many helminth species have a stage of their life cycle when they are free-living (living in the external environment). They proposed that helminths might be particularly vulnerable to desiccation during that time period, so that they might be less prevalent within the guts of mice from dry environments.

To test this hypothesis the researchers trapped mice at seven locations in Namibia and the Republic of South Africa. These regions had similar mean annual temperatures but differed markedly in annual rainfall (Figure 5.24A). The team of scientists captured 439 mice from these regions, collected fecal samples from all of the mice, and counted and identified helminth eggs from these samples. They then killed 161 of the mice, and screened the intestinal tracts for helminths.

Overall the researchers discovered a total of 13 different nematode (roundworm) species, and two cestode (tapeworm) species. The wettest regions had by far the highest prevalence of nematodes with almost 100% of the mice infected in the wettest region, and less than 25% infected in the driest region (Figure 5.24B). Species richness was also much greater in the wettest regions (Figure 5.24C), as was mean helminth abundance. All of these findings supported the hypothesis that helminth distribution and abundance were strongly influenced by moisture in the environment.

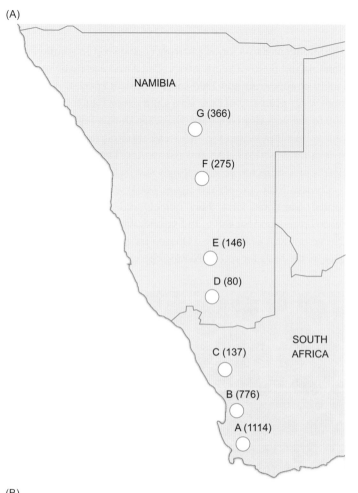

(A)

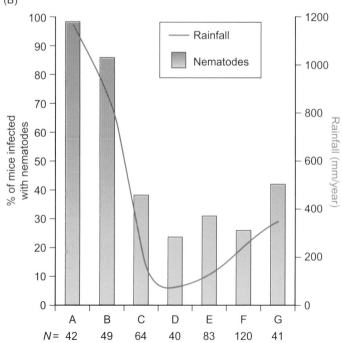

(B)

Figure 5.24 A. Seven study sites along a north/south axis in Namibia and the Republic of South Africa. Mean annual rainfall (mm) is in parentheses. B. The percentage of mice infected with nematodes at each study site. The solid line represents mean annual rainfall.

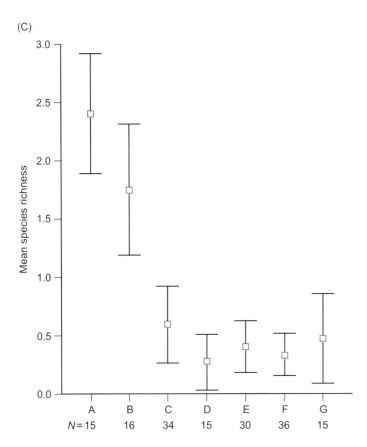

Figure 5.24 (cont.) C. Mean species richness at each site.

Thinking ecologically 5.5

The researchers propose that moisture influenced distribution because the helminths need high moisture levels to survive outside the mouse gut. Is this a valid conclusion based on the data? How might you determine the validity of this conclusion?

We are now entering an era in which temperature and rainfall patterns are expected to change quite rapidly, primarily due to the burning of fossil fuels. One consequence may be a shift in species' phenology – the relationship between climate and the timing of periodic ecological events, such as flowering, breeding, and migration. Or, as climates change, species may disperse to new environments with more appropriate climates. In the worst-case scenario, some species may go extinct. Can we see these processes occurring today?

5.6 IS EARTH'S CHANGING THERMAL AND HYDROLOGIC ENVIRONMENT INFLUENCING THE DISTRIBUTION AND ABUNDANCE OF SPECIES?

There is little doubt among climatologists and ecologists that the world, in general, is getting warmer. However, this warming is patchy – some regions are getting much warmer, while other regions are basically unchanged, and a few regions have actually become cooler over the past century (see Chapter 23).

Camille Parmesan and Gary Yohe (2003) explored the impact of climate change on natural systems. They tested two hypotheses. First, they hypothesized that global warming associated with climate change has caused an advance of phenologically related events over recent history. If true, they predicted that there would be, in general, an advancement in the date of frog breeding, bird nesting, first flowering, tree budburst, and the arrival of migrant birds and butterflies.

Parmesan and Yohe's analysis considered about 700 species of organisms, evaluating each species as one data point. In general, the analysis supports the prediction of spring advancement (Table 5.4).

Parmesan and Yohe also tested a second hypothesis – that global climate warming has shifted the distribution of species over recent history. This hypothesis generates four predictions. First, species in the northern hemisphere should expand their

Table 5.4 Phenological changes over the past 132 years for a variety of species. Modified from Parmesan and Yohe (2003).

Phenological change	Number of species or species groups	Time scale (years)	Number of species showing spring advancement	Number of species showing spring delay	Number of species showing no change
Plant budburst or flowering	461	35–132	321	47	93
Bird nesting or migration	168	21–132	78	14	76
Insect migration	35	23	13	0	22
Amphibian breeding	12	16–99	9	0	3

Table 5.5 Number of species or species groups showing changes in distribution or abundance in the direction predicted by global warming and in the direction opposite to the prediction of global warming. Modified from Parmesan and Yohe (2003).

Distribution/ abundance change	Number of species or species groups	Time scale (years)	Number of species changed in predicted direction	Number of species changed in direction opposite from predictions	Number of species with no change
Trees	9	70–1000	8	0	1
Herbs and shrubs	15	28–80	13	2	0
Lichens	165	22	43	9	113
Birds	306	20–36	157	69	80
Insects	127	98–137	67	5	55
Reptiles and amphibians	7	17	6	0	1
Fish	5	70	4	0	1
Marine invertebrates	42	66–70	39	3	0
Marine zooplankton	30	39–70	30	0	0

range in a northerly direction. Second, species in the southern hemisphere should expand their distribution in a southerly direction. Third, species on mountains should move higher in elevation. Fourth, species adapted to very high altitudes or latitudes should decline in abundance because there would be less suitable habitat available to them. In general, the data, summarized in Table 5.5, support these predictions.

In addition to supporting the predictions of their two hypotheses, Parmesan and Yohe were able to estimate the rate at which phenological or distributional changes are occurring. They estimated that, at present, phenological events are advancing at a rate of 2.3 days per decade, while range limits are moving to higher latitudes at a rate of 6.1 km per decade.

 Thinking ecologically 5.6

Tables 5.4 and 5.5 encompass a wide range of species. Do you think some species or groups of species will be more able to successfully respond to global warming? Explain your reasoning.

More recently, Allison Perry and her colleagues (2005) conducted a much narrower but more intensive study on fishes in the North Sea (between Britain, France, and Scandinavia). A survey conducted by the International Council for the

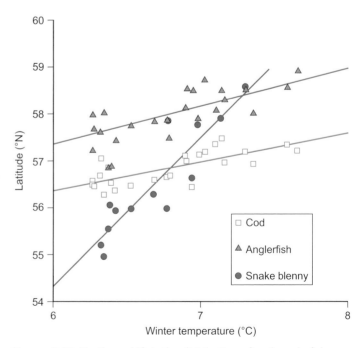

Figure. 5.25 Northern shift in the distribution of cod, anglerfish, and snake blenny in relation to increasing mean winter temperature.

Exploration of the Sea indicated that the mean bottom temperature of the North Sea had increased 1.05°C from 1977 to 2001. Perry *et al.* used data collected by the English Ground-fish Survey Programme to estimate the centers of distribution for each species (the location where the fish were most abundant each year). They hypothesized that such a rapid change in temperature should cause a change in the distribution of fish adapted to specific thermal regimes. They predicted that fish would tend to migrate in a northerly direction, or to seek out deeper water, which would provide a colder microclimate.

Figure 5.25 shows an example of some of their results. The three species of fish in the figure shifted their center of distribution in relation to temperature. Overall, 15 of 36 species showed significant changes in centers of distribution, with 13 moving north and two moving south. The mean shift in distribution was 172.3 km. Perry and her colleagues also found that 21 of the species they studied moved into deeper water, suggesting that these cold-adapted fish were seeking suitable microclimates.

The previous two studies provide substantial evidence that global climate change is influencing phenology. As we discuss in Chapter 23, humans are causing much of this change in global climate. Thus it becomes important to understand how environmental factors influence physiological processes such as thermal acclimatization, energy and nutrient intake, and water balance. Ecologists can use their understanding of the linkage between the environment and physiological performance to predict the impact of climate change on species living in extreme conditions, and to take meaningful steps to mitigate the damage that climate change causes to global ecosystems.

REVIST: Treeline communities in the Alps

Climate change is having some impact on the treeline communities of the European Alps, because the environmental shifts associated with the change are affecting organisms' physiological processes. Mean growing season air temperature increased about 1°C over the 30 years of the afforestation study there, while the mean snowmelt date decreased about 10 days over the same time period (Barbeito *et al.* 2012). The question in the European Alps, and elsewhere, is how these climatic changes will influence the survival and distribution of native trees, and the survival of other species that are adapted to these high-elevation habitats.

All three tree species survived at the Spillberg long-term research site, but *Larix decidua* was by far the most successful, with 69% of the trees still alive after 30 years. Mortality rates declined in all three species about 15 years after planting. Years with late snowmelt dates were associated with very high mortality, as was high elevation. One surprising finding was that early snowmelt was also correlated with high mortality in *L. decidua*. Once the snow melts, the trees begin growing, and the tree tissues become more vulnerable to freeze damage. The researchers suggest that early snowmelt might cause trees to begin growing unusually early in the season, only to be exposed to deep freezes that often occur in April and May in the high Alps. In addition, early snowmelt is associated with dry conditions in the spring, which could cause the trees to experience water stress.

However, late snowmelt dates are also associated with high mortality in each species. At Stillberg, late snowmelt was associated with increased tree mortality from pathogenic snow fungi. It was not clear whether late snowmelt promoted survival of the fungi, or, alternatively, weakened the physiological condition of the pines so they were more susceptible to pathogenic attack by the fungi. This study, and other ongoing long-term studies, provide unique opportunities for ecologists to gain insight into how changes in environmental conditions influence physiological performance in ways that impact entire ecosystems.

SUMMARY

Acclimatization and adaptation are two responses of organisms to a changing environment. Acclimatization is a short-term nongenetic response, while adaptation is a genetic response that occurs over many generations. Each species functions best in a particular thermal environment, and has adaptations that allow it to survive and reproduce successfully in that environment. The range of thermal conditions tolerated will vary with each species. Animals and plants face somewhat different challenges, as plants cannot move to a new environment, but there is also considerable overlap in how animals and plants deal with a variable thermal environment.

Organisms must balance their internal water and solute concentrations within tolerable levels. Terrestrial species in dry environments have adaptations for procuring and conserving water, while aquatic species are more likely to be challenged to maintain the proper solute concentration. For many species, the problem of temperature regulation and water regulation are functionally linked, and there may be tradeoffs in how these two critical processes interact. Environmental temperature, water availability, and solute concentration influence species distribution and abundance. Climate change caused by human actions is already significantly affecting species distribution and abundance, for example, shifting cold-adapted species toward the poles or to higher elevations.

FURTHER READING

Froeschke, G., Harf, R., Sommer, S., and Matthee, S. 2010. Effects of precipitation on parasite burden along a natural climatic gradient in southern Africa: implications for possible shifts in infestation patterns due to global changes. *Oikos* 119: 1029–1039.

Demonstrates a relationship between precipitation patterns and the distribution and abundance of parasites in the striped mouse along a precipitation gradient in southwest Africa. This study also introduces linear modeling as an approach to dealing with a large number of interacting variables.

Chandler, J. 2014. Ghosts of the rainforests. *New Scientist* 233: 42–45.

Nontechnical description of the plight of the white lemuroid ringtail possum, which is being forced into increasingly higher elevations in the northeast Australian rainforests as a result of global warming.

Palumbi, S. R., Barshis, D. J., Traylor-Knowles, N., and Bay, R. A. 2014. Mechanisms of reef coral resistance to future climate change. *Science* 344: 895–898.

Shows how both acclimatization and adaptation are allowing some corals to survive in the face of increased temperature.

Parmesan, C., and Yohe, G. 2003. A globally coherent fingerprint of climate change impacts across natural systems. *Nature* 421: 37–42.

A classic study of how climate change is causing significant shifts in species distribution. In addition to calling attention to this important ecological issue, this paper is a great introduction to meta-analysis as a technique for combining the findings of large numbers of studies that use very different methods.

Sheldon, K. S., and Tewksbury, J. J. 2014. The impact of seasonality in temperature on thermal tolerance and elevational range size. *Ecology* 95: 2134–2143.

Shows how environmental temperature variation influences the thermal tolerance and the distribution of beetle species across different elevations at four study sites in both hemispheres.

END-OF-CHAPTER QUESTIONS

Review questions

1. How may organisms respond to physiological challenges over different timescales? Give an example of each type of response.
2. What factors influence thermal balance in terrestrial animals? How do these factors influence body temperature in camels and kangaroo rats? What adaptations do camels and kangaroo rats have that allow them to avoid overheating?
3. What factors influence thermal balance in plants from (a) cold climates and (b) from hot climates? Use examples from the text to describe adaptations in plants that allow them to deal with these factors.
4. How do marine animals confront the challenges of thermoregulation and osmoregulation?
5. What water regulation challenges are faced by mature plants in dry and very moist environments? What are some adaptations in plants from these environments?
6. How is climate change influencing the distribution of species? Support your answer with data from the text.

Synthesis and application questions

1. In 1986, Robert Bakker wrote a book entitled *The Dinosaur Heresies*. In this book, he argued that dinosaurs maintained a constant body temperature. If so, what types of adaptations may they have had, and what type of climate might they have been living in?

2. Begin with a cube with side = 1 cm. Calculate the surface area, volume, and surface area to volume ratio. Repeat the process for cubes with sides = 2, 3, 4, 5, and 6 cm. Then graph the surface area to volume ratio in relation to the length of a side. What type of relationship is this?

3. Why do you think that so many organisms are biochemically adapted to function best at 35–40°C?

4. Tweddle *et al.* (2002) compiled a database of the desiccation resistance and dormancy patterns of over 1400 plant species. How do you think they gathered all of these data? Did they go out and study all 1400+ species, or, if not, what other sources did they use?

5. Notice that acetylcholinesterase operates efficiently over a wide temperature range in mullet and ladyfish. What might this tell you about the ecology of these two species?

6. Blubber seems like a great solution to living in cold climates because it insulates so well, and because at high body temperatures, its insulative properties can be bypassed by shunting blood to the skin capillaries. Why don't most terrestrial animals have blubber?

7. Compare and contrast the adaptations for temperature and water regulation in saguaro cacti, camels, and kangaroo rats. In what ways are they similar to each other? How are their differences in adaptations related to their respective environments and to differences in body size?

8. If Tracey and Walsberg (2002) are correct, do you think that kangaroo rats from Schmidt-Nielsen's (1997) study area would accept crickets as food? Explain your answer.

9. Which of the four phenological changes described by Parmesan and Yohe (2003) are you most confident is actually occurring?

Analyse the data 1

Tweddle *et al.* (2003) suggested that desiccation tolerance may also help protect a seed against cold temperatures. Table 5.6 shows the number of desiccation-tolerant and desiccation-sensitive species found in temperate and polar regions. Do the data support the hypothesis that desiccation tolerance offers protection against cold temperatures? Why or why not?

To answer this question you will need to go through several steps:

1. Identify the dependent and independent variables.
2. Identify the variables as qualitative or quantitative.
3. Decide on the appropriate type of test for evaluating this question.
4. Evaluate your results.
5. Discuss your conclusions.

Table 5.6 Number of desiccation-tolerant and desiccation-sensitive species found in temperate and polar regions.

Vegetation zone	Relative temperature	Tolerant	Sensitive
Northern temperate woodland	cool	16	3
Northern temperate conifer	cool	32	3
Boreal subalpine forest	cold	80	0
Southern subalpine zone	cold	7	0
Polar and alpine zone	frigid	3	0

Chapter 6
Behavioral ecology

INTRODUCTION

In his famous essay entitled "The tragedy of the commons," Garrett Hardin argues that human behavior has evolved over time so that people who overexploit resources are favored by natural selection. Within a population, individuals who take in the most essential resources should have the greatest survival and reproductive success. He also argues that powerful social forces support behavior that favors selfishness and greed at public expense. Hardin's discussion of this tragedy, and his suggestion for its resolution, ignited a storm of controversy when they were first presented in 1968, and continue to evoke tremendous response to this day.

Hardin introduces us to "the commons" – a pasture open to all herdsmen within a community. The commons has enough grass to support a herd of a particular size. Once this size has been reached, adding more cattle to the commons will cause a food shortage, which will, in an average year, cause some cattle, and by extension some herdsmen, to starve to death. As we will discuss, this argument can be applied to many human activities, and lead to the disturbing conclusion that human behavior is doomed to go down a very dark path. However, we will also learn that natural selection affects animal behavior in many complex ways, selecting for traits that enhance survival and mating success. Behavioral ecologists consider the costs and benefits of different types of behavior to address many questions about spatial distributions, foraging, and mating behavior. Different species live in different environments and thus their behavior is influenced by costs and benefits in a variety of ways. Even within species, we see tremendous behavioral variation, with some individuals cooperating and helping each other, even when it might not seem to be in an individual's best interest to do so.

KEY QUESTIONS

6.1. How does natural selection operate on animal behavior?

6.2. How can cost–benefit approaches address questions about spatial distributions, foraging, and mating behavior?

6.3. What physiological and ecological factors influence the evolution of mating systems?

6.4. How can indirect selection lead to cooperative behavior among relatives?

6.5. How do game theory models help explain the evolution of cooperation among unrelated individuals.

CASE STUDY: The tragedy of the commons

As the commons fills up with cattle, the herdsman asks himself "Should I add more cattle to my herd?" Hardin argues that from an individualistic perspective, the answer is "yes." In a community of herdsmen, there is only a very small chance that one of his herd will die from the impending resource shortage caused by his additional cow. So the rational solution is for the herdsman to maximize his return by adding another cow. In fact, he should continue to add cattle because the community is assuming most of the losses. Their loss is his gain. Unfortunately, all of the herdsmen face the same dilemma, and all are predicted to come to the same rational decision. As Hardin summarizes, "Freedom in a commons brings ruin to all" (Hardin 1968; Figure 6.1).

The power of Hardin's argument extends beyond competition for pasture. Most countries have national parks and forests that are utilized by all the citizens. All countries have fresh water to drink and air to breathe. These are resources that are held in common, and their integrity is being compromised by selfish human behavior. Humans are making the rational, but tragic, decision to pollute the environment, in the same way the herdsman is making the rational, but tragic, decision to add one more cow to his herd.

One source of the problem is overpopulation. Thousands of years ago there was no need for national parks to set aside wilderness, and no problem with waste disposal. As the human population approaches 8 billion, the human footprint on the environment continues to grow. Thus the greatest challenge articulated by Hardin is how to regulate limits on the human right to breed, and Hardin considers, but rejects, what he calls "an appeal to conscience" as being evolutionarily unstable. People will vary in their levels of morality: some will respond to the appeal to conscience by restricting both their use of resources and their reproduction, but others will pollute the environment, garner more resources for themselves and their children, and tend to have a morally unacceptable number of offspring. Assuming that variation in character traits is heritable, moral irresponsibility will increase in frequency in the human population as a result of natural selection.

You may be unaccustomed to thinking about natural selection influencing differences in behavior. But behavioral traits are like morphological or physiological traits in that their expression is a complex interaction between genes and the environment. Applied ecologists, and citizens in

Figure 6.1 The tragedy of the commons.

general, attempt to understand why humans, the most ecologically dominant animals on the planet, act the way they do. Achieving this understanding is challenging because the evolution of behavioral adaptations in animals – and particularly in humans – is influenced by so many different factors in the environment, including the behavioral interactions within and between different species. Ecologists study equivalent questions in other animals – how and why those animals employ certain behaviors to gain advantages in acquiring food, gaining territory, mating, or protecting themselves from harm. Studying the behavioral ecology of other animals may help us understand the relationship between ecology and behavior in our own species.

As students of ecology, we investigate the complex interactions between behavior, genes, and the environment. We will learn how the environment influences behavior and how behavior also influences the environment. These interactions take place across both evolutionary and ecological time. Natural selection favors behavioral adaptations that allow animals to take advantage of resources available in the environment over the long term. At the same time, natural selection also favors flexibility in behavior that allows some animals to shift their behavior over the short term, as environmental conditions and patterns of resource availability change.

6.1   HOW DOES NATURAL SELECTION OPERATE ON ANIMAL BEHAVIOR?

As we discussed in Chapter 3, adaptations are traits that enhance an individual's fitness, a measure of an individual's genetic contribution to future generations via reproductive success. As with morphological or physiological adaptations, behavioral adaptations may enhance fitness in two ways, either by increasing an individual's probability of survival or by increasing an individual's mating success. Both increased survival and mating success can ultimately increase an individual's reproductive success. Let's consider one example of each type of adaptation.

Behavioral adaptations and enhanced survival

Probably the most frequently studied types of behavioral adaptations influence foraging success and antipredator behavior. Natural selection favors the evolution of traits that improve reproductive success by enhancing an individual's success at finding and acquiring food, processing and storing food efficiently, and also not becoming someone else's food. All species of animals have behavioral repertoires that allow them to meet their nutritional needs within their environment. We'll begin by discussing foraging behavior in the Balearic lizard, *Podarcis lilfordi*.

William Cooper and his colleagues (2006) wanted to understand whether these lizards were making adaptive decisions to balance out the challenge of getting food while minimizing the

risk of being killed in the process. The researchers recognized that the lizards make tradeoffs in their foraging decisions. They predicted that lizards that foraged adaptively would take greater risks to obtain high-quality food sources, and smaller risks to obtain low-quality food sources.

To test this hypothesis, the researchers placed Petri dishes, with varying quantities of food, on the ground, withdrew to 8 m away, and stood motionless until a lizard's tongue made contact with the dish. At that point, researchers walked toward the dish at a speed of 81 m/min and stopped when the lizard fled. They recorded two responses: (1) the approach distance – how close the researcher got to the lizard before the lizard fled, and (2) the distance that the lizard ran in response to their approach (distance fled). The researchers then withdrew to their original position and recorded the return time – the amount of time that elapsed before the lizard returned to resume foraging.

The researchers varied food quality with four experimental treatments: 8, 4, 1, or 0 maggots were placed in the dish (see Dealing with data 6.1, for a discussion of how this type of experiment is designed and analysed). They predicted that, for the highest-quality resource, lizards would allow closer approach distances, would run a shorter distance from the predacious researcher, and would have a shorter return time. This was one of those very rare experiments when all of the predictions were supported by the data (Figure 6.2). Cooper and his colleagues concluded that Balearic lizards forage adaptively, minimizing risk in relation to the value of the resource.

Thinking ecologically 6.1

A challenge with many behavioral experiments is that animals don't always respond consistently or as researchers might expect. In this experiment, the lizards always fled, but they didn't always return to feed after the researcher withdrew. How would you deal with this problem when analysing the return time data?

Behavioral adaptations and mating success

In many animals, reproductive success requires finding a mate. Let's consider the firefly, which is actually a nocturnal beetle. Some of you have probably tried capturing flying fireflies at night with mixed success – they seem to turn sharply just about the same time that they flash. As you probably realize, flashing is not an adaptation to make fireflies easy for kids to catch, but is actually an integral part of the firefly mating process. Flashing has evolved as a mechanism for communication between the sexes and also for species discrimination.

Male *Photinus greeni* fireflies fly at dusk, and emit two flashes about 1.5 s apart. Females are flightless, and if responsive, flash back at a signaling male with a single flash. When he is receiving flashes from a receptive female, a male continues his flash sequences as he gradually gets closer and closer to the female. Sometimes an initially responsive female stops flashing when the

Figure 6.2 A. The Balearic lizard *Podarcis lilfordi*. B. Approach distance allowed by the lizard. C. Distance fled by the lizard in response to approach. D. Return time, the time until the lizard returned to the foraging dish, in relation to the number of maggots in the dish. Error bars indicate ±1 SE.

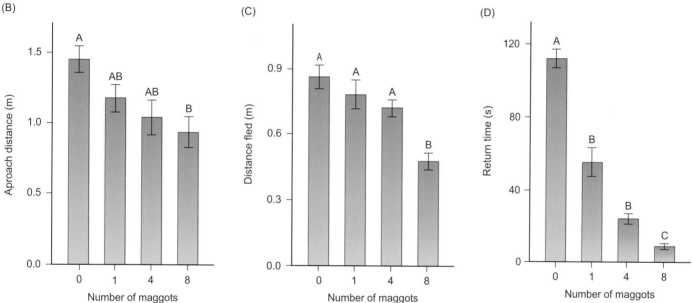

Dealing with data 6.1 Testing for treatment effects

In the Balearic lizard study, Cooper and his colleagues (2006) wanted to know whether lizards behaved adaptively to different levels of food abundance. The independent variable was the number of maggots in the Petri dish; each experiment had four different experimental *treatments* (0, 1, 4, and 8 maggots). The researchers tested to see whether there was a *treatment effect*, in which the number of maggots influenced each of three dependent variables: approach distance, distance fled, and return time. They used an **analysis of variance (ANOVA)** to answer these questions. The variance in the name of this important statistical analysis comes from the role played by the amount of variation in each treatment.

For example, let's consider the question of whether the number of maggots influenced approach distance. The research hypothesis (H_R) was that greater food abundance would cause lizards to tolerate closer approaches before they fled. The null hypothesis (H_0) was that there would be no relation between food abundance and approach distance. Table D6.1.1 shows some made-up data with very low variation among each sample within each treatment. For example, in the "0 treatment" column, all of the numbers are between 1.3 and 1.5. In contrast, Table D6.1.2 shows made-up data with much higher variation among each sample within each treatment, with values in the "0 treatment" column ranging from 0.2 to 2.4. One way to quantify the actual variation is to calculate the sample variance (see Dealing with data 2.3). ANOVA uses the differences between treatment means, and the variance within each treatment to test for significant treatment effects. Let's look at the tables to get an idea of how this works.

Table D6.1.1 Approach distances allowed by a lizard in relation to number of maggots in a Petri dish (low within-treatment variation).

Sample number	Treatment (number of maggots per Petri dish)			
	0	1	4	8
1	1.4	1.1	0.9	0.9
2	1.5	1.2	1.0	0.7
3	1.4	1.3	1.1	0.8
4	1.3	1.2	0.8	0.8
5	1.4	1.3	1.1	0.7
Mean	1.40	1.22	0.98	0.78
Variance	0.005	0.007	0.017	0.007

Table D6.1.2 Approach distances allowed by a lizard in relation to number of maggots in a Petri dish (high within-treatment variation).

Sample number	Treatment (number of maggots per Petri dish)			
	0	1	4	8
1	1.0	0.7	0.1	1.8
2	2.0	2.2	0.9	0.2
3	2.4	1.1	0.6	0.4
4	0.2	0.5	2.8	1.1
5	1.4	1.6	0.5	0.4
Mean	1.40	1.22	0.98	0.78
Variance	0.74	0.48	1.12	0.44

Intuitively, would you conclude that lizards tolerate a closer approach distance when food is more abundant? Based on Table D6.1.1, you would probably answer yes. However, the answer is much less clear based on Table D6.1.2. Interestingly, the treatment means are exactly the same for the two tables, but the within sample variation is much greater for Table D6.1.2 than it is for Table D6.1.1. An ANOVA uses an algorithm very similar to the one you intuitively used, to account for both the differences between treatment means, and the variation within treatments. The test statistic is the F-value, which is greatest when there are substantial differences between means and low variation within treatment groups (as in Table D6.1.1). A high F-value generates a low P-value, allowing us to reject H_0 and support H_R that lizards tolerate a closer approach distance when food is most abundant. We will not go into the mathematical details of how F-values are calculated.

In many cases, researchers want to know not only if there is a treatment effect, but also, which treatment means are significantly different from each other. There are several different multiple comparison tests available that accomplish this; one of the most common is the Tukey–Kramer test. Again, we won't go into the details of how they work, but we will use a convention in this text for how the results are represented. Usually, the critical value (α) is set at 0.05. When the multiple comparisons test indicates that two means are significantly different from each other ($P < 0.05$), they will have exclusively different letters over the bars of the bar graph. When the multiple comparisons test does not show a significant difference between two treatment means ($P > 0.05$), at least one letter over the two bars will be the same. Returning to Cooper's study (the real data!), from Figure 6.2B, we would conclude that the "0" and "8" treatments have significantly different mean approach distances because they have exclusively different letters above the bars (A for "0" and B for "8"). But the "1" and "4" treatments have AB above the bars, indicating that the means for those treatments are not significantly different from the means for any of the other treatments. Considering Figure 6.2D, lizards in the "0" treatment had significantly longer mean return times than did lizards in the other three treatments, In addition, lizards in the "8" treatment returned more quickly than did lizards in the other treatments. But there were no significant differences when comparing lizards in the "1" and "4" treatments to each other, though there was a trend in the predicted direction.

In some cases, researchers conduct experiments or systematic observations in which there are only two treatments. In those cases they use a **t-test** to evaluate the significance between treatment means. The test statistic is the t-value, which is conceptually and mathematically very similar to the F-value for the ANOVA. t-values are greatest when there are substantial differences between means and low variation within the two treatment groups. High t-values generate low P-values, allowing us to reject H_0 and support H_R. Because the t-test always compares only two treatments, there is no need to consider multiple comparisons.

male gets closer, effectively ending the courtship sequence. Ecologists (both amateur and professional) can observe these responsive flashing patterns as fireflies interact.

Constantinos Michaelidis and his colleagues (2006) suspected that female fireflies use male signals as a way of assessing male quality. To determine what aspects of the signal were attractive to the female, the researchers generated artificial male signals, with a computer-programmed light-emitting diode as the male flash. They could then control (1) the duration of the flash, and (2) the interpulse interval (IPI) – the time interval between the paired pulses (Michaelidis *et al.* 2006). The scientists conducted their study in a natural habitat, systematically varying flash duration and the IPI.

Results of the field experiment with the flash simulator show that both air temperature and IPI influence female response, but that signal duration does not. At low temperatures, females are more responsive to longer IPIs, while at high temperatures, females are more responsive to shorter IPIs. These differences in female responsiveness correspond to differences in the mean male IPIs at these temperatures (Figure 6.3). Males are emitting signals that correspond with female preferences at different temperatures.

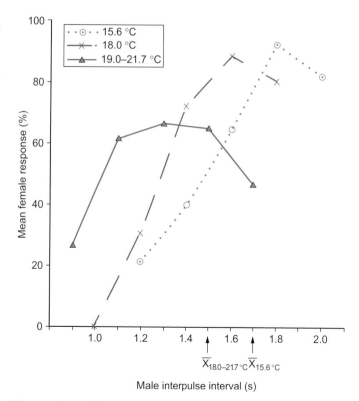

Figure 6.3 Female response to IPIs generated by a flash stimulator at three different temperatures. Arrows pointing to the bottom of the x-axis show mean male flash IPI at two different temperatures. Note the correspondence between female preferences and male performance.

Thinking ecologically 6.2

One hypothesis for why males have longer IPIs at cooler temperatures is that long IPIs require less energy than short IPIs. How might you test this hypothesis under experimental conditions? If your hypothesis is correct, what would you expect to observe?

The previous two examples demonstrate that animals have behavioral adaptations that are expected to lead to increased survival or mating success. But we really want to get a better understanding of the nature of this influence. Can we isolate different variables to understand how they might influence behavior? And what context favors particular variables as being important? For example, will the number of competitors in the environment influence whether an animal guards a resource? If so, to what extent? To answer these types of questions, we need a more quantitative approach, which behavioral ecologists have developed with a little help from their economist friends.

6.2 HOW CAN COST–BENEFIT APPROACHES ADDRESS QUESTIONS ABOUT SPATIAL DISTRIBUTIONS, FORAGING, AND MATING BEHAVIOR?

Figure 6.4 The dunnock, *Prunella modularis*.

Researchers often borrow ideas or models from different disciplines to help them address central questions within their disciplines. The field of economics has served as a fertile source of approaches for understanding many aspects of behavior and ecology.

Predictive models based on measuring the benefits and costs of behavior

The spatial distribution of resources will influence the behavior of animals that depend on those resources. Let's consider the dunnock, *Prunella modularis*, a songbird that feeds on small seeds and invertebrates in dense vegetation – primarily in flower beds and bushes (Figure 6.4). Nick Davies and Arne Lundberg (1984) were interested in the influence of food availability on the spatial distribution and ultimately on the mating behavior of these birds. If the flower beds and bushes are highly dispersed throughout a field, the dunnock will use a lot of energy going from one flower bed or bush to another. Davies and Lundberg hypothesized that natural selection will favor a dunnock that forages in an economical manner, so that energy is not wasted during the foraging process.

Let's consider a second situation in which the flowers and bushes are clumped together in a relatively small space, so that the energetic demands of foraging are greatly reduced. Have we discovered dunnock heaven? In one sense, this sounds like paradise, but a problem arises because other dunnocks will be attracted to the same resource bounty. The sharing of resources can trigger defense behaviors – individuals protecting sources of food. The question again arises: what is the most economical manner to forage? Should the dunnock eat its fill, alongside all the others that are sharing the same resource? Or should she defend this clump of space and establish a territory from which she excludes competitors, either partially or completely?

A basic model of territoriality

Jerram Brown (1964) argues that territoriality should develop when the benefits of territory defense exceed the costs. Under these conditions the territory is *economically defendable*. The challenge for behavioral ecologists interested in understanding space utilization is first to identify those benefits and costs, and second to measure them.

The most obvious benefit of territory defense is exclusive control of a certain area. This gives the territory owner access to all the resources on the territory, including food, places to hide from predators, appropriate microclimates, and access to mates that come into the territory. The most obvious costs of territory defense are the energy expended defending the territory, the risk of injury or death from fighting, and the increased risk of predation while attending to other territorial duties.

Based on Brown's concept of economic defendability, we can construct a very simple model that addresses whether territoriality should be present in a system, and if so, how big the territory should be. As Figure 6.5 shows, territoriality should be present if the benefits of territory defense are greater than the costs. The optimal territory size will maximize benefits minus costs $(B - C)$.

In the case of the dunnock, what will happen if flower and shrub density increases? The benefits curve will shift to the left because there are more resources per unit area in a territory with a high density of flowers and shrubs (Figure 6.5, bottom). It is possible that the costs curve will shift to the left as well, depending on the species and the environmental conditions. For example, if there are few high-quality territories, competition between birds may increase, which will increase the cost of defense. Whether the cost curve shifts or not, the value of $(B - C)$ will be maximized at a smaller territory size. Thus one prediction of this model is that territory size should shrink as resources become more abundant.

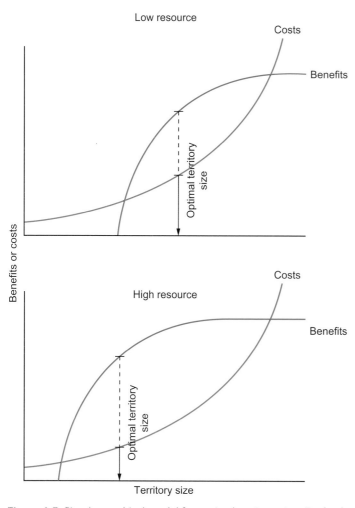

Figure 6.5 Simple graphical model for optimal territory size. For both graphs, optimal territory size is at the maximum value of (*B – C*), indicated by the dashed bracket. The benefits curve levels off because there is very little benefit gained from defending a territory that provides more resources than the animal can process. The costs curve increases exponentially, because costs rise dramatically as territory size increases. At higher resource levels (bottom graph), benefits are greater at even small territory size, shifting the benefits curve to the left, and decreasing the optimal territory size.

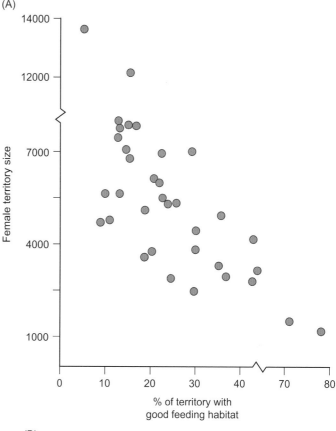

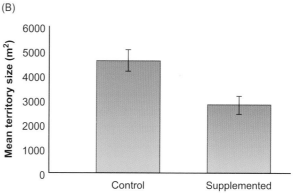

Figure 6.6 Relationship between territory quality and territory size in dunnocks. A. Observational study. B. Experimental study in which some birds received supplemental feeding. Error bars indicate ±1 SE (*P* < 0.01).

Testing the predictions of the basic territory model

Davies and Lundberg tested this prediction both observationally and experimentally. First, they captured each individual dunnock found on its respective territory, 67 in all, and put unique color bands on its legs. For the observational study, Davies and Lundberg measured territory size by repeatedly plotting each bird's position in the study site over the course of 2 years. Then they measured the percentage of the territory that had good feeding habitat, containing flower beds or bushes. As Figure 6.6A demonstrates, territory size was smaller when resources were more abundant (Davies and Lundberg 1984).

For the experimental study, Davies and Lundberg supplemented 28 of the female territories with a daily supply of oats and mealworms. They left 39 territories unsupplemented as controls. As predicted, the average territory size was significantly smaller for the supplemented birds (Figure 6.6B).

Economic approaches that consider benefits and costs can be applied to many different ecological contexts. Let's turn our attention to exploring the circumstances that favor living in groups.

Benefits and costs of group living

Many species of animals are not territorial. Some species are solitary, with broadly overlapping home ranges – areas that they traverse over the course of their lifetimes. Others may live in large social groups, such as the wildebeests we discussed in Chapter 1. There are numerous potential benefits to living in groups, as well as numerous costs. Table 6.1 outlines some of these benefits and costs.

Let's consider one potential benefit of living in groups – early detection of predators. Robert Kenward (1978) tested this hypothesis experimentally by training a goshawk to prey on pigeons under natural conditions. Pigeon flock size ranged from 1 to greater than 50. Because pigeons can generally outfly goshawks, the key for their surviving an attack is early detection. Kenward defined response distance as the distance between goshawk and pigeon at which the pigeon begins its escape flight. He made two predictions: first that pigeons in small groups would be more vulnerable to attacks than pigeons in large groups; second, that lower vigilance in small groups would result in a shorter response distance.

Kenward found support for both predictions. Goshawk attacks were much more successful against solitary pigeons or small flocks. Goshawks were successful when pigeons were slow to respond to their approach (short response distances). Response distance increased with increased flock size, indicating that pigeons in large flocks tended to flush more quickly in response to the approach of a predator

Experiments such as these can identify variables that influence the value of $(B - C)$, as a measure of the costs/benefits balance. However, there are usually several variables that influence this balance. For example, Kenward (1978) was also able to show that solitary pigeons tended to be in poorer health than pigeons in large flocks, so the increasing vulnerability that he observed in solitary pigeons could have been a result of their poor health.

While social grouping is beneficial to some species, it is costly as well (Table 6.1). Social grouping may cause competition for food among members of the group. Measuring these costs can help us predict changes in group size in relation to changes in environmental conditions. We will explore the relationship between food availability and group size in a social primate.

Optimal group size and food availability

Ring-tailed lemurs, *Lemur catta*, live in female-dominated groups of about 5–25 individuals in Madagascar (Figure 6.7A). They are omnivores, and they defend food patches or territories from members of other groups. Food availability varies with rainfall and is lowest in the dry season between May and September.

Ethan Pride (2005) studied seven groups of lemurs over the course of a year to test the basic group size model shown in

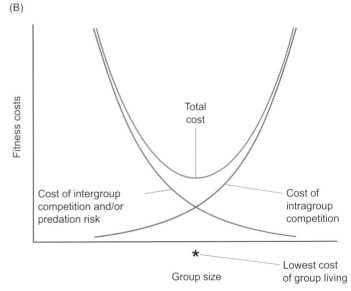

Figure 6.7 A. Group of ring-tailed lemurs. B. Pride's model showing the relationship between group size and fitness costs. The central U-shaped curve is the sum of the two fitness cost curves, which is minimized at the asterisk.

Table 6.1 Some benefits and costs of living in social groups.

Benefits	Costs
Group defense	Greater conspicuousness to outside predators
Detection of predators	Increased transmission of disease
Increased foraging success (finding food or subduing prey)	Increased competition for food
Opportunities to make alliances	Greater vulnerability to attack by individuals within the group
	Greater vulnerability to cuckoldry or reproductive interference

(A)

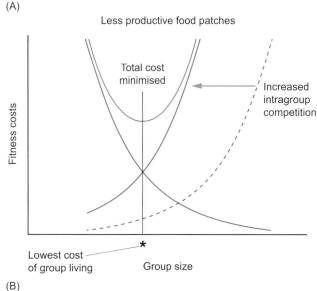

(B)

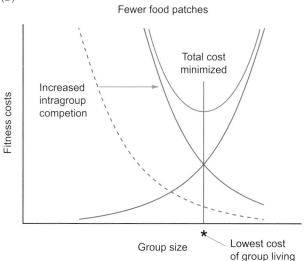

Figure 6.8 A. If food patches are less productive, intragroup competition increases and optimal group size decreases (as indicated by the bold asterisk). B. If there are fewer food patches, small groups are displaced by larger groups and optimal group size increases (bold asterisk).

Figure 6.7B. He reasoned that the costs of being in a particular social group should depend on two factors. First, lemurs in small social groups should experience very high costs due to inter-group (between-group) competition, as large groups will dominate small groups during periods of resource shortage. However, lemurs from large groups will experience high costs of intragroup (within-group) competition, being forced to share resources with members of their own group. Optimal group size minimizes the sum of the costs of both types of competition, resulting in lowest costs (and thus highest fitness) for members of intermediate-sized groups.

Furthermore, the type of food scarcity will influence the optimal group size. If the number of food patches remains the same within a group's range, but the productivity per patch (e.g.

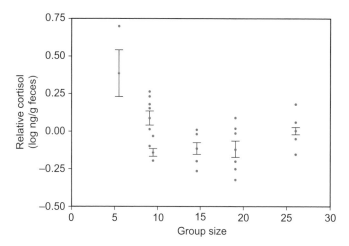

Figure 6.9 Mean cortisol levels in females from different-sized groups. The three intermediate-sized groups had significantly lower mean cortisol levels than the two smallest and one largest group.

number of fruits per tree) goes down, that change should increase intragroup competition, and reduce the optimal group size (Figure 6.8A). Alternatively, if the number of food patches goes down, that decline should increase intergroup competition, favoring larger groups that are able to outcompete smaller groups for the limited number of patches (Figure 6.8B).

Pride wanted to know if lemurs formed groups of a size that minimized costs, in accordance with his model of optimal group size. His problem was that lemurs have a complex social society, making it unlikely that group size would shift quickly in relation to food availability during the year of his study. Instead, he decided to test his model by evaluating the stress levels of individuals in different-sized groups. If his model was correct, and intermediate-sized groups experienced lowest costs, he predicted that lemurs from intermediate-sized groups should have lowest stress levels. In primates, and in some other species of mammals, cortisol levels are good indicators of the amount of stress experienced by the individuals. High blood cortisol levels are associated with high stress. Cortisol can be extracted from the feces, which lemurs are always willing to provide. Pride realized that measuring the cortisol levels of all of the lemurs would provide him with a window into the physiological health of each individual, and allow him to measure the relative well-being experienced by members of different-sized groups. His model predicted that members of intermediate-sized groups would have the lowest cortisol levels.

Pride's analyses indicated that indeed, average female cortisol levels were lowest for those in intermediate-sized groups, with females in both smaller and larger than average groups tending to have the highest cortisol levels (Figure 6.9). Males did not show this trend, but Pride explains that lemur males live on the edges of the group (they are socially subordinate to females) and that males also move from one group to another. Pride interprets these findings as supporting the basic group size model, with the fitness costs minimized in average-sized groups. It would be interesting to extend this study for an additional period of time and a larger number of groups to see whether group size ultimately changed in response to patterns of food availability.

Economic approaches to understanding behavioral ecology have many applications. We will return to questions about social groups later in this chapter. For now let's see if we can use a cost–benefit approach to understand the behavior of individuals within a population.

Predicting how long an individual should persist at a specific task

In many ecological contexts, the expected gain from a behavior decreases with the amount of time an individual has already spent doing the behavior, causing the individual to move to a new location. For example, hummingbirds sipping nectar from a patch of flowers may very quickly deplete nectar levels within the patch, and may move to a new patch. As a second example, for many insects a male's fertilization success is closely tied to how long he mates with the female. In these cases a male who copulates for a long period of time fertilizes more eggs than a male who copulates for a shorter time period. However, after an extended period, the increase in fertilization success begins to diminish, and the male may leave his female in search of other receptive females. In both cases, there are costs associated with traveling, in that it takes time and energy to find a new patch or female. The marginal value theorem makes specific predictions about when an animal should move to a new location based on its marginal rate of gain, or the rate of gain expected in the next time period.

The marginal value theorem and patch residence time
Humans are not the only animals capable of depleting resource abundance. In fact, most animals will influence resource availability as a result of their foraging behavior. Some animals are more influential than others; for example an elephant may eat up to 200 kg of vegetation per day, and desert locusts swarming in the western Sahel region of Africa may eat over 100 000 kg of vegetation per km² per day. If natural selection favors optimal foraging behavior, we would expect animals to switch locations after they have depleted the resources in a particular patch. The decision of when to switch would depend on how much time and energy is lost when searching for a new patch.

Eric Charnov (1976) developed the marginal value theorem as a quantitative approach for predicting how long an animal should continue foraging in a discrete resource patch. Intuitively, if patches are rare, we expect the animal will tend to stay in the initial patch until most of the resources are depleted. If patches are common and easily found, we expect the forager

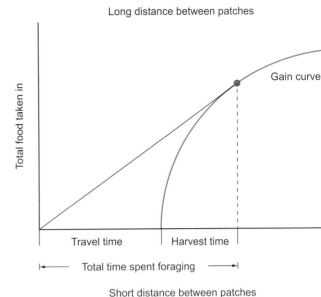

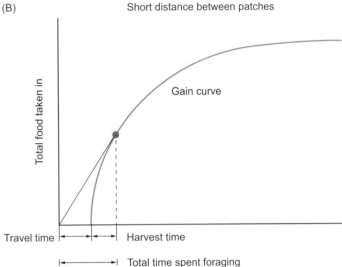

Figure 6.10 The optimal foraging time is determined by drawing a line from $T = 0$ that intersects the gain curve. Food intake rate is maximized if this line barely touches (is tangent to) the gain curve. A. Long distance between patches (long travel time). B. Short distance between patches (short travel time). Note that the harvest time (the time spent in a patch) decreases as travel time between patches decreases.

will be more likely to leave a patch before its resources are exhausted.

Quantitatively, we need to know the shape of the *gain curve*, which represents the relationship between how much energy is harvested in relation to the amount of time the forager remains in the patch (harvest time) (Figure 6.10). In most cases, the gain curve will be steep initially, meaning that a foraging animal gains a high amount of energy for the time spent foraging, but then the gain curve will level off as resources are depleted, so that extra time spent foraging leads to smaller gains in energy. We also need to know how long it will take the animal to find a new patch (travel time). Charnov's model predicts that animals should maximize their total food intake in relation to total time spent foraging. Total food intake is determined by the gain curve. Total time spent foraging is equal to travel time + harvest time. Thus the food intake rate is equal to (total food taken in)/(total time spent foraging).

One prediction of the marginal value theorem is that longer travel times will cause foragers to spend more time in the patch (or increased harvest time) (Figure 6.10A and B).

Many different researchers have tested the predictions of the marginal value theorem. Alex Kacelnik (1984) tested the prediction that starlings gathering food for offspring – traveling between nest and food sources – would forage in a way that maximized their food intake rate. Kacelnik used a long straw to drop mealworms into a feeder at a rate that generated a consistent gain curve, which the starlings were able to learn. To simulate a standard gain curve, the first mealworms were dropped one after the other, but subsequent mealworms were dropped at increasingly slower rates. The trained starlings had to decide when to go back to the nest with their loads. Kacelnik's independent variable was the distance between the nest and the feeder. He predicted that as the distance between the nest and the feeder increased (travel time increased), the parent starlings would spend a longer time at the feeder harvesting mealworms (Figure 6.11). As you can see, Kacelnik's data broadly fit the predictions of the marginal value theorem. Starlings making longer trips to the feeder were more likely to take larger loads of mealworms than were starlings making shorter trips; this strategy served to maximize their food intake rate.

Researchers have extended the marginal value theorem to other related questions as well. For example, Gordon McNickle and James Cahill (2009) used the marginal value theorem to explore the behavior of *Achillea millefolium* (common yarrow) plant roots that encountered patches of different resource quality. Based on the marginal value theorem, the researchers predicted that actively growing plant roots that encounter a particularly nutrient-rich patch of soil should slow down their growth rate until they deplete the soil of most of its nutrients. They discovered that plant roots spent more time foraging in the richer habitat, and secondly that plants allocated more root biomass to roots in particularly nutrient-rich soil.

One of the hallmarks of a useful model is its applicability to several related questions. While Charnov was developing his application of the marginal value theorem to questions about foraging, Geoff Parker was developing the same model to predict copulation duration in golden dungflies (Parker 1970).

The marginal value theorem and copulation duration

Parker (2001) describes the golden dungfly, *Scatophaga stercoraria*, as the ideal model system for behavioral ecologists studying mating behavior. The dung pile is the focus of most of the dungfly's ecology and behavior. Eggs are laid (oviposited) in the dung pile, and larvae eat the dung as they grow and metamorphose into adults. Mature males and females are more omnivorous, eating other insects and sipping nectar but also indulging in regular dung meals. Both males and females have olfactory receptors that detect fresh dung piles. Mature gravid females (females with eggs) approach a fresh pile from downwind and are usually greeted by at least one male before they even reach the pile. Because there are many more willing males than gravid females at any point in time, there is fierce competition for matings. Once a male clasps a female, the two usually retire to a nearby bed of grass and mate for at least 15 min and oftentimes much longer (Figure 6.12A). If other males spot the mating pair, they may grapple with the copulating male and attempt to push him off the female. Even if a male completes mating with a female, he must then fly with her from the bed of grass back to the dung pile and guard her for about a half hour while she oviposits. Why do dungflies mate for such long durations while many other species of flies mate for only a few seconds? Why does the male guard his mate? Wouldn't natural selection favor males who chased after other gravid females? Parker used the marginal value theorem to address these questions.

Parker had recently learned that if a female dungfly mates with more than one male, the last male to mate with her fertilizes most of her eggs (Parker 1970). In fact, the male's fertilization success rate is a function of the time spent copulating, but it levels out with increasing copulation time. Figure 6.12B shows the fitness gain curve for copulation duration. Given that most females arrive at dung piles with sperm from previous matings, it clearly benefits a male to mate with the female for a considerable duration of time, but only up to a point.

The problem for the male dungfly is that he cannot copulate with his present female and search for new females at the same time. Parker predicted that a male should maximize his fitness gained in relation to total time available for mating on the dung pile. At some point he should stop mating with the female, return her to the dung pile to lay eggs, guard her while she lays her eggs

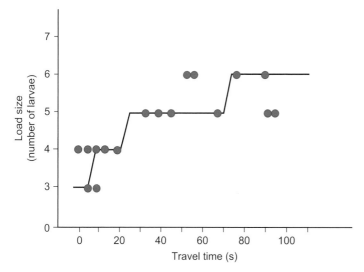

Figure 6.11 Number of larvae harvested in relation to the travel time between the nest and the feeder in European starlings.

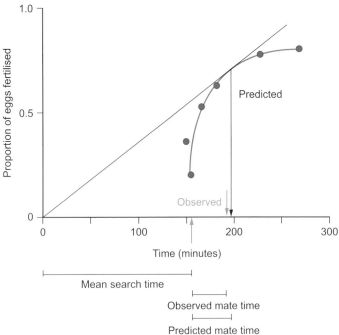

Figure 6.13 Marginal value theorem used to predict optimal copulation time for golden dungfly. Arrow pointing upwards toward *x*-axis is the mean search time for a new female = 156.5 minutes. The gain curve is from Figure 6.12. Observed copulation time is slightly less than predicted by the marginal value theorem.

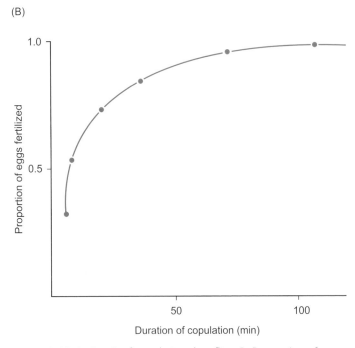

Figure 6.12 A. A pair of copulating dungflies. B. Proportion of eggs fertilized in relation to duration of copulation (min).

(fighting off all intruding males), and then search for a newly arriving gravid female. As with optimal foraging on resource patches, we can calculate an optimal copulation duration after which the male should search for a new mate.

Parker calculated that the average male took 156.5 min to find a female at the dung pile. Based on that measurement, which he obtained by lying down next to many dung piles – stopwatch in hand – over the course of several months, Parker was able to use the marginal value theorem to predict an optimal copulation duration of 41.4 min. His observed value was 35.5 min – a reasonably close fit (Figure 6.13).

Let's review our progress thus far. We've discussed examples of natural selection favoring traits that enhance survival. For example, dungflies are well adapted to locate the dung pile,

which is an essential component of its life cycle, and both the Balearic lizard and the dunnock forage adaptively. We've also discussed cases of sexual selection for traits that enhance mating success, including female fireflies choosing mates on the basis of their flash characteristics. We've developed an economic approach to understand how natural selection and sexual selection influence behavior. We're now ready to explore how natural selection and sexual selection influence the evolution of animal mating systems.

6.3 WHAT PHYSIOLOGICAL AND ECOLOGICAL FACTORS INFLUENCE THE EVOLUTION OF MATING SYSTEMS?

In birds, one egg may weigh 30% of the weight of the mother, even up to 50% in the kiwi, while a single male will store billions of sperm cells in his testes. Though the extent of the difference in gamete size varies among different animal taxa, females always invest more resources per gamete produced than do males. As a result, a female's reproductive success may be limited by the number of eggs she can produce, while a male's reproductive success is seldom limited by the number of sperm he can produce. While females get very little benefit from multiple matings, males, by virtue of their numerous (albeit tiny) gametes, have the potential to increase their fitness considerably with each new sexual partner (Figure 6.14).

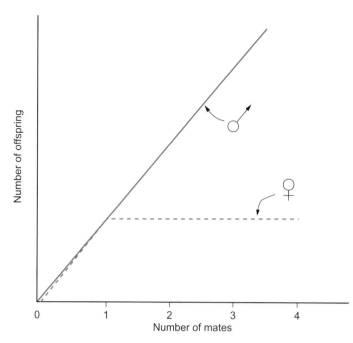

Figure 6.14 Extreme representation of potential male and female reproductive success in relation to number of matings. In this example, male reproductive success increases linearly, while female reproductive success levels off after one mating.

Table 6.2 Mating systems defined based on number of copulatory partners.

Mating system	Number of copulatory partners for males	Number of copulatory partners for females
Monogamy	1	1
Polygynandry	More than 1	More than 1
Polygyny	More than 1	1
Polyandry	1	More than 1

The two curves in Figure 6.14 help us understand the relative strength of sexual selection operating on males and females. Superficially, we might expect male fitness to increase linearly with the number of matings, with the slope of the male curve being 1.0 as I've drawn in Figure 6.14. Thus a male with two matings should have twice the fitness as a male with one mating. If this expectation were true, we would expect males to mate indiscriminately, while females would probably mate only once (particularly if multiple copulations were energetically costly to females). In this case, the strength of sexual selection would be much greater on males than females, favoring sexual dimorphism (difference in form between males and females) and intense competition between males for access to females. Males would be predicted to invest more energy in attracting mates. At the same time, females (such as the *Photinus* fireflies we introduced earlier) should be choosey over who gets to fertilize their limited collection of eggs. While this has proven to be a useful general approach to understanding mating behavior, it does not apply to all mating systems.

Both ecological and physiological factors will influence the evolution of a species' mating system. Let's define mating system as the general pattern of copulatory partners for each sex over the course of a breeding season. I intentionally use the vague term "general pattern" in this definition, because many species have considerable variation within their mating system. The four basic types of mating systems are defined in Table 6.2. We will briefly discuss each one.

Monogamy: one male mates with one female over the course of the breeding season

Monogamy is somewhat unusual in the animal world for two reasons. One we've already discussed: the basic difference in gamete investment favors multiple matings in males. The second reason is that females often benefit from multiple matings as well. We will discuss some of these benefits later.

Monogamy is most prevalent in species that require extensive parental care in order to successfully raise offspring. About 90% of birds are primarily monogamous. Biparental care is needed in many bird species because one parent may need to sit on the eggs while the other parent goes out foraging. The documentary "March of the Penguins" highlights the ridiculously cute and touching antics of emperor penguins, a species whose males sit on their one egg throughout weeks of subzero temperatures, while the females migrate more than 100 km to the ocean to forage.

But even in monogamous species, there are often cases in which individuals of both sex are more promiscuous than had been assumed – a process known as *extra-pair copulations*. Simon Griffith and his colleagues (2002) did a quantitative review of studies that used DNA analysis on offspring of more than 150 bird species that had been classified as monogamous. They discovered that true genetic monogamy was found in less than 25% of the species that were classified as monogamous. Approximately 30 species showed levels of extra-pair paternity of greater than 20% (Figure 6.15).

This surprising result raises the question, why should females in monogamous pairs accept matings from other males? There are several non-mutually exclusive hypotheses, which may apply to different species. First, females may seek out extra-pair copulations to protect themselves against the possibility that their mates are infertile. Second, females may seek extra-pair copulations to increase the genetic diversity in their offspring, which may increase the possibility that some offspring will survive. Third, females may be able to recognize high genetic quality in males,

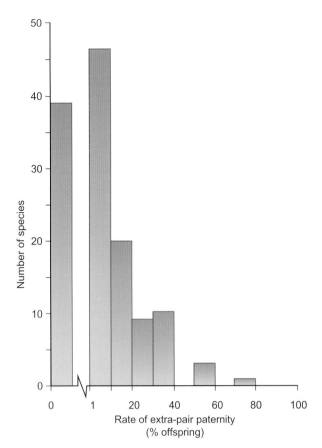

Figure 6.15 Incidence of extra-pair paternity in over 150 species of birds.

Polygynandry: males and females have multiple mates over the course of the breeding season

In the Australian toadlet, *Pseudophryne bibronii*, nest failure is very common, because survival of offspring depends on females choosing to mate with a male who has built a nest on a site with the proper amount of seasonal flooding. If these sites flood too early or too late, the tadpoles will usually die from desiccation. Phillip Byrne and J. Scott Keogh (2009) observed females mating with many different males, and hypothesized that multiple mating by females was a mechanism to reduce the chances that they would have no surviving offspring over the course of the mating season. If their hypothesis was correct, they predicted that females who mated with few males would have very high variance in the survival of their offspring, while females who mated with numerous males would have much lower variance in offspring survival. Furthermore, the researchers predicted that up to a point, females who mated with more males would have overall higher reproductive success.

Byrne and Keogh collected tissue from the toes of adults in their study population, and also collected a portion of the eggs that were laid in each male's nest. By doing a microsatellite analysis of the DNA of males, females, and fertilized eggs, the researchers identified the reproductive output of each parent. By returning to pools after flooding, they scored the survivorship of tadpoles that had successfully hatched.

Several important findings stand out from this study. First, all females mated at least twice and some mated at least eight times. As Byrne and Keogh predicted, variance in tadpole survivorship was much greater when females mated only twice than it was when females mated four or more times (Figure 6.16A). In addition, mean tadpole survivorship increased with number of nests females used for egg laying (Figure 6.16B).

In some polygynandrous systems, matings and egg laying occur within a much narrower timeframe. Odonates – dragonflies and damselflies – are good examples of polygynandrous species in which multiple matings may occur one after the other. Jon Waage (1979) and many researchers after him have studied reproductive behavior in this group of large insects. In the black-winged damselfly, *Calopteryx maculata*, males defend mating territories from which they exclude other males. Females enter the territory, mate with the territorial male, and oviposit approximately 100 eggs in the adjacent stream. Often, while a female oviposits, nearby males swarm in and attempt to mate with her. Despite the territorial male's most valiant efforts to defend his female, an intruding male may succeed and mate with the female, and the female will then resume ovipositing. This process can repeat itself several times.

Thus both females and males may mate multiple times in this polygynandrous system. Waage (1979) was able to show that male damselflies actually remove the sperm that the previous male ejaculated into the female's sperm-storage organ. Because females store sperm and mate many times, most females captured at a stream – by either a male dragonfly or a behavioral ecologist – will have a full sperm-storage organ. When Waage captured 16 females

and may seek extra-pair copulations from males with unusually good genes. Lastly, extra-pair matings by females may create a strong incentive for males in the population to cooperate with their neighbors, since there is some probability that neighboring offspring are actually theirs. Natural selection would then favor males who helped provision nests and repel predators from the neighborhood, and that showed reduced territorial aggression and infanticide (Eliassen and Jørgensen 2014). Many researchers are currently exploring the predictions of these (and other) hypotheses.

 Thinking ecologically 6.3

Imagine that as part of your research program, you decide to compare the survival of offspring of extra-pair copulations to their half-sibs, who were sired by the primary male of the "monogamous" pair. You discover that offspring from extra-pair copulations have a 5% higher survival rate. Which of the four hypothesized benefits of extra-pair copulations does this finding support?

In the remaining mating systems, multiple mating is more overt.

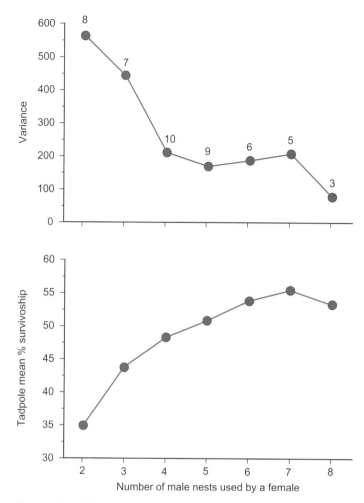

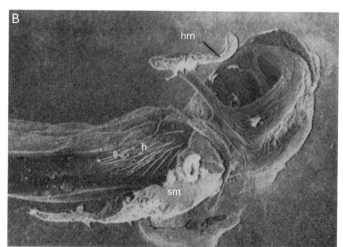

Figure 6.16 A. Variance in tadpole survivorship in relation to number of male nests used by the female. B. Mean survivorship (%) of tadpoles in relation to number of male nests used by the female.

in the precopulatory position, all of them had full sperm-storage organs, indicating they had previously mated. During the early and middle stages of copulation, males make rhythmic undulatory movements of the abdomen (Figure 6.17A). Waage captured pairs more than 15 s after undulation began, and discovered that the sperm-storage organs were empty in 18 of 24 females, and almost empty in the remaining 6 females. He concluded that males had physically removed previously introduced sperm (Table 6.3).

To complete the picture, Waage dissected the penises of males captured during the undulatory phase. Even though ejaculation had not occurred, there were numerous sperm masses on the lateral horn, and on the ventral hairs projecting from the penis (Figure 6.17B). Waage surmised that these were the sperm from a previous mating that the male had scooped out with his undulating movements.

We are accustomed to thinking of intrasexual selection favoring the evolution of weapons that allow males to fight with each other for access to females (see Chapter 3). Waage's research demonstrates that intrasexual selection can favor behavioral and morphological adaptations that allow males to compete with each other even after they have successfully secured a mate. In this example of *sperm*

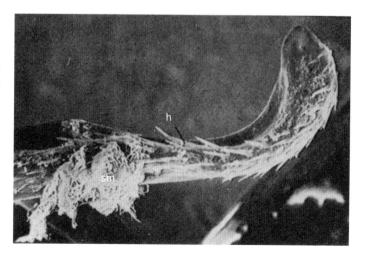

Figure 6.17 A. Copulating *Calopteryx maculata* damselflies. B. Scanning electron micrograph of dissected damselfly penis that had been inserted into a female's reproductive organ when captured. Top: lower magnification showing the lateral horn (hrn), a sperm mass (sm), and some hairs (h) that aid in sperm removal. Bottom: increased magnification of lateral horn with attached sperm mass.

Table 6.3 Mean sperm volume in female damselfly's sperm-storage organ before, during, and after copulation.

Context	N	Sperm volume (95% confidence interval)[a]
Precopulatory position	16	3.49 (0.75)
>15 s after undulation	24	0.41 (0.29)
After copulation	14	4.20 (0.59)

[a]There were no units for sperm volume in this paper.

competition, the sperm from two (or more) males are competing with each other for fertilizations. We'll now consider a mating system in which males, but not females, have multiple mates.

Polygyny: one male mates with more than one female over the course of the breeding season

In polygynous systems, some males will mate with several (or many) females, while other males may achieve no matings. Polygyny is the system associated with the greatest variation in male mating success, and thus males may experience very strong sexual selection. As we discussed in Chapter 3, this variation in mating success may lead to extreme sexual dimorphism. Polygynous males may be larger than their species' females, have weapons that allow them to dominate other males, or be more brightly or strikingly colored than the females.

In mammals, there is a long pregnancy period during which the female nourishes the developing embryos internally. In addition, females produce milk, which is the sole food for their young. Both physiological factors make it more likely that male mammals will be polygynous. Males are also less likely to provide parental care, though male mammals can help provision the young indirectly, by helping to provision the females during lactation.

Ecological factors, such as the distribution of resources, can influence the type of polygynous system animals employ. In *resource defense polygyny*, males defend resources that are essential to females during the breeding season, allowing females onto the breeding territories in exchange for matings (Figure 6.18A). In *female defense polygyny*, females form groups or clusters, and males attempt to defend the entire group (Figure 6.18B). Finally, *lek polygyny* is the closest natural phenomenon to a singles bar. Within a small area, each male defends a very small space, and displays his sexually selected traits to females who pass through the area to evaluate and choose a mate (Figure 6.18C). In lekking bird species, these displays will usually be a combination of striking plumage or complex vocalizations. In both female defense polygyny and lek polygyny, one or a few males have very high fitness, while the remaining males will have very low or no reproductive success.

Figure 6.18 A. Resource defense polygyny in the blue-headed wrasse, *Thalassoma bifasciatum*, where males defend spawning territories favored by females. B. Female defense polygyny is common in many mammals such as the moose (*Alces alces*) – a male with his harem in Alaska, USA. C. Lek polygyny is the mating system of marine iguanas (*Amblyrhynchus cristatus*) in the Galápagos Islands.

Researchers now recognize that in a surprisingly large number of cases females also benefit from multiple matings, and in some cases the benefit to females of multiple matings is as great as, or greater than, the benefit to males. These conditions may promote the evolution of polyandry.

Polyandry: one female mates with more than one male over the course of the breeding season

In some species, the relationship depicted in Figure 6.14 is reversed, and females actually benefit more than males from multiple matings. In such systems, sexual selection is actually stronger on females than on males, and females compete with each other for access to males. The result of this competition can be a polyandrous mating system, in which one female may mate with several males over the course of the mating season, while males usually mate only once.

One example of polyandry is the spotted sandpiper, *Actitis macularia*, which has been studied by Lew Oring and his colleagues for over 25 years at Leach Lake, Minnesota, USA (Oring *et al.* 1992). Female spotted sandpipers average about 50 g, while males weigh in at a relatively paltry 41 g. Females arrive earlier in the breeding season than males, females fight with each other over their territories, and males assume the primary egg incubation role. Females can lay up to five clutches of four eggs per breeding season because the lake produces breakfast and dinner in the form of copious numbers of mayflies. Thus a female's reproductive success is limited by the number of male incubators she can convince to sit on her eggs.

It is difficult to find good examples of polyandry, in part because it is relatively uncommon in nature, and in part because behavioral and microsatellite studies usually indicate that males are able to secure additional fertilizations. Over 6 years of their study of the spotted sandpiper, Mark Colwell and Lew Oring (1989) observed 174 extra-pair events, which they defined as copulations or attempted copulations in which either or both participants mated with a bird that was not their primary mate at the time. The researchers interpret these events as a means for females to establish relationships with potential mates for future clutches, and for males to obtain partial fertilization success for eggs incubated by other males.

To pursue the question of whether males actually achieve fertilization success for eggs incubated by other males, Oring and his colleagues collected blood from 77 eggs in 1991. DNA fingerprints showed that 11 of the 77 eggs could not have been fathered by the current male. But in 10 of those cases, they could have been (and probably were) fathered by the female's previous mate. As it turns out, females can store sperm for several weeks, and can use that sperm to fertilize eggs that are incubated by her new male (Oring *et al.* 1992).

Our brief discussion of mating systems highlights the influence that physiological and ecological factors can have on both male and female reproductive behavior. Let's revisit the female dunnocks, whose territory size became smaller in response to food supplementation, to see whether the reduction in territory size influenced the dunnocks' mating system.

Consequences of territory size for the type of mating system

Dunnocks exhibit amazing variation in their mating system. Davies and Lundberg (1984) discovered every type of mating system listed in Table 6.2 naturally occurring with these birds over the course of two mating seasons. One important controlling ecological factor was the distribution of females. If a female's feeding territory overlapped a male's song territory, the two would set up housekeeping together in a (mostly) monogamous relationship. But if two females' feeding territories both fell within the song territory of one male, that male would mate with both females – a polygynous mating system. This relationship was relatively uncommon, perhaps because there were more mature males than females in the population, due to higher female over-winter mortality. Alternatively, if a female spent time foraging within the song territories of two males, polyandry would likely occur. Lastly, if more than one female foraged within the overlapping song territories of multiple males, both sexes would have multiple mates – a polygynandrous mating system. In associations with multiple males, one male (the alpha [α] male) was dominant over the beta [β] male, had higher mating success, and would displace the β male if he found him attempting to mate with a female. All of the possible combinations are shown in Figure 6.19.

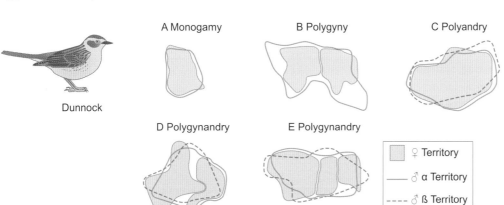

Figure 6.19 Relationship between territory size and distribution and the mating system for dunnocks. Female territories are solid brown while male territories are outlined in red.

Table 6.4 Types of mating combinations on unmanipulated and food-supplemented territories. m = male, f = female.

Mating combinations	1982		1983	
	Control	Supplemented	Control	Supplemented
Unpaired	1	0	3	0
Polyandry (3m/1f)	1	0	1	0
Polyandry (2m/1f)	9	3	5	3
Monogamy	4	7	6	4
Polygynandry (2m/2f)	1	0	1	5
Polygynandry (2m/3f)	0	1	0	1
Polygynandry (2m/4f)	0	0	0	1
Polygynandry (3m/2f)	1	0	0	0
Total #	17	11	16	14

When Davies and Lundberg supplemented the dunnocks with food, female territory size decreased but male territory size was unaffected. It may be that males defend as large a territory as possible, regardless of how much food is available, because the primary benefit of territory defense is access to females whose feeding ranges overlap their song territories. Davies and Lundberg hypothesized that supplemented females, with their smaller territories, would have reduced polyandry, and an increase in mating systems with a higher ratio of females per male. Table 6.4 shows the result of their experiment.

If you look at this table carefully, you will see a substantial reduction in polyandry from almost 50% in the controls to less than 25% for the birds that benefited from food supplementation. Davies and Lundberg argue that this reduction is a straightforward result from the reduction in female territory size that arises from the food supplementation. The dunnock system has two important lessons for us. First, mating systems are highly variable in some species. Second, ecological factors, such as the distribution of resources, can influence the type of mating system present in a population.

While both sexual selection and differences in resource distribution help us to understand the evolution of mating behavior, we need to consider that there is a second, more indirect path for individuals to pass their genes on to future generations.

6.4 HOW CAN INDIRECT SELECTION LEAD TO COOPERATIVE BEHAVIOR AMONG RELATIVES?

In *On the Origin of Species*, Charles Darwin (1859) writes

I [. . .] will confine myself to one special difficulty, which at first appeared to me insuperable, and actually fatal to my whole theory. I allude to the neuters or sterile females in insect communities: For these neuters often differ widely in instinct and in structure from both the male and fertile females, and yet from being sterile they cannot propagate their kind.

Darwin's theory proposed that natural selection operates on the reproductive success of individuals; thus it was very puzzling to observe an abundance of nonbreeders in many different species of bees, ants, and wasps. A related but more general problem is that biologists observe cooperative behavior in many species. Given that cooperation or helping can be costly to an individual, how can cooperation evolve within a species? Several different approaches have been proposed to address these problems.

Natural selection and inclusive fitness

Evolutionary biologist Bill Hamilton argued that fitness has two components. First is direct fitness, which measures the genetic contribution an individual makes to the next generation by producing surviving offspring. Second is indirect fitness, which measures the genetic contribution an individual makes to the next generation by enhancing the production or survival of genetic relatives. An individual's total fitness, or inclusive fitness, is the sum of its direct fitness and indirect fitness. Within this framework, *direct selection* is natural selection favoring traits that enhance an individual's direct fitness and *indirect selection* is natural selection favoring traits that enhance the production and survival of relatives who are not offspring (Hamilton 1964).

Hamilton's model of indirect selection

Hamilton formalized the concept of indirect selection so that researchers could understand why animals cooperated with each other even though cooperation might be costly or dangerous. He defined the following three variables:

B = benefit (in terms of reproductive success) others gain from an individual's action

C = cost to the individual for a particular action

r = **coefficient of relatedness** = the probability that two individuals share genes that are identical by descent.

To help understand what r represents, let's consider the coefficient of relatedness between you and your mom. Because you are diploid, you have two alleles for every gene (ignoring sex chromosomes). One of these alleles came from your mom, and the other from your dad. Thus half of your genes are identical by descent to those of your mom, and $r = 0.5$ ($r = 0.5$ between you and your dad as well). Table 6.5 shows values of r between relatives.

Hamilton argued that when individuals share genes by descent ($r > 0$), indirect selection favors behavior that is otherwise costly to an individual. Specifically, selection will favor helping a relative, even if the help is directly costly to the individual, if:

$$B \times r > C$$

Thus high benefits, high coefficient of relatedness, or relatively low costs will all favor the evolution of cooperative behavior. Let's see if Hamilton's model helps us understand the evolution of cooperative behavior in real animals.

Table 6.5 Values of r between relatives.

r	Direct descendants	Other relatives
0.5	Children	Siblings
0.25	Grandchildren	Half-sibs, nieces, nephews
0.125	Great grandchildren	Full cousins

Applying Hamilton's model: alarm calls

Consider the following scenario. A group of ground squirrels are out in the open, foraging on nuts and seeds. A hungry coyote craftily creeps into view and is spotted by one of the squirrels. She screams several times and squirrel bedlam ensues as all scramble for cover. Alas for the screamer, she becomes breakfast for the craven coyote.

From what we've learned thus far, coyote predation should eliminate noisy screamers from the squirrel population. But some Belding's ground squirrels, *Spermophilus beldingi*, do scream when predators approach, and Paul Sherman's research showed that screamers – or callers – suffer double the predation rates of noncallers (Sherman 1980). Sherman wanted to know why calling is present in a population – shouldn't this presumably maladaptive trait be eliminated by natural selection?

Sherman proposed that natural selection favoring parental care was partially responsible for the presence of alarm calling in ground squirrels. Male offspring usually emigrate, but female offspring usually live near their parents. Thus natural selection, specifically direct selection, would favor the evolution of alarm calls between mothers and their daughters. Sherman hypothesized, in addition, that indirect selection had also favored the evolution of alarm calling between females and other female relatives who lived nearby, despite the costly nature of the calls. He argued that the benefits to the caller from this combination of direct and indirect selection were greater than the costs incurred from higher predation rates. Even though they suffer a higher predation rate, callers would have higher inclusive fitness overall than noncallers.

How could Sherman test his hypotheses? Ground squirrel research had been ongoing for 5 years before he arrived, providing Sherman with a large database of ground squirrel genealogy. Because males emigrate, whereas females do not, resident males are unrelated to resident females. Sherman's hypothesis predicted that males will call much less frequently than females and that females without living relatives will call much less frequently than females with living relatives. Sherman and 15 students counted ground squirrel calling rates in the presence of mammalian intruders over four field seasons at Yosemite National Park and discovered that overall about 28% of the squirrels called when a predator approached. Table 6.6 compares the responses to intruding mammals of males, females with relatives, and females without relatives. Females with relatives were much more likely to call when mammalian predators approached than were females or males in the other groups studied.

We still need to know if the higher calling rate in females with relatives is simply a result of direct selection favoring parental care, or if it has also resulted from indirect selection favoring the warning of relatives who are not offspring. Fortunately, Sherman and his field assistants were able to observe 119 interactions between mammalian predators and the squirrels. These observations showed that ground squirrels favor direct descendants (direct selection hypothesis) and close relatives including parents

Table 6.6 Number of alarm calls sounded by three different classes of squirrels. $\chi^2 = 16.6$, df = 2, $P < 0.001$.

Type of squirrel	Number of times called	Number of times did not call	Rate of calling response (%)
Mature male	12	55	17.9
Mature female with relatives	75	115	39.5
Mature female without relatives	31	137	18.4

and sisters (indirect selection hypothesis). However, there is no evidence that they are more likely to sound alarm calls if their only relatives are cousins or nieces.

We can extend Hamilton's simple formula to understand other related questions in behavioral ecology, such as why organisms live in groups. We will consider one such study.

Indirect selection, ecological constraints and the formation of communal groups

The Florida scrub jay, *Aphelocoma coerulescens*, has a very small geographic range, which is limited to central Florida. The Florida scrub is a threatened ecosystem characterized by nutrient-poor soils and occasional droughts. It is populated by a variety of short trees and shrubs adapted to these conditions. The jays establish territories within the scrub and are opportunistic omnivores. Of relevance to our discussion, the jays are cooperative breeders, with some of the male young remaining within their parents' territory for several years. These helpers guard against predators and help with feeding the young. Let's investigate why many young male Florida scrub jays stay and guard the young, rather than going out to start families on their own.

Cooperative breeding in the Florida scrub jay

Glen Woolfenden and John Fitzpatrick (1984) worked on the scrub jay system for many years, and investigated the question of why helper males stay at home and help guard the young. They considered several possibilities. Direct benefits to the helpers include acquired parenting skills, reduced predation, improved foraging opportunities, and a possibility of inheriting the territory in the case of parental death. Because good breeding habitat is so uncommon, there is extreme competition between these jays for territories. Woolfenden and Fitzpatrick's research showed that groups with more helpers were able to defend larger territories, from which male helpers could bud off their own territories and begin their careers as primary breeders.

Approximately 33% of male helpers gained territories from budding, while an additional 13% inherited their natal territory following the death of one or both of the breeding pair. Thus the direct benefit of territory acquisition helps explain why males defer breeding.

Two important indirect benefits of deferred breeding in males are possible. First, helpers may benefit indirectly by increasing the survival of their parents – the breeding pair – who will then be able to produce more offspring (with high r). Second, helpers may benefit indirectly by increasing the survival of close relatives (sisters and brothers), either by defending them against predators or feeding them. Of course, if one of the parents dies or takes a different mate, that change has important implications for our calculation of ($B \times r$).

During the course of their studies, Woolfenden and Fitzpatrick were able to learn the identity and relationship of almost every individual in the Florida scrub jay population. To address whether helpers are beneficial to breeder survival, the researchers compared the survival of breeders with and without helpers. By following breeders and helpers over the birds' entire lifetimes, they observed that the survival rate of breeders with helpers was 85% per year, while the survival rate of breeders without helpers was 77% per year. Thus breeders with helpers received a substantial survival boost.

Next, they asked whether helpers receive indirect benefits by enhancing the breeders' reproductive success. Table 6.7 shows that breeders with helpers did not produce any more eggs than those without helpers, but they did have higher offspring survival rates. Woolfenden and Fitzpatrick conclude that helpers receive indirect benefits from helping.

But Woolfenden and Fitzpatrick recognized that observations such as these raise the red flag of correlation versus causation.

Table 6.7 Reproductive success of breeders and helpers in Florida scrub jays. Numbers in parentheses represent the standard deviation.

	Pairs with helpers	Pairs without helpers
Number of pairs	153	116
Nesting attempts per year	1.49	1.47
Mean number of eggs	3.39 (0.67)	3.31 (0.71)
Mean number of fledglings	2.39 (1.50)	1.58 (1.25)
Mean number of yearlings	0.77 (0.97)	0.57 (0.84)

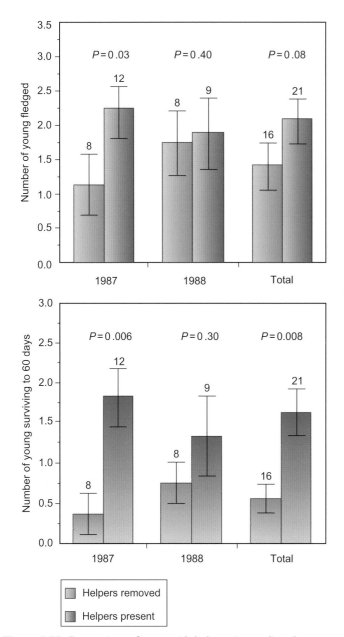

Figure 6.20 Comparison of nests with helpers (control) and nests that had their helpers removed (experimental) in 1987 and 1988. A. Mean (±SE) number of young that fledged. B. Mean (±SE) number of young that survived to 60 days old. Numbers above bars are number of nests in the study.

Are helpers actually increasing fledging success, or do helpers choose to help only in high-quality territories? If the latter is true, the increased fledging success may result from the quality of the habitat rather than the presence of helpers. In this case, helping would actually be a type of parasitic interaction.

Ronald Mumme (1992) tested this "helpers as parasite" hypothesis by removing helpers from randomly chosen nests in 1987 and 1988. He measured the mean number of offspring fledged and mean number of offspring alive after 60 days.

In both years there was a substantial decrease in offspring survival in the nests that lost their helpers (Figure 6.20). This result supports the hypothesis that helpers really help.

Thinking ecologically 6.4

As I've described Mumme's experiment, it presents some problems in design and interpretation. (a) Outline the issues that would concern you when designing and interpreting the results of this experiment. (b) How would you address these issues?

Eusociality in some Hymenoptera and in the naked mole rat

Superficially, a buzzing hive of honeybees (*Apis mellifera*) and a subterranean colony of naked mole rats (*Heterocephalus glaber*) would seem to have very little in common. Honeybees live in groups of thousands of individuals, have acute color vision, and use visual landmarks to help them navigate to and from foraging sites that are rich in nectar and pollen. Naked mole rats live in much smaller groups comprising several dozen individuals, and forage primarily on roots and tubers (Jarvis 1981). While honeybees have strict daily activity rhythms, naked mole rats are very inconsistent in their activity patterns, mixing periods of high activity with longer periods of inactivity and rest. Though their mechanisms of foraging and their activity patterns and habitats are dissimilar, the two species also share some striking resemblances in their behavior.

Both honeybees and naked mole rats are eusocial or truly social. The hallmarks of eusociality are cooperative care of young, the presence of sterile castes of non-reproductive workers, and the presence of multiple generations of individuals within the colony. Honeybees belong to the insect order Hymenoptera, which comprises over 115 000 species of bees, wasps, and ants. Eusociality has evolved independently at least 11 times in the Hymenoptera. In contrast, there are only seven different species of African mole rats, of which two – including the naked mole rat – are clearly eusocial.

It might seem surprising that natural selection would favor the evolution of non-reproductive individuals, as they have a direct fitness value of 0. Hamilton's concept of indirect selection can help us understand how helping non-descendant relatives would enhance an individual's inclusive fitness. For eusociality to be favored by indirect selection, individuals must benefit by passing on genes favoring the behavior and developmental patterns that are the hallmarks of eusociality.

Once again Bill Hamilton provided important insight that stimulated research addressing this problem. He recognized that many Hymenoptera have a haplodiploid genetic system, which can produce haploid offspring that develop into males, or diploid offspring that develop into females. The queen controls the sex of

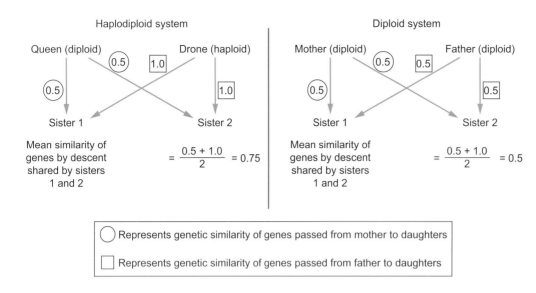

Figure 6.21 Calculating coefficient of relatedness (*r*) between sisters in a haplodiploid system with only one drone and in a diploid system with only one father.

her offspring by choosing to lay unfertilized eggs, which develop into haploid male drones, or to fertilize eggs, which develop into diploid female workers. This system has important consequences for the coefficient of relatedness (*r*). If the queen has mated with only one male, then *r* between her daughters is 0.75. This high value of *r* arises because the paternally derived chromosomes are all identical (since the father is haploid) and the maternally derived chromosomes are 50% identical (as in all diploid inheritance). Thus the average *r* between sisters in a honeybee hive will be 0.75. This compares to an *r*-value of 0.5 between siblings in an outbred population, as depicted in the diploid system (Figure 6.21). Hamilton (1964) argues that this high value of *r* is one factor favoring cooperation within the Hymenoptera.

Is there a similar genetic predisposition for eusociality to evolve within mole rats? Hudson Reeve and his colleagues (1990) estimated *r* = 0.81 between naked mole rat colony members in Kenya, based on a DNA fingerprinting analysis, even surpassing the theoretical maximum suggested for Hymenoptera! Why would a diploid species have such a high value of *r*? Reeve suggested that high inbreeding could lead to extreme genetic similarity between individuals in a population. Inbreeding could arise from offspring remaining within the natal colony and breeding with their parents, rather than dispersing (in mammals, males usually disperse). Reeve and other researchers proposed that ecological factors constrain dispersal of naked mole rats. These factors include high predation pressure, very patchy food resources, and the physical barrier caused by very hard soils that the naked mole rat must excavate. If most mole rats do not disperse, the breeders must originate within the natal colony, leading to a highly inbred population and a high value of *r*. Once established, the high value of *r* will further promote cooperative behavior among individuals in the colony, and reinforce members of the colony staying within the natal burrows.

Stanton Braude (2000) questions Reeve's analysis of the DNA fingerprint study for several reasons. First, he cites evidence that the mole rats sampled by Reeve and his colleagues derive from small, recently established colonies near the edge of the naked mole rat's geographic range. Individuals from small colonies have few alternatives to inbreeding, and as we discussed above, inbreeding will inflate the values of r among individuals of the colony. This inflated *r*-value will persist over many generations.

Braude predicted that a larger sample of colonies from a more extensive range would generate a significantly lower degree of relatedness. His student, Jon Hess (2004), analysed DNA from over 1100 mole rats representing 123 colonies. Hess discovered significantly lower values of *r* in this much larger sample, and he also found that individuals within a colony had surprisingly high levels of heterozygosity for the genes used in his analysis. High levels of hetrozygosity are associated with outbreeding, rather than inbreeding, which suggests that in most cases, the breeders are not close relatives.

Braude and his students have spent about 20 years in the field capturing, marking, and releasing over 9000 naked mole rats in Kenya and Ethiopia (Figure 6.22). He also measured and weighed every individual before release. In one study, Braude (2000) discovered 20 new colonies founded by 1–4 individuals from other colonies, including 16 that were founded by marked individuals. These dispersers, measured before they dispersed (21 females and 16 males), were much fatter than the average naked mole rat of the same length, suggesting that naked mole rats may have a special disperser morph or caste (as described by O'Riain and colleagues 1996).

Finally, Deborah Ciszek (2000) asked whether naked mole rats, if given a choice, are more likely to mate with close relatives with whom they have grown up (in accordance with the inbreeding hypothesis) or if they are more likely to mate with unfamiliar individuals who are either unrelated or distantly related. She began with two colonies captured from the field, which she allowed to breed for two or three generations. She then established six new test colonies consisting of eight individuals – each

Figure 6.22 A naked mole rat, *Heterocephalus glaber*, walking along a drift fence moments before tumbling into a pitfall trap.

had two females and males from field colony # 1, and two females and males from field colony # 2. She then systematically observed each colony, to identify the breeders.

Five of six colonies had one female breeder, and one colony had two female breeders. Three colonies had only one male breeder, while three colonies had two male breeders. Of relevance to the inbreeding hypothesis, eight of nine male breeders were unrelated to the queen (binomial test, $P = 0.019$). Ciszek concluded that in naked mole rat colonies, queens tend to mate with unrelated males, if given a choice.

If we put the findings of Braude and Ciszek together, a picture emerges that is somewhat different from the earlier description of inbred naked mole rat populations. Colonies produce disperser males and females, which either found their own colony or enter other colonies. Females are more likely to mate with unrelated males than with related males. Thus under most ecological conditions, naked mole rats will be moderately outbred.

Though our picture of eusociality in naked mole rats has changed somewhat, the basic details remain clear. Even if r between sibs is 0.5 (or 0.25 for half-sibs), as in typical outbred populations, we would still expect the evolution of eusociality if there are high indirect benefits of remaining in the natal colony

to help close relatives breed. In this case it appears that ecological factors influencing dispersal costs support most individuals' remaining in the natal colony. Only a few individuals (equipped with extra fat layers) will have the opportunity to begin new colonies, or perhaps to become breeders in other existing colonies.

Hamilton argues that natural selection favors behavior in individuals that enhances the reproductive success of relatives in proportion to r. But will natural selection favor the formation of cooperative groups among unrelated individuals, even if cooperation can be costly to the individual? Once again we will use a cost–benefit approach to address this question, this time borrowing heavily from the work of game theorists.

6.5 HOW DO GAME THEORY MODELS HELP EXPLAIN THE EVOLUTION OF COOPERATION AMONG UNRELATED INDIVIDUALS?

Game theory is a branch of mathematics and economics in which players choose solutions or strategies to maximize their success in particular situations. The Prisoner's Dilemma game was originated in 1950 by two researchers working at RAND Corporation on the problem of nuclear proliferation following World War II. At that time the United States and the Soviet Union were the two major superpowers of the world; both had significant nuclear capabilities, and both realized that an all-out nuclear war would destroy the world, but neither country trusted the other.

The Prisoner's Dilemma and the evolution of cooperation

In the Prisoner's Dilemma game, you and an accomplice have committed a crime. The police know you've both done it, but they have insufficient evidence to put you away. You and your accomplice are interrogated separately – you are told that you will be freed if you testify against your accomplice. You are a clever criminal, and you know that the police have made the same offer to your clever criminal colleague. What should you do? Should you keep your mouth shut, or should you betray your accomplice? If you both keep quiet, you'll both get off doing community service (because the police have only circumstantial evidence). If you keep quiet and your accomplice betrays you, you'll be doing big jail time, while he goes free. But if you betray him and he keeps quiet, you'll go free, while he does big jail time. Finally, if you betray him and he betrays you, you'll both do some jail time, but you'll get some time off because you did cooperate with the police by testifying against your accomplice. This Prisoner's Dilemma scenario is summarized in Table 6.8.

Table 6.8 Payoffs to prisoners in the Prisoner's Dilemma game.

	Accomplice keeps quiet	Accomplice betrays you
You keep quiet	Both do community service	You do major jail time Your accomplice goes free
You betray your accomplice	You go free Your accomplice does major jail time	You both do some jail time

Unfortunately (from the standpoint of world peace), the rational strategy is to defect (betray your accomplice). Superficially, this might seem strange, since keeping quiet has an excellent result for both players. However, regardless of your accomplice's behavior, you always have a better outcome for yourself if you defect.

Robert Axelrod realized that the solution would be more interesting, and more relevant to the human condition, if the contestants interacted repeatedly. In the Iterated Prisoner's Dilemma Game (IPDG) contestants meet again and again an indeterminate number of times, and remember what their accomplices did in their previous meeting. On two separate occasions, Axelrod invited economists and computer experts to submit solutions to the IPDG, and then ran a computer tournament to see which strategies did best in the long run.

Axelrod was quick to recognize that the IPDG had important implications for understanding the evolution of cooperative behavior and the formation of social groups, so he got together with his colleague at the University of Michigan, the ever-present Bill Hamilton, to apply this simulation to the broad question of the evolution of cooperation between individuals within or even between species (Axelrod and Hamilton 1981). Together they argue that there are many occasions in the real world in which individuals can cooperate (analogous to the keep-quiet tactic in the Prisoner's Dilemma) or defect (analogous to betrayal in the Prisoner's Dilemma). From a biological perspective, cooperation, at least in the short term, can have fitness costs. So the question becomes, can the IPDG illustrate conditions under which cooperation may develop, even if there are short-term fitness costs to the cooperator?

The answer is yes. One strategy, TIT for TAT, scored consistently high in both tournaments. TIT for TAT involves cooperating for the first interaction, and then doing whatever the other player does in future interactions. If the other player cooperates as well for the first interaction, then you cooperate the second time you meet. If the other player defects for the first interaction, then you defect the next time you meet. TIT for TAT works well, because if you interact with a cooperative

individual, you score very well in terms of fitness over a long period of time, and if you interact with a defector, then you get a truly bad payoff (in terms of the influence of his defection on your fitness) only once.

Thus Axelrod and Hamilton broadened the arena for the evolution of cooperation among unrelated individuals. Two conditions must be met. First, there must be a high probability of future interactions with the same individual. Second, there must be a way of directing the appropriate behavior to the correct individual. In organisms with developed cognitive skills, this might entail having a good memory. In organisms with poor cognitive skills, this might entail living in association with only one individual or colony.

The IPDG and the TIT for TAT solution inspired researchers in the field to seek out organisms that demonstrate behavior akin to TIT for TAT. At the same time, the IPDG also stimulated theoretical ecologists to search for more complex games that would more accurately model the complexity of human interactions. Let's first discuss a study on cotton-top tamarins that show TIT for TAT-like behavior. We will then conclude with an introduction to a more complex and richer game that has generated testable predictions about human behavior.

Cooperation between unrelated cotton-top tamarins

Cotton-top tamarins (*Saguinus oedipus*), live in small family units, feeding on fruits, insects, and tree secretions in the Colombian rainforest canopy. These small social monkeys are relatively easy to care for in captivity, yet they have advanced cognitive skills, making them good candidates for testing some of the mechanisms that could underlie cooperative behavior. Marc Hauser and his colleagues studied food-sharing behavior in tamarins; they were particularly interested in whether unrelated tamarins would cooperate with each other in a TIT for TAT-like manner (Hauser *et al.* 2003).

Hauser's experimental design was a cage with two slots big enough for a tamarin hand to fit through (Figure 6.23A). One of the slots had the handle of an L-shaped tool, oriented so that pulling on the handle would bring food toward the front of the cage, until it reached a slot and a hungry tamarin. The key to this design is that the tamarin with the tool controlled whether the food moved to the front, but could not control which slot the food came to. That was controlled by where the researchers placed the food initially. If tamarins only pulled when the food was on their side of the cage, that would indicate they only acted selfishly. Alternatively, if tamarins pulled when the food was on the other side of the cage, that behavior would indicate that they procured food for unrelated individuals.

Each test subject was initially trained with no other tamarins at the apparatus. Subjects were trained to pull when the food was in designated positions 1 or 2, but never in positions 3 or 4

A

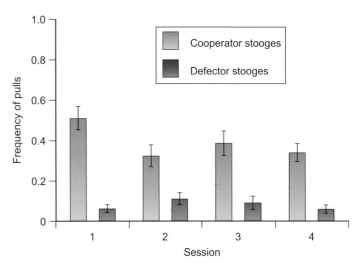

Figure 6.24 Mean (±SE) frequency of pulls by test subjects to benefit stooges that behave as cooperators and defectors.

(B)

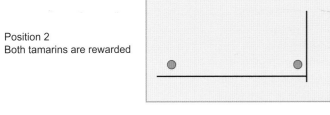
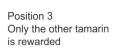
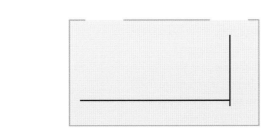

L-shaped tool

Tamarin

Food reward

Position 1
Only the puller is rewarded

Position 2
Both tamarins are rewarded

Position 3
Only the other tamarin
is rewarded

Position 4
No reward

Figure 6.23 A. Experimental apparatus with two tamarins. B. Conditions under which tamarins were trained.

(Figure 6.23B). Within 12 weeks, the tamarins learned with 100% efficiency to only pull when they would get a food reward.

After training the tamarins, Hauser and his colleagues began their experiments. For their first experiment, the researchers also trained two stooge tamarins. One stooge always pulled regardless of where the food reward was placed (and thus could be construed as playing "cooperate" in the Prisoner's Dilemma game). The second stooge never pulled (and thus could be construed as always defecting).

Did the test subjects play TIT for TAT when they were paired with either the cooperator or the defector? The researchers ran four sessions; each comprised 24 trials alternating between cooperator and defector stooges. For example, a session might begin with the cooperator stooge controlling the bar and pulling the food to the test subject; then the test subject would be given the opportunity to reciprocate and pull the food toward the cooperator stooge. The next trial would be with the defector stooge; again the test subject would be given the opportunity to reciprocate – in this case by not pulling the food toward the defector stooge. Each session thus had 12 trials with each stooge. Figure 6.24 shows that tamarins were much more likely to pull for cooperators than they were for defectors.

Thinking ecologically 6.5

As described, these results show that tamarins are more likely to pull for the cooperator stooge than the defector stooge. But does this result provide clear evidence that tamarins are playing TIT for TAT? Can you suggest alternative explanations for the increased response rate to the cooperator stooge?

Hauser and his colleagues wanted to know whether the higher pulling rate by the test subjects for the cooperator stooge in the first experiment resulted from the test subjects being aware that the cooperator stooge was pulling for them even though the stooge received no benefit from pulling. In other words, were the test subjects cognitively aware that their stooges were not benefiting, and did this awareness motivate them to pull with much higher frequency? To address this question, the researchers revised their design from the first experiment. They placed two test subjects side by side, and offered no stooges for each session. When subject 1 pulled, it received one piece of food and gave three pieces to subject 2. When subject 2 pulled, it received no food but gave two pieces of food to subject 1. Thus if both players pulled, they received a total of three pieces of food each. When subject 1 pulled, it received a direct reward, as well as providing food for subject 2. But when subject 2 pulled, it received no reward, but did provide food for subject 1. Subject 1 should, of course, always pull for selfish reasons. The real question is whether subject 2 is aware of the underlying selfish motives for subject 1 pulling food for him. If so, we expect subject 2 to pull less often for subject 1 in this experiment than the test subject pulled for the cooperator stooge in Hauser's first experiment.

The researchers discovered that the test subjects are responding as if they understand that the other test subject was behaving selfishly rather than cooperatively when providing them with food. Though subject 1 pulled 97% of the time, subject 2 only pulled 3% of the time in response. Even though subject 1 was provisioning subject 2 with food (in a manner similar to the cooperative stooge from the first experiment), subject 2 rarely pulled in return. Hauser interprets this as evidence that tamarins use complicated analyses of motivations before deciding whether to reciprocate.

Tamarins clearly have cognitive skills that allow them to make judgments of individuals within a social setting. They are able to distinguish between genuine cooperative behavior, in which an individual incurs a cost so that a second individual can benefit, and selfishly motivated behavior. Presumably, in species with complex social systems, there are fitness benefits to individuals who accurately discern the motivations of other individuals within the system. This brings us back to our opening discussion of the "tragedy of the commons." If tamarins can evaluate members of their social group, and reward individuals who show true cooperation, might Hardin's (1968) pessimism about the selfishness underlying human behavior and the resulting inevitable tragedy be a bit overly bleak?

REVIST: Tragedy of the commons

In December 2003, *Science* magazine celebrated the 35th anniversary of the publication of "The tragedy of the commons." As *Science* editor-in-chief Donald Kennedy pointed out, publication of Hardin's essay had three effects. Initially, the essay raised concern about the negative impact population growth would have on resources. A bit later, politicians and environmental activists began considering the policy implications of human population growth. More recently, a rapidly growing social science literature argued that Hardin's presentation is not an accurate portrayal of the complexity of human behavior. People routinely deal with issues that are much more complex than whether they should add more than one cow to a commons, and can evaluate the actions of other people in a much more sophisticated manner than Hardin implies (Kennedy 2003).

Martin Nowak and Karl Sigmund suggest that social systems, particularly human societies, are organized by altruistic interactions in which individuals help each other, even though helping is costly. We've discussed this in terms of direct altruism in which individuals may take turns helping and receiving help (aka TIT for TAT). Nowak and Sigmund argue that the direct model is a useful starting point, but that within human societies, much of our social existence is influenced by *indirect reciprocity*, which they define as "I help you and somebody else helps me" (Nowak and Sigmund 2005).

Let's assume that individuals meet in the roles of donor and recipient, but the same two people never meet a second time. A donor has three choices: (1) always help ("cooperate" in the Prisoner's Dilemma language), (2) never help ("defect" in the Prisoner's Dilemma language), or (3) discriminate. Discriminators make their choices to help based on the recipient's reputation. A cooperator has a higher reputation than a defector, and is more likely to be helped by discriminators, even though the discriminators had no previous interactions with the cooperator. Reputations may be built by direct observation or by transmission mechanisms such as gossip.

This type of model opens the door for more precise explanations of human social relations. For example, we can ask whether discriminators who refuse to help individuals with bad reputations should have their reputations increased or decreased. Similarly, how bad must someone's reputation be before discriminators start withholding resources? Should more resources be given to someone with a higher reputation and fewer resources be given to someone with a lower reputation? Alternatively, should there be a reputation cutoff, above which a person is always helped by a discriminator, and below which a person is never helped by a discriminator?

Reputation building is prone to error, in part depending on the size of the population. For example, estimates of past hunter-gatherer population size are generally of the order of 50–100 individuals. With populations this small, direct observation and gossip are sufficient to build a person's reputation. Nowak and Sigmund formulate an equivalent of Hamilton's rule in which C is the cost to the donor and B is the benefit to the recipient. They state that indirect reciprocity can evolve only if $B \times q > C$, where q is the probability of knowing the correct reputation of the potential recipient. Thus high benefits, low costs, and a high probability of correctly evaluating a person's reputation will favor indirect reciprocity within a population. In our current society, we are immersed in a much larger sea of anonymity than hunter-gatherers, yet our technology, for better or for worse, has created a vast social network that allows people who have never met to evaluate each other. Though it is highly prone to error, this knowledge may be one avenue for us to escape the "tragedy of the commons."

SUMMARY

As with all biological processes, animal behavior is subject to the action of natural selection, favoring individuals that behave in ways that maximize their fitness. If applied to humans, natural selection could lead to the scenario depicted by the "tragedy of the commons" in which selection favors humans who are greedy and exploitative, behaving in ways that improve their own welfare at the expense of the general good. But there are other alternatives to this grim picture that come into play when we appreciate the complex interactions between natural selection and animal behavior.

Natural selection operates on behavioral processes that promote individual survival and, ultimately, individual reproductive success. Behavioral ecologists measure the costs and benefits of alternative types of behavior to gain an understanding of basic behavioral processes such as territory defense, foraging, and mating. This cost–benefit approach allows quantitative predictions of behavior tightly tied to fitness, such as how long to guard a mate, or how long to forage at a particular location before moving on. Both physiological factors, such as the need to keep eggs warm, and ecological factors, such as the spatial distribution of resources, can influence the evolution of mating systems. Even within species, mating systems may vary in relation to the distribution or abundance of resources.

In many species, individuals cooperate with each other in procuring food or defending against predators. Hamilton's model of indirect selection is one possible explanation for the evolution of behavior favoring relatives, including the astounding degree of cooperation in eusocial animals. But in some species, cooperation is common even among unrelated individuals. Ecologists use game theory models to explain cooperation among nonrelatives; this approach can provide a framework for understanding how humans may escape the "tragedy of the commons."

FURTHER READING

Cooper, W. E. Jr, Perez-Mellado, V., and Hawlena, D. 2006. Magnitude of food reward affects escape behavior and acceptable risk in Balearic lizards, *Podarcis lilfordi*. *Behavioral Ecology* 17: 554–559.

Great demonstration of how to ask a question in behavioral ecology, and how to follow up that question with a very simple series of experiments.

Corcoran, A. J., and Conner, W. E. 2014. Bats jamming bats: food competition through sonar interference. *Science* 346: 745–747.

Provides evidence that bats compete with each other by jamming the echolocation signals of competitors, causing them to miss their insect targets.

Faulkes, C. G., and Bennett, N. C. 2009. Reproductive skew in African mole-rats: behavioural and physiological mechanisms to maintain high skew. In R. Hager and C. B. Jones (eds), *Reproductive Skew in Vertebrates: Proximate and Ultimate Causes*. Cambridge: Cambridge University Press, pp. 369–396.

A nice comparative summary of mole rate phylogeny, behavior, and ecology, which includes a good discussion of unresolved issues.

Lloyd, J. E. 1965. Aggressive mimicry in *Photuris*: firefly femmes fatales. *Science* 149: 653–654.

Female *Photuris* fireflies attract and eat *Photinus* male fireflies by mimicking the flash patterns of sexually responsive *Photinus* females, luring the males to their death. This is one of several papers that highlights the evolution of adaptations and counter-adaptations in fireflies from these two genera.

Pennisi, E. 2014. Baboon watch. *Science* 346: 292–295.

Chronicles the story of baboon research at Amboseli National Park in Kenya, begun by Jeanne and Stuart Altmann in 1963, and continued by a second, and now a third generation of researchers.

END-OF-CHAPTER QUESTIONS

Review questions
1. What is the "tragedy of the commons," and why is it a tragedy? Do you agree with Garrett Hardin (1968) that humans are doomed to make decisions that are not in their best self-interest? Support your argument with studies from this chapter.
2. Explain Jerram Brown's (1964) model of territoriality. How did Ethan Pride's (2005) study of lemurs use the concepts underlying Brown's model? What did Pride's research reveal?
3. How does dunnock behavior highlight the importance of resource distribution on the evolution of mating systems?
4. Distinguish between direct and indirect selection. Use examples to show how the concept of inclusive fitness has improved our understanding of the evolution of cooperative behavior?
5. Explain the Prisoner's Dilemma model. What is the simple solution to this dilemma for many species? What are more complex solutions to the Prisoner's Dilemma, and what are some characteristics of species most likely to use these complex solutions?

Synthesis and application questions
1. How does the "Commons" analogy help your understanding of human behavior? What are some problems associated with the use of analogy in science?
2. Hardin's (1968) controversial solution to the "tragedy of the commons" is "mutual coercion" – social arrangements that support responsible behavior, and that are agreed upon by the majority. What problems might arise from such a solution?
3. Can you think of why a firefly might turn sharply when it flashes? What might this tell you about the selection environment experienced by fireflies?
4. Returning to the Florida scrub jay system, if the female took a new mate, how might this influence whether her son would remain to help during the following year?
5. Ethan Pride's (2005) study on ring-tailed lemurs uses fecal cortisol levels as the measure of an individual's well-being. Is this a legitimate assumption? How might you test this assumption?

6. This chapter introduces several models, including the marginal value theorem, Hamilton's model of inclusive fitness and indirect selection, and the Prisoner's Dilemma. Discuss them in the context of what you learned about models in previous chapters.

7. Parker (1970) used the marginal value theorem to predict an optimal copulation duration in dungflies of 41.4 min. His observed value was 35.5 min. Just because the data are a close match to the prediction does that mean that we have now shown that male dungflies maximize their fitness by optimizing the duration of copulation? Why or why not?

8. Why might it benefit a spotted sandpiper female to fertilize eggs with sperm from her first male, in addition to sperm from her second male?

9. Choose any animal whose behavior you've observed (even casually). How would you make observations to determine whether this animal is territorial?

Analyse the data 1

Table 6.4, which presents data from Davies and Lundberg (1984), has numerous categories and relatively small numbers. Let's test the hypothesis that food supplementation causes a reduction in polyandry, and an increase in mating combinations with an equal (monogamy) or greater (polygynandry) number of females per male. To do so, you can combine the data for the 2 years, and also combine the data for both types of polyandry into one cell, and the data for monogamy and polygynandry into another cell. This will yield the following table.

Table 6.9 Polyandry and monogamy/polygynandry on unmanipulated and food-supplemented territories.

	Control	Supplemented
Polyandry	16	6
Monogamy/polygynandry	13	19

Is the reduction in polyandry in response to feeder supplementation statistically significant? In this simplified data set, I included the one case of 3m/2f with the monogamy/polygynandry group. Might it actually make biological sense to put that one case with the polyandry category? If so, what slightly different hypothesis would you be testing?

Analyse the data 2

Most studies of cooperative breeding in birds show that breeders have one of two responses to helpers who provision their nestlings with food. Breeders either (1) decrease the amount of food they deliver to compensate for the provisions supplied by the helpers, or (2) don't change their amount of food delivery. A study conducted by Juliana Valencia and her colleagues (2006) indicates a third response (Figure 6.25). What is that response, and why might some birds show this pattern?

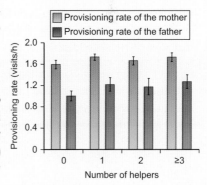

Figure 6.25

Chapter 7
Bernd Heinrich: studying adaptation in the field and the laboratory

INTRODUCTION

When I first met Bernd Heinrich he was beaming. He had just received an email message from a friend of his in Poland that included a photograph of a wasp that had resided in his memory for a very long time. This wasp was particularly significant – it was a dark bluish-black ichneumonid wasp with yellow spots forming a bright contrast on its slender abdomen – and it was so shiny that it seemed to glow. The most significant feature of this wasp was the name of the collector – Bernd Heinrich – and the date was 1945. Heinrich's father, a specialist in ichneumonid wasps, had been seeking this particular wasp since about 1925. Young Bernd found this wasp on his way back from school by peeling away some moss from the shore of a nearby creek – the wasp was already deep into hibernation. "I was 5 years old at the time I collected that wasp," Heinrich described, "and it looks exactly the way I remember it."

In this chapter we will learn about Bernd Heinrich's approach to doing ecological research. You will discover how Heinrich was influenced profoundly by members of his own family, and by his childhood experiences as a naturalist. During our conversation, Heinrich raised several issues associated with the factors that led to his success. Perhaps the most important of these were:

1. He knew his study species very well, so he was able to identify aspects that were surprising and anomalous.
2. He devoted considerable energy into trying many different approaches, and experienced great success by taking his inquiries in many different directions at the same time (he calls this "messing around").
3. He believed that chance or serendipity was an important player in his successful research program. (Note: I would argue that while serendipity was important, the key might be to be prepared to take advantage of it.)
4. He understood that successful research required a great deal of energy and motivation.
5. He worked on a variety of different animal species, carefully choosing animals that were suited to addressing specific questions.

These factors might seem basic, but they represent fundamental components of a directed approach to answering complex research questions. Not all scientists would cite these same

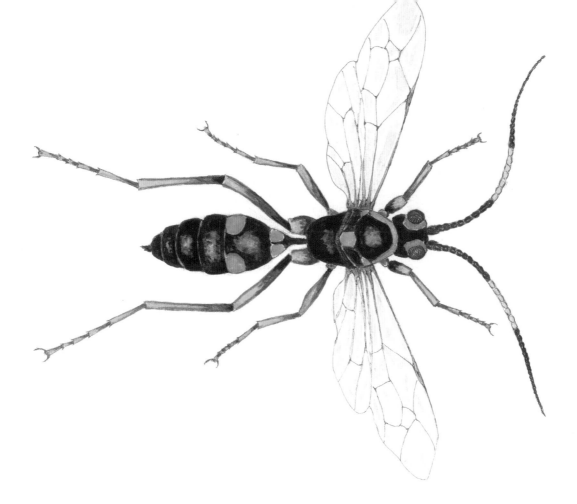

fundamental components, but all scientists who have carried out successful research would be able to explain what has worked or not worked in their respective careers, and why. In future chapters, you will have the opportunity to compare factors leading to Heinrich's success with the particular factors important to other researchers.

We begin with a brief glimpse into Heinrich's early years.

7.1 HOW DID EARLY EXPERIENCES INSPIRE HEINRICH'S APPROACH TO SCIENCE?

Heinrich's native home in Poland was annexed by Germany during World War II. Though Heinrich was fortunate to survive the war without any direct impact on his immediate family, the entire experience was devastating. But living away from a big city, his family always had nature as a refuge. Being close to nature while observing the devastation of wartime occupation helped Heinrich develop his deep appreciation for life and for its tenuousness. He treats life as a "tremendous gift," one that should be appreciated and used to the fullest possible extent.

Heinrich's father found making a living as an ichneumonid specialist to be a bit challenging in post-war Poland, and a colleague suggested emigrating to the United States. So the family moved to rural Maine in 1951. Bernd's father was also a devoted amateur ornithologist, and he allowed Bernd to raise baby birds as long as the boy took them outside once it was time for them to fledge. Bernd's friends were always impressed when some of the birds would come swooping down to visit him as he was walking home from school. Bernd's mother was also devoted to wildlife; she was the family curator of mammals, keeping pet skunks, various rodents, and raccoons. Within this naturalist backdrop, Heinrich's interest in nature flourished.

Heinrich describes his admission to graduate school as serendipitous – like many aspects of his life. In 1964 he had just completed his degree from the University of Maine, and to earn money was working part time in a University of Maine biology lab, washing dishes. Gradually he advanced to assume greater responsibility, helping his boss, Dick Cook, with experiments. Cook evidently saw some potential in his pot scrubber, and suggested that Heinrich develop their current studies of metabolic processes in *Euglena gracilis*, a unicellular protist, into his Master's thesis. This was Heinrich's first study of metabolism, and it provided him with a framework for future physiological investigations. This choice of organism typifies Heinrich's later approach of studying unusual organisms that experience unusual environmental conditions. *Euglena* is one of few organisms that function as both an autotroph and a heterotroph (Figure 7.1).

In the process of learning about the respiratory physiology of *Euglena*, Heinrich became interested in the evolution of eukaryotic organisms, and particularly in the origins of chloroplasts and mitochondria. Researchers had recently learned that both chloroplasts and mitochondria had their own DNA, a discovery that generated great enthusiasm for developing methods for extracting DNA from these two organelles. Dick Cook suggested Heinrich do his PhD at UCLA (Cook's alma mater), and off Heinrich went in 1966 to do some "real science" as a molecular biologist. Unfortunately, his approach to extracting chloroplast and mitochondrial DNA was not panning out very well, as the technology of the time was very crude. Complicating the scenario, Heinrich had a personal preference for field research, and he eventually wandered down the stairs of the biology building at UCLA into ecologist George Bartholomew's lab.

Of course, it was awkward for a beginning PhD student to switch labs, and it was also awkward for Bartholomew to accept this expatriate from the world of molecular biology. While it was not uncommon for graduate students to change advisors during their careers, it was unusual for them to switch between two disciplines as different as molecular biology and ecology. Bartholomew told Heinrich to come back in a month or two with six different potential dissertation projects. As it turned out, this daunting assignment was ultimately an act of kindness, as these six proposals helped shape Heinrich's lifetime research program (Table 7.1). Over the course of his career, he conducted research on five of his proposals, and he wrote extensively about hibernation, the topic of his sixth proposal.

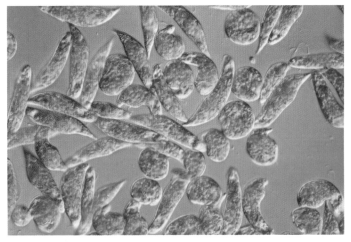

Figure 7.1 Light micrograph (125× magnification) of group of *Euglena gracilis*. *Euglena's* green color arises from photosynthetic chloroplasts, but *Euglena* can also get nutrients from organic particles in its environment.

Table 7.1 Heinrich's six questions.

Heinrich's questions	Results of Heinrich's investigations
How do sphinx moths thermoregulate?	Discussed in section 7.2.
How do desert caterpillars maintain their moisture levels?	Heinrich devoted considerable time and energy to this question, but was not able to get a satisfactory answer.
How do desert tiger beetles avoid overheating, given that they are active during the heat of the day?	The beetles have extremely shiny dorsal surfaces, which are efficient reflectors of solar radiation. In addition, they stand up tall on stilt-like legs, reducing radiative heat gain from the hot surface.
How do sphinx moth caterpillars maintain their foraging efficiency?	The caterpillars consume an entire tobacco leaf, which average about 25 cm long, without even changing position on the petiole. Heinrich evaluated the parameters of leaf morphology, so he could predict which leaf section would receive the first bite, and which would be the last part consumed.
How do bee swarms thermoregulate?	Bees on the inside of the swarm are hotter than exterior bees, which are cooled by the air. Under cool environmental conditions, exterior bees communicate their thermoregulatory needs to interior bees, who increase their metabolic rates so that exterior bees can maintain body temperatures of at least 15°C, the minimum required for flight.
What hormones control hibernation in different insect species?	Not studied by Heinrich.

As we will discuss, these topics led Heinrich in a variety of unexpected directions. As he explained, "I can't remember any circumstance in which the most interesting thing I learned from a study was something I had anticipated going in." In later chapters featuring other ecologists, we will see how some researchers take a more structured approach to their research programs.

Let's now consider Heinrich's first question, on how sphinx moths thermoregulate.

7.2 HOW DO SPHINX MOTHS THERMOREGULATE?

Growing up in Maine, Heinrich used to sneak sphinx moth caterpillars into the basement of an abandoned museum, where he allowed them to overwinter as pupae and metamorphose into adult moths. Given his childhood affinity for sphinx moths, their presence in the desert environment near UCLA, and his interest in the topic of adaptation, it is not surprising that one of his research proposals addressed the question of how these moths managed to thrive in the nearby desert environment. Sphinx moths are very large, and they use enormous amounts of energy to hover while they extract nectar from flowers (Figure 7.2A). In the mid-1960s, James Heath and Phillip Adams published two papers describing how the sphinx moth, *Manduca sexta*, maintained

a relatively constant body temperature to accomplish this energetically demanding task (Heath and Adams 1965, 1967). Heath and Adams proposed that when the sphinx moth is flying in cold temperatures, the angle of the wings is changed from the wing angle employed at warmer temperatures, to make flight more difficult. Reduced flight efficiency would result in greater energy expenditure, greater heat production, and hence high flight muscle temperature, enabling the moths to stay warm enough to fly even during the frigid desert nights. When he read Heath and Adams' papers, Heinrich was convinced that something was seriously wrong, because the researchers' measures of energy expenditures did not match up with values he estimated the moths would need, based on what he knew about moth behavior and metabolism.

For one of his first experiments, Heinrich investigated whether sphinx moths showed any significant difference in metabolic rate at low versus high temperature. He used oxygen consumption as his measure of metabolic rate, and assumed there would be a direct correlation between the two. He put moths into 10-l jars, and coerced the moths to fly for 2 min. He used an oxygen analyser to measure the amount of oxygen in the jar before and after flight. His results seemed to contradict Heath and Adams in that Heinrich found very small differences in oxygen consumption rates at a wide range of air temperatures (Figure 7.2B; Heinrich 1971a).

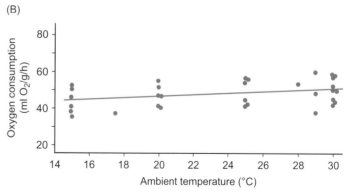

Figure 7.2 A. The hummingbird clearwing sphinx moth, *Hemaris thysbe*, extends its proboscis to sip nectar. B. Oxygen consumption rate in relation to the ambient temperature experienced by the sphinx moth, *Manduca sexta*.

Table 7.2 Mean wing beat angles and wing beat frequencies (SE in parentheses) in sphinx moths at a range of ambient temperatures. Heinrich used a minimum of five moths at each temperature. There are no significant differences between means for wing beat angle or wing beats/s at any temperature.

Ambient temperatures	15–16°C	25–26°C	30–31°C
Wing beat angle (deg)	112.2 (4.4)	112.9 (4.5)	112.1 (5.1)
Wing beats/s	25.8 (0.5)	25.7 (0.4)	27.0 (0.7)

Thinking ecologically 7.1

What else could Heinrich have measured, in addition to oxygen consumption, to assess metabolic rate?

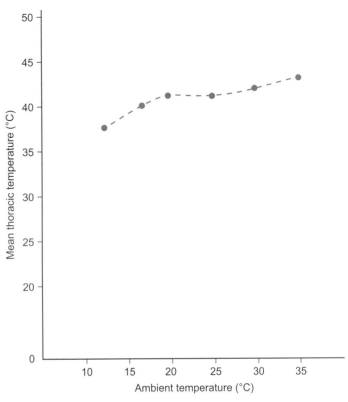

Figure 7.3 Mean thoracic temperature over a 25°C range of ambient temperature. A minimum of seven moths were tested at each ambient temperature.

To test Heath and Adams' argument that sphinx moths increase heat production by changing their flight characteristics, Heinrich measured both the angle of the wings and the wing beat frequency. The angle was measured from enlarged prints of 16 mm motion picture frames. Heinrich photographed the moths as they headed toward the camera. Wing beat frequencies were measured by shooting a strobe light at the moths. When the strobe flashed at exactly the same speed as the wings were beating, the wings would appear to be motionless. Heinrich's results indicated that over a wide range of air temperatures there was little evidence of any changes in flight characteristics (Table 7.2).

If Heath and Adams were wrong, how can the sphinx moth manage to fly over such a wide range of temperatures? From the moth's perspective there were two challenges. First, on cold nights it needed to keep its wing muscle temperature warm enough to sustain the high demand of hover flight. Alternatively, on warm nights it needed to stop its wing muscle temperature from climbing to lethal levels. Is the sphinx moth capable of such feats of thermoregulation?

Heinrich tested the sphinx moth's ability to thermoregulate by encouraging each moth to fly in a large temperature-controlled room. After a minimum of 2 min, he inserted a thermocouple (consisting of two very thin twisted wires made of different types of metal) into its thorax and measured the thoracic temperature. He discovered that thoracic temperature increased very slowly with air temperature, and even at an air temperature of 12°C, the thoracic temperature of flying moths averaged approximately 37°C. But at high air temperatures, the moths did not heat up to lethal levels and were still able to fly (Figure 7.3). Thus the secret to their ability to master the environmental challenge posed by the fluctuating desert temperatures was that they were excellent thermoregulators. Heinrich's quest became to figure out how they managed this ability.

Heinrich turned his attention to exploring how the moth's anatomy might be tied to its ability to thermoregulate (Figure 7.4). Notice the gray sheath surrounding the entire insect – this represents the scales that form a thick layer around the thorax, but only a thin layer around the abdomen. Heinrich hypothesized that these scales might insulate the moth, allowing it to fly at cold air temperatures. If the moth could retain the heat generated by its flight muscles, which are located in its thorax, the moth could keep the flight muscles warm even if the surrounding air was cold (Heinrich 1971b). Heinrich tested this hypothesis by removing the scales from some moths, and discovered that these altered moths lost a great deal of their thermoregulatory ability.

But there is more to this story than just the insulation. Somehow, the moth must be able to rid itself of excess heat when the air temperature rises. In an analogy to what happens with humans, Heinrich suspected that the moth's circulatory system might play a role. When overheated, humans open up their capillaries to increase blood flow to their extremities (we call this process flushing). Excess heat is then dumped through the skin. Perhaps the sphinx moths were doing the same thing. The problem is that insects have an open circulatory system. There is a dorsal blood vessel (also known as the heart) that extends the length of the abdomen. It then passes into the thorax, makes several loops, and ends at the base of the head.

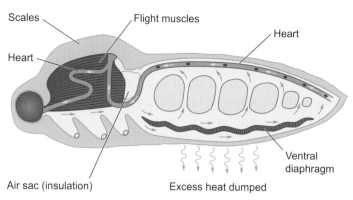

Figure 7.4 Schematic diagram of a sphinx moth, including most of the anatomy that plays a role in the sphinx moth's thermoregulatory ability.

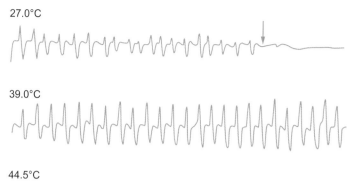

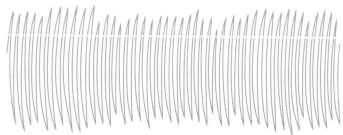

Figure 7.5 Some representative heart rates of sphinx moths at three different thoracic temperatures. Note that at 27°C, the heart stopped beating at the point indicated by the arrow.

The blood then enters the thoracic cavity, where it can still be moved by the continuous action of the dorsal blood vessel — though much more sluggishly. It is forced out of the posterior of the thorax and back into the abdomen, at which point its progress to the posterior of the moth is helped by the undulations of the ventral diaphragm. Follow the arrows in Figure 7.4 to ensure you understand this process. There is also an air sac separating the thorax and abdomen that Heinrich argued functioned as an insulator.

It is important to remember that insects don't use their circulatory system for gas exchange; they have an elegant tracheal system that exchanges oxygen and carbon dioxide for them. This allows them more freedom in controlling their circulatory system. Heinrich suggested that at cold air temperatures, sphinx moths could dramatically slow down their circulatory system, so that the heat generated by the flight muscles would stay in the thorax where it was needed, and be maintained there by the insulatory structures we've already discussed. At high air temperatures the heart would beat faster, increasing the rate of heat exchange between the thorax and the abdomen. The abdomen, with its relatively thin insulation, could dump excess heat into the environment.

This hypothesis generated several predictions. Heinrich's first prediction was that heart rate (number of beats per minute) would be determined by thoracic temperature. To measure heart rate, Heinrich implanted electrodes on both sides of the dorsal blood vessel (very carefully) and was able to record each pulsation of the vessel. He then used a microscope lamp to heat moths at a rate of about 3°C/min. He discovered that at thoracic temperatures below 35°C, heart rate was sluggish and irregular and also included intervals when the heart would not beat at all. Above 39°C, heart beats showed regular high-amplitude patterns that increased in amplitude and frequency as temperatures continued to climb (Figure 7.5).

Heinrich's second prediction was that in heated moths, the ventral portion of the abdomen (ventrum) would be hotter than the dorsal portion of the abdomen (dorsum), because blood from the thorax flows first to the ventral abdomen and last to the dorsal abdomen. Because the abdomen is poorly insulated, it loses heat along this entire path. He discovered that about 4 min after being exposed to elevated temperature, the ventrum of the second abdominal segment increased about 4°C above the temperature of the dorsum of that same segment (Figure 7.6A), and retained that higher temperature during the 25-min experiment.

Heinrich also predicted that inactivating the circulatory system (as in a dead moth) and applying heat to the thorax would cause the thorax to heat up much more rapidly, and the abdomen to show very little temperature change. By implanting temperature sensors in these regions (again very carefully), Heinrich was able to support this prediction as well (Figure 7.6B and C).

As Heinrich points out, sphinx moths experience different ecological challenges than many other animals. In order to extract nectar, sphinx moths need to hover, which is energetically very demanding. Hovering requires a mechanism for maintaining a constant high metabolic rate. Reducing blood flow works well for increasing thoracic temperature when air temperature is low. However, if air temperature is high, as it often is in the desert, sphinx moths need to be able to dump excess heat, by diverting blood to the abdomen. Heinrich wondered whether other nectar feeders that hover extensively would also face similar challenges, and whether natural selection would lead to similar adaptive solutions as those of the sphinx moth. So, in 1971 he expanded his research efforts to bees.

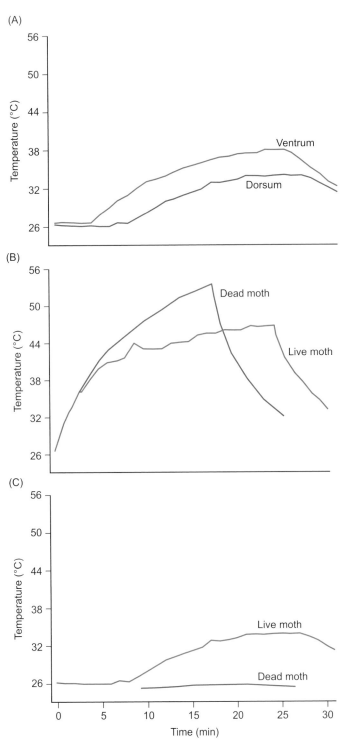

Figure 7.6 A. Temperature of the ventrum and dorsum of the second abdominal segment of a live sphinx moth whose thorax was heated by a lamp for 15 min. B. Thoracic temperature of live and dead moth. C. Abdominal temperature of the dorsal second segment of live and dead moth.

7.3   HOW DO BUMBLEBEES THERMOREGULATE?

Anatomically, bumblebees have numerous similarities to sphinx moths and also some important differences. They too are endowed with air sacs that reduce heat transfer between the thorax and abdomen. Their heart pumps blood anteriorly, and makes a loop within the flight muscles similar to the one Heinrich found in the sphinx moth. Contractions of the bumblebee ventral diaphragm help move the blood ventrally from the anterior to posterior section of the abdomen, where it can reenter the dorsal blood vessel. In contrast to sphinx moths, bumblebees are insulated with hairs rather than scales. One functionally important difference between the two insects is the very narrow petiole that separates the bumblebee's thorax and abdomen, forcing the blood that flows between these two sections through a very narrow space (Figure 7.7A).

Heinrich suspected that, as he had found in the sphinx moth, blood flow was essential for thermoregulation in the bumblebee. To test his hypothesis, he ligated (tied off) the bumblebee heart (with a piece of human hair) so that blood would no longer flow between abdomen and thorax. When he applied heat to the thorax, the temperature of the thorax increased very rapidly to lethal levels, while the abdominal temperature increased only slightly. In contrast, when he applied heat to the thorax in an unligated bee, thoracic temperature leveled off at 42°C, and abdominal temperature increased rapidly (Figure 7.7B).

We can thus return to the same question Heinrich asked for sphinx moths. How can the bumblebee thorax stay warm at low air temperatures, yet at the same time be prevented from overheating at high air temperatures? Somehow bumblebees must have mechanisms that enable them to conserve heat at low air temperatures and release heat at high air temperatures. Given that bumblebees can exist in frigid Arctic regions as well as in warm tropical regions, they are obviously up to the task.

To keep their thoracic temperatures high at cool air temperatures, bumblebees use a mechanism made famous by more glamorous animals, such as the great white shark, and various marine mammals that live in icy waters of the Arctic. These animals, and many others not so glamorous, use a *countercurrent heat exchange* system to maintain high body temperatures despite potentially enormous conductive heat loss occurring through their extremities (see Chapter 5). These adaptations allow the animals to be active over a wider range of environmental conditions, whether throughout the day, across seasons, or at a variety of locations. Heinrich showed that the bumblebee's narrow petiole provides a superb opportunity for countercurrent heat exchange. At cold temperatures, blood flows concurrently from two directions. Relatively cool blood flows from the abdomen into the thorax through the heart, while relatively warm blood, heated by the flight muscles, flows from the thorax to the abdomen, propelled posteriorly by the ventral diaphragm (Figure 7.8). As such there is a positive thermal gradient the entire length of the petiole, so that most of the heat from the thorax is returned to the thorax courtesy of the heart.

(A)

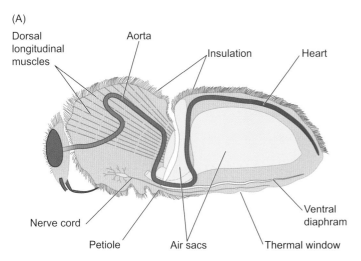

(B)

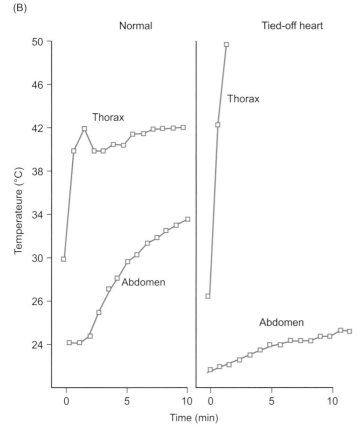

Figure 7.7 A. Anatomy of a bumblebee. B. Effect of applying heat, on the temperature of the thorax and abdomen in the normal bumblebee (left) and the bumblebee with its heart ligated to eliminate blood flow between the thorax and abdomen (right).

But what stops the thorax from overheating at high air temperatures? We know from the results described in Figure 7.7 that the circulatory system is involved. We also know that the abdomen is involved, because it heats up at high thoracic temperatures. The thorax must have a way of circumventing this countercurrent heat exchanger.

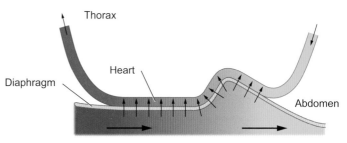

Figure 7.8 Countercurrent heat exchange in the bumblebee. Cool blood in the heart picks up heat from warmer blood flowing along the ventral surface of the thorax into the abdomen. This heat is then returned to the thorax, keeping flight muscle temperature elevated.

(A)

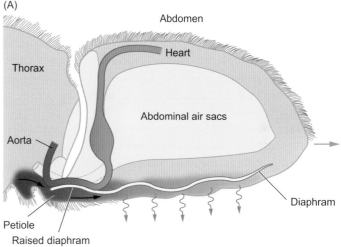

(B)

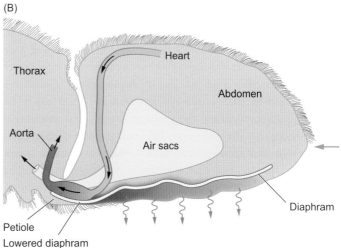

Figure 7.9 Bypassing countercurrent heat exchange in warm bumblebees. A. Expansion of the abdomen draws air into the abdominal air sacs, raising the diaphragm and allowing hot blood to flow from the thorax into the abdomen. Vertical arrows show heat escaping through the uninsulated ventral window of the diaphragm. B. Contraction of the abdomen forces air into the thorax, lowering the diaphragm and forcing blood from the abdomen into the thorax via the heart. Because blood flows first in one direction and then the other, there is no opportunity for the exchange of heat in the petiole region. Abdominal blood continues to lose heat to the environment.

Heinrich proposed an alternating current model of heat flow that could circumvent the countercurrent heat exchanger at high thoracic temperatures. Under this model, blood flow from the abdomen into the thorax would occur in alternative pulses with blood flow from the thorax into the heart. Warm thoracic blood could flow into the abdomen without making significant contact with cool abdominal blood (Figure 7.9). The bee could maintain a high rate of blood flow, and a high rate of heat flow from the thorax to the abdomen and into the environment.

One of the predictions of this model is that at low thoracic temperatures, the thorax and abdomen would be free to oscillate independently of each other. Heat is conserved efficiently if there is no flow from thorax into abdomen (in which case the heat just stays in the thorax to keep the flight muscles warm), or if there is a two-way flow (courtesy of the countercurrent heat exchanger). However, when thoracic temperatures climb to high levels, Heinrich predicted that the heart and diaphragm must oscillate in a 1:1 ratio, delivering alternating pulses of blood anteriorly and then posteriorly. To test this prediction, he recorded mechanical activity of the heart and ventral diaphragm at different thoracic temperatures (Figure 7.10). At low thoracic temperatures, the heart had primarily high-frequency, low-amplitude beats (and would sometimes even stop for brief periods), and the ventral diaphragm showed a similarly irregular pattern. However, at high thoracic temperatures, there was the predicted 1:1 ratio of high-amplitude beat of both the heart and the diaphragm.

Heinrich's bumblebee research tied together this elegant system of thermoregulation with other aspects of bumblebee natural history. For example, bumblebee queens were known to press their elongated abdomens tightly against their brood. One hypothesis was that this behavior elevated brood temperature and increased the rate of larval development. Heinrich's research showed that abdominal temperature could easily be raised by circumventing the countercurrent heat exchanger and warming up the abdomen. Heinrich measured abdominal temperature and showed that it rose sharply when a queen visited her brood, and that the queen effectively transferred heat to the brood through the thermal window on the ventral side of her abdomen.

As we've already discussed, Heinrich pursued ecological questions beyond the mechanisms of thermoregulation. He was particularly drawn to evolutionary puzzles, cases where organisms behave in ways that – from the standpoint of natural selection – appear to be maladaptive.

7.4 HOW DO SLOW-MOVING BUTTERFLIES AVOID BEING EATEN?

As a child, Heinrich collected caterpillars with his father, who was interested in the wasps that parasitize caterpillars and eat

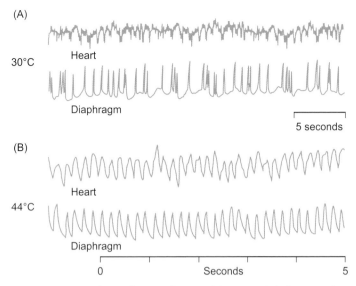

Figure 7.10 A. Independent oscillation of thorax and abdomen at low thoracic temperature, which supports countercurrent heat exchange and reduces heat loss into the environment. B. Alternate 1:1 oscillation of thorax and abdomen at high thoracic temperatures, which bypasses countercurrent heat exchange and promotes heat loss into the environment.

them from the inside out. So father and son would regularly collect caterpillars to see if they were parasitized, and to see if a wasp or a butterfly or moth would be the final product. Heinrich was particularly enamored with the fact that he could go out, put a sheet under a tree, and shake the tree vigorously, and something unexpected would always come falling down – caterpillars with new markings or a new shape, or a strange creature that was experiencing a bad hair day. And if he kept these animals alive, Heinrich could watch them metamorphose into something new, unexpected, and beautiful.

Heinrich extols the virtues of messing around. He feels that some beginning students are paralysed by doing research, not able to carry out substantive steps because they feel they will do something wrong and ruin the experiment. He argues that it is essential to do many things wrong before you can do something right. The key is spending good "messing around time" with your system – the organisms and conditions that interest you – so that you know what types of observations are meaningful, you see things that you may have missed if you were entirely systematic, and you can appreciate an observation that is anomalous. Only after some nondirected activity with your study organisms are you ready to design experiments. Thus when Heinrich saw partially eaten leaves falling from trees, he was able to appreciate that this was something unusual, because green leaves do not just fall from trees on their own accord. They generally need some coercion. And why would caterpillars toss perfectly good food into the environment? He knew there was a story there.

(A) Palatable species   (B) Unpalatable species

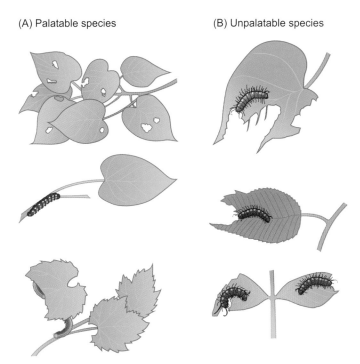

Figure 7.11 Contrasting behavior of three palatable and three unpalatable species of caterpillars. A. The palatable species remove small patches from each leaf while hiding on the leaf's ventral surface (top), or snip off portions of unfinished leaves (only the petiole remains, middle), or smooth off edges of unfinished leaves so they appear whole to a predator (bottom). B. In contrast, the unpalatable species feed in full view of predators and show no behavioral adaptations to mask their presence. Notice that the unpalatable species tend to have long setae (hairs) – a constitutive defense against predators.

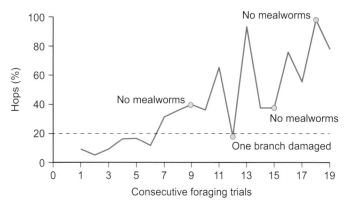

Figure 7.12 Percentage of hops by a chickadee on trees with experimentally damaged leaves. By the seventh trial, the bird had learned that trees with damaged leaves contained mealworms. Even when no mealworms were present (trials 9, 15 and 18 in this figure – with arrow), the bird was significantly more likely to hop on the damaged trees, indicating that it had learned to associate leaf damage with food. However, when only one branch had damaged leaves (trial 12), the bird showed no preference for that tree. With the exception of trial 13, there were at least 50 hops per trial.

As Heinrich describes, caterpillars are somewhat different from many foragers, because they usually have abundant food nearby. The challenge is that caterpillars need to ingest without being eaten. This is only a problem if they are palatable, and many are not, being richly endowed with toxins, or having distasteful hairs or other appendages. For species not so endowed, the food that they eat also serves as a protective cover. If such caterpillars eat their food, they are removing some of their protection against predators, and the uneaten portions of leaves serve as beacons announcing to predators that caterpillars are in the vicinity (Heinrich 1979).

Thus Heinrich predicted that natural selection operating on palatable caterpillars should favor those that behave in ways that make them less obvious to predators, including eating only a small portion of the leaf, snipping off any unfinished leaves, or rounding off the lobes of leaves so they still appeared whole and untouched. Unpalatable species of caterpillars do not experience these restrictions, thus they could perch brazenly on tattered leaves, and could forage during the day on any leaf surface regardless of its visibility to predators. Heinrich's comparative study of eight species supported this prediction that foraging strategy coincides with vulnerability as prey. Figure 7.11 shows how the contrasting behavior of six species correlates with their palatability.

While a comparative study of eight species can be used to support a hypothesis, Heinrich wanted more rigorous support of adaptive feeding behavior by caterpillars. Does leaf damage serve as a feeding cue for birds, which are caterpillars' most ecologically significant predators? If so, do birds learn to use leaf damage as a cue, do all birds use the same cues, and how much individual variation is there among birds in their foraging tactics? In 1981, Heinrich and Scott Collins began investigating these questions by studying the foraging behavior of black-capped chickadees in an artificial enclosure composed of 10 small white birch and chokecherry trees. Using a hole punch, the researchers perforated all of the leaves on two of the trees (the experimental trees), leaving eight trees with unperforated leaves. Heinrich and Collins wrapped wild rose around each tree, and impaled mealworms or caterpillars on the rose thorns on the experimental trees. The researchers counted and recorded the number of hops the chickadees made on each tree as their measure of foraging behavior. They didn't use perching time on the trees, because after finding a food item, the bird would often hop to a new tree and eat the item, before returning to the foraging mode. Given that there were two experimental trees and eight control trees, a randomly foraging bird would be expected to make 20% of its hops on the experimental tree. As can be seen from Figure 7.12, though a bird initially foraged at a frequency of less than 20% on the experimental trees, its preference gradually increased over time. In trials 9, 15, and 18, no mealworms remained on any of the trees, and still the chickadee showed a statistically significant preference for the experimental trees (Heinrich and Collins 1983).

Thinking ecologically 7.2

Why do you think Heinrich and Collins had eight control trees and only two experimental trees? Would they have been better off using five control and five experimental trees? Why or why not?

Heinrich and Collins tested numerous other chickadees and found several different feeding patterns. One way of combining these data from different chickadees is to take note of how many hops a bird makes on average before landing on a tree exhibiting leaf damage. Heinrich predicted that number would decrease significantly with consecutive foraging trials. As predicted, by the 10th trial most birds were almost immediately hopping onto trees with damaged leaves.

In the course of the chickadee study, Heinrich and Collins observed 30 different species of caterpillars that fed on trees. They classified nine species as being unpalatable by virtue of being spiny or aposematically colored (see Chapter 4). They classified 21 species as being palatable on the basis of being cryptic and nonspiny. In accordance with Heinrich's earlier data (Heinrich 1979), the researchers found that none of the unpalatable species chewed off and discarded partially eaten leaves, leaving these in view. In contrast, 20 of 21 palatable species clipped off partially eaten leaves, removing all of these from sight. This much larger data set bolstered Heinrich's argument that caterpillars feed adaptively in ways that minimize their probability of being detected by birds that use "messy leaves" as a cue for the presence of tasty prey.

Heinrich's interest in foraging and antipredator adaptations led him beyond his original research proposal to work with many other species of birds, including blue jays, golden-crowned kinglets, and red-breasted nuthatches. However, his most recent work has focused on the behavioral ecology of the common raven, which has quickly revealed itself to be one of the most uncommon animals ever studied.

7.5   WHY DO RAVENS YELL?

When Heinrich mentioned to some of his colleagues that he was preparing to embark on a study of ravens, they congratulated him for picking the most difficult study species imaginable. Their words were prophetic, but he was up to the challenge. He dedicated his book, *Ravens in Winter* (1989), to "all the raven maniacs who answered the call."

Heinrich chose ravens because of behaviors he had witnessed them displaying in nature. He was intrigued by an evolutionary puzzle – a trait that, on the surface, appears to be maladaptive. The puzzle was this: On October 28, 1984, Heinrich was strolling

Figure 7.13 Group of ravens feeding at a carcass accompanied by an equally interested group of coyotes.

through the Maine woods and heard some ravens making sounds that he described as yells. He followed the yells and came upon a moose that had been converted to a carcass by a poacher, who had covered the carcass up in a vain attempt to hide it from other hungry mouths. About 15 of these mouths belonged to ravens, who flew away and hid upon Heinrich's arrival (Figure 7.13). As he described in his book, "I was awed because I saw a paradox" (Heinrich 1989).

Ravens are rare birds and tend to be solitary or, at their most social, they travel in pairs. How did 15 of these rare, solitary birds manage to end up at the same carcass? Given all the yelling that was going on, Heinrich thought he knew – the birds were hearing the yelling from far away and coming to join the party. The paradox is that as a result of their yelling the initial finders of the carcass were now being required to share their food bonanza with numerous other birds. Recall that one of the primary tenets of evolutionary ecology is that natural selection will favor the evolution of traits that maximize an individual's survival and reproductive success (Chapter 3). Hence the paradox: why share with the world the news of its discovery, when the bird that discovered the carcass could clearly maximize its foraging success by letting soft, contented burps be the only sounds to emerge from its beak?

Over the course of several Maine winters, Heinrich tested and discarded a number of hypotheses that could explain why these ravens were showing this seemingly non-Darwinian left-wing behavior. Perhaps they were calling in other birds to help them open up the carcass? Perhaps they were helping out relatives, thereby enhancing their reproductive success courtesy of indirect selection (Chapter 6)? Alternatively, the ravens may have feared becoming some other animal's prey while feeding on the

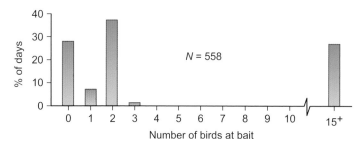

Figure 7.14 Percentage of days in which various baits had 0, 1, 2 . . . 15+ ravens at the bait. In most cases, baits were either populated by no birds, a pair of birds, or by more than 15 birds.

carcass, and if they gathered other ravens around, there would be more eyes to scan for predators.

 Thinking ecologically 7.3

The previous paragraph introduces three hypotheses for why ravens might yell loudly while feeding at a carcass. What would be a prediction generated by each hypothesis? How would you collect data to test each prediction?

Complicating the picture was the observation that recruitment only happened some of the time. Other times, a bird or a pair would fly in and feed, and come back and feed some more, and repeat this process for several days. Usually, at least with large bonanzas, a large (and raucous) group of ravens would eventually gather at the carcass, although often this would take place slowly. Heinrich would haul out a few hundred pounds of calf or moose to a potential gathering site, only to find that one or possibly two birds partook in the bounty for many days, before the large group finally moved in (Heinrich 1988). Other times, recruitment was nearly instantaneous (Figure 7.14).

When I asked Heinrich if he had a pretty good idea of the answer to his puzzle early on in the research, he replied, "No, I didn't have the slightest idea. That' s why I was so enamored with it . . . all the obvious explanations seemed unlikely." Heinrich's far-reaching (but correct) explanation recognized that there are two different classes of ravens, each of which has different foraging interests. Territorial pairs are mature adults who can be recognized by having black mouths. Juvenile birds are about the same size as adults, but they can be recognized by having red mouths. These anatomical distinctions were revealed only when Heinrich and some of his students captured a large group of birds (whom they then marked with distinct tags on their wings). The territorial adult pairs defended certain areas from all other ravens, and if they came upon a food bonanza within their territory they stayed quiet and harvested all the meat from the carcass (if they could remain undetected). However, if juveniles found a carcass, they attempted to recruit. Of course, they were instantly attacked by the territory owners, but

if they recruited a sizable number of juvenile allies quickly, they were able to harvest the carcass.

Over the course of this foraging study, Heinrich was impressed with similarities between raven and primate societies. Hence his most recent work, begun in 1990, investigates raven cognitive skills, taking Heinrich into a very different discipline than that of his prior research.

7.6 HOW SMART ARE RAVENS?

Heinrich's investigations of raven cognition focused on two questions. First, can ravens, like primates, use tools in a way that demonstrates they have a deep understanding of cause and effect relationships? Second, does an individual raven have knowledge of what other individual ravens know and don't know? It seems unlikely that any bird would have such a human-like level of intelligence and awareness. But Heinrich and Thomas Bugnyar have accumulated substantial evidence that for ravens, the unlikely is a reality.

Tool use and insight learning

Tool use is considered to be cognitively challenging to animals, because it may require them to establish causal relations between objects that are external to their own body. For his first study of potential tool use in ravens, Heinrich (1995) wanted to see whether ravens could figure out how to access meat that was dangled from a perch on a 50-cm-long piece of string. To do so, the ravens needed to execute a precise sequence consisting of at least five different steps – including reach down, grab, pull up, hold with foot, release string, hold string, and reach down again – repeated in about five sequences (Figure 7.15). Heinrich (1995) argued that quick mastery of this problem without any prior training would demonstrate that ravens can truly understand cause and effect relations.

Five of six birds were able to solve this cognitive task within 4–8 min of their first contact with the string. These birds improved dramatically with practice, with one bird completing the task in only 5.3 seconds by the 18th trial.

Having established that ravens could use cognitive skills to obtain food, Heinrich's next question was whether the birds, having solved this problem, now knew that the meat was tied to the perch, and were aware of the possible consequences of flying off with a piece of meat still securely tied to a perch (a true neck-wrenching experience). In anticipation of this question, Heinrich conducted two stages of training before running the actual experiments. In the first training stage, he gave perched birds a piece of meat, and then scared them off the perch with handclapping. In 92% of the cases the birds flew off with the meat. In the second training stage, the researchers gave perched birds a piece of meat attached to a string (but not tied to the perch), and then scared them off the perch with handclapping. In 100% of the

Meat

cases the birds flew off with the meat after the handclapping. One could argue that by now the birds had been trained to fly off with the meat. This makes the final experiment even more revealing. This was conducted after the birds had taught themselves to retrieve pieces of meat using the pull-up technique. In this experiment, the meat was tied to the perch with a piece of string. Once the birds secured the meat in their beaks, Heinrich scared these trained bids off the perch with handclapping. In contrast to the two training experiments, these ravens dropped the meat before flying 80% of the time. In the remaining 20% of the cases, they flew off with the meat in their mouths but dropped it immediately – before the string went taut in their beaks. Heinrich concluded that the ravens had established a deep understanding of the relationship between the string, perch, and meat.

From our human perspective, it is intuitively logical to pull up a string with a reward tied to it. Joined by Thomas Bugnyar, Heinrich wanted to know whether ravens were capable of solving the counterintuitive problem of pulling down on a string to raise the reward (Figure 7.16). One experimental group comprised six ravens with no string-pulling experience. The second experimental group comprised six ravens that had already mastered the string-pulling technique.

None of the inexperienced birds was able to solve this counterintuitive problem, though they tried pulling the meat that was visible through the mesh barrier screen. In contrast, the ravens that had mastered the pull-up task were able to apply their acquired knowledge to rapidly solve the pull-down challenge. The researchers argue that these successful ravens had generalized information learned from the first more intuitive task, and transferred these skills to a novel application (Heinrich and Bugnyar 2005).

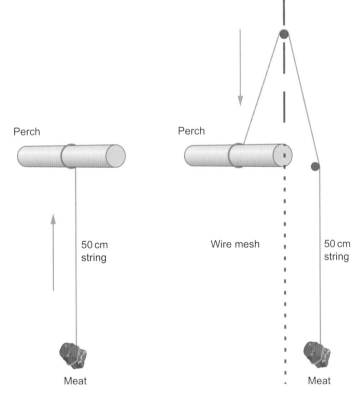

Figure 7.16 Setup for vertical pull-up task and pull-down task. Arrows indicate direction of pulling.

Given the ravens' high level of awareness of physical relationships, and their ability to solve complex problems, Bugnyar and Heinrich (2005) turned their attention to the raven social environment. They wanted to know whether ravens were aware of what other ravens in their social environment knew, and if they used that awareness to make adaptive behavioral decisions.

Do ravens know what other ravens know about them?

As Heinrich discovered in his early field studies, ravens hide extra meat in caches, which they bury in the snow or the ground (depending on the weather and the terrain). In the wild, they are sensitive to the presence of other ravens when caching, and carefully place their caches far away from other ravens, often behind objects that obstruct the view of potential food competitors. Ravens that hide caches are called "storers." If forced to hide food when competitors are in the vicinity, storers may guard their food, even though guarding prevents them from going after more food. Ravens that observe storers in action will often try to steal the cache. Bugnyar and Heinrich call these ravens "pilferers." If a potential pilferer approaches a cache, the storer will either chase the pilferer away, or dig up the cache and restore it at a new location.

Bugnyar and Heinrich (2005) wanted to know whether storers adjust their behavior to account for their perception of whether an approaching raven actually knows where their cache is stored. A potential pilferer who observed the storer in action is called an "observer" – in Bugnyar and Heinrich's study, observers were able to see the storer cache its meat through a clear window of the aviary. Non-observers were situated the same distance from the storer as observers, but were separated from the storers by an opaque window.

Even though the observers were not in the room when a storer hid its meat, it is clear that the storer knew which birds were observers, and which were not. During a 10-minute trial period, storers had a baseline of recovering about 20% of their caches when left alone in their storage room (Figure 7.17). When a

non-observer was introduced into their storage room, they still only recovered about 25% of their caches (this difference was not statistically significant). However if an observer was introduced into the storage room, storers recovered about 50% of their caches, including a nearly 30% recovery rate when the observer approached within 2 m of the cache. In contrast, storers appeared completely unthreatened when the non-observer approached within 2 m of the cache, recovering only about 3% of caches so closely approached. Clearly storers behaved as if the observer was a much greater threat than a non-observer.

The important question becomes whether the storers used their knowledge of the identity of the observer in making their discrimination between observer or non-observer. The alternative is that observers upon entering the storage room behaved differently than non-observers. Bugnyar and Heinrich noted the latency until an observer approached the cache – the amount of time the observer took to make the approach – and compared that with the latency until a non-observer approached a cache when either the observer or non-observer were alone in the storage room. They also noted the time each type of bird spent searching for the cache. The key finding is that in private (without the storer), the observer tended to approach a cache directly, and spent much less time than a non-observer searching for the cache. However, when a storer was present, the observer changed its behavior, so that there was no significant difference between its behavior and the behavior of a non-observer with the storer present (Table 7.3).

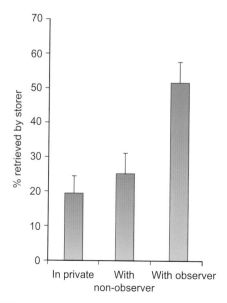

Figure 7.17 Mean percentage of caches retrieved by storers in private, in the presence of a non-observer, and in the presence of an observer.

Table 7.3 Behavior of observers and non-observers: A. when tested in private; B. when tested with storer present.

Conditions of cache search	Behavior of approaching bird*		Significance
(A) In private	Observer	Non-observer	P
Latency to approach (s)	42 (8)	116 (20)	0.06
Time spent searching (s)	5 (2)	38 (9)	0.03
(B) With storer present			
Latency to approach (s)	82 (13)	131 (31)	>0.20
Time spent searching (s)	11 (4)	26 (9)	0.16

*N = 6 for each test. Standard errors are in parentheses.

In comparison to other birds, ravens have very large brains. Heinrich speculates that their evolution of high levels of problem solving and deep understanding of connections may be related to their lifestyle. Ravens are generalists that experience an array of novel social settings. It would be impossible for them to have one special skill that allows them to cope with the variety of conditions they experience as their surroundings change, because no such specialized skill exists. The raven approach to living in its environment contrasts with other birds, such as nutcrackers, that rely almost exclusively on food caches and are endowed with phenomenal memories (Balda and Kamil 1992). Ravens in winter must make decisions under very challenging environmental conditions, which include extensive interactions with potential predators such as wolves. They must be sensitive to environmental cues of where food is, where food was (such as in caches), and perhaps most importantly, where food may be found in the future.

Perhaps the most important theme shaping Heinrich's research is his focus on uncovering the connection between the natural environment and how natural selection influences behavioral and physiological patterns. Selection favors moths and bees that can forage under a wide range of temperatures, caterpillars that forage in ways that fool potential predators, and ravens with the intelligence and flexibility to behave adaptively when facing novel challenges in their complex environment.

SUMMARY

Many factors influenced Bernd Heinrich's approach to ecological investigations. He grew up with parents who were deeply interested in the natural world, and very supportive of a son who shared that interest. As a child he raised many animals, some of which ultimately became his study species. Foremost in this group were butterflies, moths, and birds.

Heinrich's interest in evolutionary ecology led to him asking questions about how physiological and behavioral processes contributed to fitness. One approach he used with sphinx moths and bumblebees was to explore questions about how organisms could survive seemingly inhospitable conditions, or operate effectively over a wide range of environmental conditions. A second complementary approach was to explore evolutionary puzzles, including why caterpillars cast seemingly good food down onto the forest floor, and why ravens yelled when they found a food cache. Both approaches succeeded because Heinrich knew his study species very well, so he recognized interesting behavioral or physiological adaptations, and because he devoted considerable thought and energy to choosing study species that were well suited to the types of questions he was asking.

FURTHER READING

Heinrich, B., and Mommsen, T. P. 1985. Flight of winter moths near 0°C. *Science* 228: 177–179.
Short paper that demonstrates Heinrich's approach to asking questions about organisms that live in extreme conditions.

Heinrich, B. 1988. Winter foraging at carcasses by three sympatric corvids, with emphasis on recruitment by the raven, *Corvus corax. Behavioral Ecology and Sociobiology*, 23: 141–156.
The science behind the question of why ravens yell. Explores and rejects several alternative hypotheses before presenting compelling evidence for the hypothesis that ravens yell to recruit allies, so they can use a food source that is on the territory of a territorial pair.

Heinrich, B. 1989. *Ravens in Winter*. New York: Random House, Inc.
I've assigned numerous chapters from this book to introductory biology classes, because it takes the reader on a fascinating scientific journey, teaching about how problems are recognized, ideas are developed, hypotheses are posed, and predictions are generated. It is also a great read.

Heinrich, B. 1993. *The Hot-blooded Insects: Strategies and Mechanisms of Thermoregulation*. Cambridge, MA: Harvard University Press.
Heinrich summarizes his work on how and why insects thermoregulate. Interesting and informative.

Heinrich, B., and Heinrich, M. J. 1984. The pit-trapping foraging strategy of the ant lion, *Myrmeleon immaculatus* DeGeer (Neuroptera: Myrmeleontidae). *Behavioral Ecology and Sociobiology* 14(2): 151–160.

Heinrich worked on many diverse organisms over his career. Here he describes foraging strategies in the ant lion, *Myrmeleon immaculatus*, which digs pitfall traps that capture unsuspecting insects.

END-OF-CHAPTER QUESTIONS

Review questions

1. How did Heinrich's childhood help to make him a better scientist?
2. In what ways are the adaptations for thermoregulation in sphinx moths and bumblebees similar? In what ways are they different?
3. What are the different types of physiological and behavioral adaptations used by caterpillars to avoid being eaten? Which adaptations tend to co-occur in the same species?
4. What hypotheses did Heinrich test to explain why ravens yelled at a food cache? Which did he support and why?
5. What evidence is there that ravens show insight learning?

Synthesis and application questions

1. One take-home message of Heinrich's research is that both careful planning and chance events underlie successful research programs. Give an example from Heinrich's research that highlights this message.
2. Many important research questions begin with observations of evolutionary puzzles. What is an evolutionary puzzle, and what role did such puzzles play in Heinrich's research?
3. What are some differences between the approaches you have used in your science laboratory classes and the way Heinrich carries out science investigations, as described in this chapter?
4. The study by Heinrich and Collins poses an interesting evolutionary conundrum: If palatable caterpillars tend to hide evidence of their presence by clipping or trimming leaves, do chickadees still benefit by using messy leaves as a cue for the presence of palatable caterpillars? How could you evaluate this conundrum?

Analyse the data 1

Table 7.3 presents data that compare the behavior of observers and non-observers when they are introduced into a room with no other birds (in private) or with the storer present. The researchers argue that the behavior of the observer and non-observer is indistinguishable when the storer is present. Are you convinced by these data? If not, what else might you do to resolve the question of whether observers and non-observers behave similarly when introduced into a room with the storer present? What type of statistical test do you think the researchers used for this analysis?

PART III
Population ecology

Chapter 8

Life history evolution

INTRODUCTION

Only about 500 North Atlantic right whales (*Eubalaena glacialis*) remain in the world, down from estimates of approximately 10 000 individuals in the sixteenth century. Right whales earned their name because to early whalers they were the "right" whale to kill: they are slow swimmers, live in shallow water, and provide a high yield of top-quality oil. They have been protected since 1935, but remain perilously endangered. Is there something about right whales that makes their recovery so slow despite 70 years of protection?

Many endangered species, such as right whales, have life history traits that prevent a rapid population response to protection. Life history traits are adaptations that influence growth, development, survivorship, and a variety of reproductive parameters for individuals of a particular species. We begin this chapter with a discussion of how right whale life history traits make it unlikely that this species will recover quickly from the brink of extinction. We then explore how parental allocation of resources to their offspring influences some important tradeoffs between growth, development, survivorship, and reproduction. Lastly, we consider how genetically similar individuals can show variation in life history traits in response to environmental cues, which may help them survive a rapidly changing environment.

KEY QUESTIONS

8.1. How does allocation of parental resources to reproduction influence a species' life history traits?

8.2. What is the tradeoff between parental resources invested in any one reproductive event and number of lifetime reproductive events?

8.3. How does environmental variation select for phenotypic plasticity in life history traits?

CASE STUDY: Endangered species and life history traits

According to the United States Endangered Species Act (1973), an endangered species is any species which is in danger of extinction through all or a significant portion of its range. Once it is listed as endangered, a species must be treated following certain guidelines, which include designation of critical habitat essential to species survival, and establishment of a scientifically sound recovery plan. Some political leaders have suggested that the Endangered Species Act is ineffective and needs overhauling, arguing that 99% of species that were designated as endangered are still on the list and unlikely to be delisted anytime soon. Conservation biologists respond that political leaders need to consider whether a species' life history traits prevent rapid population growth following protection, before making claims that the Endangered Species Act is ineffective (Stokstad 2005).

A species' recovery rate is influenced by how long it survives, how quickly it attains sexual maturity, and how many offspring it produces over the course of its reproductive lifetime. When comparing different species, survival and reproductive rates tend to be inversely correlated, with high survival rates associated with low reproductive rates. Right whales are extremists in both regards. Historically, they have a very high survival rate – well above 90% per year – but they take an average of more than 9 years before maturing to produce their first offspring. In addition, females only produce one calf at a time, and the time interval between calves has recently averaged between 3 and 6 years. These low developmental and reproductive rates make it virtually impossible for the North Atlantic right whale to be removed from the endangered species list for many years. Thus recovery rates in right whales are constrained by their life history traits.

This species' recovery is further affected by factors beyond reproductive traits. North Atlantic right whale survival rates have recently decreased as a result of the species' behavioral life history. In one study, Susan Parks and her colleagues (2012) tagged 13 whales in Cape Cod Bay off the coast of Massachusetts with sensors that recorded how deep the whales dove for their prey, primarily copepods. Both the copepods and the whales spent the vast majority of their time within 5 m of the surface, making the whales (and presumably the copepods) highly vulnerable to being struck by ships. Indeed, five right whales are known to have been killed by collision with ships in the

Figure 8.1 Entangled right whale. The disentanglement response team used a cutting arrowhead (visible in the upper left portion of the picture), shot from a crossbow, to successfully sever the ropes that were surrounding the whale's head.

past 30 years in this one location. Even in locations where right whales do dive deeper for food, they must return to the surface to rest in between dives. Unfortunately, pregnant whales need a longer rest period to recover from a dive; hence they may be more vulnerable to being hit by ships.

Amy Knowlton and Scott Krauss (2001) compiled data on mortality and serious injuries to North Atlantic right whales. During the period of 1970–99 there was a significant increase in confirmed North Atlantic right whale mortalities. More recently, of 50 dead right whales discovered between 1985 and 2005, 19 were killed by collisions with ships, and 18 were probably killed by entanglement with fishing equipment. Approximately 75% of the victims were calves and juveniles, which make up less than 20% of the right whale population. Unfortunately, rates of entanglement with fishing equipment remain extraordinarily high, with 83% of all known right whales bearing scars from encounters with fishing ropes and traps (Knowlton *et al.* 2012) (Figure 8.1). While only a few of these entanglements are known to kill right whales in the average year, we have no way to evaluate the extent to which entanglements reduce reproductive success of surviving whales.

The good news is that North Atlantic right whales are showing some signs of recovery. At the end of this chapter, we will discuss some reasons underlying the recent population increase. However, even if we could eliminate all deaths from ship collisions and fishing gear entanglements, many years would still go by before North Atlantic right whale populations recovered to safe levels. Conservation ecologists must explain the science underlying life history traits so that the public, and their elected representatives, can make informed decisions on behalf of endangered species.

8.1 HOW DOES ALLOCATION OF PARENTAL RESOURCES TO REPRODUCTION INFLUENCE A SPECIES' LIFE HISTORY TRAITS?

Differences in reproductive life history traits can arise from differences in resource allocation – the quantity of key resources, such as energy and nutrients, that a parent can devote to reproduction. All species confront two basic issues in relation to resource allocation. First, how should parents allocate resources to offspring? For example, is it better to produce a few large offspring or many smaller offspring? Second, how much resource should parents allocate to their own growth and survival versus reproduction? Investment in their own growth and survival may allow parents to survive and reproduce for many years to come.

We can understand the tradeoff between offspring size and number by comparing reproduction in right whales to reproduction in smaller mammals, such as many species of mice. Right whales average 9 years before reproduction, whereas some species of mice mature within a few weeks of being born. Right whales have a very high probability of surviving until adulthood, whereas most mice die before reaching sexual maturity. Right whales only have one offspring at a time and a 2-year gestation period (pregnancy), and they average several years between reproductive events. In contrast, many species of mice have numerous offspring in a litter, with pregnancy lasting only a few weeks, and their litters may be separated by only a few weeks (Figure 8.2).

Ecologists have long noticed consistent differences among species in life history traits, and have devised life history

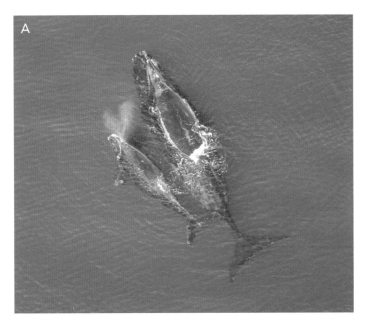

Figure 8.2 Contrasting life histories. A. North Atlantic right whale with her one immense calf. B. A house mouse with her numerous tiny pups.

Table 8.1 Ecological, behavioral, and life history traits of *r*-selected and *K*-selected species.

	r-selected	*K*-selected
Habitat	Variable, unpredictable, disturbed	Constant, predictable, undisturbed
Size of young	Small	Large
Parental care	None or low	Substantial
Survivorship of young	Often very low	Usually high
Population size	Fluctuates	Relatively constant (near carrying capacity)
Development	Rapid	Slow
Time until sexual maturity	Short	Long
Number of offspring per reproductive event	Many	Few
Life span	Short	Long
Potential population growth	Rapid	Slow

classification schemes to account for the patterns observed in nature. These schemes help ecologists consider the relationship between reproduction, development, and the environment, but they have limited application because many species do not conform to these classifications.

Life history classification

Robert MacArthur and E. O. Wilson (1967) and Eric Pianka (1970) describe two contrasting forms of natural selection: *r*-selection and *K*-selection. The letter "*r*" refers to per capita growth rate, which is a measure of how fast a population can grow (see Chapter 10). *r*-selected species, those with potentially high reproductive rates, are most successful in unpredictable environments, including those that have been recently disturbed by catastrophic events such as fires or floods. These conditions favor species with rapid development, reproduction at an early age, and the potential to produce a large number of offspring at rapid intervals. The "*K*" in *K*-selection is the symbol for carrying capacity, which is the maximum population size that can be supported or sustained by the environment (see Chapter 10). *K*-selection favors individuals that compete effectively for resources in predictable and stable environments, which tend to have populations at or near their carrying capacity. *K*-selected species tend to have greater competitive ability, which can include large body size and parents that channel resources into the production of a few large offspring that can survive and reproduce in a highly competitive environment.

An appeal of this classification scheme is that it makes intuitive sense based on the assumption that differences in reproductive life history traits arise from differences in resource allocation. Table 8.1 describes some of the ecological, behavioral, and life history traits associated with *r*-selected and *K*-selected species.

Unfortunately this dichotomy between *r*-selection and *K*-selection does not hold up across a range of species. For example, many bat species are some of the smallest mammals, yet they give birth to only one offspring at a time, whom they lavish with considerable parental care. In contrast, Pacific salmon are very large and have delayed reproduction, yet they give birth to many thousands of small offspring in one bout of reproduction. Lastly, many trees have large body size, very long lifespans, and delayed maturity. But most tree species produce many tiny offspring which receive no parental care (and oftentimes die from competing unsuccessfully with their parents for resources!).

Plant ecologists recognized that the *r* versus *K* model did not apply to plants very well, so they devised alternative classification schemes. Philip Grime (1977) recognized the importance of habitat and resource availability for plants, which he divided into three groups. Plants adapted to predictable or relatively constant habitats with abundant resources are competitive species. Where resources are limited, stress-tolerant species are favored. And disturbed sites, where plant biomass may be destroyed by abiotic factors such as fires or floods, or biotic factors such as disease, tend to favor ruderals, or rapid colonizers. The pioneer species we discussed in Chapter 5 are an example of ruderals.

Ecologists have devised many other classification schemes and they all share one important attribute: they don't always work. But they are valuable even when they fail, because they turn our attention to looking more deeply into life history questions. As one example, we may ask, why do bats defy the *r*- versus *K*-selection scheme? One obvious distinction between bats and other mammals is that they fly. It might be aerodynamically difficult for a female to fly and carry more than one developing fetus within her body.

Most important, classification schemes encourage ecologists to consider tradeoffs in resource allocation. Allocating energy to body mass diverts energy away from reproduction. Allocating energy to parental care reduces the number of offspring that a parent can successfully raise and reduces the rate at which a population can expand. Let's explore the foundation of some of these resource allocation tradeoffs.

Energy consumption and body size

Life history patterns are often studied by comparing different groups of organisms for a particular variable. One very early study done by Max Kleiber (1932) surveyed resting metabolic rates in mammals in relation to body size (Figure 8.3). Resting metabolic rate (P_{met}) is the amount of energy used by an organism over a given time period while at rest in a thermally neutral environment. Physiologists knew that larger mammals tend to use more energy than smaller mammals, but Kleiber collected data from several researchers and found the following relationship between resting metabolic rate and body mass:

$$P_{met} = 73.3M^{0.75}$$

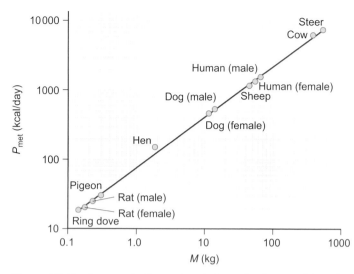

Figure 8.3 Resting metabolic rate in relation to body mass. Note that both axes are logarithmic.

Where
P_{met} = the metabolic rate in kcal/day
M = individual body mass in kg

Kleiber's analysis showed that larger animals used more energy than smaller animals, but that the exponent describing this relationship has a value of less than 1.0 (Dealing with data 8.1). An exponent of <1.0 tells us that metabolic rate *per unit mass* is actually lower in large animals than it is in smaller animals. Thus a gram of elephant tissue uses less energy to function than a gram of mouse tissue, because the mouse has a higher metabolic rate per unit mass.

As we discuss in Dealing with data 8.1, if we take the logarithm of both sides of this equation, we generate a line with the formula,

$$\log P_{met} = \log 73.3 + 0.75 \log M$$

This is the formula for the line of best fit that we see in Figure 8.3.

This type of quantitative analysis opens the door to some methodological issues. First, most of Kleiber's species live in close association with humans. Ideally, a comparative study should be unbiased in its choice of species. We should give Kleiber some allowances, as he was limited to the data that were available to him at the time. Second, it is important to use the appropriate taxonomic unit or grouping for a particular question. Some studies are best served by comparing populations within one species; some studies (such as Kleiber's) investigate patterns among different species, while other studies may compare trends among genera or even families of organisms. Perhaps most important, we should recognize that a relationship between two variables does not imply that one variable causes the other. If one species weighs ten times more than a second species, that in itself does not cause its metabolic rate to be $10^{0.75}$ (= 5.62) times higher than the second species.

Dealing with data 8.1 Power functions, logarithms, and exponents

Power functions or power laws represent the relationship between two variables, x and y, in the form of $y = kx^c$, where k and c (the exponent) are constants. Though it was probably not called a power function by your middle school teacher, one of the first power functions you learned about was the relationship between the area and radius of a circle, which is described as $A = \pi r^2$, where A is the area, r is the radius, and π is the irrational number approximately equal to 3.14. We will use Kleiber's original data to help us see how we can deal with power functions to make the relationship between resting metabolic rate and body mass or weight more easily understood. Here are the data.

Table D8.1.1

Animal	Steer	Cow	Man	Woman	Sheep	Dog (male)	Dog (fem.)	Hen	Pigeon	Rat (male)	Rat (fem.)	Ring dove
Weight (kg)	511	388	64.1	56.5	45.6	15.5	11.6	1.96	0.300	0.226	0.173	0.150
P_{met} (kcal/day)	7265	6421	1632	1349	1219.9	525	448	106	30.8	25.5	20.2	19.5

Figure D8.1.1 Kleiber's data on resting metabolic rate (kcal/day) in relation to body weight (kg).

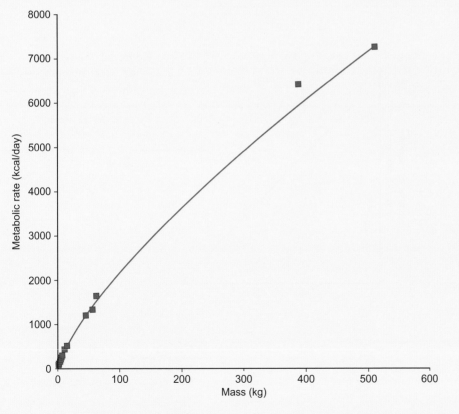

Kleiber actually found that the power function that best fits the data is $P_{met} = 73.3M^{0.74}$, but most modern versions, with the addition of extra data, use 0.75 as the exponent for this relationship. If you look at these data carefully you will see some problems with visualizing this relationship. First, there is a huge range in the data; more than three orders of magnitude (or powers of 10) for weight, and almost as much for P_{met}, which makes simple graphical representation almost impossible. The huge data range causes the metabolic rate and mass values for the smaller animals to be so close to each other that they are obscured by other data points on top of them (Figure D8.1.1). Second, though the data seem to describe some type of power function relationship between weight and P_{met}, as indicated by the curve, it is not clear (either from looking at the data, or from the graph), what the exponent of the power function is. We know the exponent is greater than 0, because P_{met} increases as weight increases, and we know it is less than 1, because the increase curves down.

Kleiber's power function relationship of $P_{met} = 73.3M^{0.74}$ is mathematically equivalent to $\log_{10}(P_{met}) = \log_{10}(73.3) + 0.74\log_{10}(M)$. Scientific notation assumes that log represents $\log_{10}$, and ln represents the natural logarithm or $\log_e$. As you will see later in this chapter, scientists sometimes use ln transformations as well. Notice that our logarithmic transformation of the equation gives us the formula for a straight line. The beauty of this transformation is that the slope of this line (0.74) is the exponent of the power function. So if we were to make a graph with the y-axis representing the $\log(P_{met})$, and the x-axis representing the log(weight), we would end up with a line with a slope of 0.74. The problem with doing this is that we are interested in the relationship between P_{met} and weight, and it is difficult to have an intuitive grasp of what $\log(P_{met})$ and log(weight) actually mean.

Fortunately, there is a wonderful solution to this problem. Rather than actually transforming the data, we can plot the actual values of P_{met} and weight on a graph that uses a logarithmic scale, in which each major increment of the axis increases by the base of the logarithm (in this case 10). This is exactly what Kleiber did in Figure 8.3. A second wonderful feature to this solution is that the data with similar low values are spread out, while the data with high values are compressed in space, so that the relationship can be easily visualized.

(A)

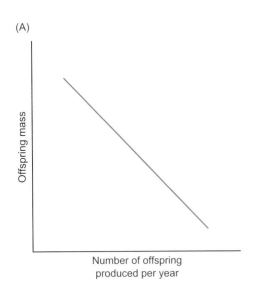

(B)

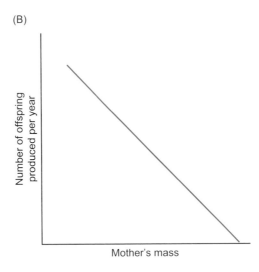

Figure 8.4 Predicted relationships based on life history theory. A. As number of offspring produced per year increases, offspring size is expected to decrease. B. As size of a parent increases, the number of offspring produced per year is expected to decrease.

This particular example also introduces many interesting ecological questions. One important life history question is how limitations in energy availability may influence how much a parent can invest in its offspring, as individual animals have limited energy budgets, and investment in offspring can be energetically expensive. A species that produces numerous offspring over the course of a year thus might be predicted to produce smaller offspring than a species of similar size that produces only one or two offspring over the course of a year (Figure 8.4A).

However, we just learned that body size increases faster than metabolic rate. Does this mean that larger animals are more limited than smaller animals in how much energy they can allocate to reproduction? If so, we might expect that large animals will have fewer offspring than smaller animals (Figure 8.4B).

Table 8.2 Relationship between number of offspring produced per year, and the weight of offspring at birth and at weaning (from Ernest 2003).

Species	Mean total # of offspring per year	Mean # of offspring per litter	Mean # of litters per year	Mean weight of offspring at birth (g)	Mean weight of offspring at weaning (g)
Trichecus imunguis (manatee)	0.5	1	0.5	12,500	67,500
Macaca mulatta (monkey)	0.75	1	0.75	476	1454
Bison bison (bison)	1	1	1	20,000	157,500
Manis temmincki (pangolin)	1	1	1	338	948
Antilocapra americana (antelope)	1.85	1.85	1	3246	8900
Lynx canadensis (lynx)	2.25	2.25	1	269.5	1860
Canis lupus (wolf)	5.49	5.49	1	250	1517
Lepus europaeus (hare)	10.58	2.82	3.75	120	390
Microtus californicus (vole)	21.93	4.54	4.83	2.83	17.67
Mus musculus (mouse)	27.74	6.07	4.57	1.25	8.95

Ecologists have developed hypotheses that explore these relationships quantitatively. Let's consider the quantitative relationship between two important life history variables: number of offspring and offspring size in mammals, in the context of the mass of the parent.

Relationship between number and size of offspring

We hypothesize that parents are limited in how much resource they can allocate to their offspring, so that they can invest either in quantity (a large number of small offspring) or quality (fewer but larger offspring). Table 8.2 shows this trend for a selected group of mammals.

Thinking ecologically 8.1

Table 8.2 records parental investment in two ways: mean weight of offspring at birth, and mean weight of offspring at weaning. Which is a better measure of parental investment? Which data do you think would be easiest to collect? Based on these data, is there a significant relationship between number of offspring per year and parental investment in individual offspring? Hint: You can make the relationship more linear by ln-transforming the data set (you can use our discussion of the relationship between body mass and resting metabolic rate as a model). Plotting out the data on log-log graph paper will help you visualize the relationship. You can then correlate ln (number of offspring per year) to ln (investment per offspring).

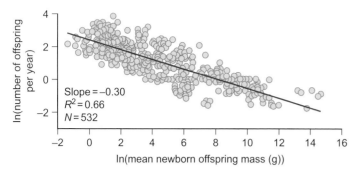

Figure 8.5 Ln(number of offspring per year) in relation to ln(parental investment per offspring) as measured by the mean newborn offspring mass.

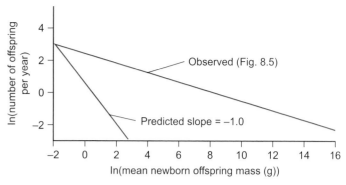

Figure 8.6 Ln(number of offspring per year) in relation to ln(newborn offspring mass). The top curve shows the line of best fit from Figure 8.5. The bottom curve shows the predicted relationship if the line of best fit had a slope of −1.0).

From our small data table, we can gather some support for the hypothesis that the number of offspring and the investment per offspring are inversely correlated. But we need more data to see if this relationship holds across a broad spectrum of mammal species. To address this question, and several other questions related to life history, Morgan Ernest (2003) created a database of 1447 mammalian species, based on the ecological literature and data from unpublished sources. Eric Charnov and Ernest (2006) collaborated to see whether a larger data set supports an inverse correlation between number of offspring per year and investment per offspring.

Figure 8.5 shows this relationship as illustrated from Ernest's much larger data set. As parental investment per offspring increases, the number of offspring decreases.

But there are a few aspects of this graphical analysis to consider. First, there is a lot of scatter in the data. As we discussed in Dealing with data 8.1, logarithmic transformations tend to reduce the apparent variation on a scattergram. Despite this data transformation, there are many species that fall well above or well below the line of best fit. Second, if parental investment per offspring is limiting the number of offspring in relation to the size of offspring, we might expect the relationship between offspring mass and offspring number to yield a slope of −1.0 (Figure 8.6). In other words, if one species produces twice as many offspring as a second species, its investment in each offspring should be half as great as the second species' investment. We would expect that the first species, with twice as many offspring, should produce offspring that are one-half as large at birth as the second species' offspring.

Instead, the data show a slope of −0.30 rather than the predicted value of −1.0 – why is this the case? This brings us to the next aspect of life history studies: once ecologists detect a pattern in life history evolution, they attempt to describe it quantitatively. It is not sufficient to say that one variable is correlated with a second variable. We really want to know precisely what that correlation is. Figure 8.5 shows a slope of −0.30 and an R^2 of 0.66. That's a highly statistically significant result, and the relationship explains 66% of the variance, but

still 34% of the variation remains unexplained. Clearly our statistical model is generating an incomplete picture.

Charnov and Ernest (2006) propose that the amount of resources a mother can allocate to her offspring will be related to the mother's size. (Obviously, whales can produce larger babies than mice.) If so, then the litter size (L) should be proportional (α) to the total resources the mother is able to allocate to her reproductive effort (R) divided by the amount of resources invested per offspring (I).

$$L \; \alpha \; R/I$$

Dividing both sides of the relationship by R allows us to rewrite it as

$$L/R \; \alpha \; 1/I$$

Then if we can figure out how the total resources the mother is able to allocate to her reproductive effort is related to the mother's mass (M), we might be able to more accurately predict the number of offspring based on offspring size. Charnov and Ernest knew from Kleiber's work that an animal's resting metabolic rate should increase with mass$^{(0.75)}$. Thus Charnov proposed substituting $R = M^{(0.75)}$ into the equation above, which gives the following relationship:

$$L/M^{(0.75)} \; \alpha \; 1/I$$

For this discussion, we will use the term *adjusted litter size* to represent $L/M^{(0.75)}$. If Charnov's model is correct, he could plot adjusted litter size against resources invested per offspring, and see two improvements in his graph. First, there should be less variation. Second, the slope of the relationship should be closer to the expected value of −1.0 than his earlier comparison. This would tell us that, after accounting for the relationship between body size and metabolic rate, species that produce larger babies produce proportionally fewer babies.

Charnov and Ernest tested this hypothesis by returning to Ernest's data set, which fortunately included data for adult

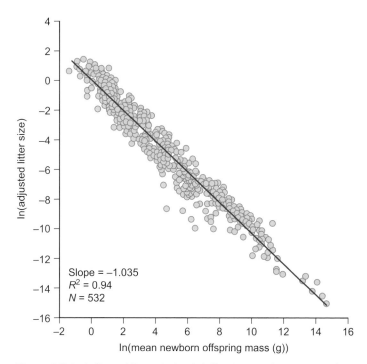

Figure 8.7 Ln(adjusted litter size) = ln(offspring number per year/$M^{(0.75)}$) in relation to ln(parental investment per offspring) as measured by the mean newborn offspring mass.

8.2 WHAT IS THE TRADEOFF BETWEEN PARENTAL RESOURCES INVESTED IN ANY ONE REPRODUCTIVE EVENT AND NUMBER OF LIFETIME REPRODUCTIVE EVENTS?

One important tradeoff is that the number of lifetime reproductive events is related to parental investment in offspring. Individuals with multiple reproductive events over a lifetime must allocate sufficient resources to keep themselves alive, thus they have fewer resources available for each reproductive event. In contrast, individuals that reproduce only once can invest all of their resources in one reproductive event.

Reproductive events: many versus one

Almost all mammals are iteroparous – having multiple reproductive events over the course of a lifetime. An exception is the marsupial mouse of the genus *Antechinus*, in which males of some species are semelparous – they have only one reproductive event over the course of a lifetime. All males that survive to reproductive age (about 10 months) die shortly thereafter, while females will survive long enough to give birth and raise the offspring. In some species of *Antechinus*, the females may live for two or even three mating seasons.

There is considerable diversity in this important life history trait across other groups of animals. For example, all five species of Pacific salmon (*Oncorynchus* spp.) are semelparous, while Atlantic salmon (*Salmo salar*) are iteroparous (Figure 8.8). One possible explanation for these differences is that Atlantic salmon may, in general, make their spawning runs up shorter or less turbulent rivers than their Pacific counterparts, and therefore expend fewer resources in the migration. This could leave some individuals with sufficient resources to support a high reproductive output yet still make a successful journey back to the ocean, and then back up the stream for a second reproductive effort. Most Pacific salmon, in contrast, migrate longer distances up very turbulent rivers, and are unlikely to survive a journey back down the river into the ocean, or to survive long enough in the ocean to accumulate sufficient resources to make a successful return journey for a second spawning. Thus semelparity has been selected for in these species.

In contrast to salmon, the diversity of life cycles in other taxa may blur the distinction between semelparity and iteroparity. For example, in temperate biomes, many insect species have annual life cycles, with distinct juvenile and adult stages. Usually the adult stage is much briefer than the juvenile stage, and it may include one or several reproductive efforts. If there is only one reproductive event, we can clearly consider the insect to be semelparous. But how should we classify an insect that sometimes lays two, three, or more clutches of eggs over a short time span? Will one female be semelparous if she only survives to lay one clutch, while her sister is iteroparous because she survives to

mass. Their analysis confirmed that, as investment per offspring increases, adjusted litter size indeed decreases. There is much less variation in the fit, and impressively the slope is −1.035 (Figure 8.7). This is very close to our prediction of −1.0.

Charnov and Ernest concluded that correcting for differences in resource allocation allows us to have a deeper understanding of life history relationships. Intuitively, we can appreciate that limitations to resource allocation will result in an inverse relation between litter size and offspring size (or investment per offspring). A relatively simple mathematical transformation gives us a more complete understanding of that relationship, and allows us to accurately predict litter size based on the mass of the mother and the offspring.

This study raises several interesting questions. First, why does resting metabolic rate increase with mass$^{(0.75)}$ across such a broad range of species? Second, does the relationship between resting metabolic rate and body size necessarily mean that resource allocation is also limited by mass$^{(0.75)}$? Finally, can we use this relationship to understand other ecological processes? Ecologists studying life history actively pursue answers to these questions.

Fundamental tradeoffs underlie life history evolution. As we discussed in Chapter 4, individuals are physiologically limited in how much resource they can take in and process. Individuals producing numerous offspring must invest less resource in each offspring, while individuals producing fewer offspring can invest more resource in each individual offspring.

Figure 8.8 A group of sockeye salmon, *Oncorhynchus nerka*, a semelparous species, jumps up Brooks Falls in Katmai National Park, Alaska, USA.

lay two clutches of eggs? There are currently no good answers to these questions.

Extreme semelparity

One group of insects that has evolved extreme semelparity is the periodical cicadas. Larvae in the genus *Magicicada* live underground for 13 or 17 years, eating root xylem sap. They emerge and metamorphose in tremendous numbers (in some cases several million individuals per hectare) to live a relatively brief adult lifespan of about 3 weeks' duration (Figure 8.9A). Adults form large groups, where males sing and mating occurs. Most females mate only once, as males provide them with copulatory plugs that block the female's reproductive organ, and presumably inhibit future matings. Males may mate several times, or not at all, with larger males in at least one species being more successful than smaller males. Females lay eggs on the twigs of trees, these eggs hatch after about 7 weeks, and the juveniles rain down to the ground, entering the soil to begin the life cycle anew. Most individuals within a population emerge within 10 days of each other in the same calendar year. Thus across 12 years there are very few adults present from a particular 13-year population (Williams and Simon 1995).

We don't have a good understanding of why this unusual life cycle has evolved within the cicadas. Predators of adult *Magicicada* include many different species of birds, mammals, reptiles, and arthropods. The larvae are eaten by moles, but probably not in huge numbers. One adaptive explanation for synchrony is predator saturation, in which cicada survival rates are high because there are simply too many cicadas present in the area for predators to eat them all (Figure 8.9B).

Richard Karban (1982) investigated whether predator saturation was a reasonable adaptive explanation for periodicity in cicadas. One prediction of the predator saturation hypothesis is that the number of cicadas eaten by predators should be similar at all cicada densities above a certain threshold. Birds discard the wings when they eat a cicada, so Karban counted the number of discarded wings as an index of predation by birds. Figure 8.10A shows that there is no relation between adult cicada density and total number of cicadas eaten. Karban suggests that even at relatively low cicada densities, the birds were eating as many cicadas as they could possibly process. Karban concluded that cicadas were saturating their bird predators, even at relatively low cicada densities.

From the standpoint of natural selection, we know that traits will be favored if they enhance an individual's reproductive success. So the real question becomes, do individual cicadas have higher reproductive success at high cicada densities? Karban counted the number of nymphs that hatched at 17 different sites in relation to the density of adult cicadas. He found that higher-density populations of cicadas had a much higher reproductive success on a per individual basis (Figure 8.10B).

Karban's research supports the predictions of the predator saturation hypothesis. However, we still don't fully understand why extreme periodicity has evolved in *Magicicada* and not in other genera. One possibility is that the root xylem sap is a very dependable food source, thus larvae eating it have high survival rates. However, root xylem sap eaten by

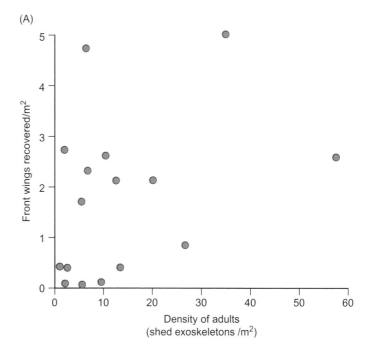

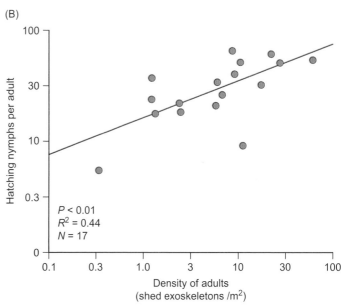

Figure 8.10 A. Cicada consumption by birds (measured by number of cicada front wings recovered) in relation to cicada density (measured by number of shed exoskeletons). Notice that the scale of the x-axis is 10 times the scale of the y-axis. B. Adult reproductive success (measured as the mean number of hatching nymphs produced per adult) in relation to the density of adult cicadas at each site (note log–log scale).

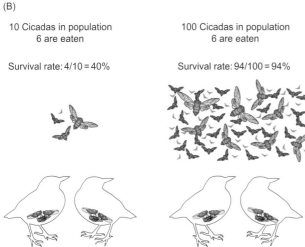

Figure 8.9 A. A recently metamorphosed cicada, perched on its shed pupal exoskeleton. B. Predator saturation. Individuals that survive to adulthood have relatively high survival rates providing they emerge in sufficiently high densities to make their probability of being eaten by predators very low. Left: If there are 10 cicadas, and 2 predators (birds) with a gut capacity of 3 cicadas each, 6 cicadas will get eaten and the cicada survival rate is 4/10 = 40%. Right: If there are 100 cicadas, and 2 predators (birds) with a gut capacity of 3 cicadas each, again 6 cicadas will get eaten and the cicada survival rate is 94/100 = 94%

Magicicada is very nutrient-poor, resulting in slow larval growth rates. It may take *Magicicada* species many years to accumulate sufficient resources to allocate to reproductive efforts. Perhaps this combination of slow development and high larval survival rates operate together to enhance the adaptive benefits gained from predator saturation.

Understanding the allocation of resources to development or reproduction is critical for understanding life history traits. Most hypotheses addressing the evolution of semelparity and iteroparity assume a tradeoff between present and future reproduction. Organisms can allocate all of their resources to one massive reproductive effort, or else they can allocate only a portion of their resources to the present reproductive effort, while retaining sufficient resources for survival and future reproduction.

Tradeoffs between present and future reproduction

Let's first see whether there is a tradeoff between present and future reproduction. If there is, we predict that semelparous species (or individuals) will have a higher reproductive output for their only reproductive event than will iteroparous species for any one of their reproductive episodes. Truman Young and Carol Augspurger (1991) tested this hypothesis by reviewing the literature that compared fecundity (average number of offspring produced) for semelparous plant species and closely related iteroparous species. In plants, fecundity is a measure of the number of seeds produced, while in animals it is the number of eggs. In all

plant studies, Young found that semelparous plants had higher mean fecundity (on a per reproductive event basis) than iteroparous species (Table 8.3). He called this the *fecundity advantage of semelparity*.

Having established that there is a fecundity advantage to semelparity, let's investigate two related hypotheses for the evolution of semelparity: the reproductive effort model, and the demographic model.

The reproductive effort model

The reproductive effort model was jointly developed by William Schaffer and Michael Rosenzweig (1977). Reproductive effort is the amount of resources an organism allocates to a reproductive event. In plants, these resources can come from leaves, stems, roots, or other storage organs. If a small or moderate reproductive effort generates relatively low reproductive success, natural selection favors plants that use all available resources for one massive reproductive effort (semelparous curve in Figure 8.11). Alternatively, if small or moderate reproductive effort generates relatively high reproductive success, and added effort increases reproductive success only slightly, then natural selection favors saving some resources for future reproductive events (iteroparous curve in Figure 8.11).

By comparing the curves in Figure 8.11, we predict that at high levels of reproductive effort, semelparous species would

Table 8.3 The fecundity advantage of semelparity is the reproductive output of a semelparous plant species (or population) divided by the reproductive output of a similar iteroparous species (or population) for one reproductive event. If a semelparous species had 20 offspring, and a comparable iteroparous species had 10 offspring, then the fecundity advantage of semelparity = 2.0. Intrageneric comparisons were among two or more species of the same genus, while intraspecific comparisons were among individuals or populations of the same species.

Species	Type of comparison	Fecundity advantage of semelparity
Temperate herbs	Very broad	2.8
Old field herbs	Very broad	1.7
Gentiana spp.	Intrageneric	2.2–3.5
Lupinus spp.	Intrageneric	2.2–3.2
Helianthus spp.	Intrageneric	1.7–4.0
Sesbania spp.	Intrageneric	2.1–2.3
Hypochoeris spp.	Intrageneric	2.4–3.7
Oryza perennis	Intraspecific	2.9–5.3
Ipomopsis aggregata	Intraspecific	1.5–2.3

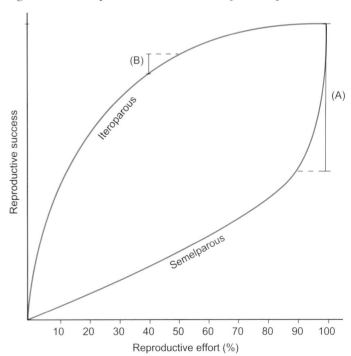

Figure 8.11 The reproductive effort model. Semelparity is favored by natural selection when reproductive success initially increases very slowly with reproductive effort, but then increases very rapidly with high reproductive effort. Iteroparity is favored when reproductive success initially increases very rapidly with reproductive effort, but then levels off at higher reproductive effort. A reduction in reproductive effort saves resources for future reproduction.

benefit more than iteroparous species from additional increases in reproductive effort. To understand this prediction, let's consider the case in which a plant allocates almost all of its available resources into one big reproductive effort. Notice that in the region of high reproductive effort (e.g. 90–100%) for semelparous species, reproductive success is increasing much faster than reproductive effort. Based on Figure 8.11, a semelparous plant that allocates 100% of its available resources has about twice the reproductive success as a semelparous plant that allocates 90% of its resources in reproduction (region A on the semelparous curve). In contrast, an iteroparous plant that allocates 100% of its resources in reproduction will have virtually the same reproductive success as an iteroparous plant that allocates 90% of its resources in reproduction. An iteroparous plant allocating a more realistic 50% of available resources in reproductive effort will still yield only a 5–10% increase in reproductive success over a plant allocating 40% of available resources in reproduction (region B on the iteroparous curve). Let's see if the data support the predictions.

Both yucca and agave plants grow in the desert and chaparral regions of western North America (Figure 8.12A). They share a similar morphology, and are taxonomically closely related. According to the reproductive effort model, in semelparous yucca species, individuals that put out very high reproductive effort should have significantly higher reproductive success than individuals putting out lower reproductive effort. In contrast, iteroparous yucca species should not show a strong relationship between reproductive effort and reproductive success. William Schaffer and Valentine Schaffer (1977) measured reproductive effort in five yucca species as the height of the flower-bearing stalk, reasoning that taller stalks represented a greater reproductive effort. They measured reproductive success in these species as the percent of flowers that developed into fruit. As predicted, the one species of semelparous yucca (*Yucca whipplei*) showed a significant positive relationship between reproductive effort and reproductive success, while none of the four iteroparous species showed a significant relationship (Figure 8.12B).

The Schaffers were concerned about drawing conclusions based on one species of semelparous yucca. Fortunately, the life history traits were the reverse in agave species that the Schaffers chose to study, with six semelparous species and one iteroparous species. Again, the researchers measured reproductive effort as the height of the flower-bearing stalk and reproductive success as the percent of flowers that developed into fruit. As predicted, each semelparous species showed a positive correlation between reproductive effort and reproductive success (the slope of the regression line was significantly greater than 0), while the iteroparous *Agave parviflora* showed no significant correlation between reproductive effort and reproductive success (Figure 8.12C). Thus the benefits of increased parental effort in yucca and agave support the predictions of the reproductive effort model; semelparous species benefit more

than iteroparous species from additional allocation of reproductive effort.

Thinking ecologically 8.2

Can you suggest a better alternative to height of the flower-bearing stalk as a measure of reproductive effort? Why do you think the Schaffers chose this measure of reproductive effort?

Let's now consider the second hypothesis for the evolution of semelparity, the demographic model.

The demographic model

Truman Young (1981) proposed a demographic model to explain the evolution of semelparity. Demography, as we will discuss in detail in Chapters 10 and 11, is the quantitative study of the size and structure of populations, and of how populations change over time. Demographers study factors that influence changes in population size, including development rates and mortality rates.

Young proposed that natural selection should favor the evolution of semelparity under two conditions. First, selection should favor semelparity in plants that require a long time to gather enough resources for even one reproductive event. Second, natural selection should favor semelparity in plants with relatively low probability of surviving to the next reproductive event. In practice, these two factors may co-vary, as plants that require a long time to gather sufficient resources for reproduction may have difficulty surviving long enough for a second reproductive effort.

Peter Lesica and Truman Young (2005) tested the demographic model in their within-species studies of a rosette-forming perennial, *Arabis fecunda*. This plant may produce one large rosette, after which it invariably dies (a semelparous pattern), or it may produce numerous smaller rosettes, after which it has a high probability of surviving (an iteroparous pattern) (Figure 8.13A).

Lesica and Young measured the fecundity advantage of semelparity, and found that it ranged between 2.0 and 3.7 based on data collected from 12 discrete populations of this species in southwest Montana, USA. They focused their demographic studies on three populations from this original group of 12, to see if there was any relationship between the frequency of semelparity, and the survival rates of individuals within each population. Figure 8.13B shows that the population with the highest survivorship (Charleys) had the lowest frequency of semelparity (about 3%), while the population with the lowest survivorship (Vipond) had the highest frequency of semelparity (about 36%). This finding provides some support for the demographic model.

The reproductive effort model and the demographic model should not be treated as two contrasting alternatives. One model may explain semelparity and iteroparity for some taxonomic groups, while the other model may apply in other groups. There may be other explanations for the evolution of these two

Figure 8.12 A. Left: *Yucca whipplei* in flower. Right: *Agave parviflora* in flower. The Schaffers defined stalk height as the measure of plant reproductive effort for their comparative studies of yucca and agave. B. Correlation between stalk height and percentage of flowers forming fruit in the semelparous *Yucca whipplei*, and in four iteroparous species of yucca. C. Correlation between stalk height and percentage of flowers forming fruit in six semelparous species of agave, and in the iteroparous *Agave parviflora*.

contrasting life history strategies. Nevertheless, both models are successful in that they direct field and laboratory researchers toward productive approaches for making relevant observations and collecting data.

The study on *Arabis fecunda* shows that even within species there can be considerable variation in life history traits associated with reproduction. Lesica and Young suggest that environmental variation may cause the variation in survival among the three populations they studied – for example, lower moisture content could lead to lower survival. Let's further explore how environmental variation influences life history variation within a species.

8.3 HOW DOES ENVIRONMENTAL VARIATION SELECT FOR PHENOTYPIC PLASTICITY IN LIFE HISTORY TRAITS?

Evaluating the influence of environmental variation on life history patterns requires comparisons across environments. Within the aquatic world, some species experience a great deal of environmental variation, while other species experience a relatively constant environment. In general, species adapted to freshwater and shallow brackish water will experience more environmental variation than will marine species adapted to deep water. In freshwater environments, species will experience variation in water depth, temperature, current, and light, while species in shallow brackish water will experience all of the aforementioned variation, plus variation in salinity. In contrast, in the deep ocean there is relatively small fluctuation in most of those variables. Thus we expect that organisms from fluctuating environments might show phenotypic plasticity – an ability to change phenotypes (expressions of developmental, physiological, or behavioral traits) in response to different environments.

Developmental phenotypic plasticity in the spadefoot toad

The spadefoot toad, *Spea multiplicata*, lives in a challenging and unpredictable environment. It spends much of its adult life buried in desert sands, but emerges during summer rains to feed on insects and to breed in temporary shallow ponds left by the rains. Fertilized eggs hatch into tadpoles in 2–3 days; tadpoles must grow quickly and metamorphose into toads before the ponds dry up. Successful toads will then use the spade-like extensions of their hind legs to dig out a retreat, where they will enter a state of torpor (estivate) until roused by next summer's rains. Many tadpoles die when their temporary ponds dry up before they metamorphose.

David Pfennig (1990) censused 37 different ponds over 3 years to study the process of spadefoot toad development. He found two morphs of tadpole in many of these ponds, an *omnivore* tadpole which fed primarily on algae and detritus, and a *carnivore* morph that fed primarily on small fairy shrimp (*Anostroca* species) and other tadpoles (Figure 8.14). Carnivores have a large broad head, powerful jaw muscles, and a sharp keratinized tooth used to subdue prey, while omnivores are smaller and have much narrower and less muscular heads, and a longer digestive tract to process their food.

Pond samples showed that carnivore frequency was highest in shallow, rapidly evaporating ponds that contained abundant shrimp populations. Pfennig also knew that carnivores

Figure 8.13 A. *Arabis fecunda* in Montana, USA. B. Percentage of semelparous *Arabis fecunda* in three populations in relation to mean annual adult survivorship.

Figure 8.14 A carnivore spadefoot toad tadpole consumes an omnivore.

developed up to 50% more quickly than omnivores, so it made sense that a rapidly drying pond might favor carnivores, which would be more likely to metamorphose before the pond dried up. But, in addition, he discovered that rapidly drying ponds also tended to have high shrimp densities, which also favors the carnivore diet. So Pfennig's challenge was to identify which of these two factors, drying rate or shrimp density, influenced the switch to carnivory.

Pfennig's approach was to establish a series of 24 artificial ponds in which he systematically controlled drying rate and shrimp density. Pfennig simulated a non-drying pond by adding water, and controlled shrimp density by adding one shrimp per tadpole every 2 days to the low-density treatment and 12 shrimp per tadpole to the high-density treatment. Conforming to Pfennig's expectations, five of six high drying/high shrimp density ponds produced some carnivores and only one of six non-drying/low shrimp density pools produced carnivores. Importantly, four of six non-drying/high shrimp density produced carnivores, while none of the high drying/low shrimp density ponds produced carnivores. Pfennig concluded that high shrimp density was the proximal cue that stimulated the development of carnivores.

Several lines of evidence indicate that these developmental differences arose from developmental plasticity rather than from genetic differences between individuals. First, in his artificial ponds experiment, Pfennig was careful to allocate similar numbers of tadpoles from each female into each pool, so there would be no genetic bias among treatments. Second, a later study showed that omnivores switched to carnivores when fed a high shrimp diet, and carnivores switched to omnivores when fed a low shrimp diet (Pfennig 1992a). Pfennig's research also showed that thyroxin, a hormone that accelerates developmental rate (and that is present in fairy shrimp), could trigger a switch from omnivory to carnivory in a laboratory population.

You may wonder why all spadefoot toad tadpoles are not carnivores. While metamorphosing before a pond dries up is essential for toad survival, it turns out that omnivorous tadpoles can build up much higher fat reserves during development than do carnivores. Pfennig also knew that omnivores could accelerate their rate of development if their pond showed signs of drying up. Pfennig hypothesized that given sufficient time to develop in a pond, omnivores, with their greater fat reserves, might have higher survival rates than carnivores after metamorphosis, but if forced to accelerate development, they might survive poorly after metamorphosis. So he set up one final experiment to test the predictions of this hypothesis.

Pfennig set up two types of pools: six long-duration pools in which water was lost only to evaporation and six short-duration pools from which he removed water so the pools would dry after two weeks (Pfennig 1992b). Each pool received either 36 omnivore or 36 carnivore tadpoles. Pfennig found similar rates of survival to metamorphosis (16–25%) for each morph in both long- and short-duration ponds. Thus some omnivores were able to accelerate development to allow them to metamorphose in rapidly drying ponds. Next he took these toads and raised them for 2 months on a restricted diet to investigate postmetamorphic survival – survival as juvenile toads. He predicted that carnivores from long- and short-duration pools would survive equally well as juvenile toads, because even short-duration pools provided ample time for carnivore development. In contrast he predicted that omnivores from long-duration pools would survive well as juvenile toads, relying on their large fat reserves, while omnivores from short-duration pools would survive very poorly as juvenile toads, because accelerating development during the tadpole stage provided insufficient time for them to build up sufficient fat reserves to survive the postmetamorphic environment.

The results of this experiment support Pfennig's hypothesis. Juvenile toad survival rates were actually slightly higher if they metamorphosed as carnivores from short-duration than long-duration pools, but these differences were not statistically significant. In contrast, juvenile toads survived extraordinarily well if they metamorphosed as omnivores from long-duration pools and extraordinarily poorly if they metamorphosed as omnivores from short-duration pools (Figure 8.15). Pfennig concluded that postmetamorphic survival was *condition-dependent* in omnivores, which survived best when given the opportunity to build up sufficient fat reserves before metamorphosis.

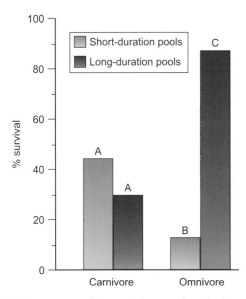

Figure 8.15 Postmetamorphic survival rates of toads that metamorphosed from 56 carnivorous and 33 omnivorous tadpoles that developed in short-duration and long-duration pools. Different letters above the bars indicate statistically significantly different survival rates among treatments.

Thinking ecologically 8.3

Given what you know about spadefoot toad survival and development, would it be adaptive for a few tadpoles to become carnivores under conditions of low (but measurable) levels of fairy shrimp density? Similarly, would it be adaptive for a few tadpoles to become omnivores under conditions of very high shrimp density? Explain your reasoning.

Pfennig's research shows that animals are adapted to different patterns of development in relation to environmental conditions. But there is much more to life history traits than developmental patterns. Are organisms also plastic in other life history traits, such as timing when to breed? Let's consider one study that addresses that question.

Environmental variation and phenological phenotypic plasticity

For birds, timing their breeding has important implications for their reproductive success. Many species of insectivorous birds achieve highest reproductive success by timing their breeding so that their offspring hatch and grow at a time that corresponds to the peak abundance of their insect prey. Timing is important for insect survival and reproductive success as well. The larvae of many leaf-eating insects rely on young shoots, which are high in nutrients and have low levels of toxins, to sustain them through development. If warmer spring temperatures lead to early budburst, and if insect abundance corresponds to the abundance of young shoots, birds that lay their eggs earlier in the season should have higher reproductive

success than birds that lay eggs that hatch too late to feast on the abundant insect prey population.

Anne Charmantier and her colleagues (2008) wanted to know whether the great tit, *Parus major*, could adaptively adjust its egg-laying date to increased spring temperature, presumably to coincide with peak prey availability for their offspring. Researchers have been studying a population of great tits in Whytham Wood, UK, since 1961, collecting reliable data on temperature, egg-laying dates, and the abundance of the birds' primary prey, larvae of the winter moth, *Operophtera brumata*. One important finding relevant to the researchers' question is that since 1980, early spring temperatures have been increasing significantly at Whytham Wood (Figure 8.16).

Two nonexclusive hypotheses could explain why birds might lay eggs at an earlier date. First, advancing egg laying could be a microevolutionary response to natural selection. If this were true, in a warming environment like Whytham Wood, birds that laid their eggs earlier in the spring would have higher fitness than birds that laid their eggs later in the spring, and over time there would be a genetic change in the population favoring early egg laying. The second hypothesis is that the great tit population may have genetically based phenotypic plasticity, so that individuals could (presumably unconsciously) adjust their egg-laying dates to the warming environment. Like the microevolution hypothesis, the phenotypic plasticity hypothesis also predicts that in a warming environment, birds that laid their eggs earlier in the spring would have higher fitness than birds that laid their eggs later

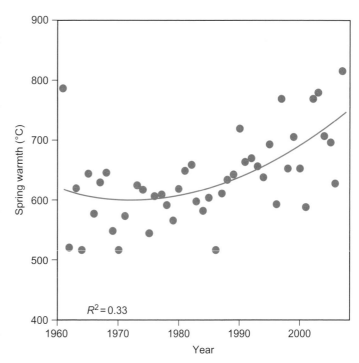

Figure 8.16 Spring warmth at Wytham Wood from 1961 until 2007. Spring warmth is the sum of the daily maximum temperatures (°C) from March 1 to April 15.

in the spring. Though the ability to respond is genetically based, this hypothesis does not predict that the population is evolving over the years of the study. Rather, birds have the plasticity to adjust to the warmer environment.

Returning to Figure 8.16, you will note that there is considerable year-to-year variation in early spring temperatures at Whytham Wood. In contrast to the microevolution hypothesis, the phenotypic plasticity hypothesis predicts that birds should be able to adjust to the highly variable year-to-year fluctuation in early spring temperatures. During unusually warm springs, birds should lay their eggs earlier, and during unusually cold springs, bird should lay their eggs later. If birds had this level of phenotypic plasticity, there would be a very good fit between mean laying date and spring warmth. The phenotypic plasticity hypothesis predicts that the fit between mean laying date and spring warmth (Figure 8.17A) would be substantially better than the fit between mean laying date and year (Figure 8.17B), which is precisely what the researchers discovered.

By working with a marked population of iteroparous individuals, the researchers could follow the egg-laying dates of females over the course of several years, which varied substantially in spring warmth. Females laid eggs earlier during warm springs and later during cold springs, regardless of their laying date in the previous year. Thus phenotypic plasticity allows the great tit population to adjust egg laying to both climate change and climate variability.

In the two studies of phenotypic plasticity we've discussed thus far, the expression of a plastic trait depends on conditions that an individual experiences in its present environment. But can the environmental conditions experienced by parents somehow influence the life history traits that develop in their offspring? *Transgenerational plasticity* (TGP), in which parents precondition the life history traits of offspring to a particular environment, would be favored in an environment where the parental environment reliably predicts the environment that will be experienced by their offspring. Let's see if TGP actually occurs in the natural world.

Transgenerational plasticity

The moor frog, *Rana arvalis*, occupies a broad range of habitats from western Europe across Asia to eastern Siberia. In central Sweden, adult frogs migrate to breeding ponds immediately after the ice melts in early spring, and females lay one clutch per year directly in the water over a period of 3–4 days. Tadpoles hatch after 2–3 weeks, and metamorphose after 2–3 months.

Alex Richter-Boix and his colleagues (2014) wanted to know whether parent frogs convey instructions to their offspring on how fast they should grow, based on when in the spring fertilization actually occurs. They reasoned that natural selection would favor tadpoles that grew more quickly if they were

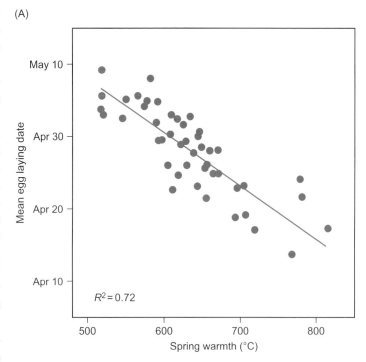

(A)

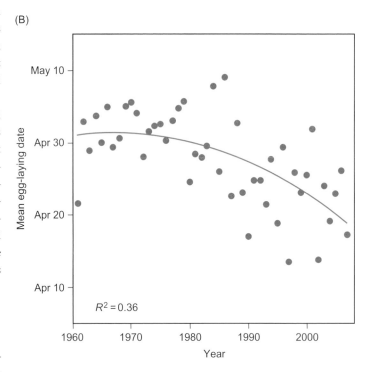

(B)

Figure 8.17 A. Mean egg-laying date in relation to spring warmth. B. Mean egg-laying date in relation to year.

born later in the spring. The researchers collected 28 pairs of moor frogs from two ponds, immediately transferred them to a laboratory, and stored them at 4°C. They established two phenology treatments: (1) delayed breeding and (2) delayed hatching. The researchers stripped eggs from 14 females within 12 h of getting them to the lab, and delayed stripping eggs from the other 14 females for an additional 12 days. Stripped eggs were immediately fertilized and half of each clutch was transferred to 16°C water which stimulated rapid development. The other half of each clutch was kept at 4°C for 6 days following fertilization, which slowed embryonic development, and mimicked a cold snap that could occur in natural conditions. Doing so allowed Richter-Boix and his colleagues to set up four experimental treatments: (1) normal breeding and hatching times, (2) normal breeding/delayed hatching times, (3) delayed breeding/normal hatching times and (4) delayed breeding/delayed hatching times.

Once the four experimental treatments were established, all tadpoles were raised at the same temperature and photoperiod, and received the same amount of food. The researchers weighed each tadpole when it hatched and followed its development until metamorphosis. Consequently, for each tadpole, they could calculate the duration of the larval period, the mass at metamorphosis, and the daily growth rate.

Both delayed breeding and delayed hatching substantially shortened the larval period (Figure 8.18A). At the same time larval growth rates increased to such a great extent that larval mass at metamorphosis was actually greatest in tadpoles which experienced delayed breeding and delayed hatching, despite the shorter larval period (Figure 8.18B and C).

Based on what you've learned thus far about phenotypic plasticity, it should not surprise you that embryos that experience a cold snap after fertilization are able to adjust to delayed development by accelerating their growth rates. However the situation is very different with delayed breeding. The embryos did not even exist at the time of the breeding delay, yet they somehow adjusted adaptively to the delay, even though there were no environmental cues to inform them that their development was behind their cohorts. Somehow the mothers must be preconditioning the eggs with information about when in the season they are being fertilized, so that their offspring accelerate development if they are produced late in the spring.

Studies of spadefoot toads, great tits, and moor frogs demonstrate considerable within-species plasticity in traits influencing reproduction and development – life history traits. Conservation biologists hope that some plasticity will, in general, promote species survival. In fact Richer-Box and his colleagues argue that transgenerational plasticity might help endangered species adjust to rapid climate change, because it allows parents to provision their offspring with an adaptive program that promotes survival and reproduction in the environment they are most likely to experience.

Let's reconsider the prognosis of the endangered North Atlantic right whales, by comparing their life history traits to a sister species, the southern right whale. This comparison enables us to understand how life history traits are important for species survival.

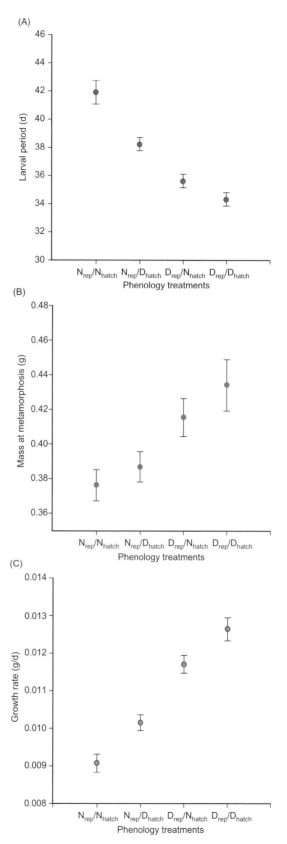

Figure 8.18 Effects of phenology treatments on moor frog larval developmental life history traits. A. Duration of larval period. B. Larval mass at metamorphosis. C. Larval growth rate. N_{rep} = normal reproduction, D_{rep} = delayed reproduction, N_{hatch} = normal hatching, D_{hatch} = delayed hatching. Error bars indicate ±1 SE.

REVISIT: The status of the North Atlantic right whale

In the past decade, the abundance of the North Atlantic right whale has climbed steadily but painfully slowly (Figure 8.19). While accurate data are more difficult to get for the southern right whale, its population has increased at a more rapid rate, and three of the largest populations are projected to double in 10–12 years. Scientists estimate the total southern right whale population exceeds 10 000 individuals. While this sounds like a lot of whales, it is still a small fraction of the estimated 70 000 that existed before commercial whaling wiped out most of the population in the nineteenth century. Let's explore why these differences exist between the North Atlantic and southern right whales, and determine whether we can use information from the rapid southern right whale recovery to understand the slow recovery in North Atlantic right whales.

Population recovery is influenced by reproductive output, survival rates, and the number of individuals immigrating into or emigrating out of the population. Unfortunately, collecting accurate data on all of these factors is very difficult in right whales. We will limit our discussion to factors influencing survival rates and reproductive output, the components of tradeoffs in life history traits. Most data come from aerial surveys of populations, which identify individuals by unique markings on their tails.

Survival rates in the northern species vary in relation to human activities, and also to unknown factors. As one example, between 2004 and 2005, observed North Atlantic right whale mortality rates tripled, including deaths of six pregnant females – four that were hit by ships, one that was entangled in fishing gear, and one that died of unknown causes. Given the low pregnancy rate, this human-induced mortality was devastating (Kraus *et al.* 2005). Since 2005, mortality rates have declined somewhat, though not steadily.

Policy changes may be playing a role in reducing North Atlantic right whale mortality. In 2008, the National Marines Fisheries Service established speed restrictions of 10 knots (18.5 km/h) on all ships greater than 65 feet (19.8 m) in length at locations with high right whale

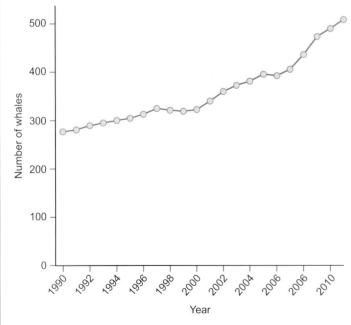

Figure 8.19 Population estimate of North Atlantic right whales since 1990.

densities. In addition, in 2009 the National Oceanic and Atmospheric Administration defined "areas to be avoided" (ATBA) off of the northeast coast of the United States. Avoiding these areas was voluntary. Initially, compliance with the speed limit restriction and ATBA was very spotty, but it has improved in recent years. Surprisingly, economic stress may be playing a role in compliance, as increased fuel costs are leading to declining ship speeds and may be reducing the number of collisions (Scott Kraus, personal communication).

Meanwhile, the number of annual births is increasing in the North Atlantic right whale, from a mean of 12 in the 1990s to between 25 and 30 since 2000. Researchers don't know why this has happened; perhaps the whales are less stressed because fewer large ships are threatening them, and perhaps the noise level has decreased and is interfering less with their ability to communicate with each other. One important life history trait, the interbirth interval (the time between births) has declined in recent years from above 5.0 years to about 3.4 years. But issues with birth rates still plague this recovering population. About 10% of the females never give birth, and a significant number of females stop being reproductively active at a very young age. Both of these trends lead to fewer calves entering the population.

The outlook for southern right whales is much brighter. Peter Best and Hirohisa Kishino (1998) determined mortality rates based on aerial photographic surveys of known individuals, and estimated a constant mortality rate of about 2% per year. This means that a southern right whale from one year will have approximately a 98% chance of still being alive one year later. Reproduction is also on the rise in southern right whales. As one example, a survey of southern right whales by Best and his colleagues off the coast of South Africa sighted 27 cows with calves in 1979, and 111 cows with calves in 1998 (Best *et al.* 2001). The interbirth interval for the southern right whale is estimated at about 3.1 years, and there have been no reports of widespread non-reproductive females in southern populations.

Both right whale species are protected, but one is responding well, while the other continues to languish. In contrast to the relatively secluded waters that are home to the southern right whale, the heavily populated and developed environment of the North American right whale interferes with its survival, reproduction, and possibly development. Unfortunately, the North Atlantic right whale seems to lack sufficient behavioral phenotypic plasticity to avoid shipping lanes and fishing lines that are important sources of mortality and stress in its environment. On the bright side, additional protective legislation for the North Atlantic right whale appears to be increasing North Atlantic right whale reproductive rates by reducing its interbirth interval to a level that is comparable to its southern right whale counterpart.

SUMMARY

A species' behavioral, developmental, and reproductive life history will influence how quickly it can recover after a population crash. Some species can recover very quickly, while others, such as the North Atlantic right whale, cannot recover quickly, because even under ideal conditions they develop slowly and have very low reproductive rates. Ecologists have described various life history classification schemes that identify important tradeoffs in resource allocation, and focus attention on interesting life history questions.

The quantitative relationship between metabolic rate and body size can help ecologists understand some life history tradeoffs, such as the relationship between number and size of offspring. There is a fundamental tradeoff between parental investment in any one reproductive event and the number of lifetime reproductive events, which in some cases can lead to a semelparous reproductive life history. Variable environments can select for phenotypic plasticity, which can lead to organisms with similar genotypes expressing alternative behavioral, developmental or reproductive life history traits. In some cases, phenotypic plasticity may help species adjust to rapidly changing environmental conditions, including climate change.

FURTHER READING

Karban, R. 2014. Transient habitats limit the development time for periodical cicadas. *Ecology* 95: 3–8.

Nice discussion of the costs and benefits of extreme semelparity in cicadas, which addresses the question of why cicadas don't take even longer to mature. It is interesting to consider the challenges of studying a species that reproduces only once every 17 years!

Richter-Boix, A., Orizaola, G., and Laurila, A. 2014. Transgenerational phenotypic plasticity links breeding phenology with offspring life history. *Ecology* 95: 2715–2722.

Uses easily understood experimental design to provide evidence for transgenerational phenotypic plasticity. Nice discussion of possible mechanisms, including a circannual clock.

Standen, E. M., Du, T. Y., and Larsson, H. C. 2014. Developmental plasticity and the origin of tetrapods. *Nature* 513: 54–58.

This paper proposes that environmentally induced phenotypic plasticity in fish promoted one group of air-breathing fish to evolve improved methods of locomotion on land, ultimately leading to the evolution of tetrapods. The researchers use an interesting experimental design in which they raise one group entirely in water and a second group entirely on land, and study differences in morphology and behavior between the two treatments.

Stokstad, E. 2005. What's wrong with the Endangered Species Act? *Science* 309: 2150–2152.

Science meets politics in this essay. A thoughtful review on how proposed changes to the Endangered Species Act would influence the prospects for recovery of endangered species. Though the details have changed since this article was written, the same political, economic and scientific perspectives appear in today's discussions.

END-OF-CHAPTER QUESTIONS

Review questions

1. What life history traits do North Atlantic right whales have that make them unlikely candidates for quick recovery following protection? Why is it important for people to understand the relation between potential recovery rates and life history traits?
2. What are two classification schemes devised by ecologists to distinguish between different groups of species. What are strengths and weaknesses of these approaches?
3. Explain in detail the relationship between number and size of offspring. Why does this relationship exist?
4. Describe two models explaining the evolution of semelparity in plants. What is the evidence supporting each model?
5. Review the data in Figure 8.18. Which findings are evidence for transgenerational plasticity, and which findings support the more standard within-generational plasticity? Explain your answer.

Synthesis and application questions

1. One hypothesis that we did not discuss in the text for the evolution of the iteroparity/ semelparity dichotomy is the bet-hedging hypothesis, which proposes that natural selection might favor iteroparity when there is a highly variable environment, so that an unsuccessful reproductive event could be replaced with another attempt. In contrast. natural selection would favor semelparity when there is a relatively constant environment. From what you learned about pioneer species in Chapter 5, does this hypothesis seem plausible to you? Why or why not?
2. Consider an Atlantic salmon female beginning her spawning run upriver to her spawning grounds. How might the following environmental and physiological factors influence her allocation of resources to reproduction: (a) distance to spawning site, (b) abundance of predators along the route, (c) size of the female, (d) age of the female? What other factors could influence her allocation of resources to reproduction?

3. Although Lesica and Young's (2005) studies support the demographic model, there is much more that could be done with this system to test the demographic model more thoroughly. What would be your next step if you were to continue their studies of life history strategies within this species?

4. Other hypotheses for periodicity being adaptive in cicadas include (a) loud calls scare away predators, or (b) females are more attracted to large aggregations. What data would you collect to test the predictions of these two hypotheses?

5. Our discussion of life history evolution considered only a few of the relationships between life history variables. Based on your understanding of life history tradeoffs, what would be your intuitive prediction about the correlations between each pair of variables listed below (positive, negative, or no correlation)? What is the biological basis for each prediction?
   (a) maximum lifespan and age at first reproduction in most organisms
   (b) seed size and seed number in plants
   (c) seed size and seed survival rates in plants
   (d) number of cones produced and the width of annual growth rings in pine trees
   (e) tree height and trunk diameter
   (f) age and parental investment per offspring in female deer.

6. Consider an iteroparous species that usually has three or four reproductive efforts per lifetime. Would you expect individuals that survive to their fourth reproductive effort to show any differences in amount of resources allocated to offspring of their fourth reproductive effort in comparison to offspring of earlier reproductive efforts? What might these differences be?

7. As presented, the study by Richter-Boix and his colleagues (2014) on transgenerational phenotypic plasticity assumes that metamorphosing at the appropriate time is adaptive for moor frogs. What are some advantages for a moor frog to metamorphose at the same time as other moor frogs? How might you test your hypotheses?

8. Do you think there are certain biomes in which phenotypic plasticity might be more prevalent? Explain your reasoning.

Analyse the data 1

Walter Koenig and Andrew Liebhold (2005) were interested in whether cicada emergences had a positive effect on bird population size. They used data from the North American Breeding Bird Survey to compare bird population size in 24 bird species in the year before a cicada emergence (year −1 in table below), the year of a cicada emergence (year 0) and the year after a cicada emergence (year +1).

	Bird species with statistically significant changes in population size comparing:	
	Year −1 to year 0	Year 0 to year +1
# of species increasing	1	8
# of species unchanged	19	14
# of species decreasing	4	2

What concerns would you need to address when using this type of data set? What factors might be responsible for why some species' population size increased in the year of cicada emergence, while others increased the year after cicada emergence? Why might some species populations decrease in size the year after a cicada emergence? Is the trend for more bird species to increase in the year after a cicada emergence statistically significant?

Chapter 9
Distribution and dispersal

INTRODUCTION

One of the most basic challenges for population ecologists is describing a population's distribution – its geographical range. Ecologists can infer past distributions of many different species in arid regions of North America by analysing the vegetation remains stored by ancient packrats (genus *Neotoma*). Packrats earn their names by scavenging pieces of wood, rock, dung, bone, feathers, and any other material from the nearby environment, which they use to create houses, and they build nests (middens) inside the houses. Careless campers who leave salt-shakers, gloves, or forks on their picnic tables are unwittingly contributing to the historical data stored in packrat middens. The materials are cemented together with urine; the evaporating urine may help keep the temperature relatively cool for the packrats, and it inadvertently benefits paleoecologists by helping to preserve the organic materials stored in the packrat's midden. Numerous middens have been dated using ^{14}C isotope decay rates, which allow age estimates back to 100 000 years ago.

Part of the challenge for determining distributions is that individuals in many species are small and difficult to locate, others live underground, while others fly, run, or swim away before they can be properly identified. A second aspect of the challenge is that all species have dynamic distributions, meaning that their distributions change over short and long time scales. Consequently, a complete study of distribution patterns must consider past, present, and future distributions. We begin this chapter with a case study that uses paleoecological information from packrat middens to evaluate historical changes in the distribution of one tree species, the pinyon pine (*Pinus edulis*) in the arid western interior of the United States. We then discuss how ecologists describe present-day distributions on a small and large spatial scale, and consider the underlying causes of present-day species distributions, which include the effects of abiotic factors such as soils and water, and the impact of biotic factors such as competition and predation. At the end of this chapter we will explore how ecologists use models to make predictions of whether a species may expand its range into an area outside of its present distribution.

KEY QUESTIONS

9.1. How are individuals distributed within populations?

9.2. How do species distribution patterns change over time?

9.3. In what ways do abiotic factors influence the distribution of populations?

9.4. In what ways do biotic factors influence the distribution of populations?

9.5. How does ecological niche theory help ecologists understand a species' current distribution and predict its future distribution?

CASE STUDY: Using packrat middens to infer past distributions of pinyon pines

Stephen Gray and his colleagues (2006) used fossil evidence in packrat middens to unravel the history of pinyon pines at their field site, Dutch John Mountain (DJM) in northern Utah, at the northern extent of the current range of pinyon pines (Figure 9.1). Note that the DJM site is isolated from the rest of the pinyon pine distribution. The researchers proposed two hypotheses to explain this disjunct distribution: (1) the DJM site was recently colonized over a very long distance, or (2) the DJM site is a surviving relic of what had been a more continuous distribution of pinyon pines in northern Utah. The second hypothesis is unlikely, because sites between DJM and sites directly south show no evidence of dead pinyon pines, which are known to persist for several thousand years in arid habitats. Let's see what packrat middens can teach us about the historical distribution of pinyon pines in this region, and how changes in climate might influence this distribution.

The researchers collected pinyon macrofossils (primarily twigs) and also pinyon pollen from 60 packrat middens at DJM that dated to the last 12 000 years. None of the 56 middens constructed before 1430 had evidence of pinyon macrofossils, while the four middens built after 1430 had substantial pinyon macrofossils. Furthermore, only trace amounts of fossil pollen were found in any of the middens until 1100, and amounts above 10% did not appear until 1400. Both macrofossil and pollen evidence support a dramatic increase in pinyon pine at DJM around 1400.

In addition, Gray and his colleagues collected tree core samples and counted the tree rings from 52 living pinyon pine trees, which gave them an accurate estimate of each tree's age. The oldest pinyon pine germinated in 1246. The next oldest pinyon pine dated from 1319 and lived nearby. By the early 1400s, pinyon pines were well established in nearby slopes, and they dominated the entire landscape by 1500. The researchers wondered why there was a lull in **recruitment** – the addition of new individuals to the population – between 1246 and 1319, and why recruitment spread so quickly after 1319. They hypothesized that dry weather could have prevented recruitment. By measuring the width of the tree rings, they could infer precipitation patterns over the last 800 years. In general,

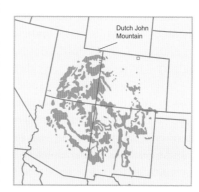

Figure 9.1 Current range of pinyon pines in southwestern USA.

wider tree rings are associated with higher precipitation totals during the period of growth (Figure 9.2A).

The ecologists' analysis of tree-ring width showed that the tree from 1246 was recruited just before a prolonged stretch of unusually dry weather, the Great Drought, which it was able to survive, in part because it germinated in a wet microhabitat. No trees were recruited during the 50 years of the Great Drought, but beginning in 1300, there was a prolonged stretch of unusually wet conditions (the fourteenth-century pluvial), which coincided with pinyon pine becoming the dominant tree in the region (Figure 9.2B)

The researchers suggest that the Great Drought might ultimately have helped pinyon pine recruitment by killing off other competitors, such as juniper trees, in the region. But they caution that a correlation between climate and recruitment at one site needs to be bolstered with numerous other studies before conclusions can be drawn about the linkage between climatic patterns and recruitment. The relationship between climate change and distribution patterns is very important for modern researchers, who are trying to make predictions about how impending climate change will influence the distribution and abundance of species in ecological communities. We begin by considering present-day distributions on a small spatial scale.

9.1 HOW ARE INDIVIDUALS DISTRIBUTED WITHIN POPULATIONS?

Population ecologists study the dispersion, or spatial pattern of distribution of individuals within a population. Over a small spatial scale, individuals in a population may have a random, uniform, or clumped dispersion. Each of these patterns may be influenced by differences among individuals within the population and by biotic and abiotic factors that influence these individuals.

Dispersion patterns: random, uniform or clumped

In a random dispersion, each individual has an equal probability of occupying any given space in the habitat (Figure 9.3A). Thus the location of one individual is independent of the location of another individual. Random dispersions are unusual in the natural world, because individuals may be attracted by resources, such as food, nutrients, or shelter, and may be attracted or repelled by individuals of the same or other species.

In a uniform (or regular) dispersion, individuals tend to avoid other individuals, with the result that they are spaced more evenly than in a random dispersion (Figure 9.3B). In animals, territorial interactions can lead to a uniform dispersion, though no dispersion is ever completely uniform. In some plant species, competition for scarce resources, such as water in desert ecosystems, may cause individuals to be dispersed somewhat uniformly.

Most commonly, organisms form a clumped dispersion within their habitat. A variety of biotic and abiotic factors will attract

Wet Dry

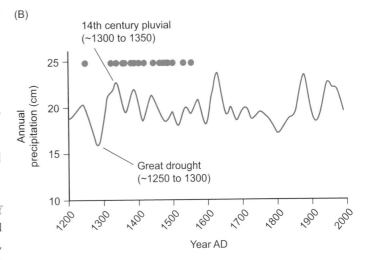

Figure 9.2 A. Variation in tree-ring width is used to infer annual precipitation. B. Age for the oldest pinyon pine tree at DJM, and the oldest pinyon pines (circles) in each of the sampled sites at DJM, in relation to the mean annual precipitation. Note the Great Drought between 1250–1300 and the fourteenth-century pluvial (period of increased rainfall) that followed.

individuals either to each other, or to the same type of habitat (Figure 9.3C).

The same species may have different patterns of dispersion at different stages of development or in response to changes in biotic or abiotic factors that can influence dispersion patterns. The South American sea lion, *Otaria flavescens*, provides a good example of how dispersion patterns may vary within and between populations.

Factors influencing dispersion patterns

The South American sea lion occupies coastal marine habitats from the southern tip of South America northward up the Atlantic and Pacific coasts, extending into tropical habitats. During the

Figure 9.3 A. Random dispersion of avalanche lilies, *Erythronium montanum*, in Olympic National Park in Washington, USA. B. Uniform dispersion within colony of king penguins, *Aptenodytes patagonicus*, on South Georgia Island. C. Clumped dispersion of blue mussels, *Mytilus edulis*, along the coast of France.

breeding season, males and females gather at traditional rooker-ies, where the females give birth, nurse their young, and – about 6 days after giving birth – mate with one or more males, to begin the cycle anew. Males defend very small territories along the high water mark, and mate with receptive females in these territories.

Claudio Campagna and Burney Le Boeuf (1988) studied a population of South Atlantic sea lions over six breeding seasons at Punta Norte, Peninsula Valdes, Argentina. In peak breeding season, territorial males defended small linear territories along the high tide line, maintaining a distance of about 4 m from each other, resulting in a regular dispersion pattern. Females attempted to approach these large territorial males, but they first needed to pass through a gauntlet of usually smaller itinerant (non-territorial) males that were dispersed randomly either lower in the tidal zone, or else higher up on the shoreline above the

territorial males. These itinerant males attempted to force arriv-ing females to mate with them, usually unsuccessfully. Some-times itinerant males chased females back into the water up to 4 km off shore.

Ultimately, females were usually able to associate themselves with a territorial male, who vigorously defended them against nearby competitors by intimidating his neighbors from approaching too closely, and coercing his small harem of females into staying nearby. This female dispersion pattern was highly clumped around the territorial male that claimed them; an average of 3.4 females associated with each territorial male during the peak breeding season in January and February. Overall, we observe three dispersion patterns in relation to individual differences: large territorial males were uniformly dispersed along the high tide line, itinerant males were ran-domly dispersed either below or above the line of territorial

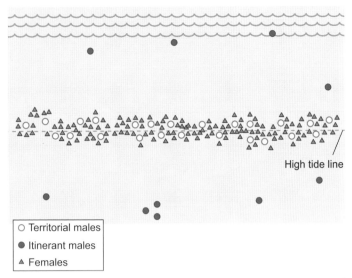

Figure 9.4 Dispersion patterns in South American sea lions. Territorial males are distributed uniformly along the high tide line. Females are clumped around each territorial male. Itinerant males are randomly distributed away from territorial males.

males, and females formed a clumped harem surrounding each territorial male (Figure 9.4).

Farther north, off the coast of Peru, Karim Soto and Andrew Trites (2011) described a spacing system that shared some similarities with the Argentinian population, but also had some substantive differences. As in the Argentinian population, males defended small linear territories, with only 2.3 m (SE = 0.1 m) separating individuals (Figure 9.5A). Air temperatures are much warmer this far north (14°S latitude) and keeping cool is a major challenge for these large, well-insulated animals. Rather than lining up along the high tide line like the Argentinian population, these males shifted position in relation to the tide, positioning themselves right along the water line, so they could be cooled by the moist sand and occasional waves that came their way.

Fights between territorial males in coastal Peru were extremely rare, while most aggression was directed toward itinerant males. The female to male sex ratio was 17.5:1 – much greater than the 3.4:1 sex ratio in the Argentina population. Females were free to move around the beach, and were never restrained or defended by males. They were most spread out during the afternoon and evening, when the western and northern cliffs shaded them from the hot sun. Even then, females were more densely packed on the north beach (Figure 9.5B). But in the morning, when the sun struck the entire beach, females clustered along the shoreline, or jumped into the water to cool off (Figure 9.5C). Exposure to sun profoundly influenced dispersion of itinerant males, who were uniformly dispersed above the shoreline when the area was shaded in the afternoon (Figure 9.5B). But when direct sun struck the beach during the morning hours, itinerant males were forced to abandon the beach and escape into the cooler water.

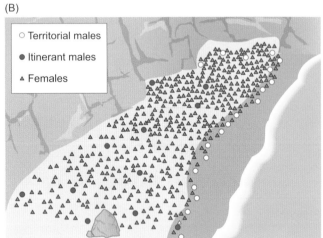

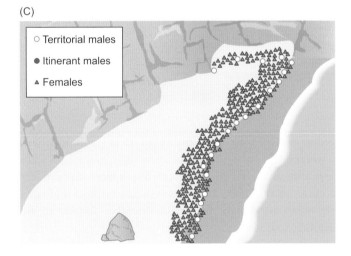

Figure 9.5 A section of shoreline showing regular dispersion of territorial males (large with dark brown bodies) along the shoreline, facing away from the water. Females and pups are clumped near the territorial males. B. Dispersion of South American sea lions during the cool shaded hours. Territorial males form a uniform dispersion along the shoreline, itinerant males form a regular dispersion between the shore and the cliff walls, while females are clumped toward the north beach. C. Dispersion of South American sea lions during the hot sunny hours. Territorial males maintain their uniform distribution, but females are compressed into the region near the shoreline, while the itinerant males are forced to abandon the beach.

Table 9.1 Examples of large groups of dispersing animals, where they were sighted, group size, and total biomass of the group (Holland *et al.* 2006).

Class	Species	Location	Number of individuals	Biomass (kg)
Insect	Dragonfly (*Aeshna bonariensis*)	Argentina	5×10^9	3.6×10^6
Insect	Monarch butterfly (*Danaus plexippus*)	Winters in Mexico	1.5×10^8	5.5×10^4
Insect	Desert locust (*Schistocerca gregaria*)	Africa, Middle East, Asia	1.0×10^{10}	1.8×10^8
Mammal	Wildebeest (*Connochaetes taurinus*)	Kenya, Tanzania	1.3×10^6	2.5×10^8
Mammal	Mexican free-tailed bat (*Tadarida brasiliensis*)	New Mexico, USA	2.0×10^7	2.7×10^5
Bird	Lesser sandhill crane (*Grus canadensis canadensis*)	Platte River	4.5×10^5	1.3×10^6

 Thinking ecologically 9.1

The Argentinian sea lion mating system is best defined as female defense polygyny, in which males defend a small harem of females and have exclusive mating rights. In contrast, the Peruvian sea lion mating system is best defined as lek polygyny, in which males defend a tiny space and females can move about freely, selecting their male. Why do you think these two distinct mating systems are found, respectively, in these two populations? Explain your reasoning.

Over a large spatial scale, populations tend to have a clumped distribution. Within the geographical range of a species there are usually hotspots with high population density, and also areas of very low population density. Both abiotic and biotic factors can influence these large-scale variations in population density. In species that can easily move around, the spatial distribution of individuals may fluctuate substantially in relation to the time of year.

9.2 HOW DO SPECIES DISTRIBUTION PATTERNS CHANGE OVER TIME?

The distribution of any species is always changing. The research at Lake Yoa (see Chapter 2) and at DJM describes distributional changes over relatively long time scales. But ecologists also study distributional changes that result from movement patterns over much shorter time scales. Dispersal is the movement of individuals from one location to another location. Migration is the back-and-forth intentional movement of individuals or populations between two locations. Thus migration is one type of dispersal.

Migration

We have already discussed the impressive migration patterns of several vertebrate species, including wildebeests in the Serengeti (Chapter 1), and European blackcaps (Chapter 3). From the perspective of sheer numbers on the move, we might be even more impressed by the migration behavior of several groups of insects (Holland *et al.* 2006) (Table 9.1). Insect migration differs from mammalian or avian migration in that, as far as we know, insects do not survive the entire journey. In some cases, for example the huge swarm of dragonflies sighted over Argentina (Table 9.1), we don't even know whether the swarm made a return journey. Were these insects truly migrating, or merely dispersing to a new environment (Martin Wikelski – personal communication)? And why were they doing so?

 Thinking ecologically 9.2

Robert Russell and his colleagues (1998) published a paper estimating the number of migrating green darner dragonflies from several swarms in the United States. One observation of 1.2 million dragonflies was made from the fourth floor office window of the Field Museum of Natural History in Chicago. How could you make such an estimate armed only with a watch and measuring tape?

Recent technological developments should help resolve some of the outstanding questions regarding insect migration. Martin Wikelski and his colleagues (2006) attached tiny microtransmitters to the thoraxes of 14 migrating green darner dragonflies, then tracked the signals from small aircraft (Figure 9.6). They learned that these dragonflies only migrate on some days, while they stay within a small area on other days. Migratory days usually follow two successive cold nights. The dragonflies' average migratory distance is about 60 km per day.

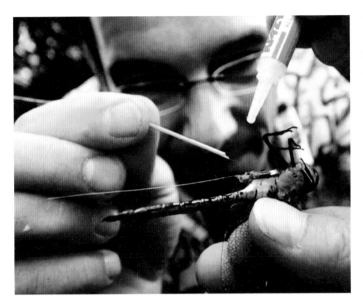

Figure 9.6. Green darner dragonfly, *Anax junius*, in the process of having a microtransmitter glued to the bottom of its thorax (top) and ready for takeoff (bottom).

Obviously there is a great deal more to learn about the proximate and ultimate cause of dragonfly migration. An ecologist studying the distribution of green darners must be aware of how distribution patterns shift in relation to time of the year. Under most circumstances, the view from the fourth floor at the Field Museum of Natural History in Chicago (see Thinking ecologically 9.2) will yield a density of 0 dragonflies per day. But every so often, that same view will yield a density of over a million dragonflies per day. Accordingly, studies of distribution of motile organisms require multiple observers and multiple observations.

Citizen scientists and the distribution of migratory species

About 70 years ago, Fred and Norah Urquhart initiated a program to study the fall migration process in monarch butterflies.

They recruited a group of volunteers to help with a *mark–recapture* study. The group captured butterflies in the northern range of their distribution, applied a unique tag to one wing of each butterfly, and then recaptured the marked insects later during migration. Of course, they recovered only a tiny fraction of the marked individuals, but each recaptured individual gave important information about the migratory route and the distances traveled during the fall migration (Urquhart and Urquhart 1978).

Several other organizations have developed a large network of citizen scientists to continue with the Urquharts' pioneering work. For example, project Monarch Watch was established in 1992 to continue the study of fall migration. Currently, over 2000 schools and organizations in the United States and Canada use the services of more than 100 000 volunteers (mostly students) to conduct mark–recapture studies of fall migrants. This vast network of data collection, coordinated by researchers at the University of Kansas, has helped clarify our understanding of monarchs' fall migration patterns (Figure 9.7A). The fall migration is relatively easy to study, because monarchs are abundant in the fall, and they tend to congregate in large clumps during their south-bound journey.

The fall monarch migration terminates in forests of the oyamel fir (*Abies religiosa*) in central Mexico's Neovolcanic Belt. These trees form a dense canopy, which helps keep the vast populations of butterflies relatively dry during occasional winter storms (Figure 9.7B). The climate at this high elevation is unique: minimum air temperatures are low but usually above freezing, and conditions are humid but usually not raining or snowing. These conditions are essential for the overwinter survival of the monarch butterflies, as low air temperatures result in low metabolic rates, which conserve lipid reserves, while high humidity allows the butterflies to conserve water. On warm winter days, butterflies can elevate their body temperatures sufficiently to forage for nectar and replenish energy supplies and water levels. A study by Karen Oberhauser and Townsend Peterson (2003) indicated that only a narrow range of temperature and moisture levels are suitable for monarch butterflies during the winter. Ideal minimum January temperatures are between 2 and 8°C, and ideal January rainfall totals are less than 1 cm.

The monarch butterfly's spring migration is more difficult to study, because of the insect's unique life history. The wintering population experiences high mortality, so the population size of the return spring migration is considerably smaller than the population size of the fall migration. A catastrophic storm in 2002 killed an estimated 273 million butterflies (about 80% of the entire migrant population) in two of the largest colonies, resulting in an unusually small population of returning butterflies the following spring (Brower *et al.* 2004). In spring, migrants exhibit very different behavior than they did during the fall migration. Spring migrants do

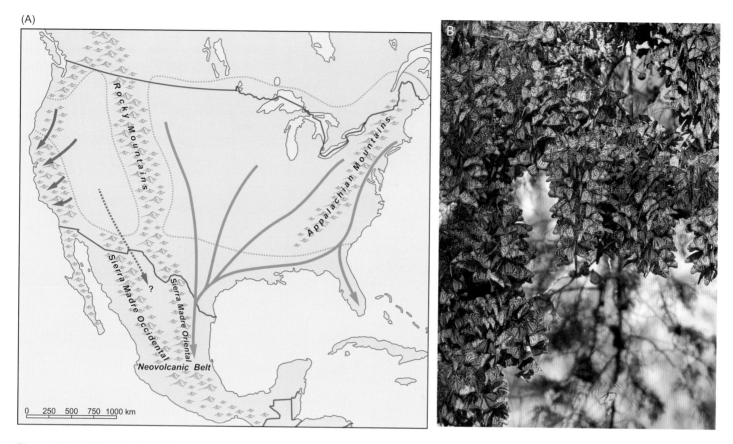

(A)

Rocky Mountains

Appalachian Mountains

Sierra Madre Occidental

Sierra Madre Oriental

Neovolcanic Belt

?

0   250   500   750   1000 km

B

Figure 9.7 A. Fall migration of monarch butterflies. The dashed line represents a track that is not well established. B. Vast populations of monarch butterflies in Sierra Chincua Sanctuary, Angangueo, Mexico.

not tend to aggregate as they work their way back north to many locations in the United States and Canada, so mark-recapture methods are even more difficult to carry out successfully then.

Journey North, an internet-based program established in 1994, is the complementary citizen scientist program to Monarch Watch. Despite a significant decrease in monarch population size, the number of monarch sightings during the spring migration has increased steadily over the years, as a result of the increase in citizen participation. About half of the participants are students in kindergarten through fifth grade (Howard and Davis 2011). When participants observe a monarch butterfly in the spring, they register this observation at the Journey North website. Collectively, these observations provide the data analysis team with the information they need to quantify the movement patterns of butterflies making the return journey (Howard and Davis 2004). Several important pieces of information have emerged from Journey North's data collection. For example, the research has supported the hypothesis that the majority of wintering monarch butterflies migrate northward to the southern United States, laying eggs along the migration routes. Their offspring then complete the journey northward to the northern United States and southern Canadian provinces (Figure 9.8A).

In addition, this research has allowed ecologists to explore important questions about the timing of the northern migration. For example, monarch butterflies depend on milkweed (*Asclepias* species) as their host plant. Ecologists wanted to know whether monarchs can time their arrival to coincide with milkweed availability, which can vary from year to year. The Journey North citizen scientists have provided some tantalizing preliminary data that addresses this question. March, 2012, was the warmest on record throughout many sections of the north central United States. Journey North's data showed that the first milkweed emerged in early April, about a month earlier than usual (Figure 9.8B). Journey North's data also showed that the spring monarch butterfly migration was much more advanced than in other years, with the first monarchs arriving in time to take advantage of early milkweed availability (Figure 9.8C). Whether early monarch butterfly arrival was a response to warm temperature, or a result of other factors, can only be ascertained by continuing to collect data in future years that vary in early spring temperature. This type of information would be very difficult to get without a large army of researchers spread across the continent.

In many cases, such as that of the monarch butterfly, migration is a regular periodic event, so a researcher has a pretty

(A)

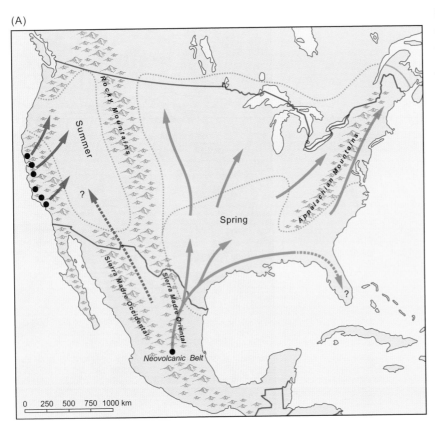

Figure 9.8 A. Northern migration paths based on data from Journey North. The breaks in the solid arrows indicate that a new generation of butterflies resumed the migration. The dashed line represents a track that is not well established B. First milkweed sightings in 2012 (warmest March on record) and 2013 (a normal March). C. First monarch butterfly sightings in 2012 and 2013.

(B)

(C)

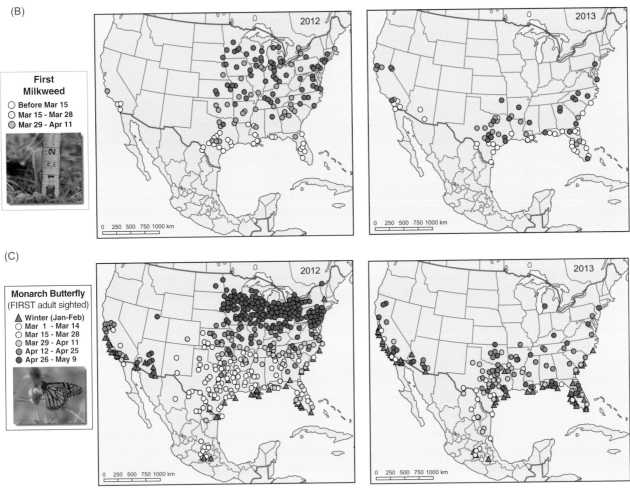

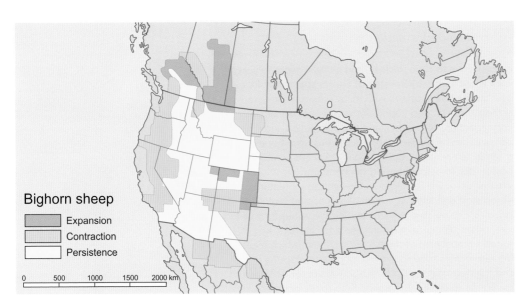

Figure 9.9 Expansion, contraction, and persistence of bighorn sheep distribution in North America.

good idea of when and where it will occur. But dispersal of a species to a new location is much less predictable, and thus more difficult to study. Let's consider one study that uses a somewhat unusual manner of data collection to address questions about large-scale changes in distribution for an important group of mammals.

Large-scale changes in mammal distribution

Andrea Laliberté and William Ripple (2004) were interested in the effects of human settlement on mammal distribution, so they investigated whether North American ungulates (hoofed mammals) had expanded or contracted their distribution or range in recent history. One problem they faced is that data on historical species distribution have not been rigorously collected. Their solution to this problem was to use range maps collected from *The Mammals of North America* by Hall and Kelson (1959), which summarized data based on actual field sightings beginning in the eighteenth century and depicted it in historical species' range maps. Map sources for current distributions were *The Smithsonian Book of North American Mammals* (Wilson and Ruff 1999) and *Mammals of North America* (Kays and Wilson 2002). Figure 9.9 shows how the distribution of bighorn sheep, *Ovis canadensis*, has changed based on these sources. Bighorn sheep have expanded their northern range along two fronts in British Columbia and Manitoba. They also have expanded into eastern Colorado. However, they have lost a lot of ground in northern Mexico, several locations in the Rocky Mountains, the Dakota Badlands, and along their former western distribution.

Laliberté and Ripple needed to quantify the data to determine whether bighorn sheep had expanded or contracted their range overall. Furthermore, they wanted to look at changes in distribution patterns in ungulates in general, so they could evaluate which species had changed the most over the past 200 years. They used GIS software to digitize the range maps from their

sources, so they could conduct comparisons within and among species (Dealing with data 9.1).

The researchers defined four variables:

% persistence = percent of former range still occupied
% expansion = percent of former range that has been expanded
% contraction = percent of former range that has been lost
% net change = % expansion − % contraction

For example, the bighorn sheep formerly occupied an area of 3262 km^2. The population added 489 km^2 based on the most recent data, primarily along its northern range. Thus the percent expansion is $489/3262 \times 100 = 15\%$. It also lost 1305 km^2 within its range, based on the most recent data, primarily in northern Mexico. Therefore the percent contraction $= 1305/3262 = 40\%$. The % net change = % expansion − % contraction $= 15\% - 40\% = -25\%$. Thus the researchers estimated that bighorn sheep distribution has contracted about 25% over the past 200 years. These data for ungulates are summarized in Table 9.2. This study highlights that many species of ungulates have smaller ranges than they did previously, while only one monitored species has expanded its range over time.

In addition to demonstrating that species distributions are dynamic, this study also highlights that we must be careful to use the appropriate spatial scale when collecting data on species distribution. Had observers confined their investigations to a few sites in British Columbia and Manitoba, they would have (erroneously) concluded that bighorn sheep were expanding their range in North America (see again Figure 9.9). It took a network of dedicated observers to compile these observations throughout the continent.

Bighorn sheep, and mammals in general, are excellent organisms for studies of distribution, because they are large, and many nonscientists want to see them and are happy to share their sightings with scientists. Many other species, such as small

Dealing with data 9.1 Geographic information systems (GIS)

A geographic information system allows researchers to store, display, and analyze *geographically referenced data* – data that describe the locations and characteristics of Earth's spatial features. The data are stored in computer databases and can be accessed with many different software applications. For ecologists, some of the most common geographically referenced data would include temperature, rainfall, soil type, solar radiation, vegetation, and the distribution and abundance of organisms. Each of these data sets can be represented in numerous ways. For example, we may represent temperature as mean annual temperature, or mean daily high temperature, or in the case of the saguaro cactus, the longest stretch of below freezing temperatures.

Many of the maps in this book were created using GIS technology. One important feature of GIS is that multiple types of data can be layered atop one another, which allows ecologists to explore spatial correlations between different variables. Ecologists can also combine the values for a variety of variables to create indices that have important biological meaning. As one example, Eric Sanderson and his colleagues (2002) created a global map of the human footprint, which measures human influence on the land surface. The researchers used four types of data to map the human footprint: human population density, land transformation (such as the conversion of grassland to agriculture, or forest to city), accessibility (as measured by the proximity to roads or other transportation media), and the extent of electrical power infrastructure. These data were collected by many other researchers beginning in the1960s; but much of the actual compilation and digitizing of the data occurred late in the 1990s or early in the twenty-first century. Sanderson and his colleagues incorporated these data onto a global map as overlaid grids of 1 km^2, and coded them into standardized scores based on their estimated contribution to human influence. Once this was accomplished, the researchers used a very simple mathematical algorithm to add the scores up for each 1 km^2 grid to create a global map of the human footprint, with higher scores corresponding to greater human influence (Figure D9.1.1).

Andrea Laliberté and William Ripple used this map as part of their analysis of range contractions of North American carnivores and ungulates. Limiting their analysis to the 17 species with the greatest range contraction (over 20%), they hypothesized that persistence of these species would be most common in areas with a small human footprint. In addition, range contraction would be most severe in areas with a large human footprint. By overlaying the range maps that we discussed in the text with the human footprint map, the researchers were able to support these two hypotheses. By using GIS technology in conjunction with data collected by many other field researchers, Laliberté and Ripple were able to help conservation ecologists identify the factors influencing range contraction, and to design approaches to address the continued range contraction of many threatened North American ungulates and carnivores.

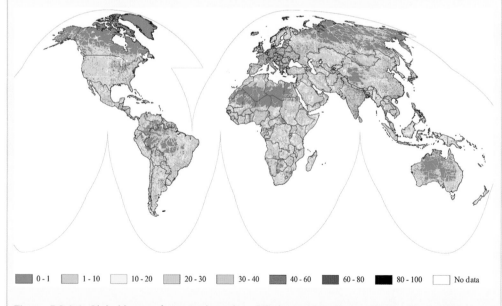

0 - 1   1 - 10   10 - 20   20 - 30   30 - 40   40 - 60   60 - 80   80 - 100   No data

Figure D9.1.1 Global human footprint based on GIS data on human population density, land transformation, accessibility, and electrical power infrastructure. Numbers are relative scores with 100 indicating the greatest human footprint.

Table 9.2 Percentage persistence (no change), expansion, contraction, and net change for 11 species of North American ungulates over the past 200 years. Note: Mean values have round-off error.

Species	% persistence	% expansion	% contraction	% net change
Elk	23	3	77	−74
Pronghorn	36	0	64	−64
Dall's sheep	36	10	64	−54
Musk ox	65	4	35	−31
Mountain goat	57	12	43	−31
Bighorn sheep	60	15	40	−25
Caribou	76	0	24	−24
Moose	85	4	15	−11
Mule deer	86	6	14	−8
Collared peccary	100	0	0	0
White-tailed deer	99	7	1	+6
Mean	66	6	33	−26

Figure 9.10 Horseshoe Falls plunges over the Niagara Escarpment in Ontario, Canada.

plants, insects, fungi, and of course microscopic organisms, have a much smaller public following, thus data on their distribution are more difficult to obtain. Many scientists estimate that fewer than 10% of Earth's existing species have even been identified, much less had their distribution described. But ecologists are not content to simply describe a species' distribution. They also want to know why a species is found in one area and not in another. Both abiotic and biotic factors influence species' distribution and abundance.

9.3 IN WHAT WAYS DO ABIOTIC FACTORS INFLUENCE THE DISTRIBUTION OF POPULATIONS?

As we discussed in Chapter 2, climate is a composite of moisture, temperature, and wind patterns that prevail in a particular region. Climate can influence the distribution of species over different time scales. In pinyon pines, the time scale for invasion in the US West was a few decades. In Chapter 2 we learned that the distributions of many types of plants in the Sahara changed dramatically over the time scale of centuries and millennia. But in bacteria, the relevant time scale might be a few hours or days. Thus an organism's life history will influence the most useful time scale for studying changes in its distribution.

In Chapter 5 we discussed adaptations in organisms to the two most important climatic variables, temperature and moisture. One example was the saguaro cactus, which is superbly adapted to the hot, dry conditions of the desert. The saguaro cactus has a tall, linear stature, which reduces the plant's absorption of radiant energy from the Sun, and its absorption of radiant and conductive heat energy from the surface. The biological tradeoff of this stature is that this cactus is unable to absorb enough radiant and conductive heat energy on subfreezing days to prevent its cells from freezing at night. Hence, it is restricted to habitats that don't drop below freezing for extended durations (Steenbergh and Lowe 1977).

Climate is arguably the most important influence on terrestrial species distribution. But there are many other abiotic factors determining species distribution. We'll investigate one example that highlights how numerous abiotic factors may vary together over a very small spatial scale, and which together exert a pronounced influence on species distribution. Our example features the plant community on the Niagara Escarpment – a very long cliff that runs through western New York, southern Ontario, northwestern Michigan, Wisconsin, and Illinois. A small portion of this escarpment is the famous cliff that forms Niagara Falls (Figure 9.10).

Abiotic factors influencing species distribution over a small spatial scale

The only tree species that grows on the face of the Niagara Escarpment is the eastern white cedar, *Thuja occidentalis*. These

Table 9.3 Comparison of abiotic conditions in the deciduous community and in the white cedar community.

Abiotic condition	Deciduous community	White cedar community
Soil depth (cm)	33.9	13.1
Maximum snow depth (cm)	20	0
Peak radiation (mol/m^2/day)	14	6
Nitrogen (mg/L)	7.2	2.2
Phosphorus (mg/L)	14.7	8.7
Potassium (mg/L)	63.9	30.3
Duration soil temperature <0°C	6 weeks	18 weeks

trees are only a few meters tall, yet tree-ring analysis shows that some of them are more than 1000 years old. Just a few meters away, on the upper plateau, we no longer find white cedars; the most common woody vegetation there by far is sugar maple, *Acer saccharum*, with red oak, *Quercus rubra*, and paper birch, *Betula papyrifera*, also represented. Thus a comparison across a spatial scale of 5 m reveals the presence of completely different types of forest community.

The Niagara Escarpment study also highlights for us that adjacent habitats can experience dramatically different climatic conditions, which can influence the species that make up the ecological community. Ruth Bartlett and her colleagues (1990) compared abiotic conditions in the plateau immediately above the cliffs (the deciduous community) to abiotic conditions on the cliff side (the white cedar community) over an 18-month period. Generally, compared to the white cedar community, the deciduous community was characterized by much deeper soils, greater nutrient availability, and the presence of deep snow (absent in the white cedar community). The amount of light striking the deciduous community was also greater, presumably because the cliff side shaded the white cedar community. Maximum temperatures were much higher in the deciduous community; this was particularly important during the winter and early spring, before the deciduous trees leafed out. Table 9.3 highlights these and other differences between the two communities.

This comparison shows major differences in abiotic conditions associated with equally significant differences in species

distribution. Although the white cedar community is continuous for many hundreds of kilometers in southern Canada, on a much smaller spatial scale, a distance of only 5 m produces a totally different distribution, one dominated by sugar maple. One challenge in this comparison is that we don't know which factors are responsible for these differences in distribution, or if several of these abiotic factors interact to cause these differences

 Thinking ecologically 9.3

How might you design further observations or experiments to resolve which abiotic factors influenced the two different distributions?

As we will now discuss, the distributions of marine species may be influenced by different factors than those affecting their terrestrial counterparts. Light plays a particularly defining role in the distribution of marine plants and algae.

Influence of light on marine species

Experiments addressing distributions of organisms in marine environments must account for the unique limiting factors determined by the abiotic factors there. For example, green, brown, and red algae tend to be found at different depths of the ocean, because ocean water – like water in other aquatic ecosystems – acts as a filter to specific wavelengths of light, filtering out red light even in very shallow water, while allowing green light to penetrate several meters deep. Brown algae, which include many species of kelp, use fucoxanthin as an accessory pigment to capture the energy of green light for photosynthesis (Figure 9.11). Kelp are a very important component of many marine ecosystems, serving as producers and often providing habitat, but different species of kelp are found at different depths in the nearshore habitat.

Christian Wiencke and his colleagues (2006) hypothesized that ultraviolet (UV) radiation might influence the distribution

Figure 9.11 The kelp, *Macrocystis pyrifera*.

of a variety of kelp species, as UV radiation can inhibit photosynthesis and damage DNA. If their hypothesis was correct, they predicted that kelp species that lived near the surface would be more immune to the toxic effects of UV radiation than would species in deeper water. The researchers focused on the zoospore stage of the kelp life cycle, because they knew from laboratory research that zoospores were particularly susceptible to UV radiation. Haploid zoospores are released by sori, structures usually located on the underside of the thallus (blade) of the kelp, and ultimately develop into small male and female gametophytes.

To test their hypothesis, the researchers first needed to evaluate the amount of UV radiation that penetrated to different ocean depths. They were particularly interested in UVB radiation, because UVB (wavelengths 290–320 nm) is higher energy and potentially more destructive than UVA radiation (320–400 nm). They discovered that about four times as much UVB radiation penetrates to a depth of 0.5 m than to a depth of 4 m. Thus ocean water is an effective screen against UVB radiation.

Scuba divers harvested three species of brown kelp, and then germinated the zoospores in the ocean at varying depths. *Zoospores* are haploid spores that are able to move about using whiplike flagella. They ultimately develop into haploid gametophytes – the stage of the kelp life cycle that produces the sperm and eggs that unite to form the zygote that develops into the large plant pictured in Figure 9.11. Wiencke and his colleagues predicted that zoospores from the species of kelp that was normally found near the surface, *Saccorhiza dermatodea*, would be less susceptible to UV radiation, and thus would have a higher germination rate at shallow depths than zoospores from *Alaria esculenta* and *Laminaria digitata*, the two species normally found at deeper depths. This was a challenging experiment to set up, requiring researchers to keep track of many thousand tiny zoospores, which had to be contained in sealed Petri dishes submerged at varying depths. This was the first such experiment done under field conditions, and the researchers could not control for variation in the clarity of the water, or the cloudiness of the day. Both of these variables influence how much UV radiation penetrates different depths over the course of a day.

To evaluate the effect of UV radiation, the researchers established three treatments. For the first treatment (Vis), the zoospores received only visible light, and were screened from UVA and UVB with light filters. For the second treatment (VisA), the zoospores were exposed to visible light and UVA, but were screened from UVB radiation. For the third treatment (VisAB), the zoospores were exposed to visible light, UVA, and UVB radiation. The researchers predicted that *Saccorhiza* zoospores would germinate as well at shallow depths as they did in deeper water, while *Alaria* and *Laminaria* would be more susceptible to the effects of high UV radiation in shallow water. If zoospores were protected from UV radiation in shallow water (the Vis treatment), researchers predicted that *Alaria* and *Laminaria* would improve their germination rates.

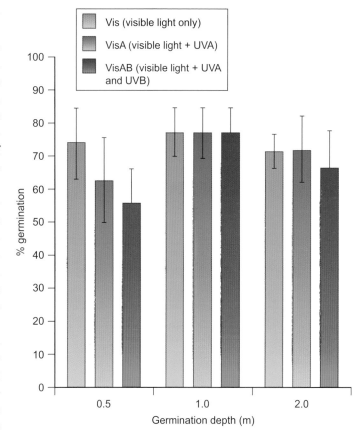

Figure 9.12 Percent germination of *Saccorhiza* zoospores at three different depths, in response to three different types of light. Error bars are 1 SD.

First let's look at the results for *Saccorhiza*. In the shallowest water (0.5 m), the Vis-treated zoospores had slightly higher germination rates than the VisA-treated zoospores, which had slightly higher germination rates than the VisAB-treated zoospores. These small differences disappeared at 1.0 m and deeper (Figure 9.12). Thus UV light has at best a very small impact on zoospore germination in this species, as predicted by the species' typical location in shallow water.

What about the other two species, normally found only at depths below 1.5 m? Are they more susceptible to UV radiation in shallow water? Both species showed very low germination rates in either of the UV treatments at 0.5 and 1.0 m. For example, the germination rate at 0.5 m for *Alaria* was approximately 60% when only visible light was permitted, but dropped to about 30% when either UVA or both UVA and UVB were permitted (Figure 9.13A). There was some improvement in germination at 1.0 m, but the VisA and VisAB treatments still germinated more poorly than the Vis treatment at that depth. At 2.0 m and below, both species showed equivalent germination rates regardless of whether the UV was screened out or not (Figure 9.13A and B). Presumably at those depths, the ocean water screened out enough UV light so that there was very little mortality at this developmental stage for any of the species.

(A)

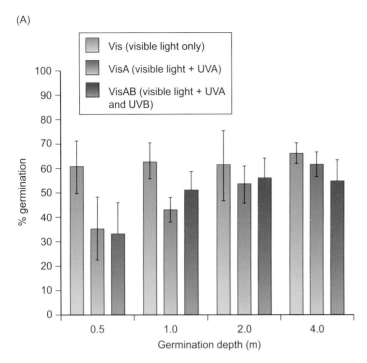

(B)

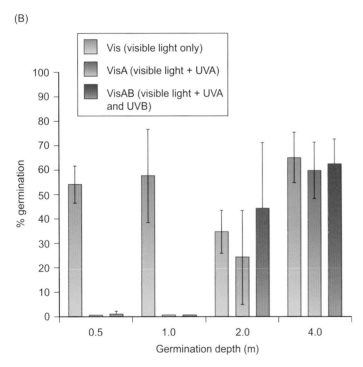

Figure 9.13 Percent germination of *Alaria* (A) and *Laminaria* (B) zoospores at four different depths, in response to four different types of light

Surprisingly, there was no difference in response to UVA and to the combination of UVA and UVB. The researchers had thought that UVB would do the most damage, as this was the result in laboratory research on related species. One possible explanation for the unexpected outcome is that the high intensity of visible light under field conditions might stimulate DNA repair mechanisms that offset some of the UVB damage to the kelp's DNA.

UV penetration of ocean water can be predicted by the underlying physical principles. We will now explore how less predictable events can influence species distribution.

Influence of disturbance on species distribution

Extreme events may limit the distribution of many species. For example, we discussed in Chapter 5 how extended frosts can kill many ectothermic species, by causing the water in their cells to freeze and bursting their cell membranes and/or cell walls. As a second example, extreme floods can cause local extinctions of invertebrate species in small streams, and local extinctions of plants and animals in associated riparian zones.

In seasonally dry biomes, such as temperate shrubland, grassland, savanna, and some temperate forests, fire plays an important role in determining species composition and distribution. Organisms in those biomes have numerous adaptations that allow them (or their progeny) to survive and reproduce even when fires recur every year or two. For example, many prairie grasses and forbs have meristematic tissue at or below the surface that can grow into new shoots following a fire. Natural selection may have originally favored this trait as protection against grazers, with the fire-protection benefits a secondary selective force. A second adaptation to fire is *serotiny*. Serotinous plants retain their seeds in the canopy for many years after the seeds mature, and release the seeds only after exposure to the heat of a fire (see Chapter 21). As of 1991, about 530 serotinous plant species had been identified – most of them are in Australia, with a smaller number in South Africa and North America.

Disturbances such as fire, floods, windstorms, and droughts are abiotic factors. However, abiotic factors such as fire can and do interact with biotic factors, such as competition with other plants or consumption by herbivores, to influence species distribution.

9.4 IN WHAT WAYS DO BIOTIC FACTORS INFLUENCE THE DISTRIBUTION OF POPULATIONS?

Biotic factors can influence spatial distribution across spatial scales ranging from very local to extremely large. We can return to the saguaro cactus for a simple example of how biotic interactions influence species distribution on a small spatial scale.

Biotic factors influencing species distribution over a small spatial scale

Many naturalists had observed an association between saguaros and so-called *nurse plants* – plants that provide a microenvironment that is conducive to saguaro survival during the early

Figure 9.14 Mesquite tree serving as nurse for saguaros.

stages of life. Three important nurse plants for saguaros are the palo verde tree, *Cercidium microphyllum*, the triangle-leaf bursage, *Ambrosia deltoidea*, and the mesquite tree, *Prosopis velutina* (Figure 9.14).

Taly Drezner and Colleen Garrity (2003) recognized that nurse plants could be providing numerous benefits to young saguaros. We've already discussed that saguaro distribution is associated with the absence of cold temperature extremes. Drezner and Garrity (and numerous other researchers) wondered whether nurse plants benefit saguaros by providing a less extreme thermal microenvironment. They hypothesized that nurse plants benefit saguaros by keeping them warmer during those frosty days and nights when saguaros are particularly vulnerable. But the researchers knew that nurse plants could also benefit saguaros by providing a shaded microenvironment with lower evapotranspiration rates, or a more nutrient-rich microenvironment arising from higher decomposition rates and greater organic matter. Their challenge was determining which of these potential benefits were most likely to underlie the association of saguaros with nurse plants.

To begin their study, the researchers tested the assumption that the microenvironment under the nurse plants *Cercidium microphyllum* and *Ambrosia deltoidea* was more thermally hospitable to young saguaros. They used data loggers to measure the air temperature during the coldest days of winter under three conditions: (1) under the north side of the nurse plant, (2) under the south side of the nurse plant, and (3) out in the open (the control). They hypothesized that the temperature under the nurse plant would be less variable than the temperature in the open over a 24-h period,

staying cooler when the Sun was out and warmer at night. Because the Sun is in the southern sky in the northern hemisphere winter, the researchers also hypothesized that the south side of the nurse plant might tend to provide a somewhat warmer microenvironment in the winter than the north side.

Figure 9.15 summarizes the results from the data loggers. Both species of nurse plants consistently provided less variable thermal environments than conditions out in the open. In addition, the south side of both species of nurse plants did not get as cold during the night as the north side of nurse plants. With this in mind, Drezner and Garrity hypothesized that if the primary benefit of nurse plants was providing a suitable thermal environment, than in colder parts of the saguaro range, we might expect to find more saguaros growing on the south side of nurse plants than on the north side.

To test the thermal environment hypothesis, Drezner and Garrity established 30 field sites throughout the saguaro range in Arizona. They hypothesized that if nurse plants were protecting saguaros against frost damage, this benefit should be primarily realized only in the coldest sites. This hypothesis predicts that in these coldest sites, we would see successful saguaro survival primarily on the south (most thermally protected) side of nurse plants, while in the warmest sites, we would observe no difference in saguaro survival on north versus south side. Table 9.4 shows the result of Drezner's and Garrity's comparisons.

These results support the prediction of the thermal protection hypothesis. In the coldest sites there were significantly more saguaros on the south side of both nurse plants, but in the warm

(A)   (B)

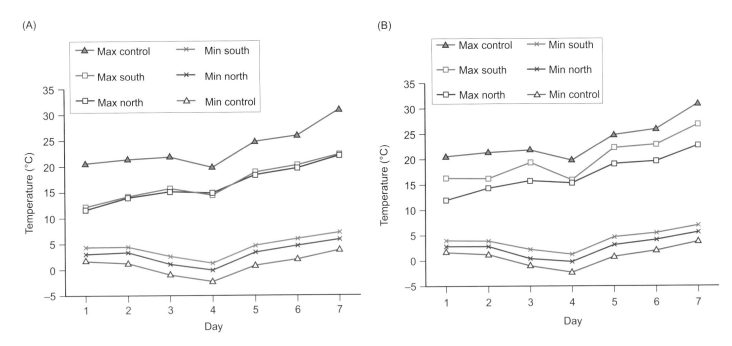

Figure 9.15 Maximum and minimum temperatures on the south and north side of the nurse plant, and for the unprotected control, on seven consecutive winter days. (A) *Cercidium microphyllum*. (B) *Ambrosia deltoidea*.

Table 9.4 Saguaro distribution on north versus south sides of nurse plants in sites at different temperatures.

Site temperature	Nurse species	# of saguaros north of nurse	# of saguaros south of nurse	χ^2	*P*-value
Coldest	*C. microphyllum*	2	11	6.23	0.01
Coldest	*A. deltoidea*	4	12	4.00	0.05
Warmest	*C. microphyllum*	15	8	2.10	0.14
Warmest	*A. deltoidea*	6	4	0.40	0.53

sites, there were actually more saguaros on the north side (although the difference is not significantly different). Drezner and Garrity conclude that nurse plants provide a thermal environment that allows saguaros to survive cold temperature extremes in the desert.

 Thinking ecologically 9.4

Because they only had access to five data loggers, Drezner and Garrity tested the temperature at only one of each species of nurse plant plus one control. Each nurse plant required monitoring with a sensor on both the north and south side. Referring to Figure 9.15, how might you redesign the sensor experiments so you would have greater confidence that nurse plants provide (a) a less variable thermal environment than conditions out in the open, and (b) a warmer environment on the south versus north side?

While the relationship between nurse plants and saguaros is clearly beneficial to the saguaros, there appears to be no benefit to nurse plants. In fact, as the saguaro gets larger, it will take up a significant portion of the available moisture, in some cases causing extreme water stress for a nurse plant. This effect indicates that the greater number of saguaros associated with nurse plants is related to the improved survival of the saguaros within the association and not to any advantage for the nurse plants. Let's now consider biotic interactions on a much larger spatial scale.

Influence of biotic interactions on species distributions on a large spatial scale

In 1845, about 20 red foxes, *Vulpes vulpes*, were introduced to Australia to provide sport for European colonists. Since that time, the red fox has spread throughout Australia, and in many areas it has become an important predator on small and medium-sized mammals. Ecologists describe the red fox as an invasive species – a non-native species that is introduced into a new habitat, and that often adversely affects numerous species in the new habitat. During the last 150 years, many species of Australian mammals have experienced range contraction and some species have gone extinct. Some researchers hypothesized that predation by red foxes was responsible for these range contractions and extinctions.

It is challenging to test the predation hypothesis for Australian prey species that have already gone extinct. However, many prey species, such as the black-footed rock-wallaby (*Petrogale lateralis*), are still present in moderate – though reduced – numbers (Figure 9.16A). If predation by red foxes is responsible for range contraction over much of Australia, we would predict that removing red foxes from regions in which they coexist with wallabies would lead to an increase in wallaby populations.

J. E. Kinnear and his colleagues (1998) tested this prediction by removing red foxes from two sites with the help of poison bait, and leaving three sites as untreated controls. They found a quick numerical response in their experimental sites, with wallaby populations increasing significantly where foxes were removed. In contrast, the wallaby population went extinct in one control site, and remained relatively constant in the other two control sites (Figure 9.16B). The researchers concluded that the red fox was important in limiting the distribution of wallaby populations.

Ideally, ecologists would like to identify all the factors that influence a species' distribution, and to understand how these factors interact. Environmental conditions such as temperature, humidity, and air or water current, and resources such as food, light, nutrients, and nesting areas, will interact to influence the survival, growth, and reproduction of individuals of a given species. Together, these conditions and resources describe a species' ecological niche, which we will define as the set of environmental conditions and patterns of resource availability, in which a species can survive, grow, and reproduce.

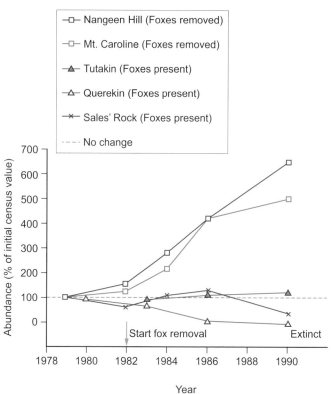

Figure 9.16 A. The black-footed rock-wallaby. B. Change in the relative abundance of rock-wallabies between 1979 and 1990.

9.5 HOW DOES ECOLOGICAL NICHE THEORY HELP ECOLOGISTS UNDERSTAND A SPECIES' CURRENT DISTRIBUTION AND PREDICT ITS FUTURE DISTRIBUTION?

Ecologists have discussed the niche concept for about 100 years. G. Evelyn Hutchinson articulated one of the more popular explanations of the niche in 1957. He described the niche as an *n*-dimensional hypervolume in which every point corresponds to a set of environmental conditions that would permit the species to exist indefinitely (Hutchinson 1957).

There are two parts of Hutchinson's description that might seem somewhat puzzling. First, you might be wondering what is an *n*-dimensional hypervolume, and what does it have to do with a niche? Second, what does Hutchinson mean when he refers to species' existing indefinitely?

Factors defining a species' niche

Let's explore *n*-dimensional hypervolumes by reconsidering the saguaro cactus. As we've already discussed, saguaros can survive freezing temperatures for only a short time (up to 36 h), but a prolonged freeze will kill them. Thus one dimension of the saguaro's niche is the set of thermal conditions defined by the line in Figure 9.17A. Second, research by Warren Steenbergh and Charles Lowe (1977) indicates that saguaro seeds need about 3 days of continuous high soil moisture conditions to germinate. Thus a second dimension of the saguaro niche is defined by this minimum. We now have defined an area that describes a saguaro's niche (Figure 9.17B). Steenbergh and Lowe also describe a third dimension of the saguaro's niche – soil texture. Saguaros prefer coarse soil types, and can't survive in clay soils. We have now defined three dimensions of a saguaro's niche. Three dimensions describe a volume (Figure 9.17C).

But several other environmental conditions and resources are essential for saguaro survival, growth, and reproduction. For example, all plants require nitrogen and a variety of other nutrients. The minimum level of soil nitrogen defines a fourth dimension of the saguaro niche. If three dimensions describe a volume, then four or more dimensions describe a *hypervolume*. Unfortunately, there is no easy way to draw a picture of a hypervolume – we must use our powers of imagination to picture all of the relevant variables in hyperspace that define the niche of a particular species.

Effect of biotic factors on a species' realized niche

But saguaro distribution is also influenced by the plant's interactions with other species, particularly predation and competition. Steenbergh and Lowe suspected that vertebrate predators were responsible for much saguaro seedling mortality. If this hypothesis was correct, young saguaros protected from vertebrate predators should have higher survival rates

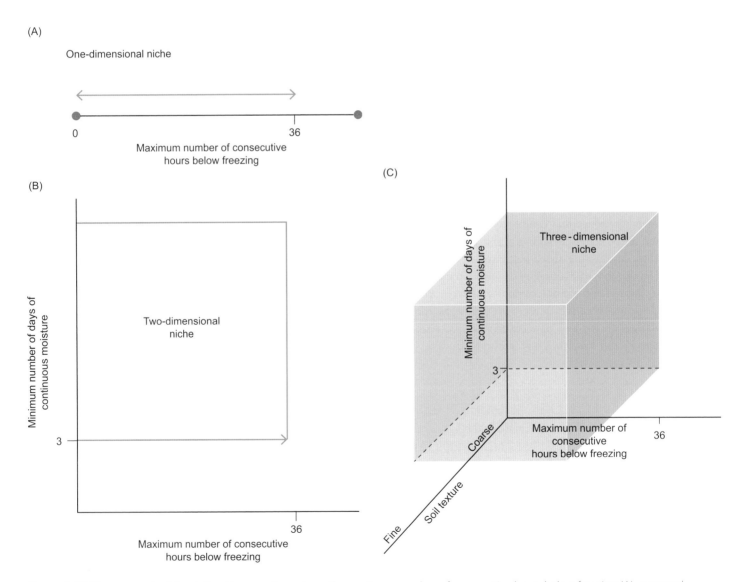

Figure 9.17 The saguaro's niche defined by one dimension – the maximum number of consecutive hours below freezing (A), a second dimension – the minimum number of consecutive days with moist soil (B), and a third dimension – soil texture (C).

than unprotected saguaros. To test this hypothesis, Steenbergh and Lowe germinated seeds and broadcast them into six plots at Saguaro National Park in Arizona. A portion of the seedlings were protected from vertebrate predation with wire mesh enclosures (the mesh was 1.3-cm squares). After 1 month, a mean of 71% of the plants were still alive at the protected sites in contrast to only 21% survival at the unprotected sites. After 1 year, 44% of the plants were still alive at the protected sites, while all the plants were dead at the unprotected sites. Steenbergh and Lowe concluded that vertebrate predators are important factors limiting saguaro survival, and ultimately saguaro distribution.

The saguaro research demonstrates that a complete picture of a species' niche should include not only environmental conditions and resources that permit a species to exist, but also biotic influences such as competition and predation. Hutchinson distinguished two types of niche. The fundamental niche is the potential set of environmental conditions and resources in which a species can survive, grow, and reproduce. The realized niche is the set of environmental conditions and resources in which a species can survive, grow, and reproduce in the presence of competitors and predators. Factors such as predation and competition exclude a species from portions of the idealized fundamental niche. The species' realized niche is thus defined by these limits.

Recall that Hutchinson described a niche as an *n*-dimensional hypervolume in which every point within the hypervolume corresponds to a set of conditions that would permit the species to exist indefinitely. What does he mean by "exist indefinitely"? We all know that species don't exist forever; ultimately the fate of every species is to go extinct or to evolve into new species. Population ecologists argue that the niche concept allows us to identify the conditions necessary for the continued survival of a species. If these conditions are maintained, a species has a much higher probability of surviving (but of course will not really exist indefinitely). If these conditions are not maintained, a species' population will decrease, perhaps to extinction.

Population ecologists also argue that the niche concept allows us to identify which species are most likely to become invasive in a particular region. If a niche is suitable for a species' survival, growth, and reproduction, and if the species can disperse into new habitats, the species may rapidly expand its distribution. The rainbow smelt is now invading numerous freshwater lakes in the northern United States and Canada as a result of its relatively broad niche, as well as its improved dispersal capabilities resulting from human activities.

The rainbow smelt's broad realized niche

The rainbow smelt, *Osmerus mordax*, is an anadromous fish, a fish that spends much of its life at sea, and moves into freshwater to breed. Its native habitat is limited to coastal streams and associated estuaries or coastal ponds and lakes that are connected to the Atlantic Ocean. The smelt migrate up rivers and streams in very large numbers to breed. Because of this historical connection to the ocean, until about 100 years ago the only established freshwater smelt populations existed in low-elevation lakes. However, in the past 100 years, the smelt's freshwater range has increased dramatically through introduction by humans, either intentionally by game wardens as forage fish for trout and salmon to feed on, or unintentionally by fishermen using smelt as bait for larger species of fish. To breed, landlocked rainbow smelt migrate from lakes into feeder streams that empty into their lakes. As a result of these introductions into new habitat, rainbow smelt are now found in many lakes in northern United States and Canada.

David Halliwell and his colleagues (2001) wanted to identify environmental conditions and resources associated with the distributions of smelt and several other species of coldwater fish in the northeast United States. To do so, they sampled the fish populations in 203 randomly chosen lakes out of 10 608 lakes with a surface area of greater than 1 ha. Sampling techniques included various types of nets and traps, and electrofishing. Rainbow smelt were native to 4% of the lakes, and had invaded an additional 14% of the lakes in the sample (Figure 9.18).

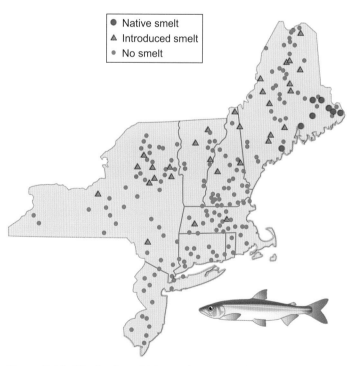

Figure 9.18. Distribution of lakes with native and non-native populations of rainbow smelt, and lakes without rainbow smelt, in northeast United States.

Thinking ecologically 9.5

Based on the map in Figure 9.18, would you describe the dispersion of (a) native rainbow smelt, and (b) all rainbow smelt, as random, uniform or clumped? Explain your answer.

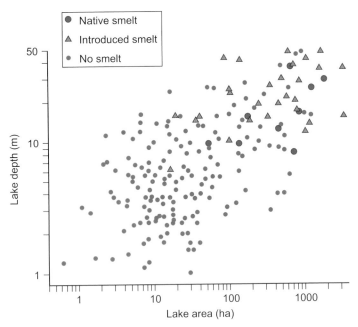

Figure 9.19 Distribution of rainbow smelt in relation to lake depth and lake surface area.

The researchers knew from studies done by other ecologists that coldwater fish distribution was most likely to be influenced by a lake's surface area, maximum depth, pH, phosphorus levels, and level of disturbance. Disturbance was evaluated based on human population density, the number of nearby roads, and the proportion of the watershed that drained urban, agricultural, and rural landscapes. Halliwell and his colleagues, and a large group of co-workers, measured these variables for each lake they sampled. Figure 9.19 shows their findings for how two variables, lake area (the *x*-axis) and lake depth (the *y*-axis) affect smelt distribution. Clearly the smelt are most commonly found in large deep lakes.

The influence of the other variables on smelt distribution is not quite as striking. Lakes with native and introduced smelt both had a median pH of about 7.0, which was similar to the median pH of all lakes in the sample. Lakes with smelt tended to have lower phosphorus levels than the median level for all lakes and had slightly lower disturbance values. We can tentatively describe the ecological niche of rainbow smelt based on lake depth, lake surface area, phosphorus levels, and disturbance, but much work remains to be done, including the consideration of other environmental factors. In addition to environmental factors, smelt distribution is influenced by historical events such as whether fishermen or games officials introduced smelt into a particular lake that had suitable environmental conditions.

Some ecologists argue that it is inappropriate to define a species' niche based on its current distribution, because, as in the case of rainbow smelt, it is very possible that a species' distribution may be expanding, and the distribution may be limited more by a species' dispersal abilities than by the unavailability of suitable niche. A second objection is that natural selection is operating continuously on individuals, and it may expand or restrict the environmental conditions and resources that define the niche, or the suite of traits associated with dispersal. A species' distribution is thus dynamic, and our evaluation of its niche based on the distribution will always be imperfect.

REVISIT: Predicting future distribution of species

In the introduction to this chapter we described how ecologists use historical data to infer past changes in distribution. We will now consider how ecologists use information based on present distributions to make predictions about future distribution patterns. This is particularly important in the case of invasive species, which upon entering an ecological community may have significant negative effects on other species in the community. One commonly used approach is to build a model based on a species distribution in its native range, and to apply the predictions generated by the model to other geographic ranges. Let's reconsider rainbow smelt to explore how this is done.

Norman Mercado-Silva and his colleagues (2006) wanted to identify which lakes in the Great Lakes region would be susceptible to invasion by rainbow smelt. Based on the research we just described (Halliwell *et al.* 2001), and on research by other ecologists, they considered five environmental variables in their model construction. Three variables were identical to those identified by Halliwell and his colleagues: lake area, maximum lake depth, and pH. A fourth variable was lake shoreline perimeter – the distance around the lake, and the last variable considered was *Secchi depth*, a measure of the water's transparency. Transparent water carries relatively few suspended particles, while turbid (cloudy or murky) water has a high concentration of tiny suspended particles. Secchi depth is obtained by lowering a patterned Secchi disk (Figure 9.20) into the water, until the pattern is no longer visible – by definition, the Secchi depth. Transparent water has greater Secchi depth than turbid water.

All of the data were collected from maps and databases, and from personal communications with other researchers (including David Halliwell). Researchers often use personal communications to obtain unpublished data, or to get clarification on published data. Mercado-Silva and his colleagues restricted their analysis to the 354 lakes that were inside the native geographical range of rainbow smelt in Maine, and they asked two simple questions. First, which lakes within the native range have smelt, and which ones don't? Second, do the lakes within the native range with smelt differ in any significant way (in regards to the five variables) from the lakes within the native range without smelt? If so, we can use these differences to make predictions about which lakes outside the native range are vulnerable to invasion, and which lakes are unlikely to be invaded by the expanding population of rainbow smelt. In the language of model building, factors that allow us to make predictions are called *predictors*.

Mercado-Silva and his colleagues used a classification tree to model the presence and absence of rainbow smelt within their native range in Maine. Classification trees use computer algorithms to partition the data set into mutually exclusive groups according to the importance of each predictor in explaining the presence or absence of smelt in each lake. The algorithm identified three predictors as important in influencing whether a lake within the native smelt range had rainbow smelt. Maximum depth was the most important predictor, with deeper lakes tending to have smelt. Surface area was also important, with lakes of large surface area more likely to have smelt. Finally, Secchi depth was also important: lakes with more transparent water were more likely to have smelt. We won't worry about the mathematics underlying how the computer did this analysis; instead we will focus on understanding the output of the classification tree (Figure 9.21).

Recall that this study was based on data from 354 lakes in the smelt native range. If you use the classification tree model in Figure 9.21, you will discover that the model correctly predicts smelt presence in 165 out of 200 lakes that have smelt. The sensitivity of a model is defined as the number of correct positive predictions (in this case for the presence of smelt) divided by the total number of positive occurrences. The sensitivity of this model was 165/200 = 82.5%. You will also find that the model correctly predicts smelt absence in 139 out of 154 lakes that lack smelt. The specificity of a model is defined as the number of correct negative predictions (in this case for the absence of smelt) divided by the total number of negative occurrences. The specificity of the

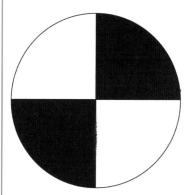

Figure 9.20 Secchi disk pattern.

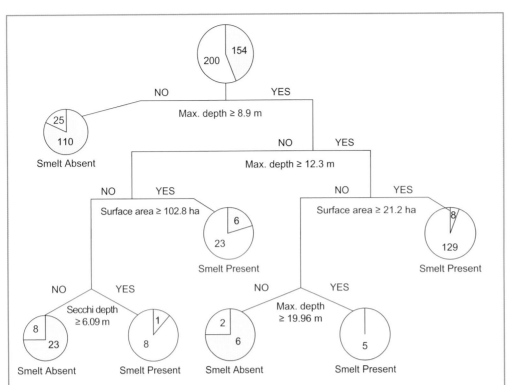

Figure 9.21 Classification tree model constructed from 354 lakes in Maine. Each of the six splits is labeled with the variable that defines the split, while the numbers in the pie chart indicate the number of lakes with (clear region) and without (shaded region) smelt based on that variable.

model was 139/154 = 90.3%. Numbers like these increase our confidence in the utility of a model. However, for a model to be useful for making predictions, it should be validated or tested.

In an ideal world, ecologists could make models based on present distributions, and use them to make predictions about future distributions. The problem is that the model should be tested or validated in some way before it is applied with any confidence. How can Mercado-Silva and his colleagues convince other ecologists that their model is actually useful for predicting smelt distributions in the future? As you might imagine, Mercado-Silva and his colleagues anticipated this objection. They argue that most of the lakes in the remainder of Maine, outside of the native distribution, had probably encountered an introduction of rainbow smelt, either intentionally by game managers or unintentionally, courtesy of bait fishermen. Thus they used their model based on lakes within the native range to see if it accurately predicted the presence or absence of smelt outside of its native range.

Outside of smelt's native range (but still in the state of Maine), the model accurately predicted smelt presence in 176 out of 221 lakes that have smelt, for a sensitivity of 79.6%. The model also correctly predicted smelt absence in 192 out of 244 lakes that lacked smelt, for a specificity of 78.7%. Thus the researchers were able to show that the model could be useful for predicting range expansion for this species. They also recognize that their model is not a perfect predictor of future reality, but that it can be used to identify lakes where monitoring, legislative, and educational efforts should be intensified. They warn that based on their model, there are a large number of lakes in Ontario and Wisconsin that are suitable for smelt, and that are likely to be successfully invaded if smelt are able to disperse into them. Similar modeling approaches can be used with other invasive species to alert conservation officials to future threats to native communities.

SUMMARY

One of the major challenges facing ecologists is determining species spatial distribution. Paleoecological studies can provide some insight into factors influencing a species' present-day distribution, and its present-day distribution can, in turn, provide some insight into its future distribution. Being able to predict future distributions is very important because climate, an important influence on species distribution, is now changing at a rapid rate.

Within a population, individuals may have a random, uniform, or clumped dispersion, though a clumped dispersion is most common because essential resources such as food, light, and undisturbed habitat are often spatially clumped. Distribution patterns change over the short term, as a result of dispersal, and over the long term from factors that influence range expansion and contraction. Abiotic factors, such as climate, soils, light availability and disturbance, and biotic factors, such as behavior, life histories and interactions with other species, can influence the distribution of species. Changes in these factors can lead to changes in distribution, including range expansion, range contraction and extinction. By quantitatively describing a species' ecological niche, ecologists can understand a species' present distribution, and may be able to make predictions about its future distribution.

FURTHER READING

Bauer, S., and Hoye, B. J. 2014. Migratory animals couple biodiversity and ecosystem functioning worldwide. *Science* 344: 54–62.

Thorough and very understandable review of how migratory species influence ecosystem functioning. If you're in a hurry to consider some of the large-scale questions in ecology, this is a good way to get there.

Laliberté, A. S., and Ripple, W. J. 2004. Range contractions of North American carnivores and ungulates. *Bioscience* 54: 123–138.

This study is a good introduction to the utility of GIS and also to somewhat unconventional ways of collecting data. The research also provides considerable quantities of data, which students can analyse.

Mercado-Silva, N., Olden, J. D., Maxted, J. T., Hrabik, T. R., and Zanden, M. J. V. 2006. Forecasting the spread of invasive rainbow smelt in the Laurentian Great Lakes region of North America. *Conservation Biology* 20: 1740–1749.

Good introduction to issues that arise when developing models and assessing their performance.

Ries, L., Taron, D. J. Rendon-Salinas, E., and Oberhauser, K.S. 2015. Connecting eastern monarch population dynamics across their migratory cycle. In K. S. Oberhauser, K. R. Nail, and S. M. Altizer, (eds), *Monarchs in a Changing World: Biology and Conservation of an Iconic Butterfly*. Ithaca, NY: Cornell University Press, pp. 268–282.

Shows how data collected by various citizen science groups are applied to drawing conclusions about patterns of monarch butterfly migration. This paper also highlights some of the challenges associated with such large-scale efforts.

Sanderson, E. W., and 5 others. 2002. The human footprint and the last of the wild. *Bioscience* 52: 891–904.

The human footprint is a critical concept in conservation biology and has penetrated into the popular press as well. This paper describes how it was measured and discusses, in detail, the underlying assumptions. This is also a nice introduction to the use of GIS technology.

END-OF-CHAPTER QUESTIONS

Review questions

1. What are three basic patterns of dispersion? Which is most common in natural populations (and why)? Give examples of each dispersion pattern in natural populations?
2. Distinguish between dispersal and migration? Why might it be advantageous for a species to migrate? Why might it be disadvantageous?
3. What factors influenced species distribution in (a) the Niagara Escarpment and in (b) kelp populations in nearshore habitat? How did the researchers identify these factors?
4. How do nurse plants influence saguaro distribution? How does saguaro influence nurse plant distribution?
5. What are invasive species? Use the red fox study by Kinnear (1998) to highlight a problem caused by invasive species.
6. Distinguish between a fundamental and realized niche. How did ecologists use the niche concept to make predictions of future distributions of rainbow smelt?

Synthesis and application questions

1. Some researchers have criticized (a) pollen cores and (b) midden macrofossils as inherently inaccurate at reconstructing vegetation history. What factors might cause either approach to be inaccurate?
2. We discussed the saguaro cactus in this chapter and in Chapter 5. Based on our discussion, what factors are most important in defining the saguaro's ecological niche?
3. Drezner and Garrity's (2003) hypothesis provides some support for the thermal environment hypothesis as an explanation of the association between saguaros and nurse plants. However, the research does not test the predictions of the alternative hypothesis, that nurse plants provide saguaros with a more nutrient-rich microenvironment from higher decomposition rates and greater organic matter. How might you test this alternative hypothesis?
4. Leonard Wassenaar and Keith Hobson (1998) measured the $\delta^{13}C$ (stable isotope) values of wings collected from different parts of the monarch butterfly's North American breeding range. They found substantial differences in the $\delta^{13}C$ values (Figure 9.22). They then used this information to test whether butterflies from different breeding ranges tended to roost together in their wintering colonies in Mexico. They collected wings from 13 different roosting sites in overwintering ranges in Mexico, and all of them showed a very narrow range of mean $\delta^{13}C$ values between -27.3 and -27.7 (SE ≈ 1.1). Based on these findings, do butterflies from different breeding locations tend to mix together when they overwinter in Mexico? Explain your answer.
5. What other predictions are generated by the hypothesis that red fox predation is responsible for the wallaby range contraction in Australia? Describe an experiment or observation that would test one of these predictions.
6. The data for Figure 9.19 show (unsurprisingly) that large lakes tend to be deeper. Thus it is difficult to evaluate whether lake size or lake depth is more relevant for determining whether a population of smelt is established. Can you eyeball the data in Figure 9.19 to help evaluate whether one factor is more important than the other, or whether both factors have equally important influences on whether a lake has a population of rainbow smelt? Explain your reasoning.
7. Based on the classification tree model (Figure 9.21), predict the presence and absence of rainbow smelt for a lake with the following

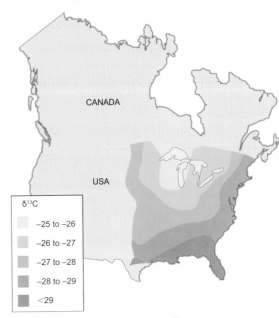

Figure 9.22. Patterns of mean $\delta^{13}C$ in monarch butterfly wings from sites across the breeding range of Eastern North America.

characteristics: (a) Maximum depth 10 m, surface area 100 ha, Secchi depth = 10 m.
(b) Maximum depth 20 m, surface area 20 ha, Secchi depth = 1 m.

Analyse the data 1

Mercado-Silva and colleagues developed a model of smelt distribution with a sensitivity of 82.5%. They then validated the model by testing it outside the smelt's natural range, and the model's sensitivity was 79.6%. We can use a chi-square test to compare sensitivity inside the natural range (where the model was developed), and outside the natural range (where the model was validated). If there is no significant difference between the two sensitivity values, that would support the model as a useful tool for identifying lakes that should be monitored for smelt invasion. Here are the data (taken from the text) for the analysis:

	Lakes with smelt	Lakes with no smelt
Lakes within natural range (model development)	165	35
Lakes outside of natural range (model validation)	176	45

(a) Is there a significant difference for sensitivity during model development and model validation?
(b) Use information from the text to construct a chi-square table comparing specificity during model development and model validation. Is there a significant difference for specificity during model development and model validation?
(c) Would you recommend this model for identifying lakes that might be prone to smelt invasion? Why or why not?

Analyse the data 2

Use a statistical analysis to compare the mean percent expansion to the mean percent contraction for ungulates, to determine whether ungulate ranges have contracted more than they have expanded over the past 200 years (see Table 9.2). Repeat the process for carnivores (Table 9.5). Interpret your findings. Would you expect to find similar patterns in ungulates and carnivores? Why or why not?

Table 9.5 Percentage persistence, expansion, contraction, and net change for 32 species of North American carnivores over the past 200 years.

Species	% persistence	% expansion	% contraction	% net change
Black-footed ferret	0	0	100	−100
Swift fox	32	8	68	−60
Grizzly bear	45	5	2	−53
Fisher	50	3	50	−47
Gray wolf	58	0	42	−42
Lynx	60	1	40	−39
Black bear	59	2	41	−39
Wolverine	61	2	39	−37

Table 9.5 (*cont.*)

Species	% persistence	% expansion	% contraction	% net change
Cougar	60	4	40	−36
River otter	75	0	25	−25
Marten	79	2	21	−19
American mink	87	1	13	−12
Polar bear	94	0	6	−6
Arctic fox	90	5	10	−5
Bobcat	90	5	10	−5
Long tailed weasel	100	0	0	0
Eastern hog-nosed skunk	100	0	0	0
Western hog-nosed skunk	97	4	3	+1
Striped skunk	99	1	1	+7
Ermine	100	1	0	+1
Eastern spotted skunk	91	11	9	+2
Ringtail	95	7	5	+2
White-nosed coati	98	5	2	+2
Gray fox	97	8	3	+5
Western spotted skunk	96	10	4	+6
Least weasel	100	8	0	+8
Hooded skunk	99	11	1	+10
Red fox	96	17	4	+13
Kit fox	88	28	12	+16
Badger	99	18	1	+17
Raccoon	100	18	0	+18
Coyote	100	40	0	+40

Chapter 10

Population abundance and growth

INTRODUCTION

Rapa Nui, also known as Easter Island, is perhaps the most remote piece of land in the world. The nearest islands are the Pitcairn Islands – themselves plenty remote – about 2000 km to its west, and the coast of Chile about 3700 km to its east. As such, Rapa Nui is an ideal model for studies of human populations because, until recently, the island has experienced very little human immigration or emigration. The most fascinating features of Rapa Nui are nearly 1000 immense evocative stone statues (*moai*) resting on even more immense stone platforms (*ahu*) that are spread around the island. Some of the moai are over 20 m tall and weigh up to 250 000 kg. Some moai were transported more than 15 km to their final positions. Who were the creators of these giant statues and what happened to their civilization? We will discuss some possible answers to these questions, and introduce some of the approaches and challenges of historical studies of populations.

In some cases, population ecologists can study patterns of population growth that extend from the past until the present, as ecologists have attempted with Rapa Nui, and extrapolate their findings into the future. For most species, under most conditions, this is a challenging endeavor, but these extrapolations are particularly important for management decisions about endangered populations, such as grey wolves in the northwestern United States and the Glanville fritillary butterfly off the coast of Finland. An accurate estimate of population size is essential to any study of past or present populations. Mathematical models depend on accurate measures of birth rates and death rates to project future population size. Numerous biotic and abiotic factors influence birth and death rates, which change as a population nears its carrying capacity. Given the importance of human impact on many other species, ecologists are using their understanding of factors that influence the growth rates of human populations to make projections of human population size on a national, regional, and global scale.

KEY QUESTIONS

10.1. How do ecologists estimate population size?

10.2. How do mathematical models project population growth rates?

10.3. How do density-independent and density-dependent factors influence birth and death rates?

10.4. How do ecologists use the carrying capacity to model density-dependent changes in birth and death rates?

10.5. What factors will influence the growth rates of future human populations?

CASE STUDY: The mysteries of Rapa Nui

The history of the island of Rapa Nui illustrates some of the factors that affect population changes. Most anthropologists agree that the first human colonists on Rapa Nui were Polynesians, based on evidence of linguistic and cultural similarities. In addition, DNA analysis of skeletons of early colonists shows a nine-base-pair deletion and three single-base-pair substitutions that are also found in the vast majority of Polynesian DNA (Diamond 2005). However, the colonists' arrival date is a focus of controversy. Most scientists suggest that first colonization occurred between AD 400 and 900. Jared Diamond, in his book, *Collapse*, proposes that the best evidence comes from radiocarbon dating of wood charcoal, and bones of porpoises collected from Anakena Beach, which most anthropologists consider to be one of the earliest colonies. Diamond's estimated arrival date of colonists is AD 800–900.

Diamond hypothesizes, using in part evidence of cultural artifacts, that the colonists increased their numbers rapidly, so that by the year 1300 there were about 15 000 individuals on the island. Approximately 3000 houses have been identified on the island. If we assume that, at the colony's peak, about one-third of these houses were occupied, with an average of 5–15 inhabitants per house, we can infer a peak population size of about 5000–15 000 people.

A second species played an important role in Diamond's historical reconstruction of the Rapa Nui mystery, the *Jubaea* palm, *Jubaea disperta* (sometimes called *Paschalococos disperta*). Pollen samples and macrofossil remains indicate that the island was heavily forested for at least the last 35 000 years, covered by millions of *Jubaea* palms and at least 21 other tree species (Orliac 2000; Mieth and Bork 2010). But sometime in the fifteenth century, tree populations began to decrease sharply, with extinction of all tree species complete by the seventeenth or early eighteenth century.

Diamond argues that a close tie-in existed between the explosion in Rapa Nui's human population and its equally rapid deforestation. *Jubaea* palms were clearly an important part of the early Rapanui culture. They may have been used to create rope and wood ladders for hauling moai from the quarry to their ahu. They were almost certainly used for canoe construction. Collections from early middens indicate that pelagic fish such as tuna were an important part of the early colonist's diet, but became less important after the thirteenth century, perhaps as a result of the unavailability of *Jubaea* wood for canoe construction. The *Jubaea* palms were certainly used for fuel, and were probably felled to make room for agriculture to feed the increasing population. Diamond argues that Rapa Nui is the most extreme example of wholesale forest destruction by humans in the Pacific. In turn, he links the forest's disappearance to the Rapanuis' collapse. Diamond describes the events on Rapa Nui as the "clearest example of a society that destroyed itself by overexploiting its own resources."

As we move toward the present, the picture on Rapa Nui becomes clearer but certainly no less dismal. The first documented European colonists (in 1722) killed about a dozen native Rapanui. By the late 1700s, the numerous European visitors had brought several plagues of smallpox to the island. Kidnapping of islanders for slaves, which began in 1805, peaked in 1862 with the abduction of 1500 Rapanui – about half of the remaining population. By 1872, the human population was down to 111. The population on Rapa Nui has rebounded sharply since that low point; the 2012 census estimated 5806 residents on the island, including a substantial number of recent immigrants.

This historical perspective highlights some of the difficulties with studies of population dynamics. Getting accurate data on population size and the timing of events is challenging. Until recently, Rapa Nui's isolation ensured that changes in human population size were results of changes in birth and death rates. But identifying the causal factors influencing birth and death rates requires understanding the dynamics of other populations, in particular the *Jubaea* palm. At the end of this chapter, we will revisit Rapa Nui to consider an alternative hypothesis to Diamond's human self-destruction hypothesis, one that infers an important role played by other species. In this chapter, we will learn how ecologists estimate population size, and use historical data – such as those available for Rapa Nui – and current estimates of population size and rates of change to project future patterns of population growth.

10.1 HOW DO ECOLOGISTS ESTIMATE POPULATION SIZE?

The simplest way to determine population size is to get out into the real world and count the number of individuals in the population. Unfortunately, this works only under a limited set of conditions.

Direct counts of population size

Table 10.1 describes six characteristics that make it easy to count population size. This can be a discouraging set of limitations for population ecologists, because there are very few species that

Table 10.1 Characteristics of populations in which direct counts of population size can be made easily.

Ideal population characteristic	Why is this important?
Distinct individuals	Possible to distinguish an individual from others
Nonmotile individuals	Individuals can't escape being counted, or won't get counted multiple times
Small geographic range	Great distances don't need to be traversed for an accurate count
Large individuals	Individuals can be seen
Small population size	It is possible to count all the individuals accurately
Friendly environment	Humans can spend enough time in the environment for accurate censusing

exhibit all six characteristics. Fortunately, there are methods for addressing each of these limitations.

Perhaps the most basic challenge illustrated in Table 10.1 is defining an individual. In unitary organisms, each individual is a unit, and presumably easy to count. If you were gazing at one human from Rapa Nui, you would never think you were looking at two individuals. In contrast, modular organisms develop an undetermined number of repeated copies of similar structures. For example, we can identify the developmental patterns for a particular species of tree in a given environment, such as when a branch will send out a branchlet, and when and where a branchlet will develop leaves and flowers. But we cannot predict precisely how many branchlets, leaves, and flowers an individual tree will produce. Complicating the picture, in modular organisms the concept of the individual becomes irrelevant. If you had been around before the extinction of the Jubaea palm on Rapa Nui, you might have noticed three trunks emerging from the soil in close proximity. Would that count as one tree or three? Some trees, including aspens, send out numerous genetically identical stems from a shared root system (Figure 10.1A), as do grasses and sedges. Some animal species, including marine corals, also have a modular growth form (Figure 10.1B). One of the largest organisms ever identified is a fungal mycelium (genus *Armillaria*) that extends its weblike modular body over an area of greater than 6 km^2, and may extend many thousands of genetically identical fruiting bodies over that area (Figure 10.1C).

Population ecologists struggle with assessing modular organisms because our definition of a population as a group of

Figure 10.1 Examples of modular organisms. A. Grove of aspen trees, *Populus tremuloides*. Many of these trunks originate from the same root system and have identical DNA. B. Colony of the modular *Sarcophyton* coral. C. Fruiting bodies of the largest known organism, the mushroom *Armillaria solidipes*, which covers 8.4 km² in eastern Oregon, USA.

interacting individuals of the same species runs into serious problems when we abandon our concept of individuals. Ecologists have developed numerous approaches for estimating population size for both unitary and modular organisms.

Estimating population sizes of unitary organisms

As Scott Mills (2007) points out, all estimates of abundance for unitary organisms boil down to the same equation mathematically:

$$\text{estimated population size} = \frac{(\text{number of individuals in the sample})}{(\text{estimated probability of detection})}$$

Or

$$N_{est} = n_s/p_{est}$$

The estimated probability of detection depends on the answer to two questions. First, what proportion of the habitat are you sampling? Second, if an organism is within your sampling area, what is your probability of seeing it?

Using the answers to these two questions, p_{est} = (proportion of the habitat you are sampling)(probability of actually detecting an organism when it exists within your sampling area).

In the simplest case, organisms will be cooperative enough that you are almost guaranteed to detect them and count them accurately.

Quadrat sampling

Quadrat sampling is used to estimate population size for easy-to-count unitary organisms. Some organisms are relatively easy to count accurately (see Table 10.1), because they are *sessile*, or fixed in one place. For example, a careful observer should be able to count the number of barnacles attached to rocks along the intertidal zone. Of course, a coastline with suitable barnacle habitat can extend for hundreds or thousands of kilometers, making counts overly tedious for even the most enthusiastic population ecologist. In this case, the ecologist would divide the habitat into equal-sized quadrats – or sampling areas, then count the number of individuals within each quadrat (Figure 10.2).

For barnacles, the probability of actually detecting the barnacle when it exists within the quadrat ≈ 1.0. So our formula for estimating population size simplifies to

$$N_{est} = \frac{\text{Number of barnacles counted}}{\text{proportion of the habitat sampled}}$$

If you counted 573 barnacles, and your quadrats took up 2.3% of the intertidal zone with suitable barnacle habitat (for a proportion of 0.023), then your estimated population size would be

$$N_{est} = 573/0.023 = 24\,913.$$

Next we discuss two approaches used by population ecologists for estimating population size of unitary organisms that are more difficult to count directly.

The mark–recapture method

The mark–recapture method is used to estimate population size when only a portion of the organisms within a habitat can be counted. It requires the researcher to capture individuals from the environment, apply a distinct mark to them, return them to the environment, and then go through a second round of capturing individuals from the same environment. If the researcher and the organisms obey a certain set of assumptions, which we will discuss shortly, then the proportion of individuals that are marked in the second sample should be representative of the proportion marked in the entire population. In mathematical terms:

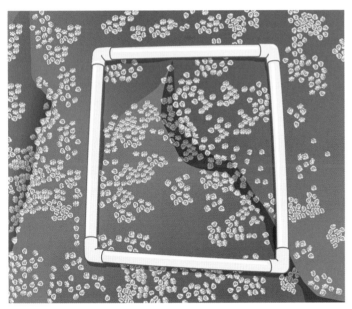

Figure 10.2 Sampling a population of barnacles within a quadrat.

Proportion marked in sample 2 = proportion marked in entire population

$$M_2/N_{S2} = M_1/N_{est}$$

M_2 = number of marked individuals recaptured in second sample

N_{S2} = total number of individuals in second sample

M_1 = number of individuals marked and released in first sample

N_{est} = estimated population size

We can cross multiply and solve for N_{est}, giving us

$$N_{est} = (M_1 N_{S2})/M_2$$

As an example, let's consider the population of the humpback whale, *Megaptera novaeangliae*, along the US Pacific coast, which was depleted by commercial fishermen until 1966 when commercial whaling for humpbacks was banned in the North Pacific. Previous analysis of mitochondrial DNA showed that this population is genetically distinct from other populations farther north along the Canadian and Alaskan coasts, and that these individuals migrate to wintering grounds in Mexico and Central America (Baker *et al.* 1990). John Calambokidis and Jay Barlow (2004) wanted to know how large this humpback whale population is, and whether it is expanding now that the animals are protected.

If you look back at Table 10.1, you will note that these whales should (and do) make difficult subjects for several reasons. They certainly move around a lot, they have a large geographic range, and the marine environment can be a very difficult place for a human to work. However, the whales also have several good features for population studies. They are huge, so they are easy to count; they surface approximately every 10 min; and they have distinct and unique markings on their tails, so that individuals can be distinguished from each other (Figure 10.3). Their

Figure 10.3 Distinct markings on a humpback whale's tail are used to identify each individual.

Table 10.2 Number of humpback whales identified photographically during the time frame analysed by Calambokidis and Barlow (2004). The number identified (# ID) is always greater than the number of unique individuals identified (# unique ID), because some whales were recorded numerous times over the course of a year.

Sample year 1 (capture year)	# ID	# unique ID (= M_1)	Sample year 2 (recapture year)	# ID	# unique ID (=N_{S2})	Number of individuals in sample year 2 which were originally photographed during sample year 1 (= M_2)
1991	668	269	1992	1023	398	188
1993	512	254	1994	402	244	108
1996	564	331	1997	382	264	104

population size can be estimated using a variation of the mark–recapture method, with the help of a camera, which records the individuals' tail markings.

Calambokidis and Barlow used data collected from a variety of sources, including boats dedicated to the whale "capture" study, but also data from whale watch boats and other commercial vessels. Table 10.2 shows some of the data from these surveys.

We can apply the formula $N_{est} = (M_1 N_{S2})/M_2$ for the period of 1991 through 1992, to estimate population size.

$$N_{est} = (269)(398)/188 = 569$$

 Thinking ecologically 10.1

Based on the data in Table 10.2, would you estimate that the whale population is increasing, decreasing, or staying relatively stable over the 7 years of the study? How confident are you of your conclusion?

In estimating population changes, it's important to remember that all models make simplifying assumptions. One important assumption of the mark–recapture method is that individuals don't enter or leave the population during the time between the two sample periods. The DNA evidence gives us some confidence that immigration and emigration are rare but, of course, whales will be born between sampling years, and whales don't live forever, so we can be confident that some mortality occurs. Estimated mortality rates are about 4% per year for humpback whales. A second important assumption is that marked whales and unmarked whales are equally likely to be recaptured during the second sample. A related assumption is that marking does not influence fitness in any way, for example, by leading to increased mortality. Population ecologists have constructed more complex models that adjust for cases when these two assumptions are not valid.

A more recent mark–recapture study of humpback whales in the entire North Pacific takes into account biases caused by new calves entering the population, and by mortality of some individuals between sampling efforts. After accounting for these (and other) violations of the mark–recapture assumptions, the researchers estimated that the humpback whale population in the entire North Pacific was 21 808, which is greater than some estimates of prewhaling abundance (Barlow *et al.* 2011).

Accumulation curve analysis

Handling dangerous, endangered, and elusive species makes mark–recapture studies problematic. Accumulation curve analysis is a second approach to estimate population size when only a portion of the organisms within a habitat can be counted. Many species will leave evidence of their genetic identity in feathers, hairs, scales, and the ever-present piles of feces. Molecular ecologists can then extract the DNA from these remains, amplify the DNA with the polymerase chain reaction (PCR), and distinguish between different individuals using several different types of molecular markers, including microsatellite markers (a technique described in Chapter 3).

One of the first studies of this type was done by Michael Kohn and his colleagues (1999) on a coyote population in southern California. They used three highly polymorphic microsatellite loci to establish the genetic identity of each coyote that created the samples. Ultimately, they were able to identify the DNA from 115 piles of feces, deposited by 30 different coyotes. Table 10.3 is a simplified version of the data collected by Kohn and his colleagues. You will note that initially each new pile is from a different individual, but that as more piles are collected, we begin to see more and more repeats.

Kohn and his colleagues constructed an accumulation curve to estimate the number of individuals in the population. An accumulation curve is a graph of the number of samples (feces

Table 10.3 Accumulation data estimated from the Kohn *et al.* coyote study (1999).

Number of piles analyzed	5	10	25	50	75	100	115
Number of unique genotypes	5	9	17	23	27	29	30

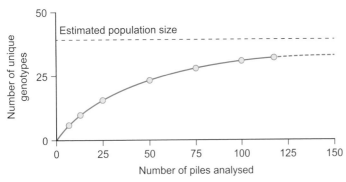

Figure 10.4 Accumulation curve that plots number of unique genotypes in relation to the number of fecal piles that are sampled. The curve is extrapolated until it reaches its asymptote, which is the estimated population size.

Figure 10.5 Cryptic morel mushrooms are easily overlooked by collectors and researchers.

in this case) on the *x*-axis and the total number of individuals (new genotypes in this instance) on the *y*-axis (Figure 10.4). After drawing the curve for the existing samples, Kohn and his colleagues used a relatively simple mathematical model to extrapolate the curve until it asymptoted, or leveled out. They estimated that the actual population size was most likely between 35 and 41 coyotes.

Estimating the abundance of modular organisms is more problematic, because individuals cannot be easily counted.

Estimating population sizes of modular organisms

Foresters typically estimate tree abundance by counting the number of stems with a certain minimum diameter (often 2.5 cm). This can be misleading, because the same individual can generate multiple stems. Estimating population size of fungi is equally challenging. If there is a distinct fruiting body, a

mycologist – a biologist who studies fungi – might estimate the abundance of mushrooms based on the number of fruiting bodies that are detected. Again, one problem is that the same individual can make numerous fruiting bodies. A second problem is that many fungi are very cryptic, and thus counts of fruiting bodies will probably be inaccurate.

Let's look at a simple example. Imagine you are given the task of estimating the abundance of morel mushroom fruiting bodies on the forest floor, but you only have time to sample 2% of the forest. Morels are very cryptic, so you estimate that you only detect 25% of them within your sampling area (Figure 10.5).

Say that you collect 19 mushrooms during your sampling session. We can then return to the basic formula we discussed at the beginning of this section to estimate the number of morel fruiting bodies:

$$N_{est} = n_s/p_{est}$$
$$N_{est} = 19/(0.02)(0.25) = 3800$$

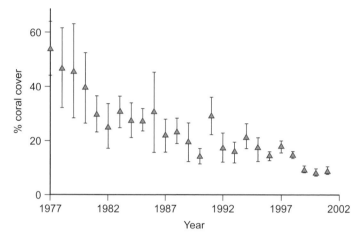

Figure 10.6. Mean percent cover (+95% confidence intervals) of coral in the Caribbean basin.

10.2 HOW DO MATHEMATICAL MODELS PROJECT POPULATION GROWTH RATES?

Now that we know how to measure population size in a variety of organisms, we will turn our attention to projecting population growth. Recall that population growth can be positive (increasing), negative (decreasing), or stable. We will develop models that correspond to two different reproductive life histories and examine how these life histories affect population growth patterns.

Discrete or continuous reproductive life histories

Population ecologists have constructed two slightly different mathematical models to describe population growth in species with different reproductive life histories. Many species have *discrete reproduction* life histories, producing offspring at discrete time intervals – usually during a specific season. For example, in temperate biomes, many annual plants will flower and be pollinated in the spring or summer, producing embryos that overwinter as seeds and resume development in the spring. Figure 10.7 shows how populations with discrete reproduction change over time.

Thinking ecologically 10.2

How often and when, over the course of a year, would you recommend sampling the population in Figure 10.7 to get an accurate estimate of how a population is changing from one year to the next?

In contrast, some species have *continuous reproduction* life histories, with individuals having the potential to reproduce at any time during the year. Humans are an example of a species with continuous reproduction. Thus reproduction is equally likely to occur for humans in January or July, or in any other month.

The dichotomy between discrete and continuous life histories is not absolute. For example, many annual plants shed their young embryos into the soil, where they may accumulate over several years before germinating (resuming growth and development), forming a seed bank. Within a seed bank, many of the individuals will die before they germinate, and many of the maturing adult plants will arise from different years or even decades. Many animals also don't fit into this dichotomy between continuous and discrete reproductive life histories. In many species, there is a distinct breeding season, but individuals may breed over the course of multiple years.

Discrete reproductive life histories and rapid growth

Let's consider the gray wolf, *Canis lupus*, as an example of a species with a discrete reproductive life history. Gray wolves

Much of the mathematics underlying estimations of population size require evaluating the fraction of organisms within your sampling area that you are able to detect.

For many modular organisms there is no obvious unit that can be counted. In these cases, ecologists may measure the *biomass* or dry weight of the organism. Unfortunately, this approach is destructive to the organism and may be infeasible for large organisms.

A noninvasive approach for assessing difficult-to-count organisms is to estimate the percent cover of an organism – the percent of its substrate that it covers as you look down at it from above. For small organisms, such as coral, ecologists may set up quadrats within the habitat, and estimate the percent of each quadrat that is covered by the organism. Digital cameras can be used to take pictures from above, and the percent cover can be accurately evaluated by computer analysis of the image of the organism(s) within the quadrat.

One study that used percent cover was a meta-analysis of coral abundance in the Caribbean Ocean between 1977 and 2001 (Gardner *et al.* 2003). The study's results highlight the sharp decline in percent cover over this time period from greater than 50% to about 10% over the 25-year period (Figure 10.6). We will discuss the complex issue of coral decline in later chapters. For now, we will explore how models developed by population ecologists describe how population size has changed in the past, and predict how it may change in the future.

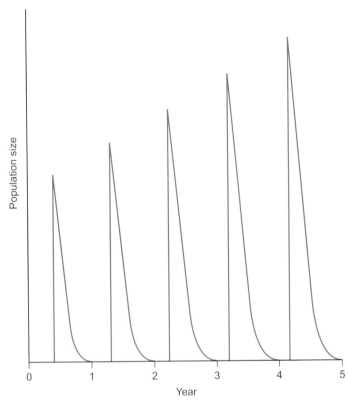

Figure 10.7 Population of an annual plant; only aboveground plants are counted. Each year there is one pulse of reproduction, followed by seasonal mortality. In species with discrete reproduction, accurate estimates of changes in population size require multiple samples. Sampling before or after the peak in population size could lead to an inaccurate estimate of the growth rate for that particular year.

are not a perfect example of a species with a discrete reproductive life history, because individuals have the potential to breed over the course of multiple years. We are using them as an example because they have a discrete reproductive season, and we have good data on them, which makes it easier to develop our model. Gray wolves generally live in a pack composed of a breeding pair, their offspring, and occasionally some unrelated adults. The average litter size for gray wolves is

about five pups per year; pups receive care from all pack members. Offspring disperse some time after 1 year, but don't mature until 2 or 3 years. In the United States, wolves were eradicated from most of their native range during the western expansion of the nineteenth century; decimation of the population was reinforced by a $20–$50 bounty granted to settlers for each dead wolf (US Fish and Wildlife Service 2013). By the time they were protected by the Endangered Species Act in 1973, gray wolves were extinct in the continental United States, with the exception of a few hundred individuals in the Upper Midwest (Figure 10.8).

A very small population of wolves immigrated to northwest Montana from Canada in the late 1970s. By 1985, Montana was host to 13 known individuals. In 1995 and 1996, the US Fish and Wildlife Service reintroduced 66 wolves from southern Canada into Central Idaho and Yellowstone National Park. Since then, the population has increased dramatically, despite continuous culling of the packs by the US Forest Service in response to wolves attacking livestock. Table 10.4 shows how the populations of gray wolves have increased since reintroduction.

The wolf populations at these two sites are unusual for two reasons. First, they are isolated by human development in the surrounding areas, so there is very limited immigration or emigration. Wolves that stray outside of protected areas are usually killed. In this simplified system, the change in population size from one year to the next is primarily the difference between births and deaths. Described mathematically:

$$N_{t+1} = N_t + B - D$$

where

$N_{t + 1}$ = population size at time $t + 1$
N_t = population size at time t
B = number of individuals born into the population between the time interval t and $t + 1$
D = number of individuals dying from the population between the time interval t and $t + 1$.

The second unusual feature of the wolf populations is that these wolves were reintroduced into a habitat with abundant resources – including some of their favorite food items, such as elk and deer – and with few competitors and predators. Under these conditions, populations can experience very high growth rates. Let's look at the numbers more carefully to derive a model that can describe population growth under these types of idealized conditions.

The geometric growth model

Note that larger wolf populations increased more rapidly than smaller populations between 1995 and 2006. For example, in 1997 there were 157 wolves, and the following year the

Figure 10.8. Gray wolf range in North America as of 1973 (green area). Range where wolves were extirpated before 1973 is shaded in pink, and range where wolves have been newly introduced is shown in pink and green. Data from Laliberté and Ripple (2004) and US Fish and Wildlife Service (2013).

Legend:
- Wolf range as of 1973
- Wolves extirpated before 1973
- Wolves reintroduced after 1973

0 500 1000 1500 2000 km

Table 10.4 Minimum fall wolf population following reintroduction in Central Idaho and Yellowstone National Park, 1995–2012 (US Fish and Wildlife Service 2013).

Recovery Area	Year	95	96	97	98	99	00	01	02	03	04	05	06	07	08	09	10	11	12
	Yellowstone	21	40	86	112	118	177	218	271	301	335	325	390	453	449	455	501	499	463
	Central Idaho	14	42	71	114	156	196	261	284	368	452	565	739	830	924	913	803	797	662
	Total	35	82	157	226	274	373	479	555	679	787	890	1129	1283	1373	1368	1304	1296	1125

population climbed to 226 wolves, an increase of 69 wolves. In contrast, in 2005 there were 890 wolves, and by 2006 the population climbed to 1129 wolves, an increase of 239. As a result of having more breeder individuals, large populations have the potential to increase more rapidly than small populations. Thus our model of population growth for species with discrete reproductive life histories must account for the effect of population size on population growth.

One approach is to assume that populations with discrete reproductive life histories increase or decrease by the same proportion every year (or whatever the most appropriate unit of time is). Using our wolf example, the number of wolves in 1998 was equal to the number of wolves in 1997 multiplied by 1.44. Population ecologists use the symbol λ (lambda) to designate the proportional change in the population from year to year. If $\lambda > 1.0$, the population is growing; if $\lambda = 1.0$, the population is stable, and if $\lambda < 1.0$, the population is getting smaller. Mathematically:

$$\lambda = N_{t+1}/N_t$$

We can rewrite this formula as

$$N_{t+1} = \lambda N_t$$

Similarly,

$$N_{t+2} = \lambda N_{t+1}$$

and

$$N_{t+3} = \lambda N_{t+2}, \text{and so on}$$

Thus the general formula for geometric population growth is

$$N_t = \lambda^t N_0$$

where

N_0 is the initial population size

Let's use $\lambda = 1.44$ from the 1997/98 data to project the growth of the total population (the bottom row of Table 10.4) under the geometric growth model. Recall that wolves were reintroduced in 1995 and 1996, so it would be misleading to include those years in our application of the model. If λ remains constant at 1.44, we should be able to project the expected wolf population from 1998 to 8 years later in 2006. We can compare the projected value to the observed value of 1129.

$$N_8 = \lambda^8 N_0$$
$$N_8 = 1.44(226) = 4178$$

Thus the model projects a value almost four times greater than is observed in the natural wolf population. There are several possible reasons for this overestimate. One explanation is that population growth may be slowing down as a result of increased crowding. A second explanation is that while population growth rates continue to be high, there is always going to be fluctuation in λ due to unpredictable factors such as weather. We can see if λ fluctuates from year to year, and if this fluctuation influences the projections of the geometric growth model, by redoing our projection using the next year of data, 1998/99, noting that there were 226 wolves in 1998, and 274 wolves in 1999.

$$\lambda = N_{t+1}/N_t$$
$$\lambda = 274/226 = 1.21$$

Notice that this value of λ is considerably lower than the value we calculated based on the previous year's data. Let's see how that influences our projection, which will now go from the year 1999 to 2006.

$$N_7 = \lambda^7 N_0$$
$$N_7 = 1.217(274) = 1041$$

This is a very slight underestimate in population size. Thus, although the geometric growth model can be used to project future population sizes, relatively small changes in λ will dramatically influence projected population size. One approach to addressing this problem is to calculate λ over more than one year, as we did in this example, to get a more realistic estimate of the population growth rate. Next we will consider population growth in species with continuous reproductive life histories.

Continuous reproductive life histories and rapid growth

Some species do have continuous reproductive life histories. Often these organisms are very small in size. Most bacteria and many protists are excellent models for studies of population growth in species with continuous reproductive life histories. Humans are an unusual example of a large species with a continuous reproductive life history.

For many decades, protistans in the genus *Paramecium* has been a model organism in several biological disciplines. These single-celled organisms move quickly, respond effectively to environmental changes, compete efficiently for resources, and show a phenomenal reproductive rate. This last feature makes them particularly interesting to population ecologists. Let's begin building our second model of population growth by taking a qualitative look at how a population of *Paramecium* may change over time.

One useful feature of paramecia is that they are easily cultured in the laboratory if provided with enough bacteria to feed on. If we start with ten paramecia in a beaker, and count the number of individuals at the same time each day, we can track the large increase in population size in subsequent days (Table 10.5), reflecting very rapid population growth.

Let's graph paramecia population size in relation to time. Notice how sampling only once per day makes the change in population size appear jumpy (Figure 10.9A). If we were to sample every 4 h, the increase in N would appear much

Table 10.5 Population size (N) for an expanding laboratory population of *Paramecium*.

Time (h)	0	24	48	72	96	120
N	10	20	40	80	160	320

(A)

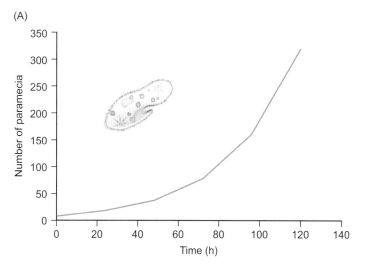

(B)

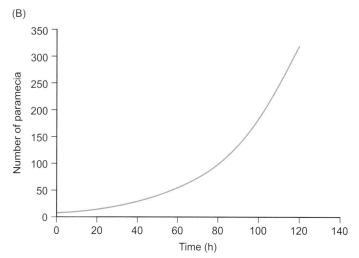

Figure 10.9 Paramecia population size in relation to time.
A. Sampling every 24 hours. B. Sampling every hour. Insert:
Paramecium aurelia.

smoother, and if we were to sample even more frequently, the curve would be even smoother (Figure 10.9B).

The exponential growth model

As we've discussed several times, in the initial stages of model construction we always make numerous simplifying assumptions. Our study of paramecia in a beaker with unlimited resources assumes by design that there is no immigration, emigration, or death due to resource limitation and interactions with other species. (Many organisms consider paramecia to be a fine meal, or at least a good appetizer.) For now, all we need to be concerned about is birth rates and death rates.

Modeling short-term population changes
The **exponential growth model** uses differences between birth rates and death rates to project changes in population size. Qualitatively, larger populations of paramecia grow more

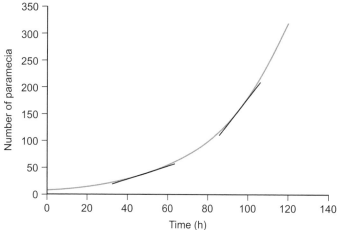

Figure 10.10 Paramecia population size in relation to time with tangent lines at 48 h and 96 h.

quickly. If we want to visualize this, we can draw a line tangent to the curve. The slope of the tangent line measures how fast the population is changing at that point in time where the tangent line contacts the curve (Figure 10.10). The tangent line at $t = 96$ h has a much steeper positive slope than the tangent line at $t = 48$ h, thus the population is growing much faster at 96 h, when N is much larger than it is at 48 h.

To measure how fast our population of paramecia is growing, we can count the number of individuals at time t (for example 96 h), and at time $t + x$, where x is a small interval of time later. Then the population growth rate is simply the number of individuals added to the population in interval x divided by the duration of interval x. We can describe this mathematically as

$$\text{Population growth rate} = (N_{t+x} - N_t)/X$$

If we enlarge the portion of the paramecia graph around $t = 96$ h, we can view how N changes in relation to t over a short time period (1 h). In this case, the population is growing at a rate of about 4.8 paramecia per hour (Figure 10.11). But if you look closely at the graph, you will note that the population is actually growing faster at the end of the time period (at hour 97) than at the beginning of the period (at hour 96). Thus an accurate estimate of how fast a continuously reproducing population is growing requires us to measure the change in N over an even smaller time period. In other words, x should approach zero. If x approaches zero, our new expression for population growth rate becomes

$$\text{Population growth rate} = \mathrm{d}N/\mathrm{d}t$$

when

$\mathrm{d}N =$ change in population size
$\mathrm{d}t =$ a very short time interval

We've established that the population growth rate, $\mathrm{d}N/\mathrm{d}t$, is the change in population size over a very short time interval.

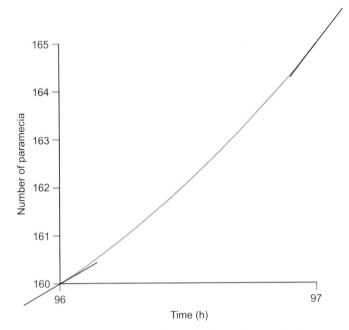

Figure 10.11 Paramecia population size in relation to time between 96 and 97 h. Notice how the tangent line at 97 h has a slightly steeper slope than the tangent line at 96 h.

We're now ready to consider the biological factors that could help explain how and why our population of paramecia is expanding under the simplified conditions we've established. If $dN/dt > 0$, the population is growing; if $dN/dt = 0$, the population is stable; and if $dN/dt < 0$, the population is getting smaller.

Factors driving continuous population growth models

Our simple continuous population growth model is driven by the difference between birth rates and death rates. For paramecia in a beaker, dN/dt is equal to the number of paramecia being born (B) minus the number of paramecia dying (D) over a small time interval.

Thus

$$dN/dt = B - D$$

The number of individuals born into a population (B) and dying within a population (D) will be directly influenced by population size. Thus a useful model will take this into account. The simplest way of forming this model is to create a new but related concept: per individual or *per capita birth rate* (b) and *per capita death rate* (d), which we can define as follows:

$$b = B/N, \ d = D/N$$

The unit for b is the number of individuals born into the population per individual over a very short time period. Similarly, the unit for d is the number of individuals dying in the population per individual over a very short time period.

Rearranging the two relationships:

$$B = bN \text{ and } D = dN$$

Substituting into our formula for population growth rate gives us

$$dN/dt = bN - dN = (b - d)N$$

Because discussions of $b - d$ are so important in population ecology, ecologists have coined a relationship as follows:

$$r = b - d$$

where r is termed the per capita growth rate.

Under idealized conditions – when birth rates are at a maximum (b_{max}), and death rates are at a minimum (d_{min}) – we will use the symbol r_m (m for maximum), which designates maximum per capita growth rate or the intrinsic growth rate when there is no competition or predation:

$$r_m = b_{max} - d_{min}$$

Thus our final model for population growth rate in a population with a continuous reproductive life history is:

$$dN/dt = r_m N$$

We can also rearrange this to estimate the value for r_m:

$$r_m = (1/N)dN/dt$$

Let's estimate r_m empirically from the paramecia data. According to Figure 10.11, there were 160 paramecia in the beaker at time 96 h and approximately 164.8 paramecia in the beaker at 97 h. Obviously you can't have eight-tenths of a paramecium, but we will use this figure as the closest approximation. Thus the overall population growth rate is:

$$(dN/dt) = 4.8 \text{ per hour}$$

and

$$r_m = 4.8/N = 4.8/160 = 0.029 \text{ per hour.}$$

This model has the advantage of describing population growth rate at any point in time. But it is difficult to make a projection of population size using the model in this form. Using the rules of calculus, we can transform this equation into a form that is more useful for making projections about population sizes at any point in time (Gotelli 2008). The somewhat more user-friendly model is:

$$N_t = N_0 e^{r_m t}$$

Where

N_t = population size at time t
N_0 = starting population size
e = base of the natural logarithm

You will notice that the product of r_m and t forms an exponent of the natural logarithm (e). As such, this model is called the exponential growth model.

Let's see if this model accurately predicts population growth in paramecia. Our observations are that there were 160 paramecia at 96 h. If $r_m = 0.029$, how many paramecia would we expect to find 24 h later (at $t = 120$ hours)? Let's put these numbers into our formula for exponential growth and solve for N_{120}.

$$N_{120} = N_{96}e^{(0.029)(24)}$$

$$N_{120} = 160e^{(0.696)}$$

$$N_{120} = 320.9$$

This is pretty close to the observed value of 320, particularly considering that we estimated an increase of 4.8 paramecia per hour by eyeballing the graph in Figure 10.11 and not through any calculation.

Both the geometric growth model and the exponential growth model make several simplifying assumptions. Let's identify these assumptions so that we can begin to make a biologically more useful model.

Simplifying assumptions of population growth models

One critical assumption of both the geometric and exponential growth models is that there is no immigration or emigration. Both immigration and emigration of wolves in Idaho and Yellowstone were largely restricted by the surrounding ranchland, and by hunters being allowed to kill wolves that venture off of restricted areas. Paramecia immigration and emigration were restricted because we confined all of the paramecia to the glass beaker. But most natural populations have considerable immigration and emigration.

Ilkka Hanski and his colleagues (1994) are studying a meta-population of the endangered Glanville fritillary butterfly (*Melitaea cinxia*) on an island in the Baltic Sea, off the coast of Finland. A metapopulation is a group of local populations inhabiting networks of somewhat discrete habitat patches. Habitat patches for these butterflies are dry meadows with populations of the goldenrod *Solidago lanceolata*, a favorite food for the caterpillars. We will now consider one aspect of the study: whether migration of individuals from one local population to another influenced the composition of local populations.

Over the course of one study, Hanski and his colleagues marked and released butterflies every day over a month-long breeding season from early June to early July. They also recaptured any butterflies they came across, so they could estimate death rates and migration rates. A butterfly that was marked and never recaptured was assumed to have died, while migration rate and distance were determined from where recaptured butterflies were discovered in relation to where they were marked.

Based on the number of marked individuals that were never recaptured, Hanski and his colleagues estimated daily death rates of 0.312 for females and 0.175 for males. They also documented

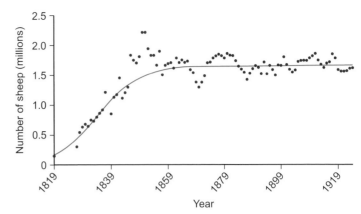

Figure 10.12 Population growth of sheep introduced into Tasmania in 1803.

very high emigration rates, particularly for females. By the ninth day of the study, 25% of surviving males and more than 75% of surviving females had emigrated to new populations. We will reconsider this study in greater detail in Chapter 11. For our present purposes, we should appreciate that migration between populations can profoundly influence the number of individuals in a population. The difference between male and female migration patterns is particularly important, because a female that migrates to a new population before ovipositing will presumably contribute all of her offspring to the new population in the following year.

A second critical assumption of the exponential growth model is that r is constant. This would be true only if resources essential for survival and reproduction were unlimited. These conditions could apply for new species moving into an area.

As an example, Europeans settled Tasmania in 1803, and quickly introduced some sheep onto the island as a source of meat and wool (Davidson 1938). Figure 10.12 shows a rapid increase in the number of sheep from 1819 and 1849. After that point, there are some significant oscillations, but the population is relatively stable around a mean of approximately 1.65 million sheep. This stability can arise from a decrease in birth rates, an increase in death rates, or both, as population size increases. Thus more realistic models must account for fluctuations in birth rates and death rates.

10.3 HOW DO DENSITY-INDEPENDENT AND DENSITY-DEPENDENT FACTORS INFLUENCE BIRTH AND DEATH RATES?

Population density is the number of individuals per unit area. As population size increases, population density must increase as well. Some factors, which we call density-independent factors, will influence population growth rates the same way regardless of population density. Examples include temperature, moisture, and unpredictable disturbances such as storms, mudslides, and rock falls. In contrast, other factors, which we call density-dependent factors, reduce population growth rates as population density increases, by decreasing birth rates and/or increasing death rates at high population densities. Common examples of density-dependent factors are disease and competition. Let's return to the Niagara Escarpment example that we introduced in Chapter 9 to discuss how density-independent and density-dependent factors have influenced the abundance of eastern white cedar, *Thuja occidentalis*.

Variable influence of density-independent and density-dependent factors on population growth

Because they are stationary, plants are often chosen as subjects for studies of population changes. Uta Matthes and Douglas Larson (2006) were interested in the period of seedling recruitment – the stage of a plant's life when it transitions from a seed to an established seedling. They wanted to identify the factors that influenced seedling recruitment of eastern white cedar, *Thuja occidentalis*, along the cliff edge and face of the Niagara Escarpment. This habitat is particularly well suited to studying seedling recruitment, in that white cedars are the only trees present on these cliffs, so competition with other tree species is not an issue.

Matthes and Larson studied the fate of all seedlings that emerged from a 25-m-wide section of cliff face between 1986 and 2003. They were particularly interested in whether seedling establishment was consistent from year to year or whether some years were better than others for seedling establishment. If there was year-to-year variation in seedling establishment, the researchers hoped to identify the factors underlying this variation. To do so, they needed to census seedlings on a fine time scale (*c.* every week), so they could observe most cases of seedling emergence and mortality.

The researchers noted great differences in the locations of successful seedling emergence. At the beginning of the study, 48.8% of the trees were found on vertical cliffs and 45.7% on horizontal ledges, with the remaining 5.6% at the cliff base. But emergence varied dramatically from year to year, and also in relation to cliff orientation. For example, in 1992 more than 200 seedlings emerged in horizontal sites, while approximately 30 emerged from vertical sites. In contrast, in 1986 fewer than 20 emerged from horizontal sites, while approximately 75 seedlings emerged from vertical sites (Figure 10.13).

The variation in total seedling emergence from year to year was influenced in part by climatic variation, a density-independent factor. Overall seedling germination was highest in years with low precipitation during the germination period, low maximum temperatures during the germination period, and lower mean temperatures during the previous summer.

Generally, seedling mortality was much higher in the summer than the winter, but the extent of the seasonal differences fluctuated from year to year. In 1993, approximately 500 trees died during the summer, while only three trees died in the winter. In contrast, in 1996, winter mortality exceeded

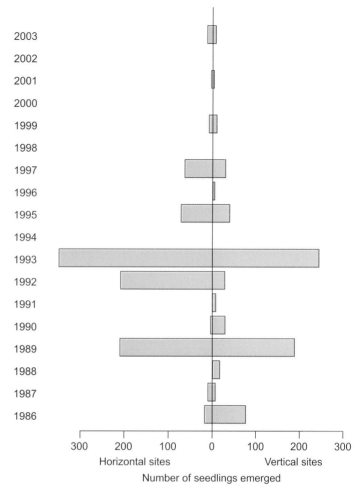

Figure 10.13 Eastern white cedar seedling emergence for the time period 1986–2003 in relation to orientation of microsite.

summer mortality. Part of the variation in mortality resulted from variation in emergence. Because so many seedlings emerged in the spring of 1993, there were more present to die in the summer of 1993.

But what caused high death rates? Unfortunately, the seedling forensic team could ascribe a cause of death to only 21% of the seedlings. The researchers discovered that the type of microsite profoundly influenced mortality in these seedlings. On ledges and crevices, more than half of the mortality was due to drought. In contrast, on vertical crevices, rock fall was a common cause of death, as falling rock exposed the shallow and vulnerable root system. Both drought and rock fall are density-independent factors. Lastly, at the cliff base, where the vertical cliff meets the horizontal shelf below, all identified mortality was due to pathogens. The researchers hypothesize that the moist microenvironment provided by the cliff base provided an ideal habitat for pathogen growth and development. Often, pathogens are assumed to have a density-dependent effect on population size. However, in eastern white cedar the researchers could not determine whether high-density seedling populations at cliff bases were more susceptible to pathogens than were low-density seedling populations. Let's consider another study that highlights the density-dependent and density-independent influence of disease on populations.

Influence of disease on population growth

The effect of disease on population size is very complex. Usually we think of disease as having a density-dependent effect on populations. We reason that as population density increases, the availability of host organisms increases as well, so disease should thrive. In many cases this is true, but as we will explore in Chapter 14, there are some cases in which the reverse is true. For example, some locusts are capable of tremendous population growth over short time periods. As they enter their rapid growth phase, locusts may produce antifungal substances that reduce the probability of fungal infection as their population density soars.

Let's consider one very early study of disease as both a density-dependent and density-independent factor. Edward Steinhaus (1958) wanted to understand the relationship between crowding and the rate of infectious disease in the alfalfa caterpillar, *Colias philodice*. He placed between 1 and 25 caterpillars in identical-sized containers, and kept temperature, humidity, and food availability constant among the different treatments. Mortality ranged from 32% for caterpillars housed individually up to 100% in the highest-density containers. There were two most common causes of death, at least as could be determined by the technology of 1958. The polyhedrosis virus killed between 10 and 25% of the caterpillars and did not appear to be influenced by caterpillar population density. In contrast, bacterial infections (of several types) became more prevalent at higher caterpillar density (Figure 10.14). This study demonstrates that disease can influence populations in a density-independent or density-dependent manner.

Many other biotic factors can influence population size in a density-dependent manner. As population size increases, the amount of food available to each individual decreases. This can result in a decrease in per capita birth rates, or an increase in per capita death rates, or both.

Food availability: a density-dependent factor influencing population size

There are many ways that food availability influences population size. As one example, let's consider the song sparrow (*Melospiza melodia*) population on Mandarte Island, a 6-ha island off the coast of British Columbia. The song sparrow population there fluctuated dramatically between 1960 and 1989, climbing above 140 individuals in 1963, 1979, and 1985, and plummeting below 40 individuals in 1980 and 1989. Peter Arcese and James Smith (1988) wanted to know what factors caused this fluctuation, and whether those factors related to differences in population density.

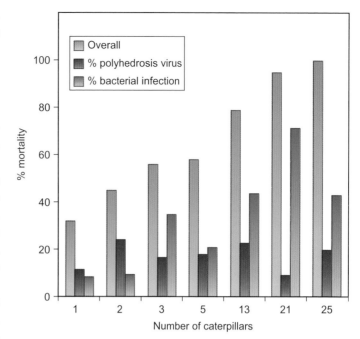

Figure 10.14 Mortality rate of *Colias philodice* in relation to population density

Because the island was so small and all of the birds were color-marked, Arcese and Smith were able to follow the reproductive history of all birds on the island. They followed the birds for the entire breeding seasons during 1975–86. The researchers were interested in six factors that could be influenced by population density. Figure 10.15 shows the results of their analysis. While there is considerable variation in the data, as population density increases, there is a significant decline in clutch size and number of independent young per female, and there is an increase in nest failure rate. Together these findings suggest that high population densities are influencing both birth rates and death rates in this population.

Arcese and Smith suspected that in years with high population density there was greater competition for food among the birds, and less food available per reproductive effort. Thus food availability was influencing population size in a density-dependent manner, reflected in between-year differences. To test this hypothesis, the researchers supplemented 18 breeding pairs with food (served in bird feeders), and compared reproductive output of the supplemented birds to the reproductive output of 40 unsupplemented breeding pairs (controls). On average, supplemented birds laid their first egg significantly earlier, had more eggs per clutch, initiated more nests, and had a greater percentage of successful nests than the controls (Table 10.6).

Predation led to complete nest failure in 45% of the control nests and only 36% of the supplemented nests. More striking, brood parasitism by the brown-headed cowbird, *Molothrus ater*, is common in this population. Female cowbirds can dump an egg

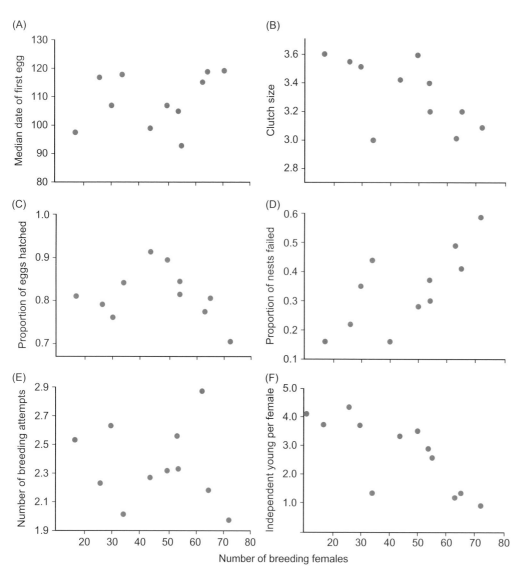

Figure 10.15 Relationship between population density (as measured by the number of breeding females) and six reproductive parameters in the song sparrow. A. Median date of first egg (Jan. 1 = 1). B. Clutch size ($P < 0.05$). C. Proportion of eggs hatched. D. Proportion of failed nests ($P < 0.05$). E. Number of breeding attempts. F. Independent young produced per female ($P < 0.05$).

Table 10.6 Comparison of reproductive success in birds supplemented with food and unsupplemented controls.

	Supplemented	Control	T-value	P-value
Mean date of first egg	April 10	April 28		
Mean number of eggs	3.31	2.77	3.45	<0.001
Mean egg mass (g)	3.08	3.03		
Proportion of eggs hatched (in nests with no predation)	78%	70%	0.90	>0.10
Mean number of nesting attempts per female	2.81	2.24	2.27	0.02
Nesting success (% that fledged at least one young)	60%	45%		

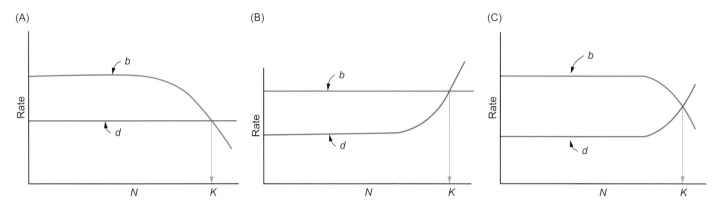

Figure 10.16 Changes in birth rates (*b*) and death rates (*d*) as population density increases. A. Increase in density leads to lower birth rate. B. Increase in density leads to higher death rate. C. Increase in density leads to lower birth rate and higher death rate. In all cases, the carrying capacity is reached at the population size where the two curves cross.

into the song sparrow nest while the song sparrow is off foraging. Often the cowbird will remove a sparrow egg in the process. Supplemented nests suffered much lower rates (18%) of parasitism than did control nests (45%). Arcese and Smith note that supplemented females guarded their nests much more diligently than did control females that were often far away looking for food. Consequently, supplemented females song sparrows could effectively chase away cowbirds and other predators and parasites.

Overall, pairs provisioned with food produced about four times as many fledglings as did control birds. Taken together, Arcese and Smith's observations and experiments provide strong support that density dependence regulates the population of song sparrows on Mandarte Island.

 Thinking ecologically 10.3

Consider all of the song sparrow evidence collected by Arcese and Smith. What is the evidence that density dependence regulates populations of song sparrows by reducing birth rates at high population density? What is the evidence that density dependence regulates population size by increasing death rates at high population density?

From a theoretical standpoint, density dependence can regulate populations in any of three ways. It can decrease per capita birth rates, increase per capita death rates, or both (Figure 10.16). The ultimate result is that as population size increases, the two curves graphed to reflect population changes – per capita birth rate and per capita death rate – will cross. The population size at which the per capita birth rate is equal to the per capita death rate is the carrying capacity of the population.

10.4 HOW DO ECOLOGISTS USE THE CARRYING CAPACITY TO MODEL DENSITY-DEPENDENT CHANGES IN BIRTH AND DEATH RATES?

Ecologically, the carrying capacity (signified with the letter K) is the maximum population size that can be supported or sustained by the environment. It encompasses all of the density-dependent factors that influence population growth rate. A more realistic model of population growth rate should incorporate the carrying capacity into its formulation. Our realistic model would show near-exponential growth at low population size, but it would reflect that growth should slow down and eventually stop as the population size approached the carrying capacity.

The logistic growth model

To accurately assess population changes, this concept of how populations approach carrying capacity must be incorporated into models. As population size increases, we wish to decrease the value of dN/dt, so that population growth rate slows as population size nears the carrying capacity, and population growth rate stops ($dN/dt = 0$) when population size reaches the carrying capacity (Figure 10.17A). Fortunately, there is a relatively easy way of achieving these desired dynamics, which is to multiply the right side of the exponential growth formula by $(1 - N/K)$. This gives us the logistic growth equation or logistic model of population growth.

$$dN/dt = r_{m}N(1 - N/K)$$

It may be helpful to think of $1 - N/K$ as the unused portion of the carrying capacity (Gotelli 2008). At very low population size, the unused portion of the carrying capacity is very high; mathematically, it is close to a value of 1. Thus the population growth rate will be nearly exponential at low population sizes. As population size increases, the unused portion of the carrying capacity decreases, and the population growth rate slows down. When a

(A)

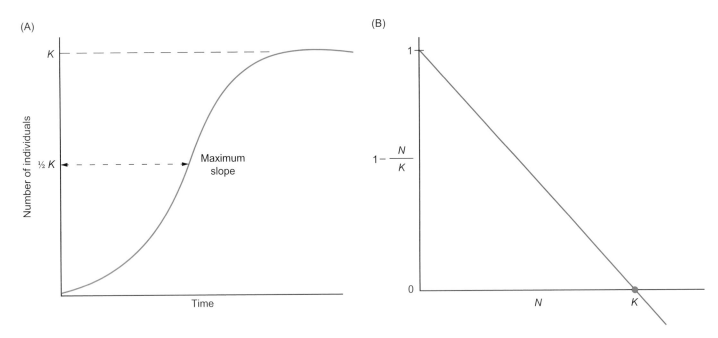

(B)

Figure 10.17 A. A more realistic representation of population growth. Note how the slope of the curve is at a maximum when $N = 0.5K$. B. At very low population size, the unused portion of the carrying capacity is close to 1. But as population size approaches the carrying capacity, $(1 - N/K)$ approaches 0. Notice that if population size exceeds the carrying capacity, then $(1 - N/K)$ is a negative number, which will cause the population to decline.

population reaches its carrying capacity, then $N = K$ and $(1 - N/K)$ is equal to $(1 - 1) = 0$. This achieves the desired result of reducing the population growth rate to zero when the population reaches its carrying capacity (Figure 10.17B). Thus population size (N) is stable when $N = K$.

 Thinking ecologically 10.4

Assume $K = 100$. Draw a graph of $(1 - N/K)$ (y-axis) versus N (x-axis) for several values of N. Describe the relationship between population size and the unused portion of the carrying capacity.

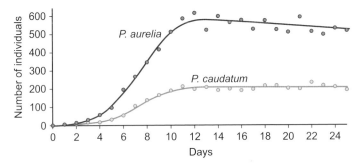

Figure 10.18. Population size of *P. aurelia* and *P. caudatum* when they are grown in separate culture vials.

Logistic growth in the laboratory and the field

G. F. Gause (1934) pioneered experimental population ecology. One of his favorite research organisms was *Paramecium caudatum*. Using carefully regulated media, which had to be changed daily, and taking great care to feed each culture vial with the same quantity of bacteria, he was able to show that *P. caudatum*, and a second species, *Paramecium aurelia*, both showed approximately logistic growth when grown separately in their own culture vials (Figure 10.18). Note that the carrying capacity for *P. aurelia* is about three times as great as for *P. caudatum*. Gause explains that this threefold difference is probably a result of *P. caudatum* being approximately 2.6 times as massive as *P. aurelia*.

One of the luxuries of working in the laboratory is that researchers can define their starting conditions. Thus Gause

could begin each experiment with the same number of individuals, and raise them under identical conditions. Of course there always is some experimental error even in the laboratory. But studying populations in the field is much more challenging, because it is usually difficult to define the starting conditions, and to control resource availability.

Saara DeWalt (2006) studied changes in the population size and the population growth rate of an invasive shrub, *Clidemia hirta*, on the Island of Hawaii. This plant is also known as Koster's curse in honor of the man who accidentally introduced it to Fiji, where it successfully competed with coffee plants. DeWalt's goal was to identify approaches to control the spread of this invasive plant, but first she needed to describe how its population was changing over the 4 years of her study. She studied two different forest populations, and was able to show that the low-density population was increasing rapidly while the high-density population was leveling off (Figure 10.19A).

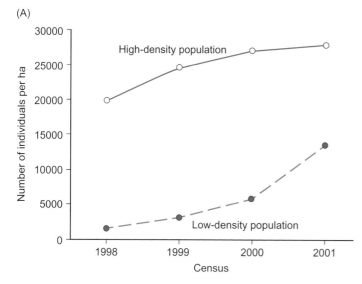

Figure 10.19 A. Population censuses over the four years of the invasive shrub study. B. Values of λ over the four years of the study.

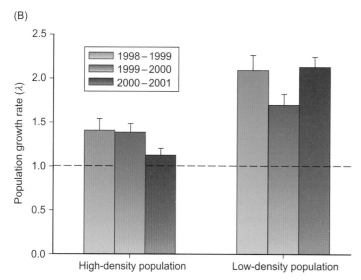

Based on the number of individuals alive at her census date for each year, she was able to show that λ was significantly above 1.0 in the low-density population, and was only a little above 1.0 in the high-density population (Figure 10.19B). Thus the high-density population may be approaching its carrying capacity.

We will discuss biological invasions in more detail in Chapter 11. For now, we will consider the population biology of

Earth's most dominant invasive species, *Homo sapiens*. Over the past millennium, human distribution and abundance have increased in dramatic fashion. These increases in both distribution and abundance are particularly impressive considering that humans do not seem like good candidates for rapid increases based on their reproductive life history traits. Unlike most invasive species, humans have unusually long generation times, long gestation periods, and small litter size. Despite these disadvantages, humans have proliferated in almost every biome, presumably because humans can transform their environments in ways that increase their birth rates and decrease their death rates.

10.5 WHAT FACTORS WILL INFLUENCE THE GROWTH RATES OF FUTURE HUMAN POPULATIONS?

As with all populations, human population growth rates are determined by a variety of factors such as fertility, death and migration rates, and age distribution. Globally, the human population growth rate is beginning to decrease (Figure 10.20). This trend is the result of the interplay of enhanced economic development and family planning education, which both tend to decrease fertility rates. In some regions, this slowdown is tragically supported by increased death rates from HIV/AIDS. Let's look at the data more carefully, so that we can analyse these trends.

Global slowdown in human population growth

Over the past 50 years, annual human population growth rates have declined from a peak of 2.22% in 1962 to about 1.11% in 2012. Table 10.7 shows the importance of distinguishing between net population increase (i.e., recruitment) and population growth

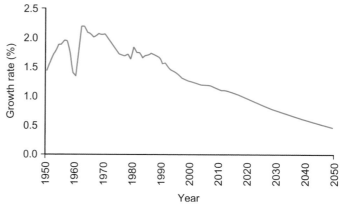

Figure 10.20 Human population growth rates since 1950, including projections to 2050. The decline in the late 1950s was a result of the communist Great Leap Forward in China, when natural disasters and decreased agricultural output associated with major social reorganization led to very high mortality rates.

Table 10.7 Human population growth since 1952 from the US Census Bureau Database. N_t is the number of individuals estimated in the annual survey (for example 1951 for the first row of data), and N_{t+1} is the number of individuals estimated in the next year of the survey (which is 1952 for the first row of data). The net population increase (N_i) = $N_{t+1} - N_t$. The annual growth rate is $N_i/N_t \times 100$.

Year	N_t	N_{t+1}	Net population increase (N_i)	Annual growth rate
1952	2 636 732 631	2 681 994 386	45 261 755	1.72
1957	2 891 211 793	2 947 979 287	56 767 494	1.96
1962	3 139 919 051	3 209 631 895	69 712 844	2.22
1967	3 490 051 163	3 562 007 503	71 956 340	2.06
1972	3 866 158 404	3 941 664 971	75 506 567	1.95
1977	4 231 636 519	4 303 675 842	72 039 323	1.70
1982	4 613 830 568	4 694 935 057	81 104 489	1.76
1987	5 025 796 394	5 113 007 284	87 210 890	1.74
1992	5 455 057 523	5 537 583 721	82 526 198	1.51
1997	5 858 582 659	5 936 039 484	77 456 825	1.32
2002	6 243 351 444	6 319 822 330	76 470 886	1.23
2007	6 629 668 134	6 708 196 774	78 528 640	1.18
2012	7 017 543 964	7 095 217 980	77 674 016	1.11

rates. Large populations can achieve a higher net increase than small populations, even if the large populations have a relatively low annual growth rate. For example, 1962 had the highest annual growth rate of any year in the survey (2.22%) and added over 69 million to the human population. In contrast, 2012 had the lowest annual growth rate of any year in the survey (1.11%), yet it added over 76 million people to the human population.

Much of the decline in human population growth rates stems from a global decline in number of offspring. In 1950–5, the average number of offspring was 4.97 children per woman; by 2005–10, this was down to 2.53 (United Nations 2013). But number of offspring is highly correlated with level of economic development, as family planning efforts increase with improved standards of living and higher levels of education. Though birth rates have fallen in most countries, they remain much higher in the poorest countries.

Some of the decline in human population growth rate results from increased mortality from HIV/AIDS, which has its greatest impact in southern (sub-Saharan) Africa. For example, in Botswana, about 36% of the population is infected with HIV, and the average life expectancy has declined from 62 years in 1980–5 to about 47 years as of 2010 (United Nations 2013).

AIDS has influenced mortality patterns disproportionately in different age classes. Before AIDS was widespread in sub-Saharan Africa, mortality was highest in young and old age classes, and relatively low among young adults. As a result of the AIDS pandemic, by the early twenty-first century, almost 50% of the population in sub-Saharan Africa died between age 20 and 40 (Figure 10.21). This has the potential to significantly change the age structure of the population – the proportion of individuals in different age groups.

Age structure and population growth

Rapidly growing human populations generally have more young people, who are likely to reproduce, while numerically declining

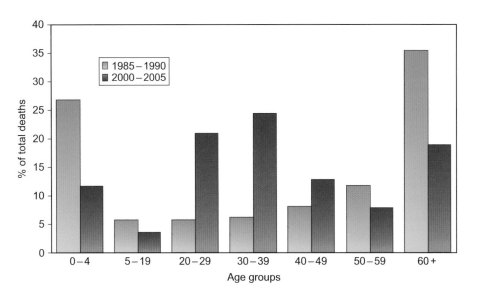

Figure 10.21. Percent distribution of deaths by age group in sub-Saharan Africa before the spread of AIDS (1985–1990), and after AIDS became widespread (2000–2005).

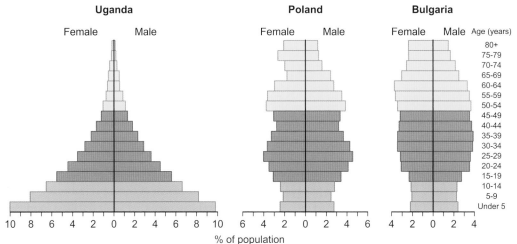

Figure 10.22. Age structure diagrams of the rapidly growing population in Uganda, the stable population in Poland, and the slowly declining population in Bulgaria. Data from United Nations (2013).

populations tend to have a higher proportion of older people, who are beyond their reproductive years. Age structure diagrams can show these differences, and can be used to make quick projections of future population growth. Let's compare the age structure of three countries: Uganda, which is growing exceptionally quickly; Poland, which is relatively stable; and Bulgaria, which is slowly declining in population.

One striking feature of this comparison is that almost half of Uganda's population is primarily pre-reproductive (less than 15 years old), while only about 15% of Poland's and 13% of Bulgaria's population are primarily pre-reproductive. On the other side of the age spectrum, less than 8% of Uganda's population is primarily post-reproductive (older than 50 years old), in contrast to 35% for Poland and 40% for Bulgaria (Figure 10.22). Thus we are not surprised to learn that Uganda is growing more rapidly than Poland and Bulgaria, and we might suspect that Bulgaria, with its large post-reproductive population, might actually be shrinking in population. But looking at an age structure diagram does not allow

us to make accurate projections into the future; for example, we don't know whether Uganda's small over-40 population is a consequence of a high reproductive rate, or a reflection of a short life expectancy. Thus age structure diagrams are useful for picturing the distribution of people in different age groups, and provide some insight into future changes in population size. However, a more complete picture needs more information.

Projecting future population growth based on birth, death, and migration patterns

Based on our discussion of age structure, we might expect that in Uganda, per capita birth rates would be greater than per capita death rates, because there are so many reproductive people, and there are relatively few old people in the population. In contrast, Poland might have similar birth and death rates, and in Bulgaria, with its high frequency of old age classes, the death rate might be predicted to exceed the birth rate. Table 10.8 supports this expectation.

Table 10.8 Annual per capita birth rates (*b*), death rates (*d*) and migration rates (*m*) for human populations in Uganda, Poland, and Bulgaria in 2005–10. A positive migration rate means that immigration > emigration, while a negative migration rate means that emigration > immigration (United Nations 2013).

	Birth rate (*b*)	Death rate (*d*)	*b* − *d* = *r* (per capita growth rate)	Migration (*m*)	*b* − *d* + *m*
Uganda	0.046	0.011	+0.035	−0.001	+0.034
Poland	0.010	0.010	0.000	0.000	0.000
Bulgaria	0.009	0.015	−0.006	−0.002	−0.008

One important message from these data harkens back to the introduction to this chapter, where we point out that one feature that simplifies studying Rapa Nui is the absence of migration after the initial colonization event, at least until more modern times. Population growth is influenced by the difference between per capita birth and death rates (the column headed by (*b* − *d* = *r*) in Table 10.8), and in addition by the difference between per capita immigration and emigration rates (*m*). The final column in Table 10.8 (headed by (*b* − *d* + *m*)) gives a more accurate picture of population change, because it calculates the interaction of per capita birth, death, and migration rates for each country.

The numbers in this table are small and inconvenient, but remember that they are per capita (per individual) rates of change, and that they are calculated on an annual basis. Thus (*b* − *d* + *m*) is a description for the annual per capita growth rate for a population that experiences migration. Because humans have continuous reproductive life histories, we will modify the formula for exponential growth to include the per capita migration rate. This modified formula is

$$N_t = N_0 e^{(r+m)t}$$

If we assume that Uganda maintains its current rates of birth, death, and migration (*r* + *m* = 0.034), we can project the number of individuals we would expect to find in Uganda in 2050 based on the United Nations estimate of *N* = 33 987 000 in 2010. This assumption means that we are not anticipating any influence from density dependence.

$$N_{40} = 33\ 987\ 000 e^{(0.034)(40)}$$
$$= 33\ 987\ 000 e^{1.36}$$
$$= 132\ 419\ 922$$

However, the United Nations actually projects a much smaller population of 104 078 000 for Uganda in 2050. This smaller projection arises because population ecologists working for the United Nations expect that Uganda's birth rate will drop considerably by 2050. Like all projections, these population estimates make several simplifying assumptions.

Assumptions of population growth projections

As we've discussed previously, all models make assumptions, and all assumptions introduce some error into a model. Sometimes we are forced to make assumptions because good data are simply not available. Other times we make assumptions because they help keep the model relatively simple. Perhaps the simplest assumption of all population projections is that the population sizes for current and past years are correct. We know this assumption is false, but our only reasonable choice is to go with the best available data.

The United Nations actually makes six different projections of human population growth between 2010 and 2050, each of which has its own set of assumptions. Each of these projections is called a variant. All of the variants assume that high fertility countries that until now have experienced no reduction in fertility will go through a process in which fertility begins to decline, and ultimately levels off at a much lower level. This transition is associated with increased industrialization and economic development. There are many assumptions that can be incorporated into these models; for example, the high fertility variant assumes that fertility in developing countries drops to 2.6 children per female, the medium variant drops to 2.1 children per female (approximately replacement level), while the low fertility assumption drops to 1.6 children per female. The medium variant is considered most likely, so we will explore some of the other assumptions associated with this model.

A second assumption of the medium variant is that death rates will decline in all countries, and will decline most quickly in countries that currently have the highest death rates. The projected decrease in death rates is based on how death rates in other countries have decreased in the past. Clearly this assumption is highly flawed. For example, in 1980, very few population ecologists could have predicted the devastating effect AIDS would have on human death rates in sub-Saharan Africa. A third assumption of the medium variant is that future migration patterns will be similar to past migration patterns, with the exception that future migration patterns were adjusted for recent or anticipated political changes in migration policy for each country. This assumption is also incorrect in some cases; for example, Bulgaria's population declined dramatically in the early 1990s due to high emigration rates associated with the fall of the Soviet Union. Finally, the medium variant assumes that improved antiretroviral treatments will increase the life expectancy of patients suffering from AIDS.

Most population analyses, even those of current populations, contain some inaccuracies, which depend on the quality of data collection and on the accuracy of underlying assumptions. In Rapa Nui new data are surfacing regarding past events. Similarly, projections of human population growth are constantly being revised as new data describe current and recent trends. Let's now revisit Rapa Nui to explore how new data have caused some scientists to revise their perspectives of what actually happened to the human populations there after colonization.

REVISIT: Rapa Nui reconsidered

In the introduction to this chapter, we discussed Jared Diamond's (2005) hypothesis that Polynesian colonizers arrived at Rapa Nui at about AD 900, increased in population at a 3% annual rate, harvested all of the *Jubaea* palm trees, and ultimately were responsible for the crash of their own population. Terry Hunt and Carl Lipo (2006) have proposed an alternative hypothesis that argues that there is very little evidence for colonization before AD 1200. They reject most of the pre-1200 fossil material as having been contaminated by older carbon, which caused inaccurate age estimates. They also reject all analyses that were based on only one sample collection, arguing that it is essential to have more than one sample from a stratigraphic layer to evaluate the variation in the age estimates. To test their hypothesized colonization date, Hunt and Lipo collected multiple samples of charcoal, and plant and animal remains from a beautifully preserved stratigraphic column from Anakena (Figure 10.23).

Figure 10.23 Stratigraphic column from Anakena. The different colors and textures are a result of different soil composition; for example layer III averages 18% clay while layer II has no clay. The deepest layer was radiocarbon-dated to 1250–1410. Layers IV–VII had prehistoric artifacts. Remember that, in general, artifacts from deeper in the column are older than artifacts from nearer the surface.

The animals included the now-extinct Polynesian rat (*Rattus exulans*), which arrived with the first colonizers, and the plants included the extinct giant *Jubaea* palm, which we discussed earlier. Farther down the stratigraphic column there is a very distinct transition point, below which there is no evidence of any human-related artifacts. Six of eight charcoal samples from just above that transition had an estimated date of AD 1280–1400. Two samples were estimated from AD 1130. Accepting some error in even the most carefully collected samples, Hunt and Lipo argue for a settlement date at Rapa Nui of about AD 1200.

Hunt and Lipo argue that the later colonization date does not allow enough time for the population explosion envisioned by Diamond. If we begin with 50 colonists, and the population increases at the rate of 3% per year (which is a very rapid growth rate for humans), the population by AD 1350 will only be about 4500 individuals. Hunt (2006) estimates that the population leveled off at about 3000–4000 individuals in the fourteenth century, and stayed at that level until European colonization in the eighteenth century.

Both Diamond and Hunt agree that the fossil record indicates that the Polynesian rat was a major staple of the colonizer's diet. Hunt and Lipo (2011) estimate that the rat population was probably above 3 million. But Hunt suggests that the Polynesian rat, not deforestation by humans, was responsible for the extinction of the *Jubaea* palm. Some palm nuts discovered in caves have rodent tooth marks. Hunt also argues that the *Jubaea* palm's decline in the pollen record began before the human population increase, but after the rat population increase. To support his argument, he cites archeological research by Stephen Athens and his colleagues (2002) on the Hawaiian Island of Oahu that links deforestation of a major section of the island to the spread of the Polynesian rat prior to human settlement.

The debate over the recent history of Rapa Nui highlights some of the challenges of reconstructing past populations. Some of these challenges will be resolved as more data and more technology become available. For example, at Rapa Nui, recently discovered fossil remains of large expanses of charred stumps and roots from *Jubaea* palms indicate that humans cut down and burned large stands of forest, in order to create agricultural fields. In addition, some recent finds show that less than 10% of the palm nuts were gnawed by rats (Mieth and Bork 2010).

Equally daunting challenges apply to efforts to predict future population growth on a national and global basis. For example, until recently the United Nations World Population Prospects data assumed that developing countries transition from high to low fertility in a similar manner. More recent data indicate that different countries experience different, but predictable, patterns of fertility decline in association with development (United Nations 2013). By understanding these patterns of fertility decline, population ecologists can make more informed projections of future population growth that are unique to each country.

Regardless of which assumptions are closer to reality, there is no doubt that humans are the most dominant species on the Earth. Patterns of resource use and environmental disturbance profoundly influence the world's ecosystems on a local level, as on Rapa Nui, and on a more global level. As a result of our activities, many species have already gone extinct, and numerous species are going extinct every day. Chapter 11 will use some of the concepts developed in this chapter to explore the science of conservation biology – a science that has sprung into existence as a result of human dominance.

SUMMARY

Population ecologists work in three time frames: the past, present, and future. Research in each time frame has its own set of challenges, tools and assumptions. Historical studies of populations often use fossil evidence to make inferences of past distributions and abundance of populations of different species. In Rapa Nui, and other studies of human populations, ecologists also use cultural remains to help with their inferences.

Only rarely can ecologists accurately count the numbers of individuals within a present-day population. Instead they rely on a variety of tools and techniques to estimate population size and population growth rates. Ecologists have identified density-independent factors, such as temperature, rainfall, and disturbance, and density-dependent factors, such as competition and disease, that influence population growth. Once growth rates are estimated, ecologists can apply mathematical models to make projections of future population size, which are particularly important for making management decisions about endangered species. Population models are also applied to human populations, allowing planners to anticipate resource needs in regions of the world that will experience substantial changes in population size in future decades.

FURTHER READING

Calambokidis, J., and Barlow, J. 2004. Abundance of blue and humpback whales in the eastern North Pacific estimated by capture–recapture and line-transect methods. *Marine Mammal Science* 20: 63–85.

Difficult article to understand completely, because some of the methods are unfamiliar. If you can get past that problem, this is a good article for using mark–recapture and line-transect approaches to estimate whale population abundance. Nice discussion of sources of uncertainty and of assumptions made by the researchers.

Campbell, A. H., Vergés, A., and Steinberg, P. D. 2014. Demographic consequences of disease in a habitat-forming seaweed and impacts on interactions between natural enemies. *Ecology* 95: 142–152.

Very nice study of impact of disease on populations of red seaweed. Demonstrates that the impact of disease goes well beyond a simple impact on survival, and that many factors need to be considered to understand population growth rates.

Hollowed, A.B., and Sundby, S. 2014. Change is coming to the northern oceans. *Science* 344 (6188):1084–1085.

Describes the changes in the abundance of cod and pollock in northern oceans, and discusses the influence of global warming on these changes.

Morell, V. 2014. Science behind plan to ease wolf protection is flawed, panel says. *Science* 343: 719.

This brief article addresses updates the status of the grey wolf in the United States. This is a nice introduction to an important issues we will soon address: what is a species?

Ripple, W. J., and 13 others. 2014. Status and ecological effects of the world's largest carnivores. *Science* 343: 1241484.

Far-ranging discussion of changes in abundance and distribution of the world's largest carnivores. This is a great introduction to questions such as why the average citizen should care about carnivore populations, and to what the future holds for these species.

END-OF-CHAPTER QUESTIONS

Review questions

1. What are the two hypotheses for the historical reconstruction of changes in Rapa Nui's population over the past 12 years? What type of evidence is used to support each hypothesis?

2. Your ecology class conducts a mark–recapture study on grasshoppers in an abandoned field near campus. At the beginning of lab, your class marks and releases 75 grasshoppers. An hour later, you recapture as many grasshoppers as you can, and find 12 marked individuals out of a sample of 120. How many grasshoppers do you estimate were in the population? What assumptions are you making with this estimate?

3. Start with a population of 10 paramecia in a beaker. Assume $r_m = 0.029$ per hour. How many paramecia will there be in 10 days, assuming exponential growth. What are the assumptions of the exponential growth model, and which are likely to apply to your paramecia?

4. Based on reading this chapter, list as many density-independent and density-dependent factors as you can. Then discuss whether any of these factors could be both density-independent and density dependent. Explain your reasoning.

5. Explain how human population growth rates are declining steadily, yet each year still has a large increase in the number of people added to the population?

Synthesis and application questions

1. There is good evidence that morel mushroom abundance is greatest in forests that have been recently burned. Assume you are asked to estimate morel fruiting body abundance in a forest of 10 000 ha, of which 10% had been recently burned. Would you focus your sampling efforts more on burned or unburned areas? Justify your answer.

2. Arcese and Smith's research on song sparrows (1988) showed that food-supplemented birds, on average, initiated their first clutch of eggs significantly earlier than unsupplemented controls. However, the unsupplemented birds were much more variable in date of first egg than the supplemented birds, with several nests initiated on the same schedule as the supplemented birds. How would you determine whether there is something special or different about the unsupplemented birds that initiated their nests as early as the supplemented birds? Include in your answer what you would measure or compare.

3. The differences between the age structures of Poland and Bulgaria are somewhat subtle. Based on Figure 10.22, how would you quantitatively describe the differences between the age structures of those two countries?

4. Return to the accumulation curve analysis used by Kohn and his colleagues (Figure 10.4). Suppose your boss gave you the task of counting and identifying the invertebrates in 10 mud samples from a lake bottom. How might you use an accumulation curve approach to decide when you should quit looking through one mud sample and move on to the next one?

5. Hunt and Lipo (2006) argue that if Rapa Nui were colonized by 50 humans in AD 1200, that there was simply not enough time for the population to grow to 15 000 by the year AD 1350, even assuming a 3% annual growth rate. How many years would it take the human population to reach 15 000 under these assumptions? Let's assume that the colonizing party was somewhat larger – for example 100 humans? How long would it then take the population to reach 15 000?

Analyse the data 1

Beginning with the information that human population growth rate peaked in 1962 at 2.22% per year, use the data from Table 10.7 to test the hypothesis that there has been a linear decline in growth rates since that peak in 1962. Then explain why it would be unreasonable to use those data to project future population growth rates.

Analyse the data 2

The population of Bulgaria in 2010 was estimated at 7 389 000. Use the data from Table 10.8 to predict the population of Bulgaria in 2050. How close was your prediction to the UN's prediction in the latest revision of World Population Prospects?

Chapter 11
Conservation ecology

INTRODUCTION

One sad episode in the history of conservation ecology is the decline of the ivory-billed woodpecker, *Campephilus principalis*, which is presently so rare that experts cannot agree if it still exists. In the preface to his book *The Grail Bird*, Tim Gallagher (2005) writes,

> But every so often this obsession would flare up again, and I would start dreaming about swamps and imagining what it would be like to find an ivory-bill. The bird is so iconic: big, beautiful, mysterious – a symbol of everything that has gone wrong with our relationship to the environment. There is such a finality about extinction. I thought that if someone could just locate an ivory-bill, could prove that this remarkable species still exists, it would be the most hopeful event imaginable: we would have one final chance to get it right, to save this bird and the bottomland swamp forests it needs to survive.

Since the last documented ivory-bill observation in Louisiana in 1944, numerous sightings have been reported, but no one has been able to convince the scientific community that ivory-bills still exist. Most twenty-first-century ornithologists believed that the ivory-billed woodpecker was extinct. In June 2005, *Science* magazine published an article entitled "Ivory-billed woodpecker (*Campephilus principalis*) persists in continental North America" (Fitzpatrick *et al.* 2005). The title says it all, and the scientific community was dumbfounded, ecstatic, and skeptical.

Unfortunately, as we discuss in this chapter, even if this report is correct, conservation ecologists will face a major challenge to restore this tiny remnant population of ivory-billed woodpeckers to a sustainable size. Small populations are at a greater risk of extinction than large populations due to low levels of genetic diversity and fluctuations in population size arising from disturbance and other chance events. To project future population growth and predict whether a species is likely to go extinct in the near future, conservation ecologists can use mathematical models as part of a population viability analysis. High immigration rates can sustain populations and help them increase in size, while reduced immigration rates can lead to extinctions. And lastly, as happened with the ivory-billed woodpecker, human-mediated changes to essential habitat can cause a species to become endangered, or to go extinct.

KEY QUESTIONS

11.1. Why do very small populations have a high extinction rate?

11.2. How do ecologists use mathematical models to predict population viability?

11.3. How can immigration of individuals from nearby populations maintain species richness and high population size.

11.4. How can human-mediated changes to habitats cause species to become endangered, or to go extinct?

CASE STUDY: Conservation ecology and the decline of the ivory-billed woodpecker

The Society for Conservation Biology has over 5000 members worldwide, including students, faculty members, and working scientists at all levels. Its mission is "to advance the science and practice of conserving Earth's biological diversity" (Society for Conservation Biology 2010). Biological diversity has different meanings in different contexts. Genetic diversity is a measure of differences in nucleic acid composition among individuals within a population or species. Recall from Chapter 3 that populations may be polymorphic – have more than one allele – for each gene in the genome. Populations that are polymorphic for a higher proportion of their genes have greater genetic diversity than populations that are polymorphic for a smaller proportion of their genes. Species diversity is a measure of the number of species, and the relative abundance of these species, within a particular area. Ecosystem diversity is the number of different ecosystems within a larger area – either a landscape or a region. Conservation ecologists recognize that all three types of biological diversity must be conserved. In this chapter, I will begin discussing genetic and species diversity, but defer a more extensive discussion of species and ecosystem diversity until later chapters.

One approach to maintaining species diversity is to prevent the extinction of individual species, such as the ivory-billed woodpecker. Much of what we know about ivory-billed woodpecker ecology comes to us from James Tanner, a graduate student working with Arthur Allen, who had done some preliminary research on one of the few remaining nests of ivory-bills in 1935. Tanner was with Allen on that expedition to northeastern Louisiana, and proved his mettle as a tireless and insightful researcher. He also had an engaging personality that inspired confidence in the local people, who were happy to share their knowledge of the environment and of woodpeckers they had seen or heard about. When the National Audubon Society came up with funding to study these birds, Allen recommended Tanner for the job. Tanner spent 3 years in Louisiana studying a few birds at their nest sites, and traveling around the southern United States investigating claims of other nest sites within the shrinking ivory-bill habitat.

In the 1700s, ivory-bills were found along much of the Mississippi River and its tributaries, from southern Louisiana into southern Illinois, with extensive populations in southern Mississippi and Alabama, throughout Florida, and up the eastern seaboard into North Carolina. The common element of all ivory-bill habitat is large trees, which furnish nest cavities for raising young, and food in the form of insects, which bore under tree bark, and are excavated by woodpeckers with their specialized beaks.

Tanner (1942) reported a maximum density of one pair of ivory-billed woodpeckers per 15 km^2. Other species of woodpeckers reach a much higher density, and it is not clear why ivory-bills were present at such a low density. Table 11.1 shows that the population in Tanner's major study site in Arkansas declined over the 3 years of his study (1937–9) in comparison to censuses conducted in previous years by other researchers. The area adjacent to his study area had been recently logged, and one of the areas within his study area was logged in 1938, destroying the habitat of one of the breeding pairs.

We have no way of knowing how many ivory-bills there were before the dramatic decline that began in the mid-nineteenth century. After the American Civil War, the price of soybeans shot through the roof, and great profits could be made by clearing the vast lowland forests and using them to grow soybeans. As a result, ivory-bills suffered severe range contraction, and the total population size of the species plummeted to approximately 24 individuals in 1939 (Tanner 1942). The final nail in the coffin was that ivory-bill skins became more valuable as their population declined. During the early twentieth century, many of the remaining ivory-bills were shot and stuffed. The unfortunate reality is that even if the reported sighting of an ivory-billed woodpecker in 2005 is correct, at best the remnant population is tiny. And under the best conditions, restoring small populations to a sustainable size is a daunting task.

Table 11.1 Censuses of ivory-bills from 1934 to 1939 in Tanner's study site.

Year	# of pairs	# lone males	# young	total
1934	7	0	4	18
1936	6	0	6	18
1937	5	1	2	13
1938	2	2	3	9
1939	1	3	1	6

11.1 WHY DO VERY SMALL POPULATIONS HAVE A HIGH EXTINCTION RATE?

Large populations usually harbor more genetic variation than do small populations. In Chapter 3 we discussed how marine populations of sticklebacks were polymorphic for alleles influencing the number of dorsal plates. Individuals with fewer plates who migrated into freshwater systems were at a selective advantage. Genetic diversity provided the sticklebacks with the evolutionary potential to survive and reproduce more successfully in the novel freshwater environment. In contrast, small populations, such as the ivory-billed woodpecker in 1934, have very low genetic diversity, and are usually less successful in novel environments.

Small populations and low genetic diversity

Small populations have low genetic diversity by virtue of having a very small gene pool. But a related problem is that small populations tend to lose their genetic diversity through genetic drift – a change in allele frequency from chance events. For example, if only two individuals within a population have a particular allele, and those individuals by chance don't breed, or don't pass those alleles to their surviving offspring, then that allele will become extinct. Genetic drift operates in populations of all size, but is strongest in small populations.

A second concern with small population size is the increased probability of inbreeding depression – a reduction in survival or viability of offspring produced when two relatives mate with each other. In the case of a brother–sister mating, the coefficient of relatedness (r) is 0.5, which means that there is a 50% chance that the two siblings will have the same allele as a result of being close relatives (see Chapter 6). If each sib is heterozygous for the same recessive lethal allele that they've inherited from one of their parents, then 25% of their offspring will be homozygous for the lethal allele, and die. Thus inbreeding tends to increase the overall homozygosity of individuals within a population. Small populations tend to have more inbreeding, leading to lower offspring viability and a higher incidence of spontaneous abortions and birth defects.

A final problem related to small population size is that the genetic diversity of a population is generally smaller than might appear. Conservation ecologists define the genetically effective population size (N_e) as the average number of individuals in a population that effectively contribute genes to succeeding generations. N_e is usually smaller than actual population size for several reasons. In polygynous or polyandrous mating systems, some individuals will not mate, and therefore will not contribute genes to the next generation. Even in monogamous mating systems, some individuals contribute more of their genes than others. In addition, if two individuals are related, their effective contribution to population size is reduced in relation to their degree of relatedness. In the most extreme case, two identical twins that breed within the same population are counted as one individual when calculating N_e. Finally, if a population has gone through a *population bottleneck* – a temporary reduction in population size – there will be a significant reduction in N_e in future generations. Conservation ecologists have developed quantitative methods for estimating N_e in these cases. Of particular interest to our discussion is whether low N_e reduces the viability of a population.

Effect of low N_e on the viability of a plant population

Dara Newman and Diana Pilson (1997) set up 32 experimental populations of the evening primrose *Clarkia pulchella* to test whether low genetic variation reduces the viability of a population (Figure 11.1A). Sixteen of the populations started with

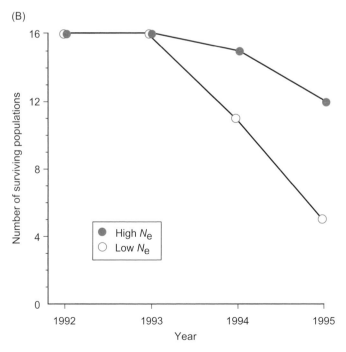

Figure 11.1 A. The evening primrose, *Clarkia pulchella*. B. Number of surviving evening primrose populations in each generation of the experiment.

12 unrelated plants (high N_e), while the other 16 populations started with 12 closely related plants (low N_e). Though 12 plants is not a huge initial population size, the N_e of the high-variation group was 12.0, while N_e of the low group was 7.3. The researchers predicted that the 16 high-N_e populations would have higher viability than the 16 low-N_e populations.

Between 1992 and 1995, only 5 of 16 low-N_e populations survived, while 12 of 16 high-N_e populations survived over the same time period (Figure 11.1B). This increased survival resulted from two sources. First, 21.1% of high-N_e seeds germinated, in contrast to only 8.7% of low-N_e seeds. Second, 98% of the high-N_e plants with flowers set fruit, while only 50% of the low-N_e plants with flowers set fruit. Both of these comparisons had statistically significant differences ($P < 0.02$). At least in the evening primrose, we have good evidence that high N_e of a genetically variable population leads to higher population viability. Conservation ecologists can use information about N_e as part of a general approach for predicting the viability of populations.

11.2 HOW DO ECOLOGISTS USE MATHEMATICAL MODELS TO PREDICT POPULATION VIABILITY?

Conservation ecologists have a toolbox of approaches for projecting population growth. As we discussed in Chapter 10, populations tend to increase when they have low mortality rates and high birth rates, so ecologists measure birth rates, mortality rates, and migration rates within natural populations. These data are entered into mathematical models to project the dynamics of population growth.

Population viability analysis

Conservation ecologists are concerned with two important aspects of populations that are at risk. First, they evaluate whether the population is likely to go extinct in the near future. If extinction is likely, then they identify management tools that can reduce the probability of extinction. Population viability analysis (PVA) is a quantitative analysis of extinction risk that allows ecologists to recommend management options to improve the prognosis for continued survival of a population.

There are several different types of PVAs, that are suitable for different species and different types of available data (Groom *et al.* 2006). Single-site PVAs project the population growth and probability of extinction for a single population, while multi-site PVAs consider multiple populations of the same species. Within these two broad categories there are several approaches that can be used. *Count-based PVAs* use some of the techniques we discussed in Chapter 10 for estimating population size at several different points in time, and then project the total number of individuals in a single population at some time in the future. The key to count-based PVAs is that all individuals are assumed to be equivalent to each other in characteristics such as age and the

Figure 11.2 Male chimpanzee holding a red colobus monkey he has just captured. Other chimpanzees are joining him in hope that he will share some of the prey.

probability of surviving and reproducing. In contrast, *demographic PVAs* take into account the *population structure*, which are the differences between individuals within a population in factors such as age, size, and social status that influence survival and reproductive success. This information is difficult to gather for long-lived species, requiring extended time in the field. But collecting these data allows for more accurate projections about future population growth. We will discuss in Chapter 12 how a 40-year study of the chimpanzees of Gombe Stream Research Center in Tanzania is only beginning to yield enough data to generate projections about the future viability of the population.

Let's turn to a less famous primate, the red colobus monkey (*Procolobus rufomitratus*) at Gombe Stream, as our introduction to using population structure to project future population growth. We will see that the viability of the red colobus population is linked to its interactions with the chimpanzees of Gombe.

Projecting the viability of red colobus monkey populations

Jane Goodall (1986) stunned the world with her vivid descriptions of cooperative hunting by chimpanzees (Figure 11.2). Males who are successful hunters achieve higher rank and consequently better access to fertile females (Stanford, 1998). Curiously, some populations of chimpanzees hunt very little, but we don't know why populations vary so much in their hunting

behavior. Here we introduce life tables as an approach for evaluating the future viability of colobus monkey populations subjected to different levels of chimpanzee hunting.

Using life tables to project a population's future growth
Marc Fourrier and his colleagues (2008) used data collected by Goodall, Stanford, and many other researchers to construct life tables that summarized survival and reproductive rates of colobus monkeys at Gombe at different ages. Ecologists can use the data on survival and reproductive rates to understand current population trends and to project whether population size is likely to increase, remain the same or decrease in the future. Some colobus groups are inside chimpanzee hunting ranges at Gombe, while others are outside the hunting ranges. Thus we can use life tables to see how hunting influences colobus mortality rates at different ages (Table 11.2).

There are several aspects of this data set. First, notice that for computational ease both populations start with $N = 500$. Stanford (1995) estimated that the actual colobus populations were 450–550, so these are reasonable and convenient starting numbers. Population ecologists will often make proportional adjustments to life tables so that the numbers are easier to work with. Most importantly, we can see from this life table that hunting chimpanzees severely depress colobus survival at Gombe. Only 102 of an original group of 500 newborn monkeys were estimated to survive to 5 years in a habitat frequented by chimpanzees, while 250 of an original group of 500 were

Table 11.2 Partial life table showing number of individual colobus monkeys surviving from a starting population of 500 newborn individuals, in relation to the presence or absence of a population of hunting chimpanzees.

Age interval (years)	Age class (x)	Hunted population		Nonhunted population	
		Number of survivors at beginning of age class x	Proportion of original population alive at beginning of age class x (l_x)	Number of survivors at beginning of age class x	Proportion of original population alive at beginning of age class x (l_x)
0–1	0	500	1.0	500	1.0
1–2	1	241	0.482	349	0.698
2–3	2	152	0.304	283	0.566
3–4	3	116	0.232	256	0.512
4–5	4	103	0.206	250	0.500
5–6	5	102	0.204	250	0.500

estimated to survive to 5 years in a chimpanzee-free habitat. In addition, mortality rates changed in relation to age, with very high juvenile mortality in the areas hunted by chimpanzees. Colobus are extremely aggressive when attacked, and a chimpanzee is very likely to get injured if attacked by several adult colobus. Consequently chimpanzees favor attacking juveniles, particularly when there are no adult colobus to defend them. Finally, notice that the data set ends at age 5 years, with much of the original colobus population still alive. Colobus monkeys can live to 16 years, but most research programs do not last that long. Thus long-lived species, including most primates, are difficult subjects for life-table studies.

We might be tempted to conclude that the hunted populations are doomed to extinction. That conclusion would be invalid for several reasons. First, in most primates, females tend to migrate out of their natal population. So even if juvenile mortality is high, new migrants will be constantly entering from other populations. In addition, many of the females born into the population will migrate out when they mature. Second, Table 11.2 does not present any information on fecundity – the number of offspring produced by females of a particular age class. High fecundity could compensate for high mortality, and produce a viable population.

Estimates of fecundity: projecting a population's future growth

Colobus fecundity is the number of annual births divided by the number of reproductive-age females in the population, which in Stanford's (1998) analysis came to 0.625 offspring per year. In life-table analyses of sexual species, demographers often keep track only of the females. This works out conveniently, because male reproductive output is often very difficult to determine, though the mean for males should equal the mean for females. Assuming an equal male/female sex ratio, the average number of female offspring per year was 0.625/2 = 0.3125. Table 11.3 summarizes all of the information we've discussed so far about survival and fecundity of hunted and nonhunted colobus monkeys.

Fourrier and his colleagues were forced to make some assumptions in their handling and analysis of the data. Assumptions about mortality and fecundity values allowed the researchers to make projections about the viability of the hunted and nonhunted populations. But the projections are only as good as the data. In this case, the researchers very clearly articulated their assumptions, so that other scientists and managers could evaluate the quality of their projections. For example, a future researcher could test (with a great deal of work) whether fecundity is constant from age class 3 through 16, as is assumed in Table 11.3. Let's combine our understanding of survival rates and fecundity to project changes in colobus population size.

Calculating the net reproductive rate to project a population's future growth

We will use these life-table data to evaluate whether the colobus populations at Gombe are viable. First we will calculate the net reproductive rate (R_0) – the mean number of

Table 11.3 Life table with estimates of survival and fecundity for colobus monkeys. The data are modified from Fourrier et al. (2008).

Age interval (years)	Age class (x)	Hunted population		Nonhunted population		Fecundity (b_x)
		Number of survivors at beginning of age class	Proportion of original population alive at beginning of age class x (l_x)	Number of survivors at beginning of age class x	Proportion of original population alive at beginning of age class x (l_x)	
0–1	0	500	1.0	500	1.0	0
1–2	1	241	0.482	349	0.698	0
2–3	2	152	0.304	283	0.566	0
3–4	3	116	0.232	256	0.512	0.3125
4–5	4	103	0.206	250	0.500	0.3125
5–6	5	102	0.204	250	0.500	0.3125
6–7	6	102	0.204	250	0.500	0.3125
7–8	7	102	0.204	250	0.500	0.3125
8–9	8	102	0.204	250	0.500	0.3125
9–10	9	102	0.204	243	0.486	0.3125
10–11	10	99	0.198	217	0.434	0.3125
11–12	11	87	0.174	173	0.346	0.3125
12–13	12	66	0.132	119	0.238	0.3125
13–14	13	41	0.082	66	0.132	0.3125
14–15	14	19	0.038	29	0.058	0.3125
15–16	15	6	0.012	9	0.018	0.3125
16$^+$	16	0	0	0	0	0

female offspring produced by a female over the course of her lifetime. If $R_0 > 1.0$, the population will increase exponentially. If $R_0 < 1.0$, the population will decline to extinction. If $R_0 = 1.0$, the population will be static, with births balancing mortality each generation.

Let's focus on the survival data for the hunted population in Table 11.3. Survival (l_x) is always measured to the beginning of the age class, so the proportion of newborns surviving to the beginning of age class 0 (l_0) is by definition 1.0. Only 241 of 500 newborns survive to 1 year, which is the beginning of age class

1. This gives a l_1 value of $241/500 = 0.482$. Similarly only 152 individuals survive until 2 years (the beginning of age class 2), yielding $l_2 = 152/500 = 0.304$. These calculations are continued for each age class.

Let's next turn to the fecundity estimates (b_x) from Table 11.3. Females in age classes 0, 1, and 2 produce no offspring. Beginning with age class 3, surviving females average 0.3125 daughters per year. Mathematically, b_3 (and beyond, for each age class) = 0.3125.

Our task is to put these numbers together to calculate R_0. This is not as daunting as it may appear initially. The average number of offspring produced by each age class is the probability of surviving to that age class (l_x) multiplied by the fecundity of individuals who survive to that age class (b_x). Mathematically, the reproductive output for each age class is = $l_x b_x$. The net reproductive rate (R_0) is then simply the sum of the reproductive outputs of each age class = $\Sigma l_x b_x$.

Finally, we will create a new table (Table 11.4) to help calculate R_0 for the hunted population, using the values of l_x and b_x from Table 11.3 to make a final column which is $l_x b_x$. Adding all the values of $l_x b_x$ together gives us $\Sigma l_x b_x = R_0 = 0.654$. This indicates that, on average, each female will produce 0.654 female offspring. Because R_0 is so much less than 1, we project our population of hunted colobus to go extinct relatively rapidly.

Let's look at the nonhunted population to see if they enjoy brighter prospects (Table 11.5).

Going through the same steps as above gives us $\Sigma l_x b_x = R_0 = 1.476$. This indicates that, on average, each female will produce 1.476 female offspring. According to these data, our population of nonhunted colobus will increase rapidly.

Thinking ecologically 11.1

In some mammals, females will come into estrous more quickly if their offspring die while still nursing. How might this influence our projection of the viability of the hunted colobus population at Gombe? Check out your intuition by substituting different fecundity values into Table 11.4, and recalculating R_0.

Is the hunted population of red colobus monkeys in Gombe really doomed? Will the nonhunted population in Gombe explode in abundance? To answer these questions we need to explore some of the simplifying assumptions we made when doing this analysis.

Life-table analyses: simplifying assumptions about survival rates and fecundity

Though our analysis paints a bleak picture for the hunted population of colobus monkeys at Gombe, there are several reasons for holding out hope for the future viability of the population. First, the mortality estimates are based on one data set collected by Stanford in 1995. In reality, it would be very surprising if

mortality were constant from year to year. As one example, we could easily imagine that smaller colobus population have reduced mortality due to decreased infectious disease and decreased competition between colobus monkeys for food. Thus it is unreasonable to expect that mortality rates will be the same for different population sizes.

Second, we assumed a constant fecundity schedule of 0.3125 female offspring per year, beginning in age class 3 and

Table 11.4 Calculation of R_0 for the hunted population of colobus at Gombe.

Age class	l_x	b_x	$l_x b_x$
0	1.0	0	0
1	0.482	0	0
2	0.304	0	0
3	0.232	0.3125	0.0725
4	0.206	0.3125	0.064375
5	0.204	0.3125	0.06375
6	0.204	0.3125	0.06375
7	0.204	0.3125	0.06375
8	0.204	0.3125	0.06375
9	0.204	0.3125	0.06375
10	0.198	0.3125	0.061875
11	0.174	0.3125	0.054375
12	0.132	0.3125	0.04125
13	0.082	0.3125	0.025625
14	0.038	0.3125	0.011875
15	0.012	0.3125	0.00375
16	0	0	0
		$R_0 = \Sigma l_x b_x = 0.654$	

Table 11.5 Calculation of R_0 for the nonhunted population of colobus at Gombe.

Age class	l_x	b_x	$l_x b_x$
0	1	0	0
1	0.698	0	0
2	0.566	0	0
3	0.512	0.3125	0.16
4	0.5	0.3125	0.15625
5	0.5	0.3125	0.15625
6	0.5	0.3125	0.15625
7	0.5	0.3125	0.15625
8	0.5	0.3125	0.15625
9	0.486	0.3125	0.151875
10	0.434	0.3125	0.135625
11	0.346	0.3125	0.108125
12	0.238	0.3125	0.074375
13	0.132	0.3125	0.04125
14	0.058	0.3125	0.018125
15	0.018	0.3125	0.005625
16	0	0	0
		$R_0 = \Sigma l_x b_x = 1.476$	

continuing until death. Fourrier and his colleagues very clearly state that the assumption of constant fecundity for different age classes is untested, but is used in the absence of any evidence to the contrary.

To address these assumptions, Fourrier and his colleagues used six different mortality assumptions and three different fecundity assumptions in their overall analysis. This gave a total of 6 × 3 = 18 different analyses, of which we've discussed only

two. We will not take the time to go through them all, or even to discuss the reasoning underlying the different assumptions. However, the researchers conclude from these 18 analyses that the hunted population is viable over the long term, only if females compensated for losing their babies by immediately going into estrous and becoming impregnated. This corresponds to an annual fecundity of 0.6667 per year as opposed to 0.3125, which is highly unlikely in this species.

Fourrier and his colleagues conclude that the colobus monkeys at Gombe have historically been part of a metapopulation – which we defined in Chapter 10 as a group of local populations inhabiting networks of somewhat discrete habitat patches. Until recent deforestation, the entire region around Gombe had been a forested network, with female colobus monkeys migrating into and out of different populations. Presumably, the high mortality rates from hunting in the Gombe population of colobus could be compensated for by high immigration rates of new females, and their juveniles, into the Gombe habitat. Unfortunately, most of the forests surrounding Gombe have been cut down (see Chapter 12), resulting in a much lower immigration rate of colobus monkeys into Gombe.

For many species studied by conservation ecologists, we need to consider immigration and emigration processes. In some cases, although the local population may go extinct, other individuals immigrating from nearby populations may reestablish it. Let's consider a study that explores the relationship between immigration rates and extinction rates on a very small spatial scale.

11.3 HOW CAN IMMIGRATION OF INDIVIDUALS FROM NEARBY POPULATIONS MAINTAIN SPECIES RICHNESS AND HIGH POPULATION SIZE?

Moss landscapes have animal communities dominated by small arthropods – primarily ticks and mites (subclass Acari) and springtails (Order Collembola) (Figure 11.3A and B). By focusing on very small arthropods, Andrew Gonzalez and his colleagues (1998) could study the distribution and abundance of numerous species in a very small area, and track the animals' movements. They tested the hypothesis that a reduction in immigration rates among different patches of moss habitat would cause a decline in population size, and cause some species to go extinct.

Effects of reduced immigration rates

To test the hypothesis that reduction in immigration rates causes population size to decline, in some cases to extinction, Gonzalez and his colleagues established four different experimental moss habitat treatments on a background of bare rock. The moss was suitable habitat for the small arthropods, while the rock was relatively inhospitable. Figure 11.3C is a schematic representation

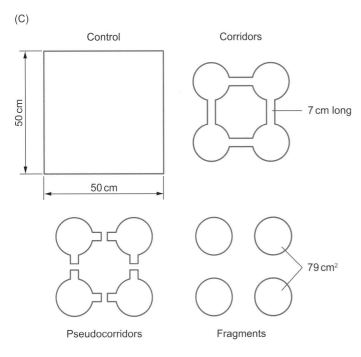

Figure 11.3 A. Group of mites traveling on a beetle's back. B. Small group of the hairy ground springtail, *Orchesella flavescens*, on a rock. C. Schematic representation of Gonzalez's experimental design.

of the four treatments, which vary in the extent to which they are fragmented. The control treatment was a continuous moss landscape, so it was completely unfragmented. The corridor treatment had four identical moss habitats joined by 7-cm-long immigration corridors. The pseudocorridor treatment had four identical fragments joined by 7 cm long corridors that were broken by thin (1 cm wide) strips of bare rock, which is an area long enough to prevent these tiny animals from immigrating between fragments. Finally, the completely fragmented treatment had four identical fragments separated by 7 cm of bare rock.

The researchers predicted that the completely fragmented treatment would have higher extinction rates in comparison to equivalent-sized and -shaped samples taken from the unfragmented controls. They also predicted that the immigration corridors would reduce the extinction rates of the fragments, but that the pseudocorridor treatments would have extinction rates similar to the fragmented treatments. After 6 months, the researchers surveyed these treatments, and as predicted, discovered that extinction rates were greatest in the completely fragmented and pseudocorridor treatments. Adding corridors substantially reduced extinction rates, and the unfragmented controls suffered no extinctions at all (Figure 11.4A).

The researchers also predicted that the mean population size of species that did not go extinct should be higher if immigration corridors were provided, in comparison to fragmented treatments. Figure 11.4B compares the population size before and after fragmentation for species that survived the corridor and pseudocorridor treatments. As predicted, the population size of most species was higher for the corridor than for the pseudocorridor treatment.

These results are important to conservation ecologists, because they emphasize that endangered species are more likely to recover from low numbers if they are provided with immigration corridors. In essence, immigration corridors create a metapopulation, in which some populations may decline to low numbers, or even go extinct, but other populations may be recolonized and reestablished following extinction.

Metapopulations

Recall from Chapter 10 that Ilkka Hanski and his colleagues (1994) are studying metapopulations of the endangered Glanville fritillary butterfly, *Melitaea cinxia*, on an island in the Baltic Sea off the coast of Finland. These butterflies have become a model system for asking questions about population ecology (Ehrlich and Hanski 2004). In Chapter 10, we demonstrated that both sexes showed high immigration rates into new populations, with females migrating more than males. Let's now investigate, using data on the Glanville fritillary butterfly, how these high levels of migration translate into maintaining the stability of metapopulations.

(A)

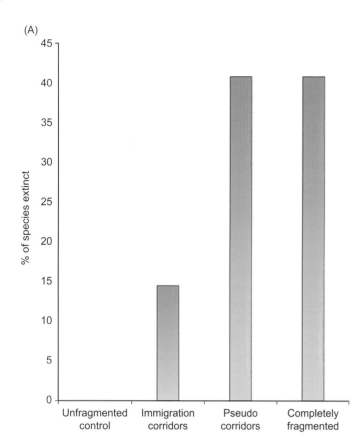

(B)

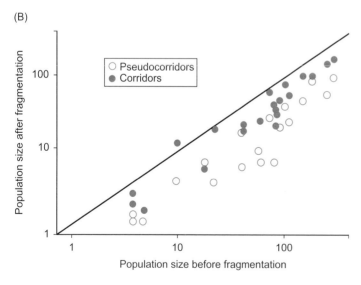

Figure 11.4 A. Arthropod extinction rates for four experimental treatments. B. Arthropod population size before and 6 months after fragmentation for the corridor (filled circles) and pseudocorridor (open circles) treatments. Points above the line had increased population size 6 months after treatment, while points below the line had decreased population size 6 months after treatment. Overall, across species, the corridor treatments had greater mean population size 6 months after the treatment than did the pseudocorridor treatment (paired $t = 4.31$, $p < 0.001$).

Stability of metapopulations versus local populations

In Finland, where Hanski and his colleagues have conducted much of their work, *M. cinxia* inhabits dry meadows. Females lay *c.* 150 eggs on the leaves of two host plants, *Plantago lanceolata* and *Veronica spicata* (Figure 11.5A, B). On Aland, an island complex in western Finland, there are approximately 4000 habitat patches that are suitable habitat for these butterflies. Nearby patches form a network, which if occupied by at least one local population (and usually many more) is considered a metapopulation. Within the Aland complex, there are many existing metapopulations, though the number is constantly changing as some metapopulations go extinct and new metapopulations are founded. We might expect that metapopulations, being made up of several local populations, would be less likely to go extinct than individual local populations. To test this hypothesis, Marko Nieminen and his colleagues (2004) summarized extinction rate data for *M. cinxia* over a 9-year time period. Between 1993 and 1994, about 50% of the existing local populations had gone extinct, and by 2001 more than 90% of the originally surveyed populations were extinct. But by 2001, only about 30% of the metapopulations went extinct (Figure 11.5C).

But while some local *M. cinxia* populations are going extinct, others are being recolonized. By intensively sampling all of the habitat patches on the Alands, Hanski (2011) tallied both the number of extinctions and the number of colonizations. As Figure 11.6 shows, both extinctions and colonizations are common events, but the rates vary considerably from year to year.

 Thinking ecologically 11.2

Based on Figure 11.6, can you identify a pattern that describes the relationship between the number of extinctions and colonizations. Draw a rough graph that shows this relationship. Propose a hypothesis for why this relationship might exist.

The research team wanted to identify the factors that influenced extinction and colonization rates. We will begin our discussion by considering only two of the many factors that could influence extinction rates: patch size and patch density. We will then expand our discussion to consider other factors as well.

Relatively low extinction rates of populations occupying large habitat patches

Hanski and his colleagues reasoned that larger patches would have lower *M. cinxia* extinction rates for a number of reasons. First, by virtue of being bigger, larger patches would tend to have more space for a greater number of butterflies. Second, larger patches are more diverse in the types of structure, food plants, and microclimates they contain. Finally, other research showed that the emigration rate from large patches is lower than

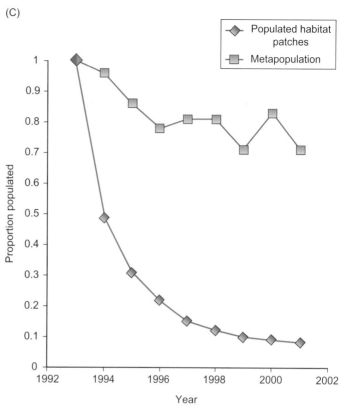

Figure 11.5 A. A mating pair of *M. cinxia*. B. *M. cinxia* egg cluster on the underside of a leaf. C. Proportion of populated habitat patches and metapopulations in 1993 that were populated in subsequent years.

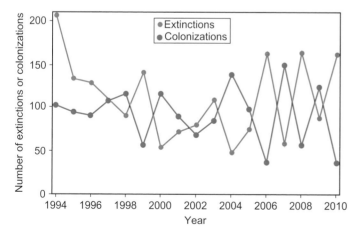

Figure 11.6 Number of extinctions and colonizations of habitat patches by *M. Cinxia* in 1994–2010 on Aland.

the emigration rate from small patches (Petit *et al.* 2001). Data collected by Hanski and his colleagues (1995) support the hypothesis that larger patches have lower extinction rates (Figure 11.7A)

The researchers also reasoned that a high patch density, or high number of patches per unit area, would be associated with lower extinction rates, because populations that were surrounded by other nearby patches would be likely to have their populations supplemented by immigrants arriving from these other populations. The data support this hypothesis as well. Populations surrounded by many other patches had lower extinction rates than populations surrounded by relatively few patches (Figure 11.7B).

The effect of patch size and patch density on extinction rates raises many important questions. Are large patches less prone to extinctions because they host larger populations? Or are large patches less prone to extinction for some of the other reasons we have already discussed? Earlier in this chapter we discussed two reasons why small populations tend to go extinct, including low genetic diversity and chance fluctuations in population size. A third reason for why small populations tend to go extinct is that small populations tend to have a low population density, which may further suppress population growth due to the Allee effect.

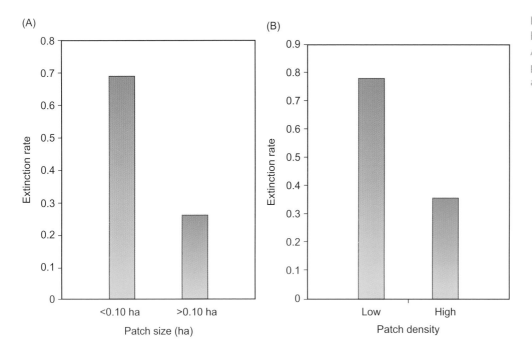

Figure 11.7 Extinction rates between 1991 and 1993 comparing A. small ($N = 19$) and large ($N = 23$) patches and B. low-density ($N = 9$) and high-density ($N = 23$) patches.

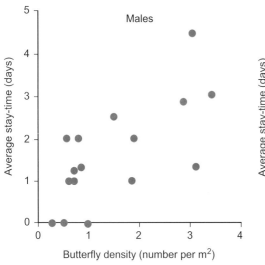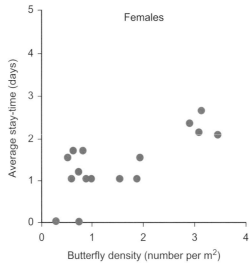

Figure 11.8 Effect of butterfly density on the amount of time a (left) male or (right) female butterfly stayed in a patch after being captured and marked. Note that a long stay-time indicates a low emigration rate.

The Allee effect predictions: low-density populations with low per capita growth rates

W. C. Allee (1931) proposed that very low-density populations should have reduced per capita growth rates (r) because they have reduced defense against predators, decreased foraging efficiencies, difficulty finding mates, or higher emigration rates. Low per capita growth rates at low population densities will reduce population growth rates and may increase the probability that low-density populations go extinct.

Mikko Kuussaari and his colleagues (1998) tested the hypotheses that mating success would be reduced and that emigration rates would increase in small populations of *M. cinxia*.

From their field research, the scientists knew that small populations also tend to have lower density than large populations. The researchers used number of larval groups as their measure of population size. The larvae tend to stay together after they hatch, so the researchers can count each group, rather than disturbing the caterpillars and counting each individual in the group. The researchers also collected mature females from the field, and dissected them to see whether they contained spermatophores – packets of sperm and nutrients deposited by males during the mating process. Less than 70% of the females from the smallest populations had spermatophores, while more than 90% of the females in the larger populations had spermatophores. As predicted by the Allee effect, low-density populations had lower mating success.

Kuussaari and his colleagues did a field experiment to see whether smaller populations had higher emigration rates. They captured, marked, and released 882 butterflies into 16 habitats at different population densities. They then recaptured the butterflies either at the release patch, or in newly colonized patches over the course of several days. The researchers found that, in general, individuals released in low densities remained in the patch for only a short time before emigrating, while butterflies released at high densities tended to remain in the patches for substantially longer (Figure 11.8). As predicted by the Allee effect, low-density populations had higher emigration rates.

Of course, while some populations are going extinct, others are being colonized. Let's now explore the factors that influence the colonization of new habitat patches.

High colonization rates of habitat patches with good connections to existing populations

We can approach the problem of colonization of new habitats by using our scientific intuition. Compare Figures 11.9A and B. In Figure 11.9A, an unoccupied habitat patch has one occupied patch 500 m away. In Figure 11.9B, an unoccupied habitat patch is separated from three occupied patches by 100 m. Intuitively, we expect that focal patch B is much more likely to get colonized than focal patch A ("focal" patch is shorthand for the patch we are talking about). Hanski and his colleagues have quantified this relationship by defining connectivity as a measure of the distance between a focal habitat patch and nearby populations, weighted by the sizes of the nearby populations and their distances to the focal patch. Using this definition, focal patch A has low connectivity, and focal patch B has high connectivity. We will not develop the mathematical formulation of connectivity.

Just how important is connectivity to colonization rates? Saskya van Nouhuys and Hanski (2002) showed that patches with low connectivity have almost no colonization, but that colonization rates increase exponentially at higher connectivity (Figure 11.9C). Interestingly, overall occupancy rates show a similar relationship to connectivity because a highly connected patch receives a steady influx of immigrants from nearby populations.

Metapopulations persist if there is a balance between local extinctions and the establishment of new populations. Hanski's research team found that metapopulations do go extinct but also can become reestablished when one or more habitat patches become colonized. But, of course, suitable habitat must be available for colonization. The Glanville fritillary went extinct in mainland Finland by the 1980s, because people converted most dry meadows into agricultural land that no longer supported the host plants.

On a global level, extinction of the Glanville fritillary in mainland Finland is not too problematic, because there are existing metapopulations on the Aland islands, and in other

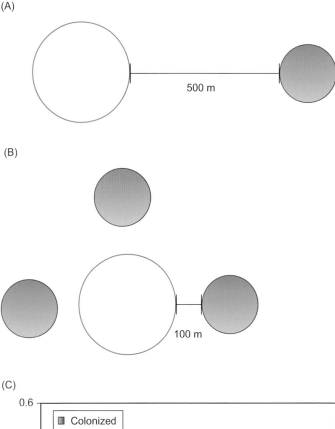

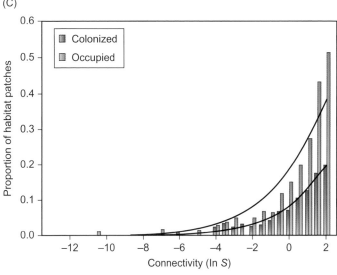

Figure 11.9 Focal patch with A. low connectivity, and B. higher connectivity. C. Proportion of habitat patches colonized (green bars) and occupied rates (blue bars) in relation to connectivity. Ln(S) is the researchers' measure of connectivity of the focal patch, which is based on the number of populated local patches, the distance to these local patches, and the size of the local patches.

regions of southern Europe. However, preventing local extinctions helps slow global extinction. In their concern about global extinction of species, conservation ecologists seek to minimize human impacts on species that are threatened on a local level. In recent history, human expansion and habitat manipulation have led to one of the greatest extinction events in the history of life.

11.4 HOW CAN HUMAN-MEDIATED CHANGES TO HABITATS CAUSE SPECIES TO BECOME ENDANGERED, OR TO GO EXTINCT?

Historically, humans have changed habitats in numerous ways. Perhaps the most damaging to species survival is habitat degradation or destruction, in which the quality of an area is reduced for a particular species, so that its population is no longer viable in that habitat. A second human-mediated change is habitat fragmentation, in which a large, continuous habitat is broken down into small isolated patches. Human overexploitation of a species – using a species as a resource beyond its ability to replenish itself numerically – may also lead to species extinction. Less directly, humans may intentionally or inadvertently introduce novel species, sometimes called invasive species, into an area. These species may compete with, parasitize, or eat the native species, causing them to go extinct. We will briefly consider each of these human-mediated changes to habitats.

Human-mediated effects: habitat degradation and destruction

The possible global extinction of the ivory-billed woodpecker and the extinction of the Glanville fritillary butterfly from mainland Finland were both caused by habitat destruction. Human-caused habitat destruction of southern US forests and of dry open pastures in the Finnish mainland was motivated by the desire for more agricultural fields. Unfortunately, these are not isolated examples. As human populations and resource consumption continue to increase, many of our terrestrial and marine biomes are shrinking dramatically.

Historically, the most degraded biomes are Mediterranean and temperate forest, which were colonized early in human history. As Figure 11.10 shows, conservation ecologists project that tropical forests will be most severely degraded over the next 35 years (Millennium Ecosystem Assessment 2005). The least-affected biomes will be temperate forests and woodland, which are actually predicted to expand slightly, and boreal forests and tundra, which have very low human population density, and low accessibility to developers. Ironically Mediterranean biomes are also projected to experience only minimal loss, presumably because most Mediterranean habitat has already been degraded.

Even if habitat is not completely degraded, its ability to sustain biological diversity may be at risk. Breaking habitat into small fragments can also threaten the structure and functioning of ecosystems, and reduce both species and ecosystem diversity.

Human-mediated effects: habitat fragmentation

Habitat fragmentation can lead to extinctions in many different ways. First, small fragments contain fewer types of habitats, and thus fewer ecological niches. Second, small fragments support smaller populations, which can lead to low genetic diversity. Third, small fragments have a much larger portion of their area near the edge of the habitat. Conditions near the edge of a habitat are very different from conditions in the interior, including differences in climate, rates of predation, and interactions with non-native competitors. We will discuss this edge effect in more detail in Chapter 22. The reduction in biological diversity from fragmentation can change the dynamics of competitive, parasitic, predatory, and mutualistic interactions within each fragment. A predator may go extinct because its preferred prey is missing from the fragment, or a mutualist may go extinct because its mutualist partner is missing or present in reduced numbers.

As one example of how habitat fragmentation can disrupt a mutualism, let's consider the small subcanopy tree *Oxyanthus pyriformis pyriformis*, which is now restricted to two small fragments within the city of Durban, South Africa. This tree previously had a much wider distribution, which has contracted dramatically as the human population of the area exploded over the past century. *Oxyanthus* is pollinated by hawkmoths – large moths with very long tongues that can penetrate about 10 cm to the nectar reward awaiting them at the bottom of the floral tube (Figure 11.11). In the process of drinking nectar, the hawkmoths pick up pollen, which they may transfer to other flowers (*Oxyanthus* does not self pollinate). Thus far, this interaction has the makings of a successful pollination mutualism.

The problem is that hawkmoths are very scarce in these two remaining fragments. In 121 h of observation during prime pollination time, only 16 hawkmoth foraging bouts were observed in the two habitats. As a result, fewer than 1000 of over 200 000 flowers actually set fruit in the two natural populations. To determine whether pollen transfer was the limiting factor, Steven Johnson and his colleagues (2004) hand-pollinated 26 flowers, leaving 28 flowers as an unmanipulated control. Nineteen of 26 flowers that were hand-pollinated successfully set fruit, while none of the 28 unmanipulated flowers set fruit. The researchers concluded that pollination rates were extraordinarily low in these two fragments, presumably because of low hawkmoth abundance.

One of the challenges with working on endangered species is that the species are endangered, and thus rare. In this case, it would be very useful to know what the pollination rates are like in unfragmented populations, but the problem is that there are no unfragmented populations.

The effects of fragmentation on species survival may be subtle, particularly when, as we've just discussed, the effects arise by disturbance of species interactions. It is much easier to understand how species may become threatened or endangered when humans exploit them for their own needs.

Human-mediated effects: overexploitation

Humans have a long history of causing extinctions of other species from overexploitation. One of the most spectacular

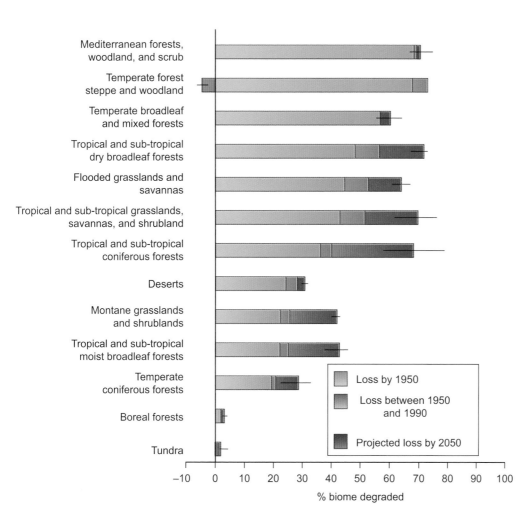

Figure 11.10 Habitat degradation in the past, and projected degradation into the future for the 14 biomes described by the Millennium Ecosystem Assessment.

examples was the extinction of the passenger pigeon, *Ecto-pistes migratorius*, which was the most abundant bird in North America, with a population of about 4 billion individuals in 1800. Just over 100 years later, the last known individual died in the Cincinnati Zoo. In this and many other cases of extinction or endangerment, the life history of the species played an important role in its vulnerability to overexploitation.

As described by Dave Blockstein and Bud Tordoff (1985), there was a maximum of about 25 flocks of passenger pigeons at their peak, but each flock had many millions of birds. This was an excellent example of protection by predator saturation, a topic we introduced in Chapter 8. In such a large flock, the chance of an individual bird getting eaten by a predator was close to zero. The nut-producing trees in passenger pigeon habitat used predator saturation as a life history strategy as well. Pigeons feasted on enormous crops of nuts that were produced in some years by forest trees. By unpredictably producing these huge nut crops, populations of trees could saturate their avian predators, so that a large proportion of the nuts would germinate and develop into juvenile trees. Large flocks of pigeons scoured the habitat for large masses of nuts,

and thus were able to feed their huge populations. Because the pigeons were so mobile, they were able to cover vast swaths of forest in search of these nuts.

Despite the passenger pigeons' immense flock size, each pair produced a maximum of one offspring per year (Figure 11.12A). If nut production was low, the pair did not nest. Thus the per capita growth rate was relatively low. Their low per capita growth rate made the pigeons vulnerable to changes in their environment that arose from two developments in technology: the railroads and the telegraph. Railroads gave hunters quick access to the shifting nesting colonies, and to distant markets around the country. The telegraph gave hunters access to information on the location of each flock (Figure 11.12B). The last observed colonial nesting was in 1887.

Each case of extinction or endangerment from overexploitation has its own story, which usually includes some aspects of the species' life history that make it vulnerable to human activities. As a second example, the Chinese giant yellow croaker, *Bahaba taipingensis*, has declined to less than 1% of its former abundance in the early 1930s (Sadovy and Cheung 2003). Like many vulnerable species, this marine fish has a

Figure 11.11 The hawkmoth *Coelonia mauritii*, sipping nectar from an *Oxyanthus* flower. Both the floral tube and moth's tongue are about 9 cm in length.

Figure 11.12 A. Passenger pigeons and nest with one egg. B. Hunters shooting into vast flocks of passenger pigeons.

narrow distribution (Figure 11.13). During their spawning runs, *Bahaba* aggregate into huge schools in shallow waters, which makes them particularly vulnerable to trawlers (fishing vessels) using large nets. Fishermen can easily locate the schools, because the fish emit a loud drumming sound to communicate with each other. Finally, the croaker's swim bladder is prized for medicinal purposes, and in a market-driven economy, the swim bladder has increased in value as the fish nears extinction.

Franck Courchamp and his colleagues (2008) proposed that humans may accelerate extinction rates of endangered species by causing an Allee effect, or by enhancing an already existing Allee effect at low population densities. As we just saw with the value of the yellow croaker's swim bladder, and as we discussed earlier with the ivory-billed woodpeckers, the value of a resource can increase significantly as it becomes rarer. If human exploitation reduces the per capita growth rate below 1, the species will continue to get rarer, and therefore more valuable. This could increase the hunting pressure even further, resulting in the extinction of an endangered species.

The final human-mediated threat to species survival that we will discuss is more indirect, in that humans are not directly removing habitat or individuals. Instead, humans are introducing non-native species into the habitat, which then compete with or prey on the native species, or even more indirectly, remove resources that are essential for the survival of native species within their habitat.

Human-mediated effects: introduction of non-native species

In 1954, the Uganda Game and Fisheries Department probably introduced a few Nile perch, *Lates niloticus*, into Lake Victoria in hopes that the perch would multiply within the lake, and provide a valuable commercial and sport fishery. I use the word "probably" because the introduction was not officially sanctioned, and there is controversy over when it occurred and who was actually responsible (Pringle 2005). In 1960, local fishermen began catching the perch in small numbers. But fishermen did not catch Nile perch in large numbers until the late

Figure 11.13 Huge specimen of *Bahaba* caught off the coast of Hong Kong in 1993.

1970s in some regions, and the middle 1980s in other regions (Figure 11.14A).

Mass extinction of native fish species caused by intentional introduction of the Nile perch into Lake Victoria

Unfortunately, the increase of the Nile perch is associated with the decline and extinction of a huge number of other fish species in Lake Victoria. Before the Nile perch introduction, Lake Victoria had over 500 species of cichlid fish, including many species that were *endemic* to the lake – found in Lake Victoria and nowhere else. Many of these species were in the genus *Haplochromis*, and had evolved and gone through an adaptive radiation (see Chapter 3) within the relatively isolated confines of the lake (Figure 11.14B).

Kees Goudswaard and his colleagues (2008) went through catch records of commercial fishermen, and through catch records of scientific expeditions to document how the abundance of Nile perch correlated with the decline in the haplochromines. They used fish catch rate (kg/h) as their measure of fish abundance. In all of six locations described, there was a dramatic increase in the Nile perch that coincided with a dramatic decrease in the haplochromine cichlids. The most complete data are from the Mwanza Gulf, where, between 1982 and 1987, the catch rate

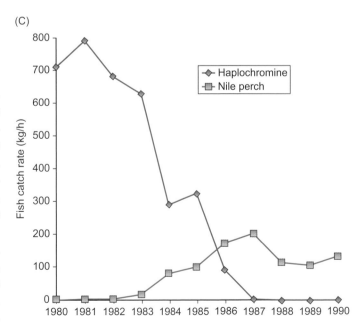

Figure 11.14 A. Fishermen with Nile perch along the shore of Lake Victoria. B. Haplochromine cichlids from Lake Victoria. C. Mean catch rate of Nile perch and haplochromine cichlids in the Mwanza Gulf of Lake Victoria between 1980 and 1990.

of Nile perch jumped from 4 to 203 kg/h, while at the same time, the catch rate of haplochromine species plummeted from 682 to 4 kg/h (Figure 11.14C).

There is a problem with ascertaining causality in extinction events. In cases where one species increases dramatically and another species goes extinct, can we be sure that the increasing species caused the extinction? We must always consider the alternative that another factor – or several other factors – led to the decline of one species and the increases in the other species. For example, Lake Victoria has become more eutrophic in the past 40 years and has lower oxygen levels than in the past. Perhaps eutrophication has led to the demise of the haplochromines, and the subsequent increase in the Nile perch.

There are several pieces of evidence that argue against that alternative hypothesis. First, the data show the rapid increase in Nile perch and decline in haplochromines co-occurring in many different locations in the lake at different periods of time. In Kenya, this process occurred around 1979, while in several other locations it did not even begin until several years later. But the eutrophication and oxygen depletion did not occur any earlier in Kenya than elsewhere in affected areas. Second, as a result of intensive fishing on Nile perch, there has been a significant decline in their abundance in recent years. In association with this decline, many of the species of *Haplochromis* that had declined, but not gone extinct, have shown a substantial recovery in recent years. In 1987, haplochromines comprised about 0.2% of the catch weight hauled in by trawlers in Mwanza Gulf of Lake Victoria. By 1997, this number had increased to greater than 20% (Witte *et al.* 2007).

The Nile perch is an example in which a species was deliberately introduced by biologists into a novel environment, with hopes of creating a new and more financially rewarding fishery. We are not too surprised when a voracious predator such as the Nile perch can lead to the extinction of a variety of prey species. But many of the most damaging invasive species are plants. I have a very small wild woodlot in my backyard, which in recent years was overrun by the tree of heaven, *Ailanthus altissima*, an import from central China. If I drive into lower elevations near my house, I am immediately confronted by hectares of trees and shrubs encased in coffins of kudzu, *Pueraria lobata*, which arrived from southern China and Japan (Figure 11.15). Conservation ecologists note that most invasive plants are much less dominant in their native range than they are in their recently invaded, or naturalized, range. These scientists have proposed many hypotheses for why some plant species become invasive in their naturalized range, while others fail to get established. We will explore three of these hypotheses.

Hypotheses explaining the success of invasive species

The *enemy release hypothesis* proposes that invasive species are successful in their naturalized range if they can disperse without

Figure 11.15 A. Three young trees of heaven (light-coloured bark) dominate a meadow in my home town of Radford, Virginia. B. Kudzu overgrowing shrubs and trees in North Carolina, USA (right).

many of the natural enemies that keep them in check in their native range. The *evolution of increased competitive ability (EICA) hypothesis* proposes that invasive plants, because they have fewer herbivores attacking them, can allocate fewer resources to defense, and thus can allocate more resources to survival and

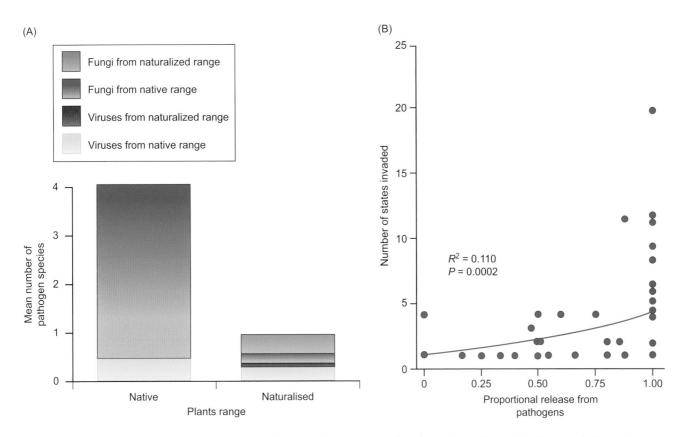

Figure 11.16 A. Mean number of pathogen species and origins of pathogens within the native and naturalized range of plants that were classified as invasive in the United States. B. Number of states invaded in relation to proportional release. Proportional release = (number of pathogens in Europe – number of pathogens in United States)/number of pathogens in Europe. Thus a species with 10 pathogens in Europe and only 2 in the United States would have a proportional release = (10 − 2)/10 = 0.80. (R^2 = 0.110, P = 0.0002).

reproduction in their naturalized range. This gives them a competitive edge in their naturalized range when competing with native species for resources. Lastly, the *resource hypothesis* argues that most invasive plants succeed because they are already extraordinarily well adapted to the new environment. The combination of temperature, nutrients, moisture, and soil types that they experience in their naturalized range may be better suited to their successful survival and reproduction than what they experience in their native range.

THE ENEMY RELEASE HYPOTHESIS. The enemy release hypothesis makes several predictions – we will consider two of them. One prediction is that invasive plants will have more pathogens in their native range than in their naturalized range. To test this prediction, Charles Mitchell and Alison Power (2003) consulted the United States Department of Agriculture database (http://plants.usda.gov), which has a comprehensive listing of plants that have invaded the United States from Europe. They then referenced two different databases and numerous publications to extract information on incidence of fungal pathogens and viruses affecting each plant in the native range (Europe) and the naturalized range (United States). In accordance with their prediction, naturalized plants had about four times as many pathogens in their native range than in their naturalized range (Figure 11.16A)

A second prediction of the enemy release hypothesis is that the plants that were most released from their native pathogens would have a greater naturalized range in the United States. As Figure 11.16B shows, there was a trend for species that were most released from native pathogens to successfully invade more states. Note that there is a great deal of scatter in the data, resulting in a low R^2 value, suggesting that other factors, in addition to enemy release, also influence how large a range these invasive plants occupy within the United States. The ability to reallocate resources from defense to growth and reproduction may be one factor that promotes the success of invasive plants.

THE EICA HYPOTHESIS. Let's next turn to predictions of the EICA hypothesis using the Chinese tallow tree, *Sapium sebiferum*, as our test organism (Figure 11.17A). This tree arrived in Georgia in the late eighteenth century, and has spread its range throughout the southern United States, and also along the western US coast. One prediction of the EICA hypothesis is that Chinese tallow trees from Asia will have effective but costly anti-herbivory adaptations that protect them against herbivores in their local habitat in Asia. But upon arriving in Georgia, where these herbivores are absent, *Sapium* with lower levels of these costly adaptations would have been at a selective advantage over *Sapium* with higher levels of costly anti-herbivore adaptations.

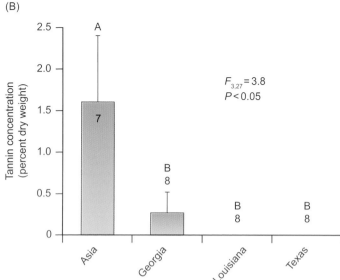

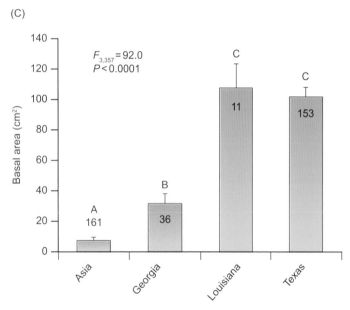

Figure 11.17 A. Chinese tallow tree showing dark seed pods from the previous year and new flowers formed in the spring. B. Tannin levels in plants grown from seeds harvested from four locations. C. Basal area of plants grown from seeds from four locations. Numbers in or above bars indicate sample size. Different letters above the bar indicate significant differences between the treatments in pairwise comparisons.

To test this prediction, Evan Siemann and William Rogers (2001) planted a field of *Sapium* in Galveston, Texas, from seedlings collected in the species' native range in Taiwan, and from seedlings collected in its naturalized range in Georgia, Louisiana, and Texas. After 14 years the researchers measured the tannin levels in leaves from trees from each source. Recall from Chapter 5 that tannin is a very common and effective defensive compound produced by trees, to protect them against herbivores. As predicted by the EICA hypothesis, tannin levels were much higher in the trees from Taiwan than in the trees from the United States (Figure 11.17B).

The EICA hypothesis assumes that it is costly to plants to produce and maintain high tannin levels. Because they allocate more resources to defense, the trees from Taiwan should grow or reproduce more poorly than the trees that are naturalized in the United States. Figure 11.17C confirms this prediction; the trunks of trees from Taiwan had, on average, a much smaller basal area than did the trunks of trees from Louisiana and Texas, while the basal area of trunks from Georgia was intermediate.

 Thinking ecologically 11.3

Recall that *Sapium* arrived in Georgia about 200 years ago. Its arrival in Texas and Louisiana is much more recent. Use this information to construct a plausible hypothesis for the intermediate tannin and basal area values that Siemann and Rogers found in trees from Georgia. How would you test this hypothesis?

In a separate series of experiments, Siemann and Rogers (2003) tested the effect of grasshopper herbivory on *Sapium* seedling growth. They imported seeds from China and collected seeds from Texas, which they germinated and grew together in pots

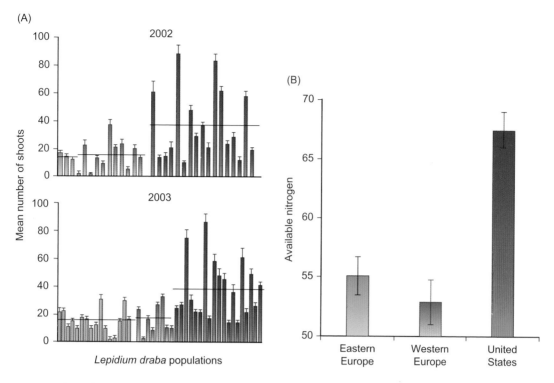

Figure 11.18 A. Mean number of *L. draba* shoots per quadrat (0.5 × 0.25 m) in eastern Europe (green bars), western Europe (blue bars) and the United States (red bars) in 2002 and 2003. Each bar represents one site at which researchers established approximately 30 quadrats. Error bars are ±1 SE. Horizontal lines are means for each range. B. Available nitrogen in each range. Available nitrogen = (fine particulate organic matter/total particulate organic matter) × 100.

in a greenhouse. Each pot had one seedling from China and one from Texas. Grasshoppers fed more avidly on seedlings from Texas. In contrast to the results from the previous experiment, which was conducted on 14-year-old trees, seedlings from either location have no detectable tannins. Thus some other substance – not yet isolated – was influencing the preference of the grasshoppers for the Texas seedlings. The protection enjoyed by the China seedlings is costly, as they grew more slowly in the field than did the Texas seedlings, when they did not suffer predation by grasshoppers.

The first two hypotheses consider the direct or indirect effects of herbivory on plant invasions. Let's conclude this discussion with a hypothesis that addresses the role of abiotic factors.

THE RESOURCE HYPOTHESIS. The resource hypothesis predicts that invasive plants experience a more ideal combination of resources in their naturalized habitat than they do in their native habitat. Hariet Hinz and her colleagues (2012) wanted to know whether differences in resource quality might be one important factor influencing the success of the invasive hoary cress, *Lepidium draba* in the western United States. *L. draba* is native to eastern Europe, but expanded into western Europe at least 300 years ago, and was introduced into the western United States in the mid-1800s. This small mustard is toxic to livestock,

and is designated as noxious in 16 western states and three Canadian provinces.

The first question the researchers asked was whether *L. draba* was indeed more invasive in western United States and Canada than it was in Europe. They used several measures of invasiveness including shoot density, plant biomass, percent cover and plant reproductive success. In general, the sites in the United States produced larger, and reproductively more successful, *L. draba* plants than did the sites in eastern or western Europe (Figure 11.18A). The researchers then asked whether there was anything different about abiotic conditions in the western United States where *L. draba* is found, in comparison to the European sites. As we discussed in Chapter 4, nitrogen is often a limiting nutrient for plants, so Hinz and her colleagues hypothesized that perhaps western US soils had more available nitrogen than their European counterparts. Samples taken from each site supported this hypothesis, with substantially greater nitrogen levels in US soils compared to soils from either eastern or western Europe (Figure 11.18B).

This finding supports a prediction of the resource hypothesis, but is very far from proving that the higher nitrogen levels caused the increased invasiveness of *L. draba* in the United States. For example, recall that the enemy release hypothesis predicts that invasive plants will suffer higher rates of pathogenic and herbivore attack in their native than in their naturalized range. In accordance with this prediction, the researchers did find that European *L. draba* suffered from more herbivory than did US *L. draba*.

The three hypotheses for why species become invasive are clearly not mutually exclusive. All of the studies described remind us that populations are part of an ecological community composed of many different species, with which they may have important interactions. We will pick up this thread in the next unit.

Of course, the most dominant invasive species is us, *Homo sapiens*. Let's return to the ivory-billed woodpecker to evaluate its current status in its habitat that has been dramatically altered by humans.

REVISIT: The (apparent) resurrection of the ivory-billed woodpecker

As we discussed earlier, bird enthusiasts have been on the trail of the ivory-billed woodpecker since its last documented observation in 1944. Many birders believed this species still existed in small remnant populations, based on observations made by trusted friends and fellow birders. Other bird enthusiasts believed the species to be extinct, pointing to the absence of any verifiable documentation. Fitzpatrick and colleagues' announcement in 2005 divided the scientific community into three camps: some scientists believed the evidence to be solid, others were somewhat skeptical and awaited further evidence, and still others found the science to be very weak. In a rebuttal paper, Jerome Jackson (2006) argues that considering Fitzpatrick's evidence as proof "is delving into faith-based ornithology and doing a disservice to science."

Figure 11.19 Woodpeckers in flight viewed from below (arrow points to very blurry image of the bird). A. Image of ivory-billed woodpecker captured from video (left) and drawing that highlights the pattern of the flight feathers (right). Note the triangular white patches on underside, and the absence of posterior black border to the wing feathers. B. Similarly obscured pileated woodpecker captured on video (left) and drawing that highlights the pattern of flight feathers (right). Note the distinct posterior black border to the wing feathers.

What is the evidence that Fitzpatrick's team presented in their *Science* paper? First, trained ornithologists made seven sightings between April 2004 and February 2005. Second, one of the sightings resulted in a brief (and unfortunately) low-resolution video of a bird taking flight from its perch on a tupelo tree (http://www.birds.cornell.edu/ivory/multimedia/videos/aoupresentation.htm). The researchers did a frame-by-frame analysis of the video, which convinced them that the bird was an ivory-billed woodpecker, and not the relatively common pileated woodpecker. Some evidence from the video included considerable morphological similarity in size and plumage pattern with the ivory-bill, and morphological differences with pileated woodpeckers (Figure 11.19).

Wing beat frequencies obtained from the video were much faster than the pileated wing beat frequency, and identical to one ivory-bill recorded in 1935.

David Sibley and his colleagues (2006) argued that the bird in the video was actually a large pileated woodpecker. His video analysis indicated a black trailing edge on the upper wing in several frames, similar to what a pileated woodpecker would reveal. Unfortunately, the quality of the images is poor enough that with present technology it is difficult to be certain about who is correct. Somewhat surprisingly, Sibley and his colleagues argued that conservation efforts on behalf of the ivory-billed woodpecker should continue. The US government agreed, and allocated substantial funding to help draw up a conservation plan for this species, and for researchers at Cornell University to continue the search. Unfortunately, five more years of searching across the southern Unites States yielded no credible ivory-billed woodpecker sightings.

Why support funding for a bird whose existence is in question? From a conservation standpoint, ecologists usually take a highly conservative approach, and are willing to risk spending time and money to save a species that may already be extinct. The alternative, of failing to save a species that is highly endangered, is not acceptable. But some ecologists argue that the millions of dollars allocated to ivory-bill conservation could have been better spent on other endangered species, whose existence is more clearly documented. In response, conservation ecologists point out that the land acquisitions in Arkansas and Florida that are restoring ivory-bill habitat are providing habitat for many other species that are also highly endangered. Thus conserving ecosystems is a viable approach to conserving endangered species.

SUMMARY

Biological diversity should be viewed through the lens of genetic diversity, overall species diversity, and on a broader scale, ecosystem diversity. Small populations have very low genetic diversity, and have high probabilities of extinction. Ecologists use various types of population viability analyses to predict the probability of extinction. Field ecologists collect population data on survival of young and fecundity of females to construct life tables that help with making projections of future population growth.

Immigration of individuals from nearby populations can maintain population viability and species diversity. Metapopulations are most viable when they are large and well-connected to numerous subpopulations, so they experience high immigration rates. Humans have caused the decline or extinction of many populations and species by degrading or destroying habitats, by fragmenting habitats, by overexploiting species, and indirectly by introducing non-native (invasive) species to a novel environment. Habitat destruction for the purposes of agriculture, and direct exploitation of a naturally small population led to the precipitous decline of ivory-billed woodpeckers in the early twentieth century.

FURTHER READING

Côté, I. M., and 6 others. 2014. What doesn't kill you makes you wary? Effect of repeated culling on the behaviour of an invasive predator. *PLoS One*, 9(4), e94248.
Lionfish are invasive predators that feed on coral reef fish in the western Atlantic Ocean and Caribbean Sea. This paper shows how removal efforts by conservation ecologists are leading to behavioral changes in lionfish that make them more difficult to capture or kill.

Edgar, G. J., and 24 others. 2014. Global conservation outcomes depend on marine protected areas with five key features. *Nature* 506: 216–220.
Far-reaching article that identifies five important features that marine protected areas should have in order to achieve their conservation goals.

Gonzalez, A., Lawton, J. H., Gilbert, F. S. Blackburn, T. M., and Evans-Freke, I. 1998. Meta-population dynamics, abundance, and distribution in a microecosystem. *Science* 281: 2045–2047.

Shows the importance of immigration corridors for maintaining population viability and species diversity. This paper nicely demonstrates how simple small-scale experiments allow researchers to draw inferences about large-scale processes.

Li, Z., Wang, W., and Zhang, Y. 2014. Recruitment and herbivory affect spread of invasive *Spartina alterniflora* in China. *Ecology* 95: 1972–1980.

Shows how an invasive species is able to spread to new habitats despite suffering intense seed predation and herbivory. This process is converting much of the south China shoreline from mangrove forests to grassland.

Mace, G. M. 2014. Whose conservation? *Science* 345: 1558–1560.

Very clear statement of how conservation biology has evolved in the past 50 years from a focus on species, wilderness, and protected areas to a broader consideration of extinctions, ecosystems, and environmental change.

END-OF-CHAPTER QUESTIONS

Review questions

1. Why did the ivory-billed woodpecker decline, and what is the evidence for and against its persistence today?
2. Even if the ivory-billed woodpecker does persist, why is it likely to go extinct in the near future?
3. Use the example of the red colobus monkey in Gombe Stream National Park to explain how ecologists use life tables to predict the viability of populations.
4. What factors led to the persistence of metapopulations of Glanville fritillary butterflies on Aland? Explain the evidence for the importance of each factor that you list.
5. What human-mediated effects can cause a decline in population abundance and species diversity? Give an example of each effect.
6. What are three hypotheses for why invasive species become established in new habitats? Discuss evidence that supports each hypothesis. Are these hypotheses mutually exclusive? Why or why not?

Synthesis and application questions

1. Tanner (1942) suggested that the decrease in population size of ivory-billed woodpeckers over the course of his study may partially result from the initial 1934 population being above carrying capacity. What factors might cause the population in 1934 to rise above carrying capacity?
2. Propose a hypothesis to test whether environmental conditions or cultural traditions influence how much hunting is done by a particular population of chimpanzees. How would you test this hypothesis?
3. Given that PVAs are used to make projections into the future, how can we know if they are any good?
4. We could imagine that hunting-induced mortality in colobus monkeys could be either positively or negatively density dependent. Give a plausible explanation for each type of density dependence.
5. Why did Gonzalez *et al.* (1998) require the pseudocorridor treatment? Why not simply compare unfragmented, corridor, and completely fragmented treatments? What would be an advantage and a disadvantage of only having three treatments (rather than four)?
6. Nieminen *et al.* (2004) describe several factors that could conceivably influence the rate at which empty habitats are colonized by Glanville fritillary butterflies. For each factor,

indicate how it would influence the colonization rate of an empty habitat. Justify your answer.

(a) number of females ovipositing within the habitat simultaneously
(b) size of habitat patch
(c) number of host plants
(d) grazing by herbivores
(e) abundance of nectar plants for adults.

7. Environmental conditions on habitat patches will vary from year to year. Some environmental conditions may be correlated from habitat to habitat within a metapopulation. For example, if it is unusually hot and dry in one population, it will probably be hot and dry in any nearby population. Other conditions may be less correlated. For example, an infectious disease may affect only one or a few habitat patches, leaving the remaining populations unaffected. Will correlated conditions tend to increase or decrease the long-term viability of a metapopulation? Will uncorrelated conditions tend to increase or decrease the long-term viability of a metapopulation? Explain your answer in terms of variation in numbers of populations over time.

8. If there are more pathogens and herbivores in Europe that eat *L. draba* than there are in the United States, it is possible the EICA hypothesis could also explain the greater invasiveness of *L. draba* in the United States. Propose greenhouse experiments that could test this hypothesis.

9. Apply the concepts you've learned about in this chapter to recommend to the United States government how it should work with conservation ecologists to promote the recovery of the ivory-billed woodpecker.

Analyse the data 1

Return to Newman and Pilson's (1997) population viability data shown in Figure 11.1. Are the differences in population viability statistically significant?

Chapter 12

Jane Goodall and Anne Pusey: researching the chimpanzees of Gombe

INTRODUCTION

I arrived in Tanzania's Gombe Stream National Park in March 2008 for a very brief visit. My excuse for missing 2 weeks of classes was the opportunity to visit the famous chimpanzees of Gombe, *Pan troglodytes schweinfurthii*, with Anne Pusey, my former professor, present-day colleague, and then director of the Jane Goodall Institute's Center for Primate Studies at the University of Minnesota. I had just begun writing this book, and I knew that I wanted to write about these legendary chimpanzees and the researchers who have helped make them famous. These chimpanzees have now been studied continuously for over 50 years, and they continue to provide researchers with new insights into their behavior, ecology, and evolution.

Unfortunately, human activities threaten the viability of the Gombe chimpanzee population. During the boat ride on Lake Tanganyika from Kigoma to Gombe, Anne pointed to the devastated landscape on the lake's eastern side that climbs very steeply to heights almost 1000 m above the lake's surface. There were no chimpanzee-sustaining forests, and few trees of any size. Instead, I saw a partially eroded grassland that is planted periodically with cassava, which can be cultivated on marginal soils and provides an excellent source of carbohydrates (Figure 12.1A). Though part of the chimpanzees' former range, it was clear that the eastern shore of Lake Tanganyika, south of Gombe, was not going to sustain chimpanzee populations anytime soon, courtesy of the explosion in human populations in East Africa.

Because these chimpanzees have been studied for over 50 years, ecologists can answer a variety of questions that are usually very difficult to address in populations of long-lived species. For example, researchers now understand how some biotic and abiotic factors influence the distribution and abundance of the three social groups of chimpanzees that live within the park. Together these three social groups make up the Gombe Stream National Park population of chimpanzees: the Mitumba group in the north, the centrally located Kasekela group, and the southern Kalande group, each named after a stream that runs through its respective range (Figure 12.1B). Within these social groups, the results of interactions are influenced by rank,

KEY QUESTIONS

12.1. Why did Jane Goodall study chimpanzees?

12.2. What did Goodall learn during her years at Gombe?

12.3. How did environmental activism change Goodall's career?

12.4. How have Anne Pusey and other researchers collaborated in chimpanzee research?

12.5. What ecological factors threaten the viability of the Gombe chimpanzee population?

age, social affiliations, and other factors that we will discuss. Researchers also understand how dispersal patterns between groups influence population dynamics. In addition, they know a great deal about the population demography, and are using that information to predict the viability of the population. We will discuss all of these population-level relations throughout this chapter, and explore how biologists from a variety of disciplines have worked together to learn about these amazing animals.

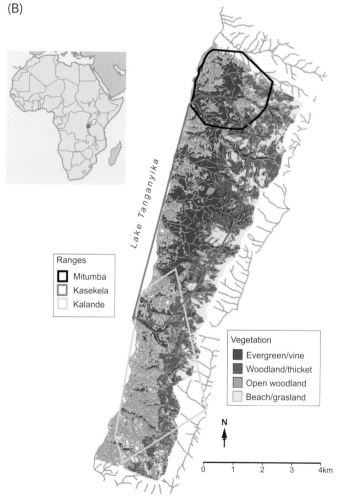

Figure 12.1 A. Mostly deforested eastern shore of Lake Tanganyika, south of Gombe Stream National Park. B. Gombe Stream National Park and ecosystem.

Let's begin by considering why chimpanzees are worthy of study and why Jane Goodall was chosen to study them.

12.1 WHY DID JANE GOODALL STUDY CHIMPANZEES?

The chimpanzees' evolutionary relationship to humans motivated Jane Goodall's first study, begun in 1960. Louis Leakey, the world-renowned anthropologist, sent her there for two reasons. First, as an anthropologist, Leakey wanted to understand human behavior. He reasoned that because chimpanzees are our closest relatives, any patterns of behavior that humans shared with chimpanzees might have been present in the common ancestor to both species. So, studying chimpanzee behavior would provide a window for viewing and understanding the behavior of ancestral humans.

The second reason Leakey sent Goodall was more personal. She was employed as his secretary, had already traveled with him to Oldavai Gorge to work on an archeological dig, and had demonstrated both her toughness and resourcefulness. In addition, Leakey believed that two of Goodall's traits that the scientific community held against her – that she was untrained as a scientist, and that she was a woman – were actually assets. He reasoned that an untrained observer would have no expectations based on existing scientific theory, so Goodall's observations were likely to be accurate and unbiased. He also argued that women tended to be more patient observers, and less threatening to wild animals, and consequently less likely to provoke aggressive behavior from the chimpanzees.

I later asked Goodall whether she agreed with Leakey's assessment of the relative skills of women versus men as observers, and she did, up to a point. "I think [women have] patience and the ability to understand the needs of a non-speaking primate, which is what human babies are [. . .] also the traditional female role [is] of protecting a child by having a good eye for recognizing discord and potential conflict in a family." She added,

> The feminist movement hadn't started, a man has got to make a living, so if he went out in the field he hadn't the luxury of just staying and staying. Whereas [. . .] it was just the beginning of women getting jobs, so me, Birutė, and Dian [Galdikas and Fossey, both chosen by Leakey to study orangutans and gorillas, respectively], we had the luxury of staying out there, without worrying about making a living.

Perhaps noting the look of concern on my face, Goodall then reassured me that men can be good researchers as well.

Our boat arrived at Gombe in the evening, and I was trying to get settled before dark. Meanwhile, the Sun was treating me to the most glorious sunset of my life – I won't even try to describe it, because no description could do it justice (Figure 12.2A). But perhaps an even more glorious sight awaited me the next morning. Anne Pusey had already warned me how difficult it was to get near the chimpanzees, and that we would have to take some

Figure 12.2 A. Sunset over Lake Tanganyika. B. A group of chimpanzees visits camp.

very long and difficult hikes to see them. Even then, they would be atop lofty trees and difficult to see. So the next morning, Pusey and I were talking in the Goodall house, and suddenly there was a loud chorus of vocalizations. I looked at Anne, and asked "Baboons?," and she said, "No, those are chimpanzees." So off we ran, and were treated to a parade of about half of the Kasekela social group strolling along the lake shore, including Kris, the alpha male, and Frodo, who was the alpha male about 10 years previously and is a feature creature in several of Goodall's writings (Figure 12.2B). Anne looked at me and said "I can't believe how lucky you are, Fred. This never happens."

Of course, this did not happen to Goodall upon her arrival at Gombe Stream Game Reserve. It took many months of crawling through the forest, and sitting still for hours at a time, before the chimpanzees would allow her to get within 50 m. She begins her first book, *In the Shadow of Man* (1971), with a description of a long crawl through the woods, hoping to get to a vantage point for viewing four feeding chimpanzees. But after 10 more minutes of crawling, she arrives at her goal, and as had happened so many times in the past 6 months, the chimpanzees had vanished without a trace. She writes,

> The same old feeling of depression clawed at me. Once again the chimpanzees had seen me and silently fled. Then all at once my heart missed several beats. Less than 20 yards away from me, two male chimpanzees were sitting on the ground staring at me intently. Scarcely breathing, I waited for the sudden panic-stricken flight that normally followed a surprise encounter between myself and the chimpanzees at close quarters. But nothing of the sort happened. The two large chimps simply continued to gaze at me. Very slowly I sat down, and after a few more moments, the two calmly began to groom one another [. . .] Without any doubt whatsoever, this was the proudest accomplishment I had known. (Goodall 1971)

That was the beginning of a long series of accomplishments in Goodall's life with chimpanzees, but without that moment, none of the others could have happened. Once the chimpanzees accepted Goodall as a peculiar, but harmless, great ape, it seems as though they could not wait to reveal to her their impressive behavioral repertoire.

12.2 WHAT DID GOODALL LEARN DURING HER YEARS AT GOMBE?

Even after this initial contact, acceptance by the chimpanzees did not come easily. One of the two males at Gombe was the famous David Greybeard, who later provided Goodall with her first observations of several stunning types of chimpanzee behavior (Figure 12.3). As we will learn, chimpanzees have unique foraging adaptations, which include a sophisticated tool-making technology, and have an equally diverse repertoire of social interactions.

Chimpanzee foraging behavior

Chimpanzees are opportunistic feeders, because most of their major types of food are available in large quantities at certain times of the year, and not at others. When diet components are broken down by months of the year, it is clear that there are seasonal components to the time individuals spend eating particular food types (Figure 12.4). For example, seeds were a major

Figure 12.3 The aptly named David Greybeard.

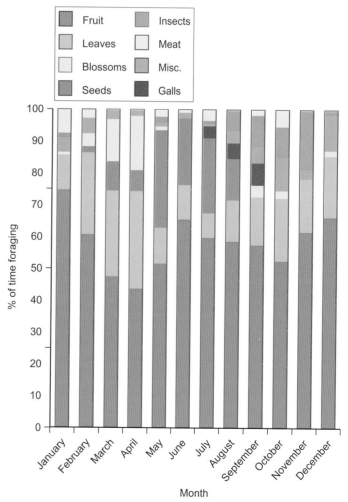

Figure 12.4 Percentage of time chimpanzees spent foraging on different types of food in 1978.

part of the chimpanzee diet during May–July 1978, while insects became important in September–November (Goodall 1986). Also evident is that on an irregular basis, chimpanzees will eat meat.

Carnivory in chimpanzees

Goodall's first observation of meat-eating in chimpanzees was provided by David Greybeard, who she observed eating a bushpig baby high up in a tree. At the base of the tree were the bushpig mother and her three remaining piglets. Up in the tree with David Greybeard were a female chimpanzee and her young child, who were begging for and receiving pieces of meat from David. Upon dropping a piece of meat, the youngster went down to retrieve it, only to be furiously attacked by a rather upset mother pig. The youngster scampered back up the tree and resumed his begging. Goodall did not observe the actual capture of that piglet, but subsequent research has shown that chimpanzees at Gombe capture and eat a diverse array of animal prey; the most prevalent are the red colobus monkey, *Colobus badius*, which accounted for more than 60% of the prey in the 1970s, the bushpig, *Potamochoerus porcus*, and the bushbuck, *Tragelaphus scriptus*.

Most hunts are done in groups, and male chimpanzees are more common in hunting parties, though females participate as well. Upon spotting a troop of colobus monkeys, the chimpanzees appear to assess the situation from the ground. Richard Wrangham (1975) noted that chimpanzees seldom hunt colobus passing through an unbroken canopy, which presumably affords the monkeys more escape routes. Often, chimpanzees will attack mothers with infants or juveniles, whose screams may attract adult male colobus to the scene. Colobus males are formidable defenders; they can usually drive off one or several attacking chimpanzees, and may injure them in the process. Nonetheless, chimpanzees were successful in about 50% of their hunts, though the success rate of individual hunters was much lower – about 25% (see Chapter 11 for an analysis of how chimpanzee hunting affects the viability of colobus populations).

A lone successful hunter will usually eat silently, while the scene following a successful group hunt is pure bedlam. Rank in chimpanzee groups influences many of their interactions, including access to food following a group hunt. A low-ranking captor may have the meat immediately grabbed by a higher-ranking male. Often several males will grab large portions of the prey for themselves. But begging is a chimp's most successful strategy for getting part of the kill; one female was successful in over 70% of her begging efforts. Unsuccessful beggars are described by Goodall (1986) as throwing temper tantrums, which, in at least one case, convinced the male to share his prey. Ian Gilbey (2006) provided evidence that in many situations, chimpanzees share meat because it is too difficult and costly to defend their meat in the face of continuous harassment by beggars (Figure 12.5). But successful hunters will sometimes share some of their meat without being begged, most commonly at the end of a long meat-eating session.

Figure 12.5 A group of chimpanzees share a colobus monkey.

Insectivory and tools

Observing chimpanzees feeding on insects was not that surprising, but the methods chimpanzees used to obtain their insects captured the imagination of human society. David Greybeard was the hero of this story, providing Goodall with her first observation of chimpanzee tool use. He was squatting beside a termite mound, and she watched him push a long grass stem into a hole in the mound. He appeared to be eating termites; this was confirmed in numerous subsequent observations (Goodall 1964). In addition, he was using a tool and, most exciting of all, he fabricated his termite-fishing tool by stripping off the leaves from twigs, so they had the proper shape for the job (Figure 12.6A). Previously, humans had been considered the only tool-making animal, and, in fact, tool use was often a characteristic employed by anthropologists to define the essence of humanity. When Goodall wrote to Leakey to describe her discovery, he famously responded, "Now we must redefine 'tool,' redefine 'man,' or accept chimpanzees as humans" (Peterson 2006).

Inspired by this finding, other researchers soon discovered other examples of tool use and tool-making at Gombe. For example, William McGrew (1974, 1992) described how a chimpanzee may modify a stick from a branch that it breaks off from a tree or shrub, then pulls off the leaves and strips off the bark with its teeth, generating a smooth, straight, tapered tool that averages 66 cm in length. The tool-bearing chimpanzee then digs a hole into the nest of the highly aggressive driver ant, *Dorylus nigricans*, and inserts this tool into the hole, behavior McGrew named "ant dipping." Ants swarm the tool in large numbers, the chimpanzee removes the tool from the hole, and slides its hand over the stick to remove a mass of about 300 ants, which it quickly transfers to its mouth and chews frantically, presumably before the mass of ants can inflict painful bites on the inside of its mouth (Figure 12.6B). Sometimes the chimpanzee will suspend itself from a tree next

Figure 12.6 A. A group of Kasekela chimpanzees fishes for termites. B. Female chimpanzee is ant-dipping.

to the nest, so it can dip with relative impunity, though McGrew described one unfortunate chimpanzee whose supporting tree limb snapped, dropping the chimpanzee into the mass of swarming – and no doubt very annoyed – ants. McGrew estimated a mean intake of about 20 g of ants per feeding session.

Observations of hunting, tool use, and tool fabrication were changing human conceptions of chimpanzees as peaceful vegetarians with limited cognitive skills. But perhaps most damaging to the chimpanzees' reputation were reports of war making among neighboring social groups.

Chimpanzees and war making

In 1972, some chimpanzees from the Kasekela social group broke away to form the new Kahama social group along Kasekela's southern border. This was not a sudden event; by 1966 the researchers had already recognized northern and southern subgroups within Kasekela, which, though they had friendly relations, still tended to spend substantially more time with members of their own subgroup. By 1972, members of the two social groups stopped associating, and had only a very limited zone where their home ranges overlapped.

Conditions deteriorated in 1974 when the Kasekela males began a systematic annihilation of the Kahama chimpanzees, which culminated in the killing of all of the mature males and at least one female. Attacks were consistently brutal and prolonged, and even more astounding to our human perspective was that some of the victims had been close allies with some of the killers. By 1977, all of the Kahama males were dead, and the daughter of the one killed female transferred her affiliation to the Kasekela social group.

However, to the victors did not belong all of the spoils. As it turns out, the Kahama social group was a buffer against the Kalande social group situated farther to the south (see Figure 12.1B). With the annihilation of Kahama, the Kalande social group pushed its boundaries farther to the north. One Kasekela female and her infant son were severely wounded, presumably by Kalande chimpanzees, when they traveled to the southern part of the range. One Kasekela male disappeared; his skull was later found at the eastern portion of the range. Meanwhile the Kasekela home range continued to shrink to about half of its 1977 maximum. With only five mature males, the Kasekela social group was in a precarious position.

These militaristic outings were not aberrant, but appear to be part of the chimpanzee behavioral repertoire, as evidenced by similar descriptions of warfare in other populations (Wilson *et al.* 2014). Though details of chimpanzee brutality captured the attention of a fascinated general public, daily chimpanzee activities are dominated by peaceful feeding and grooming, which indicate that social relationships are of paramount importance in their social group.

Chimpanzee social and family relationships

I asked Goodall what her most exciting discovery was, expecting her to respond with some of the findings we have just discussed. She surprised me with her response:

What's always fascinated me is this long-term bond between family members [...] just realizing how long the period of childhood really was – nobody had any conception that there would be 5 years between births. When I started, I was not allowed (by the scientific community) to talk about childhood – that was a human concept.

Adolescence was a totally human culturally determined stage of the life cycle, but oddly enough I was allowed to use the word baby [...] it was very obvious to me that there were babies, infants, children, and adolescents, and I was fascinated by working out the subtle communications that derived from the changing mother–child interaction.

In contrast to the picture of chimpanzee behavior that I've described thus far, most chimpanzee interactions are what we would describe as friendly, if we were applying them to humans. The strongest friendships are between mothers and their grown daughters, who often form lifelong alliances. Relationships between mothers and their grown sons can also be quite friendly, but less reliably so. But in actuality, relations between males are the most social, though the alliances may shift over the course of an adult's lifetime. Some brothers, in particular, will often have very close relationships and may support each other as they attempt to climb the dominance hierarchy. Other brothers do not form close relationships as adults.

Social grooming is the most common type of chimpanzee social behavior. Grooming can occur in almost any social context. In some cases, one chimpanzee may approach and start grooming a second chimpanzee without any invitation. Or one chimpanzee may solicit grooming by raising an arm, or presenting any body part that needs to be groomed. Grooming may occur when two chimpanzees greet each other after a prolonged period of separation, or following any type of stressful interaction. Males tend to groom for longer periods of time than females; the average duration of grooming session between two males was 25.9 min, between a male and female was 13.5 min, and between two females was 6.5 min (Goodall 1986). There was a slight tendency for males to spend more time grooming males than females, while adult females were more likely to groom males than females. However, adult females with offspring directed most of their grooming efforts toward their offspring.

Grooming is common after a period of separation. Surprisingly, in the majority of these cases, the higher-ranking individual will groom the lower-ranking individual. Goodall interprets this as indicating that grooming, in this context, is used to reassure the lower-ranking individual (Figure 12.7).

When Goodall started her research, DNA had only recently been identified as the genetic material, and there was no using DNA to learn about family relationships. The chimpanzees had to be followed continuously to see who gave birth to whom, and to identify siblings. Goodall compiled all of her data first in notebooks, and then on standardized data sheets that are used to this day.

Long-term data and tragedy

Perhaps the most important data are *follows* of focal individuals (Goodall 1986). Researchers follow one chimpanzee from the earliest possible time in the morning (usually when the chimpanzee leaves its nest), to the latest possible time at night, usually

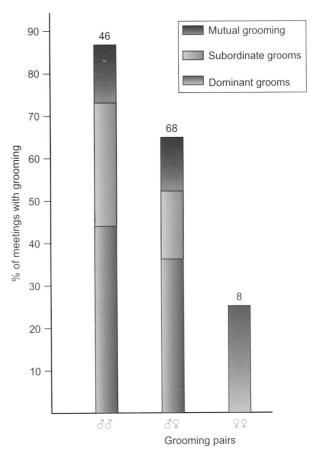

Figure 12.7 Percentage of meetings after a period of separation that resulted in grooming between two males (left bar), between a male and female (center bar), and between two females (right bar). Note that the dominant individual was often the groomer. Numbers above bars are numbers of observations.

when the chimpanzee constructs and enters a nest high up in a tree. During the course of a follow, the pair of observers records the following data: the frequency of feeding, resting, and traveling, and a large variety of interactive activities including grooming, fighting, hunting, mating, playing, and vocalizing. Of course, many of these activities occur simultaneously, which is why two observers are needed for each follow. The researchers also note the animals' travel route by marking the focal chimpanzee's position on a map every 15 min. In addition, the researchers collect data on social associations by noting all chimpanzees within 100 m of the focal chimp.

Between 1968 and 1975, follows were done primarily by undergraduate and graduate students, and also by postdoctoral researchers from the United States and Europe. By 1971, the Tanzanian field assistants were helping with the follows, and assuming greater responsibility for the travel route and association data. By the spring of 1975, 18 students were doing research at Gombe, but then tragedy struck. A boat carrying approximately 40 heavily armed intruders who were members of the Parti de la Révolution Populaire entered the camp at night

Figure 12.8 Hilali Matama kept track of chimpanzees for 36 years.

and kidnapped four American students. As Dale Peterson (2006) describes, the invaders' goal was to overthrow the government in Zaire, and replace it with a Marxist society, using the students as valuable hostages for securing funding for this operation. Ultimately, the kidnappers exchanged their hostages for money and the release of several political prisoners. Twenty-two years later, the rebel leader, Laurent Mobuto, became president of Zaire, which he renamed the Democratic Republic of the Congo.

As you might imagine, this event was catastrophic to the ongoing research at Gombe, in that all of the remaining students were recalled by their universities. However, the now fully trained Tanzanian field assistants insisted on continuing the data collection. One researcher, Hilali Matama, worked at Gombe for 36 years beginning in 1968; his research findings are cited 26 times in Goodall's 1986 book (Figure 12.8).

Though the kidnapping put a serious damper on collaborative research, by 1982 some of the "old timer" researchers returned to Gombe, for visits or to begin new research programs. Some of the former graduate students, including Anne Pusey (about whom we will soon have much to say), were now professors, and collaborative research resumed with increased vigor on topics as diverse as factors influencing the distribution and abundance of the social groups, population demography, dispersal, hunting, analyses of male and female reproductive success, and further explorations of tool use. Some of these studies have been largely independent of the follows, while others have relied heavily on the continuous data set. But since 1986, most of the research has been done without Jane Goodall's direct involvement, as by that time she was responding to a new calling.

12.3 HOW DID ENVIRONMENTAL ACTIVISM CHANGE GOODALL'S CAREER?

Relatively early in her research program, Jane Goodall (1967, 1971) published two popular books on chimpanzee behavior

and ecology that were targeted at lay audiences. Though she had published several articles in scientific journals, it was not until 1986 that Goodall published her first book aimed at a scientific audience, *The Chimpanzees of Gombe: Patterns of Behavior*. To celebrate this event, the Chicago Academy of Sciences sponsored a 3-day conference entitled "Understanding Chimpanzees," which was attended by primatologists from all over the world, including field researchers, laboratory scientists, and zoo researchers.

Though intended to celebrate Goodall's science, this conference effectively ended it. There were two sessions, one on conservation and one on captive conditions. As Goodall described to me, "Everybody had the same grim story [...] the picture of destruction across Africa, the huge drop in chimpanzee population numbers, slide after slide showing habitat destruction, feet caught in wire snares, and the beginning of the bushmeat trade." Deeply shocked by what she heard, Goodall abandoned her career as a scientist for a new career as a conservation activist.

Advocating for chimpanzees in captivity

Initially, Goodall's concerns focused on issues that directly influenced chimpanzee welfare (Peterson 2006). Chimpanzees had been routinely captured for research by hunters, who killed the mothers and gathered up surviving infants. Usually the infants died during processing, but about 10% survived, and many of these survivors made it to research laboratories in the United States sponsored by the National Institutes for Health (NIH). The Endangered Species Act in 1975 reduced chimpanzee import into the United States, but, concerned about possible procedural loopholes, Goodall successfully lobbied for an amendment to the NIH appropriations bill that prevented the NIH from funding any laboratories that used wild-caught chimpanzees for their research activities. The NIH was not enthused about imposing any restrictions, because its top priority was funding research, and chimpanzees – because of their genetic similarity to humans – were the best model for medical research. In 2011, the Institute of Medicine of the National Academies issued a report that concluded that chimpanzees had been a valuable research model in the past, but that they were no longer necessary for most current biomedical research. The use of chimpanzees could only be considered if three conditions were met. First, the knowledge gained must advance public health. Second, there must be no other suitable research model available, and the research could not be ethically done with humans. Lastly, the chimpanzees must be maintained in natural habitats or in an ethologically appropriate physical and social environment – an environment that is conducive to normal behavior (Altevogt *et al.* 2011). Given the stringency of these requirements, future research activities on chimpanzees are expected to decline substantially.

Goodall was also deeply concerned about inhumane conditions suffered by captive chimpanzees in research laboratories. For

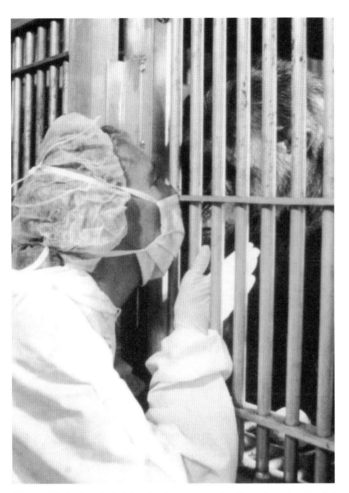

Figure 12.9 Goodall visits the Laboratory for Experimental Medicine and Surgery in Primates in New York, USA, where chimpanzees are kept in stacked, small cages.

example, one research laboratory routinely housed their chimpanzees in isolation in tiny sealed boxes that measured 1.0 × 0.79 × 0.66 m ostensibly to prevent the spread of airborne viruses. The chimpanzees could not even lie down in these cages, and received no diversions nor exercise (Figure 12.9). Goodall went public with her condemnation of these conditions, describing the despair experienced by these highly intelligent animals as a result of sensory and social deprivation (Goodall 1987). She concluded an op-ed piece in the New York Times magazine as follows, "Chimpanzees have given me so much in my life. The least I can do is to speak out for the hundreds of chimpanzees who, right now, sit hunched, miserable, and without hope, staring out with dead eyes from their metal prisons. They cannot speak for themselves."

Ecological consequences of the human–chimpanzee interface

Meanwhile, the physical environment surrounding Gombe continued to deteriorate for chimpanzees, as a result of deforestation

caused by the ever-expanding human population. Gombe National Park was now mostly surrounded by habitat that was unsuitable for chimpanzees. The upshot was that chimpanzees rarely entered or exited the park. In the long term, curtailing immigration and emigration in this way is expected to reduce chimpanzee genetic diversity.

A second problem for some chimpanzee populations was that bushmeat became very fashionable table fare for some of the more wealthy African people who could afford to pay exorbitant prices for some fresh chimpanzee meat. When female chimpanzees were killed, their babies, which were too small for human consumption, were sometimes picked up and sold as pets. Remarkably cute, chimpanzees become less so as they mature, and their owners release them. Thus there was a large population of domesticated chimpanzees, unable to survive in the wild, in urgent need of permanent caretaking. This problem was particularly acute in the country of Burundi, just to the north of Gombe, and also in Zaire and the People's Republic of the Congo. Fortunately, the local human population surrounding Gombe has no history of eating chimpanzees.

By the late 1970s and through the 1980s, Goodall was keeping research at Gombe afloat, in part, by earning money from speaking engagements, which she funneled to the research station through the Jane Goodall Institute (JGI). Partly in response to the problems caused by deforestation and the bushmeat industries, JGI helped found the Lake Tanganyika Catchment Reforestation and Education Project (TACARE) in 1994, a project designed to address poverty and promote sustainable development in the villages around Lake Tanganyika. TACARE engages local residents in the process of promoting conservation values in the areas of community development, forestry, agriculture, health, and environmental education for young people. As Goodall described to me, "The whole idea behind TACARE was how can we even try to protect the chimps [. . .] while the people living around are in such dire straits? All the trees are gone and there are too many people for the land to support, so the idea was to improve the people's lives and persuade them to become our partners in the conservation effort."

TACARE teaches local residents how they can meet their personal needs while becoming responsible stewards of the forest ecosystem. Here are some examples. Local residents within the villages are educated about how they can improve their own standard of living by using fuel-efficient stoves that generate more heat while burning less fuel, which reduces the demand for wood fuel. Residents establish tree nurseries, which grow trees for forest restoration, provide needed wood fuel, and include important cash crops such as coffee and palm trees, for palm oil. Coffee has turned into an incredibly successful cash crop, substantially improving the standard of living for farmers, while using only about 1 ha of land per coffee farm. Palm tree cultivation has a greater downside, as it requires substantial irrigation, and the large plantations may replace native trees.

Farmers and peer educators are taught to plant vegetation that prevents soil erosion across steep landscapes. And TACARE has educated local citizens about birth control, which, if successful, has the potential to substantially reduce the need for converting forest into cropland. TACARE's approach is very non-coercive, and each village can agree or refuse to be part of the project. After seeing the benefits derived from participating, most villages east of Lake Tanganyika have embraced the project.

Global conservation education

Goodall recognizes that regardless of what local residents do, chimpanzees throughout their range will not be protected unless there is a global awareness of conservation issues. She firmly believes that the future lies with educating children to become stewards of the environment, which led JGI to found a youth environmental organization called "Roots and Shoots." Its mission is "to foster respect and compassion for all living things, to promote understanding of all cultures and beliefs and to inspire each individual to take action to make the world a better place for people, animals and the environment." The Roots and Shoots model proposes that knowledge of the environment – particularly when gained at an early age – generates compassion for life, which promotes action to conserve life and the environment, which in turn generates more knowledge of the environment (Figure 12.10A).

As a global organization, Roots and Shoots recognizes that different localities have very different environmental issues and concerns. So Roots and Shoots, which has active chapters in over 120 countries, is flexible in how it educates its students. For example, in some regions of the world with large Islamic populations, Roots and Shoots has collaborated with *madrasas* – Islamic parochial schools – to develop and teach environmental lessons based on writings in the Koran. Across Roots and Shoots chapters, students service their communities with environmentally friendly activities including planting native trees, establishing community gardens, recycling, and organizing environmental cleanups (Figure 12.10B).

Roots and Shoots has had extraordinary success globally, including the establishment of over 900 chapters in mainland China. In 1998, Goodall was invited to China for the first time to explain her organization to the Chinese vice minister for the environment. She approached this task by pretending to be a low-ranking female chimpanzee who comes across a high-ranking male in the forest. She made the appropriate submissive chimpanzee vocalization and told the vice minister that if he approved he would reach out and pat her on the head, at which point she grasped his hand, causing the audience of state officials to gasp in astonishment, and gently patted her head with his hand. After a moment of astonishment, the vice minister roared with laughter, the audience relaxed, and Goodall was given the opportunity to describe

(A)

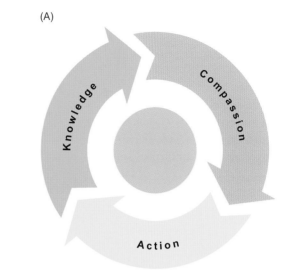

Figure 12.10 A. The Roots and Shoots model. B. Roots and Shoots kids in action.

Roots and Shoots to a television audience of 50 million Chinese people (Peterson 2006).

Jane Goodall the activist has completely replaced Jane Goodall the scientist. Fortunately, research at Gombe has continued in her absence, with the Tanzanian researchers doing the bulk of the daily data collection, and a network of researchers from across the world conducting special projects that interest them.

12.4 HOW HAVE ANNE PUSEY AND OTHER RESEARCHERS COLLABORATED IN CHIMPANZEE RESEARCH?

Oftentimes, when a principal investigator moves on to pursue different interests, his or her research program effectively ends. However, by the early 1990s, Gombe was home to a strong community of researchers who were able to continue collecting data on the chimpanzees, and to further Jane Goodall's work. Anne Pusey was one important member of this research community.

Anne Pusey's early days at Gombe

After completing her undergraduate degree at Oxford University, Anne Pusey came to Gombe in 1970 as Goodall's research assistant, studying mother–infant relations in chimpanzees. Pusey was fascinated by chimpanzee behavioral development, and in particular, the period of adolescence. She enrolled in the graduate program at Stanford University in 1972, and continued research on adolescent development for her doctorate. The student kidnapping happened near the end of Pusey's planned stay at Gombe, but she was fortunate in three respects. First, she was not there at the time of the kidnapping, because she was seeking medical attention for hookworm. Second, since she was approaching her departure from Gombe, she had all of the data that she needed even though she had to terminate her fieldwork a bit prematurely. And, third, the Tanzanian field assistants at Gombe continued to collect data, which she was able to use for her dissertation and for subsequent research.

Returning to Stanford, Pusey began to write her dissertation, which focused on changes in chimpanzee behavior during adolescence, and the implications of these changes for population structure and functioning. Initially, she investigated how the relationship between mother and offspring changed over the course of development.

Developmental changes in chimpanzee behavior

One way Jane Goodall initially earned the trust of the local chimpanzees at Gombe was by providing them with a supply of bananas at a provisioning station (Figure 12.11A). While the chimpanzees approved of this setup, Goodall and other researchers recognized that artificial provisioning was disrupting the natural social group dynamics, so over time provisioning was reduced, and finally eliminated in 2000. But provisioning gave researchers easy opportunities to observe the chimpanzees. For example, Pusey wanted to know how much time mothers were spending with their children, and how this changed over the course of development for male and female children. To do this, she simply calculated the percentage of time that each child arrived at the provisioning station with its mother over the course of its childhood.

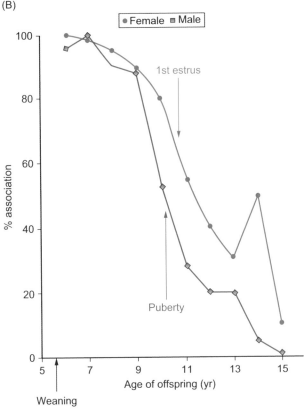

Figure 12.11 A. Infant takes bananas from her mother at Gombe. B. Mean percentage of time mothers associated with their male and female offspring at the feeding station in relation to offspring age. The means were usually based on two male chimpanzees and three or four female chimpanzees.

Males and females showed somewhat similar patterns, always arriving at the feeding station with their mothers until they were weaned, which usually occurred between 5 and 6 years of age (range = 50–86 months). After being weaned, most males reduced association with their mothers at the feeding stations with the onset of puberty, while females reduced association with their mothers at the time of their first estrous period (Pusey 1983). By the time they were full-grown, males rarely associated with their mothers, while females were more variable in their association patterns (Figure 12.11B).

Pusey described behavior away from the feeding station based on over 700 h of follows of four male and four female adolescents between 1972 and 1975. Between the ages of 5 and 12, female adolescents spent most of their time near their mothers (defined as within 15 m), while males spent considerably less time with their mothers as they approached maturity (Figure 12.12A). Males are much more social than females, and even during adolescence, males left their mothers, or attempted to lead them toward groups of chimpanzees (Figure 12.12B). Adolescent females showed no tendency to lead their mothers in the direction of chimpanzee groups.

In summary, Pusey observed a prolonged period of dependence for offspring, even for males, but especially for females. Subsequent studies showed that in some cases mothers and daughters maintained a close relationship throughout their lives, although about 50% of Kasekela females disperse from their natal social group, effectively ending their relationship with their mother (Kasekela is the largest and best-studied social group). Most males stay in their social group, and often maintain friendly relationships with their mothers.

Females' remaining in the social group can have important consequences for population dynamics. Chimpanzees are highly promiscuous, and in some cases, females mate with most or all of the males in the social group while in estrus. Recall from Chapter 11 that mating with close relatives can cause inbreeding depression as a result of increased homozygosity causing deleterious recessive alleles to be expressed. In such a promiscuous mating system, females have the potential to mate with their brother and even their fathers, whose identity they presumably could not even know!

Inbreeding avoidance in chimpanzees

Chimpanzees are unusual among primates in that the males almost always remain in their natal group, while the females often transfer out. At Gombe, however, many females do remain in the group for their entire lives. Pusey wanted to know whether females could somehow bypass the costs associated with conceiving and rearing a child who was sired by a close relative.

Pusey (1980) hypothesized that females avoid the costs of inbreeding by changing their pattern of association once they mature and became fertile. To test this hypothesis, Pusey studied association patterns between females and males in the period of time before and after females first entered estrus. She knew that adolescent females commonly associated with their brothers, and predicted that they would avoid associating with them once they became fertile. Because she did not have relationship data on all of the females in her study, Pusey tested a related prediction that females, upon achieving their first estrus, would switch their

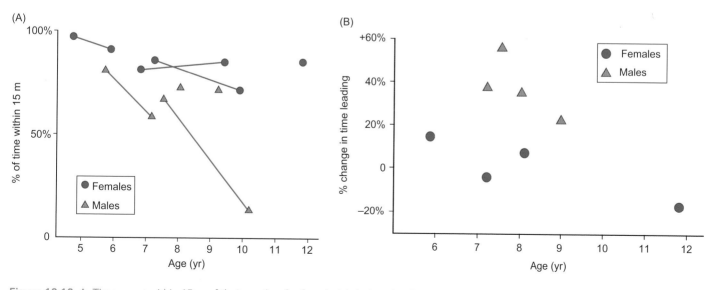

Figure 12.12 A. Time spent within 15 m of their mother for female (circles) and male (triangles) adolescents. Lines connect the same individual. Adolescent daughters had a higher association frequency than did adolescent sons. ($P < 0.03$). B. Percentage increase in time spent leading their mothers toward groups when other chimpanzees were nearby. $P < 0.03$ for male/female comparison.

Table 12.1 Associations and copulations between females and their pre-estrus favorite males (POFM).

	Focal female				
	Pooch	Gigi	Miff	Gilka	Fifi[a]
Number of mature males in Kasekela	15	15	14	8	14
POFM's mean rank[b] before estrus	1.0	1.0	2.0	1.3	2.85
POFM's mean rank after estrus	8.5	9.7	9.0	4.8	4.4
Median number of times focal female copulated with a male during estrus	9	6	8	4	41
Actual number of times female copulated with her POFM	0	1	1	0	5.5

[a]Fifi had two brothers who were her POFMs. For ease of interpretation, I have given the average values for associations and copulations, which has the unfortunate effect of her copulating with her POFM 5.5 times.

[b]Each male was given an association rank for six 3-month periods before estrus and six 3-month periods after estrus, based on the frequency with which he arrived at the feeding station with the female. The mean rank was the mean of those six ranks. The POFM had the lowest mean rank of all males that associated with the focal female.

association from preferring one male before estrus (the *pre-estrus favorite male*) to avoiding that same male. Natural selection would favor such a switch if it reduced the probability of a female mating with her brother.

Pusey gave each male an association rank based on the frequency with which the focal female associated with each male at the feeding station. She calculated the association rank of each male for the six 3-month periods before the female achieved her first estrus. The pre-estrus favorite male had the mean rank (based on the six periods) closest to 1.0. In most cases, this male was known to be the focal female's brother, but in one case, the relationship between the female and the pre-estrus favorite male was unknown. Pusey also recorded all copulations that occurred at the feeding station. In support of Pusey's prediction, after achieving estrus, females tended to associate much less frequently with their pre-estrus favorite males, and also mated with them much less frequently than they did with other males in the group (Table 12.1). Given that in four of these cases, the pre-estrus favorite males were definitely the females' brothers, this avoidance would dramatically reduce the probability of inbreeding.

Anecdotal behavioral observations also support the inbreeding avoidance hypothesis. Pre-estrus favorite males usually did not court their favored females, and when they did court them, the females usually did not respond to their courtship. In the few copulations observed with pre-estrus favorite males, the females usually screamed extensively. Similarly, when sexually approached by males old enough to be their fathers, four of six

females in the social group usually retreated while screaming. Pusey suggests that avoiding older males might be a way for females to avoid inbreeding (mating with their fathers). However, she also discovered that after females got older and more attractive to males, their brothers try to mate with them more frequently.

 Thinking ecologically 12.1

Natural selection has favored female chimpanzees who avoid mating with close relatives. Would you expect natural selection to also favor males who discriminate between relative and non-relative sexual partners? Why or why not?

Pusey's success, and the success of other researchers at Gombe, resulted from having a continuous long-term database for these very long-lived animals. By knowing family relations, Pusey could test the hypothesis that close relatives avoided mating with each other. But in 1978, Pusey began a new research project in the Serengeti with her husband, Craig Packer, studying the behavioral ecology of lions. The two investigated processes such as cooperation and competition, the importance of kinship, and mother–offspring relationships, which bore striking parallels to Pusey's studies into chimpanzee societies. Though Pusey abandoned active participation in the world of chimpanzees to other researchers, her interest in chimpanzees never flagged, and she returned to chimpanzee research in the 1990s.

Saving the data

While in the Serengeti studying lions, and also a postdoctoral associate at the University of Chicago, Pusey kept in touch with Goodall, discussing their mutual research interests in chimpanzees. All of the Gombe chimpanzee data, on data sheets, were being stored in Goodall's house in Dar es Salaam, Tanzania's largest city, where she did much of her writing, and Pusey offered to take responsibility for it. As Goodall described the data and Pusey's offer, "It was in my house in Dar es Salaam, it was being eaten by rats, and I was now moving into my other phase [environmental activism]. So how exciting that she came and rescued all of those moldy bits of paper, and now they are all being computerized."

Pusey moved the data to the University of Minnesota in 1995 under the auspices of the JGI's Center for Primate Studies. As director, Pusey coordinated a team of ecologists and computer scientists to organize and digitize the data. This involved transcribing actual data sheets, written descriptions, slides, films, and videotapes into databases that could be accessed by the research community to help answer a variety of questions about the chimpanzees at Gombe. Pusey estimates that there are over 350 000 pages of data, which are still being collected on data sheets by the Tanzanian research assistants. In 2010, Pusey, and the data, moved to Duke University, where she directs the JGI Research Center.

Meanwhile, western researchers were beginning to return to Gombe, as travel restrictions eased and political conditions in Tanzania became less threatening. Pusey, too, returned for brief visits to help coordinate research conducted by her graduate students. Let's explore some of these more recent studies.

Chimpanzee rank, reproduction, and survival

Rank plays an important role in chimpanzee society, with higher-ranking individuals often displacing lower-ranking individuals from food and from mating opportunities. Rank determination is complicated; chimpanzees born to high-ranking females tend to achieve high rank as they get older, but rank is also influenced by relationships with other members of the social group. The relationships are complex, in part due to the chimpanzees' unusual social structure, described as a fission–fusion social system. Within each social group, chimpanzees break up into much smaller subgroups, which are unstable associations that may persist for a few hours or days. Thus within each subgroup, access to females will be influenced, in part, by the rank of each male that happens to be present.

Rank and male reproductive success

When two chimpanzees meet, the lower-ranking chimpanzee will often emit a *pant-grunt* vocalization to the higher-ranking individual. Thus rank is easily evaluated by the direction of pant-grunts.

Emily Wroblewski and her colleagues (2009) wanted to know whether rank was correlated with male reproductive success. While rank is relatively easy to score, reproductive success is more complex because of the diversity of mating tactics used by male chimpanzees. *Opportunistic* matings occur when a female presents herself to most or all available males in the subgroup, and males mate with the female, apparently indiscriminately. *Possessive* matings occur when one male (usually the highest-ranking male in the subgroup), disrupts matings between the female and other males in the subgroup. *Consortships* occur when the female and her mate retire to the edge of the social group's range during her estrous period and mate exclusively with each other. Because of their secretive nature, consortships are very difficult to observe.

To determine paternity, the researchers matched DNA extracted from fecal samples of adult males and offspring that survived until 2 years old within the Kasekela social group during 1984–2005. In general, the highest-ranking males sired the most offspring, but middle-ranking and even low-ranking males had considerable reproductive success (Figure 12.13A). One reason for the relatively high reproductive success of low-ranking males is that they showed a strong tendency to engage in consortships (often with lower-ranking females). Because these consortships gave them

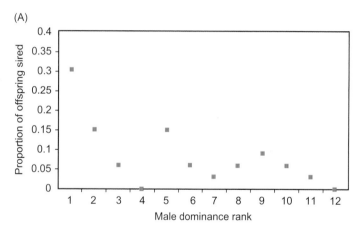

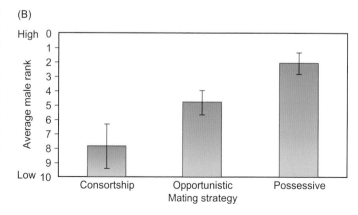

Figure 12.13 A. Proportion of offspring in the social group sired in relation to rank of adult male. B. Mean (±1 SE) rank of males that sired offspring through each mating tactic

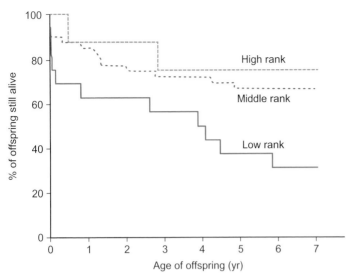

Figure 12.14 Survival rate of offspring (% still alive) born to females of high rank (dashed line), middle rank (dotted line) and low rank (solid line). Notice that by 6 years old, only 30% of offspring born to low-ranking females were still alive).

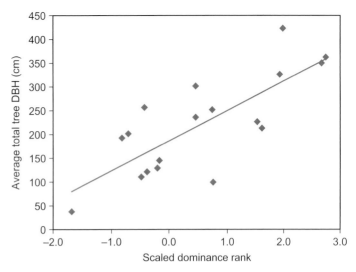

Figure 12.15 Quality of core area in relation to female rank. DBH is tree diameter at breast height, a standard measure of tree size.

nearly exclusive rights to females, these low-ranking males could be successful reproductively (Figure 12.13B).

Rank and female reproductive success

Rank is more difficult to determine in females, because mature females, at least when not in estrus, are often accompanied only by their offspring. Nonetheless, mature females do spend substantial time with other females, which enabled Anne Pusey and her colleagues (1997) to classify females as either low-, middle-, or high-ranking based on the directions of their pant-grunts. They found that offspring of high-ranking females had a slightly higher survival rate (to 7 years old) than offspring of middle-ranking females, and much higher survival rate than offspring of low-ranking females (Figure 12.14).

High-ranking females have been observed to kill the offspring of lower-ranking females, a behavior that explains in part why offspring of high-ranking females enjoy higher survival. But there are clearly nutritional benefits to having a high-ranking mother. Daughters of high-ranking females have greater rates of weight gain. Perhaps related to the nutritional benefits, daughters of high-ranking females mature much more quickly, which should increase their lifetime reproductive success considerably.

Rank and foraging success

Carson Murray and her colleagues (2006) wanted to know how rank influenced female foraging success. They included both pant-grunt data and the results of actual conflicts between females to calculate rank, which ranged from −2.0 for the lowest possible rank to +3.0 for the highest possible rank. The researchers then evaluated the quality of each female's core foraging area by first identifying the 10 tree species most commonly used by females as food sources. Because previous research had indicated

that larger trees bear more fruit, Murray and her colleagues used the size of commonly used trees within each female core area as a measure of habitat quality.

The researchers discovered that higher-ranking females occupied substantially higher-quality core foraging areas (Figure 12.15). Lower-ranking females spent considerably more time foraging and ate lower-quality foods, such as leaves and pith. Each increase of one unit in female rank was associated with a 2.6% decrease in the amount of time spent eating leaves and pith. The researchers concluded that high rank benefits females by providing them greater access to high-quality food.

In a related study, Murray and her colleagues (2008) investigated whether males inherited their mothers' core foraging areas, and thus whether male foraging was influenced by their mothers' ranks. They used data from the long-term database during 1971–91 to map out the mother's core foraging area. They then directly collected data on each female's son during 2001–4, but only when the sons were off by themselves (Figure 12.16A). By restricting data to times when sons were foraging by themselves, the researchers avoided the complications that could arise from movements being influenced by social interactions. They predicted that male foraging core area would tend to overlap with their mother's core area more than it would with the core area of randomly selected females from the social group.

Analysing spatial data can be quite complex, and this case is no exception. The researchers quantified core area overlap by measuring (using a map derived from *follows* data) how frequently the mother and son foraging areas were within 400 m of each other, and comparing that value to a similar statistic using the same males and randomly selected females from the social group. Even though, in many cases, the mothers had died many years before the analysis was conducted, males still tended to forage in a core area that overlapped their mother's

(Figure 12.16B). Even alpha males whose mothers had poor-quality core areas did not move to better foraging areas when they became alpha. Murray and her colleagues conclude that site fidelity allows males to forage in areas where they have extensive

experience with resource availability, allowing them to minimize the time they spend searching for high-quality food.

To summarize thus far, high social rank in chimpanzees is associated with higher survival and reproductive success in males and females. However, even low-ranking individuals can be successful in the complex Kasekela social group. Let's next look at the big picture for the population at Gombe, which includes the Kasekela, Mitumba, and Kalande social groups. We will have less to say about Kalande, because its members are less habituated to humans and observations of that group are difficult to obtain. Mitumba has been studied since 1985, so we have more data on population-level issues at Mitumba than at Kalande.

(A)

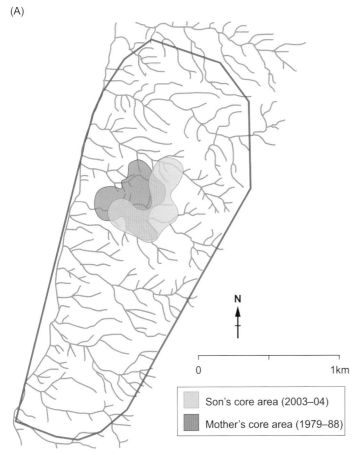

12.5 WHAT ECOLOGICAL FACTORS THREATEN THE VIABILITY OF THE GOMBE CHIMPANZEE POPULATION?

Recent estimates of the total populations of the three Gombe social groups are 14–18 individuals for Kalande (Rudicell *et al.* 2010), 61 for Kasekela, and 26 for Mitumba (Anne Pusey, personal communication). These are obviously very small numbers, but Kasekela and Mitumba have been increasing recently, while Kalande is experiencing a sharp decline (Figure 12.17). Let's look more closely at factors influencing population sizes within each social group, so we can begin to glimpse what may lie ahead for the Gombe population.

Causes of mortality

In many cases it is difficult to ascribe a definite cause of mortality in Gombe chimpanzees, as individuals may disappear from the social group and never be seen again. This is particularly true for females who may leave their natal social group and emigrate to another group. In addition, decomposition and scavenging both

(B)

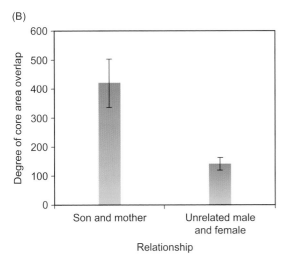

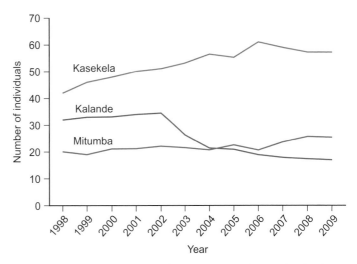

Figure 12.16 A. Foraging core area of one male measured 25 years after the core area of his mother was measured. B. Mean overlap of son and mother core areas in comparison to mean overlap of male and unrelated female core areas. A score of 1000 indicates complete overlap, while a score of 0 indicates no overlap.

Figure 12.17 Population size of Kasekela, Mitumba, and Kalande social groups.

Table 12.2 Causes of death of male and female chimpanzees.

	Illness	Killed by chimpanzees	Injury	Orphaning	Maternal disability	Poaching
Number of males (%)	30 (59)	12 (24)	4 (8)	1 (2)	3 (6)	1 (2)
Number of females (%)	20 (61)	5 (15)	2 (6)	5 (15)	1 (3)	0

Numbers of males and females killed by various causes. Number in parentheses are the percentage of deaths of known cause due to each factor. An additional 17 males and 21 females died from unknown causes.

occur very rapidly after death, so if a body is not discovered quickly, it may be impossible to analyse. Despite these difficulties, Jennifer Williams and her colleagues (2008) were able to identify the cause of death for 67% of Gombe chimpanzees that died between 1960 and 2006 (Table 12.2).

Illness was the primary cause of death, with 50% of the illness-related deaths occurring during epidemics of polio (in 1966), respiratory diseases (in 1968, 1987, and 2000) and mange (in 1997). Polio infections may cause no symptoms, so it is difficult to know what proportion of chimpanzees in 1966 were actually infected but asymptomatic. We do know that 18% of the population suffered paralysis (73% of the paralysed were males) and six individuals died (all males). Respiratory epidemics killed mostly older chimpanzees, while mange killed the very young and very old, mostly sparing the middle-aged chimpanzees. Lastly, wasting disease, defined as noticeable weight loss and weakness before death, caused an additional 28% of illness-related mortality. Illness-caused deaths is probably a conglomerate category including intestinal diseases, parasitic infections, and simian immunodeficiency virus (SIV)-induced infections.

An estimated 20% of mortality was caused by chimpanzees killing other chimpanzees. This is probably an underestimate, because there were several cases of chimpanzee infants disappearing without cause, and they were likely victims of infanticide. Within the Kasekela social group, one mother–daughter team was guilty of at least four infanticides. There was one case of one adult male being killed by a second, and another case where a deposed alpha male was attacked and nearly killed by his successor. Aggression between different social groups sometimes resulted in mortality, usually claiming adults as victims. The bloodiest event occurred when the Kahama social group split off from Kasekela, prompting the Kasekela males to systematically attack and kill all males and one female within the Kahama social group.

Aggression between females is common when one female attempts to enter a different social group. Eventually, immigrant females may be accepted, particularly if they are large and are able to gain assistance from the resident males. Males are most likely to defend immigrant females in estrus (Pusey *et al.* 2008). The reason for aggression between adult females may be tied to resource competition. As we have already described, higher-ranking females have access to better resources, gain weight more

quickly, and reach reproductive age more quickly than other females. Presumably, a female who enters a new social group will compete with resident females for resources. Thus, even though undernourished chimpanzees have not been seen at Gombe, reduced food competition may be driving aggression between females.

Differences in mortality between social groups

Returning to Figure 12.17, you will note that Kasekela and Mitumba populations have increased slightly, while the Kalande population declined sharply between 2002 and 2009. The increase in Kasekela and Mitumba resulted from female immigrants from Kalande. Conversely, part of the decline in Kalande is a result of losing these same females to the other two social groups, but part is a result of higher mortality rates in the Kalande social group.

Why are Kalande chimpanzees dying? Let's first look at some data collected by Brandon Keele and Beatrice Hahn and their colleagues (2009) on SIV affecting chimpanzees (SIVcpz) in the Kasekela and Mitumba social groups. The researchers hypothesized that SIVcpz was an important cause of disease-related death of Gombe's chimpanzees. They followed 94 individuals from Kasekela and Mitumba over a 9-year period, collecting both urine and fecal samples at least once per year. They used western blot analysis to detect SIVcpz antibodies, and a reverse transcription PCR to screen for viral RNA. Over the course of the study, 17 chimpanzees with SIVcpz infection were detected.

Did the infected chimpanzees have higher mortality, and did they show evidence of compromised immune function? The answer to both questions is yes. During the study, 41% of infected chimpanzees died, in comparison to 14% of uninfected chimpanzees. Infection also affected the next generation. All 4 infants born to infected mothers died before their first birthday, while only 6 of 30 infants born to uninfected mothers died before their first birthday.

One Kasekela female died 3 years after becoming infected with SIVcpz. She had no injuries, but became very weak and lethargic shortly before dying. Her autopsy revealed many of the symptoms we normally associate with end-stage AIDS in humans, including skeletal muscle atrophy, multiple abdominal abscesses, and depleted lymphocyte abundance. Researchers tested her for CD4-T-cell depletion, and also tested two other SIVcpz-infected

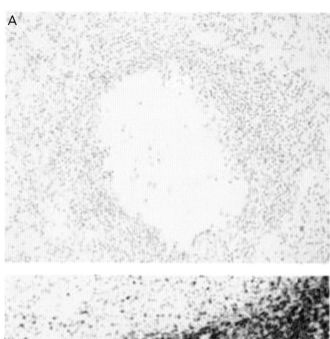

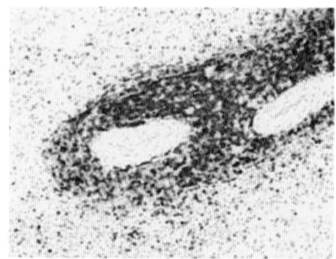

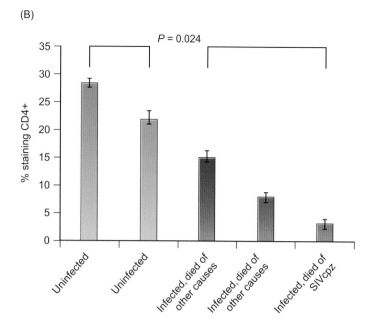

Figure 12.18 A. The researchers assayed CD4-T-cell abundance in the spleen and lymph nodes by staining tissue with antibodies that turn CD4-T-cells a dark blue, and then measuring the percentage of the tissue that took up the stain. The top photo is from the infected female, while the bottom photo is from a chimpanzee who died of other causes. Note the absence of CD4-T-cells in the infected female. B. CD4-T-cell abundance calculated by measuring the area of the spleen that took up CD4-T-cell-specific stain. This area is then divided by the area of the spleen that has the potential to contain CD4-T-cells to calculate the percentage staining CD4+.

chimpanzees who died of other causes (Figure 12.18A). They predicted that the chimpanzee with AIDS-like symptoms would have the lowest CD4-T-cell levels, while the two who died of other causes would show some evidence of CD4-T-cell depletion, in comparison to two uninfected controls. As predicted, all three infected chimpanzees had low CD4-T-cell levels, and the chimpanzee who died with AIDS-like symptoms displayed the lowest CD4-T-cell levels of all (Figure 12.18B).

Next, Rebecca Rudicell and her colleagues (2010) investigated whether Kalande chimpanzees have higher SIVcpz infection rates than chimpanzees from Kasekela and Mitumba. Remember that Kalande populations are not nearly as habituated to humans as are the other two social groups, so sampling of chimpanzee feces and urine was more sporadic in Kalande. The researchers found that Kalande chimpanzees had a consistently much higher rate of SIVcpz infection than did chimpanzees from Kasekela and Mitumba (Figure 12.19).

Given the promiscuous mating system of chimpanzees at Gombe, researchers are very concerned about the future effects of SIVcpz infections on the population growth rate. But diseases of all types also reduce birth rates, which can

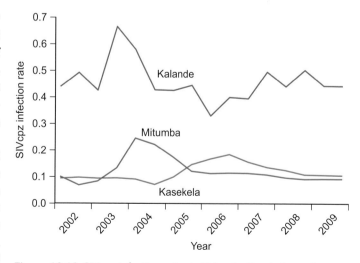

Figure 12.19 SIVcpz infection rates in Kalande, Kasekela, and Mitumba social groups.

also depress population growth. Let's first see what we know about chimpanzee birth rates, then we will quickly look at how SIVcpz may be affecting birth rates of chimpanzees at Gombe.

Chimpanzee birth rates at Gombe and the effects of disease

The rate at which new individuals are born into the population is determined by a female's chance of surviving to maturity, the number of years it takes her to mature, her expected lifespan, the mean number of offspring per birth, and the interbirth interval – the amount of time between births. As we've already discussed, high-ranking chimpanzee females mature at an earlier age than low-ranking females, and thus can contribute more individuals into the population.

Chimpanzee populations have very slow growth potential for several reasons. First, females usually give birth to one baby at a time; of 119 live births examined by Holland Jones and his colleagues (2010), only two were twins. Second, as noted previously, females are slow to mature; most females don't give birth until they reach about 14 years old. Third, the interbirth interval is unusually long.

James Holland Jones, Mike Wilson, and their colleagues (2010) wanted to know what factors influence interbirth intervals of the Gombe chimpanzees. They hypothesized that a female's physiological state would influence her interbirth interval. This generated several predictions. For example, the interbirth interval should be shorter if the previous offspring died while still an infant, if the mother was high ranking, and if the previous child was a female. The researchers reasoned that losing an infant would cause a female to resume her estrous cycle more quickly, reducing the length of her interbirth interval. They expected that high-ranking females should be in better physical condition than low-ranking females, and thus could resume their estrous cycle more quickly after giving birth. Similarly, raising a female offspring requires fewer resources than raising (the usually) larger male offspring, so a mother should resume cycling more quickly after giving birth to a female. In contrast, the first baby may be more physiologically demanding on a mother, given her absence of parenting experience, so the researchers expected the first interbirth interval to be longer than subsequent intervals. However, they expected very old females, with their somewhat diminished physiological condition, to have longer birth intervals.

Most of these predictions were supported by the data, with one exception. There was no significant effect of the sex of the previous offspring on the interbirth interval ($P = 0.210$). But the interbirth interval was much greater after the female's first birth, and when the previous infants survived (Figure 12.20). Female rank was important, with high-ranking females having shorter interbirth intervals than either middle- or low-ranking females, but middle- and low-ranking females had similar interbirth intervals.

Disease can also influence birth rates. SIVcpz infection reduced birth rates in mature cycling females to 0.13 births per year, in comparison to 0.32 offspring per year in uninfected females (Keele *et al.* 2009).

These data show us that the Gombe chimpanzee population has a limited capacity for rapid growth. Overall, mortality is quite low,

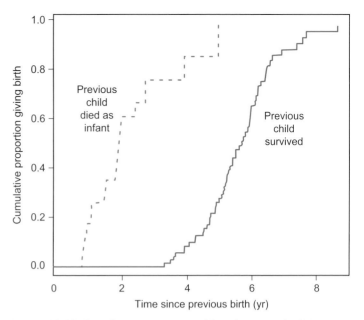

Figure 12.20 Cumulative proportion of females giving birth in relation to time since their previous birth, when their previous child died in infancy (dashed line), or survived (solid line). Note that about 60% of females gave birth within 2 years of their child dying in infancy.

but there is considerable death from disease and from lethal aggression among chimpanzees. Birth rates are very low, with most females not giving birth until they are 14 or 15 years old, and then averaging about 5 years between births. (The champion baby maker at Gombe produced only nine offspring over her long life, of which seven survived to independence.) This low birth rate raises the final question of what lies ahead for Gombe chimpanzees.

Demographic analysis: an uncertain future for Gombe's chimpanzees

Researchers have been collecting demographic data at Gombe for well over 50 years. Based on these data, Anne Pusey, Anne Branikowski, and William Morris have constructed a life table that describes the population growth rate. In Chapter 11, we described how Marc Fourrier and his colleagues (2008) estimated R_0 based on survival and fecundity estimates for red colobus monkeys at Gombe. We have a much more complete data set for chimpanzees, from which we can calculate R_0, and then calculate the per capita growth rate (r) for populations, such as chimpanzees, which have a continuous reproductive life history. Recall that a population is predicted to increase over time if $r > 0$.

As we discussed in Chapter 11, $R_0 = \Sigma l_x b_x$, where l_x is the proportion of the original population alive at the beginning of age class x, and b_x is the mean fecundity of individuals that survive to age class x. As we did with red colobus monkeys, we will simplify the analysis by only measuring female survival and female reproduction, which yields a very long life table (Table 12.3). When we add up all of the values in the $l_x b_x$ column, we get $R_0 = \Sigma l_x b_x = 1.02919$.

Table 12.3 Life table of the chimpanzees of Gombe.

Age class[a]	Female l_x	Female b_x[b]	$l_x b_x$	$x l_x b_x$
0	1	0	0	0
1	0.829787234	0	0	0
2	0.731337901	0	0	0
3	0.716711143	0	0	0
4	0.671445176	0	0	0
5	0.641267865	0	0	0
6	0.626354659	0	0	0
7	0.611788271	0	0	0
8	0.584896479	0	0	0
9	0.571294236	0	0	0
10	0.571294236	0	0	0
11	0.530487504	0.0627	0.033261567	0.365877232
12	0.516885261	0.03935	0.020339435	0.24407322
13	0.506649909	0.05515	0.027941742	0.363242652
14	0.506649909	0.0811	0.041089308	0.575250307
15	0.506649909	0.1886	0.095554173	1.433312593
16	0.497914566	0.1271	0.063284941	1.012559061
17	0.497914566	0.10935	0.054446958	0.925598282
18	0.488339286	0.0329	0.016066362	0.289194525
19	0.478373178	0.0662	0.031668304	0.601697783
20	0.468085583	0.09365	0.043836215	0.876724296
21	0.468085583	0.09825	0.045989408	0.965777578
22	0.456806412	0.11665	0.053286468	1.172302295
23	0.456806412	0.13275	0.060641051	1.394744177
24	0.456806412	0.04305	0.019665516	0.471972385
25	0.420261899	0.05505	0.023135418	0.578385438
26	0.420261899	0.08405	0.035323013	0.918398328
27	0.408254416	0.20355	0.083100186	2.243705033
28	0.371140378	0.08055	0.029895357	0.837070009

Table 12.3 (*cont.*)

Age class[a]	Female l_x	Female b_x[b]	$l_x b_x$	$x l_x b_x$
29	0.371140378	0.06345	0.023548857	0.682916853
30	0.330652337	0.0508	0.016797139	0.503914162
31	0.289320795	0.05745	0.01662148	0.51526587
32	0.259646867	0.05665	0.014708995	0.470687841
33	0.245222041	0.17155	0.042067841	1.388238759
34	0.230797215	0.03255	0.007512449	0.255423278
35	0.201947563	0.08925	0.01802382	0.630833701
36	0.201947563	0	0	0
37	0.183588694	0.05445	0.009996404	0.369866962
38	0.166898813	0.2319	0.038703835	1.470745717
39	0.13351905	0	0	0
40	0.09537075	0.125	0.011921344	0.476853751
41	0.09537075	0	0	0
42	0.09537075	0.125	0.011921344	0.500696438
43	0.09537075	0	0	0
44	0.09537075	0.30725	0.029302663	1.289317171
45	0.0762966	0	0	0
46	0.05722245	0	0	0
47	0.0381483	0	0	0
48	0.0381483	0	0	0
49	0.0381483	0.25	0.009537075	0.467316676
50	0.0381483	0	0	0
51	0.0381483	0	0	0
52	0.0381483	0	0	0
			$\Sigma l_x b_x =$ 1.02918867	$\Sigma x l_x b_x =$ 24.2919624

[a] Each age class is a one-year interval that begins at that age class number. So age class 0 begins at birth and extends until a chimpanzee's first birthday.

[b] b_x is the mean number of offspring produced by a female in age class x divided by 2. For example, a female who survives to 20 years old will, on average, produce 0.09365 female offspring between the time she is 20 and 21 years old.

Remember that our goal is to use these data to calculate the per capita growth rate (r) for these chimpanzees. For populations with a continuous life history,

$$r = \ln R_0 / T$$

where T is the mean generation time, or the mean age of a chimpanzee's mother when it is born:

$$T = \left(\sum x l_x b_x\right) / R_0.$$

We will not derive either of these formulas, but we will apply them to the life-table data. The far right column of the life table shows the age class (x) multiplied by $l_x b_x = x l_x b_x$. The bottom right cell of the life table is simply the sum of all of those $x l_x b_x$ values $= \Sigma x l_x b_x = 24.29196$.

$$\text{Then } T = \left(\sum x l_x b_x\right) / R_0 = 24.29196 / 1.02919$$
$$= 23.603 \text{ years.}$$

It may be surprising that chimpanzee mean generation time is so long, but remember that females give birth even when they are very old. There is no evidence of any menopause-like stage in chimpanzees, and, for example, one female had her last child when she was about 49 years old!

Then it follows that

$$r = \ln(1.02919) / 23.603 = 0.0012$$

Because the per capita growth rate is slightly positive, but very close to 0, we can conclude that the Gombe chimpanzee population is, at present, relatively stable.

 Thinking ecologically 12.2

Refer to Table 12.3. What aspects of the data on the table do you think would change the most if data were collected at Gombe for another 100 years? Explain your reasoning.

The status of the chimpanzee population at Gombe National Park is precarious, at best. The good news is that the per capita growth rate is slightly positive; thus the short-term picture for population survival is relatively good. But several factors conspire to make the long-term picture worrisome. One important concern is population size, which hovers around 100 individuals. A population this small will tend to lose genetic diversity relatively rapidly from genetic drift. In addition, the population is surrounded by very large areas of habitat that are unsuitable for chimpanzee dispersal, so no new genes are expected to enter from gene flow. Female emigration opportunities are limited to the two nearby social groups within Gombe, which may explain why female emigration rates are lower than found in other populations across Africa. These low emigration rates could contribute to inbreeding depression. The upshot is that loss of genetic diversity is likely for the foreseeable future.

The strong social bonds among individuals and the promiscuous mating system make chimpanzee populations good vectors for disease, which has the potential to sicken or kill a large number of individuals in a population that has low genetic diversity. The research program begun by Goodall, and continued by Pusey and her colleagues, has helped to identify how social interactions, dispersal patterns, and the spread of disease influence population dynamics. The numerous conservation education projects that we have discussed are helping to build public support for chimpanzee conservation, and a greater understanding of why this population is so valuable both locally and globally. But if the chimpanzees of Gombe are to survive, the local human population must embrace chimpanzee conservation as a worthwhile project. Rather than competing with each other for resources, humans and chimpanzees must shift their interaction from competition to mutualism.

SUMMARY

A long-term study of the population of chimpanzees at Gombe Stream National Park has provided researchers with numerous insights into the ecology and behavior of our closest relative. Initial studies by Jane Goodall startled the scientific and lay community with discoveries of very human-like behavior including close social relationships, tool fabrication, cooperative hunting, and war. Destruction of chimpanzee habitat, declining chimpanzee populations and increase in the bushmeat trade inspired Jane Goodall to shift her focus to chimpanzee conservation, and the founding of the TACARE and Roots and Shoots.

In Goodall's absence, the Tanzanian researchers continued collecting data, and Anne Pusey and many other researchers from around the world joined them to continue ongoing projects and to begin new ones. Pusey's research answered some questions about how long offspring are dependent on their mothers and about how chimpanzees reduce the probability of inbreeding. More recent studies highlight the importance of rank on reproductive and foraging success, and demonstrate the important role played by disease in limiting the growth of the Gombe population. Demographic analysis shows that the Gombe population is currently stable; nonetheless, with about 100 individuals, the population is threatened by low genetic diversity, and by the loss of suitable habitat nearby which curtails immigration and emigration.

FURTHER READING

Markham, A. C., Lonsdorf, E. V., Pusey, A. E., and Murray, C. M. 2015. Maternal rank influences the outcome of aggressive interactions between immature chimpanzees. *Animal Behaviour* 100: 192–198.

Shows that juveniles from high-ranking females are much more likely to win aggressive interactions than are juveniles from low-ranking females. Juveniles from high-ranking females have this advantage even though their mothers seldom intervene in any aggressive interactions.

Murray, C. M., and 6 others. 2014. Early social exposure in wild chimpanzees: mothers with sons are more gregarious than mothers with daughters. *Proceedings of the National Academy of Sciences*, 111(51), 18189-18194.

This study showed that mothers with sons spent more time in larger social groups and in male-dominated social groups than did mothers with daughters. The authors suggest that these differences provide sons with opportunities to learn from observation and experience sex-typical social behavior.

Pusey, A. E., Pintea, L., Wilson, M. L., Kamenya, S., and Goodall, J. 2007. The contribution of long-term research at Gombe National Park to chimpanzee conservation. *Conservation Biology* 21: 623–634.

Clear summary of the benefits to conservation of continuous long-term research programs in general, and at Gombe in particular.

Wilson, M. L. and 29 others. 2014. Lethal aggression in *Pan* is better explained by adaptive strategies than human impacts. *Nature* 513: 414–417.

Analysis of 152 cases of lethal aggression at 18 different chimpanzee communities indicates that incidents of chimpanzees killing other chimpanzees were most common in communities with large numbers of males, and were usually directed at males from other communities rather than at males from the killers' community. Importantly, communities with extensive exposure to humans actually showed somewhat reduced levels of killing.

END-OF-CHAPTER QUESTIONS

Review questions

1. Why did Jane Goodall study chimpanzees?
2. What were Goodall's most significant discoveries?
3. What factors motivated Goodall's conversion to chimpanzee activism? What has been her approach to conservation?
4. How is social rank important in chimpanzee interactions and chimpanzee fitness?
5. What ecological factors are of greatest concern to the viability of the Gombe population?

Synthesis and application questions

1. JGI has established a program called ChimpanZoo, in which students from around the world can study captive chimpanzees, collecting data and adding them to a growing chimpanzee database. If you were given the opportunity to study captive chimpanzees, what would you do, and how would you go about conducting the research?
2. Some of Pusey's data came from follows of focal animals, which in this case comprised four males and four females from the Kasekela population. How might this influence her conclusions and what could she do to make her findings more robust?
3. Newly mature females frequently disperse to a new social group after leaving their mothers. How might the sexual behavior of these females differ from the sexual behavior of females

who stay in their natal social group?

4. Why don't high-ranking males engage in consortships?

5. The chimpanzees of Gombe are only one population of chimpanzees that are being studied across Africa. If you had the opportunity to study numerous populations across Africa, what questions would you ask, and how would you collect the data to answer those questions?

6. Collecting data needed for creating a chimpanzee life table is very challenging for a variety of reasons. What types of problems do you think the researchers needed to address when deriving the l_x and b_x values?

Analyse the data 1

Compare these data for percentage of time spent eating various food types in 1979 (Figure 12.21) to similar data for 1978 (Figure 12.4). What similarities do you see, and what differences do you see? Provide a hypothesis for why you might observe differences in diet, and also how you might test your hypothesis.

Analyse the data 2

Return to Table 12.1. Are the differences in rank of pre-estrus favorite males before and after the female achieves her first estrus significantly different? Do females mate significantly less frequently with their pre-estrus favorite males than they do with other males in the population?

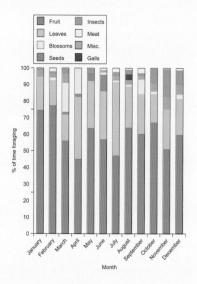

Figure 12.21 Percentage of time chimpanzees spent foraging on different types of food in 1979.

PART IV
Community ecology

Chapter 13
Interspecific competition

INTRODUCTION

The country of Tokelau in the South Pacific Ocean may seem like an unusual venue for studies of competition – interactions between individuals that have a negative effect on the fitness of the interacting individuals. The country comprises three low-lying atolls – coral islands that surround shallow lagoons – with each atoll forming numerous islands. The greatest elevation is less than 5 m above sea level, and the total land area of this country is a mere 12 km^2. About 1400 humans live on these isolated islands, but of interest to us are some other types of interacting animals, the invasive yellow crazy ant, *Anoplolepis gracilipes*, and three species of native hermit crab in the genus *Coenobita*.

Islands are excellent venues for studies of ecological processes, because islands may have relatively low species diversity, which can lead to interactions involving fewer species than in a more species-rich habitat. The numerous islands on Tokelau are a natural laboratory for studying competition between the ants and the crabs. Ants have invaded some islands, but not others, so researchers can study crab foraging and habitat preferences in the presence and absence of ants. Even so, on Tokelau and elsewhere, it is often difficult to establish conclusively that competition is occurring, and to identify precisely what the interacting species are competing for. In this chapter, we discuss the possible outcomes of competition. When two species occupy very similar ecological niches, competition can result in the superior competitor excluding the inferior competitor from an area. Alternatively, ecologically similar species may coexist by changing their resource use in regions where they co-occur. The Lotka–Volterra competition model describes the conditions under which two competing species can coexist, based on the strength of competition and the carrying capacity of each species within a given environment. But competition in natural systems often involves multiple species that can interact with each other directly, or may interact with each other indirectly via their effects on other species.

KEY QUESTIONS

13.1. What types of resources do organisms compete for?

13.2. How can interspecific competition lead to competitive exclusion?

13.3. How do theoretical models and empirical studies identify conditions promoting the coexistence of competing species?

13.4. How do indirect effects and asymmetric interactions operate in natural communities?

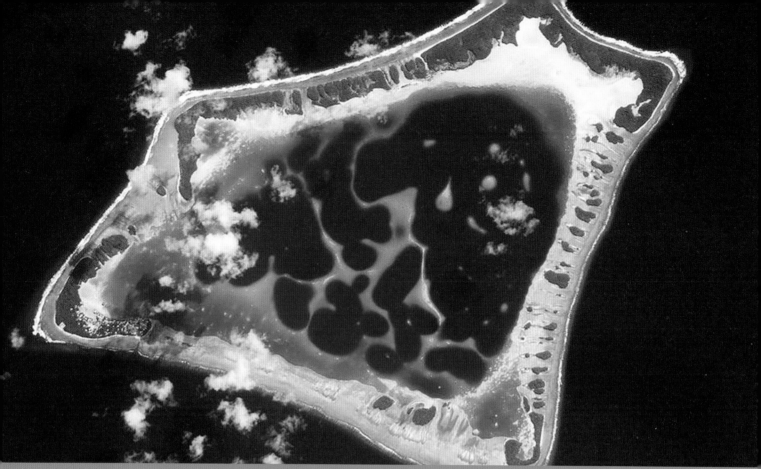

CASE STUDY: The ants and crabs of Tokelau

Researchers on Tokelau observed that yellow crazy ants and three species of native hermit crabs tended to eat similar types of food. This set up the potential for *interspecific competition*, or competition between individuals of different species. Alice McNatty and her colleagues (2009) used several approaches to investigate whether competition truly existed between these two very different animals in Tokelau. One prediction of the interspecific competition hypothesis is a negative correlation between the distribution of the two species; if one species is common, then the other species should be rare. To measure hermit crab abundance, the researchers established 2 m × 2 m quadrats in various locations on islands where both species were present. Then they simply counted the numbers of hermit crabs within these quadrats. To measure ant abundance the researchers loaded 10 cm × 10 cm bait cards next to these quadrats with 2 g of canned tuna, and measured the number of ants attracted to the bait. As predicted by the interspecific competition hypothesis, there was a negative correlation between crab and ant abundance ($r = -0.555$, $P = 0.003$).

But is the difference in distribution a result of competitive exclusion, where one species outcompetes a second species for a limiting resource, and excludes it from a habitat? An alternative explanation for this study's results is that ants may prefer one type of habitat, while crabs prefer another, and that there is no actual competition between them. Ants tended to be more abundant in the interior of islands, which provide them with a rich assortment of plant and animal products (Figure 13.1A). If ants are competitively excluding crabs, then crabs should be less abundant in the interior of islands. But crabs might be less abundant in the interior, independent of any competitive exclusion by ants, simply because they prefer to live along the shore so they can glean carrion brought in by the waves. Fortunately, some Tokelau islands have no yellow crazy ants on them. The researchers reasoned that if ants are competitively excluding crabs from interior habitats, then islands with ants should have fewer crabs living in the interior than islands without ants. When the researchers tested this prediction, they discovered almost no crabs living in interior habitats on islands with ants, but a mean of almost five crabs per quadrat in interior habitats on islands without ants (Figure 13.1B).

We are still left with the problem of showing that these two species are competing for food, rather than for something else, such as space or cover from the elements. McNatty and her colleagues observed the ants chasing crabs from food baits with a spray of formic acid – an irritating compound used both offensively and defensively by many ant species. To determine the overall effect of ant behavior on crab feeding, the researchers created exclosures, which were tuna bait stations surrounded by water. These afforded access to the crabs but excluded the ants. They compared crab feeding rates with ants (unexclosed) and without ants (exclosed). Crab feeding rates increased dramatically when ants were excluded in the shore areas, where the crabs were commonly found, and increased slightly in the interior areas, where the crabs were much less common (Figure 13.2).

Traditionally, ecologists distinguish between two types of competition. Resource competition occurs when species compete for a limiting resource such as food, space, or nest sites. Interference competition occurs when species engage in direct aggressive acts. In many cases, including the crab/ant interactions, both types of competition co-occur. While we are confident that *Coenobita* crabs and *Anoplolepis gracilipes* ants are competing for food on Tokelau, they may also be competing for habitat, or some other resource as well. Let's explore the diversity of resources that can underlie competition.

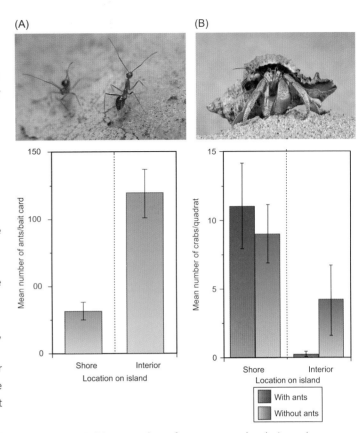

Figure 13.1 A. Mean number of ants attracted to bait cards near shore vs. the interior portion of the islands. B. Mean number of crabs per quadrat near shore and interior of islands with ants and without ants.

13.1 WHAT TYPES OF RESOURCES DO ORGANISMS COMPETE FOR?

One major problem for ecologists is identifying exactly which limiting resource(s) organisms are competing for. Recall from Chapter 4 that a limiting resource is any resource that limits population size. Limiting resources for motile organisms include food, water, nutrients, and suitable habitat. Nonmotile organisms may also compete for ways to move their gametes, such as pollinators, while photosynthetic organisms may also compete for access to light. Competition for these limiting resources can influence the distribution and abundance of species.

Competition for space between barnacle species

As a graduate student in 1947, Joe Connell was required to lead a zoology seminar that discussed a published paper of his choice. His paper referred to a little-known series of controlled field experiments under natural conditions conducted by a French researcher (Hatton, 1938). Connell had not known that controlled field experiments were possible; leading this seminar and reading Hatton's paper convinced him to do his PhD work on a natural community of organisms inhabiting the rocky shores of Scotland. Connell's doctoral investigations (Connell 1961) became one of the classic studies of community ecology.

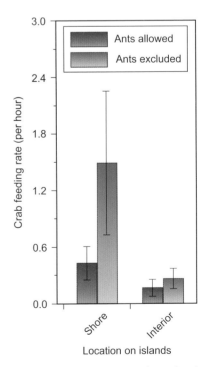

Figure 13.2. Crab feeding rate (mean number of crabs feeding at a bait station per hour) along the shore and in the island's interior when ants were allowed and excluded.

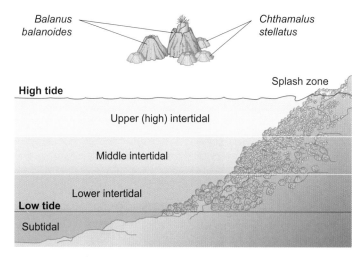

Figure 13.3 *Chthamalus stellatus* and *Balanus balanoides* in the intertidal zone along the coast of Scotland. Note that *Balanus balanoides* was recently reclassified as *Semibalanus balanoides*.

The intertidal zone is the region between the highest and lowest tides along the ocean's shore. Oceanographers divide it into many regions, but for simplicity's sake, we will divide it into only three – upper, middle, and lower. Importantly, the upper zone may go for many hours without water, while the lower zone is submerged much of the time. As larvae, barnacles settle on intertidal zone rocks, metamorphose into adults, and never move again, though they continue to grow larger. Two species were common at Connell's site off the coast of Scotland, *Chthamalus stellatus*, which dominated the upper zone and decreased dramatically in the lower zone, and *Balanus balanoides*, which dominated the lower zone and decreased dramatically in the upper zone (Figure 13.3). Connell hypothesized that this distribution arose from differences in sensitivity to drying out (desiccating), and from interspecific competition for space.

Connell chose barnacles because they are small, abundant, easy to handle, and easy to transplant to new habitats. Connell knew from previous research that *Chthamalus* could withstand constant immersion for up to 2 years; thus he suspected that its scarcity from the lower intertidal zone resulted from competition with *Balanus*. A chance event allowed him to deduce that *Balanus* was rare in the upper intertidal zone because it desiccated easily. A prolonged drought and very unusual low tides in April 1955 led to the ocean level dropping considerably. During this time period, *Balanus* in the upper intertidal zone suffered a mortality rate of about 70%, while the corresponding *Chthamalus* mortality rate was in the order of 30%. Thus moisture level appears to limit the distribution of *Balanus*.

But does *Balanus* outcompete *Chthamalus* in the lower intertidal zone? Several lines of evidence support this hypothesis. First, Connell observed *Balanus* growing over, and completely covering *Chthamalus* individuals so they were unable to feed, and also undercutting *Chthamalus* individuals, ultimately dislodging them from the rocks to be smashed by the ocean's waves. *Balanus* grew about twice as quickly as *Chthamalus*, which probably gave the species this competitive edge. Second, Connell established eight 0.1 m² quadrats on rocks that contained both *Chthamalus* and *Balanus*. He removed *Balanus* individuals from half of each quadrat, leaving them in place on the other half of the quadrat. He then monitored *Chthamalus* survival every 4–6 weeks. At the end of the experiment, seven of the eight quadrats had more *Chthamalus* individuals counted where *Balanus* was removed than where it was present. In most quadrats the differences were very dramatic (Figure 13.4)

Figure 13.4 *Chthamalus* survival with and without *Balanus* in two of Connell's eight quadrats.

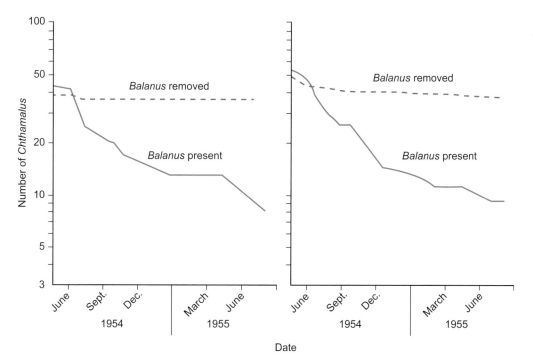

Finally, Connell transplanted rocks with *Chthamalus* on them into the lower intertidal zone, where *Balanus* was abundant. Again he established eight 0.1 m^2 quadrats, and removed *Balanus* individuals from half of each quadrat as described in the previous experiment. In all eight quadrats *Chthamalus* survival was considerably higher in the half quadrat that had no *Balanus* competitors. Taken together, these experiments provide elegant evidence that *Balanus* outcompetes *Chthamalus* for space in the lower intertidal zone.

Thinking ecologically 13.1

Connell's evidence for *Balanus* outcompeting *Chthamalus* in the lower intertidal zone is quite strong, but the evidence for *Balanus* being rare in the upper intertidal zone because of the threat of desiccation is considerably weaker. Can you think of other hypotheses for *Balanus* rarity in the upper intertidal zone? How might you test your hypothesis versus the desiccation hypothesis?

Because adult barnacles are completely sedentary, they need to contend with the problem of moving their sperm from one individual to another. One adaptation is an ultra-long penis – barnacles are endowed with the greatest penis length/body mass ratio of any animal group. Other sedentary groups of organisms face the same challenge of moving sperm and have different adaptations that in some cases lead to interspecies competition. For example, flowering plants send tiny male gametophytes that fly through the wind or are carried by pollinators to female gametophytes. Following arrival, the gametophytes ejaculate their sperm, which, if the timing is right, will fertilize the female's egg. Under some conditions there may not be a sufficient supply of pollinators, setting up conditions whereby flowering plants may compete with each other for pollinators.

Competition for pollinators between dandelion species

In urban and suburban Japan, a native dandelion species, *Taraxacum japonicum*, is being replaced at a high rate by an invasive dandelion species, *Taraxacum officinale*. Ikuo Kandori and his colleagues (2009) hypothesized that *T. officinale* may be outcompeting *T. japonicum* for pollinators. This hypothesis makes two predictions. First, pollinator visitation rates to *T. japonicum* should be reduced in plots with *T. officinale*. Second, seed set (the proportion of flowers that produce seeds) in *T. japonicum* should be reduced in plots with *T. officinale*.

To measure pollinator visitation rates, the researchers grew plants of both species, and trimmed them so that each plant had the same number of capitula – floral displays composed of a large number of tiny flowers. They established three treatments: (1) 30 pots of *T. japonicum*, (2) 15 pots of *T. japonicum* and 15 pots of *T. officinale*, and (3) 30 pots of *T. japonicum* and 30 pots of *T. officinale*. Treatments 1 and 2 had the same overall number of plants, while treatments 1 and 3 had the same number of

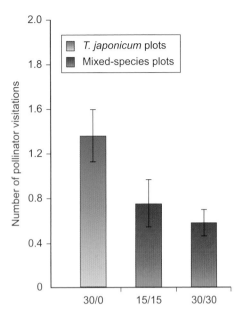

Figure 13.5 Mean pollinator visitation rates ($\pm$SE) per 30-minute observation period. 30/0 = 30 *T. japonicum*; 15/15 = 15 *T. japonicum* and 15 *T. officinale*; 30/30 = 30 *T. japonicum* and 30 *T. officinale*. $F_{1,39} = 17.21$. $P < 0.001$ when comparing visitation rates to *T. japonicum* in pure *T. japonicum* plots vs. mixed-species plots

T. japonicum. They then counted the number of pollinators (mostly copper butterflies, *Lycaena phlaeas*) that visited each plot over a 13–17-day period. Pollinator visitation to *T. japonicum* was much lower in mixed-species plots (treatments 2 and 3) than in pure *T. japonicum* plots (Figure 13.5).

But does this reduction in pollinators cause a reduction in *T. japonicum*'s reproductive success? The researchers measured seed set under the same three conditions. They found a significant reduction in the proportion of *T. japonicum* flowers that produced seeds in plots where both species were present (Figure 13.6A). To investigate whether this reduction also occurred in natural dandelion communities, the researchers chose seven widely separated communities, each of which had at least one pure site of *T. japonicum* and one mixed site with high proportions of both species. They then compared the proportion of *T. japonicum* capitula that set seed in pure *T. japonicum* sites and in mixed sites. Figure 13.6B shows that *T. japonicum*'s seed set was higher in pure sites for all seven populations.

Thus far we've discussed organisms competing for food, space, and pollinators. For many years ecologists have recognized that a species' distribution and abundance may be influenced by many different limiting resources, and by other abiotic and biotic factors. Understanding how these resources and factors together influence a species' distribution and abundance can help us describe a species' *ecological niche*, which we defined in Chapter 9 as the set of environmental conditions and patterns of resource availability, in which a species can survive, grow, and reproduce. We will now explore how competition between species with similar ecological niches may result in the extinction of one or more species within a community.

(A)

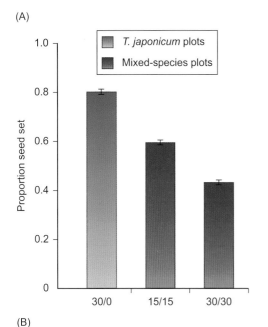

(B)

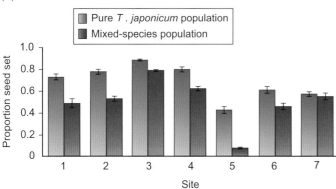

Figure 13.6 A. Proportion *T. japonicum* seed set for pure and two mixed-species plots ($F_{2,26}$ = 225.51, P < 0.001). B. Proportion *T. japonicum* seed set for pure and mixed-species natural communities at seven different sites ($F_{6,973}$ = 127.29, P < 0.001).

13.2 HOW CAN INTERSPECIFIC COMPETITION LEAD TO COMPETITIVE EXCLUSION?

When two or more co-occurring species occupy very similar niches, we would intuitively expect competition to be very strong. Under these conditions, we might expect the best competitor to drive the inferior competitors to extinction – a process ecologists call competitive exclusion. We will discuss several approaches ecologists use to test for competitive exclusion.

Checkerboard species distribution pattern and competitive exclusion

The Bismarck Archipelago off the coast of New Guinea hosts several species of cuckoo doves – arboreal fruit-eating pigeons that live in shaded forests (Figure 13.7A). Two species are closely related, having speciated from a common ancestor relatively

recently. These two species occupy very similar ecological niches as well; both eat mainly small fruit and are almost exactly the same size.

Jared Diamond (1975) studied the distribution pattern of these two species on this archipelago, and found 14 islands with only *M. mackinlayi*, 6 islands with only *M. nigrirostris*, and 11 islands with neither. Importantly, he found no islands hosting both of these ecologically similar species (Figure 13.7B). This checkerboard distribution pattern, in which two or more ecologically similar species have mutually exclusive distributions, suggests that one species competitively excluded the other species from invading the island. Diamond found similar checkerboard patterns with other species groups he examined on the archipelago, as well.

But as Diamond points out, a checkerboard distribution by itself is only weak evidence for competitive exclusion. One problem is explaining the islands that have neither species, even though both species are good colonizers and the habitat seems suitable. A second problem is we don't understand the types of interactions that have led to this mutually exclusive distribution.

Removing a competitively dominant species and competitive exclusion

While taking a summer field ecology course in Costa Rica, Krushnamegh Kunte (2008) elected to investigate whether two very common species of butterflies, the banded peacock, *Anartia fatima*, and the white peacock, *Anartia jatrophae*, were competitively excluding other species of butterflies from a simple plant community. The two major nectar-producing plants were *Lantana camara* and *Wedelia trilobata*. Kunte suspected competitive exclusion simply because the peacock butterflies were so numerically dominant. Surveys conducted over a 1-week period identified 80.2% peacock butterflies on *Lantana*, and 64.4% peacock butterflies on *Wedelia*.

Kunte then removed almost all of the peacock butterflies from his study site, to investigate whether they were competitively excluding other species. If peacock butterflies were competitively excluding other species, he predicted that species diversity should increase following their removal. Table 13.1 shows the substantial increase in numbers of species on both plants that occurred following peacock butterfly removal. In particular there was an increase in the number of large butterflies. He attributes this to greater nectar abundance; following peacock butterfly removal, morning measures of nectar volume per flower increased from 0.003 μl to 0.63 μl.

 Thinking ecologically 13.2

Kunte points out in his paper that you could argue that species richness would have gone up over the time period even without peacock butterfly removal. He then discusses why this is unlikely. How might you design an experiment that controls for this potentially confounding variable?

(B)

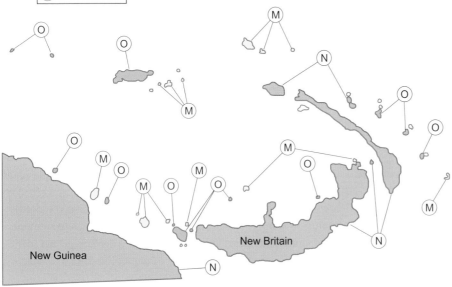

⊙ M *M. mackinlayi*
⊙ N *M. nigrirostris*
⊙ O Neither

New Guinea

New Britain

Figure 13.7 A. *Macropygia mackinlayi* (left) and *Macropygia nigrirostris*. (right). B. Checkerboard distribution of cuckoo doves on Bismarck Archipelago.

Table 13.1 Butterfly abundance and species richness on *Lantana* and *Wedelia* before and after *Anartia* removal. SE is in parentheses.

	Butterflies on *Lantana*			Butterflies on *Wedelia*		
	Pre-removal	Pre-removal excluding *Anartia*	Post-removal	Pre-removal	Pre-removal excluding *Anartia*	Post-removal
Total number of butterflies	1126	223	399	1052	375	703
Mean number of butterflies per survey	75 (19.0)	15 (5.1)	27 (3.3)	70 (7.1)	25 (5.7)	47 (6.9)
Total number of species	23	21	39	19	17	35
Mean number of species per survey	10.5 (2.6)	8.5 (2.6)	12.8 (2.4)	10.9 (2.3)	8.9 (2.3)	14.7 (2.2)

Kunte's research demonstrates that we can see the effects of competitive exclusion on a community by removing the dominant competitor. A second approach is to increase the abundance of the limiting resource driving the competition and see how that influences species richness. If adding more of the resource creates an environment more suitable for the dominant competitor, we might expect a resultant decrease in species richness.

Increasing the level of competition and competitive exclusion

Over the past 50 years, the amount of nitrogen and phosphorus available to plants has approximately doubled, mostly due to human activities such as intensive farming and the production of synthetic fertilizers. Many studies have shown two effects of this fertilization process – an increase in primary productivity, and a decrease in species richness within plant communities. Why does increasing a limiting resource lead to a decrease in species richness? One hypothesis is that some plant species are better competitors for nitrogen and/or phosphorus and thus can outcompete other species for these resources and competitively exclude them from the community. This competition can be above ground, in that taller plants can shade out the shorter plants, or below ground, via competition between roots for resources. Yann Hautier and his colleagues (2009) designed an experiment to get at the root of this problem.

The researchers grew 32 grassland communities for 4 years in the field, before transferring them to a greenhouse. They designed three experimental treatments and a control. The treatments were added fertilizer, added light, and added fertilizer + light. The controls received no fertilizer and normal light. One clever feature of the researchers' design is that the light was added only to the lowest levels of the grassland, so it would only benefit plants growing in the understory – presumably the species that would otherwise be outcompeted for light by the dominant competitors (Figure 13.8A).

After 2 years, the fertilized communities in normal light conditions showed a significant increase in productivity, with the greatest increase enjoyed by the communities that received both fertilizer and light (Figure 13.8B). The community that received only fertilizer had much lower light levels in the understory than the other communities, because the increased biomass blocked out more than half of the natural light (Figure 13.8C). Importantly, understory light levels in the fertilizer + light treatment were almost exactly the same as understory light levels in the controls. The most important finding was that the fertilized community without supplemental light lost more than a third of its species richness compared to the other three communities. By restoring understory light to the fertilized community, species richness was equivalent to the control (Figure 13.8D). Hautier and his colleagues conclude that

competitive exclusion of poor light competitors is responsible for the loss of species richness that is associated with adding nutrients to communities.

Competitive exclusion of poor light competitors can operate in two ways. First, it can cause species present in the community to go extinct at higher than normal rates. Second, it can prevent the recruitment of new species into the community. The researchers observed similar extinction levels for all four treatments but much lower recruitment levels in the fertilized communities that received no supplemental light.

Taken together, the previous two studies – Costa Rican butterflies and grassland communities – provide experimental evidence of competitive exclusion in ecological communities. But these studies also show that competing species do coexist within communities. Even though species richness was greater without the dominant butterfly species, there were still many species of competing butterflies when the dominant species were present. And even though species richness was greater when normal understory light levels were restored to fertilized plant communities, there were still several species of competing plants in the understory at low light levels. Thus our task now becomes understanding the conditions that promote the coexistence of competing species.

13.3 HOW DO THEORETICAL MODELS AND EMPIRICAL STUDIES IDENTIFY CONDITIONS PROMOTING THE COEXISTENCE OF COMPETING SPECIES?

A species' niche is influenced by many biotic and abiotic interactions. Initially we will ignore this complexity, and limit our discussion to two species competing with each other in the absence of the other components of each species' niche. The reason we do so, with caution, is that it is easiest to design and carry out studies with only two species. This is particularly true when deriving models; we start out with the simplest possible case, and then introduce complexity to make the models more realistic. Complexity comes at a cost, because it makes a model more difficult to understand, and may require researchers to collect more data to account for a greater number of variables. Thus complexity should only be added when it improves a model's predictive value.

The Lotka–Volterra model: competition between two species

Recall from Chapter 11 the logistic formula for population growth.

$$dN/dt = r_m N(K - N)/K$$

(A)

Upper grassland canopy with natural light and glasshouse lamps

Supplementary light added to grassland understory

Plastic box

60 cm

40 cm

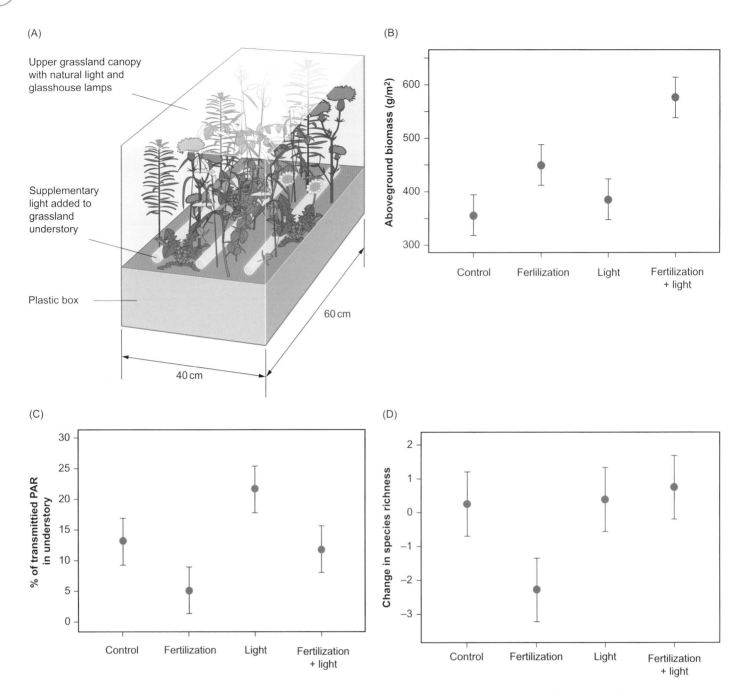

(B)

(C)

(D)

Figure 13.8 A. Hautier's experimental setup to light the understory. B. Mean (±SE) aboveground biomass for control and three treatments. C. Mean (±SE) photosynthetically active radiation (PAR) for control and three treatments. D. Mean (±SE) change in species richness for control and three treatments.

where

dN/dt = the population growth rate,

r_m = the maximum per capita growth rate

N = population size,

K = the carrying capacity.

We were able to show that the population at low levels would show exponential growth, but that the population growth rate would level off as N approached K. For example, if you substitute the same values for N and K into the formula, then $(K-N)$ is equal to $1-1 = 0$, and $dN/dt = 0$, and the population remains at carrying capacity indefinitely, neither growing nor shrinking. We will now modify the basic logistic model to explore competition between two species, creating a model known as the Lotka–Volterra competition model.

Derivation of the Lotka–Volterra competition model

For simplicity, we will assume that competition is only between two species, and they are only competing for one limiting resource. For example, we can consider a simplified version of Kunte's study, with two butterfly species competing for one limiting resource (nectar). If individuals in species 1 were larger than individuals in species 2, they would need more nectar, and we could imagine that on a per capita, or per individual, basis, species 1 would have a stronger (negative) competitive effect on species 2 than species 2 would have on species 1. The symbol α designates the competition coefficient – the per capita competitive effect of one species on the other.

α_{12} = the per capita competitive effect of species 2 on species 1
α_{21} = the per capita competitive effect of species 1 on species 2

Based on our description in the previous paragraph, because species 1 consumed more nectar than species 2, then $\alpha_{21} > \alpha_{12}$.

We will abandon butterflies competing over nectar for now, to consider a general model developed by Alfred Lotka (1925) and Vito Volterra (1926), who extended the basic logistic model for one species into a model for competition between two species over one limiting resource. If N_1 is the population size of species 1 and N_2 is the population size of species 2, then

$$dN_1/dt = r_1 N_1 (K_1 - N_1 - \text{overall competitive effect of species 2 on species 1})/K_1$$

and

$$dN_2/dt = r_2 N_2 (K_2 - N_2 - \text{overall competitive effect of species 1 on species 2})/K_2.$$

r_1 = the maximum per capita growth rate of species 1
r_2 = the maximum per capita growth rate of species 2

The overall competitive effect of species 2 on species 1 is equal to the per capita competitive effect (α_{12}) multiplied by the population size of species 2 (N_2), which equals $\alpha_{12} \times N_2$.

Similarly, the overall competitive effect of species 1 on species 2 is equal to $\alpha_{21} \times N_1$.

Substituting these values gives us the Lotka–Volterra competition equations:

$$dN_1/dt = r_1 N_1 (K_1 - N_1 - \alpha_{12} N_2)/K_1$$

and

$$dN_2/dt = r_2 N_2 (K_2 - N_2 - \alpha_{21} N_1)/K_2$$

Our next task is to understand the predictions of this model.

The Lotka–Volterra model: predicting outcomes of competitive interactions

The Lotka–Volterra model has two possible solutions or outcomes: (1) competitive exclusion, or (2) stable coexistence. The actual solution depends on the carrying capacity and competition coefficients for each species.

To generate these predictions, we will draw graphs of N_1 versus N_2. We'll focus first on what happens to species 1 (Figure 13.9A). Remember the two species are competing with each other. If you start out with an environment with a large

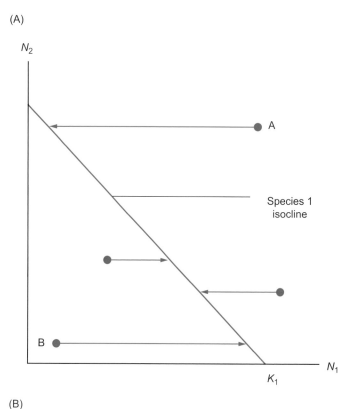

(A)

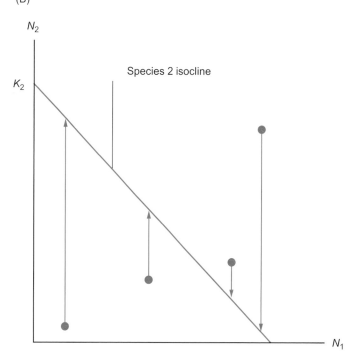

(B)

Figure 13.9 A. Population size of species 1 is attracted to its isocline. In this diagram all changes are parallel to the N_1 axis, because we assume species 2 is not changing. B. Population size of species 2 is attracted to its isocline. In this diagram all changes are parallel to the N_2 axis, because we assume species 1 is not changing.

number of both species (point A on the graph), you would expect species 1 to decrease from competition with species 2 (as shown by the arrow from point A). However, if you start out with relatively few individuals of species 1 and species 2 (point B on the graph), you would expect species 1 to increase (arrow from point B on the graph). In the extreme case, with no species 2 in the environment, species 1 should increase to its carrying capacity (K_1 on the N_1-axis).

In fact, we can draw a line of equilibrium, called a zero growth isocline, from K_1 on the N_1 axis to a point (derived later) on the N_2 axis that describes the dynamics of competition. Any combination of species 1 and species 2 to the right of the line will cause species 1 to decline (go to the left) until it reaches the isocline. Any combination of species 1 and species 2 to the left of the line will cause species 1 to increase (go to the right) until it reaches the isocline. Once species 1 reaches its isocline, it has no tendency to increase nor decrease. Similarly, species 2 will have its own zero growth isocline, from K_2 on the N_2 axis to a point on the N_1 axis (also derived later) (Figure 13.9B). Species 2 will either increase (move up) or decrease (move down) until it reaches its isocline.

The species 1 zero growth isocline (henceforth called the species 1 isocline) intersects the N_1-axis at K_1. That tells us that in the absence of species 2, species 1 is at equilibrium at its carrying capacity. We knew this already from the logistic model. To describe the rest of the isocline, we first recognize that the isocline is a curve (in this case a line) that defines conditions of no change in population. Mathematically, along the isocline, $dN/dt = 0$. Returning to the Lotka–Volterra model, for species 1 this gives us

$$0 = r_1 N_1 (K_1 - N_1 - \alpha_{12} N_2 / K_1)$$

If we multiply both sides of the equation by $K_1/r_1 n_1$, we get $K_1 - N_1 - \alpha_{12} N_2 = 0$ along the isocline. This is the formula for a line (to assure yourself of this, you can substitute X for N_1 and Y for N_2). Using standard graphing techniques, you can convince yourself that the N_1 intercept $= K_1$, as follows:

$$K_1 - N_1 - \alpha_{12} N_2 = 0$$

Set $N_2 = 0$, which by definition is where the line intercepts N_1

$$K_1 - N_1 = 0$$

$$N_1 = K_1$$

But we need two points to draw our isocline. Let's determine the N_2 intercept by setting $N_1 = 0$ for the isocline line $K_1 - N_1 - \alpha_{12} N_2 = 0$.

$$K_1 - \alpha_{12} N_2 = 0$$

$$\alpha_{12} N_2 = K_1$$

$$N_2 = K_1 / \alpha_{12}$$

Our species 1 isocline, where $dN_1/dt = 0$, is a line that intercepts the N_1 axis at K_1, and intercepts the N_2 axis at K_1/α_{12} (Figure 13.10A). You should convince yourself that the species 2 isocline intercepts the N_2 axis at K_2, and the N_1 axis at K_2/α_{21} by going through the same procedure with $r_2 N_2 (K_2 - N_2 - \alpha_{21} N_1)/K_2 = 0$ (Figure 13.10B).

To determine whether each species will increase or decrease when allowed to interact with each other, we must put both isoclines on one graph. There are four possibilities or outcomes of

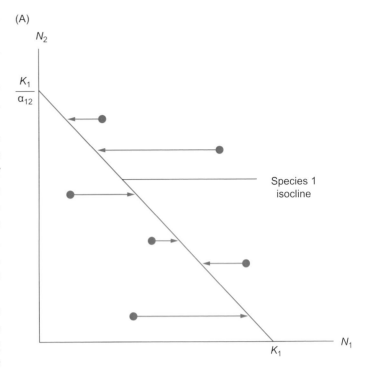

(A)

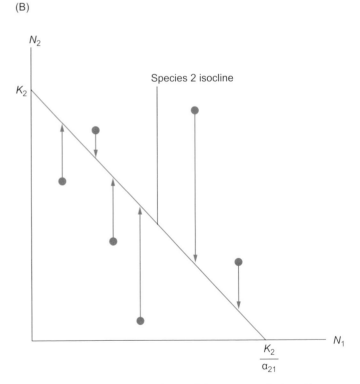

(B)

Figure 13.10 A. Species 1 isocline with intercepts calculated by setting $dN_1/dt = 0$. B. Species 2 isocline with intercepts calculated by setting $dN_2/dt = 0$.

the interactions that are distinguished by the relative positions of the two isoclines. Our task will be to draw arrows (actually vectors) that describe the direction each species will increase or decrease. For our purposes here, we will take the average of the two arrows to indicate the predicted direction in population size for both species considered together.

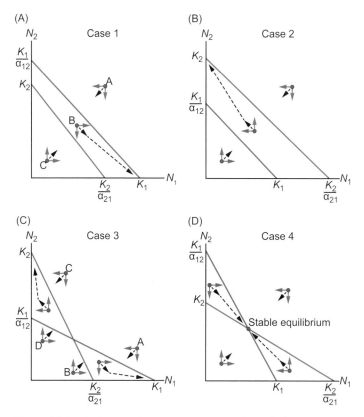

Figure 13.11 A. Case 1. Species 1 competitively excludes species 2. B. Case 2. Species 2 competitively excludes species 1. C. Case 3. One species competitively excludes the other depending on starting population size of each species. D. Case 4. Stable coexistence.

In Case 1 the two isoclines do not intersect each other (Figure 13.11A). If we begin with large populations of both species (for example, at point A), both species will tend to decline, driving the abundance of both species into the middle (trapezoidal) region of the graph. Once this happens (at point B), species 1 will tend to increase toward the species 1 isocline, while species 2 will still tend to decrease toward the species 2 isocline. There will be a gradual increase in species 1, and a gradual decrease in species 2, ultimately resulting in species 2 going extinct and species 1 increasing to K_1. If we begin with small populations of both species (for example, at point C), both species will increase, again driving the abundance of both species into the middle region of the graph, at which point, the exact same dynamics will prevail, resulting in species 2 going extinct, and species 1 increasing to K_1. Case 2 is similar to Case 1, except the species 2 isocline is on top (Figure 13.11B). In case 2, species 1 goes extinct and species 2 increases to K_2. In both cases the species whose isocline is on top and to the right competitively excludes the other species. Once this happens, the successful competitor increases to its carrying capacity.

In Case 3, the two isoclines intersect (Figure 13.11C). If we begin with a large number of species 1 in relation to species 2 (points A or B in the graph), the trajectory of population change will move species abundance into the lower right triangle. In that zone,

species 2 will decrease to extinction, while species 1 will increase to K_1. Conversely, if we begin with a large number of species 2 in relation to species 1 (points C or D in the graph), the trajectory of population change will move species abundance into the upper left triangle. In that zone, species 1 will decrease to extinction, while species 2 will increase to K_2. Where the two isoclines cross is an equilibrium point, with both species on their isocline. Theoretically, neither population should change at that equilibrium. However, this equilibrium is unstable, and random events are sure to drive the population away from that equilibrium. Once that occurs, the two populations are predicted to move in accordance to which of the two triangles they next enter.

In Case 4, the two isoclines also intersect (Figure 13.11D). As in Case 3, the trajectory of population change will move species abundance into one of the two triangles. Once within the two triangles, in contrast to Case 3, the population abundance of both species will converge on the intersection point of the two isoclines, which forms a stable equilibrium. Even when random events drive the populations away from that equilibrium point, they will tend to return to that point over time. Thus Case 4 defines conditions for stable coexistence of two competing species.

Mathematically, from the relative values of K_1, K_2, K_1/α_{12}, and K_2/α_{21}, we can predict the outcomes of all possible interactions (Table 13.2).

Let's translate the simplified mathematical conditions into biological concepts. $K_1\alpha_{21}$ and $K_2\alpha_{12}$ represent the potential strength of interspecific competition. If there is the potential for a large number of species 1 individuals (K_1 is high) and the per capita effect of species 1 on species 2 (α_{21}) is substantial, then the product ($K_1\alpha_{21}$) indicates a strong negative effect of species 1 on species 2. Similarly a large value for $K_2\alpha_{12}$ indicates a strong negative effect of species 2 on species 1. Using the same logic, $K_1\alpha_{11}$ and $K_2\alpha_{22}$ represent the potential strength of intraspecific competition. The per capita effect of a species on itself (α_{11} for species 1 and α_{22} for species 2) always, by definition, $= 1$. Thus the overall effect of species 1 on itself is equal to $K_1(1)$, and the overall effect of species 2 on itself is equal to $K_2(1)$. So if K_1 is high, there is a strong negative effect of species 1 on itself (strong intraspecific competition), and if K_2 is high, there is a strong negative effect of species 2 on itself.

Thus Case 1 predicts species 1 competitively excluding species 2 when the effect of species 1 on itself is greater than the effect of species 2 on species 1, and the effect of species 1 on species 2 is greater than the effect of species 2 on itself. In other words species 1 is a strong interspecific competitor, while species 2 is a weak interspecific competitor. The reverse is true for Case 2. In Case 3, interspecific competition is stronger than intraspecific competition. Case 4 predicts stable coexistence when both species have less of a competitive effect on the other species than they do on themselves – in other words, interspecific competition is relatively weak. Using the Lotka–Volterra model, we can make predictions on the outcome of competition if we know the values

Table 13.2 Mathematical summaries and simplifications of conditions for each outcome of competition for the Lotka–Volterra model.

Case	Mathematical conditions	Simplified mathematical conditions	Predictions of results of competition
1	$K_1/\alpha_{12} > K_2$ and $K_1 > K_2/\alpha_{21}$	$K_1 > K_2\alpha_{12}$ and $K_1\alpha_{21} > K_2$	Competitive exclusion – species 1 wins
2	$K_2 > K_1/\alpha_{12}$ and $K_2/\alpha_{21} > K_1$	$K_2\alpha_{12} > K_1$ and $K_2 > K_1\alpha_{21}$	Competitive exclusion – species 2 wins
3	$K_2 > K_1/\alpha_{12}$ and $K_1 > K_2/\alpha_{21}$	$K_2\alpha_{12} > K_1$ and $K_1\alpha_{21} > K_2$	Competitive exclusion – unstable equilibrium
4	$K_1/\alpha_{12} > K_2$ and $K_2/\alpha_{21} > K_1$	$K_1 > K_2\alpha_{12}$ and $K_2 > K_1\alpha_{21}$	Coexistence – stable equilibrium

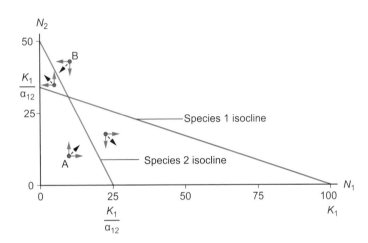

Figure 13.12 Using the Lotka–Volterra competition model to predict the outcome of competition between two species. Let's take a simple hypothetical example using two species of butterflies competing over nectar. Assume $K_1 = 100$, $\alpha_{21} = 2$, $K_2 = 50$, and $\alpha_{12} = 3$. What will be the results of competition? To solve this problem you need to draw the two zero growth isoclines. The species 1 isocline intersects the N_1 axis at K_1, and the N_2 axis at $K_1/\alpha_{12} = 100/3 = 33.3$. The species 2 isocline intersects the N_2 axis at K_2, and the N_1 axis at $K_2/\alpha_{21} = 50/2 = 25$. As the figure shows, when you draw the isoclines, you can see that they cross as in Case 3, yielding an unstable equilibrium. One species will competitively exclude the other. If you start with 10 individuals of each species, the trajectory of population change (dashed arrow from point A) will move species abundance into the lower right triangle. In that zone, species 2 will decrease to extinction, while species 1 will increase to K_1. Thus species 1 competitively excludes species 2. In contrast, If you start with 10 individuals of species A and 40 individuals of species B, the trajectory of population change (dashed arrow from point B) will move species abundance into the upper left triangle. In that zone, species 1 will decrease to extinction, while species 2 will increase to K_2. Thus species 2 competitively excludes species 1.

of K and α. Case 3 is unique in that it also requires us to know the starting values of N_1 and N_2 (Figure 13.12).

As with all models, the Lotka–Volterra model makes simplifying assumptions. Some assumptions are mathematical conveniences, while others raise important conceptual issues.

The Lotka–Volterra model's many simplifying assumptions

One reason the Lotka–Volterra model appears prominently in most ecology textbooks is that it is relatively simple. A second feature of the Lotka–Volterra model is that it builds directly on the logistic model, adding the concept of interspecific competition to a model that was designed to show the dynamics of intraspecific competition or density dependence. This highlights an important feature of mathematical models – that they often build upon previously derived models.

Three of the essential components of the model, r, K, and α, are assumed to be constants. To varying degrees this can be problematic. For example, we know that a species' per capita growth rate can vary, and carrying capacities of the environment can also vary for a particular species. We also know that the competition coefficient of a 30-m-tall oak tree is very different from that of a sapling. One solution to these problems is to make the model more realistic, but of course that also makes it more complex.

A more serious problem articulated by David Tilman (1987) and many other researchers is that the Lotka–Volterra model does not consider the mechanisms of how species compete in the natural world. It assigns one value, a competition coefficient, to each species in a two-species interaction, but it does not consider the richness of competitive interactions. Within a community, a species might influence another species directly, by interference or resource competition, or indirectly, by influencing a third species within the community. Species also may adapt genetically, changing their pattern of resource use. Finally, species may also change their pattern of resource utilization, shifting their realized niche as a response to competition. Tilman argues that these concerns should stimulate ecologists to develop new, more robust models that take into account the mechanisms underlying competition. Several important models have been developed to incorporate these factors, but they are beyond the scope of our discussion.

Many recent ecological studies highlight the dynamic complexity of competitive interactions. This complexity paves the way for the coexistence of competing species in the natural world.

Factors favoring species coexistence in natural systems

The simplified world of Lotka–Volterra predicts the coexistence of competing species if interspecific competition is relatively weak between two species. We will shift our attention to the more complex natural world to identify other factors that can influence the long-term coexistence of competing species.

Environmental variation

Several species of mosquitofish have been introduced to many countries worldwide in an attempt to control the mosquitoes that causes malaria. Unfortunately, mosquitofish have also been implicated in the local extinction and endangerment of many native fish and amphibia. Mosquitofish are dominant species in many freshwater habitats but are less abundant in moderately and highly saline habitats. The Mediterranean toothcarp, *Aphanius fasciatus*, a small endangered native fish formerly common in freshwater habitats, is now restricted primarily to saline environments. Carles Alcaraz and his colleagues (2008) suspected that the eastern mosquitofish, *Gambusia holbrooki*, competitively excluded toothcarp from freshwater, but not saltwater, habitat. The researchers proposed two competition mechanisms. The interference competition hypothesis proposes that saline environments provide toothcarp with a refuge from attacks by eastern mosquitofish. The resource competition hypothesis proposes that toothcarp compete more effectively with eastern mosquitofish for food in saline environments.

The interference competition hypothesis predicts that aggression by eastern mosquitofish to toothcarp should decrease at higher salinity values. To test this prediction, the researchers established three colonies of fish that were acclimated to three salinity treatments: 0‰ (freshwater), 15‰, and 25‰ (ocean water averages about 35‰, or a salt content of 35 parts per thousand). They then put various combinations of toothcarp and mosquitofish together in aquaria at these different salinities, and observed for aggressive behavior. We will restrict our discussion to the simplest case in which there was one individual of each species. The results of this experiment show a small reduction in mosquitofish aggression directed against toothcarp between 0‰ and 15‰ salinities, and a substantial reduction in aggressive acts at 25‰ salinity (Figure 13.13A).

The resource competition hypothesis predicts that toothcarp will have higher feeding rates at higher salinity values. The researchers placed three of each fish species, toothcarp and eastern mosquitofish, into aquaria of different salinities, allowed the fish to acclimate, and then introduced 10 fly larvae to each aquarium. Mosquitofish ate 100% of the food in pure water but a significantly lower percentage of the food in the saline environment, when toothcarp feeding rates increased (Figure 13.13B). The reduction in feeding success occurred because mosquitofish took longer to find their prey at the higher salinity treatments. The researchers speculate that the decrease in aggression and

decrease in feeding rates are results of a decrease in mosquitofish metabolic rate under the stress of relatively high salt concentrations. Consequently in salt water, eastern mosquitofish do not competitively exclude Mediterranean toothcarp.

There are two important conclusions from this study. First, the results of competition are profoundly influenced by the environment. Second, both interference competition and resource competition can operate together, as they do in the interaction between eastern mosquitofish and toothcarp, to influence the distribution of species.

The experimental results of this study, and the present restriction of toothcarp to moderately saline habitat within the Mediterranean region, support the hypothesis that mosquitofish have competitively excluded toothcarp from freshwater habitat. But species under competitive pressure, such as the toothcarp, may avoid competitive exclusion from a habitat by reducing overlap in habitat use.

Niche shift

The American mink, *Neovison vison*, became established in Britain about 100 years ago as a result of escaping from mink farms. Concurrently, two native species, the Eurasian otter, *Lutra lutra*, and the European polecat, *Mustela putorius*, were hunted and poisoned to near extinction. Both of these native species are now recovering within their native ranges.

Lauren Harrington and her colleagues (2009) suspected that as otters reestablish themselves in their former habitats, they may be outcompeting the invasive minks. Otters are much larger than minks, they have been observed stealing food from minks, and some recent studies have shown mink abundance declining with increasing otter abundance. Very little is known about the interactions between polecats and mink. But minks are still relatively abundant in many areas where otters have recovered significantly. The researchers hypothesized that niche shift – an adaptive change in a species in one or more of its niche dimensions, underlies the continued coexistence of mink in areas recolonized by otters and polecats.

Their study site was a 20-km stretch of the River Thames near Oxford, England. This site had a healthy population of mink during the 1990s, but no otters or polecats. Mink populations remained abundant in the mid-2000s, while otters and polecats returned and reestablished themselves. The researchers considered three niche dimensions that might have changed in minks, in response to the return of otters and polecats: (1) diet, (2) use of space, and (3) timing of daily activity.

Because otters spend almost all of their time in water, the researchers hypothesized that the mink diet had become more terrestrial with the return of otters. The researchers analysed mink scat collected from the 1990s and 2000s and found a significant decrease in fish consumption and a significant increase in insect and bird consumption during the 2000s (Figure 13.14A). To test the use of space hypothesis, the researchers radio-tracked minks to see whether they used the main river less frequently in

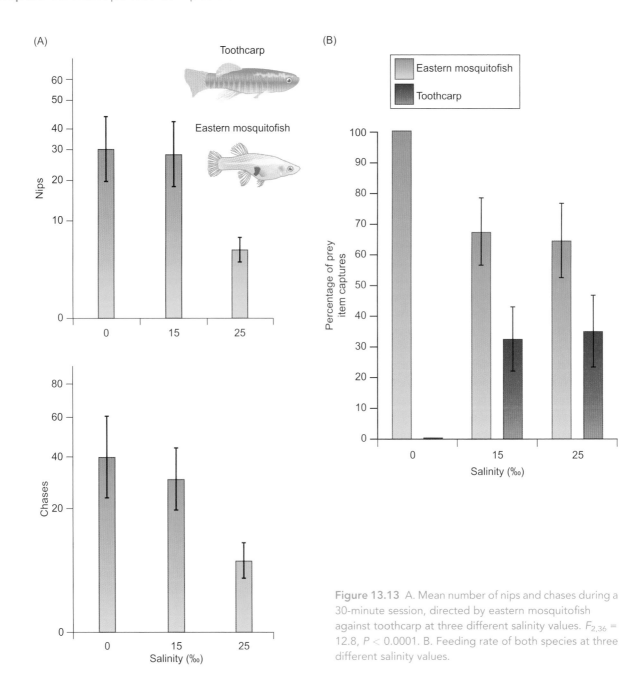

Figure 13.13 A. Mean number of nips and chases during a 30-minute session, directed by eastern mosquitofish against toothcarp at three different salinity values. $F_{2,36} = 12.8$, $P < 0.0001$. B. Feeding rate of both species at three different salinity values.

the 2000s, in response to the return of the otters. River use did decline from about 70% to about 60%, but these differences were not statistically significant.

Lastly, the researchers analysed their radio-tracking data to see if minks changed their activity patterns to minimize their interactions with highly nocturnal otters. As you can see, mink activity shifted dramatically from primarily nocturnal in the 1990s to primarily diurnal in the 2000s (Figure 13.14B).

Minks shifted their niches in at least two dimensions in association with the return of the native species. But the change in timing of activity was much more significant than the change

in diet – all individuals tracked in the 1990s were primarily nocturnal, while 9 of 10 individuals from the 2000s were primarily diurnal. As a cautionary note, the authors remind us that they have not shown that niche shift was a direct result of competition – factors other than the reestablishment of otters could have caused the changes in mink behavior. The investigators encourage research on other sites to further test the niche shift hypothesis.

Niche shift is an extreme example of niche partitioning – a process in which ecologically similar species coexist by using different resources or specializing on different factors in the

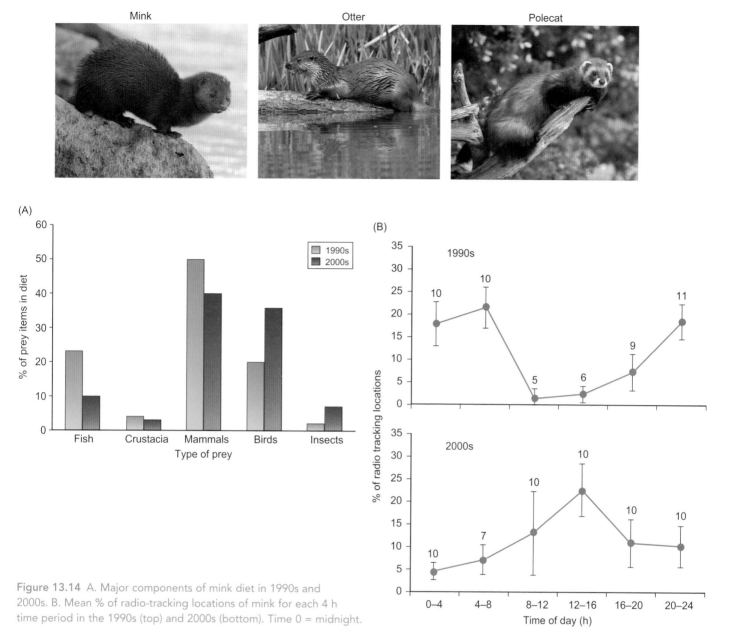

Figure 13.14 A. Major components of mink diet in 1990s and 2000s. B. Mean % of radio-tracking locations of mink for each 4 h time period in the 1990s (top) and 2000s (bottom). Time 0 = midnight.

environment. In Chapter 3 we discussed one example of niche partitioning in which each species of seed-eating finches on the Galápagos specialized on different-sized seeds. We will turn our attention to another group of species, mosquito larvae living in bromeliad pools.

Niche partitioning

If you've eaten pineapple, you've had the pleasure of interacting with bromeliads, a very diverse family of over 3000 plant species found almost exclusively in North, Central, and South America. In addition to producing fruit for our palettes, bromeliads also provide habitat for many small organisms that live within the tiny pools that are formed by their leaves. One such group is mosquito larvae, which feed off small detritus particles suspended in bromeliad pools (Figure 13.15A).

Benjamin Gilbert and his colleagues (2008) used turkey basters to collect all the mosquito larvae living within 71 bromeliad pools. The four most common species were *Anopheles neivai*, *Culex rejector*, *Wyeomyia melanopus*, and *Wyeomyia circumcincta*. If mosquito larvae partition their niches, they could do so along many different niche dimensions, including the size of the mosquito larva, their position in the water column, the size of bromeliad, and the species of bromeliad.

The researchers investigated relative position in the water column by placing the larvae into a 50 ml centrifuge tube (with detritus to feed them) and allowing the larvae to settle in the absence of other species. *A. neivai* settled very near the surface, and *W. circumcincta* near the bottom, while *C. rejector* and *W. melanopus* inhabited the middle regions of the water column (Figure 13.15B). Along this niche dimension,

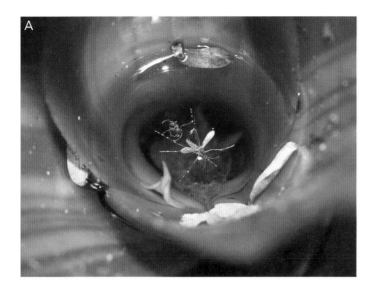

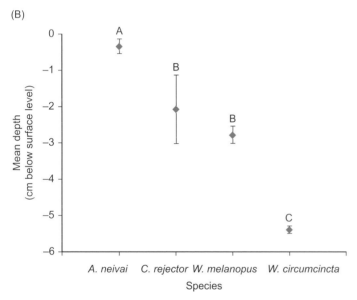

Figure 13.15 A. Adult mosquito on the surface of bromeliad pool B. Mean (±SE) depth in water column for larvae of four mosquito species. Different letters above the data points indicate statistically significant between-species differences in mean water column depth.

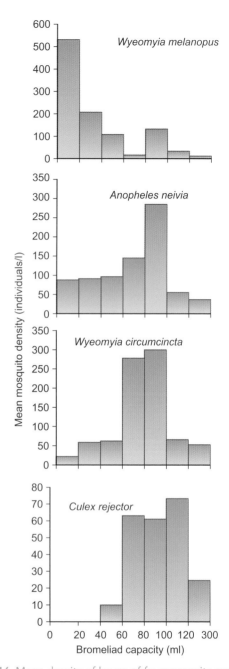

Figure 13.16 Mean density of larvae of four mosquito species in different-sized bromeliad pools.

C. rejector and *W. melanopus* had the most interspecific competition, because they had the greatest overlap in their resource utilization (position in the water column). In contrast, there was almost no competition between *A. neivai* and *W. circumcincta*.

Gilbert and his colleagues reasoned that if mosquito larvae were actually partitioning water depth to reduce interspecific competition, then strongly competing species such as *C. rejector* and *W. melanopus* should be found together only rarely within an individual plant, while weakly competing species such as *A. neivai* and *W. circumcincta* should be found together much more commonly. Their surveys support this line of reasoning;

less than 15% of the pools contained both *C. rejector* and *W. melanopus*, while more than 50% of the pools contained both *A. neivai* and *W. circumcincta*.

The question then becomes, what mechanism causes *C. rejector* and *W. melanopus* to rarely co-occur in natural bromeliads? Is there a battle for spatial supremacy within the contested middle of the bromeliad water column, or is there some other mechanism for partitioning resources? Surveys of bromeliad pools showed strong differences in species abundance in different-sized pools, with *W. melanopus* abundance greatest in the smallest pools, and *C. rejector* abundance greatest in the largest bromeliad pools (Figure 13.16).

These mosquito species partition niches along at least two dimensions: depth in the water column and pool size. But we still don't know what determines pool size distribution in the two species that show overlapping preferences in pool depth. Is *W. melanopus* a better competitor in small bromeliads and *C. rejector* a better competitor in large bromeliads? Or do *W. melanopus* females lay their eggs in small pools, and *C. rejector* females lay their eggs in larger pools?

 Thinking ecologically 13.3

How might you design observations or experiments to distinguish between the two suggested mechanisms explaining the distribution of *W. melanopus* in small pools and *C. rejector* in large pools?

The two logical outcomes of competitive interactions are competitive exclusion or species coexistence. But even in simple two-species systems, competition may be difficult to define and characterize. The situation becomes even more complex in natural communities, composed of multiple species with a variety of interactions.

13.4 HOW DO INDIRECT EFFECTS AND ASYMMETRIC INTERACTIONS OPERATE IN NATURAL COMMUNITIES?

The effect of one species on another is often indirect, in that the outcome is influenced by the interaction of one or both species with other species. These indirect interactions are common in biological communities (see Chapter 16 for a more complete discussion of indirect interactions). Underscoring the complexity of interspecific interactions are cases that are neither interference nor resource competition, yet both species have a negative effect on each other's abundance.

Apparent competition

If one predator has two prey species, then an increase of one prey species may cause an increase in the predator species. Similarly, an increase in the other prey species may also cause an increase in the predator. But an increase in the predator may harm both species. Thus each prey species is indirectly having a negative effect on the other, courtesy of their positive effect on the predator (Figure 13.17). This process is called apparent competition.

Heather Johnson and her colleagues (2013) wanted to know whether apparent competition was limiting the recovery of the endangered Sierra Nevada bighorn sheep, *Ovis canadensis sierrae*. Cougars, *Puma concolor*, prey on the population of 360 sheep, but their primary prey is a population of 19 000 mule deer,

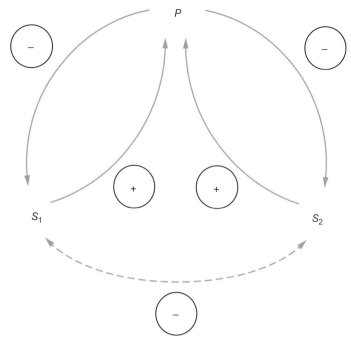

Figure 13.17. Apparent competition. Each prey species (S_1 or S_2) serves as food and thereby directly increases the abundance of the predator (P), as represented by the solid arrows with + sign. Conversely, predators directly decrease prey abundance (solid arrows with − sign). By increasing predator abundance, each prey species is indirectly decreasing the abundance of the other prey species (dashed arrows with − sign).

Odiocoileus hemionus. Mule deer generally favor lower elevation and less rugged terrain than bighorn sheep, but there are extensive areas of overlap between the two species (Figure 13.18A). The researchers hypothesized that mule deer, by increasing the survival rate of cougars, were reducing the population growth rate of the bighorn sheep.

We will discuss two predictions generated by the apparent competition hypothesis. One prediction is that sheep mortality caused by cougars should be greatest in the populations with the most spatial overlap with mule deer. Researchers radio-collared 162 of the sheep, and were able to track and find sheep soon after they died, and ascribe a cause of death to 39 animals, which included 22 events of known cougar predation. In accordance with their prediction, the researchers found a strong correlation between the percent overlap of the winter range and the annual sheep mortality due to cougar predation. The Mono Lake population, which did not overlap mule deer at all, experienced no cougar-induced mortality, while the Baxter population, which overlapped with mule deer in 67% of its range, had the greatest annual mortality rate from cougar predation (Figure 13.18B)

A second prediction of the apparent competition hypothesis is that within each of the three sites that have both sheep and deer, most of the predation on sheep should occur in the areas where sheep overlap with deer. At Wheeler, where 58% of the sheep range also has deer present, 24 of 27 cougar-killed sheep occurred

(A)

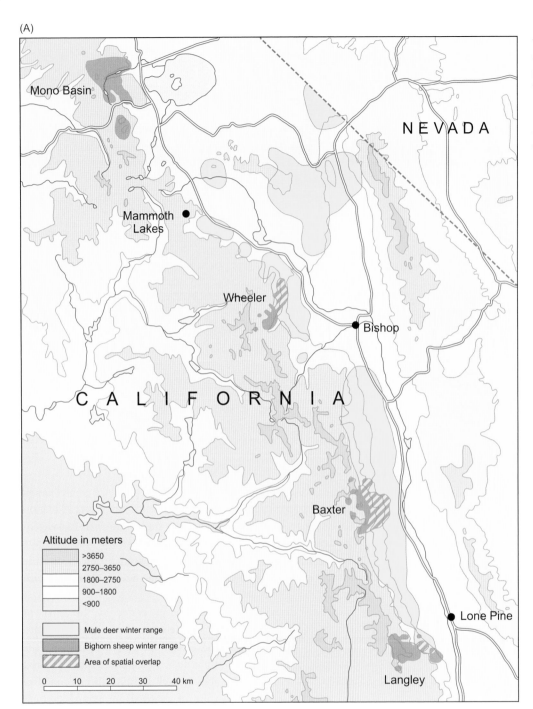

(B)

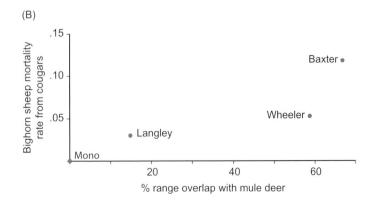

Figure 13.18 A. Winter ranges of the four populations of bighorn sheep and the much larger populations of mule deer (Figure 1 from Johnson). B. Annual mortality rate from cougar predation in relation to percent of spatial overlap of the bighorn sheep winter range with that of mule deer.

in the overlap range. At Baxter, where 67% of the sheep range has deer, 17 of 18 cougar-killed sheep occurred in the overlap range. Lastly, the three deer that were killed by cougars at Langley were from the overlap range, which comprised only 14% of the sheep winter range. These results support the hypothesis that mule deer are negatively affecting the survival of the Sierra Nevada bighorn sheep via the mechanism of apparent competition.

In this case, and many others described in this chapter, the researchers described the negative impact of one species on the other but did not describe or even investigate the negative impact of the second species on the first. We have strong evidence that mule deer are negatively affecting the Sierra Nevada bighorn sheep, but is the reverse also true? In this case, the effect, if it exists, is probably very small, given that deer density is so much greater than sheep density. In many interspecific interactions, two species have very different effects on each other. As we discuss in Chapter 16, measuring the strength of interspecific competition in both directions is essential for understanding interactions between multiple species.

Many of the cases we've discussed in this chapter involve competition between invasive species and native species. One reason for this focus is that we can actually see the competitive interactions occurring in real time. Many competitive events in the past are historical events whose effects are no longer visible. If the result of competition were competitive exclusion, we would only see the successful competitor existing within the community. And if the result is coexistence, competition may be difficult to measure, because niche shift or niche partitioning may have already occurred, which weakens the interspecific interactions, and makes them more difficult to detect (Connell 1980). But invasions are ongoing processes, and they can be studied in regions where two or more species have recently come together.

For example, the red imported fire ant, *Solenopsis invicta*, has spread into North America and wreaked havoc on biological communities in the southern United States. In an attempt to curb the spread and reduce the abundance of these fire ants, ecologists have imported parasitoids – organisms that live attached to or within a host organism, killing it in the process. The results of this biological control experiment highlight many of the principles we've developed in this chapter.

Biological control: direct and indirect competition

Fire ants are highly polymorphic, with the largest workers more than 20 times heavier than the smallest workers. To control fire ants, two species of parasitoid fly were released in or near Austin, Texas: *Pseudacteon tricuspis* in 2000, and *Pseudacteon curvatus* in 2004. Both species were used because they show resource partitioning in their native South American habitat, with *P. tricuspis* parasitizing the larger workers, and *P. curvatus* parasitizing the

smaller workers. However, there is some overlap, with intermediate-sized fire ants suitable hosts for both *Pseudacteon* species.

Biologists hoped that these parasitoids would be able to decimate a fire ant colony by virtue of parasitizing workers from all size classes. Because small ants are much more numerous, they expected that ultimately *P. curvatus* populations would be much more abundant than *P. tricuspis* populations.

Population samples demonstrated that coexistence of significant numbers of both parasitoid species was simply not happening (LeBrun *et al.* 2009). Rather, as *P. curvatus* spreads into habitat previously inhabited only by *P. tricuspis*, it is competitively excluding *P. tricuspis*. As examples, three sites colonized by *P. curvatus* in 2005 all showed dramatic increases between 2006 and 2007 in *P. curvatus*, and significant declines during the same time period for *P. tricuspis* (Figure 13.19A). Similar results were found throughout the surveyed regions.

The researchers asked, why is *P. curvatus* outcompeting *P. tricuspis*? One explanation is that there are many more *P. curvatus* because they parasitize the much more abundant small-sized ants. Perhaps *P. curvatus* is directly outcompeting *P. tricuspis* for medium-sized ants, which both species can parasitize. Data to address this direct competition hypothesis would ordinarily be very difficult to get, were it not for a fluke in the *P. tricuspis* life history. *P. tricuspis* that parasitize medium-sized ants produce only males, while *P. tricuspis* that parasitize large ants produce primarily females. If *P. curvatus* is directly outcompeting *P. tricuspis* for medium-sized ants, there should be fewer male *P. tricuspis* produced in regions that have had both parasitoid species present for at least 2 years, in comparison to regions that have recently been colonized by both species. The reasoning is that where both species have coexisted for 2 years, *P. curvatus* numbers have increased to where they can outcompete *P. tricuspis* for the medium-sized ants, effectively reducing the number of adult male *P. tricuspis* to emerge from medium-sized ants, and thus increasing the female/male sex ratio in those regions. Figure 13.19B shows strong support for this prediction; seven mounds where both species were present for 2 years show a much higher *P. tricuspis* female/male sex ratio than seven recently colonized mounds. The researchers concluded that *P. curvatus* is directly outcompeting *P. tricuspis* for medium-sized ants.

But behavioral observations indicated to LeBrun and his colleagues that indirect competition might be important as well. Once ants detect parasitoid flies, they tend to scurry underground. The researchers hypothesized that the first few flies (usually *P. curvatus*) present scare ants away from the surface of their mound, so they are no longer vulnerable to parasitoid attack. Thus *P. curvatus* is indirectly outcompeting *P. tricuspis* by causing the abundance of hosts to decrease. To test this hypothesis, the researchers scraped open the tops of fire ant mounds, inducing numerous ants to emerge to defend the

(A)

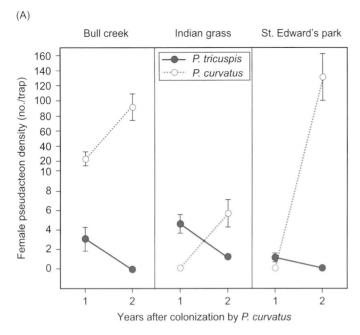

(B)

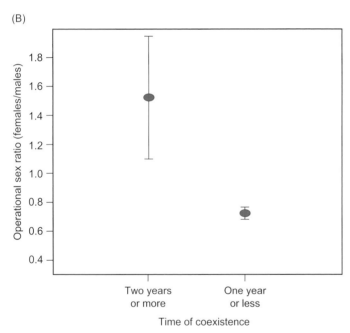

Figure 13.19 A. Abundance of *P. curvatus* and *P. tricuspis* species 1 year and 2 years after colonization by *P. curvatus*. B. *P. tricuspis* female/male sex ratio in regions that have had both parasitoid species for at least 2 years vs. regions where *P. curvatus* is a recent colonizer.

mound. On the top of each mound, the researchers placed a white tile, against which ants and parasitoid flies were clearly visible. Control mounds received no further treatment, while experimental mounds had 25 *P. curvatus* or *P. tricuspis* added to the tiles.

After 25 min, more than 80% of the ants retreated back into the control mounds. But a mean of about 95% of the ants retreated back into the mounds attacked by either parasitoid. Interestingly, small ants retreated slightly (but not significantly) more than large ants when *P. curvatus* was the attacker, while both sizes of ants retreated equally when *P. tricuspis* attacked, even though the small ants were immune to *P. tricuspis*. Thus ants can distinguish parasitoid flies as a threat, but they do not appear to be able to evaluate which size class of ant is threatened by a particular species. The outcome of attack by either parasitoid species is that all sizes of ants retreat, thereby reducing ant availability for later-arriving parasitoids. The researchers concluded that indirect competition between the two parasitoid species was reducing the availability of fire ant hosts.

In this system, niche partitioning reduces competition between the parasitoid species but has not completely eliminated it. Competition is both direct and indirect. Understanding the niche dimensions may allow ecologists to design better methods for biological control. For example, both parasitoid species detect their hosts by tracking the fire ant pheromones from disturbed nests. Researchers are now seeking a parasitoid species that can follow fire ants underground, which would allow the parasitoids to take advantage of a much larger population of susceptible hosts.

As we discussed earlier, one of the greatest challenges confronting ecologists who study competition is that once competitive exclusion occurs, it is difficult to demonstrate that two or more species were actually competing with each other. This is why many field researchers study invasive species, such as the fire ants discussed above, or the yellow crazy ants that compete with native crabs on Tokelau. In addition, these same researchers can turn to new technology to help them answer questions that were previously very difficult to address. Let's revisit our introductory case study to see how stable isotope analysis allowed McNatty and her colleagues (2009) to address the question of whether native crabs change their diets in response to competition with these ants.

REVISIT: Do invasive ants on Tokelau cause native crabs to shift their diet?

Stable isotope analysis allows researchers to infer the type of food eaten by a particular organism, or group of organisms. Recall from Chapter 4 that ^{15}N is the relatively rare and heavy stable isotope of nitrogen, while ^{14}N is the common and lighter stable isotope of nitrogen. Fortunately for researchers, an organism preferentially takes up ^{15}N during protein synthesis, which causes it to have a higher $\delta^{15}N$ level in its body tissues than is found in the tissues of its food. Thus animals tend to have higher $\delta^{15}N$ levels than plants, and carnivorous animals tend to have higher $\delta^{15}N$ levels than herbivores.

McNatty and her colleagues hypothesized that competition with ants forced crabs to reduce the amount of meat and fish they were able to scavenge, and to replace these food sources with plant matter. If true, the researchers predicted that crabs on islands without ants would have a more carnivorous diet and higher $\delta^{15}N$ values than crabs on islands with ants, which would be forced to a more herbivorous diet due to competition. Their surveys of eight islands confirmed this prediction for three different species of hermit crabs. All three hermit crab species had higher $\delta^{15}N$ levels on islands without ants (Figure 13.20).

In this system, an invasive ant species is profoundly influencing the structure of the ecological community by successfully competing with other very different species. In many systems, the effects of invasive species on community structure are more apparent through predation and parasitism, the topics we explore in the next chapter.

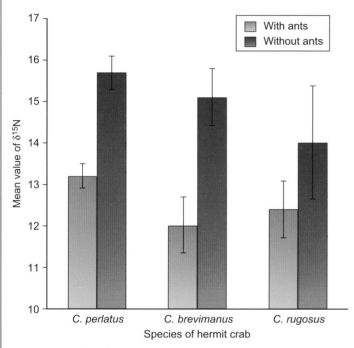

Figure 13.20. $\delta^{15}N$ for three species of hermit crabs on island invaded by ants, compared to uninvaded islands. See Dealing with data 3.3 for a discussion of how $\delta^{15}N$ is calculated.

SUMMARY

Organisms may compete for a great variety of limiting resources, such as food and habitat and, in the case of plants, light and pollinators. Direct mechanisms of competition, as highlighted by interactions between yellow crazy ants and hermit crabs on Tokelau, include resource and interference competition, while indirect mechanisms of competition that are mediated by other species are also widespread in ecological communities. Introductions of species into novel environments allow ecologists to study competitive interactions in real time.

Interspecific competition can lead to competitive exclusion when two or more species occupy similar niches. A variable environment, niche shift, and niche partitioning can promote species coexistence. Theoretical models, such as the Lotka–Volterra competition model, help identify conditions in which two or more competing species can coexist. When conservation ecologists introduce two or more species as biological control agents, they must consider potential competitive interactions among the introduced species, keeping in mind the factors that promote the coexistence of the introduced species.

FURTHER READING

Hanna, C., Foote, D., and Kremen, C. 2014. Competitive impacts of an invasive nectar thief on plant–pollinator mutualisms. *Ecology* 95: 1622–1632.

Shows how an invasive wasp outcompetes native pollinators by stealing nectar from a dominant Hawaiian tree species. When wasps were removed from the habitat, competition between the remaining flower visitors decreased and visitation rates of effective pollinators increased, resulting in increased fruit production by the tree.

Kandori, I., Hirao, T., Matsunaga, S., and Kurosaki, T. 2009. An invasive dandelion reduces the reproduction of a native congener through competition for pollination. *Oecologia* 159: 559–569.

Clearly laid-out experimental design and accessible statistical analyses make this an excellent choice for generating discussion on hypothesis testing, analysis, and drawing conclusions.

LeBrun, E. G., Jones, N. T., and Gilbert, L. E. 2014. Chemical warfare among invaders: a detoxification interaction facilitates an ant invasion. *Science* 343: 1014–1017.

Very fun article showing how the invasive tawny crazy ant, *Nylanderia fulva*, is outcompeting the invasive fire ant, *Solenopsis invicta*, in the southern United States, by virtue of its ability to inactivate the fire ant's chemical arsenal.

Orrock, J. L., Witter, M. S., and Reichman, O. J. 2008. Apparent competition with an exotic plant reduces native plant establishment. *Ecology* 89: 1168–1174.

Nice example of apparent competition between two plant species that is mediated by an animal seed consumer. The authors do a great job describing the experimental design and articulating the predictions generated by each treatment.

Radville, L., Gonda-King, L., Gómez, S., Kaplan, I., and Preisser, E. L. 2014. Are exotic herbivores better competitors? A meta-analysis. *Ecology* 95: 30–36.

Straightforward meta-analysis that addresses the question of whether exotic (invasive) herbivores are better competitors than are native herbivores. This study nicely points out how coevolutionary history with their host plant influences the foraging success of competing herbivores.

END-OF-CHAPTER QUESTIONS

Review questions

1. What is the evidence of competition between yellow crazy ants and hermit crabs on Tokelau? Is this an example of resource competition, interference competition or both? Explain your answer.
2. In your own words, explain the reasoning for the three different experimental treatments established by Kandori and colleagues (2009) in their dandelion studies.
3. What are the assumptions of the Lotka–Volterra competition model? Which seem most problematic to you? Explain your reasoning.
4. What processes have been shown to lead to species coexisting within a community? Describe a study that demonstrates each process.
5. How does the study by LeBrun and colleagues (2009) demonstrate direct competition between introduced parasitoids, and how does it demonstrate indirect competition?

Synthesis and application questions

1. How might you design a study to measure the competitive effect of *Coenobita* crabs on *Anoplolepis gracilipes* ants?
2. Regarding Diamond's research, given the number of islands with each species of cuckoo dove, how many islands would you predict to have both species present, if there were no competitive exclusion?
3. It is somewhat counterintuitive that increasing the supply of a limiting resource can reduce species richness. Explain why this can happen, using Hautier's study as your supporting evidence. Do you think that increasing the supply of a limiting resource reduces species richness in other systems as well? Explain your reasoning.
4. In Harrington *et al.*'s research, minks shifted both their diets and the timing of peak activity in a way that would reduce competition with otters. Can you think of alternatives to reduction of competition that might underlie this niche shift? How might you test for your idea?
5. Assume the following conditions: $K_1 = 150$, $K_2 = 120$, $\alpha_{12} = 0.2$, $\alpha_{21} = 0.6$. What will be the most likely outcome if you start out with 100 individuals in species 1 and 150 individuals in species 2? Explain your answer.
6. How is apparent competition similar to direct competition, and how does it differ?

Analyse the data 1

Kandori *et al.* wanted to know whether pollinators preferred one species of dandelion to the other. They set up four mixed-species plots with 30 *T. japonicum* plants and 30 *T. officinale* plants, and measured pollinator visitation rates (Table 13.3). What can you conclude from these data?

Table 13.3 Number of pollinator visits in mixed-species plots at four sites.

Location	Visits to *T. japonicum*	Visits to *T. officinale*
A	70	114
B	348	446
C	225	341
D	175	126

Analyse the data 2

Based on Table 13.1, are the increases in species richness associated with removing *Anartia* butterflies statistically significant? Justify your answer.

Chapter 14

Predation and other exploitative interactions

INTRODUCTION

The cockle, *Austrovenus stutchburyi*, is an important member of the biological community off the New Zealand coast. This bivalve mollusk grazes on algae and suspended organic materials, provides food and housing for numerous species, and influences the abiotic environment by extracting nutrients from the water and by forming mounds on the seafloor with its burrows, about 2 cm below the surface of the seafloor. Its impact is so great that it is considered to be an ecosystem engineer – a species that substantially influences the availability of essential resources within an ecosystem by causing important changes to the biotic or abiotic environment. While most of these cockles live and develop unseen in burrows just below the seafloor, some live right on the surface of the seafloor, while others are only partially buried. The seafloor surface is a dangerous place for slow-moving, non-aggressive, and tasty animals, as many species prey on them. So why do some cockles abandon the safety of their burrows for the dangerous world of the seafloor surface?

It turns out that there are several species involved in the seemingly suicidal cockle behavior. In this chapter, we consider the different types of species that benefit at the expense of other species: predators, parasites, grazers, and parasitoids. These exploiters can regulate the survival and abundance of their hosts or prey directly, and may have indirect effects on the survival or abundance of other species in the community. In turn, prey and hosts have adaptations to protect them against being consumed or parasitized. There is an evolutionary dynamic in the relationship between exploiters and their prey that changes over evolutionary time. We introduce two models that predict the outcomes of exploitative interactions involving two species, but most exploitative interactions, including the complex world of cockles and their exploiters, involve multiple species acting in diverse ways in different contexts.

KEY QUESTIONS

14.1. What are the different types of exploitative interactions?

14.2. How do exploiters regulate the abundance of their prey or hosts?

14.3. How do prey and hosts defend themselves?

14.4. How does the interaction between exploiters and their prey or hosts evolve over time?

14.5. How do theoretical models in association with empirical studies describe the outcomes of exploitative interactions?

CASE STUDY: The complex world of cockles and their exploiters

Wading birds such as the oystercatcher, *Haematopus ostralegus*, prey heavily on cockles. Also, many fish eat them, or take bites out of their meaty and exposed foot, a behavior described by researchers as cropping. So why do cockles come to the surface at all? The answer is that these cockles are under the influence of many parasites, including the trematode, *Curtuteria australis*, and an unnamed trematode in the flatworm genus *Acanthoparyphium* (Leung and Poulin 2007).

Curtuteria australis has a complex life cycle, which requires it to parasitize three different hosts and to move from one host to another in a variety of ways (Allison 1979). The adult trematode lives within a shorebird, such as the South Island oystercatcher (*Haematopus finschi*), where it lays eggs that pass out of the bird in its feces. These eggs hatch and develop into free-swimming miracidia, which enter the first *intermediate host*, usually a gastropod such as the whelk, *Cominella glandiformis*. Intermediate hosts support immature stages of a parasite's life cycle. Within the whelk, the miracidium develops and generates a population of rediae, which feed actively on the whelk, developing into a population of free-swimming larval forms called cercariae. These cercariae swim to their second intermediate host, in this case a cockle, entering through the cockle's siphon. Each cercaria sheds its tail, transforms into a metacercaria, which forms a cyst within the cockle's foot. When a shorebird eats an infected cockle, it also ingests metacercariae, which then emerge from their cysts transforming into mature adult trematodes. The shorebird is the *definitive host* – the organism in which a parasite matures and produces eggs (Figure 14.1).

There is obviously a great deal of randomness in this process. Most miracidia never enter a gastropod, most cercariae don't successfully colonize a cockle, and most metacercariae die before being eaten by a bird. Thus there is strong selection pressure favoring parasites that are able to journey to the appropriate host. Ecologists recognized that cockles on the seafloor would be more accessible to oystercatchers than would cockles buried away in burrows. Perhaps the metacercariae are manipulating the cockle to make it more likely to get eaten by oystercatchers, allowing the trematode to complete their lifecycles. We will explore this question in more detail at the end of this chapter.

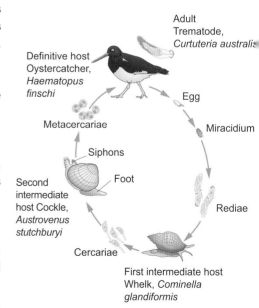

Figure 14.1 Complex life cycle of the trematode, *Curtuteria australis*, and its three host species.

This case study introduces you to several different types of exploitative interactions – interactions in which an organism of one species benefits by consuming all or part of an organism of another species, or by diverting resources that would otherwise nourish the organism. It also reminds us that understanding adaptation requires evaluating the costs and benefits of traits to each interacting organism. Let's begin our discussion with a systematic approach to understanding the different types of exploitation.

14.1 WHAT ARE THE DIFFERENT TYPES OF EXPLOITATIVE INTERACTIONS?

To most people, the definition of exploitative interactions brings to mind vivid images of orcas eating sea lion pups, or lions killing and eating wildebeest. Both of these examples of predation are lethal to the prey. Equally lethal are the interactions between *Pseudacteon* parasitoids and their fire ant hosts that we discussed in the previous chapter. Recall that parasitoids, like predators, kill their hosts, but they differ from predators in that they spend a significant part of their lives living within or attached to their hosts. In contrast to parasitoids, predators generally do not have an intimate relationship with their prey; they kill them, eat them, and move on. But predators are not only large fierce animals; mice that eat seeds, and shorebirds that eat cockles are equally lethal to their prey.

Many exploiters don't kill their source of food under most circumstances. Grazers only eat portions of their prey but generally don't kill them, at least not immediately. In addition to common grazers like cows, grasshoppers, and most moth larvae, we would also consider aphids sucking sap, mosquitoes sucking blood, and fish cropping cockles as legitimate examples of grazers. Grazers are similar to predators, in that they too lack any sort of intimacy with their prey or hosts. Parasites, like the parasitoids we mentioned earlier, have a highly intimate relationship with their hosts, living within or attached to their hosts. Unlike parasitoids, most parasites don't usually kill their hosts. Thus we can organize the four basic types of exploitative interactions on a two-axis graph of intimacy and lethality (Figure 14.2)

Of course, this two-axis approach does not capture the full flavor of exploitative diversity. Many seed predators are also grazers, while some species of aphids attach to only one plant over the course of a lifetime, and thus may be more accurately classified as parasites. In addition, parasitic organisms may be lethal if they infect hosts with compromised immune systems, and large population of grazers can devastate a population of plants. Furthermore, many aquatic ecologists classify zooplankton as grazers, even though they often consume the entire organism. This two-axis system gives a useful but imperfect picture of how exploiters can influence the population of their prey or hosts.

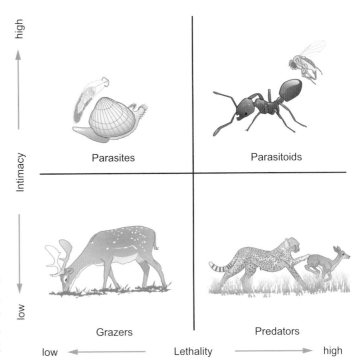

Figure 14.2 Two-axes visualization of the range of exploitative interactions.

14.2 HOW DO EXPLOITERS REGULATE THE ABUNDANCE OF THEIR PREY OR HOSTS?

From a population standpoint, exploiters have a negative effect on the abundance of their prey or hosts, while the prey or hosts have a positive effect on the abundance of their exploiters. Let's consider some examples of how these effects on abundance play out.

Grazing deer and the viability of ginseng populations

American ginseng, *Panax quinquefolius*, is one of the most important and controversial medicinal plants harvested by humans from forests in the United States. It is purported to be useful in stress reduction, nutrition, treatment of diabetes, and enhancing sexual performance. The root has high levels of ginsenosides, steroid-like compounds proposed to be the major active compound in the plant. Mean plant size and mean population size have declined over the past century, and biologists are concerned that several causes may be leading to the extinction of this important plant.

James McGraw and Mary Ann Furedi (2005) hypothesized that deer browsing was the most important factor causing the decline in ginseng abundance. They censused seven ginseng populations in West Virginia, USA, over a 5-year period to measure each population's geometric growth rate (λ) (see Chapter 10). The mean λ was 0.973, which indicates that the mean population size was declining by 2.7% per year. The researchers measured deer

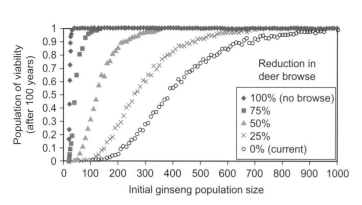

Figure 14.3 Projected ginseng population viability in relation to population size at five levels of deer browse.

browse rates and browse-induced mortality over this period, and based on these figures estimated that λ would have been 1.021 in the absence of deer browsing, or an annual increase of 2.1%.

The researchers used these estimates as part of a population viability analysis (PVA – see Chapter 11) that assessed the probability that a ginseng population would avoid extinction within a 100-year time frame, based on ginseng population size and amount of deer browsing. The results are sobering. At current browsing levels, a ginseng population would need to be composed of about 800 individuals for a 95% probability of viability after 100 years, and about 400 individuals for a 50% probability of viability (Figure 14.3). If browsing rates are reduced by 50%, about 300 individuals are needed for 95% probability of viability, and about 100 individuals are needed for a 50% chance of viability. Unfortunately, the mean population size of 36 surveyed ginseng populations was 93 individuals, and the largest population had 408 individuals. Consequently, ginseng abundance is projected to decrease sharply over the next century, victims of Bambi's appetite.

Of course, many factors could influence the accuracy of these projections. For example, the PVA assumes that human consumption of ginseng will remain constant. If new medicinal products are discovered to replace ginseng, human impact on ginseng populations would be reduced, leading to an increase in ginseng population growth rates. Alternatively, if the demand for ginseng increases over the next century, population viability would be even lower than McGraw and Furedi's projection.

Similar to grazers, many parasites have a small negative impact on the population growth rates of their host species. Some parasites compete with their hosts for resources; for example, an intestinal tapeworm uses nutrients that its host would otherwise allocate to growth and reproductive effort. Others, including some parasites that cause infectious disease, can regulate populations by killing a significant number of their hosts.

Parasites and host species abundance

Ecologists consider two types of parasites. *Microparasites* are tiny, generally live within cells, and multiply directly within their hosts. Viruses, bacteria, some protistans, and simple fungi are microparasites. *Macroparasites* are visible to humans without a microscope, and live outside of the host's cells. Ticks, nematodes, trematodes, and some fungi are common macroparasites. We will show how each type of parasite can regulate a population of its host species.

West Nile virus and the decline in North American bird populations

West Nile virus is an emerging infectious disease – an infectious disease that is projected to become much more prevalent in the near future. Mosquitoes are vectors for the virus, transmitting it between hosts, which are primarily birds of many different species. Within birds these microparasites can increase to very high numbers, and then be transmitted by mosquitoes to a new host. Humans can be hosts as well, but the virus usually doesn't reproduce very well within humans. Consequently, it is very difficult for mosquitoes that bite infected humans to become infected, and most humans who become infected show very minor symptoms. In the United States in 2007, there were 3630 cases of the most severe type of West Nile virus infection, which caused 124 deaths. Though it has been present in Africa for about a thousand years, its first North American appearance was in New York City in 1999, and it has spread through most of North and Central America.

Shannon LaDeau and her colleagues (2007) tested the hypothesis that West Nile virus caused significant population declines in a broad range of bird host species across North America. The researchers acknowledge that demonstrating that a virus caused a population decline is very challenging, because only a very tiny sample of dead birds were tested for infection. LaDeau and her colleagues also were challenged by the difficulties of measuring bird population sizes. Fortunately, the North American Breeding Bird Survey is conducted annually by citizens across the United States and Canada during bird breeding seasons. Each observer participating in the survey notes all of the birds observed over a 40-km stretch of a specified route. Currently, there are over 3700 routes that are surveyed every year. This effort has provided a rich continuous database of bird distribution and abundance in North America since 1966.

The researchers reasoned that if West Nile virus was causing bird populations to decline, then there should have been a significant drop recorded in the Breeding Bird Survey after 2002 – a year of peak virus epidemic in the United States. But this drop should only be detectable in birds that are susceptible to the effects of the virus. LaDeau and her team tested this prediction with a mathematical model that incorporated three variables associated with susceptibility: (1) mortality in laboratory tests where birds were injected with live virus under controlled conditions, (2) mosquito exposure patterns based on habitat use in natural conditions, and (3) *seroprevalence* – the percentage of individuals who show antibodies to the virus. The researchers used a formula based on these three variables to predict the relative impact (high, moderate, or low) of West Nile virus on the population sizes of 15 species of North American birds deemed susceptible (Table 14.1).

High-impact species were predicted to decline in population size following the 2002 outbreak of West Nile virus, while low-impact species were not expected to decline. Fewer moderate-impact species were expected to decline than were high-impact species. In

Table 14.1 Predicted impact of West Nile virus based on the researchers' model, and observed population decline after a peak virus epidemic in 2002, based on the North American Breeding Bird Survey.

Species	Predicted impact	Observed population decline?	Year of minimum abundance[a]
American crow	High	Yes	2004
Blue jay	High	Yes	2004
Fish crow	High	No	2005
Tufted titmouse	High	Yes	2004
American robin	Moderate	Yes	2005
House wren	Moderate	Yes	2003
Chickadee	Moderate	Yes	1996
Common grackle	Moderate	No	2003
Northern cardinal	Moderate	No	1997
Song sparrow	Moderate	No	2004
Downy woodpecker	Low	No	1996
Gray catbird	Low	No	1997
Mourning dove	Low	No	1997
Northern mockingbird	Low	No	1997
Wood thrush	Low	No	2003

[a]The period of 1996–2005 was considered for years of minimum abundance. Observed population decline was scored as "Yes" if the observed population was below the 95% confidence interval of the expected population if there had been no West Nile epidemic in 2002.

accordance with these predictions, three out of four of the high-impact species and three out of six of the moderate-impact species showed a significant population decline after the 2002 West Nile epidemic, based on Breeding Bird Surveys, while none of the species with predicted low impact showed a significant population decline after 2002. In addition, most of the impacted species had their year of minimum abundance after West Nile virus became established.

It is difficult to justify research on emerging infectious diseases that could cause additional hosts to become infected. Consequently, LaDeau and her colleagues could not infect birds with the virus to study the population-level consequences of infection. Instead, they limited their analysis to investigating whether there was a correlation between potential impact based on three variables, and actual impact based on natural observations. But we always need to be careful to recognize that correlational studies are by their nature, uncontrolled.

This limitation is less of a factor for many other studies of parasites. We will turn our attention to a parasitic plant that has been studied extensively under controlled and natural conditions.

Parasitic plants: regulating the abundance and diversity of plants in a community

The yellow rattle, *Rhinanthus minor*, is a *hemiparasite* – a plant that is parasitic on other plants but can also carry out photosynthesis (Figure 14.4A). *R. minor's* fitness may be enhanced by a factor of 10 or more if it associates with an optimal host (Westbury 2004). It has a modified root, or hausterium, that can form connections to the xylem of over 50 species of host plant(s) and remove nutrients and sugars. Under natural conditions, hausteria of an individual *R. minor* plant may parasitize several host plants.

R. minor is widespread in England, and generally found in low- to medium-fertility grasslands. Richard Bardgett and his colleagues (2006) wanted to know whether *R. minor* could regulate the abundance and diversity of plants within a community. They set up an array of mesocosms, each planted with six grasses and five forbs commonly found in grassland communities. Each mesocosm then received either 0 (control), 30, or 60 *R. minor* plants, and the experiment was run for 3 years.

Three important results emerged by the end of the experiment. First, *R. minor* substantially decreased the biomass of the plant community in comparison to the control mesocosm (Figure 14.4B). Second, species richness was substantially higher in the *R. minor* mesocosms (Figure 14.4C). Finally, there was a higher percentage of forbs in the *R. minor* mesocosms, reflecting the parasite's suppression of the competitively dominant grass, *Lolium perenne*, which enabled several species of forbs to become more abundant (Figure 14.4D).

Interestingly, *R. minor* had potentially important influences on soil processes as well. The *R. minor* mesocosms had a higher proportion of bacteria in relation to fungi in the soil. Usually, high bacteria concentrations are associated with a higher rate of nitrogen mineralization, which is the process that converts organic

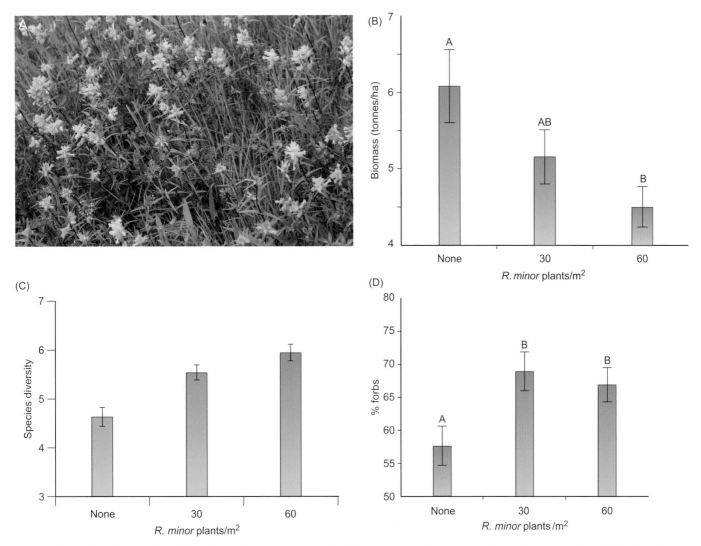

Figure 14.4 A. The yellow rattle, *Rhinanthus minor*, in a meadow. B. Mean biomass of mesocosms in relation to number of *R. minor* plants. C. Mean species diversity (number of species) of mesocosms in relation to number of *R. minor* plants. D. Mean percent of forbs in mesocosms in relation to number of *R. minor* plants. Nonoverlapping letters above bars indicate significant differences between treatments.

nitrogen from decaying organisms into inorganic nitrogen that can be taken up by plants. In this case, nitrogen mineralization rates were substantially higher in the *R. minor* treatments, which increased usable nitrogen to the plants. This raises the question of why biomass was lower in the treatments with more available nitrogen. The researchers argue that the direct effect of parasitic plants suppressing biomass was stronger than the indirect effect of nitrogen mineralization increasing biomass.

A recurring theme of this unit on community ecology is that ecologists must consider both the direct and indirect effects of all interactions. We will emphasize this theme by turning to a study of predation by birds and bats in a tropical forest.

Birds and bats: direct and indirect effects on prey populations

Many species of bats and birds prey upon arthropods (such as insects and spiders) in tropical forests (Figure 14.5A). By independently excluding either bats or birds from plants, it is possible to compare the effects of these two groups of predators on arthropods. If there is an effect, we can then see how reduction in predation influences the rest of the biological community.

To exclude birds and bats from forest plants, Margareta Kalka and her colleagues (2008) covered common understory plants in a Panamanian tropical forest with mesh exclosures that permitted access to arthropods but eliminated access for their predators. But birds are diurnal and bats are nocturnal, so exclosures in place during the day excluded birds but allowed access to bats, while exclosures in place during the night excluded bats but allowed access to birds. The researchers visually inspected each leaf six times during the 10-week study to search for arthropods. Comparing the number of arthropods in these two treatments versus an unexclosed control demonstrated that excluding either type of predator increases arthropod abundance, but that excluding bats had a stronger effect (Figure 14.5B).

Does increasing the number of arthropods increase the amount of herbivory? To answer this question, the researchers collected

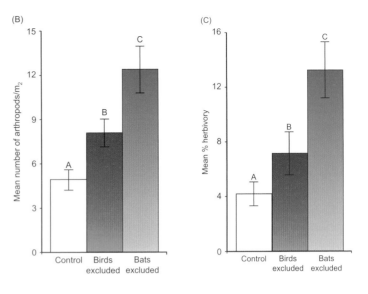

Figure 14.5 A. A bat (*Micronycteris microtis*) eats a katydid at the field site at Barro Colorado Island, Panama. B. Mean number of arthropods (±SEM) per m^2 in controls, treatments that excluded birds, and treatments that excluded bats. C. Mean percent (± SEM) of leaf area eaten for each treatment. N = 43 controls, 42 bat exclosures, 35 bird exclosures. Nonoverlapping letters above bars indicate significant differences between treatments.

every leaf at the end of the study and used analysis of digital photographs to calculate leaf damage area. As you might expect, the bat-exclosure plants, which had the greatest number of arthropods, suffered the greatest amount of herbivory; the bird-exclosure plants suffered an intermediate amount of herbivory; while the control plants suffered the least herbivory (Figure 14.5C).

The researchers concluded that bats and birds both directly regulate the abundance of their prey, but bats do so more than birds in the understory of this tropical forest. Both of these predators also indirectly reduce herbivory, by removing herbivores from the community.

 Thinking ecologically 14.1

The exclosures also protected carnivorous arthropods that eat herbivorous arthropods from bat and bird predators. How does this influence interpreting the results of this experiment?

Grazers, parasites, and predators clearly can regulate populations of their prey or hosts. However, prey and hosts don't just sit there and take it. Natural selection favors prey or hosts that can avoid being consumed, at least until after they have had the chance to reproduce.

14.3 HOW DO PREY AND HOSTS DEFEND THEMSELVES?

We can group defensive traits into two categories: **constitutive defenses** are expressed continuously, while induced defenses are expressed in response to an injury or a threat (see Chapter 4).

Constitutive defenses

Some defenses prey species use against predators or grazers are so much a part of their morphology or behavior that we don't really think about them as being defenses. For example, if I were to ask you to describe an adaptation in a giraffe, you might start talking about its long neck that allows it to access leaves high up in trees, or, if I prodded you, you might discuss the long neck as an adaptation in males that allows them to compete for access to females. But you would probably not immediately think about large body size in giraffes as an antipredator adaptation that makes giraffes immune from predation.

But large body size is a constitutive defense against predation in the Serengeti ecosystem (Sinclair *et al.* 2003b). Small herbivores suffered predation from many more species of predators than did larger herbivores. Each of the three smallest ungulate species (hooved mammals) was preyed upon by seven species of carnivores, while the three largest ungulate species suffered virtually no predation. Almost all mortality in small ungulates resulted from predation, while larger ungulates such as giraffes, rhinoceroses and hippopotamuses suffered very few losses to predators.

Tony Sinclair and his colleagues were able to use a natural experiment to test the hypothesis that large body size is a constitutive defense in the Serengeti ecosystem. Beginning in 1980, most carnivores were removed from the Serengeti through poaching and indiscriminate poisoning. In the adjacent Mara Reserve, there was almost no predator removal. If large body size is a constitutive defense against predators, then predator removal should cause an increase in the abundance of small-bodied, but not large-bodied, species. There were sufficient data for analysis of abundance for five small-bodied ungulate species and one large-bodied ungulate species, the giraffe (Figure 14.6). As predicted the small ungulate abundance increased dramatically with predator removal in the Serengeti, but did not increase during the same time period in the Mara, where predators were still present. In contrast giraffe abundance did not increase with

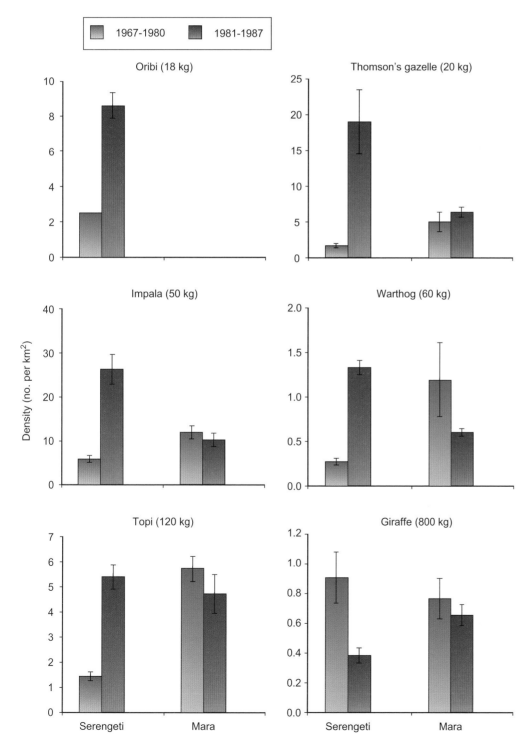

Figure 14.6 Mean population density of six ungulate species before predators were removed (1967–80) and after predators were removed (1981–87) in the Serengeti, and during the same time period in the adjacent Mara Reserve. Poaching in the Serengeti caused predator removal, but this did not occur in the Mara Reserve. No oribi lived in the Mara Reserve.

predator removal; instead it actually declined substantially in the Serengeti. Like the other five ungulates, its population remained relatively constant in the Mara Reserve over the same time period. The researchers concluded that predators were regulating the populations of small, but not large ungulates (Figure 14.6).

Thinking ecologically 14.2

Propose two hypotheses for why giraffe populations actually declined when predators were removed in the Serengeti. How would you test the predictions of each hypothesis?

Many constitutive defenses are clearly antipredator adaptations. Some organisms have spines or thorns to deter predators or herbivores. A recent study by Adam Ford and his colleagues

(2014) highlights the adaptive nature of thorns as a constitutive defense. Impalas are browsers that prefer open woodland habitat where they can see and escape predators, and shy away from closed-in woodland where they are vulnerable to ambush predators. Using global positioning telemetry, the researchers were able to show that the open woodland habitat favored by impalas was dominated by *Acacia* trees with very long thorns, while habitat avoided by impalas was dominated by much more poorly defended *Acacia* trees. The researchers suggest that low predation risk in open woodlands increases impala abundance, thereby selecting for thorny *Acacia* trees that are more protected against damage caused by browsing impalas.

There are many other examples of constitutive defenses (Figure 14.7). Some species have a cryptic appearance or coloration that makes them almost invisible within their environment.

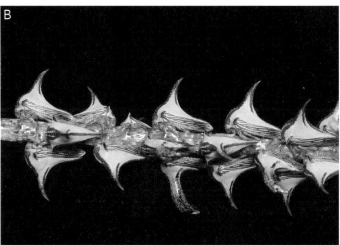

Figure 14.7 Constitutive defenses. A. Long-tailed tit perches gingerly in a well-protected black the three, *Prunus spine*. B. The treehopper, *Umbonia crassicornis*, has a thornlike body that can fool predators. C. Poison ivy, *Toxicodendron radicans*. D. Large body size makes adult elephants (*Loxodonta africana*) virtually immune against all predators, even this group of lions.

Others, such as the monarch butterfly, possess warning coloration that makes them more conspicuous and advertises that they are richly endowed with large quantities of toxic substances that they get from the milkweed they eat as caterpillars. Many plants have high levels of secondary compounds, molecules that are not essential for growth, development, or reproduction, but that function as constitutive chemical defenses. Some examples of secondary compounds that ironically are consumed by humans, though not necessarily to their advantage, are caffeine, cocaine, and morphine. As I write this, I sport a nasty festering rash, courtesy of a brush with urushiol toxin secreted by poison ivy, *Toxicodendron radicans*. Though I am an omnivore, and enjoy leafy vegetables, I would never consider eating poison ivy, even if I were starving, because uroshiol is a highly effective constitutive defense.

We need to be careful to distinguish between the effects of a trait, and an assumption that a trait was a result of natural selection favoring the trait because it has those effects. Returning to body size in ungulates, we see that larger ungulates are relatively immune from present-day predators. Does that mean that large bodies arose because they afforded protection against predators? Maybe, but maybe not! It is equally likely that herbivores with larger and more complex digestive systems could process the abundant low-quality forage that is available in the Serengeti. A large, complex digestive system needs a large body to carry it around, a large circulatory system to provide nutrients and oxygen to the cells, and a large head to take in the large quantities of food. For other traits such as quills, thorns, or cryptic morphology, it is much easier to argue that natural selection favored these traits because they afforded protection against predators. This is particularly true for induced defenses, which are expressed by the prey or host in response to the actions of a predator or parasite.

Induced defenses

Antipredator and anti-parasite adaptations are generally costly to the organism possessing them, for many reasons. To construct quills or thorns, an organism must divert energy and nutrients that could otherwise be allocated to growth and reproduction. Chemical defenses may be costly to construct, requiring extra genetic information and metabolic pathways. In many cases, the toxin must be compartmentalized, so it doesn't harm the organism that produces it. One way to reduce these costs is for an organism to only produce defenses when it senses the presence or threat of exploitation.

Plants' synthesis and release of toxic chemicals in response to herbivore attack

Herbivore attack can induce chemical defenses in many species of plants. Wild tobacco, *Nicotiana attenuata*, increases production of the toxic alkaloid nicotine in its roots, in response to leaf wounding by several mammalian and insect herbivores. It also increases production of several other defense compounds, including proteinase inhibitors. Proteinases are digestive enzymes, such as pepsin and trypsin that help digest proteins; consequently proteinase inhibitors interfere with an herbivore's ability to digest proteins in its diet. Both nicotine and proteinase inhibitors are also present in lower levels in plants that have not been attacked, so they function as both constitutive and induced defenses.

Do higher levels of these molecules actually benefit the plant? Previous research had indicated that jasmonic acid is a wound hormone that increases the synthesis of several defense chemicals. Ian Baldwin (1998) showed that exposing plants to 500 μg of jasmonic acid methyl ester increased *N. attenuata* nicotine levels by about 31%. When grasshoppers, *Trimerotropis pallidipennis*, were given access to both jasmonic-acid-treated and untreated control plants, they removed over 10 times more leaf area from the control than from the treated plants. Baldwin concluded that increased nicotine production was an induced chemical defense that was benefiting the plants.

More recent research has identified the importance of proteinase inhibitors to *N. attenuata*'s defense against herbivory. In one series of experiments, Jorge Zavala and his colleagues (2008) silenced the expression of the gene that produces the proteinase inhibitor that severely reduces the effectiveness of the digestive enzyme trypsin in the gut of the herbivore. They incorporated this gene into the *N. attenuata* DNA, and fed it to larvae of the tobacco hornworm, *Manduca sexta*, a very damaging *N. attenuata* herbivore.

The researchers created a population of 24 wild-type plants, and 24 transformed plants that had the silenced gene for trypsin proteinase activity. If gene silencing actually works in this system, the researchers predicted much higher trypsin proteinase inhibitor (TPI) activity in the wild-type plants than in the plants with the silenced gene. Figure 14.8A shows a highly significant reduction in TPI activity in the silenced population.

But does the high TPI activity present in wild-type plants have a negative effect on the herbivores? Caterpillars fed on wild-type *N. attenuata* had about 40% less mass after 7 days than caterpillars fed on plants with the silenced gene (Figure 14.8B).

 Thinking ecologically 14.3

From the research described thus far, we still need to show that induced production of proteinase inhibitors is actually beneficial to *N. attenuata*. How might you design experiments to show this?

Earlier we argued that induced defenses benefit a plant by reducing the costs of producing and maintaining high constitutive levels of toxic chemicals. This argument assumes that chemical defenses are costly to a plant. Let's see whether this assumption is valid.

(A)

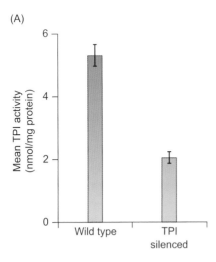

(B)

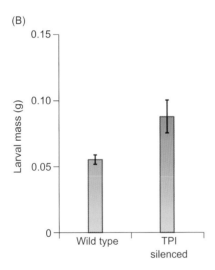

Figure 14.8 A. Mean TPI activity in wild-type plants and plants with silenced TPI genes. $P < 0.0001$. B. Mean larval mass of caterpillars fed wild-type *N. attenuata* vs. those fed plants with silenced TPI genes. $P = 0.04$.

Costs of anti-herbivore defenses

In an earlier study, Zavala and other colleagues (2004) used genetic engineering techniques to create *N. attenuata* plants with genetically determined differences in proteinase inhibitor levels. In the absence of herbivores, assuming anti-herbivore defenses are costly to a plant, we would expect that plants with lower

levels of proteinase inhibitors would be more successful than plants with higher levels. Zavala's measure of success was the number of seed capsules produced by plants.

To test this hypothesis, the researchers placed two plants with different proteinase inhibitor levels together in a pot; this arrangement simulated conditions in the field, where *N. attenuata* tend to grow closely together. The researchers replicated this procedure multiple times, with differing TPI levels in each plant. They predicted that plants with similar TPI levels should produce similar numbers of seed capsules, while a plant that had lower TPI levels than its neighbor should produce substantially more seed capsules. Their results confirm this prediction. In all cases, the plant with the lower TPI level produced the greater number of seed capsules. The difference between seed capsule production between plants was greatest when the two neighboring plants had very different TPI levels.

N. attenuata's defensive repertoire is even more vast than that we've described thus far. Some induced defenses are used by the plant to call in other defenders.

Attacked organisms: using other species to help with defense

When attacked by an herbivore, *N. attenuata* also releases a group of volatile organic compounds. Andre Kessler and Ian Baldwin (2001) wondered why a plant might release a volatile compound that would not directly inhibit an herbivore foraging on the plant. They hypothesized that these volatile compounds may function to summon predators that could attack the herbivore or its offspring on the plant.

To test this hypothesis, the researchers measured the quantity of each volatile compound that was released following an herbivore attack. They then manually applied this quantity of compound to *N. attenuata* leaves. Altogether they tested five different volatile organic compounds, with each plant receiving only one compound. They glued 10 fertilized eggs of the herbivore, *Manduca sexta*, to the undersides of two of the leaves of each plant. One treatment was a control, which received 10 eggs but no volatile organic compound. The researchers compared egg predation by the bug, *Geocoris pallens*, on the five experimental treatments and the control. Three of the five treatment groups that received application of a volatile organic compound showed lower *M. sexta* survival, a result of attack by the predacious *G. pallens* (Figure 14.9).

Given the higher predation rate suffered by hornworms from *N. attenuata's* induced defenses, Kessler and Baldwin hypothesized that hornworm moths might be less likely to oviposit – lay their eggs – on plants that were emitting volatile organic compounds. They measured hornworm oviposition rates on plants treated with linalool, the volatile organic compound which was associated with the lowest egg survival in the previous experiment, and discovered that oviposition was reduced by almost 60% in comparison to untreated control plants.

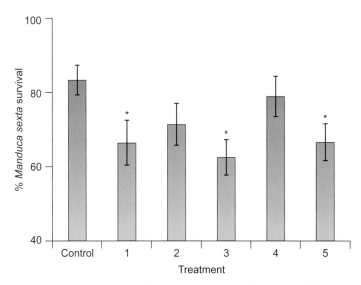

Figure 14.9 Mean percent hornworm egg survival on control plants in comparison to experimental plants treated with five different volatile organic compounds hypothesized to call in egg predators. * indicates significantly lower survival rates than controls.

Taken together, these experiments demonstrate that *N. attenuata* has a complex system of induced defenses that discourages herbivores by reducing herbivore growth rates, and increasing herbivore mortality. One adaptive response by the herbivores is reducing their oviposition rates on plants that exude organic volatile molecules in response to herbivory. But plants are not the only group with induced defenses. Many species of animals also have induced defenses in response to predation cues. In fact, even within a species, some populations of animals can respond with induced defenses, while other populations of the same species may lack the induced defense.

Natural selection and differences in induced defenses within a species

Mussels have an array of constitutive defenses, including strong shells that protect them from predators, and powerful adductor muscles that prevent most predators from prying these shells open. But some crab species can break through these shells and eat the underlying flesh. Aaren Freeman and James Byers (2006) wanted to know whether experience with particular species of crabs would induce an adaptive defensive response in the mussel, *Mytilus edulis*. Two crab species were of particular interest, as they are recent invaders into the range of *M. edulis*. The green crab, *Carcinus maenas*, arrived from Europe in the nineteenth century and quickly spread up and down the east coast of North America. Thus mussels throughout New England have been exposed to predation by *Carcinus* for about 100 years. In contrast, the Asian shore crab, *Hemigrapsus sanguineus*, was first found off the New Jersey coast in 1988 and is still expanding its range both north and south. Thus mussels in southern New England have a few years of exposure to *Hemigrapsus*, while

mussels in central Maine have still never experienced predation by *Hemigrapsus*.

Previous work identified an induced defense in mussels – they thicken their shells in response to predators. This defense works pretty well; a 5–10% increase in shell thickness can require so much more work for predators that they often give up trying to open the mussels (Stokstad 2006). Freeman and Byers wanted to know whether exposure to either species of crab would induce the shell-thickening response in mussels from populations taken from southern and northern New England. They hypothesized that shell thickening might be present in both southern and northern mussels in response to *Carcinus*, a predator for about a century. They did not know whether mussels from either population would show any response to *Hemigrapsus*, the much more recent predator of southern mussels that has never encountered the northern population. Part of the solution would depend on whether both crab species secrete the same chemical signal that induces shell thickening in mussels.

The researchers collected mussels from six northern and six southern sites, and established three experimental treatments: exposure to (1) water that had drained through a bucket of *Carcinus*, (2) water that had drained through a bucket of *Hemigrapsus*, and (3) water that had drained through an empty bucket (control). After 3 months, the researchers measured mussel shell thickness. Exposure to *Carcinus* induced a substantially thicker mussel shell than did the control treatment. Exposure to *Hemigrapsus* induced a substantially thicker shell in the southern population, but not in the northern population that had never experienced predation by *Hemigrapsus*. (Figure 14.10) The researchers concluded that mussels from the southern population had evolved an induced response to predation by *Hemigrapsus* in less than 15 years, and that the induction signal exuded by *Hemigrapsus* was different from the signal exuded by *Carcinus*.

The research on induced defenses highlights the interplay of ecological and evolutionary processes. Induced defenses occur over a short time period – over the lifetime of an individual. But the ability to generate induced defenses is a genetically influenced process arising from natural selection favoring the survival and reproductive success of individuals with genes that code for induced defenses. The tobacco plant study shows how complex these defenses can be, while the mussel study demonstrates that these defenses can evolve over a relatively short time period. Both studies highlight how exploitation and defense are dynamic processes.

14.4 HOW DOES THE INTERACTION BETWEEN EXPLOITERS AND THEIR PREY OR HOSTS EVOLVE OVER TIME?

Studying the fossil record makes it clear that predators and their prey change over evolutionary time. If you visit a natural

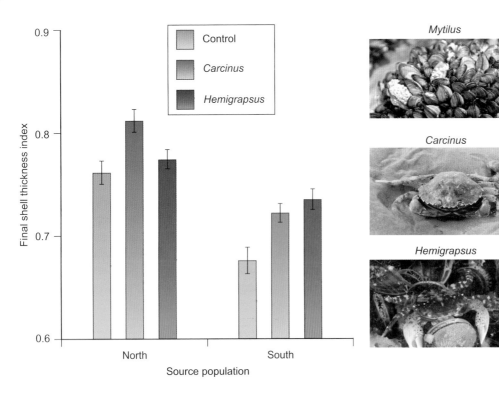

Figure 14.10 Shell thickness index of mussels from northern and southern parts of their geographic range. Experimental mussels were exposed to water that had drained through either *Carcinus* or *Hemigrapsus* predators, while control mussels were exposed only to water. (Figure 3 from Freeman and Byers, 2006.)

history museum, you may encounter a fierce-looking dinosaur sporting dagger-like teeth attacking an herbivorous dinosaur five times its size munching on vegetation growing in a swamp. Dagger-like teeth are an adaptation in the predator that allow it to subdue a much larger animal, while large body size can be an effective counteradaptation in the herbivore. Perhaps not highlighted in the museum diorama are adaptations in the plants, such as the ones we've just discussed, that make them less likely to be eaten. Similarly, herbivores have adaptations that allow them to be more successful plant consumers, or to avoid eating highly toxic plants. All of these adaptations have one feature in common; they are responses to changes in a dynamic biotic environment.

Leigh Van Valen's Red Queen hypothesis (Van Valen 1973) is a metaphor inspired from the Red Queen's comment in *Through the Looking Glass* that "here, you see, it takes all the running you can do, to keep in the same place" (Carroll 1871). Applied to predator–prey systems, the hypothesis is that over time selection favors predators that can run fast enough to capture their preferred prey, and prey that can run fast enough to avoid capture. This is an example of coevolution – a process in which interactions between species lead to reciprocal adaptation.

Jakob Bro-Jørgensen (2013) tested this coevolutionary hypothesis in a study of five predator species and 14 prey species that live in the African savanna. The predators were the cheetah (*Acinonyx jubatus*), hyena (*Crocuta crocuta*), wild dog (*Lycaeon pictus*), lion (*Panthera leo*), and leopard (*Panthera pardus*). Cheetahs, hyenas, and wild dogs are pursuit predators, using a combination of speed and endurance to run down their prey, while lions and leopards are ambush predators relying on surprise to capture prey. These differences in predator foraging strategies

are important, because we would predict that selection for fast sprint speed would be stronger on the prey preferred by pursuit predators, while selection for fast sprint speed would be much weaker, or even nonexistent, on the prey preferred by ambush predators. Bro-Jørgensen's measure of prey preference, which he called prey vulnerability, was the proportion of a particular prey species in a predator's diet in relation to how many of that prey species are found within the predator's habitat.

This coevolutionary hypothesis predicts a positive correlation between the vulnerability of the prey of pursuit predators and prey sprint speed. The alternative hypothesis is that pursuit predators will prefer the slowest prey, as they are easiest to catch. This alternative hypothesis predicts a negative correlation between prey vulnerability and sprint speed. The data support the prediction of the coevolutionary hypothesis. For example, cheetahs are the fastest predator, with a mean sprint speed of *c.* 110 km/h, and their most vulnerable prey tended to be the swiftest prey species on the savanna (Figure 14.11A). Similar patterns were found for the prey of other pursuit predators. Somewhat surprisingly, rather than no correlation, which we might have expected, there was actually a negative correlation between prey vulnerability to ambush predators and prey sprint speed (Figure 14.11B). Bro-Jørgensen suggests that the best defense against ambush predators such as lions and leopards is not swiftness, but rather a large body, robust build and defensive weaponry. Thus the favorite targets of ambush predators are likely to have evolved these defensive traits, which represent an evolutionary tradeoff associated with reduced sprint speed (see Chapter 3 for a discussion of evolutionary tradeoffs).

A more complete test of this coevolutionary hypothesis would require demonstrating that the sprint speed of pursuit predators traits evolved by parasites (Decaestecker *et al.* 2007; Spotti-swoode and Stevens 2012). A prediction of this hypothesis is a cycle of adaptation and counteradaptation in traits that influence the fitness of the parasite and its host.

The coevolution of fast running speed in pursuit predators and their prey, and the coevolution of parasitic capabilities and parasite resistance in parasites and their hosts, are examples of genetic adaptation over an evolutionary time scale. Our previous discussion of induced defenses showed how some prey or host species exhibit changes in morphology, physiology, or behavior over the course of their lifetimes in response to predation or parasitism. Both long-term and short-term responses to exploitation will influence the abundance of exploiters and their hosts or prey. We will now explore a mathematical approach to understanding how predator and prey abundance changes over time in exploitative interactions.

(A)

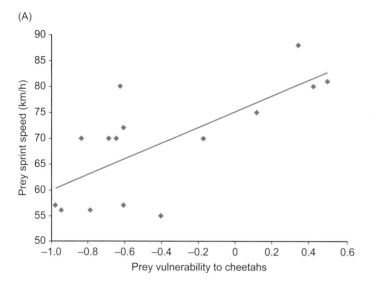

(B)

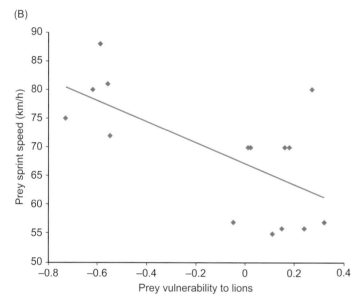

Figure 14.11 Prey sprint speed in relation to: prey vulnerability to cheetahs (A), and prey vulnerability to lions (B). Prey vulnerability (range of −1.0 = never eaten, to +1.0 = always eaten) is measured as the proportion of prey in the predator's diet in relation to how abundant that prey is in the predator's habitat.

14.5 HOW DO THEORETICAL MODELS IN ASSOCIATION WITH EMPIRICAL STUDIES DESCRIBE THE OUTCOMES OF EXPLOITATIVE INTERACTIONS?

As with our exploration of competition, we have both Alfred Lotka (1925) and Vito Volterra (1926) to thank for some of the most influential early work on understanding exploitative interactions. Lotka was impressed with observations by entomologists that predatory insect populations and their hosts or prey tended to fluctuate together in repeated patterns or cycles. He was also motivated by efforts of mathematicians to explain the spread and time course of malaria epidemics. Volterra's entry into predator–prey ecology was precipitated by research done by his daughter's boyfriend, which showed that predacious fish in the Adriatic Sea increased in abundance during the World War I – a time when there was almost no commercial fishing (Kingsland 1985). Both researchers developed similar models that describe changes in populations of predators and their prey, or parasites and their hosts.

The Lotka–Volterra predation model

To explore changes in population growth rates of predators/prey or parasites/hosts, we are most interested in how two variables change over time: N_h, which is the number of hosts or prey, and N_p, which is the number of predators or parasites (or more broadly, the number of exploiters). If we borrow from the exponential growth model that we derived in Chapter 10, then:

$$dN_h/dt = r_h N_h = \text{the rate of host population change}$$

where

$$r_h = \text{the host per capita growth rate}$$

is subject to strong selection as a consequence of the increased sprint speed of their prey. This may prove difficult to test, particularly given the small number of predator species on the African savanna. Bro-Jørgensen also points out that understanding this coevolutionary process would be helped by studies that investigate the endurance and acceleration of prey species, but that these data are not currently available.

One important feature of the Red Queen hypothesis is that it can be used to understand other types of exploitative interactions. For example, applied to parasitic infections, the Red Queen hypothesis predicts that over time selection favors novel traits in parasites that enhance their parasitic capabilities, and novel traits in hosts that make them more resistant to the novel

This tells us that in the absence of exploitation, hosts or prey will increase exponentially. But we know that this exponential growth is unsustainable and will be limited by density-dependent factors. In Chapter 13, we derived formulas that describe how interspecific competition regulates population growth – here we derive formulas describing how predation or parasitism regulates population growth. Thankfully, there are many similarities linking the Lotka–Volterra competition and Lotka–Volterra predation models.

Derivation of the Lotka–Volterra predation model

Intuitively, we know that exploiters will reduce the growth rate of their hosts or prey. For the purposes of this model, I will use the term "predator" to refer to either predators or parasites, and "hosts" to refer to prey or hosts. Three factors should be most important: the number of predators (N_p), the number of hosts (N_h), and the efficiency with which a predator captures or infects its host (c). The product of these three factors measures the number of hosts removed from the population. Subtracting this product from the exponential growth formula gives the differential equation describing the rate of host population change.

$$dN_h/dt = \text{number of hosts born} - \text{number of hosts killed by predators or parasites}$$

$$dN_h/dt = r_h N_h - c N_h N_p$$

Let's next consider the predator population. We will assume that its growth rate is influenced by two factors: how many predators are born into the population, and how many predators die within the population.

The number of predators born into the population will be the number of hosts removed from the population as described above ($c N_h N_p$), multiplied by the efficiency with which these hosts are converted into predator babies (f). Thus the number of predators born into the population = $f c N_h N_p$.

The number of predators dying = $d_p N_p$, where d_p = the predator death rate.

The differential equation describing the rate of predator population change is

$$dN_p/dt = \text{number of predators born} - \text{number of predators dying}$$

$$dN_p/dt = f c N_h N_p - d_p N_p$$

The host and predator population growth formulas can be combined to generate predictions about how both populations will change under a variety of assumptions.

The Lotka–Volterra predation model's predictions of outcomes of exploitative interactions

In analogy to the Lotka–Volterra competition models, we can draw zero growth isoclines that describe the dynamics of exploitative interaction. For this model, the x-axis of our graph

describes the host population (N_h), while the y-axis describes the predator or parasite population size (N_p). Setting $dN_h/dt = 0$ describes the host zero growth isocline: the conditions under which the host population is stable, neither increasing nor decreasing. Setting $dN_p/dt = 0$ describes the predator zero growth isocline: the conditions under which the predator or parasite population is stable. Let's first see what happens when we set $dN_h/dt = 0$.

If

$$dN_h/dt = r_h N_h - c N_h N_p = 0$$

$$r_h N_h = c N_h N_p$$

solving for N_p gives us

$$N_p = r_h/c$$

Notice that N_p (the number of predators) is equal to the constant r_h/c, which means that the host isocline is a straight line parallel to the x-axis that intersects the y-axis of our graph at r_h/c. When N_p falls below the host isocline, the host population will tend to increase, and when N_p rises above the host isocline, the host population will tend to decrease (Figure 14.12A).

Let's next consider the predator or parasite population.

If

$$dN_p/dt = f c N_h N_p - d_p N_p = 0$$

$$f c N_h N_p = d_p N_p$$

solving for N_h gives us

$$N_h = d_p/fc$$

The predator isocline is a straight line parallel to the y-axis that intersects the x-axis of our graph at d_p/fc. When N_h (the number of hosts) falls to the left of the predator isocline, the predator population will decline due to lack of hosts. But if N_h increases, moving to the right of the predator isocline, the predator population will increase because there is an ample supply of prey or hosts (Figure 14.12B)

If we combine the predator and host isocline on one graph, we can generate a set of predictions on how N_h and N_p interact with each other. When hosts are abundant, predators increase, and when hosts are scarce, predators decrease. Conversely, when predators are abundant, hosts decrease, and when predators are scarce, hosts increase (Figure 14.13A). This generates the important prediction that predator and prey populations will tend to go through population cycles under these simplified conditions.

Points A–D on Figures 14.13A and B represent four sequential periods of time in this cycle. At point A, the host population is at its maximum, while the predator population is at its average value. At point B the host population declines to its average value, and the predator population climbs to its maximum. At point C the host population falls to its minimum value, while the predator population declines to its average value. At point D the host population recovers, increasing to its average value while

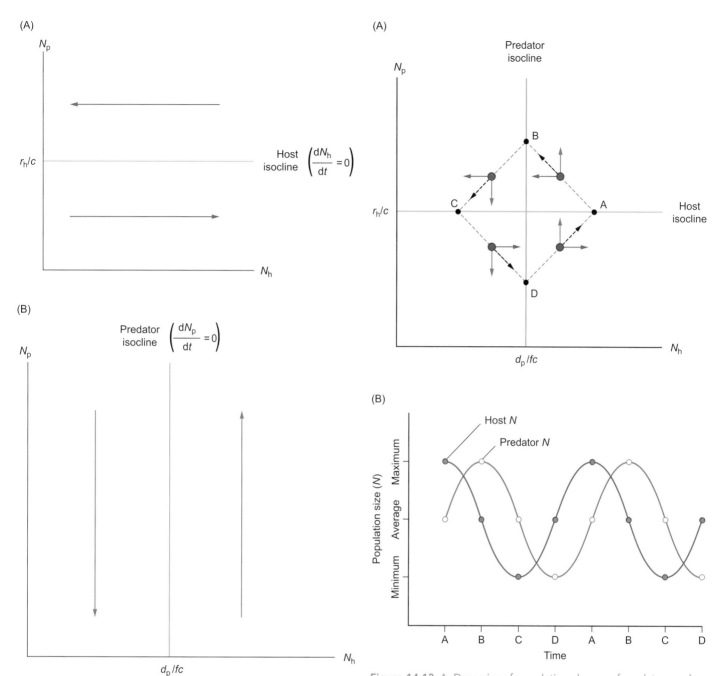

Figure 14.12 A. Prey or host isocline. B. Predator or parasite isocline.

Figure 14.13 A. Dynamics of population change of predators and hosts in relation to abundance of predators and hosts. B. Abundance of predator and host populations in relation to time.

the predator population continues to fall to its minimum value. The next cycle begins with the host population continuing to climb to its maximum value, while the predator population begins its recovery, reaching its average value again at point A (Figure 14.13B).

An important prediction is that predator and host populations will cycle over time, but the predator population peaks and valleys will lag behind the host population peaks and valleys. Researchers have worked diligently to investigate cycles of predators and hosts, both in the field and in the laboratory. The data base of lynx and snowshoe hare populations in Canada

that dates back to 1850 is probably the best-known long-term study of predator–prey cycles under natural conditions.

Lynx and snowshoe hare: periodic 10-year cycles in abundance

Snowshoe hares, *Lepus americanus*, have historically been a staple food for several Native American populations across northern and central Canada. People who depend on these animals for their sustenance have long known that theses hares go through periods of time when they are remarkably abundant, and other

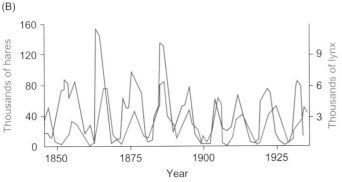

Figure 14.14 A. Canada lynx hunts down snowshoe hare. B. The lynx–snowshoe hare population cycle.

are low, hare populations increase rapidly, given hares' notoriously high reproductive rates. Once hares become very abundant, lynx populations begin to climb, due to high consumption rates of the now very available hares. High consumption rates cause the hare population to fall rapidly. When hares decline to low levels, lynxes begin to experience food shortages, and their populations also decline. In accordance with the Lotka–Volterra model, changes in predator abundance lag behind changes in prey abundance.

Though the data beautifully match the predictions of the model, the model does not necessarily provide a complete explanation for the dynamics of population cycles in this and other examples of population cycles. Researchers are becoming increasingly aware that nonconsumptive effects can profoundly influence population dynamics in exploitative interactions.

Influence of nonconsumptive effects on exploitative interactions

Nonconsumptive effects are changes in the behavior, growth, development, or reproduction of prey or hosts in response to the presence or absence of predators or parasites (Peckarsky *et al.* 2008). For example, prey or hosts may emigrate when predators become abundant, or they may forage in more protected – and less productive – areas within the habitat, reducing their rate of growth or development. Alternatively, prey or hosts may reduce their reproductive rates in response to high predator or parasite abundance. Thus high abundance of exploiters may reduce prey or host abundance, even in the absence of consumption.

Several avenues of research have highlighted that nonconsumptive effects of lynx on hare populations may influence hare population cycles. One possible nonconsumptive effect addressed in two different studies is whether reproductive output – the number of young produced per female per year – varies with stage of the cycle. About 15 years separated the end of the first study (Cary and Keith 1979) and the beginning of the second study (Stephan and Krebs 2001), which were conducted more than 1500 km apart. Despite these differences in space and time, both studies showed that a decline in reproductive output already had begun by the time the hare population reached its peak. For the second study, researchers retrapped female hares several times over the course of the breeding season, and identified three factors that contributed to this pattern. First, females only had two litters during the years of lowest reproductive output – 1991 and 1992 – as opposed to a mode of three litters in all other years of the study. Second, the second litter of 1991 exhibited by far the highest stillborn rate of any year studied. Lastly, the offspring also tended to be significantly smaller in 1991 and 1992 (Stephan and Krebs 2001).

Researchers wondered whether nonconsumptive effects could be causing these declines in reproductive rates. In particular, high stress levels caused by high predator abundance could

periods of time when they are extremely scarce. Snowshoe hares are also a staple food for Canada lynx, *Lynx canadensis* (Figure 14.14A). Like the human Native American populations, survival is relatively easy for the lynxes when hares are abundant, and very challenging for lynxes when hare populations decline. D. A. MacLulich (1937, p. 101) in his tome on snowshoe hare cycles, describes the observations of one field researcher during a winter of low hare abundance, "I met with a dozen lynxes that were dying of starvation – mere walking skeletons – and in the silent woods found a dozen shriveled corpses."

Many researchers, including MacLulich, and Charles Elton and Mary Nicholson (1942), pored through historical records to measure these fluctuations in abundance. One important source of information was the Hudson's Bay Company, which had archives of numbers of both hare and lynx pelts brought to them by trappers dating back to the eighteenth century. These records revealed a 10-year boom and bust cycle, in which hare abundance built up to a peak and then crashed over about 2 years to less than 10% of peak abundance, with some crashes substantially lower (Figure 14.14B). Lynx populations followed the same pattern, crashing usually about 1 year behind the hare crashes.

We can use the Lotka–Volterra predation model to understand the dynamics of this population cycle. When lynx populations

Table 14.2 Mean cortisol and testosterone levels for hares collected in 1991 and 1992 (decline phase), and 1994 (very end of low phase).

Year	Cortisol (nmol/l)	Testosterone (nmol/l)
1991	278.3 (40.4)	5.86 (0.81)
1992	301.8 (47.4)	2.76 (1.03)
1994	166.5 (30.6)	14.68 (3.80)

A minimum of 12 hares were sampled for each test. Year effect for cortisol $P < 0.0001$, for testosterone $P = 0.003$.

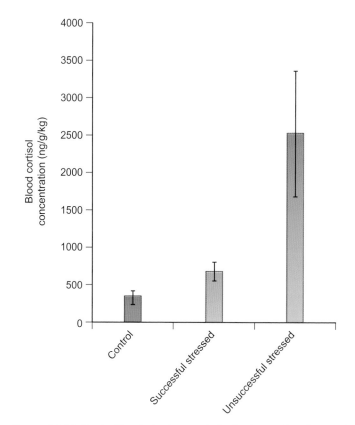

Figure 14.15 Cortisol levels (ng/g sample/kg body weight) of control hares, those that were exposed to a dog and gave birth (successful stressed) and those that were exposed to a dog and failed to give birth (unsuccessful stressed).

reduce reproductive rates and the health of the offspring. Recall in Chapter 6, we learned that chronic stress leads to high levels of glucocorticoids (primarily cortisol in hares, lemurs, and humans), which can cause an array of symptoms, including low fertility rates, inhibition of growth, and long-term changes in brain structure and function. Rudy Boonstra and his colleagues (1998) investigated whether hares from the decline phase (the years 1991 and 1992) showed evidence of a physiological response to high stress levels caused by exposure to high numbers of predators.

One important finding is that cortisol levels in the blood were almost twice as high during 1991 and 1992, the years of lowest reproductive output, than they were in 1994, which was the very end of the low phase, when reproductive output had climbed considerably (Table 14.2). Testosterone levels in males were much higher in 1994 than in either of the low reproductive output years. Taken together, these findings suggest a hormonal response to chronic stress that could potentially reduce reproductive output.

In the vertebrate stress response system, the pituitary gland releases corticotropin (ACTH), which causes the adrenal gland to release cortisol. Boonstra and his colleagues proposed that chronic stress from high predator levels during the decline phase (1991 and 1992) caused an increase in the responsiveness of the adrenal glands to ACTH, so that the hares could respond more rapidly to predators. If so, injection of decline-phase hares with ACTH should cause more of a rise in cortisol levels than would injection of low-phase hares from 1994. They found that following injection of ACTH, free cortisol levels in decline-phase hares increased to about twice the concentration as the free cortisol levels of low-phase hares. The researchers concluded that decline-phase hares had higher free cortisol levels because high stress from predators caused their adrenal glands to be more responsive to ACTH.

But were high predator levels causing the stress? Michael Sheriff and his colleagues (2009) investigated this question by live-trapping pregnant snowshoe hares and exposing 14 females

to a dog who was trained to cause no harm to the hares, but to be present in their pen for 1–2 min ever other day, for the 15 days before they gave birth. Twelve pregnant females were kept as an unstressed control and encountered no dogs. The researchers then compared reproductive output in the unstressed versus stressed females.

Eleven of 12 unstressed females gave birth, in comparison to 9 of 14 stressed females. Thirty hours after giving birth, the cortisol levels of the successful stressed females (the nine who gave birth) were about twice as high as the control, while the unsuccessful stressed females, measured at the same time, had cortisol levels almost 10 times greater than the controls (Figure 14.15). The successful stressed females had slightly smaller litter size, and their offspring were substantially smaller at birth than the offspring of unstressed controls. The researchers concluded that exposure to predators substantially elevates blood cortisol levels, and causes diminished reproductive output in snowshoe hares.

One puzzling observation is, given their renowned reproductive potential, why don't hare populations immediately recover once the predation threat is reduced? The low phase of the population cycle generally lasts about 2 years (sometimes

3 years), which is longer than would be predicted by a short-term response to stress. One possibility is that chronic stress can have long-term effects on snowshoe hares by reducing their muscle mass, interfering with proper development, and reducing cognition. If so, population recovery may not occur until mothers that have not been chronically stressed produce young that are free from the long-term negative effects of chronic stress. Thus the effect of the chronic stress on the females is permanent.

Tony Sinclair and his colleagues (2003a) tested this hypothesis by establishing laboratory colonies created from hares collected either from the decline phase or from the low phase. They maintained hares under identical laboratory conditions, and measured their reproductive output over the course of each hare's lifetime. They discovered that even under ideal laboratory conditions, decline-phase female reproductive output was less than one-half that of low-phase females. This supports Boonstra's hypothesis that the effect of chronic stress is permanent, prolonging the low phase of the snowshoe hare cycle.

To summarize this discussion, the lynx–snowshoe hare cycle appears to follow the classic dynamics predicted by the Lotka–Volterra predation model. Predation by lynx on hares increases well before hares reach maximum density, and survival rates of both juvenile and adult hares decreases sharply even as the hare population approaches its peak level. Hare survival rates continue to decline during the decline phase, but survival rates rebound sharply during the low phase. Hare reproductive output declines sharply as its population increases; this decline is influence by nonconsumptive effects. But there is much more to this story. For example, there are many other predators, such as coyotes, goshawks, and owls, that eat snowshoe hares and they too have a pronounced impact on hare survival rates. So this is much more than a two species system. In addition, this cycle is quite (but not absolutely) synchronous across all of boreal North America. Clearly there is a great deal left to understand about this system.

A consideration of nonconsumptive effects takes us beyond the simple predictions of the Lotka–Volterra predation model. But the model itself can be modified to deal with some of the realities of ecological complexity in exploitative interactions. We will very briefly explore some of these modifications.

Extending the Lotka–Volterra predation model

One limitation of our basic Lotka–Volterra predation model is that until now we have not accounted for the carrying capacity. Doing so will bend the host isocline down, intersecting the x-axis at the host carrying capacity. The effect of this modification is to cause the host population to decrease whenever it exceeds its carrying capacity regardless of predator abundance (e.g. point A on the graph, Figure 14.16). In addition, as the host population approaches its carrying capacity, it will have more of a tendency

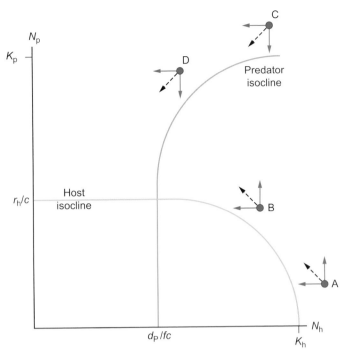

Figure 14.16 Predator–prey isoclines with carrying capacity. At points A and B prey abundance declines even though predators are relatively scarce. At points C and D predator abundance declines even though prey are abundant. The differences between these predictions, and the predictions of the simpler model in Figure 14.14, arise from incorporating the carrying capacity into the model.

to decrease even at low to moderate predator populations (e.g. point B on the graph). Similarly, we can account for the predator carrying capacity by bending the predator isocline toward the right, ultimately running parallel to the x-axis at the predator carrying capacity. The effect of this modification is to cause the predator population to decrease whenever it exceeds its carrying capacity regardless of host abundance (e.g. point C on the graph). As the predator population approaches its carrying capacity, it too will have a tendency to decrease, even a moderately high host populations (e.g. point D on the graph) (Figure 14.16).

There are many ways to extend this model to make it more realistic. For example, we could consider the Allee effect (Chapter 11), which, for predator–prey systems, results in reduced per capita growth rate at low population densities of predators and prey. A second extension is that predators may form a *search image* – a pattern of sensory input that describes the prey's important features. Upon encountering the same prey multiple times, the predator will learn to recognize a particular prey species by its appearance, smell or pattern of movement, and may hunt more effectively. Lastly, some hosts develop immunity to parasites, and this immunity may vary over time. All of these factors can be incorporated into the basic Lotka–Volterra predation model. Each factor will modify the isoclines, and predictions of the model, in different ways. As with all models, each modification has the potential to make the model more biologically relevant in describing outcomes of populations involved in exploitative interactions. But the downside is that each modification also makes the model more complex and more difficult to understand.

Thinking ecologically 14.4

One modification of the basic Lotka–Volterra model is to account for the presence of hiding places or refuges for prey (or hosts). For example, there may be a small number of crevices or burrows for rodents to hide in. Once the crevices fill up, the rodents are exposed to predators. How might you change the prey isocline to account for the refuge effect, and how would this influence the predictions of the model?

One assumption of the basic Lotka–Volterra model is that host capture rate (c) is a constant. Recall that the number of hosts removed from the population – the consumption rate – in the Lotka–Volterra model is equal to cN_hN_p. If c is a constant, and we hold N_p at a certain level, this tells us that for predator–prey systems, the prey consumption rate will increase linearly with N_h. If we double the number of prey, we double prey consumption. If we quintuple the number of prey, we quintuple prey consumption. Can predators really do this? The answer to this question rests on the predator's functional response to prey population density.

The predator's functional response

The functional response is the relationship between prey density and predator per capita consumption rate. C. S. Holling (1959a) described three types of functional response. In a type I response, the number of prey consumed increases linearly with prey density. In a type II response, the increase in consumption rate slows down as prey density increases, and reaches an asymptote at a level that may be determined by the predator's appetite or by the time it takes to handle each prey item. In a type III response, consumption rate is initially very low, then increases with prey density, and then asymptotes at a level determined by the predator's appetite or by the prey handling time (Figure 14.17A).

Obviously a type I functional response can only persist within certain bounds of prey density – predator per capita consumption rate cannot increase indefinitely. In a type I response, the consumption rate increases linearly and asymptotes abruptly, while in a type II response the asymptote is approached more gradually. Type II responses are probably most frequently encountered in natural systems. Holling's research shows a nice example of type III responses in small mammals responding to changes in the density of the predaceous sawfly larvae (Holling 1959b). All three species of mammals showed evidence of a type

III functional response, though one could argue that the error bars are large enough so that a type II response would be consistent with the data (Figure 14.17B).

Three factors can lead to the low consumption rate at low prey population densities that typify a type III response. First, a predator may need to develop an effective search image before it can efficiently capture prey. Second, there may be a limited number of refuges available to prey within the habitat, so that at low populations all prey have suitable refuges, while at higher populations some prey are forced to occupy more vulnerable

(A)

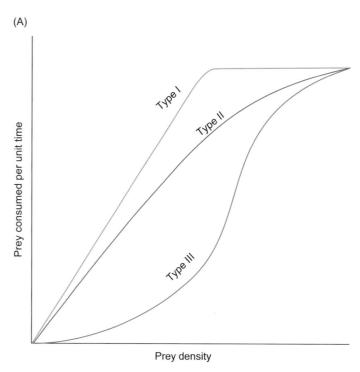

(B)

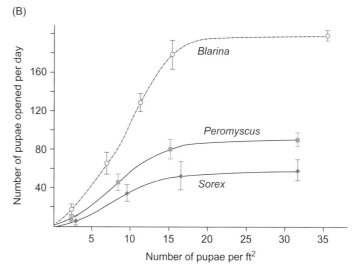

Figure 14.17 A. Three types of functional responses. B. Type III functional response of three genera of small mammals consuming sawfly larvae (Holling 1959b).

habitats. Lastly, predators may have other prey options available, and may switch to a more abundant type of prey when the first type is scarce.

Given the relative scarcity of type I functional responses in natural systems, does it make sense to assume a type I functional response in the Lotka–Volterra predation model? If we were to treat prey capture rate as a variable, rather than a constant (c), we would have a more precise, but more complex model. Resolving this tradeoff between precision and simplicity requires modelers to choose simplifications that don't compromise the qualitative predictions of their models.

As we discussed earlier, Lotka's interest in exploitative interactions was piqued by his observation of cycles in insect populations and in outbreaks of diseases. For many years ecologists have recognized that infectious diseases can cycle through populations of animals, including humans. But infectious disease has some important attributes that distinguish it from predator–prey systems. Some potential hosts may be immune to a disease or may acquire (or lose) immunity. Some diseases are more easily transmitted than others. Consequently we can use a different modeling approach to understand how some disease organisms exploit their hosts.

A simple model of how infectious diseases spread in host populations

The measles virus is a highly contagious airborne pathogen that is transmitted in respiratory secretions. Unvaccinated people living in the same house are very likely to spread the disease to each other. Before vaccination programs, measles tended to show cycles of about 2 years in human populations. Once populations became even partially vaccinated, overall prevalence declined, and the periodicity tended to disappear.

In a series of important articles published more than 30 years ago, Roy Anderson and Robert May argued that the spread of infectious disease can be understood using an ecological perspective (Anderson and May 1979, 1982; May 1983). They presented a simple descriptive model of disease caused by microparasites such as viruses. Hosts are born into the population with rate b, and die naturally with rate d. Susceptible individuals become infected with rate β and recover with rate v, at which point they are assumed to be immune. But immune individuals lose their immunity at rate γ. The death rate of infected individuals is the natural death rate (d) + the death rate from the disease (α) = $d + \alpha$ (Figure 14.18)

Given these factors, Anderson and May argued that we could predict the future of a disease in a host population by estimating its value of R_0 – the net reproductive rate, which is the mean number of offspring produced by an individual over the course of its lifetime (see Chapter 11). Applied to microparasitic infection, R_0 is the mean number of hosts infected by a host organism carrying the disease. According to Anderson and May's model, R_0 is equal to the product of the number of susceptible host organisms in the population (S), the transmission coefficient that measures how easily the infection is transmitted to a new host (β), and the amount of time the host is infectious (L). Mathematically this gives us

$$R_0 = S\beta L$$

R_0 is often called the *basic reproductive rate* when applied to infectious disease. By this formula, the basic reproductive rate increases as the number of susceptible hosts increases.

As we discussed in Chapter 12, if R_0 is greater than 1, the population will tend to increase, and if R_0 is less than 1, the population will tend to decrease. This allows us to define a threshold population size (S_T – sometimes called the critical community

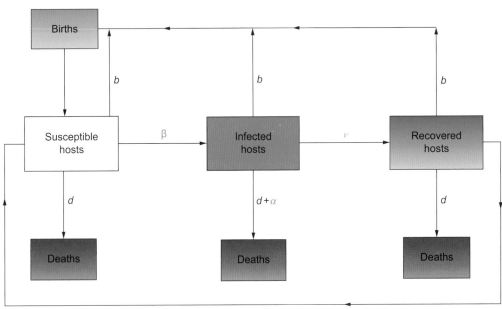

Figure 14.18 Diagrammatic flow chart of Anderson and May's (1979) model of microparasite infection.

size) as the threshold number of susceptible hosts needed to make R_0 greater than 1. Substituting 1 for R_0 and S_T for S in our formula gives:

$$1 = S_T \beta L \text{ or } S_T = 1/\beta L$$

Let's explore two of the implications of this formula. Pathogens vary in how easily they are transmitted; some, such as the HIV virus, require intimate physical contact, while others can be transmitted with casual contact, and still others, including many influenza viruses, are transmitted even more easily in droplets or aerosols. If other factors are held constant, easily transmitted pathogens (high value of β) should have low values of S_T. But hosts are infectious for highly variable periods of time. Someone infected with HIV will be infectious for many years, while a host with measles or influenza may be infectious for only a few days. Thus pathogens such as the HIV virus, those that are not easily transmitted but are infectious for a long period of time within an individual host, will have moderate values of S_T.

Let's see how these considerations can influence the progression of a disease in a population. If we choose a disease with $\beta = 0.001$ per day, and $L = 5$ days, then $S_T = 1/(0.001 \text{ per day} \times 5 \text{ days}) = 200$ individuals. Let's introduce an infected host into a population of 1000 susceptible individuals (Figure 14.19). This gives us $R_0 = 5$ for our starting conditions, because

$$R_0 = S\beta L = 1000 \times 0.001 \text{ per day} \times 5 \text{ days} = 5.0$$

We expect the disease to spread rapidly, infecting most individuals in the population. As the number of infected individuals increases, the number of susceptible individuals (S) decreases, ultimately dropping below S_T (200 individuals). When that happens, the disease will be much less likely to encounter susceptible hosts, and R_0 declines below 1. Ultimately, according to this model, the disease may go extinct in the population (Figure 14.19).

What actually happens in real populations can be much more complex. Some infected hosts may be very social, and may be more likely to spread the disease, even when the number of susceptible individuals falls below S_T. This may allow a disease to persist at population sizes below S_T. After the disease has declined and R_0 drops below 1, the number of susceptible individuals may begin to increase from three sources: births, immigrants, and loss of immunity at rate γ. If the disease has gone extinct in the population, the number of susceptibles will continue to climb until a new infected host enters the population. At that point, the infection cycle can begin again. Alternatively, if the disease is still present in the population, when the number of susceptibles climbs above S_T, there may be a much milder outbreak that drops the population below S_T again.

We can apply the concept of a threshold population size to developing a logical approach for disease control. First we can reduce transmission rate (β) by using good hygiene. For easily

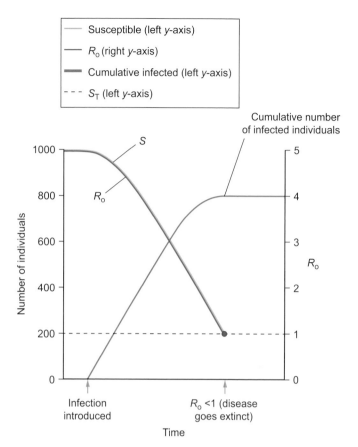

Figure 14.19 One possible response of a population to infectious disease. In this case, the infectious disease goes extinct when R_0 drops below 1, which occurs when S drops below S_T.

transmitted diseases such as influenza, this can involve washing hands, avoiding contact with other individuals, and sneezing into tissues. Second, we can effectively reduce the number of susceptible hosts by reducing the number of individuals an infected individual comes in contact with. This is the idea underlying quarantine.

Lastly, epidemiologists develop vaccination programs, which seek to reduce the number of susceptible hosts below the threshold population size, and to keep it below that level so that the pathogen will die out in that population. This was achieved with smallpox, which had a relatively high value of S_T. Measles, though infectious for only a few days, has a very high transmission rate, and epidemiologists estimate that completely eradicating measles from the population will require about a 93% vaccination rate at current population levels. More disturbingly, malaria, which kills approximately 2 million people per year, will be particularly difficult to eradicate, because it is infectious for a very long period of time (very high L) and is transmitted very effectively by its mosquito vector (very high β).

These models are abstract, difficult to understand, and imprecise. However, they are essential steps for developing an understanding of how disease spreads within a population.

Epidemiologists use more complex versions of these models to develop strategies for preventing the spread of diseases, ultimately saving many lives and reducing human suffering.

One theme of this chapter is that exploitative interactions are complex, often involving species affecting one another in different ways, both directly and indirectly. Unraveling the complex web of interactions is challenging, exciting, and often puzzling. Let's return to New Zealand to see how researchers have unraveled the complex interactions between cockles, their parasites, and their predators.

REVISIT: Unraveling the mystery of cockles

The cockle manipulation hypothesis proposes that trematodes cause cockles to expose themselves to oystercatcher predators, so that the trematodes can complete their life cycles by parasitizing the oystercatchers. One prediction of this hypothesis is that oystercatchers should eat more surface-dwelling cockles than burrowing cockles. Frederic Thomas and Robert Poulin (1998) tested this prediction in a field setting. They collected 102 cockles from the surface and 102 cockles from burrows, marked them, and attached them to metal stakes driven into the ocean floor near the shore, so they would be less likely to be lost from the ocean's current. They knew from previous work that oystercatchers usually shatter the cockles' shells, so they scored a broken shell still attached to the string as predation by oystercatchers. The experiment only went on for 7 weeks because oystercatchers left the coastline to breed, but the researchers still observed seven documented cases of oystercatcher predation on the surface-collected cockles, and only one case of oystercatcher predation on cockles collected from burrows.

A second prediction of the cockle manipulation hypothesis is that surface cockles will have higher levels of metacercariae (encysted trematodes) than burrowing cockles. This is difficult to address based on field data, because a substantial number of surface-dwelling cockles have their feet cropped by fish, which of course reduces metacercariae numbers. When Kim Mouritsen and Robert Poulin (2003) addressed this prediction by restricting their analysis to uncropped cockles, they discovered almost twice as many metacercariae in surface-dwelling cockles than in fully buried cockles, while partially buried cockles had intermediate values (Figure 14.20A).

A third prediction of the cockle manipulation hypothesis is that habitats with overall higher rates of parasitism should have a higher proportion of cockles living at the surface. Mouristen and Poulin surveyed 12 different sites and found a strong positive relationship between mean parasitism rate, and the proportion of cockles living at the surface (Figure 14.20b).

Do these three results taken together make a compelling case that natural selection has favored trematodes that bring their cockle hosts to the surface? To address this question, we need to understand the mechanism underlying this process. Mouritsen knew from previous observations and research that highly infected cockles could remain in their burrows if undisturbed, but if they were dislodged by waves or burrowing animals, they could not effectively rebury themselves. He suspected that high infection rates reduced the foot muscle contraction rate. To test this hypothesis, Mouritsen (2002) measured foot length in 37 cockles, tapped the foot to stimulate muscular contraction, and remeasured foot length immediately after contraction. In support of muscle contraction hypothesis, he found that contraction efficiency decreased as infection intensity increased (Figure 14.20C). He speculates that the high number of metacercariae in the tip of the foot may immobilize the transverse muscles needed for burrowing.

There is still more to this story. For example, cockles on the seafloor surface are also eaten by some species of fish; as far as we know this abruptly ends parasite transmission. In addition, there appears to be more than one species of trematode parasitizing these cockles. Many exploitative interactions are clearly much more complex than a simple two-species predator–prey interaction. The same holds for all types of ecological interactions, including mutualisms, the subject of our next chapter.

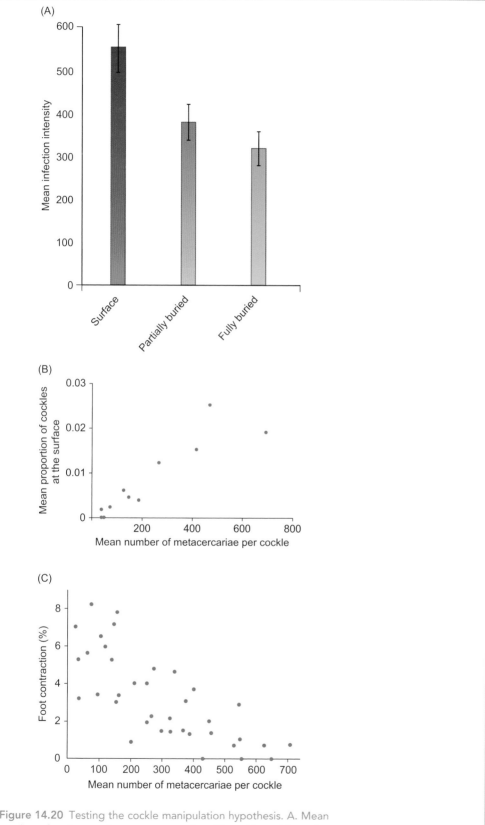

Figure 14.20 Testing the cockle manipulation hypothesis. A. Mean number of metacercariae on uncropped cockles collected from the surface, in comparison to cockles that were partially buried and fully buried. B. Mean proportion of cockles at the surface in relation to mean number of metacercariae per cockle. C. Percent cockle foot contraction in relation to mean number of metacercariae per cockle.

SUMMARY

Exploitative interactions can be understood in terms of their lethality and intimacy. Predators and parasitoids cause highest lethality, parasites and parasitoids have highest intimacy with their hosts, while grazers are low on both scales. Exploiters can regulate the populations of their hosts directly by killing or injuring them, or through nonconsumptive processes such as increasing their prey's stress level and thereby reducing reproductive rates, as has been implicated for the snowshoe hare. Exploiters can also regulate community processes indirectly; for example bats and birds eat arthropods in the forest, which reduces leaf damage by herbivorous arthropods. Prey and hosts use constitutive defenses, such as thorns in plants, and large body size in Serengeti grazers, against exploiters. Some species have evolved induced defenses; for example some plants release toxic chemicals following herbivore attack.

The Red Queen hypothesis proposes that exploiters and their prey or hosts are engaged in a coevolutionary process, in which interactions among species lead to reciprocal adaptations. The outcomes of these exploitative interactions can be predicted by the Lotka–Volterra predation model, which, in its most basic form, predicts that the relative abundance of predators and prey will cycle. A simple model of disease transmission can explain how disease spreads in host populations based on the ease of transmission, the amount of time the host is infectious, and the population size of the host. Both models make numerous simplifying assumptions. Ecologists can incorporate biological complexity into these models, which makes them more realistic, but also more difficult to understand and apply.

FURTHER READING

Connolly, B. M., Pearson, D. E., and Mack, R. N. 2014. Granivory of invasive, naturalized, and native plants in communities differentially susceptible to invasion. *Ecology* 95: 1759–1769.

Clear demonstration of how seed predation by small mammals influences the establishment of invasive and native plant species within forested and steppe communities.

Decaestecker, E., Gaba, S., Raeymaekers, J. A., Stoks, R., Van Kerckhoven, L., Ebert, D., and De Meester, L. 2007. Host–parasite "Red Queen" dynamics archived in pond sediment. *Nature* 450: 870–873.

Very clever experiment. Using the water flea, *Daphnia magna*, and its bacterial parasite, *Pasteuria ramosa*, collected from various sediment levels at the bottom of a pond, the researchers show that there is a coevolutionary cycle of the evolution of high infectivity in the parasite, and parasite resistance in the *Daphnia* host.

Peckarsky, B. L. and 10 others. 2008. Revisiting the classics: considering nonconsumptive effects in textbook examples of predator–prey interactions. *Ecology* 89: 2416–2425.

Thoughtful consideration of four classic studies of predator–prey systems that should be reconsidered in the light of nonconsumptive effects. The authors argue that nonconsumptive effects often play a critical role in many different aspects of predator–prey systems.

Stastny, M., and Agrawal, A. A. 2014. Love thy neighbour? Reciprocal impacts between plant community structure and insect herbivory in co-occurring Asteraceae. *Ecology* 95: 2904–2914.

Shows that the presence of other plant species can either increase or decrease a particular species' susceptibility to insect herbivory. High levels of herbivory reduce overall plant productivity and increased colonization by other plant species.

Sugimoto, K. and 14 others. 2014. Intake and transformation to a glycoside of (Z)-3-hexenol from infested neighbors reveals a mode of plant odor reception and defense. *Proceedings of the National Academy of Sciences*, 111: 7144–7149.

Remarkable demonstration that tomato plants can pick up chemical signals from other tomato plants, and reconstruct those same chemicals into defensive compounds.

END-OF-CHAPTER QUESTIONS

Review questions

1. What are the different types of exploitative interactions and give an example of one type of each interaction as described in this text. How do the different types of interactions differ in relation to intimacy and lethality?
2. How did the study by LaDeau and her colleagues (2007) test the hypothesis that West Nile virus was regulating populations of songbirds across North America? What were the strengths and weaknesses of this study?
3. How did the study by Kalka and her colleagues (2008) show direct and indirect effects of predation on community processes?
4. How do constitutive and induced defenses differ? Give an example of each.
5. What is the Red Queen hypothesis, and how does Bro-Jørgensen's research test its predictions?
6. What are the variables in the Lotka–Volterra predation model and what are the constants? Describe, in your own words, how the model leads to a prediction of population cycles.

Synthesis and application questions

1. An alternative to the cockle manipulation hypothesis is that the behavioral change in highly infected hosts is not the result of natural selection favoring manipulative trematodes, but rather is a non-adaptive side effect of trematode infection. Do you think this is a valid argument? If so, what types of data might you collect to address this objection? If not, why is this objection invalid?
2. Can you think of environmental, behavioral, or technological changes that might increase ginseng population viability over the next century?
3. After 1987, predators in the Serengeti made a gradual recovery, as anti-poaching laws were enforced. What would you predict happened to the abundance of the six ungulates in Sinclair's 2003 study?
4. Ian Baldwin's research showed that herbivory rates were 10 times higher in control plants than in plants that had elevated nicotine levels induced by jasmonic acid. Why should you be concerned about concluding that induced nicotine levels reduce herbivory based on this experiment alone? What other research could you do to make you more confident that induced production of nicotine reduces herbivory?
5. In the lynx/snowshoe hare system, lynx data depended on pelts brought in by trappers to the Hudson's Bay Company. These trappers ate hares as an important food over the winter. In what ways could a scarcity of hares influence the number of pelts? Is this concern a problem for our estimates of lynx abundance?
6. Can you imagine a situation where the functional response curve decreases with increase in prey density?
7. Human populations are very open, with people immigrating and emigrating at high rates. What are the implications of this mobility for our understanding of how well vaccination programs will work?
8. Bardgett *et al.* (2006) were surprised to find high nitrogen mineralization rates in soil with the hemiparasite *R. minor*, because high nitrogen mineralization rates are expected to increase plant biomass, which is contrary to their findings. The researchers argue that the direct effect of parasitic plants suppressing biomass was stronger than the indirect effect of nitrogen mineralization increasing biomass. How might you use systematic observations or experiments to evaluate this hypothesis?
9. Refer back to Figure 14.6 and note that the density of giraffes actually decreased following predator removal. Propose a hypothesis that could account for this finding. How could you test your hypothesis?

10. The study by Ford and his colleagues (2014) shows a strong correlation between predation risk, the presence of impala, and the presence of thorny *Acacia* trees. But showing a causal relationship among these three factors is challenging. What types of observations or experiments might you conduct to show this connection?

Analyse the data 1

Are you convinced that oystercatcher predation is higher on surface cockles than burrowing cockles, based on Thomas and Poulin's results?

Analyse the data 2

Mouritsen wondered whether cockles surface for reasons other than parasite load, and while cockles are on the surface, whether parasites can more easily infect them. To address this concern, Mouritsen collected 60 cockles from burrows, measured the number of metacercariae in 20 of the cockles, and transferred the remaining 40 cockles to small cages submerged in a sand flat near the shoreline. Twenty of these cages had a mesh bottom that forced the cockles to remain at the surface, while the other 20 cages permitted burrowing. The results of this experiment are shown in Figure 14.21. What should Mouritsen conclude?

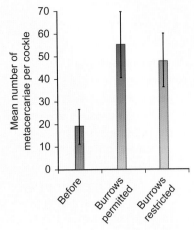

Figure 14.21 Mean number (±SEM) of metacercariae per cockle before the experiment, and after cockles were placed in cages and permitted to burrow or restricted from burrowing.

Chapter 15
Facilitation

INTRODUCTION

Perhaps more than any other insect, the honeybee, *Apis mellifera*, swarms in facilitative interactions – interactions between two organisms, or two species, that benefit at least one of them and harm neither. Mutualism occurs when both benefit, while commensalism occurs when only one benefits. Pollination of flowers by honeybees is a mutualistic interaction, in which flowers receive pollen, which they need to develop seeds and fruits, while honeybees obtain both nectar and pollen, which are critical foods for them. Farmers use wild and domesticated bees to pollinate their crops, with some very large farms depending almost exclusively on domesticated colonies. This demand increases the strain on domesticated hives, which are sometimes called into service during the winter when they would normally be quiescent, and may be hauled thousands of kilometers over the course of the season. Beekeepers are accustomed to losing a small fraction of their colonies over the course of the winter due to a variety of causes, including disease, bad weather, and excessive travel stress on the bees.

Our opening case study discusses how the human food chain is threatened by a recent sharp decline in domestic honeybee populations from an emerging disease. We then explore developing threats to other facilitative interactions, focusing primarily on mutualisms, but considering commensalism as well. Mutualisms benefit the interacting species in many different ways, providing food, nutrients, pollen, dispersal services, suitable habitat, and protection against predators and herbivores. In most cases, there is a cost to being involved in a mutualistic interaction, so we would expect mutualisms to evolve only when the benefits of the association exceed the costs to the interacting species. As we discovered with competitive and exploitative interactions, the positive effect of one species on another may be mediated by the actions of one or more other species.

KEY QUESTIONS

15.1. What are the consequences of disrupting mutualisms?

15.2. What are the benefits of facilitative interactions?

15.3. What conditions favor the evolution of facilitative interactions?

CASE STUDY: Honeybees under attack, but who is the enemy?

Recall from Chapter 6 that honeybees are eusocial or truly social. They care cooperatively for their young, and have sterile castes of non-reproductive workers. These workers are mutualists in the sense that they cooperate with each other to build the hive, raise the brood, defend the hive, and forage for resources. They use elaborate dances to explicitly inform other hive members where they should go to find food, and how much food they are likely to find.

Related to this behavior is the interspecific mutualism between honeybees and flowering plants. Honeybees visit flowers primarily in search of nectar – the raw material used to create honey, which feeds the bee colony throughout the year. In the process of collecting nectar, the bees also collect pollen grains, which are in reality very small male plants or male gametophytes (see Figure 15.1A). Some of these protein-rich pollen grains are brought back to the hive to feed the brood. However, of great importance to flowering plants, when the pollen-carrying bees visit a second flower of the same plant species, some of these pollen grains may be transferred to the stigma of the second flower – a process that we call pollination. If the process works properly, the pollen grain then transfers its male gamete into an ovule (the female gametophyte) located within the plant's ovary, where it fertilizes an egg cell that will then develop into a new seed. The fruit, which develops from the second flower's ovary wall, subsequently encases and protects one or more seeds. Later on we will discuss other mutualisms that allow the seeds to disperse into environments that are suitable for seed germination and development.

Pollination is usually mutualistic, as bees obtain food and flowering plants get pollinated. Humans benefit tremendously from this interaction, as pollinated flowers develop into fruits and seeds that are important parts of their diet. In the United States, honeybees are essential pollinators of almost 100 species of fruits and vegetables consumed domestically and exported to other countries around the world. From an economic standpoint, researchers estimate that pollination by insects increases the worldwide value of agricultural crops by about US$215 billion annually (Gallai *et al.* 2009).

The agricultural community became very concerned in 2006 when commercial beekeepers began noticing that honeybee workers in many of their previously healthy colonies were simply

Figure 15.1 A. Honeybee worker sipping nectar and receiving additional pollen grains. Notice the pollen basket attached to her hind leg. B. Frame from (top) a normal beehive, and (bottom) a hive suffering from CCD. The brood and a few workers are still evident in the CCD hive.

disappearing over a few-week time period, leaving the colonies with only the queen, her brood, and almost no workers. There was no evidence of any dead worker bees either within or near the affected colonies. Pests such as hive beetles and wax moths were noticeably absent, and bees from neighboring colonies did not move in to steal the honey. This syndrome, **colony collapse disorder (CCD)**, was even more of a problem in the winter of 2007–8, which was partially responsible for extraordinary overwinter mortality rates above 35% in the United States (vanEngelsdorp *et al.* 2008; Figure 15.1B). Since then, overwinter mortality rates have fluctuated, but have remained substantially above the historic levels of 10–15%. Though reliable data are difficult to get, it appears that much of the northern hemisphere is experiencing increased hive mortality from a variety of causes. Recently, the first documented case of CCD outside the United States was discovered in Switzerland (Dainat *et al.* 2012).

Despite heroic attempts by many researchers, we still don't know what is causing CCD, or what the future holds for honeybees as our world's most important animal pollinators. Many researchers are actively searching for clues by comparing variables influencing affected and unaffected colonies, but thus far the results, while suggestive, are inconclusive. Later in this chapter we will explore the progress that researchers have made to unravel this critical puzzle.

This example highlights how important facilitative interactions are in our world, and how disruption of mutualisms can have dire consequences. Disruption of facilitative interactions may extend beyond keeping humans fed, by threatening the continued existence of entire biomes or ecosystems.

15.1 WHAT ARE THE CONSEQUENCES OF DISRUPTING MUTUALISMS?

One consequence of extraordinarily high plant species diversity in the tropics is that the population size and density of each individual species is usually very low. Under conditions of low population density, wind pollination is rarely a viable option, as there is a low probability of the pollen reaching its target. Consequently, tropical plants with low population density tend to rely more on animal pollination than they do on wind pollination. Thus tropical rainforests, the most species-rich biome, are greatly threatened by the potential loss of pollinator services.

Threat of pollinator loss to tropical species diversity

Pollinator systems run the gamut from completely specialized, where one pollinator species services one plant species, to highly generalized, where many pollinator species service many plant species (Figure 15.2A). Steven Johnson and Kim Steiner (2000) argue that natural selection favors specialization by a plant on one pollinator when the most effective pollinator is abundant

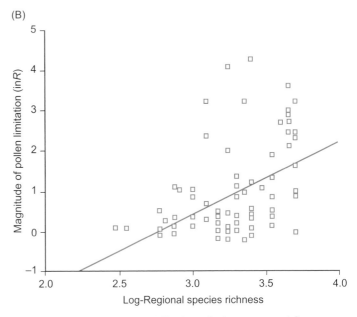

Figure 15.2 A. Extreme specialization of a long-tongued fly (*Moegistorhynchus longirostris*) pollinator adapted to a flower with an equally long floral tube. B. Pollen limitation increases with species richness of the region ($P < 0.0001$).

year after year. They also propose that long-lived plants can afford to specialize on one pollinator, because missing 1 or 2 years of reproduction does not have a catastrophic effect on those individuals' fitness. Lastly, rare plants are more likely to specialize, because specialization reduces the amount of pollen lost to generalist pollinators that are unlikely to land on another plant of the same species.

Jana Vamosi and her colleagues (2006) wanted to know whether plants in hotspots of biological diversity, such as tropical rainforests, were more likely to suffer from a shortage of pollen. They reasoned that high plant species richness would be associated with a low mean number of individuals of each plant species, and thus pollen limitation could be a serious problem. They surveyed the literature for studies that quantified pollen limitation across the world, and tested to see whether there was a positive correlation between pollen limitation and species richness.

In all studies included in this meta-analysis, researchers studied pollen limitation by comparing fruit set in plants given supplemental pollen (experimental plants $= E$) to the fruit set in plants that were pollinated by their natural pollinators (control plants $= C$). They calculated pollen limitation as the ln response ratio ($\ln R$),

$$\ln R = \ln(E) - \ln(C)$$

Plants that were pollen limited had higher values of $\ln(E)$ than $\ln(C)$, and thus greater response ratios. Based on this extensive meta-analysis, plants in regions with high species richness had a significant tendency to be more pollen limited than plants from areas with lower species richness (Figure 15.2B). Thus plants in biodiversity hotspots are more likely to be pollen limited.

If pollen limitation is greater in biodiversity hotspots, and specialist pollinator systems are more common in those areas as well, conservation ecologists must be vigilant about preserving specialist pollinator systems in those regions. Specialist pollinators are most vulnerable when ecosystems are degraded or destroyed, because their specialization often makes them poor competitors for alternative sources of nectar. For example, the long proboscis on *Moegistorhynchus longirostris* (Figure 15.2A) is presumably not efficient at extracting nectar from more traditional flower architecture. On the other hand, if its host plant goes extinct, and the pollinator goes extinct as well, the ecosystem- and biome-level impact will be much less catastrophic than the impact of losing an important generalist pollinator such as the honeybee. The important question to pursue is whether other examples of facilitation play equally important functional roles within biomes. The short answer to this question is "yes." Let's turn our attention to a different facilitative interaction with biome-level consequences.

Mycorrhizal associations

Most terrestrial plant species form mycorrhizal associations – symbiotic and usually mutualistic associations between their roots and various types of fungi (organisms that live in close associations like this are often called *symbionts*). Ectomycorrhizal associations are very common in taiga and temperate forest biomes, but they are found throughout the biosphere. The fungal hyphae form a sheath around the exterior of roots, and also penetrate into the spaces between individual root cells. These hyphae have an enormous surface area, which allows the plant to greatly increase its uptake of essential nutrients, particularly nitrogen and phosphorus. In exchange, the plant provides the fungus with carbohydrates.

Arbuscular mycorrhizal associations are between plant roots and fungi of the phylum Glomeromycota, in which the hyphae grow between the root cells, ultimately entering some of them and forming finely branched arbuscles. About two-thirds of all land plants form these associations, most commonly in tropical and subtropical biomes but also in woody and nonwoody vegetation throughout the biosphere. The fungal partner often extends itself into the soil, extracting nutrients and water for the plant in return for carbohydrates..

Given how widespread mycorrhizal associations are, we might expect that they play an important role in terrestrial biomes worldwide. One approach to testing this assumption is to study systems that are deficient in mycorrhizae to see if there is a reduction in plant diversity, abundance, or growth. On a large scale, several managed pine forests have failed from a lack of mycorrhizal fungi. On a somewhat smaller scale, as demonstrated by Martin Nuñez and his colleagues (2009), many species of trees have failed to become established on Isla Victoria, Argentina – also because of a lack of mycorrhizal fungi.

Isla Victoria was cleared for ranching in the early twentieth century, but it became part of a national park in 1934. In 1925, the Argentine government planted over 100 species of trees, including 18 species of invasive pines, in an effort to establish a vigorous and productive forest on the island. Most plantings of invasive trees failed, with the exceptions of those planted close to tree plantations that already had established networks of existing invasive pines. The researchers hypothesized that most of the ectomycorrhizae were host specific, and thus there were insufficient mycorrhizal fungi adapted to the invasive pine species that were planted far from the plantations.

One prediction of the mycorrhizae deficiency hypothesis is that invasive pines grown in soil collected near existing plantations should grow better than invasive pines grown in soil collected far from the plantations. A second prediction is that adding living ectomycorrhizal fungi (an inoculum) to the soil near existing plantations will not increase invasive pine growth, because presumably the fungi are already present in the soil. However, adding an inoculum to the soil collected far from the plantations should significantly increase invasive pine growth by providing the deficient ectomycorrhizal fungi for the trees.

Greenhouse studies: the importance of ectomycorrhizal associations

To test these predictions, the researchers conducted greenhouse experiments on three species of invasive conifers: *Pseudotsuga menziesii* (Douglas fir), *Pinus ponderosa* (ponderosa pine), and *Pinus contorta* (lodgepole pine). They planted seeds in 1-l pots, and grew the trees for 9 months. The researchers collected soil either from within 100 m of existing plantations (near soil treatment) or from greater than 1000 m (far soil treatment). They established five treatments by adding to some of the soil a 25 cm^3 mycorrhizal inoculum as follows: (1) near soil + inoculum, (2) near soil + no inoculum, (3) far soil + inoculum, (4) far soil + no inoculum, and (5) sterilized far soil + no inoculum, as a control.

For all three species, the best performers were both of the near soil treatments, and the far soil treatment with the inoculum of ectomycorrhizal fungi added. On average, adding the inoculum to the far soil tripled the growth rate over the 9 months of the experiment. Plants grown in sterile soil had similar growth as plants grown in far soil without inoculum (Figure 15.3). The researchers concluded that adding ectomycorrhizal fungi increased the growth of trees grown in soil collected far from the plantations.

Field studies: confirming the greenhouse findings

But Nuñez and his colleagues wanted to extend this study to field conditions. They planted the same three species at 50 different locations either close to (c. 100 m) or far from (>1000 m) plantations. Each location had one treatment (25 cm^3 inoculum addition) and two controls. The first control was a sterilized 25 cm^3 inoculum, while the second control was the addition of 25 cm^3 of soil from the same location as the planting. The researchers protected the plots from seed predators by surrounding them with wire mesh. Only about 11% of the plots had seedlings by the end of the experiment; presumably a pine's life in the field is more difficult than life in a cushy greenhouse, where resources were controlled and supplied at a steady rate.

Despite the reduced survival, the field experiments supported the findings of the greenhouse experiments. In general, adding inoculum did not seem to benefit trees in near sites but substantially benefited trees grown at far sites. All trees in near sites, and inoculated trees in far sites had substantially greater aboveground biomass, and higher survival rates, than the control trees at far sites (Figure 15.4).

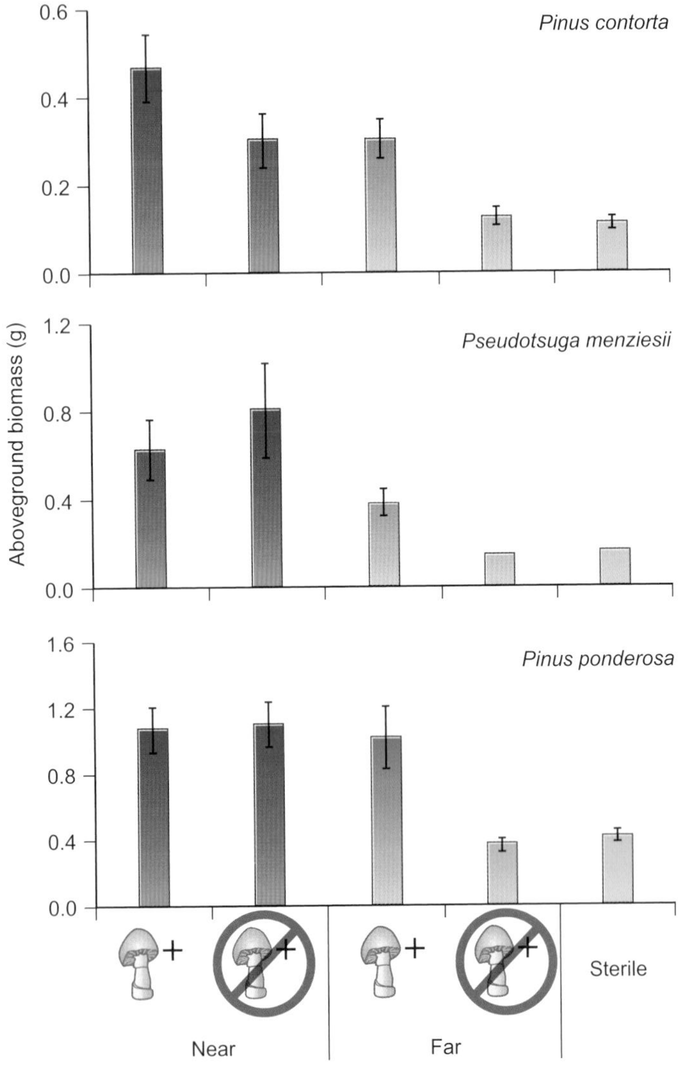

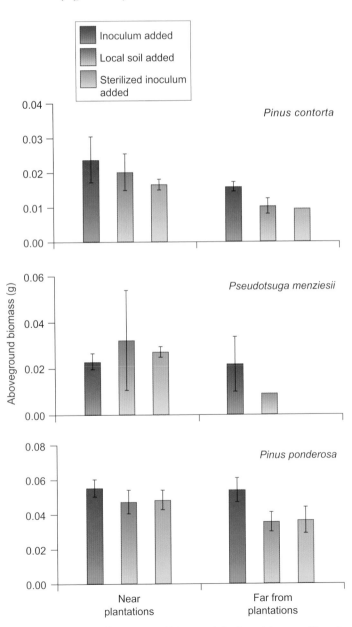

Figure 15.3. Aboveground biomass of three tree species under the five treatments described in the text. There were 10 replicates of each treatment. From left to right: (1) near soil + inoculum, (2) near soil + no inoculum, (3) far soil + inoculum, (4) far soil + no inoculum, and (5) sterilized far soil + no inoculum. ($F = 38.15$, $P < 0.001$).

Figure 15.4 Mean aboveground biomass (g) of surviving seedlings in the field experiments.

When the researchers harvested the trees, they inspected the roots for the presence of mycorrhizal fungi. They discovered the greatest number of mycorrhizal fungi in near trees. Importantly, they also discovered many more mycorrhizal fungi in far trees that had received the inoculum in comparison to far trees that had not received the inoculum. They concluded that inoculating the far trees with mycorrhizal fungi increased tree aboveground biomass and survival rates.

 Thinking ecologically 15.1

Why did the researchers use two controls for their field study? Which control was more important? Would they have been better off to use only one control and to have a larger sample size?

Many other studies highlight the essential role played by facilitative interactions in terrestrial biomes. Let's turn to marine biomes, which also are highly dependent on facilitative interactions.

Coral reefs: the symbiotic mutualism between coral and its zooxanthellae

In Chapter 2, we discussed the similarities between tropical rainforest and coral reef biomes. Some of their striking similarities include extraordinary species richness, high productivity, complex layering of the dominant species, and low nutrient availability. A coral reef's diversity depends on the mutualism between the coral hosts and the symbiotic zooxanthellae algae that live within the coral's endoderm cells. All of these algae are in the genus *Symbiodinium*, which has eight clades (A though H). A clade is a set of species descended from a particular common ancestor. Consequently, all species in one clade (for example clade A) are more closely related to each other than they are to any species in the remaining clades (B through H). Systematists have still not worked out whether the different algal symbionts are distinct species, or subspecies, so we will simply refer to them as types. No doubt more types of symbionts will be discovered as researchers continue working to understand this mutualism.

Coral reef environments: conditions favorable for photosynthesis and nutrient retention

The coral polyps feed on zooplankton that float in the water, but that only provides the coral with a small fraction of their energetic needs. They get the remaining 90% of their energy from the zooxanthellae, and return the favor by providing the zooxanthellae with a place to live, and a generous allotment of nitrogenous waste. Recall that fixed nitrogen is a limiting resource for many organisms, particularly autotrophs. Coral animals use calcium that is abundantly available in shallow and warm salty water to build their skeletons, while the zooxanthellae are limited to shallow clear waters so they can photosynthesize. Other algae also thrive in these ideal conditions, so productivity is very high.

The coral animals are sessile, so they live on top of previous generations of coral, forming reefs that may be thousands of meters thick. This structural complexity provides diverse habitat for over 4000 species of fish, and hundreds of thousands of invertebrate species. Because the water is shallow, nutrient-rich waste products and decomposition products remain available to members of the community, rather than sinking out of areas of high production as occurs in the deep ocean.

The threat of coral bleaching to coral reef ecosystems

The coral–algal mutualism forms the foundation of the coral reef biome. Unfortunately, coral bleaching – a breakdown in the coral–algal mutualism – has become a serious problem in the past 25 years. Though we know that coral bleaching occurs when the coral loses its algal pigmentation, we still don't know exactly what causes this loss, either physiologically or environmentally. Physiologically, we can ask: what actually happens within the coral tissues when they lose most or all of their pigmentation? From an environmental perspective, we can ask: what biotic or abiotic factors lead to coral bleaching?

Researchers have proposed that high light and high temperatures together prevent the algae from processing the excited electrons generated during photosynthesis. This "roadblock" leads to the buildup of oxygen radicals, which can damage or destroy both host and symbiont tissues. When this occurs, either the symbionts may be expelled from the host, or else the endoderm cells may die and be sloughed off.

While the physiology underlying coral bleaching is somewhat murky, researchers have amassed substantial evidence that high water temperatures are the most important environmental cues underlying bleaching. In the summers of 1983, 1987, and 1998, the ENSO (see Chapter 1) caused substantial increases in sea surface temperatures (SSTs) at two of the world's largest coral reefs – the Great Barrier Reef off the Northeast Australian coast, and the East African reefs off the coasts of Kenya, Tanzania, and Comoros (Figure 15.5). These significant increases in SSTs, known

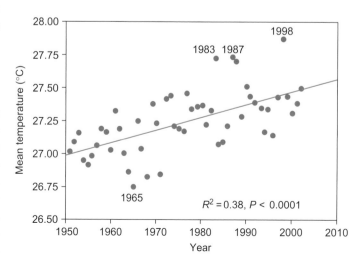

Figure 15.5. Mean sea surface temperatures of coral reefs off the East African coast.

as thermal anomalies, coincided with extensive coral bleaching events. For example, of 29 reefs surveyed by Timothy McClanahan and his colleagues (2007) off the East African coast, 28 had reduced coral cover after the 1998 event. In 27 of these sites, the reduction in coral cover was at least 10%.

Coral adaptation to thermal stress

More recent investigations provide a glimmer of hope for the future of this essential mutualism. Jeffrey Maynard and his colleagues (2008) studied bleaching at 14 sites on the Great Barrier Reef during the great 1998 thermal anomaly (Figure 15.6A). Some of these sites experienced much greater thermal stress than others. For three important genera of coral, there was a very strong linear correlation between thermal stress and bleaching severity at a particular site. Based on this relationship, the researchers now had a model to apply to the next major thermal anomaly (Figure 15.6B).

The researchers returned for the next thermal anomaly in 2002, to discover that the response to thermal bleaching was much less than they had anticipated based on their 1998 model. The model predicted bleaching severity above 70% for all three genera, but the observed bleaching severity values averaged below 20%. Maynard and his colleagues interpret the lower level of bleaching as evidence that all three coral genera had increased their thermal tolerance between the two bleaching events. One possible explanation for this increase in thermal tolerance is *symbiont shuffling* – that coral may be shifting to more heat-stress-tolerant symbionts in response to higher SSTs.

Is there any evidence for symbiont shuffling in response to thermal anomalies? One approach to studying this question is characterizing the symbionts in corals before and after bleaching events. A prediction of the symbiont shuffling hypothesis is that coral will shift from stress-sensitive to stress-resistant symbionts in response to thermal stress.

Alison Jones and her colleagues (2008) tested this prediction on a section of the Great Barrier Reef. Characterizing the *Symbiodinium* clades requires DNA amplification, cloning, and sequencing, and does not detect rare symbionts. The researchers tagged and characterized 460 coral colonies in 2004/5 before the bleaching event. They found that 93.5% of the coral harbored exclusively *Symbiodinium* clade C2, 3.5% harbored clade D, and 3.0% had a mixture of both clades.

Almost 90% of the colonies were severely bleached from the thermal anomaly of 2005/6. When Jones and her colleagues resampled the survivors 3 months later, symbiont composition had changed dramatically. The majority of the surviving C2 colonies switched to D or C1, a previously undetected type. The majority of the surviving D colonies remained D, while a small number switched to C1. And all of the colonies that had harbored a mix of C2 and D were now exclusively D (Table 15.1). Clearly symbiont shuffling occurred after the thermal anomaly of 2005/6.

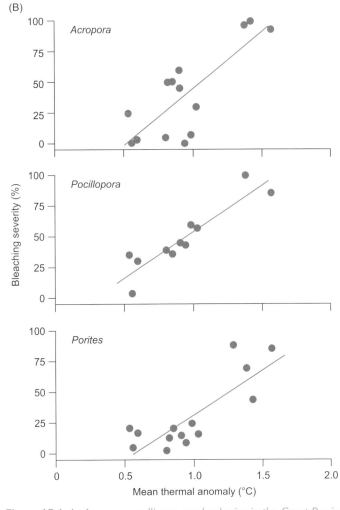

Figure 15.6 A. *Acropora mellipora* coral colonies in the Great Barrier Reef. White corals are bleached, while darkly pigmented corals still contain their symbionts. B. Bleaching severity in relation to mean thermal anomaly for three coral genera at the Great Barrier Reef. Each point represents one reef system within the Great Barrier Reef. Bleaching severity is defined as the percentage of colonies either bleached or recently dead. Mean thermal anomaly is the mean increase in sea surface temperature above the historical mean for each day, over the course of the entire summer.

Table 15.1 Percentage of colonies of each type after the bleaching event, in relation to their initial type before the bleaching event.

Initial type	Sample size	% of each type post-bleaching			
		C2	D	C2/D mix	C1
C2	58	29.3	51.7	0	19.0
D	15	0	80	0	20
C2/D mix	6	0	100	0	0

These results need to be interpreted with caution. Laboratory studies have confirmed that Clade C1 is heat tolerant, but they are somewhat less consistent about the relative heat tolerance of clade D. On a more global level, while these results give some reasons to be optimistic that coral ecosystems may survive the anticipated major rise in sea temperatures of the next century, we still don't know what the thermal limits are for the continued success of this mutualism that is the foundation of the ocean's most diverse biome.

In the mutualism between coral and *Symbiodinium* both species exchange food or nutrients, but these are not the only positive outcomes of mutualisms. We will now explore the diversity of types of benefits associated with facilitative interactions.

15.2 WHAT ARE THE BENEFITS OF FACILITATIVE INTERACTIONS?

Our discussion of the global consequences of mutualisms has introduced us to three different types of benefits available to partners in a mutualistic interaction: food, habitat, and services. Let's look at each of these types of benefits.

Providing mutualistic species with food or nutrients

Of mutualisms discussed thus far, at least one partner has benefited by increasing the quality or quantity of food or nutrients it receives. In plant–pollinator mutualisms, the pollinator usually is rewarded with energy-rich nectar. In a mycorrhizal association, the host plant increases its nutrient uptake, while its fungal symbiont is rewarded with the carbohydrates produced by the plant. Lastly, coral provides its symbiotic algae with nitrogen, and presumably other nutrients as well, while the algae return the favor with carbohydrates from photosynthesis.

But in many cases, the mutualist is also being provided with a better habitat, or the ability to inhabit a previously inaccessible habitat. In mycorrhizal associations, both the plants and fungi may occupy habitats that might otherwise be unsuitable for them, and the same is true for coral and their algal symbionts. Let's look at a mutualism that is clearly based on expanding a species' range of usable habitat.

Providing mutualistic species with access to different types of habitats

One of the most important variables defining a species' niche is environmental temperature. In most cases, if the temperature is unsuitable for survival, the species will move away, if movement is an option, or become dormant, or else die. But some mutualisms expand the range of habitats that are suitable for each partner.

Like all plants, the panic grass *Dichanthelium lanuginosum* lives in association with fungi. But panic grass has a unique relationship with the fungus *Curvularia protuberata,* an *endophyte* that grows inside its leaves and roots and also on its seed coat. Regina Redman and her colleagues (2002) noticed panic grass thriving in soils heated by geysers and underground magma sources in Lassen Volcanic and Yellowstone National Parks in the western United States, at temperatures up to 57°C – well above the thermal tolerance levels of most plants. They analysed the contents of 200 plants and discovered large amounts of *C. protuberata* in each plant, but none living in the soil. They suspected that the presence of the fungus might be responsible for the success of the plant at these elevated temperatures, and conversely, that panic grass may provide a microenvironment enabling the fungus to survive.

In laboratory cultures between 25 and 35°C, fungal spores showed 100% germination rates, while at 40°C, fungal germination rate dropped to 14%, and no germination occurred above 40°C. The researchers concluded that fungal survival at elevated temperatures was only possible inside the panic grass plant. To test whether panic grass required the fungal endophyte in hot soils, the researchers compared the growth and survival of panic grass grown with and without the symbiont at elevated temperatures. At soil temperatures of 50°C for 3 days, symbiotic plants thrived, while nonsymbiotic plants became yellowed and shriveled. In fact, symbiotic plants withstood intermittent temperatures of 65°C with only minor damage (Figure 15.7). When Redman and her colleagues transplanted symbiotic and nonsymbiotic plants to soil near Amphitheater Springs in Yellowstone National Park, growth rates of nonsymbiotic plants were severely depressed in 40°C soil, and all nonsymbiotic plants died in 45°C soil. Meanwhile, symbiotic plants continued to grow under these field conditions.

These results beg the question of what makes both the plant and the fungal species tolerant of high temperatures. Luis

S (50°C) NS (50°C) S (65°C) NS (65°C)

Figure 15.7 Symbiotic (S) and nonsymbiotic (NS) plants grown at 50 or 65°C under laboratory conditions.

Marquez and his colleagues (2007) knew that viruses were important in some other plant–fungi symbioses, so they decided to see if the fungus *C. protuberata* was routinely infected with any particular virus. Fungal viruses usually have double stranded RNA, which is rare in uninfected fungal cells, making it possible to more easily detect and isolate viruses from fungal cells. The researchers successfully isolated a new virus from *C. protuberata*, which they descriptively named *Curvularia* thermal tolerance virus (CThTV).

To see whether this virus conferred thermal tolerance to *C. protuberata*, the researchers cultured fungi with and without the virus. They then inoculated 25 panic grass plants with CThTV-infected fungus, and another 25 plants with virus-free fungus, and left 25 plants uninoculated. When subjected to a 14-day regimen of 10 h/day at 65°C, and 14 h/day at 37°C, all of the uninoculated plants and all of the plants inoculated with virus-free fungus either became very yellow or died outright. In contrast, all of the plants inoculated with CThTV-infected fungus survived, and only four plants showed signs of yellowing (Figure 15.8).

Researchers are now trying to understand how the virus interacts with the endophyte to make the plant heat tolerant. There is some evidence that the virus induces the fungus to increase its production of proteins that help the plant and fungus survive osmotic stress by stabilizing cell membranes, and heat stress by increasing production of heat-shock proteins (Rodriguez and Roossinck 2012). But much more remains to be learned in this system.

This case demonstrates how a complex mutualism can greatly expand the thermal tolerance, and resulting habitat range, of two species, using a third species to mediate the interaction. While facilitative interactions can make new habitats available to different species, they can provide other services as well.

Other services to mutualists

We've already discussed one important service provided by facilitative interactions; pollinators increase plants species' reproductive success. While we've focused primarily on insect pollinators, many animal species and groups are important pollinators. But for a plant to be successful!, pollination must be followed by fertilization, seed development, seed germination, and growth of the new offspring. Germination is problematic, because plants must be dispersed to habitats that are suitable for growth. Plants have evolved numerous mechanisms to increase the probability that their seeds will be dispersed to appropriate habitats.

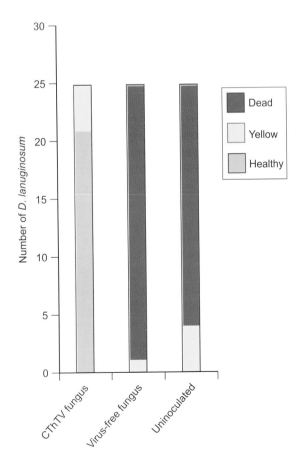

Figure 15.8 The number of dead, yellow, and healthy *D. lanuginosum* after the experiment.

Seed dispersal facilitated by animals

Animal dispersal of seeds may be highly beneficial to plants. In some cases plants encase their seeds in fleshy fruits, which provide a nutritious meal for animals that then move the seeds to new locations that may be favorable for germination. This process can be costly to the plant as well. Fleshy fruits are energetically expensive to produce, and consume a large portion of a plant's energy budget. Many seeds may be destroyed by the digestive process, or they may be dispersed to unsuitable habitats. In addition, many seeds are simply eaten by seed predators with only a tiny fraction actually surviving the journey to a new habitat.

Some trees produce seeds in such large numbers that seed predators cannot eat all of them, and instead the animals store their seeds in underground caches for future retrieval. Fortunately, from the trees' perspective, many of these caches are never retrieved, converting would-be seed predators into tree-planters. Jennifer Briggs and her colleagues (2009) undertook an 8-year study to see whether forest rodents cache Jeffrey pine seeds in a manner that facilitates germination and growth of new seedlings, a process they describe as *directed dispersal*.

The researchers placed individuals of four different rodent species into test enclosures containing 150 radiolabeled seeds. After each trial they excavated the caches and measured the characteristics of each cache for each species. There were significant between-species differences in cache depth, size, soil preference (mineral soil, light litter, or heavy litter), amount of understory cover, and amount of overstory cover (Table 15.2).

There is a great deal of variation in these findings that has important implications for seedling survival. For example, previous research indicated that seeds buried deeper than 40 mm had low germination rates. Thus golden-mantled ground squirrels are burying seeds deeper than is optimal for the plant, but the large standard error indicates enough variation so that some of their caches will be at a suitable depth for seeds to germinate.

The important question then becomes: which cache choices favor successful emergence by the plant? Briggs and her colleagues planted artificial caches at 0, 5, and 25 mm deep. Soil was either mineral soil or light litter, as they had previously established that heavy litter offered poor conditions for emergence. Understory cover was either open or under bushes. The researchers then followed seed progress in the caches to see which conditions were most favorable for emergence. One clear finding was that emergence was very low (4.7%) when seeds were set at the surface, and much higher when seeds were buried (27.6% at 5 mm and 33.5% at 25 mm). Across all treatments combined, emergence was much greater in mineral soil (27.2%) than in light litter (6.7%).

But there were some pretty interesting interactions between cover and soil type for buried seeds at 5 and 25 mm. The probability of emergence was greatest in mineral soil, under shrubs, with minimal canopy cover. As canopy cover increased, emergence success dropped quite rapidly. But caches that were out in the open, without shrubs to protect them, were not as adversely affected by a high canopy cover, indicating that emergence benefits from some shading, either from the understory or the overstory (Figure 15.9). Presumably the shade maintains higher water availability during the seedlings' first year.

Table 15.2 Cache preferences of four different rodent species. Standard errors are in parentheses.

Peromyscus maniculatus *Tamias amoensus* *Tamias quadrimaculatus* *Spermophilus lateralis*

Species	Common name	Mean cache depth (mm)	Number of seeds per cache	Soil preference	Understory preference	Overstory preference
Peromyscus maniculatus	Deer mouse	3.4 (3.1)	1.4 (0.8)	light litter	random	open
Tamias amoensus	Yellow-pine chipmunk	5.7 (2.6)	3.7 (2.2)	mineral soil	under shrubs	open
Tamias quadrimaculatus	Long-eared chipmunk	9.7 (8.7)	7.2 (4.1)	light litter	under shrubs	random
Spermophilus lateralis	Golden-mantled ground squirrel	49.1 (37.3)	17.3 (12.0)	NA	NA	NA

NA – data were not available for habitat preference for *S. lateralis* due to small cache sample size.

(A)

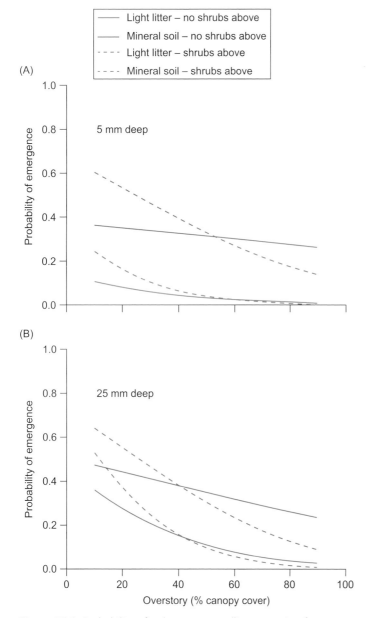

Light litter – no shrubs above
Mineral soil – no shrubs above
Light litter – shrubs above
Mineral soil – shrubs above

5 mm deep

Probability of emergence

(B)

25 mm deep

Probability of emergence

Overstory (% canopy cover)

Figure 15.9 Probability of at least one seedling emerging from artificial caches of three seeds in four different microsites in relation to canopy cover. A. Caches planted 5 mm deep. B Caches planted 25 mm deep.

If we combine rodent behavior with plant emergence data, we can see that both species of chipmunks create caches that are suitable for Jeffrey pine seedling emergence. The caches are small enough so that there won't be too much competition, deep enough so that emergence success will be reasonably great, and shaded enough to reduce water stress. Briggs and her colleagues concluded that at least the two chipmunk species are providing directed dispersal services for Jeffrey pine trees.

 Thinking ecologically 15.2

Compare Figures 15.9A and B. What are two differences in emergence of seedlings planted at 5 mm and at 25 mm?

Defense is another type of service animals can provide for plants. The best-known plant protectors are ants.

Ants protecting plants

I will let Dan Janzen describe to you his moment of discovery of ants defending plants, in Central America, as he described to me in an email message.

1962. I am walking across a Veracruz pasture and there are two acacias about two meters apart, each with a big healthy ant colony. I swing my machete and cut both down, intending to take both back to the house to dissect. For reasons I have zero idea of, I only picked up one and walked off with it. Six weeks later, I just happen to walk through the same pasture and boom! Right in front of me are the two stumps. The one where I left the crown (and colony) has a one-meter gorgeous sucker shoot with full perfect leaves and swarming with the ant colony (which obviously had survived the cut off crown, and then moved into the growing sucker shoot), and the other is a stump with a couple of ratty defoliated 10–15-centimeter trashed shoots with no intact growing point. Bingo. Any field biologist confronted with that lucky coincidence of things would have come to the same conclusion. My thesis was to repeat this with about 2500 pair of acacias growing in every imaginable circumstance for about 1.5 years. I also figured out later to remove the ant colony with parathion – nearly killing myself in the process – thereby getting the same effect without trashing the tree structure.

Janzen describes the ant–acacia interaction as a "fully developed interdependency" (Janzen 1966, p. 252). The queen ant lands on an unoccupied acacia plant, lays her first eggs in a swollen thorn, drinks nectar produced by the nectaries, and eats protein-rich solid food in the form of beltian bodies (Figure 15.10). The growing colony expands to occupy all of the acacia thorns. While it is quite clear that the ants benefit from both food and accommodations provided by the acacia, Janzen wanted to confirm that the acacia was also benefiting from housing and feeding the ants.

BENEFITS ACACIA ANTS PROVIDE THEIR HOST PLANTS. There were several pieces of observational evidence supporting the hypothesis that ants were benefiting the plants. Janzen could see that there were many fewer herbivorous insects on acacia trees with ants than on acacia trees without ants. He also observed that there was very little vegetation surrounding acacia trees with ants. One hypothesis is that ants provide herbivore removal and competitor removal services for the acacias. But an alternative hypothesis is that queen ants may be attracted to trees with low levels of herbivores and surrounding vegetation.

To distinguish between these two hypotheses, Janzen cut down a group of acacia trees, leaving the stumps to regenerate. Vegetative reproduction is actually very common in these

Figure 15.10 Fully developed interdependency between ants, *Pseudomyrmex ferruginea*, and the bullhorn acacia, *Acacia cornigera* (now *Vachellia cornigera*). A. Acacia tree in a cleaning with surrounding vegetation suppressed by its symbiotic ants. B. *Pseudomyrmex* ants tending larvae inside a thorn. C. An ant can be seen entering a thorn next to a group of nectaries.

acacias, because there is a high incidence of fires in the tropical forest and also a great deal of tree-cutting by humans. Janzen removed the ants from some stumps, and allowed them to remain on other stumps. He then compared the success of occupied (with ants) stumps to unoccupied stumps (without ants). As Table 15.3 demonstrates, ant occupation increased the growth and survival of regenerating stumps, supporting the hypothesis that this interaction benefits both species.

All of these findings support the hypothesis that ants are providing protection to acacia trees, ultimately enhancing the plants' growth and survival. There are numerous other examples of ant–plant mutualisms.

"DEVIL'S GARDENS" CREATED BY AGGRESSIVE ANTS WITHIN TROPICAL FORESTS. In a second ant–plant system, Megan Frederickson and two colleagues (2005) were interested in

Table 15.3 Growth, survival, and herbivorous insect infestation of regenerating occupied and unoccupied stumps (SD in parentheses). Data from Janzen (1966).

Variable	Occupied	Unoccupied
Mean new branch growth (cm)	72.9 (37.4)	10.2 (9.8)
Mean new branch wet weight (g)	579.9	43.9
Mean number of leaves	108.1	52.4
Mean number of swollen thorns	103.9	39.3
Survival (for 9.5 months)	28/39	30/69
% of shoots with herbivorous insects (day)	2.7	38.5
% of shoots with herbivorous insects (night)	12.9	58.8

why one small tree species, *Duroia hirsuita*, forms a monoculture within Amazonian rainforests. These monocultures are known as "devil's gardens," which, according to legend, are maintained by evil forest spirits (Figure 15.11A). The researchers considered two hypotheses. First, perhaps *D. hirsuita* is *allelopathic* – exuding chemicals that restrict other plants from growing nearby. Alternatively, perhaps *D. hirsuita's* ants, *Myrmelachista schumanni*, aggressively keep other plants out of the area.

The researchers tested the ant aggression hypothesis by planting saplings of a common cedar tree, *Cedrela odorata*, both inside and outside of "devil's gardens." They set up four experimental treatments: (1) cedar tree planted inside devil's garden, ants not excluded; (2) cedar tree planted inside devil's garden, ants excluded; (3) cedar tree planted outside devil's garden, ants not excluded; and (4) cedar tree planted outside devil's garden, ants excluded. In support of the ant aggression hypothesis after one day, cedar tree leaf necrosis (localized tissue death) was much greater when it was planted inside a devil's garden with ants present (Figure 15.11B). In addition, after 5 days the cedar tree shed a much greater percentage of its leaves when it was planted inside a devil's garden with ants present in comparison to the other three treatments (Figure 15.11B). Lastly, the researchers could see the ants attacking the cedar tree and appearing to inject a substance into the base of its leaves.

When the researchers analysed the ant's poison glands, all they found was formic acid, which is commonly used by many ant species for aggression and defense. When they injected the cedar leaves with formic acid, they observed the same necrotic pattern in all the leaves they tested. They concluded that ants were killing competitors as a service for their *D. hirsuita* hosts.

In this section, we've discussed how mutualisms provide food, habitat or a variety of services to each mutualistic partner. Analogously, the benefiting partner in a commensalism also receives a broad diversity of benefits.

Diversity of benefits to commensalists

As in mutualistic interactions, the benefiting partner in a commensalism may receive more or better food. For example, many tropical seabird species depend on small fish as a mainstay of their diets. The problem is that these birds cannot dive deeper than a few meters, so their prey are often inaccessible. However, predators such as tuna and swordfish that also eat these small fish motivate them to attempt escape by surfacing or by jumping into the air, making these small fish easy prey for the seabirds. The predaceous fish suffer no cost because these schools of small fish are so abundant that a few lost potential prey don't affect them, but the seabirds are rewarded with a meal. This commensalism is so common, that conservation biologists are using the presence of large flocks of seabirds as an indicator of where they should focus their conservation efforts, reasoning that if seabirds abound, then fish species diversity and abundance are likely to be very high in that area as well (Le Corre *et al.* 2012)

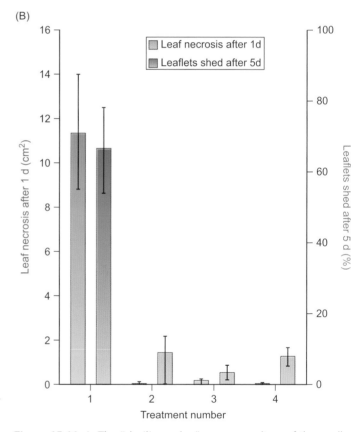

Figure 15.11 A. The "devil's garden" – a monoculture of the small tree *Duroia hirsuita* with a diverse tropical forest in the background. B. Mean *C. odorata* leaf necrosis (cm² of leaf surface) after one day and percentage of *C. odorata* leaflets shed after five days in response to the experimental treatments. See text for description of treatments.

Perhaps the most common benefit enjoyed by commensal species is a place to live. Many different microorganisms are currently residing on you and in you, taking advantage of suitable temperature, dependable moisture, and available food. Most of them are not harming you, or providing much benefit, but are simply using you as a suitable habitat. Some species are of course parasitic, while others are mutualists (Dunn 2011).

When you walk through an unplowed field in the autumn, you are participating in a commensalism with many species of flowering plants. After your journey, you may find yourself covered with burrs, which are seeds or fruits that have hooks or teeth that allow them to adhere to animals. You, and other animals that pass through, are dispersing seeds and fruits produced by various plants. In most cases the disperser is unaffected by the interaction, while the seeds may be moved to a new location that is more suitable for germination.

Many of the examples of facilitative interactions we've discussed thus far involve plants extracting some type of service from animals or fungi. One possible explanation for this is that plants are very common, and thus would be expected, by chance, to be involved in the majority of the well-studied facilitative interactions. A second possibility is that plants, in general, cannot move around very much, so they benefit greatly from services provided by other, more mobile species. They also can benefit other species courtesy of the large supply of carbohydrates they can amass from photosynthesis. If we understand the benefits gained by each species through a facilitative interaction, can we predict when mutualisms are likely to evolve in natural systems?

15.3 WHAT CONDITIONS FAVOR THE EVOLUTION OF FACILITATIVE INTERACTIONS?

I hope you are convinced by now that facilitative interactions are widespread in the natural world, and that species can provide a wide diversity of services for each other. In some cases, including most commensalisms, and some mutualisms such as directed dispersal of plant seeds by rodents, or pollination of plants by honeybees, each interaction is very brief, and the costs of the interaction may be very low or nonexistent for at least one species. Presumably honeybees are not overly burdened by the pollen load they carry from plant to plant. But in other cases the interactions are very intimate, with both species living together and dependent on each other for continued existence. In these cases, the costs of the interaction may be very high as well. For example, many zooxanthellae have substantially lower growth rates than free-living algae, presumably because they are contributing carbohydrates to their coral hosts. If one species is providing a costly service to a second species, natural selection should favor the evolution of traits in the first species that ensures that the second species rewards the first species' for its efforts. Otherwise natural selection would favor individuals of the second species who did not reciprocate, ultimately leading to the collapse of the mutualism. Consequently, for the mutualism to be stable, the first species must be able to punish the second species for cheating.

Punishing cheating mutualists

Studies of several different mutualisms have demonstrated that one partner has the potential to punish the other partner if it does not provide the appropriate benefit. One of the well-studied examples of this interaction is the mutualism between host legume plants and their symbiotic *Rhizobium* bacteria. Plants provide the rhizobia with root nodules to live in, and carbohydrates to use for growth and reproduction. The bacteria are in turn able to fix nitrogen from the air, and convert it to ammonia with the following reaction:

$$N_2 + 3H_2 + 16ATP \xrightarrow{\text{nitrogenase}} 2NH_3$$

Recall that fixed nitrogen is a limiting resource for plants, so supplying fixed nitrogen is a wonderful service provided for the legumes. But this is also an energetically very costly process for the rhizobia, as indicated by the 16ATPs that fuel the reaction. Toby Kiers and his colleagues (2003) wondered why the rhizobia did not cheat by either reducing or eliminating nitrogen fixation once they became established within the snug confines of the plant nodules (Figure 15.12A).

To investigate this puzzle, the researchers manipulated the atmosphere by replacing nitrogen with argon, an inert gas. They left enough N_2 in the atmosphere so that the bacterium *Bradyrhizobium japonicum* could fix about 1% of their normal amount, enough to easily meet its own nitrogen needs, but with little left over to share with its host plant, the soybean, *Glycine max*.

Retribution was swift and terrible. The number of rhizobia per nodule was substantially lower in plants whose nodules were forced to cheat (Figure 15.12B), and the number of rhizobia released into the surrounding sand in the next generation as part of the *Rhizobium* reproductive cycle declined by about 50% (Figure 15.12C). In a second experiment, the researchers controlled the atmosphere experienced by individual nodules on the plant root, so that one nodule experienced a normal atmosphere, and a nearby nodule received the nitrogen-poor atmosphere. In this experiment the number of rhizobia per nitrogen-deficient nodule declined by over 50% as compared to nodules experiencing a normal environment (Figure 15.12D). Thus host plants can identify which nodules are cheating, and can direct sanctions against the offending nodules. The underlying mechanism appears to be a reduction in oxygen supply to the bacteria; within 5 h of the bacteria cheating, oxygen levels within nodules declined by about 50%. *Rhizobium* with reduced oxygen levels may be unable to efficiently carry on cellular respiration.

Kiers and his colleagues concluded that intimate mutualisms, with high costs, can be stabilized if species have mechanisms to reduce the incidence of cheating by their mutualistic partners. Presumably these mechanisms themselves are costly to the host species, and would be expected to evolve primarily in systems where the costs of being cheated are very high. But there are

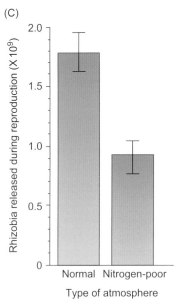

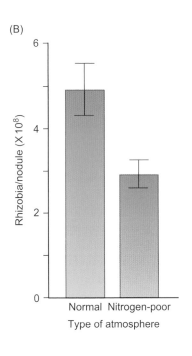

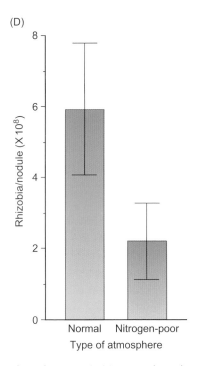

Figure 15.12 A. Root nodules formed by soybean plant to house *Rhizobium* bacteria. B. Mean total number of rhizobia per nodule in plants experiencing either normal atmosphere or nitrogen-poor atmosphere. C. Mean number of rhizobia released into sand during *Rhizobium* reproduction. D. Mean number of rhizobia per nodule when individual nodules receive either normal or nitrogen-poor atmosphere.

many examples of facilitative interactions in which some species do cheat. As a simple example, many plant–pollinator mutualisms are exploited by *nectar robbers*, animals that steal nectar from flowers while bypassing the anthers. In many cases, nectar robbers show much lower rates of pollen transfer, although some studies show them to be moderately effective pollinators.

Because provisioning another species is often very costly, we might expect facilitation in general to be more common in environments where the benefits of facilitation are also very high. For example, we might expect benefits to be greater in inhospitable environments, when a species is living at the extreme boundary of its geographical range. In general, we might expect facilitative interactions to be more common under conditions that are stressful to a particular species.

Environmental conditions favoring facilitative interactions

Ragan Callaway and her colleagues (2002) proposed that alpine plants growing at relatively high elevation experience more physiological stress than the same species at much lower elevation. They reason that at high elevation, plant success tends to be limited by abiotic factors such as extreme temperature, wind scouring, or soil instability, rather than access to light or nutrients. In contrast, in the more benign environmental conditions experienced at low elevation, plant success is more likely to be limited by competition for light or nutrients. Under this assumption, high-elevation plants are more likely to benefit from association with other plants. This hypothesis predicts that facilitative interactions will be more common at high altitudes, and competitive interactions will be more common at low altitudes.

The researchers recognized that any findings would be more convincing if they did their studies at numerous alpine sites around the world, so they conducted research at five locations in North America, four in Europe, and at one site each in South America and Asia. At each location, researchers established low-elevation and high-elevation sites that were dominated by herbaceous vegetation.

At each site researchers removed all vegetation within 10 cm of a target plant, and identified a second plant of the same species, which had no surrounding vegetation removed, as a control. They carried out approximately 10 replicates of this treatment on 3–10 species per location. At the end of the next growing season, the researchers harvested each plant and measured its dry weight. The relative neighbor effect (RNE) measures the effect of removing the surrounding plants; a positive RNE indicates that the control plants with no vegetation removed grew better than target plants, which had nearby vegetation removed. In contrast, a negative RNE indicates that target plants grew better than control plants. Thus a positive RNE indicates facilitation, and a negative RNE indicates competition.

In support of the hypothesis that physiological stress favors facilitative interactions, plants had a significantly positive value of RNE at 9 of 11 high-elevation sites, and a significantly negative value of RNE at 7 of 11 low-elevation sites. At the other sites, the RNE values were not different from 0 (Figure 15.13). Thus interactions between a plant species and its neighbors shift from facilitation to competition when the physiological stress experienced at high altitudes – temperature extremes, wind scouring and soil instability – is reduced or eliminated at lower altitudes.

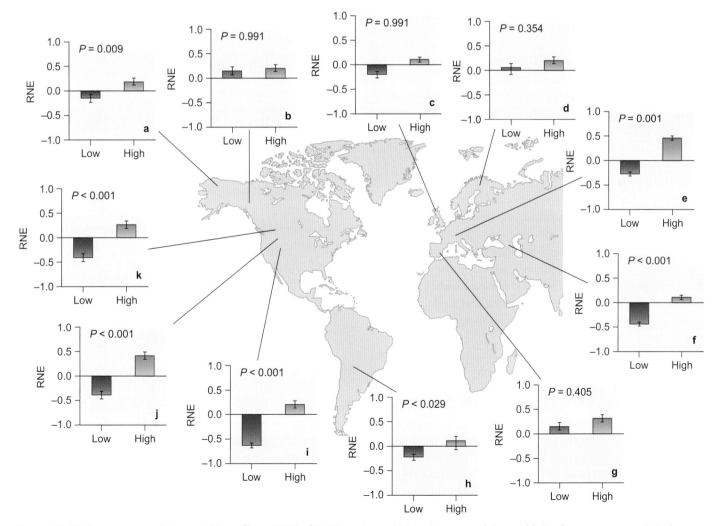

Figure 15.13 Comparative relative neighbor effects (RNE) of 3–10 species at 11 locations around the world. P-values compare statistical significance between low- and high-elevation sites at each location. A positive bar indicates facilitation, and a negative bar indicates competition.

These findings suggest that facilitation may be conditional – in this case dependent on abiotic factors. But facilitative interactions can be conditional on biotic factors as well.

Conditional facilitation

Brood parasites lay eggs in the nests of other birds. The adoptive host parents raise the chicks, provisioning them with food, warmth, and protection against predators. In most cases, there is substantial cost to the host species, as this parasitic interaction can lead to complete nest failure. So why discuss this study in a chapter on facilitation, shouldn't we have explored this interaction in the previous chapter on exploitation? As it turns out, some species benefit from brood parasitism, converting a seemingly parasitic interaction into a mutualism.

Daniela Canestrari and her colleagues (2014) explored the effect of a brood parasite, the great spotted cuckoo *Clamator glandarius*, on its host species, the carrion crow *Corvus corone corone*, in a series of observations and manipulation experiments. Based on 16 years of data, parasitized and non-parasitized nests had similar probabilities (about 75%) of surviving to hatching. Of nests that produced hatchlings, 76.4% of parasitized nests and only 53.8% of non-parasitized nests produced at least one fledgling crow (fledglings are young that survive to leave the nest, $P = 0.003$). However, among successful nests, non-parasitized nests produced a mean of 2.564 fledgling crows in comparison to a mean of only 2.073 fledgling crows in parasitized nests ($P = 0.008$). Combining these two findings, parasitized nests produced an average of about 0.2 more crow fledglings, but these differences were not statistically significant.

To further explore the effect of parasitism, Canestrari and her colleagues transferred one or two cuckoo hatchlings from parasitized crows' nests into non-parasitized crows' nests. Some parasitized crows' nest were left as unmanipulated parasitized controls. Nest success (the number of nests that produced at least one fledgling crow) was significantly greater in the two parasitized treatments than in either non-parasitized treatment (Figure 15.14). The researchers concluded that having a cuckoo young growing up in the nest somehow protected against nest failure.

Annual carrion crow nest failures range from 21 to 78%, usually from predation by a vast community of mammalian and avian predators. The researchers knew from personal experience that cuckoos voided a malodorous cloacal secretion when handled; perhaps that secretion was protecting nestlings of both species. Chemical analysis indicated that the secretion was a highly volatile mix of caustic and repulsive compounds. When they added this compound to cooked chicken, it was regularly rejected by feral cats that were otherwise delighted to eat chicken without this added secretion. Thus the researchers concluded that this toxic compound was the most likely mechanism for brood protection.

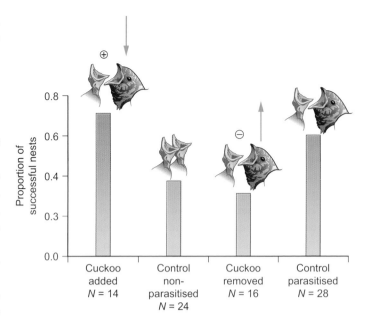

Figure 15.14 Proportion of successful nests (that fledged at least one crow) in relation to experimental treatments.

Recall that though nest success was greater in parasitized nests, overall reproductive success of parasitized and non-parasitized nests was similar because non-parasitized nests that escaped predation averaged about 0.5 more crow fledglings per nest. Thus the benefit of protection depends on the intensity of predation. In years with high predation rates, crows benefited by sharing a nest with cuckoos, but in years with low predation rates, the crows suffered a cost in reduced fledgling success (Figure 15.15). Above the zero line in Figure 15.15, carrion crows are receiving a reproductive benefit from sharing a nest with cuckoo young. Consequently the relationship is mutualistic in years with high predation and parasitic in years with low predation.

This example highlights that in the natural world, facilitation and other interactions may be very complex, involving direct and indirect interactions among several species. Interactions may be exploitative in one context and mutualistic in another. Unraveling these indirect effects requires a deep understanding of natural history. In some cases these indirect effects can occur between seemingly isolated ecosystems.

Indirect facilitative interactions across ecosystems

Seemingly, fish and terrestrial flowering plants would have little opportunity for interaction, except for the occasional fish whose misguided leap lands him on shore where his decaying body can provide nutrients for nearby terrestrial plants. But research conducted by Tiffany Knight and her colleagues (2005) provides evidence that fish in northern Florida lakes are able to indirectly increase the reproductive success of terrestrial plants growing near the shore without any misguided leaps. They hypothesize that fish eat dragonfly larvae, thereby reducing the abundance of dragonfly

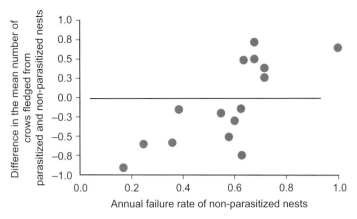

Figure 15.15 The difference in the mean number of crows fledged from parasitized and non-parasitized nests per year, in relation to predation pressure (the annual failure rate of non-parasitized nests).

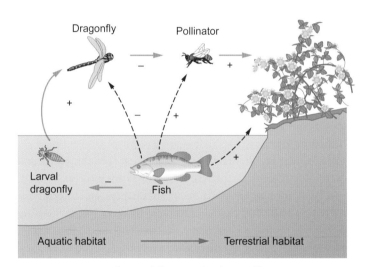

Figure 15.16 Hypothesized direct and indirect effects among species in this complex system. Solid arrows are direct effects, while dashed arrows are indirect effects.

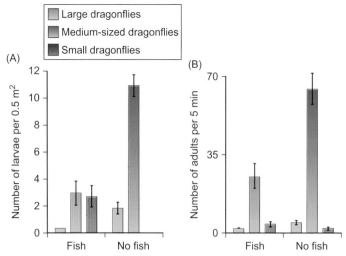

Figure 15.17 A. Mean abundance of larval dragonflies in four ponds with fish and four ponds without fish. B. Mean abundance of adult dragonflies near same eight ponds under same conditions as A.

As they predicted, both larval and adult dragonflies were much more common in and around ponds with no fish. In addition, ponds with no fish had more large and medium-sized dragonflies, and fewer small dragonflies (Figure 15.17).

 Thinking ecologically 15.3

Why do you think that there were more small dragonflies in ponds with fish? How would you test your hypothesis?

Effects of fish on plant pollination rates

To determine whether fish have a positive effect on plant pollination rates, the researchers chose a common plant, *Hypericum fasciculatum* (St. John's wort), because it is mostly pollinated by bees – a favorite dragonfly prey item. They watched each plant for 20 min, and counted the number of visits by bees, flies, and moths. They were able to document significantly higher visitation rates of all three types of pollinators to this plant species when it grew near ponds with fish in comparison to ponds without fish. The number of visitations by bees, in particular, was much greater near ponds with fish.

Dragonflies can reduce visitation rates by pollinators either by eating them, or by scaring them away (nonconsumptive effects). Unfortunately, predation by dragonflies is very abrupt and therefore unusual to observe, but nonetheless the researchers did observe several cases of dragonflies eating pollinators. To explore nonconsumptive effects, Knight and her colleagues surrounded *H. fasciculatum* plants with cages that had mesh large enough to admit most pollinators, but small enough to contain dragonflies (which are somewhat easily contained because they can't fold up their wings). They chose pairs of *H. fasciculatum* of similar size, placed a cage over both plants, and placed a large

adults. Large adult dragonflies reduce the abundance of pollinators by eating them, and also by scaring them away from plants. This is an example of a nonconsumptive effect of an interaction, which we discussed in Chapter 14. By reducing the numbers of large dragonflies, fish are indirectly increasing pollinator abundance, and subsequent plant reproductive success (Figure 15.16).

The researchers tested this hypothesis by breaking it into smaller pieces. Let's go through the steps in the process.

Effects of fish on dragonfly abundance

To address whether fish had a negative effect on dragonfly abundance, the researchers selected four ponds with fish and four ponds without fish as their study sites. These ponds had similar surface area, sunlight, and vegetation structure surrounding them. The researchers then surveyed the number of larval dragonflies of each species in each pond, and the number of adult dragonflies of each species flying along the shore of each pond.

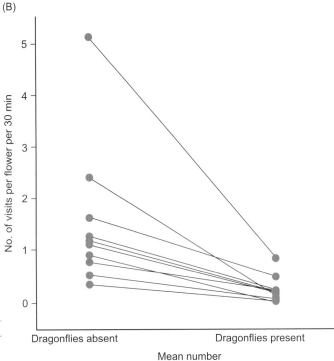

Figure 15.18. A. Dragonfly, *Erythemis simplicicollis*, eats a bee-fly pollinator (*Bombylius* species). B. Mean pollinator visitation rates over a 30 min time period in paired experiments with dragonflies absent or dragonflies present.

dragonfly into one cage. Pollinator visitation rates were much greater when there was no dragonfly in the cage (Figure 15.18).

Lastly, it doesn't really matter if fewer pollinators visit a plant, if the plant still has all of its flowers pollinated. So the biologically relevant question is whether *H. fasciculatum* is more pollen limited when it lives near ponds with no fish. At each pond, the researchers chose 10 plants, and carried out supplemental pollination on all flowers of one branch by rubbing each flower with anthers from another *H. fasciculatum* plant, and left a second branch as an unmanipulated control. They defined pollen limitation as the difference in the number of seeds produced by pollen-supplemented flowers and the number of seeds produced by the unsupplemented controls. The researchers discovered that pollen limitation was more than twice as great in *H. fasciculatum* near ponds with no fish in comparison to ponds with fish. Taken together, these findings

support the web of interactions that formed the hypothesis outlined in Figure 15.16.

This study highlights several important features of natural systems. First, many facilitative interactions are indirect. Second, as we will discuss in more detail in Chapter 16, interactions between species can cascade through food webs, influencing the success of species far removed from an initial interaction. Lastly, though ecologists discuss ecosystems as separate entities, they recognize that ecosystems interact with each other in important ways.

Along a similar vein, the impact of CCD highlights that interactions between species can cascade through food webs, and that ecosystems from different regions of the world interact with each other in important ways. Let's return to this mysterious case to see if researchers have identified any of the factors that are associated with this puzzling process.

REVIST: Colony collapse disorder

If researchers could identify what causes CCD they would be in a much stronger position to treat it. The problem is that the collapse usually happens very quickly, with little indication that the bees are about to die or abandon the hive. Thus it is challenging to get information on sick colonies, because the transition from "apparently healthy" to "sick and mostly abandoned" is so abrupt. Because CCD has such significant global implications, researchers have put forth considerable expense and effort to follow hives over extended time periods, so they have data on colonies right before they collapse. Their comparisons of colonies with CCD to colonies without CCD provide intriguing, but not compelling, indications of what underlies this disorder.

Diana Cox-Foster and her colleagues (2007) extracted, amplified, and sequenced ribosomal RNA from bees in CCD colonies and healthy colonies, which they compared to ribosomal RNA sequences of 18 different parasites and pathogens known to live within honeybees. Four of these pathogens were found more commonly in the 30 CCD colonies than in the 21 healthy colonies. One virus, Israeli acute paralysis virus, which in early stages causes wing-shivering, and which ultimately leads to paralysis and death, was found in 83% of CCD colonies and only 5% of healthy colonies. Other pathogens found significantly more frequently in CCD colonies were Kashmir bee virus, which also causes paralysis and death, and two species of *Nosema*, a microsporidian fungus that leads to dysentery-like symptoms.

Dennis vanEngelsdorp and his colleagues (2009) conducted a survey of 19 types of disease organisms known to afflict honeybees in 91 colonies located in 13 apiaries in Florida and California, USA. Bee specialists classified each apiary as either CCD or nonsymptomatic (controls). They also classified each colony within the apiary as either strong, weak, or dead. About 150 bees were removed from each surviving colony and screened for the presence and quantity of each disease organism. These bees were also scrutinized morphologically, to see if any physical characteristics were present that distinguished CCD colonies from controls. In addition, a sample of wax was taken from each colony to see if there were differences in pesticides or chemical used to treat some bee pathogens. Overall, the researchers considered over 200 variables in their comparisons.

The researchers found that CCD colonies tended to have a clumped dispersion – if one colony had CCD, it was likely that adjacent colonies would also be affected. This finding suggested to them that an infectious agent was causing CCD. In one analysis, the researchers compared parasite and pathogen prevalence in adults from colonies that were dying from CCD to adults from colonies that were classified as strong. There was a higher incidence of three parasites and pathogens out of 20 that were tested. Similar to the Cox-Foster (2007) study, Kashmir bee virus and the microsporidian fungus *Nosema* were more common in CCD colonies than in control colonies. These and other studies suggest that CCD does not arise from one single causative agent.

Ecologists remind us that part of the underlying problem is that modern agriculture is creating demands on ecosystems that are likely to disrupt important mutualisms. Huge monoculture plantations of almonds in California require the pollination services of approximately 1.6 million honey bee hives, which is about 60% of the commercial hives in the United States (USDA 2013; Figure 15.19)

These large monocultures are problematic in several ways. First, developing these plantations destroys native communities, so that native pollinators either decrease in abundance or go locally extinct. A recent study has shown that almond orchards that retain a native pollinator community have higher fruit set than orchards that depend exclusively on honeybees (Brittain *et al.* 2013). Second, there is good evidence that the stress associated with the long journey to far-away orchards makes the bees more susceptible to various pathogens. Lastly, many large plantations use herbicides and pesticides to help produce healthy crops. Many growers make efforts to avoid using chemicals at levels that have been shown to harm bees. However, recent studies have demonstrated that low, and previously presumed harmless, doses of several commonly used chemicals can have a strong negative impact on honeybee health and behavior. Neonicotinoid pesticides are commonly used to protect crops against aphids and other sap-sucking insects. In one study exposure to low doses of a widely used neonicotinoid increased honeybee susceptibility to *Nosema* infection (Pettis *et al.* 2012). A study on bumblebees, *Bombus terrestris*, showed that low doses of the same neonicotinoid resulted in much slower weight gain, and an 85% reduction in queen production (Whitehorn *et al.* 2012). Finally, low doses of a related neonicotinoid interfered with the ability of foragers to find their nests (Henry *et al.* 2012).

Figure 15.19 Vast almond orchard in the Sacramento Valley in California, USA.

The mutualism between honeybees and plants is complex, and its success is influenced by the presence or absence of numerous other species in the community. Next chapter we will begin to explore how one or a few species within a community can profoundly influence how a community functions.

SUMMARY

Facilitative interactions include mutualisms, in which both species benefit, and commensalisms, in which only one species benefits and the second species is unaffected by the interaction. The commercially important pollination mutualism between honeybees and plants is under assault by an emerging disease, CCD. Researchers have identified some potential causative agents, but are still unable to determine exactly what causes this disease. Mutualistic species play critical roles in biological communities, including coral and their algal symbionts that are the foundations of coral reef communities, and the mycorrhizal association between plant roots and their fungal symbionts that is essential for most plant communities.

A facilitative interaction can benefit species either directly, or indirectly by its effect on another species. Benefits include food, nutrients, pollination services, seed dispersal, access to habitat, and refuges from predators and aggressive actions against competitors. There is usually some cost to each mutualistic species; thus mutualism is most likely to evolve if the benefits exceed the costs, and if each species can ensure that its mutualistic partner provides the appropriate benefit. For example, legumes can ensure that their *Rhizobium* symbionts are providing enough fixed nitrogen, by reducing oxygen available to noncompliant symbionts. Facilitation may be more common in stressful environments, where the benefits of facilitation are greater than they might be in more benign environments. Some facilitative interactions, such as the interaction between the great spotted cuckoo and the carrion crow, are beneficial under some conditions and detrimental under other conditions.

FURTHER READING

Dixson, D. L., Abrego, D., and Hay, M. E. 2014. Chemically mediated behavior of recruiting corals and fishes: a tipping point that may limit reef recovery. *Science* 345: 892–897.

    Clear experimental design showing that coral juveniles are attracted to chemical cues exuded by healthy coral reefs and repelled by chemical cues exuded by sick or degraded coral reefs. This finding has important implications for applied ecologists attempting to restore sick or degraded coral reefs.

Hanna, C., Foote, D., and Kremen, C. 2014. Competitive impacts of an invasive nectar thief on plant-pollinator mutualisms. *Ecology* 95: 1622–1632.

    This study used a manipulative experiment to explore how an invasive nectar-thief wasp is disrupting plant-pollinator mutualisms in an endemic Hawaiian tree species. This disruption leads to a reduction in overall pollinator effectiveness and a reduction in tree fruit set.

Mascarelli, A. 2014. Designer reefs. *Nature* 508: 444–446.

    Good summary of researchers' efforts to artificially select for traits in coral that will make them more heat-resistant and able to survive increasing ocean temperatures associated with climate change.

Stachowicz, J. J., and Whitlatch, R. B. 2005. Multiple mutualists provide complementary benefits to their seaweed host. *Ecology* 86: 2418–2427.

    Nicely designed series of observations and experiments shows how a red alga receives complementary benefits from the presence of two different snail species, which, in turn, receive refuge from attacks by predatory crabs within the protective confines of the red alga.

Zhong, Z., and 5 others. 2014. Positive interactions between large herbivores and grasshoppers, and their consequences for grassland plant diversity. *Ecology* 95: 1055–1064.

    Indirect mutualism between two very different organisms is based on their feeding behavior. Grasshoppers consume a dominant grass, which makes a forb preferred by sheep more available, while sheep prefer the forb, which makes the grass more available to the grasshopper. There are also some interesting community-wide effects of this mutualism.

END-OF-CHAPTER QUESTIONS

Review questions

1. Distinguish between mutualism and commensalism, and give two examples of each.
2. Why are plants in species-rich communities more likely to have specialized pollinators? What is the evidence that supports this claim?
3. Why did certain tree species fail to become established on Isla Victoria? Support your answer with observational or experimental evidence.
4. What are five different types of services that are associated with facilitative interactions? Give an example of how each operates within an ecological community.

5. Sometimes the connection between two species involved in a facilitative interaction can be quite remote. Which study in this chapter best highlights the point that facilitative interactions can have strong indirect effects? Describe all of the interactions in your example.

Synthesis and application questions

1. Plant–pollinator systems are susceptible to coextinction – the loss of one species resulting from the loss of a species it depends on (Dunn *et al.* 2009). Do you think coextinction will be more of a problem for plant–pollinator systems that are specialized or for those that are generalized? How might you collect data to answer this question?

2. Can you think of other hypotheses besides thermal shuffling to explain the increase in thermal tolerance in coral?

3. The research on dispersal of Jeffrey pine seedlings by rodents indicates that chipmunks seem to do a good job of dispersing seeds to locations that are suitable for germination. Do you think this was favored by natural selection, or simply an accident of what's best for the chipmunk is best for the tree? How might you go about supporting your hypothesis?

4. Dan Janzen writes "Ledyard Stebbins commented to me when I defended my thesis that removing the ant colony from an acacia tree is like removing the nicotine from a tobacco plant." What does Stebbins mean by that? If this analogy is accurate, can you make any predictions about constitutive levels of defense compounds in acacia trees with ants, in comparison to related plant species that don't have a mutualism with ants?

5. In reference to the "devil's garden" study by Frederickson and her colleagues, what would be the predictions of the allelopathy hypothesis for the four experimental treatments established by the researchers?

6. Reconsider our discussion of the realized niche as the portion of the fundamental niche occupied by a species. We argued that the realized niche was a subset of the fundamental niche, because biotic interactions such as competition and predation restricted the distribution of a species to only a portion of its fundamental niche. How does our discussion of mutualism influence our understanding of a niche?

7. vanEngelsdorp and colleagues (2009) show that 55% of dying bee colonies are infected by three or more viruses, in comparison to 29% of strong colonies. They argue that multiple infections might weaken bee colonies, leading to a CCD event. Can you think of an alternative to this hypothesis for the correlation between multiple infection and CCD?

Analyse the data 1

Jones *et al.* (2008) report that 54 of 147 clade C2 coral colonies died from the 2005/6 thermal anomaly in comparison to 1 of 15 clade D coral colonies. Are these differences statistically significant? How do these findings influence your interpretation of how coral responds to thermal stress?

Chapter 16

Complex interactions and food webs

INTRODUCTION

In the year 2000, Angela Doroff and her colleagues flew over the fractured Aleutian Islands coastline in southwestern Alaska, repeating a procedure that had been followed in 1992 and 1965. Their mission was to conduct an accurate census of the sea otters, *Enhydra lutris*, that lived in the subtidal zone along those broken shorelines. Their findings were sobering: Over the 8-year interval between 1992 and 2000, sea otter populations had declined an average of 17% per year. What caused this sharp decline in sea otter abundance, and how did sea otter decline influence the rest of the ecological community?

We begin this chapter by introducing the species that interact with sea otters within their ecological community. We discuss how these interactions can be explored in different spatial scales from meters to thousands of kilometers. This leads us to consider how communities are structured or put together, and introduces the concept that some species are much more important than others in influencing community dynamics. The feeding interactions between species within a community can be summarized by a food web, which gives a useful but incomplete picture of community structure. Depending on biotic and abiotic conditions, consumers or producers may have a more important influence on how food webs are structured and how they function.

KEY QUESTIONS

16.1. How do ecologists explore community processes across scales of space and time?

16.2. How are ecological communities structured?

16.3. How do food webs describe community structure?

16.4. What factors influence community structure and functioning?

Sea otter declines were much more modest between 1965 and 1992 than they were between 1992 and 2000. Based on some mathematical extrapolations, Doroff and her colleagues (2003) estimated that the sharp decline did not begin until the mid-1980s. This is not the first time that sea otter populations have suffered declines; they were hunted to near extinction by fur traders in the eighteenth and nineteenth century, until the International Fur Seal Treaty protected the few surviving colonies in 1911. Subsequently, the sea otter population increased rapidly over much of its previous range, including the Aleutian Islands.

From our discussion of species interactions, you will recognize that a decline in large animal abundance in a community is likely to have major consequences for other members of the community. Sea otters are large endothermic homeotherms with a very high metabolic rate (see Chapter 5), so they consume substantial resources, about 20–25% of their body weight per day. Their preferred food in the Aleutian Island community is sea urchins, *Strongylocentrotus polyacanthus*, and fish. James Estes and his colleagues (1978) realized that the rapid changes in distribution and abundance of sea otters in the Aleutian waters provided a natural experiment to test hypotheses about the impact of these large predators on the structure of the community. They hypothesized that sea otters feed on sea urchins, reducing their numbers considerably. Sea urchins, in turn, feed on several species of kelp. Thus when sea otters are abundant, sea urchins are relatively rare. Under these conditions, kelp can thrive, growing to high densities in the subtidal offshore waters.

Kelp, sea urchins, and sea otters are part of the subtidal Aleutian Island food web. A food web describes the feeding relationships among organisms within a community. In this food web, numerous species of macroalgae, including kelp, form the first feeding or **trophic level**, which is made up of producers, organisms that produce their own energy-rich organic compounds by photosynthesis or chemosynthesis. Because sea urchins eat kelp, they are primary consumers (consumers of producers), and occupy the second trophic level. Sea otters, by virtue of eating sea urchins are secondary consumers and occupy the third trophic level. In almost all food webs there are many species at each trophic level, and in most cases, the distinction between trophic levels is very blurry. For example, barnacles are filter feeders that eat a variety of plankton including tiny

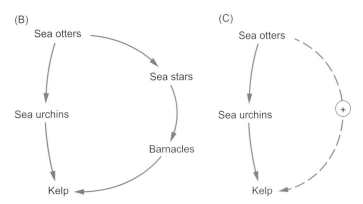

Figure 16.1 A subtidal food web. A. (Top) Sea otter munches on a sea urchin. (Bottom) Sea urchins forage in a kelp forest. B. A small portion of the offshore Aleutian Island food web. C. Trophic cascade hypothesis. Solid arrows indicate consumption. Dashed arrow with + sign indicates indirect facilitation.

pieces of broken-off kelp that float by. As such they are primary consumers. Sea stars eat barnacles, so they are considered secondary consumers. But sea otters are happy to eat sea stars, so from this perspective they are now tertiary consumers (consumers of secondary consumers), occupying the fourth trophic level by virtue of these linkages (Figure 16.1A, B).

According to the hypothesis proposed by Estes and his colleagues, the sea otter–sea urchin–kelp portion of the food web is so fundamental to the community that it creates a trophic cascade, in which the effects of consumption cascade down the food web from higher to lower trophic levels. Trophic cascades have two important attributes. First, the structure of the community is influenced primarily by the action of predators. Second, there are significant indirect effects at least two trophic levels removed from the action of the predator (Figure 16.1C).

Food webs and trophic cascades are conceptual models used by community ecologists to understand community structure, or how a community is put together and how the different species within the community interact with each other. Important elements of community structure include species composition and diversity, and the distribution and abundance of each species within the community. Ecologists recognize that community structure needs to be studied across diverse scales of space and time.

16.1 HOW DO ECOLOGISTS EXPLORE COMMUNITY PROCESSES ACROSS SCALES OF SPACE AND TIME?

Sometimes, one community process might occur over a small spatial scale, but a very different process might occur over a larger spatial scale. For example, if an ecologist obtained evidence that a trophic cascade was important at one particular site, it would be bad science to infer that this same trophic cascade was important at all sites that contained this same group of species. It would be equally unwarranted to conclude that trophic cascades, in general, were important parts of community structure across ecosystems. Thus researchers have investigated the sea otter–sea urchin–kelp trophic cascade across several different scales.

Small-scale studies of species abundances

In 1976, there were no sea otters in Torch Bay in southeast Alaska. The trophic cascade hypothesis predicts that the absence of sea otters would be associated with abundant sea urchins and very low kelp abundance, which was indeed the case. But a causal link needed to be demonstrated to increase confidence that a trophic cascade was responsible for this pattern. To test the linkages between sea otter predation, sea urchin herbivory, and kelp abundance, David Duggins (1980) functioned as a predacious sea otter by continuously removing sea urchins from three locations in Torch Bay, leaving two locations as unmanipulated controls. All five 50-m^2 quadrats in the study had no kelp at the beginning of the study. However, by the end of the first year, kelp density increased to an average of over 300 individuals/m^2 in the removal quadrats, while remaining at zero in the two controls. This provided further support that sea otter predation on sea urchins indirectly leads to abundant kelp populations.

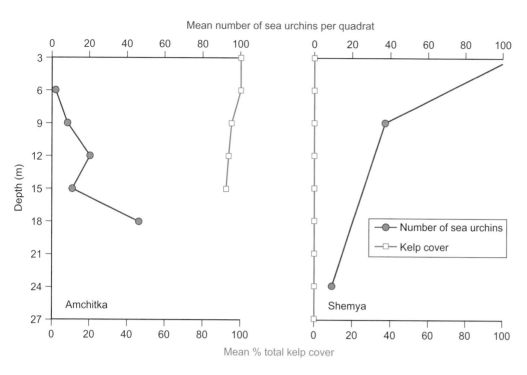

Mean number of sea urchins per quadrat

Figure 16.2 Mean percent total kelp cover (bottom x-axis) and mean sea urchin abundance (top x-axis) in relation to water depth at Amchitka Island, which has been recolonized by sea otters, and Shemya Island, which has no sea otters.

Expanding the spatial scale somewhat, James Estes and his colleagues (1978) conducted their research at two island chains in Alaska that had very different sea otter abundances. Amchitka Island had a large sea otter population, while Shemya Island had no sea otters. Otherwise these two islands were similar structurally and climatologically. The trophic cascade hypothesis predicts that Amchitka, with its large sea otter population, should have a relatively low sea urchin population and high kelp abundance. Conversely, Shemya should have a much higher sea urchin population and a much lower kelp population. The researchers counted, weighed, and measured all of the kelp and sea urchins from 0.25-m^2 quadrats established at varying depths off the shores of both islands. Their findings supported the trophic cascade hypothesis. Almost the entire subtidal surface of Amchitka was covered by a luxuriant kelp population, but urchins were relatively rare, particularly at shallow depths. In contrast, there was almost no kelp off the Shemya Island shore at any depth, and very high densities of sea urchins, particularly in shallow water (Figure 16.2).

Estes and his colleagues argue that sea urchins favor shallow water at Shemya because kelp grows best in shallow water where more light penetrates (see Chapter 5). Once they consume most of the kelp, sea urchins can still subsist on other less preferred producers, such as diatoms and various types of algae besides kelp, that also live in shallow water. However, at Amchitka the sea urchins are forced to forage in deeper water, where the sea otters are less likely to prey on them. Thus kelp cover approaches 100% to depths where there is sufficient light for photosynthesis.

Estes and David Duggins (1995) are quick to point out that ecologists must be careful about extending conclusions derived

from small-scale comparisons (for example, of two nearby islands) to larger geographical scales. Thus they expanded the scale of their research effort.

Large-scale studies of species abundances

Sea otters, sea urchins, and kelp are found in many locations along the western coast of North America. Estes and Duggins (1995) identified 23 research studies of this food web along the western North American coast. In almost all studies of sea otter predation, sea otters significantly reduced the abundance of sea urchins. In studies that investigated whether sea otter predation increased kelp abundance, the answer was usually that it did. The one geographical exception was off the central California coast, where some researchers argue that kelp abundance is influenced by many other factors in addition to sea otter predation on sea urchins.

To further address the issue of geographical scale, Estes and Duggins (1995) conducted an extensive comparison of the sea urchin/sea otter/kelp communities in southeastern Alaska to those they had already studied in the Aleutian Islands, including several additional Aleutian Island study sites. Most of these data were collected in 1987 and 1988, which also allowed a comparison with the earlier study (Estes *et al.* 1978). These two major study locations were separated by more than 1000 km. Within the sites, they sampled from a large number of habitats and collected specimens from many different quadrats over the course of 15 years. They observed most of the same patterns in both geographical locations. Where sea otters were long established, sea urchin abundance was greatly reduced, and kelp

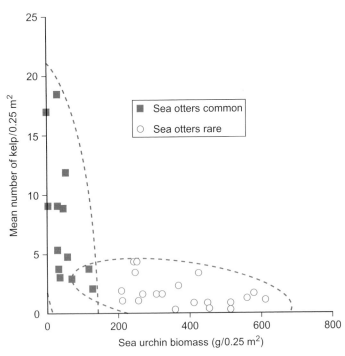

Figure 16.3 Relationship between sea urchin biomass and kelp density off islands where sea otters are common (more than 6/km² – closed squares) or rare (fewer than 6/km² – open circles).

cover was much higher than in nearby equivalent sites that only differed in having no sea otters. And, in general, sites where sea otters were recently introduced showed intermediate values.

This relationship between sea otter abundance, sea urchin biomass, and kelp density is consistent over relatively long time scales. More recently, Estes and his colleagues (2010) surveyed 463 sites off of 19 islands across the 1900 km extent of the Aleutian Island chain over a 20-year period. Islands with abundant sea otters had low sea urchin biomass and high, but somewhat variable kelp density. Islands with few or no sea otters had high sea urchin biomass and low kelp density (Figure 16.3). The researchers concluded that evidence for this particular trophic cascade is strong over relatively large scales of space and time.

The consistency of this trophic cascade is particularly useful for conservation ecologists who must address the issue of whether the diminishing populations of Aleutian sea otters should continued to be protected by the Endangered Species Act. The researchers argue that a great deal of information about sea otter abundance can be obtained by surveying the kelp populations across ecosystems. These data can supplement much more difficult to obtain data from direct sea otter observations to make informed decisions about the status of sea otter populations.

Returning to the airplane flying over the Aleutian Islands, what has happened to all of the Aleutian Island sea otters? James Estes and his colleagues (1998) propose that killer whales, *Orcinus orca*, are eating them and devastating their populations. Other researchers believe that there may be a different cause for their

decline. We will return to this discussion at the end of this chapter, when we consider the consequences of removing sea otters from the Aleutian subtidal community. But first we will explore how communities of organisms are structured by breaking them down into their basic units, and seeing how these units interact with each other.

16.2 HOW ARE ECOLOGICAL COMMUNITIES STRUCTURED?

On one level, species are the basic unit within a biological community. But a list of the species within a particular area does not prompt an appreciation of the interactions that occur within a biological community any more than a list of the naturally occurring elements generates an appreciation for the richness of chemical reactions. Furthermore, constructing a complete species list of a community is nearly impossible, because most communities have so many species. Thus community ecologists often lump together similar species when describing communities.

Grouping similar species

When we introduced this chapter with the Alaskan subtidal food web, we discussed kelp as if it were only one species. In fact, there are numerous species of kelp within this community. We combined the different species of kelp to simplify our description of the food web. In this particular case, all of the kelp species were taxonomically related within the order Laminariales, and functionally related as producers in the subtidal zone.

But in many cases closely related species may have different functions within a community. A **guild** is a group of species that depend on the same resource for survival and reproduction. A **functional group** is a group of species that performs the same function within the ecological community. Sometimes species are in both the same guild and functional group, but in other cases species from the same guild may occupy different functional groups, and vice versa.

As one example, consider the life of a typical dung beetle. An egg is laid within a ball of dung, a brood ball, located in a nest built by one or both parents. The nest provides protection to the larva while it consumes the dung generously provided by its parents (ultimately originating from a large herbivore or omnivore digestive tract). This benign and protective environment results in dung beetle larvae having much higher survival rates than most insects (Figure 16.4A). The larva forms a pupation chamber, metamorphoses into an adult, and usually emerges after seasonal rains soften the brood ball enough so it can escape the ball that has been both its home and its prison. It then emerges to find itself a mate, with whom it collaborates to create new brood balls and new babies.

This description ignores the astounding diversity of dung beetle life histories. There are over 7000 species of dung beetles,

Figure 16.4 Dung beetle development and behavior. A. Larva developing within cut-open brood ball. B. Roller dung beetle couple in action.

which by virtue of depending on the same resource – dung – for survival and reproduction, all belong to the same guild. However, researchers have identified anywhere from three to seven functional groups within this guild. Dwellers set up housekeeping and go through their entire developmental process within one unmodified dung pat. In contrast, tunnelers drag brood balls well below the soil surface, and females lay one egg per ball. After hatching, tunneler larvae feed on the dung, but they must then abandon the ball to find new dung as a food source as they continue to mature. Lastly, rollers cut off a portion of the dung pat and roll it to a new location where the male and female mate, and the female lays her egg in the ball, which is buried in a subterranean nest (Cambefort and Hanski 1991). In some roller species, parents defend the nest from predators for an extended period of development. These different functional groups have important effects on the ecological processes within the community. Both tunnelers and rollers harvest the dung in a way that reduces its accessibility to other members of the community such as flies, predatory beetles, numerous species of ants, and a

diverse array of parasites. Tunnelers and rollers play an important role in nutrient cycling by dragging the dung to deeper levels of the soil horizon. Rollers disperse the dung to new locations, and thereby spread nutrients throughout the ecosystem (Figure 16.4B).

The role of a guild or functional group is an important part of community structure. But sometimes one individual species may play a very critical role.

High-impact species

In this chapter, we have discussed how trophic interactions – i.e. different types of consumption – influence community structure. But as you might imagine, non-trophic interactions have an important impact on community structure as well. In some communities, some species may be so prevalent that they dominate the community both in biomass and in how they influence other species.

Dominant species

Dominant species, also called *foundation species*, influence community structure primarily by virtue of their great abundance or biomass. Depending on geographical regions, spruce trees, genus *Picea*, will form near monocultures over vast swaths of the taiga biome. These trees dominate the community with their great biomass, but also because they have a tremendous influence as the major sources of production, providing food for seed-eating mammals and birds, browse for herbivores, and a variety of food opportunities for arthropods. In addition these trees offer a relatively benign microclimate, which many species of all types use for shelter. Finally, as they drop needles and branches and ultimately die, these trees form the basis for a community of decomposers.

Keystone species

In contrast to dominant species, keystone species have a much greater impact on the community than would be expected by measuring their abundance or biomass. In the architectural world, a keystone is the structural piece at the crown of an arch, which locks the other pieces into position. Thus a keystone species has a central function in maintaining the structural integrity of a community. This term was coined by Robert Paine (1969) to describe the role of the sea star, *Pisaster ochraceus*, in Mukkaw Bay off the northern Washington State coast.

A CLASSIC STUDY OF KEYSTONE SPECIES. Paine studied three different food webs in the marine rocky intertidal zone, but we will only explore one of those. As we discussed in Chapter 13, living space is a limiting resource for many species in these communities. *Pisaster* occupies the top trophic level in the Mukkaw Bay food web, while the whelk, *Nucella*, is one trophic level down, because it too is occasionally eaten by

Pisaster. Numerically, acorn barnacles are the primary foods for both carnivores, but *Pisaster* gets most of its nutrition from eating the much larger bivalves and chitons (Figure 16.5). The producers of this system are four species of algae, which are primarily eaten by the chitons and limpets. One species of sponge, *Haliclona*, feeds on algae and suspended particles and may be eaten by the nudibranch, *Anisodoris*.

To identify the role of *Pisaster* in this food web, Paine removed *Pisaster* from an 8-m stretch of shoreline and compared its community composition over a 1-year period to that of a similar stretch of unmanipulated shoreline. By repeatedly visiting the area, he was able to effectively exclude *Pisaster* from this experimental community. The results were striking. After *Pisaster* removal, juvenile *Balanus* acorn barnacles moved in and occupied about 70% of the available space. Several months later, these acorn barnacles were crowded out by competitively superior *Mytilus* mussels and *Mitella* gooseneck barnacles. Most of the

algae disappeared (presumably being poorer competitors for space), as did their herbivores – the chitons and limpets. The sponges and their nudibranch predators also were displaced by the increasing numbers of mussels and gooseneck barnacles. After one year, the 17-species community was reduced to eight species (Paine 1966).

Paine concluded that *Pisaster* was a keystone predator in this system, removing competitively superior species from the community and enabling other poorer competitors to become established. Thus keystone species may indirectly influence other species through non-trophic interactions. Many researchers suggest that we need to use an interactions web model to understand community structure, which includes not only feeding relations among species from different trophic levels, but also competitive and mutualistic interactions that may occur among species that occupy the same trophic level (Figure 16.6).

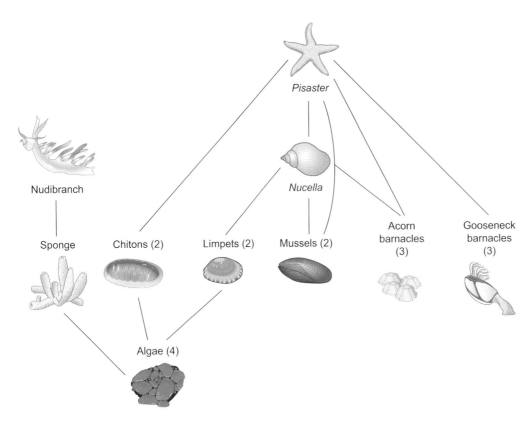

Figure 16.5 The food web at Mukkaw Bay. The number in parentheses indicates the number of species Paine identified in each taxonomic group that included more than one species.

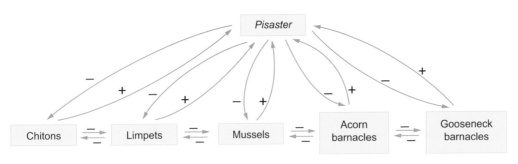

Figure 16.6 Greatly simplified interactions web based on Paine's food web indicating positive and negative interactions among some of the species. The negative interactions are occurring among all five taxonomic groups at the bottom of the interactions web.

But not all intertidal zone communities along the Washington coast are structured the same way. *Pisaster* is a keystone species under some conditions, yet relatively unimportant under other conditions.

CONTEXT DEPENDENCY OF KEYSTONE SPECIES. Bruce Menge and his colleagues (1994) suspected that *Pisaster*'s status within a community might vary in relation to habitat. For example, if *Pisaster* was very uncommon in its natural habitat, it could not eat enough mussels and gooseneck barnacles to allow competitively inferior species to move into the habitat. The researchers knew from reviewing the existing literature that *Pisaster* tended to be much more common in habitats that were exposed to strong wave action. So they established two research locations along the Oregon coast that had local sites that were exposed to or protected from wave action. Initial surveys of multiple quadrats established that *Pisaster* was more abundant at the Strawberry Hill location than it was at Boiler Bay. But at both locations, *Pisaster* was much more abundant at exposed sites than it was at protected sites.

Menge's team tested to see whether *Pisaster* could function as a keystone predator even at protected sites where it is relatively uncommon. As Robert Paine had done three decades previously, they continuously removed *Pisaster* from half of their study plots, and compared mussel survival and recruitment at plots with and without *Pisaster* (Figure 16.7A). They worked primarily in the low intertidal zone, where *Pisaster* tends to be most abundant. Their observations tested two predictions. First, if *Pisaster* is a keystone predator at both exposed and protected sites, the scientists predicted that mussel density would be greater at all *Pisaster*-removal sites. But if *Pisaster* is a keystone predator only at exposed sites, then removing *Pisaster* should cause mussel density to be higher at exposed sites than at protected sites. They discovered that after 15 months, removing *Pisaster* had no effect at the protected sites where *Pisaster* was rare. Somewhat surprisingly, removing *Pisaster* also had no effect at exposed sites at Boiler Bay but had a strong effect at exposed sites at Strawberry Hill; after 15 months mussel cover averaged about 40% without *Pisaster*, and less than 10% with *Pisaster* (Figure 16.7B).

The researchers also tested a second prediction. They reasoned that if *Pisaster* is indeed a keystone species at these locations, then removing *Pisaster* from the lower intertidal should induce mussels to expand from the middle intertidal where they are more common, into the lower intertidal where they are very rare. In support of this prediction, after 3 years of continuously removing *Pisaster*, the lower edge of the mussel bed dropped substantially at all four exposed sites (mean vertical drop of 0.47 and 0.63 m at Strawberry Hill, and 0.30 and 0.40 m at Boiler Bay) while there was no measurable change at the four protected sites. In addition, where *Pisaster* was not removed, there was no change in the lower edge of the mussel bed in any of the exposed or protected sites.

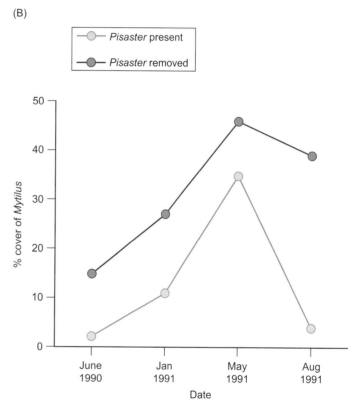

Dependent variable: % cover of *Mytilus*

Figure 16.7 A. Experimental design of Menge and his colleagues. B. Percent cover of *Mytilus* in response to removing *Pisaster* from the exposed Strawberry Hill sites. *Pisaster* feed very little in cold water, so *Pisaster* removal would have little effect until water temperatures increase in late spring.

The researchers were concerned that they had no idea how quickly to expect a community response to *Pisaster* removal, so they used a different approach to test for the intensity of *Pisaster* predation in exposed and protected sites. They transplanted 20 × 20-cm clumps of mussels to the lower intertidal zone, below the natural mussel zone. *Pisaster* is normally found in these regions, so as before, Menge's team removed *Pisaster* from half of the sites, and allowed them to remain in the other

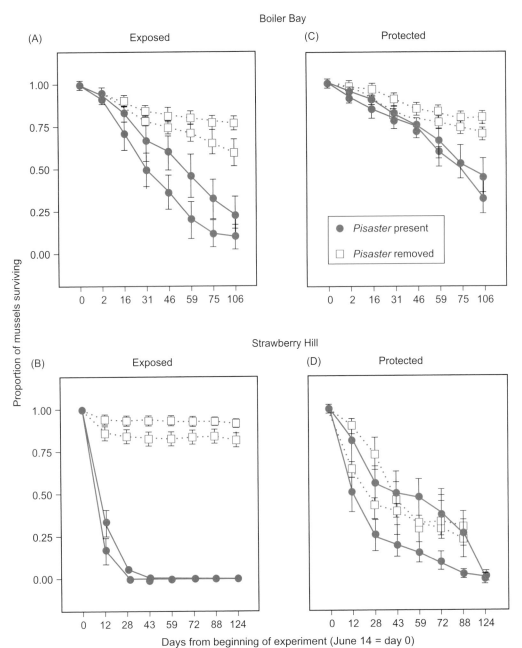

Figure 16.8 *Pisaster* removal experiment. Mean proportion survival ($\pm$SE) of mussels transplanted into the lower intertidal zone in exposed (Figures A and B) and protected (Figures C and D) sites at Boiler Bay and Strawberry Hill. At each location, two sites had *Pisaster* removed, and two sites had normal levels of *Pisaster* present.

half of the sites. Two replicates of each treatment were conducted at both exposed and protected sites at both locations. If *Pisaster* is a keystone predator, the researchers expected much higher mortality of these transplanted clumps in the *Pisaster* sites.

In exposed sites at Boiler Bay and Strawberry Hill, *Mytilus* survival was very low if *Pisaster* was allowed to remain. But when *Piaster* was removed from these exposed sites, *Mytilus* survival increased sharply (Figure 16.8A and B). At the protected locations, which had naturally lower *Pisaster* abundance, *Mytilus* survival with *Pisaster* present was somewhat greater than it was at the exposed sites. When *Pisaster* was removed

from protected sites at Boiler Bay, *Mytilus* survival showed a moderate increase (Figure 16.8C) Interestingly, the protected sites at Strawberry Hill showed very little effect of *Pisaster* predation, with *Mytilus* mortality relatively high whether or not *Pisaster* was present (Figure 16.8D). At this site mussels were killed when wave action buried them in a thick layer of sand for a prolonged period of time. Taken together, this extensive series of observations and experiments supports the hypothesis that *Pisaster* is a keystone predator at exposed sites off the Oregon coast, but not at protected sites, where other factors, including abiotic factors such as sand burial, are important in structuring communities.

Inspired by Paine's work, researchers have questioned whether keystone species are common in communities, or if they are somewhat unusual. The jury is not yet in for this debate, but researchers have come together informally, and at formal conferences and workshops, to sharpen the definition of a keystone species, and to explain why keystone species are important for conservation ecology.

RE-EXAMINING KEYSTONE SPECIES. In 1995, a group of ecologists met at a research conference – the Keystone Workshop – in Hilo, Hawaii. One important conclusion of this workshop was a working definition of a keystone species as "one whose effect is large, and disproportionately large relative to its abundance" (Power *et al*. 1996). This definition is critical, because it alerts researchers that they must identify ways of measuring the effect of a species on its community. The workshop participants then defined a measure, *community importance*, which is the change in the community trait per unit change in the abundance of the proposed keystone species. Examples of community traits might be species richness, productivity of the producers in the community, or the diversity or abundance of functional groups. The *total impact of a species* is its community importance multiplied by its proportional abundance or biomass in the community. We can use this framework to graphically visualize keystone species. Species whose total impact is proportional to their biomass would fall along the line in Figure 16.9. Keystone species must be significantly to the left of the line of proportionality, and must exceed a certain threshold of total impact on the community.

Thinking ecologically 16.1

As we will discuss shortly, viruses such as rinderpest, which affects buffalo and wildebeest populations in the Serengeti (see Chapter 1), can be considered keystone species. In what ways would it be challenging to fit rinderpest into the framework created by the Keystone Workshop? How would you deal with these challenges?

The Keystone Workshop group also articulated why conservation ecologists care about keystone species. First, the preservation of an endangered species may depend on the presence of a keystone species even though there may be no direct interaction between the keystone species and the endangered species. Many keystone effects are indirect. Second, and related, is that the loss of a keystone species, particularly a top carnivore, may cascade through food webs, affecting seemingly unrelated guilds and functional groups.

The food web model continues to be very useful for understanding community structure. Equally important, it also provides a framework for understanding ecosystem-level processes such as energy flow and nutrient flow. But our typically simple model of positive and negative effects can produce a distortion of the complexities of community structure.

16.3 HOW DO FOOD WEBS DESCRIBE COMMUNITY STRUCTURE?

There is no hard and fast rule about what type of species grouping should be represented in food webs. Most food webs will use a combination of higher taxonomic groupings (for example, sea urchins or kelp each treated as one group) or guilds or functional groups, as a way of simplifying their interactions and reducing the number of links. Researchers make these decisions based on the types of feeding relations they are exploring. Even in very simple communities, each species is usually not represented, in part because researchers usually bias their observations in the direction of species they can actually see.

Some species don't fit neatly into the food web trophic structure

Parasites are probably the most overlooked group in food web analyses. But excluding parasites from food webs can severely hamper our understanding of food web dynamics. Recall from Chapter 1 the simple food web that links lions and hyenas as secondary consumers of buffalo, wildebeest, and zebras, all of which are primary consumers of many grass species. Until Tony Sinclair's research included the rinderpest virus in the food web, researchers could not understand why buffalo and wildebeest

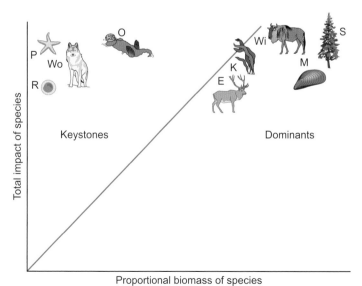

Figure 16.9 Keystone species have a high total impact on community structure despite relatively low abundance. Dominant species have a high total impact on community structure and high overall abundance. Keystones: rinderpest (R), *Pisaster* (P), grey wolves (Wo), and sea otters (O). Dominants: spruce tree (S), *Mytilus* (M), wildebeest (Wi), kelp (K), and elk (E).

(A)   (B)

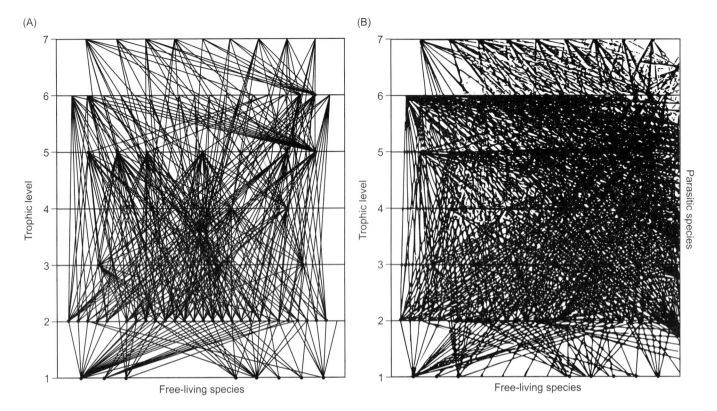

Figure 16.10 A. The relatively simple (but still overwhelmingly detailed) predator–prey web at Carpinteria. B. The complete web including parasites (note that parasites enter from the side).

populations increased sharply in the 1960s and zebra populations remained stable.

While the rinderpest virus was important in limiting wildebeest and buffalo populations in East Africa, there are many other parasites that also influence the food web. For example, my most recent journey to the Global Mammal Parasite Database disclosed that wildebeest harbor 53 parasite species, which are taxonomically broken down into 20 arthropod, 2 bacteria, 14 helminth, 2 protozoa, and 15 viral species (Nunn and Altizer 2005). If we were to include all of these species, and also include all of the parasites of lions, buffalo, and zebras, and all of the producers they consume, we would have an unwieldy and unhelpful mess. In addition, parasites are very problematic because many have complex life cycles with different developmental stages parasitizing different host species.

This complexity issue is highlighted by a study by Kevin Lafferty and his colleagues (2006b) on the 93-ha salt marsh food web near the city of Carpinteria, along the California coastline. When all of the nonparasitic species are included in the food web, we have a boggling amount of detail, including 12 first-trophic-level groups, which form a basis for 29 invertebrate, 8 fish, and 38 mammal and bird species (Figure 16.10A). But adding the 47 parasite species increases apparent complexity almost beyond comprehension (Figure 16.10B). Even this web includes many simplifications. For example, all macroalgae in this food web (including green algae and red algae) are grouped together,

even though they belong to at least three genera, and numerous distinct species. Ecologists use many approaches to deal with this amount of detail; we will briefly discuss a few of these approaches.

Simplifying and quantifying food webs

One way of dealing with food web complexity is to break a complete food web down into several subwebs. Lafferty and his colleagues broke the Carpinteria food web down into four subwebs: (1) the standard predator–prey web (Figure 16.10A), (2) a parasite–host subweb, (3) a subweb that included predators that feed on parasites, and (4) a subweb that included parasites of parasites. By breaking the full web into components, the researchers were able to gain a clearer picture of how each class of interaction was contributing to the complete web of interactions.

Ecologists have tools they can use to describe food webs quantitatively. They can then use these quantitative descriptions to compare the structure of different communities.

Measuring webbiness

At a glance, the two very simple food webs in Figure 16.11 differ in many respects. Web A has more species, but fewer trophic levels. But we also intuitively feel that Web B is webbier, that the species are more tightly linked. Biologically this means that

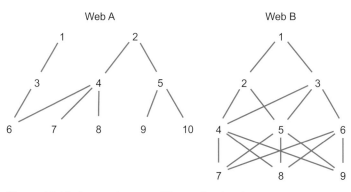

Figure 16.11 Comparing two different food webs.

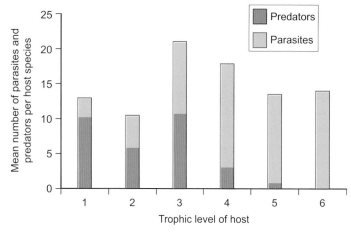

Figure 16.12 Mean number of parasites and predators affecting different trophic levels of the Carpinteria food web.

the average species in Web B is more of a generalist feeder than the average species in Web A.

Ecologists use three very simple measures of webbiness. *Linkage density* is the number of links divided by the number of species. In effect it measures the average number of feeding links per species. Web A in Figure 16.11 has a linkage density = 9/10 = 0.9. *Directed connectance* is the number of links in a food web divided by the maximum possible number of links. If we assume that species can commit cannibalism, and that each species can eat every other species, the maximum possible number of links is equal to S^2, where S = the number of species in the web. Thus directed connectance = #links/S^2. Web A has a directed connectance of 9/100 = 0.09. Lastly, *connectivity* is the number of links in a food web divided by the maximum possible number of links, assuming that links are only in one direction, and that species don't commit cannibalism. The formula for connectivity is $C = 2(\#links)/[S(S - 1)]$. Thus Web A has a connectivity of $2(9)/10(9) = 18/90 = 0.20$. Ecologists may multiply directed connectance and connectivity by 100 to generate a value that is easier to work with.

 Thinking ecologically 16.2

What are the linkage density, directed connectance, and connectivity of Web B in Figure 16.11? Do these numbers fit your mental picture of which web is webbier?

Let's return to the Carpinteria salt marsh to see how adding parasites to food webs changes our understanding of how parasites influence community structure. Some findings will be obvious. For example, adding parasites increases the number of species from 87 to 134, and increases the number of links from 505 to 2313. We can use these numbers to see what happens to the overall webbiness. At Carpinteria, linkage density increases from 5.8 to 17.3, while directed connectance increases from 6.7 to 12.9. Our intuitive picture drawn from looking at Figure 16.10 is confirmed; adding parasites makes food webs considerably webbier. (Lafferty *et al.* 2006a).

But where are these new links on the food web? This is an important question that helps us understand community

structure. The location of these links tells us which trophic levels are most vulnerable to parasites. The researchers found that a relatively small number of first-trophic-level species were afflicted by parasites. But parasite infection tended to increase with each trophic level. In contrast, the number of predators per prey species was at a maximum at intermediate trophic levels (Figure 16.12).

Based on several studies showing that parasites have important influences on community structure, ecologists are beginning to include parasite–host interactions in food webs, even though they add complexity. As with all biological models, complexity increases biological realism but decreases the model's utility. One approach to simplifying food webs is to recognize that not all links are created equal – some are much stronger than others.

Strong versus weak species interactions

Returning briefly to the sea otter–sea urchin–kelp trophic cascade, sea otters have a strong effect on sea urchins, and sea urchins have a strong effect on kelp. There are many ways to measure interaction strength – the strength of the link between two species in a food web. One relatively easy, though somewhat unpleasant, approach is to open up a predator's stomach and calculate the percent representation of each prey species it has consumed. If an individual predator has a relatively large amount of particular prey species in its stomach, then we can argue that the two species make up a strong predator–prey interaction. An advantage of this approach is that no experimental design is required; animals are simply captured and then analysed. However, the community-wide effects of this interaction will depend on the abundance of both species in the community, which this approach can't determine; if there are very few predators or prey, then the overall effect will be small.

A better but more time-consuming approach for studying interaction strength is to exclude a predator from a community, and to observe if there are significant changes in prey abundance following predator exclusion. Robert Paine (1992) used this approach to study the interaction strength between seven different grazers and several species of macroalgae at Tatoosh Island along the Washington coast. Paine observed the community under natural conditions to measure the natural density of each species of grazer. He then created enclosures to measure the density of brown algal sporelings (juveniles) with and without natural densities of grazers over a 7.5-month period. Per capita interaction strength was simply $(P - A)/A$, where P represents sporeling density in the presence of grazers, and A represents sporeling density in the absence of grazers. Maximum per capita interaction strength is -1.0. A value close to zero indicates no

significant interaction strength, while a value above zero indicates a mutualistic interaction.

Interaction strength values for the Tatoosh Island predator removal experiment are summarized in Table 16.1. Both the sea urchin and the chiton, *Katharina tunicata*, had relatively strong negative per capita interactions with brown algae. The overall interaction strength within the community is the per capita interaction strength multiplied by the consumer's natural abundance in the community. Because they had the highest natural abundance, both the sea urchin and *K. tunicata* have very strong overall community effects, while other grazers have much weaker effects on community structure. Two of the weakly interacting species, the chiton *Tonicilla lineata* in the first series of experiments, and the limpet *Acmaea mitra*, actually had slightly positive effects on sporeling density. Paine explains that these species eat coralline algae, which compete effectively with the brown algae for space. By reducing coralline algae, these two grazer species are actually increasing the abundance of brown algae.

There is no magic value of interaction strength that is considered a strong interaction. From the standpoint of understanding food webs, we should probably focus our efforts on identifying which interactions are strongest among the hundreds or thousands of links in most webs. This is particularly challenging given the many indirect effects that are difficult to identify and quantify.

Ecologists have developed many quantitative descriptions of food webs that allow us to compare various aspects of different webs. Let's see if we can use the few measures we have discussed to visualize similarities and differences between different webs.

Comparing food webs and community structure

Most work on food webs has looked at either terrestrial communities, or shallow aquatic and marine communities. One reason for this focus is that it is much easier to collect data in these environments than in the deep ocean. The well-named Jason Link (2002) presented findings that compare the food web of the marine shelf ecosystem in the northeastern United States to food webs characteristic of these better-studied terrestrial or shallow aquatic and marine communities.

Link used data on fish stomach contents from part of a survey by the National Marine Fisheries Service. In most cases he, along with other researchers, evaluated the stomach contents of at least 500 individuals per species to reconstruct the diet of these fish. Species eaten that were in the lowest trophic level were impossible to distinguish, so Link used the broad functional group, *phytoplankton*, as the only producer, and *detritus*, the remains of dead organisms, as the other group in the lowest trophic level (Figure 16.13). This aggregation reduced the number of links connected to the first trophic level. Other species were also aggregated into mostly taxonomic groups – for example, one group is toothed whales and porpoises, while another group is

Table 16.1 Effects of grazer removal on brown algae sporeling density.

Taxon/species	Natural density (number/m²)	Per capita interaction strength	Overall interaction effect on sporeling density[a]
Sea urchin/ *Strongylocentrotus purpuratus*	25.4	−0.031	−78.7
Chiton/*Katharina tunicata*	28.0	−0.025	−70.0
Chiton/*Mopalia hindsii*	1.9	−0.003	−0.6
Chiton/*Tonicella lineata* (first series)[b]	12.3	+0.005	+6.2
Chiton/*Tonicella lineata* (second series)	12.3	−0.002	−2.5
Limpet/*Acmaea mitra*	4.3	+0.017	+7.3
Limpet/*Lottia painei*	3.2	−0.001	−0.3
Limpet/*Lottia ochraceus*	0.1	−0.007	−0.1

[a] Overall interaction effect = natural density × per capita interaction strength.

[b] Paine (1992) reports two sets of data for *T. lineata* that were conducted at different densities.

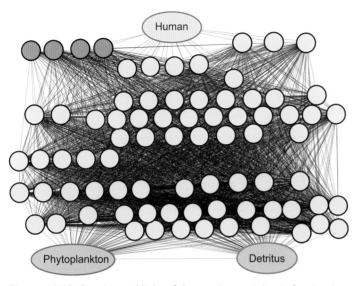

Figure 16.13 Species and links of the northwest Atlantic food web.

food webs. Intuitively, there is some limit to the number of species that an animal can feed on, so a larger web should have fewer links per species, and hence lower connectivity. When Link examined this pattern, he discovered that the northwest Atlantic food web had dramatically greater connectivity than other food webs of similar size (Figure 16.14). He argued that marine food webs may be fundamentally different from food webs of other biomes, and he encouraged other researchers to investigate whether other marine food webs also have unusually high linkage density and connectivity.

Another measure ecologists use to describe communities is *stability*, which is the tendency of a community to retain its current structure and functioning even when it is disturbed by biotic or abiotic forces. Community ecologists argue about whether stability increases or decreases with the number of species in a group, and with the number of links and the strength of the interactions. Some ecologists propose that a species-rich community with numerous weak connections, like the northeast United States marine shelf, should be very stable, while other researchers argue that such a community is inherently unstable. Given the number of perturbations faced by marine communities, in part from the actions of the species on the top of Link's food web, this question is of paramount importance. We will explore this question in Chapter 17 when we discuss biological diversity.

Predation and parasitism are obviously important factors influencing community structure. But producers may play an equal or even greater role than consumers in structuring ecological communities.

birds. In total, the web comprised 81 groups: 33 invertebrate and 46 vertebrate groups, as well as the two first-trophic-level groups.

Despite these aggregations, and the absence of any parasites, you can see that this is a very webby food web, and that there appear to be a large number of groups in intermediate trophic levels. Link compared the linkage density of this web to 14 other published webs, and discovered that this web has by far the highest linkage density. It also has a high percentage of intermediate trophic level groups, though there are other webs with an equivalent percentage of intermediate trophic level groups.

Several researchers have noted that connectivity is generally much lower for species-rich food webs than it is for species-poor

16.4 WHAT FACTORS INFLUENCE COMMUNITY STRUCTURE AND FUNCTIONING?

Let's consider a simple community with three trophic levels: producers, grazers, and predators. The trophic relationships are

Figure 16.14 Connectivity in relation to number of species in a food web. Each mark represents one food web described in the scientific literature.

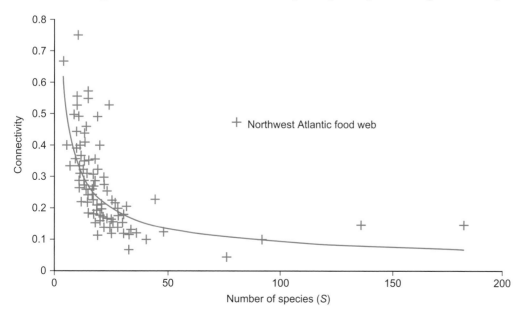

(A)

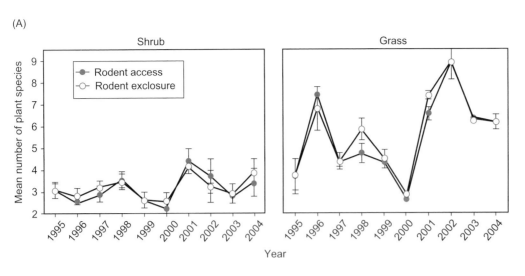

Figure 16.15 New Mexico desert community structure. A. Mean plant species richness in relation to rodent access over a 10-year period. B. Mean rodent density in relation to previous summer precipitation in 1989–2004 in shrub (top) and grassland (bottom) communities.

(B)

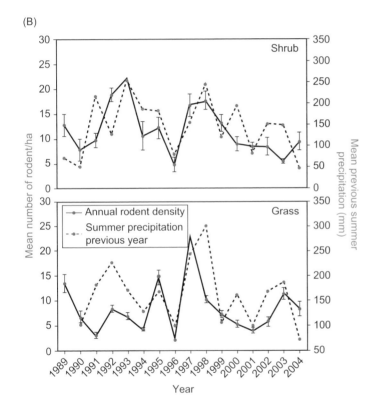

that predators eat grazers and grazers eat producers. We now know that all communities have structure, which includes species composition, richness, and abundance of populations of different species, the presence of dominant and keystone species, guilds and functional groups, and the interactions that link all of these species and groups. The question we are asking is, within this simple three-trophic-level community, which level has the greatest influence on community structure? The bottom-up control hypothesis argues that producers are primarily responsible for community structure. The top-down control hypothesis argues that predators are primarily responsible for community structure. We will begin by considering a simple

two-trophic-level community that shows almost exclusive bottom-up control.

Bottom-up control of community structure in low-production communities

Selene Baez and her colleagues (2006) were interested in whether a desert grassland community in New Mexico, USA, was primarily influenced by bottom-up or top-down effects. In arid habitats, moisture levels usually limit plant production, so the researchers kept track of rainfall over the course of this 10-year study. To investigate how consumption by rodents influenced this community, the researchers set up permanent 36 × 36-m exclosures without rodent access, with nearby 36 × 36-m control plots that allowed rodent access. There were four exclosure–control pairs in the grassland, and another four exclosure–control pairs in shrub habitat.

If there were significant top-down effects, the researchers expected that plots without rodents should have higher production, which they quantified as percent plant cover, than would control plots. They also expected to see differences in species richness between exclosures and controls, and differences in heterogeneity – a measure of how different the species composition was in each sample. Alternatively, if there were significant bottom-up effects, the researchers expected that percent cover, species richness, and heterogeneity would primarily be influenced by rainfall, and that rodents would have little effect on these three measures of community structure.

After 10 years, the researchers saw no effects of rodents on community structure (Figure 16.15A). In contrast, rainfall had strong effects on both the plants and consumers in these communities. For example, high winter rainfall increased cover and species richness in the shrub habitat. And high summer rainfall increased cover, species richness, and heterogeneity in the grassland. High previous-year rainfall also was associated with increased rodent densities in both habitats (Figure 16.15B).

Baez and her colleagues conclude that in this system there is very little evidence of consumer regulation of community structure. Rather, bottom-up effects cascade upward through the community, so that plants respond to recent pulses of precipitation, and rodents respond to increased plant production a year after the precipitation pulse.

The researchers considered whether this case of bottom-up control is unusual, or is a general finding for grassland communities. They pointed out that their study site has unusually low annual rainfall and primary production for a grassland community. Thus rodents are not abundant enough to exert major top-down effects on the plant community. Baez and her colleagues cited studies conducted on higher-productivity grasslands, which boast higher rodent density and show considerable evidence of top-down effects.

Other factors besides productivity may influence whether a community is controlled by bottom-up effects. For example, interactions between producers, or between producers and consumers, can influence soil nutrient levels, which can then have positive cascading effects up the food web.

Interactions between species and soil nutrient levels

Let's leave North America and head back to Africa, which is suffering from numerous environmental crises. In some regions elephant populations are increasing rapidly. Elephants are voracious feeders that can reduce tree abundance over a large geographical scale. Other factors, such as fires – some set by humans, and harvesting of trees for charcoal production, all are acting together to reduce tree abundance in many African savannas.

But trees are not the major food item for most savanna animals – grass is. The grass feeds the large herds of wildebeest and zebras, which migrate into Tarangire National Park in northern Tanzania during the dry season. Over the past 20 years, the elephant population within the park has increased sharply as a result of habitat destruction outside the park, forcing more elephants into its friendly confines. Fulco Ludwig and his colleagues (2008) wanted to know whether destruction of the dominant *Acacia tortilis* trees by these elephants was influencing the savanna food web by changing the soil nutrient levels. They hypothesized that reduction in nutrient levels would reduce grass quality, and possibly have serious consequences for the large herbivores that eat that grass.

Previous work identified three herbaceous vegetation zones in association with large acacia trees: (1) under the canopy, (2) just outside the canopy projection (which the researchers call "around the canopy"), and (3) open grassland more than 50 m from any tree. Researchers collected grass samples from each of these zones during peak wildebeest migration season and analysed nutrient levels, calorie content, protein content, fiber content, and digestibility of

organic matter. They found substantially higher levels of many important nutrients and lower levels of fiber in grass growing under trees and around the canopy in comparison to grass from open grassland.

 Thinking ecologically 16.3

The researchers did not discuss in their paper why grasses under trees were more nutritious. Help them out by suggesting several reasons why nutrient levels might be higher under trees. How might you set up experiments or observations to test your hypotheses?

The researchers wanted to know whether the decline in tree abundance would have serious implications for wildebeest survival. They knew that wildebeests had minimum daily requirements of nutrients and calories. If they did not take in enough of both, they would lose weight and die of starvation. You might reasonably ask, why can't they just eat more low-quality grass to compensate for the lower nutritional value of grass from the open grassland? The problem is a forage intake constraint, or limitation, that is determined by how easily the food is digested, and how long it takes to pass through the animal's rumen. The forage intake constraint is influenced by the grass's content of neutral detergent fiber – mostly indigestible fibrous material such as lignin and cellulose. Grass with high fiber content slows down digestion so that the wildebeest cannot ingest as much over the course of a day.

To evaluate whether the loss of trees was important for the herbivore food web, Ludwig and his colleagues constructed a linear programming model that evaluated how much grass from the three different vegetation zones a wildebeest must eat over the course of the day. They used published data to estimate minimum daily energy and nutrient needs for wildebeests, and maximum amount of fiber that a wildebeest could ingest. Plotting the results of this linear model showed that wildebeests need grass from under or around acacia trees for survival (Figure 16.16).

A summary of this system reveals a complex interaction web. This study showed that grass under and around acacia trees had higher nutrient levels than grass from open grassland, presumably, in part, from higher soil nutrient content, though other factors are probably important as well. Thus herbivory by elephants is causing an overall reduction in grass quality, which may have serious long-term consequences for wildebeests. On one level, this study supports the hypothesis that this community is controlled by bottom-up processes, with soil nutrient levels influencing grass quality, which in turn influences wildebeest abundance. But on another level, this study also introduces some top-down processes as well, with herbivory by elephants reducing tree abundance. Interestingly, there is also a horizontal process at work in which one producer, acacia trees, facilitates a

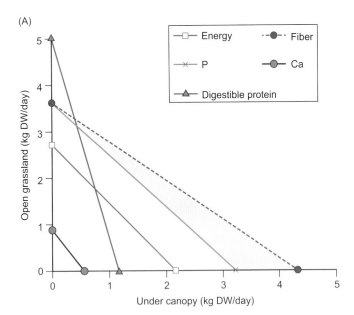

(A)

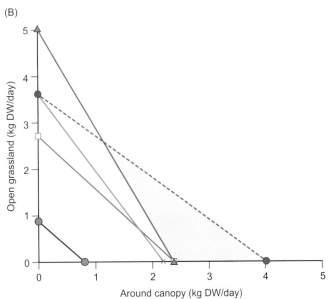

(B)

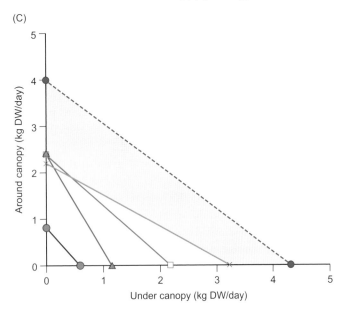

(C)

Figure 16.16 Output of linear model evaluating wildebeest energy and nutrient needs and constraints. The solid lines represent the minimum energy, digestible protein, phosphorus, and calcium that a wildebeest must take in over the course of a day. The dashed line represents the maximum neutral detergent fiber a wildebeest can take in over the course of the day. All measures are in terms of kilograms dry weight per day (kg DW/day). The brown area represents the proportion of grass from two vegetation zones that meet the wildebeests needs, but still fall below the forage intake constraint (as determined by the neutral detergent fiber maximum). A. The x-axis represents forage from under canopy, while the y-axis represents forage from open grassland. Thus for example, if a wildebeest foraged exclusively from open grassland (look at the y-axis), it would need a minimum of 0.9 kg to meet its calcium needs, 2.8 kg to meet its energy needs, 3.7 kg to meet its phosphorus needs, and 5 kg to meet its digestible protein needs. Unfortunately for the wildebeest, the neutral detergent fiber content limits it to ingesting 3.7 kg of grass from the open grassland, which means that the wildebeest falls about 25–30% short of its protein needs when foraging exclusively from the open grassland. But the wildebeest can meet its protein needs by adding under canopy grass to its diet of open grassland grass. B. The linear model output for a wildebeest that mixes grass from open grassland and around canopy. C. The linear model output for a wildebeest that mixes grass from around canopy and under canopy.

second producer, the grass growing beneath it, in effect linking these top-down and bottom-up processes.

There are many systems in which top-down processes play a more prominent role. Recall that in Chapter 10 we discussed the Greater Yellowstone Ecosystem, which provided an excellent opportunity to study the population growth rates of wolves after they were reintroduced into the community. As you might imagine, ecologists have been following changes to the entire community in response to wolf reintroduction. One finding is that these wolves are influencing the abundance and distribution of their prey in several different ways.

Top-down effects at Yellowstone National Park and at other ecosystems

Reintroduction of a species that has been absent from a community for an extended period of time is often highly controversial. Before considering introducing wolves back into the Yellowstone community, researchers attempted to predict the impact of wolves on other members of the community. Of greatest interest was the potential impact on elk, as all of the models predicted that elk would be the wolves' primary prey. Most of the models agreed that elk abundance would decline by anywhere from 5 to 30%, though some predicted sharper declines. But researchers also considered the impact on bison, a larger and more dangerous prey item. An additional consideration was how wolf predation would influence other predators. For example, would wolves eliminate or reduce the abundance of coyotes? And how would wolves influence the endangered grizzly bear? One model

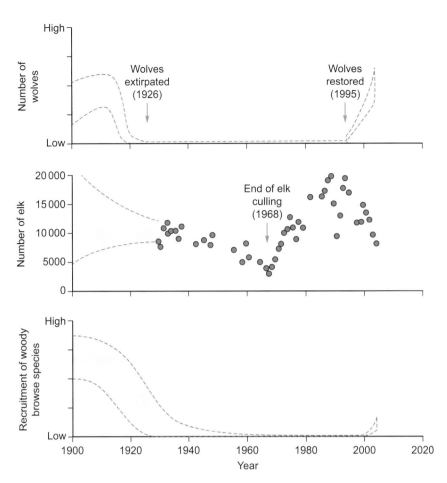

Figure 16.17 Changes in northern range of Yellowstone Park since 1900. Brown area in the graph represents range of estimates based on different data sets. A. Number of wolves. B. Number of elk. C. Recruitment of woody browse species such as aspen, cottonwood, and willow.

predicted that although wolves would occasionally kill a bear cub, that overall, wolves would have a positive impact on grizzly bears, as the much larger grizzlies would be able to chase wolves away from the prey the wolves had killed and eat it themselves. And of course, ranchers were concerned that wolves would not stay within the friendly confines of the national park, and instead help themselves to the easy pickings afforded by relatively defenseless domestic cows and sheep (Smith *et al*. 2003).

If wolves do exercise strong top-down control on the Yellowstone herbivore community, this would have important implications for lower trophic levels as well. If herbivore abundance were reduced, we would expect producer abundance to increase. Bill Ripple and Robert Beschta (2004) describe the past century of Yellowstone management history as "the Yellowstone Experiment." Very briefly, Yellowstone National Park was established in 1872, but wildlife was not protected until the US Army assumed control in 1886, and protected ungulates, bears, and beavers during their administrative period, which ended in 1918. During the US Army period, cougars and wolves were hunted, and continued to be hunted after the National Park Service assumed control in 1918. Ultimately, wolves went extinct in the park in 1926 (Figure 16.17A).

Park managers recognized that elk were devastating the woody plants, particularly those in their winter range in the northern section of the park. So in addition to allowing elk hunting, they also routinely culled the herd, keeping elk population at 8000–11 000 in the best-studied northern range until about 1950, and reducing them to 4000–8000 until 1968, when culling ceased (Figure 16.17B). The Park Service was disappointed that their culling did not seem to increase recruitment, or establishment, of woody plants – specifically aspen, cottonwood, and willow – that elk were overbrowsing during the harsh winters (Figure 16.17C). In 1968, the Park Service opted for a policy of natural regulation, which resulted in a dramatic increase in the elk herd until wolves were reintroduced in 1995. Since then, elk populations have declined substantially. Several factors underlie the decline of the elk in association with the return of the wolves.

Effects of wolves at Yellowstone on elk abundance

Wolves at Yellowstone reduced elk abundance in many different ways. Elk have declined dramatically since 1995, but we need to recognize that "the Yellowstone Experiment" has a sample size of one (as in one predator–prey association). So although elk decline is associated with wolf introduction, we should critically scrutinize all claims of a causal link between wolf introduction and elk population declines. With that in mind, let's consider some of the best-studied mechanisms by which wolves have reduced elk abundance.

TRENDS IN NUMBER OF ELK KILLED BY WOLVES. L. L. Eberhardt and his colleagues (2007) used data collected by several other research groups to estimate kill rates by wolves and humans between 1996 and 2004 in Yellowstone. The researchers found a steady increase in the number of elk killed by wolves per year over this time period, and a general decline in the number of elk killed by humans. Since then, the wolf population has leveled off (see Chapter 10), and actually declined in some areas, including the northern range, which harbors the largest elk population in Yellowstone National Park (Peterson *et al.* 2014). Nonetheless elk numbers continue to decline sharply (Figure 16.18). Consequently, killing of elk by wolves and humans cannot account for all of the decline in the elk herd. Other factors must be important as well.

INFLUENCE OF DENSITY-INDEPENDENT FACTORS ON PREY POPULATIONS. Between 1996 and 1998 the elk herd estimates declined from about 23 000 to about 15 000. There were not nearly enough wolf and human kills over that time period to account for that decline. One consideration is that elk have other predators besides wolves, including grizzly bears and cougars, which are both increasing in abundance. But another consideration is that the winter of 1996–7 was unusually harsh, with deep snow in January, followed by rain, followed by unusually cold temperatures, which formed a difficult-to-penetrate seal covering the winter forage. Many elk probably starved, or became so deprived of nourishment that reproduction was reduced the following year.

But the elk herd has continued to decline since that severe winter. Several research groups have shown that the risk of predation, which increased dramatically in 1995, can cause important top-down effects even beyond the direct impact of predation.

HOW PREDATION RISK INFLUENCES ELK BEHAVIOR AND PHYSIOLOGY. One way to avoid getting eaten is to detect a predator before it gets too close. Another approach to avoid getting eaten is to run away successfully after being detected. Researchers have observed several behavioral changes among Yellowstone elk in the years following wolf introduction.

John Laundré and his colleagues (2001) measured the mean percentage of time Yellowstone elk spent vigilant – raising their heads to survey the area – and the amount of time spent foraging. They hypothesized that females with calves, as the most vulnerable group, would be most likely to show increased vigilance and decreased foraging as wolves colonized Yellowstone National Park. Furthermore, these differences should be more pronounced in areas with wolves in comparison to areas without wolves. Vigilance times in females with calves increased markedly over the 5 years following wolf introduction (1996–2000), but the increases were, as predicted, greater in areas with wolves than at sites without wolves (Figure 16.19A). In addition, the time

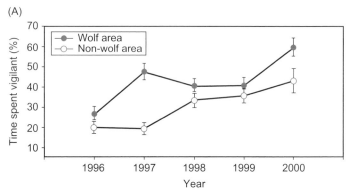

(A)

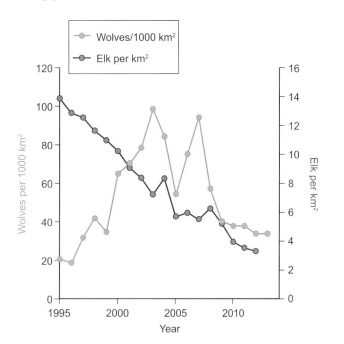

Figure 16.18 Estimated abundance of wolves and elk in the northern range of Yellowstone National Park.

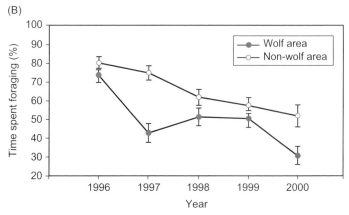

(B)

Figure 16.19 Behavior of females with calves in areas with and without wolves, between 1996 and 2000. A. Time spent vigilant. B. Time spent foraging.

spent foraging declined markedly over the same 5-year period, with a greater decline in areas with wolves (Figure 16.19B). Researchers have indicated other changes in antipredator behavior. For example, some studies have shown that elk are less likely to forage in areas that have barriers in the form of downed trees or high ledges that block escape routes. Consequently, the elk miss out on some high-quality winter food.

This suite of behavioral changes can have physiological consequences extending beyond missed foraging opportunities. The hormone progesterone plays an important role in maintaining pregnancy in mammals. Ecologist Scott Creel and his colleagues (2007) suspected that predation risk could reduce progesterone level and thus reduce pregnancy rates. This could happen two ways: Continued predation risk could increase blood cortisol levels, which could reduce progesterone levels, or continued predation risk could reduce nutrient intake, which could reduce progesterone levels. Their investigations confirmed a link between predation risk and nutrient intake, but not between predation risk and cortisol levels.

Creel and his fellow researchers followed five elk herds and monitored predation risk, which they quantified as the ratio of elk to wolves. A high elk–wolf ratio indicated low predation risk, while a low elk–wolf ratio indicated high predation risk. The researchers collected fecal pellets from 1495 elk over the 5 years of the study, and measured the progesterone levels in the pellets, which other research has shown is correlated with blood progesterone levels. High progesterone levels were associated with high calving rates across all sites (Figure 16.20A). In support of the ecologists' hypothesis, high predation risk (low elk–wolf ratios) was associated with low progesterone levels (Figure 16.20B). And finally, high predation risk was associated with low calving rates (Figure 16.20C). Taken together, these results provide an important link between high predation risk, poor nutrition, low progesterone levels, and reduced reproductive rates.

If wolves have such varied and substantial effects on the elk herd, we might expect that wolves have an indirect top-down effect on the elk's food. Can we see any responses in the plant communities to wolf reintroduction?

Top-down effects of wolves on the Yellowstone plant community

Yellowstone National Park managers were disappointed that culling the elk herd between 1930 and 1968 did not restore much of the vegetation that was being destroyed by elk grazing and browsing. Thus researchers were very interested in whether wolf-induced herd reduction would restore any parts of the plant community.

WOLF REINTRODUCTION AND WILLOW TREE ESTABLISHMENT. Predation risk for elk varies in relation to many different factors. Wolves are highly territorial, thus elk that

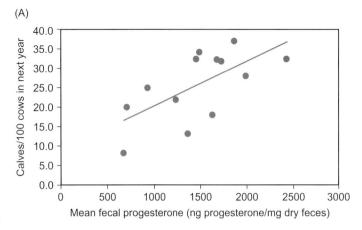

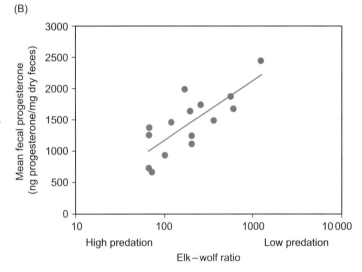

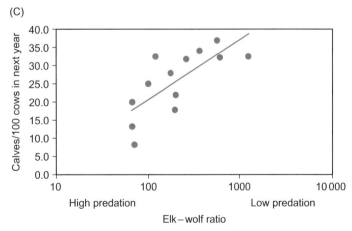

Figure 16.20 A. Calving rates in relation to progesterone level in five elk populations between 2002 and 2006. B. Progesterone levels in relation to predation pressure. Low elk–wolf ratio indicates high predation pressure. C. Calving rates in relation to predation pressure.

forage outside of a wolf pack's home range will have lower predation risk. Elk have a higher chance of escaping a wolf attack if they don't have impediments in their way, and if they have a clear view of incoming wolves. Several studies have also shown that uplands, particularly open coniferous forests that have a clear view and few impediments, afford elk a high probability of escaping predaceous wolves. Thus Bill Ripple and Robert Beschta (2006) hypothesized that elk browsing on willows along valley-bottom riparian sites (along rivers) would experience higher predation risk than elk browsing in upland riparian sites (more than 20 m above the rivers). They predicted that willows along the valley bottom would show much greater recovery following wolf introduction than willows in upland sites.

Testing recovery can be challenging, because it involves going back in time to evaluate the pre-recovery condition. Ripple and Beschta could do so, thanks to a large library of photographs of willow stands from a 20-year time period prior to wolf introduction (Figure 16.21). Based on these photos, they were able to identify 16 upland and 26 valley-bottom sites, and to estimate the height of the tallest willow trees at each site

A 1949

B 1988

C 2004

Figure 16.21 Time series of willow growth in a valley bottom of the northern range of Yellowstone National Park. The small shrubs in the foreground in 2004 are willows.

before wolf introduction. Prior to wolf introduction, tallest tree height at both upland and valley-bottom sites was in one of two size classes: 0–1 m, or 1–2 m. When they remeasured willows in 2004, Ripple and Beschta found no change in the height of upland willows. However, the riparian willows had grown substantially, with tallest trees in eight of the riparian sites now averaging 2–3 m, and the tallest trees at an additional 10 sites now in the 3–4 m size class. Overall, none of the upland sites increased in tree size class, while 22 of 26 riparian sites increased in tree size class. These increases in tree height at valley sites were due to reduced browsing. Only 49% of valley willows were browsed in comparison to 94% of upland willows.

Why are elk browsing less at some valley sites, while still browsing at high rates at other valley sites? The researchers were able to show that tree growth was negatively correlated with two factors: the average view distance, and the average impediment distance. Trees grew best in areas that gave elk a poor view of the surrounding habitat ($r = -0.46$, $p = 0.02$), and in areas where impediments to elk escape were nearby ($r = -0.52$, $p = 0.01$). The researchers concluded that predation risk keeps elk out of these high-risk sites, allowing willows to recover.

Researchers wanted to know whether other trees were also showing patchy recovery in Yellowstone following wolf introduction. For some tree species, the answer is "maybe."

ASPEN ABUNDANCE FOLLOWING WOLF INTRODUCTION. Prior to wolf introduction, the last century was particularly grim for many tree species at Yellowstone. Joshua Halofsky and Bill Ripple (2008) documented such trends in their study of the age structure of aspen trees near the Gallatin River in Yellowstone. In 1945 park personnel established two elk exclosures, which they have maintained since then. The researchers used increment cores to estimate tree age and found remarkably different patterns of age distribution. Within the exclosures, most of the trees originated in the last few decades. In the natural system, all of the trees were old, with relatively few originating after 1940, and even fewer after 1960. But Halofsky and Ripple found that a small number of new aspens are becoming established within the natural system, providing evidence for a top-down controlled trophic cascade, where increased wolf abundance decreases elk browsing (via several mechanisms) leading to increases in aspen abundance.

But a study by Matthew Kaufman and his colleagues (2007, 2010) casts doubt on this implied trophic cascade. The researchers created a predation risk map in the northern part of Yellowstone National Park based on the actual locations where elk were documented as killed by wolves (Figure 16.22A). They then investigated the trophic cascade hypothesis by testing the prediction that aspen should be browsed less in high-risk areas,

(A)

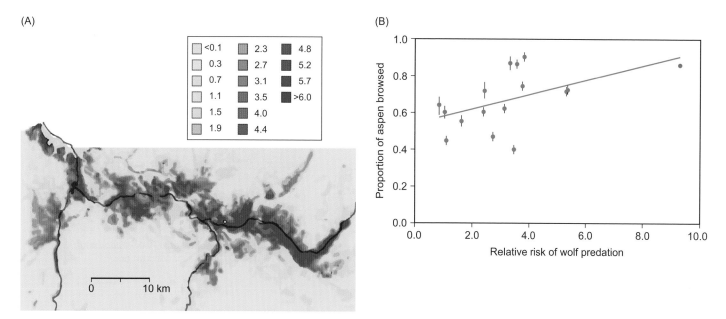

(B)

Figure 16.22 A. Relative risk of wolf predation on the northern range of Yellowstone National Park (deep purple indicates highest risk). B. Proportion of aspen browsed in relation to the risk of wolf predation.

and more in low-risk areas. The researchers found no evidence in support of this prediction, and in fact they found a slight tendency for aspen being browsed more in high-risk areas (Figure 16.22B).

The question remains: is there a trophic cascade operating within Yellowstone National Park? Beschta and Ripple (2013) point out that Kaufman and his colleagues' study was based on only 16 sites in one section of Yellowstone National Park. Kaufman and his colleagues (2013) respond that the increase in aspen growth described by Ripple and Beschta was very minimal, and should not be considered a genuine recovery. Data collected in future decades will answer the question of whether, and to what extent, a trophic cascade is an important part of community structure in the Yellowstone ecosystem.

Let's turn to the question of whether top-down control of community structure by large carnivores is common in other communities.

Prevalence of strong top-down effects of large carnivores in some communities

Researchers continue to discuss the question of whether bottom-up or top-down forces structure animal communities. Most researchers agree that both influences are important, but they want to know if there are any generalizations about when we might be more likely to see important bottom-up effects, and when we might see important top-down effects. Unfortunately, that discussion is beyond the scope of this book. But

we can focus our inquiry on whether the lessons of "the Yellowstone Experiment" apply to other western US communities.

To answer this question, Beschta and Ripple (2009) surveyed five national parks in the western US where top carnivores had been removed. They evaluated historical records to determine when these removals occurred, how the ungulate populations responded, and how the tree species changed. Tree species were chosen based on whether they were ecologically important, long-lived, and easily aged via increment cores, and whether their leaves were palatable to ungulates. The researchers also did field measurements of existing tree stands, including increment core studies, to characterize the age structure of each population.

For each park, there was a strong reduction in tree establishment associated with the extirpation or major reduction of the top carnivores in the community. For example, at Wind Cave National Park, no cottonwood recruitment has occurred for over a century (Figure 16.23A). In other cases, there has been much lower than expected recruitment of ecologically important tree species. Yellowstone was the only park in this group where carnivores were restored, and also the only park where some trees are becoming reestablished (Figure 16.23B).

Our discussion of bottom-up and top-down effects on community structure has only considered a small fraction of the species in each community. Let's return to the sea otter–sea urchin–kelp trophic cascade, and add a few species to this interaction to get a broader perspective of how this community may actually operate.

Winds Cave National Park – Plains cottonwood

Yellowstone National Park – Black and narrowleaf cottonwoods

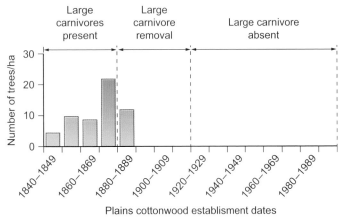

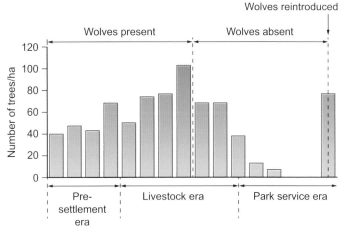

Figure 16.23 Age structure diagram of A. plains cottonwood trees at Wind Cave National Park. B. Black and narrowleaf cottonwoods at Yellowstone National Park. Without excessive browsing by herbivores, we would expect to find a continuous increase in the number of trees with more recent establishment dates.

REVISIT: Aleutian Islands food web

Recall that we suggested that predation by killer whales might underlie the recent decrease in sea otter abundance. If so, then our trophic cascade expands into four trophic levels, with predatory killer whales causing sea otter declines, which cause sea urchin increases, resulting in kelp declines. In an earlier study, Estes and his colleagues (1998) summarized the evidence that killer whale predation was one more link in this trophic cascade. First, an increase in observed killer whale attacks on sea otters after 1990 coincided with the sharp decline in sea otter abundance. Before 1990, there were no observed attacks in 3405 person-days of observation. After 1990, there were 10 observed attacks in 4005 person-days (the probability of this increase occurring due to chance was 0.006). In addition, the researchers compared trends of sea otter abundance in two adjacent sites, one that had killer whales and one that was too shallow to be accessible to killer whales. At the site with killer whales, sea otter abundance declined 76% over 4 years, while sea otter abundance remained stable at the site without killer whales.

The researchers considered numerous other hypotheses but were unable to find any support for alternative explanations for sea otter declines. They ruled out disease, toxins, or starvation because there was no evidence of any increases in beach-cast carcasses. Estes and his colleagues argue that these findings leave killer whales as the most likely driver of a four-level trophic cascade.

But what has changed the dynamics of this community? Why have killer whales suddenly started eating sea otters in the Aleutian Islands? In the initial report, Estes and his colleagues describe how both Steller sea lions and harbor seals had recently declined sharply, and argued that killer whales have simply switched to sea otters to compensate for losing these important food sources. A more recent report expands on this hypothesis by proposing that the observed decline in sea otter abundance is the most recent stage in an ongoing legacy of industrial whaling (Springer *et al.* 2003). The argument is that the decimation of the great whales by commercial whalers during the twentieth century forced killer whales to turn to alternative food sources. Killer whales with their voracious appetites depleted each species that they shifted to, which resulted in a sequential decline in abundance of each prey species. Harbor seals were the first to decline, followed by harbor seals, fur seals, sea lions, and finally sea otters (Figure 16.24)

This hypothesis is controversial for two reasons. First, we don't know how important the great whales were in the killer whale diet. Second, though the decline of great whales correlates with a sequential collapse of other marine species, there were other factors that correlated with the collapse of seals, sea lions, and sea otters. Andrew Trites and his colleagues (2007) propose that the Steller sea lion decline was caused by changes in the ocean climate, rather than changes in predatory behavior of killer whales. The researchers point to substantial warming of the SST that began in 1977, which was associated with changes in barometric pressure, wind patterns, and ocean currents in

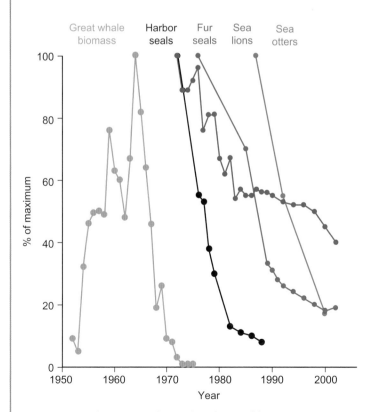

Figure 16.24 Changes in relative abundance of five marine mammal species since 1950.

the North Pacific. Their models predict a reduction in upwelling, and presumably in nutrient availability, in the regions where sea lion populations crashed. Consequently, a bottom-up process that begins with nutrient limitations may underlie the decline in sea otter abundance.

Trites and his colleagues (2007) suggest that both bottom-up and top-down forces could be interacting. For example, bottom-up forces reducing availability of high-quality food could cause important shifts in marine mammal foraging behavior, forcing the animals to forage in areas frequented by killer whales, and weakening them so that they are less likely to escape a killer whale attack.

We have introduced numerous species into what had been an artificially simple food web for the Aleutian Islands. Even so, we have merely scratched the surface of both species diversity and interaction diversity within this community. Chapter 17 will look at community structure from a broader perspective, as we consider factors influencing the biological diversity of a community. We will explore whether communities with high biological diversity are inherently more or less stable than communities with low biological diversity.

SUMMARY

The strong interactions among sea otters, sea urchins, and kelp are the basis of a trophic cascade, a community-level process that is influenced primarily by the actions of predators, and which has indirect effects at least two trophic levels removed from the actions of the predator. The trophic cascade in the introductory case study describes how sea otters prey on sea urchins, reducing their abundance, which in turn increases the abundance of kelp, which are sea urchins' preferred food. In effect, predation by sea otters on sea urchins indirectly increases the abundance of kelp. Ecologists explore community processes across scales of space and time to investigate whether trophic cascades such as these are common, and whether they are conditional on a particular set of environmental conditions.

Ecological communities are structured by species composition and diversity, by the distribution and abundance of each species within the community, by the interactions among species, and by the presence of guilds and functional groups. In addition, community processes are strongly influenced by the actions of dominant species, such as Sitka spruce in the taiga, and by keystone species, such as the sea star *Pisaster* in the rocky intertidal zone. Ecologists use food webs to construct a useful but incomplete picture of community structure. By quantifying certain attributes of food webs, such as linkage density and connectance, ecologists can compare food webs to identify unique features of community structure in a particular web.

Food web structure and functioning are influenced by both consumers and producers. Ecologists are attempting to identify conditions under which bottom-up versus top-down processes are more likely to influence community structure. Wolf reintroduction to the Greater Yellowstone Ecosystem shows top-down effects on elk abundance, but ecologists argue about whether these effects cascade down to the plant communities. Along similar lines, there is conflicting evidence over whether the sea otter decline is mediated by a top-down process in which predation by killer whales reduces sea otter abundance, or by a bottom-up process in which climate change leads to a reduction in ocean upwelling, reducing nutrient availability and thus reducing kelp growth. According to this bottom-up hypothesis, low kelp abundance causes low sea urchin availability, reducing sea otter populations.

FURTHER READING

Barton, B. T., and Ives, A. R. 2014. Species interactions and a chain of indirect effects driven by reduced precipitation. *Ecology* 95: 486–494.

Neat demonstration of apparent competition between spotted aphids and pea aphids that is mediated by a predaceous ladybeetle. Experimentally simulated drought shows how abiotic and biotic factors can work together to influence the abundance of several different species.

Beschta, R. L., and Ripple, W. J. 2009. Large predators and trophic cascades in terrestrial ecosystems of the western United States. *Biological Conservation* 142: 2401–2414.

Demonstrates an association between top-predator removal and the loss of dominant plant species via a trophic cascade at five National Parks in the United States. This study suggests that reintroducing top predators should be considered a viable policy to help restore the plant communities at these locations.

Laundré, J. W., and 9 others. 2014. The landscape of fear: the missing link to understand top-down and bottom-up controls of prey abundance? *Ecology* 95: 1141–1152.

Well-conceived study that makes clear predictions regarding the circumstances in which bottom-up vs. top-down control should be most important regulators of the population size of kangaroo rats in the Chihuahuan Desert. The researchers argue that nonconsumptive effects help explain the results of experiments and observations that had previously yielded seemingly inconsistent results.

Treu, R., and 8 others. 2014. Decline of ectomycorrhizal fungi following a mountain pine beetle epidemic. *Ecology* 95: 1096–1103.

Shows how increased herbivory by the mountain pine beetle in western Canada is leading to a decline in ectomycorrhizal fungi diversity and abundance in association with the destruction of lodgepole pine. This has important management implications, because the loss of ectomycorrhizal associations could interfere with forest recovery.

Beschta, R. L., and Ripple, W. J. 2013. Are wolves saving Yellowstone's aspen? A landscape-level test of a behaviorally mediated trophic cascade: comment. *Ecology* 94: 1420–1425.

Beschta, R. L., Eisenberg, C., Laundré, J. W., Ripple, W. J., and Rooney, T. P. 2014. Predation risk, elk, and aspen: comment. *Ecology* 95: 2669–2671.

Winnie Jr, J. A. 2012. Predation risk, elk, and aspen: tests of a behaviorally mediated trophic cascade in the Greater Yellowstone ecosystem. *Ecology* 93: 2600–2614.

Winnie Jr, J. 2014. Predation risk, elk, and aspen: reply. *Ecology* 95: 2671–2674.

These four articles highlight part of the controversy over whether the behaviorally mediated trophic cascade hypothesis applies to the Greater Yellowstone ecosystem and the "recovery" of aspen populations. Clearly the jury is out!

END-OF-CHAPTER QUESTIONS

Review questions

1. What is the trophic cascade involving sea otters, sea urchins, and kelp? What evidence did researchers collect supporting it? What is the evidence that sea otter decline is based on top-down versus bottom-up processes?
2. Distinguish a guild from a functional group and give an example of each.
3. How did *Pisaster* act as a keystone species? Under what conditions is it not considered a keystone species?
4. Distinguish between food webs and interaction webs. How do ecologists measure interaction strength?

5. Describe all of the interactions in the food web studied by Fulco Ludwig and his colleagues (2008). What parts of the food web indicate top-down control, and what parts indicate bottom-up control?

6. What is the evidence that predation risk influences the structure of plant communities in Yellowstone National Park?

Synthesis and application questions

1. Refer to the food web pictured in Figure 16.1B. Barnacles are not restricted to eating floating plant debris, they also eat a large quantity of small marine invertebrates that float or swim by. How does that information influence the possible trophic levels occupied by sea stars and sea otters?

2. One could argue that sea otters are keystone species in subtidal communities. In what ways are sea otters similar to *Pisaster*, and in what ways are they different in their effect on the community?

3. Return to Figure 16.12. What ecological and evolutionary factors might be responsible for the trends observed in the Carpinteria food web?

4. Draw a four-species food web that has the maximum connectance.

5. How would adding parasitoids to food webs influence linkage density and connectance?

6. Lake trout were introduced by humans into Yellowstone Lake in the late 1980s and are now outcompeting the native cutthroat trout. Previously, grizzly bears sustained themselves on large numbers of cutthroats. Lake trout prefer deeper water than cutthroats, so they are less available to grizzly bears, which have shifted their behavior to preying on baby and juvenile elk. Draw an interactions web that demonstrates the direct and indirect effects of these species on each other, and on the abundance of aspens.

7. Look carefully at the output of Ludwig's linear model (Figure 16.16). If a wildebeest is foraging from both open grassland and under canopy, what is the minimum proportion of its grass that must come from under the canopy? If a wildebeest is foraging from both open grassland and around the canopy, what is the minimum proportion of its grass that must come from around the canopy?

Analyse the data 1

At exposed sites, Menge *et al.* (1994) observed a mean vertical drop in the mussel zone of 0.47 and 0.63 m at Strawberry Hill, and 0.3 and 0.4 m at Boiler Bay. What is the 95% confidence interval for the mean drop of the mussel zone? Can we conclude that this drop is statistically different from the no change that was observed in protected sites?

Analyse the data 2

Jason Link's (2002) study summarizes data for the number of species and number of links in 12 different food webs. Use the data in Table 16.2 to answer the following questions:

1. Is there a relationship between number of species in the web and directed connectance?

2. Is there a significant difference in linkage density between terrestrial and aquatic species? Make sure you use statistical tests to evaluate your conclusions.

Table 16.2 Number of species and number of links in 12 food webs in different biomes.

Biome	Number of species	Number of links
Lake	12	36
Pond	22	110
Desert	30	409

Table 16.2 (*cont.*)

Biome	Number of species	Number of links
Stream	34	109
Stream	40	80
Tropical forest	44	218
Swamp/stream	75	514
Grassland	77	126
Marine	81	1562
Estuary	92	409
Tropical forest	136	1322
Lake	182	2366

Chapter 17

Biological diversity and community stability

INTRODUCTION

In *The Giving Tree*, Shel Silverstein (1964) describes a lifetime of love between an apple tree and a boy. As the title implies, the tree is the primary donor in this love affair, giving leaves which the boy makes into crowns, a trunk to climb, branches to swing by, apples to eat, and shade to sleep in. As the boy matures, his interest in the tree wanes, but he sells her apples so he can move to the city. Still later, the man/boy returns and cuts her branches for a house, and later, her trunk for a sailboat. Much later, a tired old man returns, and the tree's final gift is a stump that is "good for sitting and resting."

In many terrestrial communities, flowering plants such as apple trees provide a wide variety of services for a diverse assemblage of species. Many ecologists propose that the tremendous biological diversity of some terrestrial communities is based on the numerous interactions between flowering plants and their predators, competitors, and mutualists. Even one plant species will have many different predator species eating its fruit, leaves, stem, and roots, and many mutualists moving its pollen in exchange for nectar or increasing its nutrient uptake in exchange for carbohydrates. Our opening case study compares the biological diversity of terrestrial communities and marine communities, and suggests some reasons for the apparent differences between them. We then describe several ways that ecologists measure biological diversity, and discuss studies of marine fish, vascular plants, and insects in tropical forests, that use these different measures. Biotic and abiotic factors contribute to the biological diversity of a community both directly and via indirect interactions. Many studies have shown that diverse communities tend to be stable, retaining their basic structure and functioning even when disturbed by biotic or abiotic forces, but the relationship between diversity and community stability is complex, and is an exciting focus of ecological research.

KEY QUESTIONS

17.1. How do ecologists describe and measure biological diversity?

17.2. How do biotic and abiotic factors influence community species diversity?

17.3. What is the relationship between species diversity and community stability?

CASE STUDY: Biological diversity in terrestrial and marine communities

Given the complexity of the interaction web based on one plant species such as an apple tree, we quickly realize that a diverse plant community will be the basis for a very rich community of interacting species. Community ecologists have begun unraveling some of this complexity. As one example, Katja Poveda and her colleagues (2007) were interested in the linkages between below- and aboveground processes associated with the wild mustard plant, *Sinapis arvensis*. This flowering plant participates in numerous interactions that benefit the invertebrate animal community both above and below the surface. Some of these interactions are direct, such as providing nectar for a community of pollinators and phloem sap for two species of aphids. Some of these interactions are indirect, for example, sap gained by aphids benefits aphid parasitoids. Not surprisingly, decomposers such as the earthworm, *Octolasion tyrtaeum*, also benefit the aphids, by increasing nitrogen mineralization rates in the soil, ultimately increasing protein content in the phloem. Somewhat surprisingly, click beetle larvae (*Agriotes* species) that eat the mustard plant's roots actually benefit these same aphids, by reducing water uptake by the plant so that the phloem has less water and a higher nutrient concentration. In addition, through mechanisms that are not yet worked out, root herbivory by click beetles actually increases the attractiveness of the flower to pollinators (Figure 17.1).

The situation is somewhat different in the ocean, which has a very limited diversity of flowering plants – mostly mangroves and seagrasses. Overall, ecologists estimate that the oceans contain between 4 and 15% of the global species diversity, despite encompassing 71% of Earth's surface area (May and Godfrey 1994; Benton 2009). Some ecologists argue that marine species diversity is so low because flowering plants are so poorly represented, and there is no opportunity for the evolution of the complex community interactions that occur between plants and invertebrates, such as those we described in Chapter 16. Other ecologists argue that there is a bias in these estimates caused by the difficulty in doing research in the ocean in comparison to terrestrial environments.

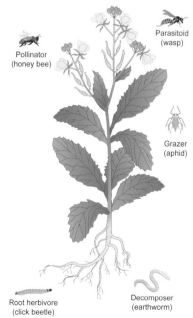

Figure 17.1 Complex ecological community based on the flowering plant, *Sinapis arvensis* (wild mustard).

Figure 17.1 (*cont.*)

In support of the estimate bias hypothesis, William Whitman and his colleagues (1998) described and analysed the diversity and abundance of prokaryotes in sediment cores that were collected by drill ships at varying depths below the ocean's floor. Prokaryote abundance was greatest just below the surface, but prokaryotes were abundant to depths of 600 m, and were probably present to at least 3000 m. Despite their tiny size, these subsurface microorganisms account for about 30% of the total biomass of life on Earth! Analysis of 16S ribosomal RNA revealed that most of the organisms were new to science (Inagaki *et al.* 2006).

Despite these findings on prokaryotes, there is no doubt that the diversity of larger organisms is greatest in terrestrial biomes. Later in this chapter, we will consider whether the ability to persist in locally small populations, in other words to survive although rare, underlies the diversity of terrestrial species. Before exploring patterns of species diversity in more detail, we need to recognize that species diversity can have different meanings in different contexts.

17.1 HOW DO ECOLOGISTS DESCRIBE AND MEASURE BIOLOGICAL DIVERSITY?

In my backyard is a patch of corn. A few days ago I spent 15 minutes counting and identifying (limited by my feeble identification skills) all of the organisms I could see living there. More recently, I spent 15 minutes doing the same exercise in one of my favorite riparian communities (communities along the shorelines of rivers) along a tributary of the New River, just a few blocks from my house. I was inspired to do so by observing a mink in that exact location, though he failed to grace me with his presence while I did my species counts. I'm guessing he was not particularly interested in hanging with a threatening megamammal like myself. Table 17.1 shows the results of both surveys.

Table 17.1 15-minute surveys of garden community and riparian community.

Garden community		Riparian community	
Species (if known)	Abundance	Species (if known)	Abundance
Corn	143	Box elder saplings (*Acer negundo*)	8
Wild mustard	16	Horsetail (*Equisetum*)	22
Unidentified flowering plant 1	2	Wingstem (*Verbesina alternifolia*)	27
Unidentified flowering plant 2	1	Unidentified flowering plant 3	10
Ants (*Formica* species)	2	Johnson grass (*Sorghum halepense*)	36
Slug	1	Isopods (roly-poly bugs)	20
Small flies	4		

Species richness is a simple count of the number of species in a community. The garden survey had a species richness of 7, while the riparian community had a species richness of 6. Is it reasonable to conclude that the garden patch had higher species diversity?

Species diversity indices

In the garden community, if you were to close your eyes, spin around randomly, and then open your eyes, the first organism you would notice would probably be corn. In addition to being larger than most of the other species there, corn is much more abundant, comprising 85% of the individuals in the community. In contrast, in the riparian community there is no dominant species, and the most abundant species is Johnson grass, which comprises only 29% of the community. Thus the riparian community has much greater evenness, or much less variation in the distribution of abundances of each species, than the garden community.

Because the riparian community has greater evenness, it feels much more diverse to us. Several indices of species diversity take evenness into account when measuring species diversity. Probably the most common diversity index is the Shannon index, which uses the following formula:

$$H' = -\sum_{i=1}^{S} (p_i \ln p_i)$$

H' = Shannon diversity index

S = the number of species in the community

p_i = the proportion of species i in the community. Thus if there are 50 individuals in a community, and there are four individuals of species 1, then $p_1 = 4/50 = 0.08$.

$\ln$ = the natural logarithm

Let's use this formula to calculate the Shannon index for both communities (Table 17.2).

Summing the values of $p_i \ln p_i$ for species 1–7 in the garden community gives us $\Sigma p_i \ln p_i = -0.619$. But $H' = -\Sigma p_i \ln p_i = -(-0.619) = 0.619$. Similarly H' for the riparian community = 1.677. Notice that H' for the riparian community is considerably higher than H' for the garden community. This corresponds nicely with our intuition that the riparian community has a higher diversity, even though it has slightly lower species richness than the garden community.

The interaction of richness and evenness can be studied on a large scale as well. In some cases, regional or global patterns of species richness are very different from regional or global patterns of evenness. As one example, Rick Stuart-Smith and his colleagues (2013) were interested in the relationship between global patterns of species richness and species evenness. They deployed volunteer divers on almost 2000 reefs throughout the world, establishing groups of 10 × 50-m quadrats, with each diver trained to use consistent methods to identify each fish to species, thereby providing consistent measures of both richness and evenness. Their findings revealed one expected and one surprising pattern (Figure 17.2).

As expected, species richness was greatest in warm ecosystems, and decreased at higher latitudes. However, evenness showed a very different pattern, with maximum evenness consistently found at higher latitudes. These findings have important management implications. For example, since species diversity is influenced by both richness and evenness, perhaps it would be important to shift some conservation efforts into more temperate marine communities. Furthermore, as we will discuss later in this chapter, ecologists want to understand the relationship between species diversity and community stability. These data on diversity could be used as a baseline to explore this fundamental question in community ecology.

Table 17.2 Using the Shannon index to calculate biological diversity for two communities.

Garden community					Riparian community				
Species	Abundance	p_i	$\ln p_i$	$p_i \ln p_i$	Species	Abundance	p_i	$\ln p_i$	$p_i \ln p_i$
1	143	143/169 = 0.846	−0.167	−0.141	1	8	8/123 = 0.065	−2.733	−0.178
2	16	16/169 = 0.095	−2.357	−0.223	2	22	22/123 = 0.179	−1.721	−0.308
3	2	0.012	−4.437	−0.053	3	27	0.220	−1.516	−0.333
4	1	0.006	−5.130	−0.030	4	10	0.081	−2.510	−0.204
5	2	0.012	−4.437	−0.053	5	36	0.293	−1.229	−0.360
6	1	0.006	−5.130	−0.030	6	20	0.163	−1.816	−0.295
7	4	0.024	−3.744	−0.089					
				$\Sigma p_i \ln p_i = -0.619$					$\Sigma p_i \ln p_i = -1.677$

Note: For simplicity I have numbered the species sequentially in the table.

(A)

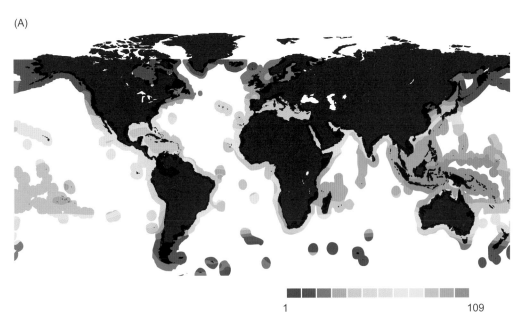

1　　　　　　　　　　　　　　　109

(B)

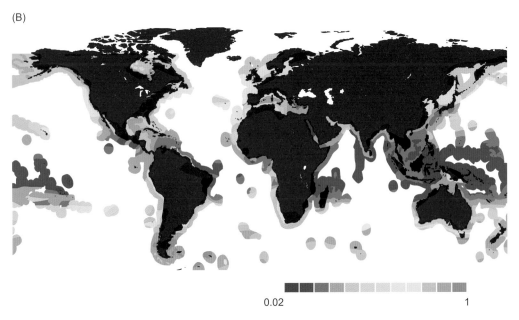

0.02　　　　　　　　　　　　1

Figure 17.2 Global patterns of fish diversity based on A. species richness (number of species/quadrat) and B. evenness (0.02 = low evenness, 1 = maximum evenness).

Several other diversity indices are also a function of richness and evenness. You may come across some of these other indices as you read the ecological literature. But ecologists may go beyond richness and evenness in their evaluation of species diversity.

Measures of community diversity

Robert Whittaker (1956) recognized that the species composition of a community varies in response to how large an area is sampled, and in response to habitat heterogeneity – a measure of how different the habitat is over space or time. He explained that species composition in many communities changes along environmental gradients – changes in an abiotic factor through space or time. In fact, these environmental gradients may define the boundaries of different communities. For example, in his

study of vegetation in the Great Smoky Mountains National Park in the southeastern United States, Whittaker was able to define seven different forest communities based on two environmental gradients, elevation and geological structure (Figure 17.3). Later (1960, 1972), Whittaker and other researchers defined three measures of diversity within and across habitats.

We have already described alpha diversity at length. It is simply the species richness within a defined area. Beta diversity, as conceived by Whittaker, is a measure of species turnover, or the change in species composition, along an environmental gradient. Over time it has come to represent species turnover between places in general, with or without a defining environmental gradient. More recently, beta diversity and species turnover are also used to describe a change in species composition within a community over time (see Chapter 22). Lastly, gamma

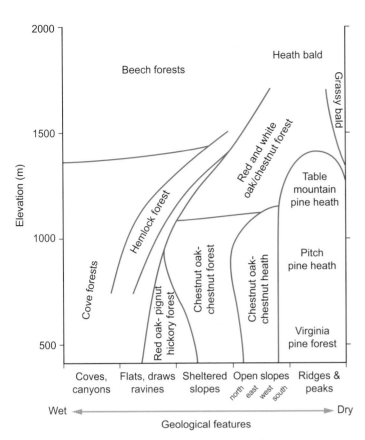

Figure 17.3 Forest communities in the Great Smoky Mountains in relation to elevation and geological features.

Table 17.3 Whittaker's survey of trees off Bullhead Trail along an environmental gradient from a very protected wet site (site 1) to a very exposed dry site (site 8). The numbers represent the percentage of trunks with diameters > 2.5 cm of each species. The species are numbered to simplify the table (modified from Whittaker 1956).

Tree species	Site number from wettest to driest (# 1 is wettest)							
	1	2	3	4	5	6	7	8
1	11	2	4	3				
2	29	24	25	24	1			
3	9	3	9					
4	9	14	33	25				
5	16	27	13	13				
6	10							
7	9							
8	1							
9	<0.5							
10	<0.5							
11	<0.5	1	2					
12	1		1					
13	<0.5	11	3	1				
14	1	<0.5			6			
15	4		<0.5	<0.5	21	54	20	<0.5
16		1	1					
17		1	2					
18		1	1	4				
19		14	1	11	6			
20		1	4	7	8	10		
21			3	8	28			

diversity is the combined species richness of all the communities that are being evaluated. Let's use a small portion of Whittaker's research to more clearly define these three measures of diversity.

On a relatively small scale, Whittaker wanted to know how tree species diversity varied in relation to a moisture gradient in the Smoky Mountains. As part of one study, he identified a sample of at least 50 trees at eight sites along the Bullhead Trail in the Great Smoky Mountains National Park, at an elevation of about 950 m, that varied along a moisture gradient from very wet in a deep valley cove, to very dry with a southwest exposure. Table 17.3 gives Whittaker's data.

With these data, we can evaluate tree diversity along Bullhead Trail. Alpha diversity is simply species richness at a particular site. Thus alpha diversity at Site 1 equals 15, and at Site 2 equals 13. Gamma diversity is total species richness at all sites and equals 32. Beta diversity is a bit more complex to calculate because researchers use several different measures.

One measure of beta diversity is Sorensen's Index, which measures how similar two sites are in species composition (Whittaker 1960). The formula is:

$$\beta = 2S_b/(S_1 + S_2)$$

Where β = Sorensen's beta diversity index,
S_b = the number of species shared by both samples,
S_1 = the number of species in sample 1, and
S_2 = the number of species in sample 2.

Table 17.3 (cont.)

Tree species	Site number from wettest to driest (# 1 is wettest)							
	1	2	3	4	5	6	7	8
22					9	4		
23						2		
24			1	<0.5		4	2	<0.5
25				3	9	10	2	1
26					6	8	38	4
27					4	10	5	1
28					1	<0.5	16	2
29							5	1
30							<0.5	2
31							11	12
32							2	77

Let's calculate β to compare Sites 1 and 2. Sites 1 and 2 have eight species in common with each other. Thus

$$\beta = 2(8)/(15 + 13) = 16/28 = 0.57$$

Sorensen's Index has a maximum of 1.0 when both communities have identical species composition (no species turnover), and a minimum of 0, when both communities have no species in common (complete species turnover). Thus beta diversity is greatest at low values of β.

Returning to Whittaker's data set, you can see that there is moderate beta diversity when comparing two adjacent sites, but that there is very high beta diversity when comparing two distant sites. Thus if we compare Site 1 and Site 8, β = 0.08, reflecting that these sites share only one tree species in common (# 15).

Thinking ecologically 17.1

How might the sample size influence the values of alpha, beta, and gamma diversity?

Whittaker's Bullhead Trail data set was collected from transects only 25 m apart. Let's increase our perspective to view biological diversity on a larger scale, using some of the measures of diversity we've introduced. We will first consider species richness on a global scale and then discuss large-scale patterns of beta diversity.

Global patterns of species richness in terrestrial communities.

In Chapter 2, we introduced tropical forests as the most diverse biome, so it is no surprise that tropical forests have the greatest number of plant species. Wilhelm Barthlott and his colleagues (1996) divided the world into a grid system of 100 × 100 km, and used more than 1400 published studies to create a global map of vascular plant species richness. They identified five areas with highest vascular plant species richness: (1) Costa Rica–Choco, (2) Andes–Amazonia, (3) East Brazil, (4) northern Borneo, and (5) Papua New Guinea (Figure 17.4).

In a related study, Barthlott and his colleagues (1999) proposed that vascular plants are good indicators of species richness of other groups of organisms within a community. For example, in communities across the globe, there is a strong positive correlation between vascular plant and tetrapod (four-legged vertebrate) richness ($r = 0.86$) and between vascular plant and insect richness ($r = 0.97$). These patterns generated many questions, including: (1) Why do rain forests have so many species of animals? (2) How do geographical features influence species diversity? (3) Do rain forests have unusually high animal diversity using other measures of species diversity? (4) What is the relationship between animal diversity and vascular plant diversity in rain forests? We will reconsider the first two questions in Chapter 22. For now we will turn to a recent study that addresses the third question of whether rain forest diversity is high when estimated with other measures of species diversity.

Beta diversity of tropical insect communities

Vojtech Novotny and his colleagues (2007) recognized that many biological diversity studies in tropical forests have shown high species turnover rates (high beta diversity) in cases where the forests had gradients associated with elevation changes, differences in moisture levels, or disturbance events. This finding is similar to Whittaker's early studies of high beta diversity along a moisture gradient on the Bullhead Trail, and at many other sites. Novotny and his team wondered whether a tropical forest lacking these gradients might have substantially lower beta diversity. So they turned their attention to a lowland forest in Papua New Guinea, where they sampled about 75 000 caterpillars at eight communities separated by as much as 500 km.

Each research site was adjacent to a village, and most of the researchers were recruited from the local village (Figure 17.5A).

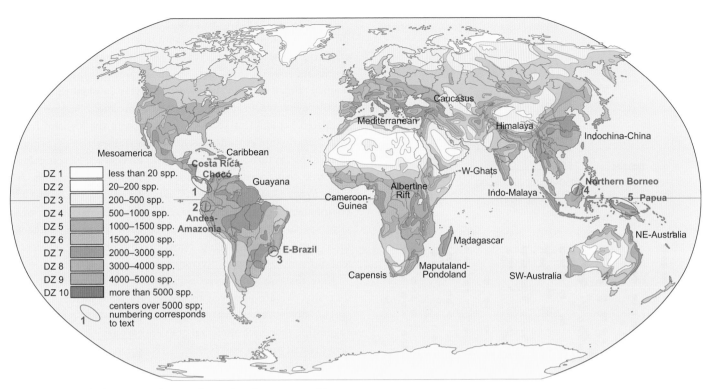

Figure 17.4 Global species richness of vascular plant communities.

Figure 17.5 The Novotny team research site in Papua New Guinea. A. Map of villages that served as sampling sites within the lowland forest. Ohu is a village in Madang province. B. Research team sorting and raising caterpillars.

At each site, the research team established 50, 20 × 20-m quadrats, gathered about 1500 m² of leaves from each focal tree species, and identified all of the insects foraging on the leaves. In many cases the researchers had to raise the insects to sexual maturity so they could identify them to species (Figure 17.5B). The focal tree species came from four large genera: *Ficus*, *Macaranga*, *Psychotria*, and *Syzygium*.

Forests with low beta diversity will have low species turnover among communities. Novotny and his colleagues measured beta diversity in two ways. First, they calculated the geographical range of each caterpillar species based on their sampling sites and discovered that most species had a geographical range of greater than 500 km (Figure 17.5C). Second, they used a modification of the Sorensen Index to calculate similarities in caterpillar species composition at all of the sites. They discovered that, for each tree genus, species composition of the caterpillar community was similar, even when the trees compared were separated by a distance of

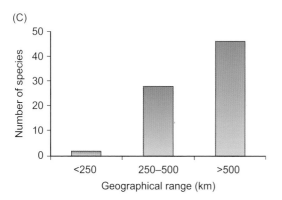

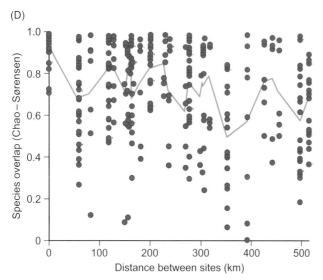

Figure 17.5 (*cont.*) C. Geographical range of Lepidoptera species in surveys of eight villages D. Species overlap in caterpillar community composition at eight sites separated by distances ranging from 50 to 500 km. High Chao-Sorensen values indicate high species overlap, or low beta diversity. Solid line represents the mean value at each distance.

500 km (Figure 17.5D). For example, the caterpillars on *Ficus* in one site were mostly the same species as the caterpillars on *Ficus* trees more than 500 km distant. However, *Ficus* and *Macaranga* did host very different caterpillar communities, even at the same site. The researchers concluded there was very low beta diversity in caterpillar communities across this forest.

Novotny and colleagues conducted similar experiments on two other guilds of insects. Insects from one guild – Ambrosia beetles – dig tunnels in dead or dying wood, which they inoculate with a fungus that serves as their food source. The other guild consists of fruit flies (family Tephritidae) native to southern and southeastern Asia, which the researchers collected by setting out traps baited with a chemical fruit fly attractant. For these studies the maximum distance separating the sites was 950 km, but there were fewer intervening sites. The results were very similar, with species composition almost identical in communities separated by 950 km.

These results point to surprisingly low beta diversity of three different insect guilds in this lowland tropical forest. The researchers interpret the low beta diversity as arising from an absence of any significant environmental gradient in the area that they surveyed. Annual rainfall and elevation were relatively constant throughout the surveyed area. In contrast, when Novotny and colleagues (2005) compared Lepidoptera species composition in the lowland forest with a nearby uplands site in central New Guinea, there was much greater species turnover, or higher beta diversity. The researchers conclude that forests with strong environmental gradients are likely to have high beta diversity, while forests lacking significant environmental gradients may have low beta diversity, even over large spatial scales. They point to the need for more studies to understand the relationship between environmental gradients, and different measures of diversity.

Now that we have some background on measuring species diversity, we are prepared to investigate what factors promote species diversity. We've already introduced this question by suggesting that environmental gradients are associated with high beta diversity in tropical forests. Let's explore this question more fully in other communities.

17.2 HOW DO BIOTIC AND ABIOTIC FACTORS INFLUENCE COMMUNITY SPECIES DIVERSITY?

Interactions between species can influence species diversity in many different ways. In some cases these effects are direct, while in other cases biotic influences on species diversity are indirect and involve several interacting factors.

Biotic factors: direct influence on species richness

I'm hoping that if I were to ask you what biotic factors might influence the diversity of the Lepidoptera (butterflies and moths) that you would immediately start thinking about host plants. You would point out that the caterpillars need leaves to eat, and the adults need nectar to drink. Juha Pöyry and his colleagues (2009) were interested in understanding how features of the habitat, including the host plants, influence Lepidoptera diversity. As conservation ecologists, they were particularly interested in whether declining Lepidoptera species – those that are becoming rare – showed different patterns than relatively abundant species.

The researchers conducted their study in 48 grasslands in two geographical regions in southwest Finland. Grasslands ranged in size from 0.25 to 6.0 ha. The ecologists censused each site seven times, identifying 51 species of Lepidoptera out of almost 15 000 individuals. Based on other census data, 34 of the Lepidoptera species were classified as abundant, and 17 were classified as declining. The researchers considered nine independent variables that might influence species richness, and used a method of analysis that allowed them to distinguish which variables were most strongly affecting Lepidoptera species richness.

In the 34 abundant species, species richness was most strongly correlated with the abundance of nectar-producing plants (Figure 17.6A). In the 17 declining species, species richness was also correlated with the abundance of nectar-producing plants, but the variable with the strongest correlation was mean plant height (Figure 17.6B). Somewhat surprisingly species richness was not correlated to the size of the patch (of grassland habitat), or to the connectivity of the patch to other patches. However, butterfly density was correlated with connectivity, but only in abundant species, with a greater butterfly density in more highly connected patches.

It is interesting that habitat variables influenced declining and abundant species to different degrees. These findings remind us that endangered species may respond differently to changes in their environment than do abundant species. Consequently, conservation ecologists should tailor recovery plans to the unique niche occupied by each declining species.

While the biotic environment directly influences Lepidoptera species richness in Finland, there are many ecological settings where important biotic factors indirectly influence species richness.

Biotic factors: indirect influence on species richness

In African savannas, mammals and birds may influence the abundance and richness of other species in a variety of ways. For example, mammals and birds may compete with each other for resources such as seeds or insects. Some bird species are important prey items, both for mammals and for other bird species. Both groups of animals can have important influences on plant diversity and abundance, by directly consuming plants, by dispersing fruits and seeds, or by providing nutrients for the plants. In return, plants may benefit both groups, by providing them with food, shelter, or nesting areas. Community ecologists are particularly interested in unraveling how these and other mammal/bird/plant interactions influence community structure.

As one example, Darcy Ogada and her colleagues (2008) wanted to know how the presence of large herbivorous mammals influenced the species richness of the bird community at the Mpala Research Center in central Kenya. Their approach was to create 4-ha replicates of communities with and without mammalian herbivores. They suspected that very large mammals (giraffes and elephants) might have a different influence than more moderate-sized mammals, such as zebras, buffalo, and several species of antelope. For one treatment, the researchers excluded all mammals from some exclosures with a 2.3-m-tall electrical fence. For a second treatment they hung a single electrical strand 2 m off the ground, which allowed all but the very tallest mammals (giraffes and elephants) to pass through. A total of six exclosures allowed no mammals, six allowed all mammals except for giraffes and elephants, and six control areas allowed all

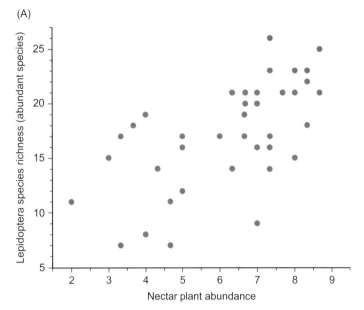

(A)

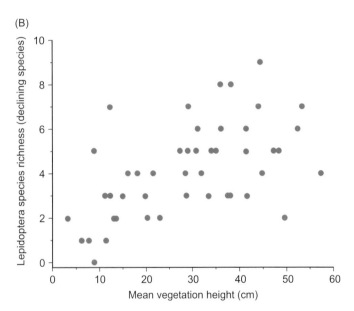

(B)

Figure 17.6 Biotic factors affecting Lepidoptera diversity. The researchers considered nine variables: nectar abundance, plant species richness, plant diversity (Shannon index), plant height, management history, patch area, connectivity between patches, temperature, and study region. A. Relationship between nectar plant abundance and species richness of abundant Lepidoptera species. Nectar plant abundance was scored on a 0–10 scale, with 0 = lowest abundance and cover, and 10 = greatest abundance and cover. B. Relationship between mean vegetation height and species richness of declining Lepidoptera species. $P < 0.05$ for both correlations.

mammals. Over a 2-year period, the researchers surveyed all of the birds and arthropods in these exclosures on a regular basis.

Bird species richness was significantly reduced when all mammals, very large and moderate sized, were allowed into the exclosures, but only slightly – and insignificantly – reduced when only moderate-sized mammals were allowed into the

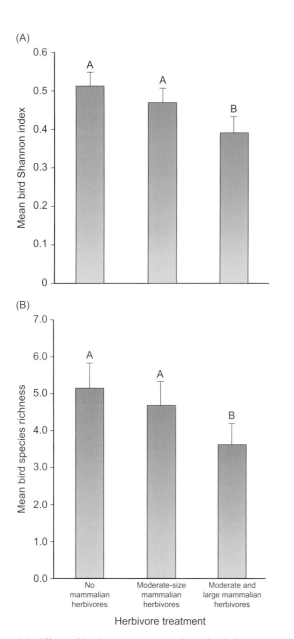

Figure 17.7 Effect of herbivorous mammals on bird diversity in the African savanna, as measured by A. Shannon index and B. total bird species richness for the exclosure experiment. Nonoverlapping letters above bars indicate significant differences between treatments.

exclosure (Figure 17.7). Of 61 bird species observed in all treatments, 45 species were observed in the exclosure that allowed only moderate-sized mammals, and only 31 bird species in the exclosure with all mammals. With these results in hand, the challenge for Ogada and colleagues became to figure out how giraffes and elephants were reducing bird species diversity.

Giraffes and elephants can have numerous impacts on a community. They are prodigious feeders and can remove large quantities of vegetation very quickly. They also generate large quantities of waste products, which can add nutrients to the soil. Their ultra-heavy footsteps compress the soil, which can reduce the soil's ability to exchange gases such as oxygen and carbon dioxide, and may reduce permeability to water. These changes to

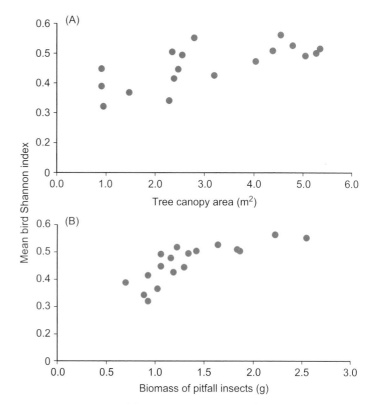

Figure 17.8 Proposed factors underlying the influence of large mammals on bird diversity. A. Bird Shannon index in relation to tree canopy area ($R^2 = 0.76$, $P < 0.001$). B. Bird Shannon index in relation to insect biomass ($R^2 = 0.76$, $P < 0.02$).

the soil could affect both seed and insect abundance, which could in turn influence the richness or abundance of birds in the community.

The researchers measured two of these potential effects: the impact of mammals on vegetation, and the impact of mammals on the insect community. They discovered that giraffes and elephants reduced subcanopy tree coverage by more than 50%. Giraffes and elephants had a somewhat more modest impact on the insect community, reducing the number of insects captured in traps by about 30%.

But are differences in these two variables associated with differences in bird species diversity? The correlation analysis showed that both tree canopy area and insect abundance were correlated with bird species diversity (Figure 17.8). The researchers concluded that mammalian herbivores were reducing bird species diversity indirectly, by reducing tree canopy area and by reducing insect abundance.

 Thinking ecologically 17.2

This study provides evidence that giraffes and elephants decrease bird species diversity by removing subcanopy trees and reducing the abundance of insects. What is a weakness in this study, and how might you follow up on this study to address this weakness?

Our two examples – butterflies in Finnish grasslands and birds in the African savanna – provide evidence that biotic factors can influence species diversity either directly or indirectly. Let's increase our scale now, and return to tropical forests to address an unresolved question about biotic influences on species diversity.

Biotic factors: a controversy in the tropics

Ecologists agree that there are many more folivorous (leaf-eating) insects in the tropics than in temperate forests, even though they don't know how many species there are, because most have not been identified. However, ecologists disagree on what factors underlie this difference in folivorous insect species richness between these two biomes; they offer three major hypotheses. The simplest hypothesis is that the only difference between tropical and temperate communities is host tree diversity (a host tree provides leaves for a folivorous insect to eat). Because there is greater host tree diversity in the tropics, there is greater folivorous insect diversity in that biome. This hypothesis predicts that an individual tree species from a tropical community should, on average, have the same folivore richness as a tree species from a temperate forest.

A second hypothesis to explain the differences in folivorous insect diversity is that trees in tropical forests serve as hosts to a greater diversity of insects than do trees in temperate forests. This hypothesis predicts that an individual tropical tree species should, on average, host a richer community of folivores than would an individual temperate tree species.

Finally, the third hypothesis is that insects in tropical forests have greater host specificity, with each species of insect, on average, specializing on fewer tree species. In essence, according to this hypothesis, the feeding niche of tropical folivores should be narrower than the feeding niche of temperate folivores. In practice, distinguishing between these three hypotheses is challenging. But recent studies have helped unravel some of the complexity.

Evidence for a simple correlation between host tree diversity and folivorous insect diversity

Vojtech Novotny and his colleagues (2006) tested these three hypotheses by comparing patterns of folivore diversity and host specificity in a lowland tropical forest in Madang, Papua New Guinea, and a lowland temperate forest in Moravia, Central Europe. The researchers were concerned about making this comparison, because they recognized that differences in folivore community richness or folivore host specificity could arise from differences in the phylogenetic relationships of the trees that were sampled. For example, they expected that two closely related trees would host a similar community of folivores, while two distantly related trees would host very different folivore communities.

To explore this connection between phylogeny and folivore diversity patterns, the researchers chose 14 of the most common tree species from the temperate Moravia site. They then matched this community to 14 tree species from the much more diverse tropical forest from Madang (Figure 17.9A). These species were

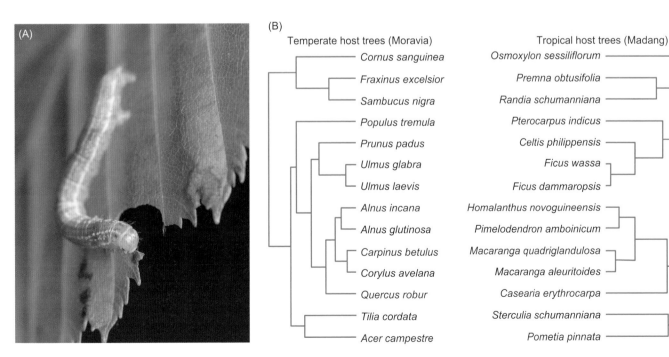

Figure 17.9 A. Caterpillar (*Operophtera brumata*) eating leaves. B. Phylogenies of 14 temperate tree species from Moravia and 14 tropical tree species from Madang. Notice how similar the patterns of branching are when comparing the two communities. The length of each branch indicates the number of DNA base sequence differences separating two species. For example, in the Madang community, *Ficus wassa* and *F. dammaropsis* have the shortest branches, indicating that their DNA base sequences are most similar, while *Sterculia schumanniana* and *Pometia pinnata* have much longer branches, indicating that their DNA is much less similar to each other.

matched based on the differences in the DNA base sequences of several genes, using sequence information available in Genbank. The goal was for the community of trees in Madang to have a similar phylogenetic structure as the community in Moravia. Thus the branching order and branch length of the phylogenetic trees representing the two communities were almost identical (Figure 17.9B).

Novotny and his colleagues compared the folivore communities on each pair of tree species. For example, in the temperate forest, they compared folivores on *Cornus sanguinea* to the folivores on *Fraxinus excelsior*. Then they compared the folivores on *C. sanguinea* to the folivores on *Sambucus nigra*, and so on down the tree, testing all possible pairwise comparisons with *C. sanguinea*. They repeated the process with all possible *F. excelsior* comparisons, ultimately yielding 13! (13 factorial) pairwise comparisons for each forest. The researchers discovered that at both forests, closely related trees tended to host very similar folivore species, while more distantly related trees hosted more dissimilar folivore communities (Figure 17.10). The fitted curves are so similar that they cannot be distinguished. Based on these findings the diversity patterns of tropical and temperate forest were very similar.

To further test the predictions of the first two hypotheses, the researchers scrutinized 150 m^2 of leaves from all 14 tree species in Moravia, and 1500 m^2 of leaves from all 14 tree species in Madang. They raised all larvae to adulthood to help with species identification. They surveyed a larger number of leaves in Madang, because they needed data for other experiments. To account for this difference in sampling effort, Novotny and colleagues normalized the results of their survey by using the average number of species, or individuals, that were found on a 100-m^2 sample at each site. They found some interesting differences among the major groups of insects. For example, there were many more Lepidoptera larvae per 100 m^2 in temperate forests, and many more Coleoptera (beetle) adults per tree species in tropical forests. Overall, trees in temperate forests averaged 29.0 species of folivorous insects, while trees in tropical forests averaged 23.5 species of folivores. These differences were not statistically significant. These findings support the prediction of the first hypothesis that higher tropical insect diversity is simply a reflection of higher tropical forest tree diversity, but rejects the second hypothesis that tropical trees, on average, host a greater diversity of insect species.

To test the third hypothesis that tropical folivores have greater host specificity, the researchers summarized the data for the number of tree species used by insect species as hosts. In both communities, tropical and temperate, most folivorous insect species were found on only one tree species, though a substantial number of insect species used two or three hosts. But importantly, the pattern of usage is almost identical in each community. In contrast to the prediction of the host specificity hypothesis, there is no evidence that folivorous insects in tropical communities feed on a narrower range of hosts.

Research on these two communities shows no evidence for greater host specificity in tropical forests. Novotny and his colleagues warn us that we need to investigate patterns in numerous communities before making any generalizations regarding why insect diversity is so high in tropical forests. They also concede that their surveys could have missed rare folivores, which conceivably could be more prevalent in tropical forests.

These findings are compelling, but research conducted in North, Central, and South America suggests that folivorous tropical insects may have greater host specificity than temperate insects.

Evidence for greater host specificity in tropical folivorous insects

Lee Dyer and his colleagues (2007) used a different approach than that used by the Novotny team and came up with a different conclusion. They addressed the question of whether host specificity was greater in the tropics by sifting through eight very large data sets compiled along a latitudinal gradient in North, Central, and South America. In one analysis, the researchers compared the mean number of plant host species used by caterpillars in the four tropical sites to the number of plant host species used in the four temperate sites. For the four largest caterpillar families, temperate caterpillars fed on a much larger number of plant species than did tropical caterpillars (Figure 17.11). These general findings apply to broader taxonomic scales as well. For example, in Canada, 89% of the caterpillar species fed on more than one genus of plants, in comparison to only 36% in Costa Rica that fed on only one genus of plants.

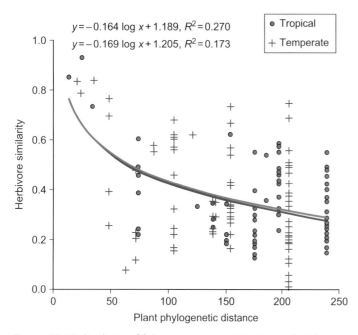

Figure 17.10 Similarity of folivorous communities (using the Chao–Sorenson Index) supported by pairs of host plant species, in relation to plant phylogenetic distance (how closely related the two host plant species are), based on the number of DNA base pair differences between the two host plants.

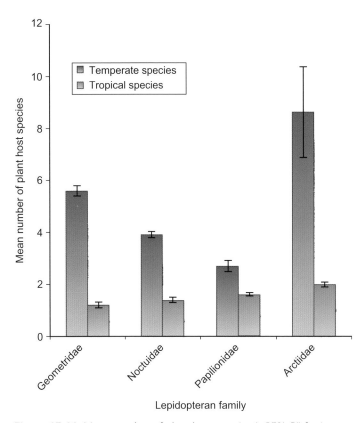

Figure 17.11 Mean number of plant host species (+95% CI) for insects in the most abundantly sampled lepidopteran families in four temperate vs. four tropical forests.

How can we reconcile these differences in host specificity identified by the Novotny and Dyer research teams? One approach is to collect more data. Dyer and his colleagues suggest that there may be real differences in folivore host specificity between the Americas and Eurasia. They also propose that the absence of host specificity differences in the Moravia/Madang study could be an artifact of reducing the data set to 14 tree species. They argue that in their study, host specificity may be greater in the tropics as a result of trophic interactions. In particular, tropical trees may have more distinct and potent defensive compounds, which could reduce a folivore's ability to detoxify leaves from a diversity of potential host species.

Let's now turn our attention to understanding how abiotic factors can influence species diversity. Initially we will explore processes that can influence diversity on a large scale, and then we will shift our attention to smaller-scale processes.

Abiotic factors and species diversity

Ecologists try to understand diversity patterns on many different scales, from global to extremely small. Many researchers study small-scale patterns with hopes that the insight they gain from their observations and experiments will apply to larger scales as well. Researchers can manipulate variables on a small geographical scale that they are unable to manipulate on a very large scale. For

example, they can introduce nutrients to garden plots, or they can create habitats that are identical except for differences in geological features such as slope, aspect, or elevation. Researchers who study global patterns cannot (and obviously should not) manipulate variables affecting entire regions or biomes. We will begin exploring differences in scales of diversity patterns by looking at two studies of abiotic factors influencing a large-scale pattern, and then moving to much smaller-scale studies.

Abiotic factors: land versus water

Earlier in this chapter we described how species diversity was much greater in terrestrial biomes in comparison to marine biomes. Geerat Vermeij and Richard Grosberg (2010) propose that the ability to persist in locally small populations, in other words to survive although rare, underlies the diversity of terrestrial species.

Vermeij and Grosberg propose that differences between air and water as media underlie high animal and plant diversity. Water is thicker or more viscous and has much higher specific heat than air. Because air is less dense, gases tend to diffuse much more rapidly and efficiently in air than in water, which results in oxygen having a diffusion coefficient approximately 10 000 times greater in air than in water. Vermeij and Grosberg argue that small animals such as arthropods can navigate more easily and more rapidly, expending less energy as they pass through the relatively unresistant air. Animals will lose much less heat to air than to water, allowing them to maintain a high body temperature and consequently a high metabolic rate. In addition, the high diffusion coefficient in air makes oxygen more available in air than in water, which also increases a terrestrial animal's potential metabolic rate (Table 17.4). Other molecules, such as pheromones, also diffuse much more rapidly in air, allowing terrestrial arthropods to communicate with each other, and to receive messages from flowering plants, as we discussed in Chapter 14. Lastly, light attenuates more rapidly in water, making it difficult for marine animals to see long distances. Therefore, small and dispersed populations of animals are more likely to survive in terrestrial than in marine environments (Figure 17.12).

Table 17.4 Important physical differences between water and air that may influence species diversity (from Vermeij and Grosberg 2010).

	Sea water	Air
Density (g/cm^3)	1.02	1.2×10^{-3}
Viscosity (g/cm/s)	1.1×10^{-2}	1.8×10^{-4}
Specific heat (J/g)	4.10	1.01
Diffusion coefficient of O$_2$ (m^2/s)	2.10×10^{-9}	2.03×10^{-5}

Figure 17.12 Poor long-distance visibility and low diffusion coefficients reduce the potential for long-distance visual and chemical communication among diverse species found in a coral reef (A) in contrast to greater communication potential in terrestrial ecosystems such as the fynbos (B) in South Africa.

Light attenuation in the ocean also prevents photosynthetic plants from establishing themselves across most of the ocean floor. In contrast, most of the terrestrial surface has an adequate light regime for a diversity of photosynthetic plant species, which can specialize on different light levels. A high diversity of plant species allows animal consumers to specialize on rare, widely scattered food sources. Rare consumers specializing on rare plants may also attract fewer predators. But perhaps the most important interactions between rare consumers and rare plants are pollination mutualisms. Plants separated by very long distances can still exchange pollen, if a suitable pollinator can locate them, and if the plants can lure the appropriate pollinator to pick up their pollen. As we discussed above, pollinators can see farther and travel longer distances over land than in water. In addition, pollinators may be more motivated to secure the food reward offered by terrestrial plants than that in marine environments, because there are very few suspended food particles in air, in contrast to a relatively high abundance of suspended food particles in the highly buoyant water.

The evolutionary process shaping the relationship between pollinator and plant operates through direct and indirect interactions. Mobility of consumers allows stationary producers to be rare, which supports the evolution of relatively rare consumer and mutualist specialists. In addition to pollinators, mutualists may include mycorrhizae and dispersers of seeds and plant propagules. The expansion in diversity of mutualists was set in motion by the explosion in angiosperm diversity about 110 million years ago. That event begs the question of why angiosperms became so successful at that point in time – a question to revisit at the end of this chapter.

The basis of Vermeij and Grosberg's argument is abiotic – the difference between water and air as media for physical processes. But their argument quickly leads to a consideration of biotic factors as well; in this case interactions among mutualists. Let's

next consider how a second abiotic factor – environmental heterogeneity – influences terrestrial species diversity.

Abiotic factors: geological heterogeneity

In our earlier discussion of tropical species richness, we showed a map that highlighted five regions of the world with unusually high plant diversity. In a later paper, Wilhelm Barthlott and his colleagues (2005) describe these five global centers of plant diversity in more detail. Each center has a species richness in excess of 6000 species per 10 000 km², and also exhibits a very high rate of *endemism*, with at least 30% of their species found in that region and nowhere else. But each center is extremely geologically heterogeneous, with elevations that range a minimum of 2800 m. These elevation differences cause differences in moisture, temperature, soils, and many other dimensions that define a species' ecological niche. Thus regions with a great deal of geological heterogeneity are likely to support considerable species diversity.

Let's now turn to a smaller-scale study that investigates how nutrient heterogeneity influences species richness.

Abiotic factors: nutrient levels

Recall that in Chapter 9 we defined a niche as an *n*-dimensional hypervolume, where each dimension is the level of one environmental variable that a species requires for survival and reproduction. Within a community, a species' realized niche is determined by species interactions, with competition being the best-studied interaction in relation to niches. There is good evidence that some species have evolved traits that allow them to compete effectively when a particular resource, for example nitrogen, is limiting or in short supply. In most cases, a species that competes well when one resource is limiting will not compete as well when a different resource is limiting. The *niche dimension hypothesis* proposes that there are more niche dimensions when there are a greater number of limiting resources within a community, because each limiting resource opens up opportunities for a species that competes well when that particular resource is in short supply. This hypothesis predicts that decreasing the number of limiting resources will decrease species diversity.

To test this hypothesis, Stanley Harpole and David Tilman (2007) systematically supplemented a series of grassland plots at Sedgwick Reserve in California, USA, with one or more of the following: (1) nitrogen, (2) phosphorus, (3) cations (K^+, Ca^{2+}, Mg^{2+}), and (4) water. They established 16 treatments, with six replicates of each treatment, for a total of 96 plots. Four treatments received one additional resource, six treatments received two additional resources, four treatments received three supplemental resources, and one treatment received all four supplemental resources. One treatment was an unsupplemented control.

The researchers designed this experiment to vary the number of dimensions in the niche; presumably with greater supplements there are fewer limiting resources, resulting in fewer niche dimensions, and consequently reduced species richness. In accordance with this prediction, the researchers found that species richness declined with the increasing number of supplemental resources (Figure 17.13A). In particular, the treatment with four added resources had substantially fewer species than the control plots or the plots with only one supplemented resource. When all four resources were supplemented, the invasive grass *Bromus diandrus*, an excellent competitor for light, increased from 24% to 56% relative biomass.

To compare their experimental results with natural environmental heterogeneity, the researchers established 215 unmanipulated plots in the same reserve, and measured natural levels of nitrogen, phosphorus, potassium, and magnesium. Each plot was categorized as high or low for each resource, based on whether it was above or below the median level of that resource in the 215 plots. Of course, natural environmental differences between the plots were much smaller than the differences between the artificially supplemented plots. Nonetheless, the researchers found that plots with high levels of all four nutrients tended to have lower species richness than plots with lower levels of these four nutrients (Figure 17.13B).

Harpole and Tilman remind us that human activities are adding large quantities of nutrients to communities and thereby reducing niche dimensionality. If we extrapolate from a small scale to a large scale, resource supplementation may reduce species diversity within communities and across the globe.

Both of the studies discussed above argue that differences in abiotic factors can influence the number of species that fit into an ecological community. In effect niche space is increased by greater geological heterogeneity or by increased number of limiting resources. This perspective focuses on the community in an equilibrium state; we can predict species diversity by knowing how resources and environmental factors vary spatially within the community. However, the natural world is not so simple, and no serious community ecologist would argue that

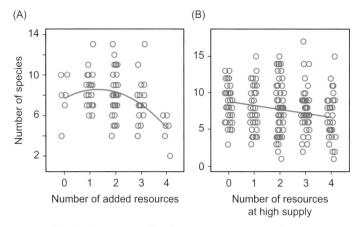

Figure 17.13 High nutrient levels can reduce species diversity. A. Number of plant species in relation to number of supplemented resources in experimental plots. B. Number of plant species in relation to the number of resources in high supply (above the median value) at 215 natural sites.

we can accurately predict species diversity simply by knowing how resources and environmental factors vary spatially at a particular time. For example, disturbances can disrupt the structure of a community by changing either the physical environment or the resources it contains.

Abiotic factors: disturbance

Ecologists recognize that communities are dynamic, and that conditions are changing over the course of time. Relatively unpredictable disturbances such as volcanoes, storms, fires, and floods, and more predictable seasonal climatic variation, as well as complex biotic interactions, will move a community away from an equilibrium species diversity. We will discuss disturbance in more detail when we explore in Chapter 21 how disturbance influences ecosystems. For now, we will consider one example of disturbance influencing species diversity within a community, and explore one general principle of how disturbance has its effect.

TREEFALL GAPS: AN EXAMPLE OF DISTURBANCE AFFECTING SPECIES DIVERSITY. A large tree that falls in a tropical forest creates immediate changes to the local environment. The most dramatic change is the increase in light that reaches the canopy floor. Understory plants suddenly have much greater access to photosynthetically active radiation (PAR). Other important environmental changes can include increased temperature, decreased moisture, and higher rates of evaporation. Biotic changes may include higher transpiration rates in the understory plants, and changes to the species composition of the entire community.

Many researchers have studied how the plant community responds to treefall gaps – openings in the canopy formed when an overstory tree falls down. Ecologists propose that treefall gaps help to maintain plant species diversity in the forest, by creating an environment conducive to the growth and development of shade-intolerant pioneer species that would otherwise be unable to compete in a more homogeneous environment. Some ecologists also propose that treefall gaps can increase the diversity of shade-tolerant species as well, by creating a variable light environment that can be partitioned by tree species that thrive under a range of light conditions.

Stefan Schnitzer and Walter Carson (2001) wanted to know whether the environmental heterogeneity associated with treefall gaps actually increased plant species diversity. They compared species diversity in 17 gap and non-gap (control) sites in a tropical forest in Baro Colorado Island (BCI), Panama. The researchers only studied gaps greater than 5 years old, because they were most interested in the long-term effects of gaps on species diversity. They hypothesized that the density and species richness of shade-tolerant trees, pioneer tree species, and lianas – woody vines that climb trees – would all be greater in treefall gaps. When a tree falls, so do its lianas, and because many different lianas form a tangled network across the forest, a tree-fall can cause many lianas to move into the treefall gap (Figure 17.14A). Until this study most researchers had ignored the influence of gaps on lianas, which are notoriously difficult to identify to species.

Supporting Schnitzer and Carson's hypothesis, both liana richness and pioneer tree species richness were significantly greater in treefall gaps than in the control sites (Figure 17.14B). In fact, the researchers never found any pioneer tree species outside of treefall gaps. However, there was no effect of gaps on species richness of shade-tolerant trees. Previous studies on younger gaps on BCI revealed an increase in shade-tolerant tree species in gaps, so presumably this effect is only temporary and disappears as gaps get older. The researchers point out that the woody flora at BCI is composed of 28% liana and 15% pioneer tree species, so the impact of treefall gaps on species diversity is biologically significant.

If treefall gaps increase plant species richness, we might expect they also increase the richness of animal species that interact with those plants. In one study investigating that hypothesis, Sylvia Pardonnet and colleagues (2013) surveyed fruit-feeding nymphalid butterflies in treefall gaps and the surrounding closed canopy within a Peruvian rain forest. The researchers found no differences in butterfly abundance, but they did find 42% greater species richness in treefall gaps in comparison to the surrounding closed canopy. There were also major differences in species composition with 39% of the species present only in gaps, and 13% of the species present only in the surrounding closed canopy.

Do all disturbances increase species diversity? The answer to this is clearly "no," as evidenced by the complete loss of species diversity in nearby communities following major disturbances such as the eruption of Mount St. Helens. Let's discuss one general principle that relates species diversity to the intensity of disturbance.

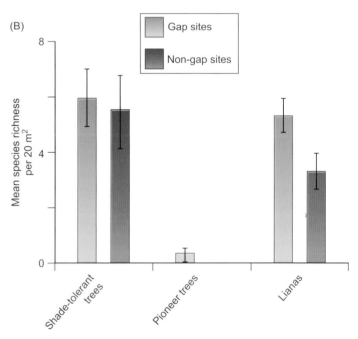

Figure 17.14 A. Downed tree, entangled with lianas, forms a treefall gap in a tropical forest. B. Mean species richness for different types of woody vegetation at gap and non-gap sites. $P = 0.03$ for pioneer species, $P = 0.0003$ for lianas.

HOW INTENSITY OF DISTURBANCE INFLUENCES SPECIES DIVERSITY. Joe Connell (1978) explored six hypotheses for how high species diversity is maintained in natural communities. We will discuss the intermediate disturbance hypothesis, which proposes that highest species diversity is maintained at intermediate levels of disturbance. The rationale is that frequent or intense disturbance will favor only pioneer species that have good dispersal powers and that reach maturity quickly. But as disturbance becomes less frequent or intense, other species that are poorer dispersers and are slower to mature can enter the community. As disturbance is reduced further, species diversity will decline because species that are good competitors will outcompete the pioneer species for resources (Figure 17.15A).

As evidence, Connell presented data from a coral reef community off Heron Island in Queensland, Australia. The reef was struck by a hurricane in 1972, which damaged much of the coral on the outer slopes of the reef, but did less damage to coral in more protected areas. He surveyed coral diversity 4 months after the hurricane and found low species richness in the protected areas, dominated by a few species of competitively superior staghorn coral, which outcompetes its neighbors by shading them out. He also found low diversity in the highly disturbed areas, where most of the coral was killed and only a few pioneer species had begun recolonizing. But diversity was highest in areas of intermediate disturbance, where 10–50% of the coral survived the storm (Figure 17.15B). Connell suggested that

diversity was highest in this area of intermediate disturbance because some of the competitive species survived and regenerated new growth, but plenty of habitat remained available to be colonized by propagules of pioneer species.

Connell's coral reef data provided weak support for the intermediate disturbance hypothesis. But Connell's hypothesis spurred other researchers to test the hypothesis, which continues to be modified and extended even in the past few years. As part of his doctoral dissertation, one of Connell's students, Wayne Sousa, tested the intermediate disturbance hypothesis on an intertidal boulder field along the southern California coast. Sousa reasoned that boulders in the ocean are disturbed by waves when they are moved into new positions. He predicted that easily moved boulders experience high wave disturbance, while difficult-to-move boulders experienced low wave disturbance. Boulders are mostly protected from extreme wave action by the Channel Islands, except during winter, when northwest winds bypass the islands and bring very high waves to the boulder field.

Sousa measured disturbance for each boulder by attaching a spring balance to each boulder and attempting to pull the boulder toward shore. The amount of force required to successfully dislodge the boulder indicated the expected level of disturbance. Boulders that required less than 49 N (newtons) of force to dislodge were classified as highly disturbed, those requiring 50–294 N were classified as experiencing intermediate disturbance, and those requiring 295 N and above were classified as experiencing low disturbance. Once he had classified the boulders, Sousa conducted surveys every 5 or 6 months and counted the number of species on each boulder. Colonizing species were mostly red and green algae and barnacles.

As predicted by the intermediate disturbance hypothesis, species richness was greater for boulders in the intermediate disturbance class at each census date (Table 17.5). It is particularly noteworthy that boulders in the intermediate disturbance

(A)

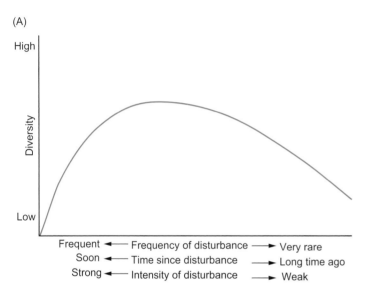

(B)

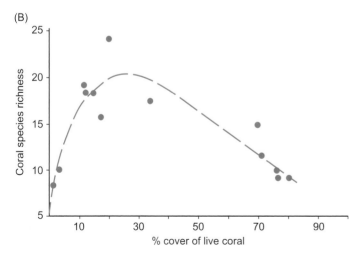

Figure 17.15 The intermediate disturbance hypothesis. A. According to this presentation of the hypothesis, species diversity is maximized when disturbance occurs with intermediate frequency, after an intermediate period of time since the last disturbance, and in association with disturbances of moderate intensity. B. Coral species richness 4 months after a hurricane in relation to storm-induced mortality on a coral reef. Percent cover of live coral is a measure of disturbance intensity.

Table 17.5 Species richness (±SE) at three disturbance levels from four censuses taken over a two-year period. N = Newtons.

	Level of disturbance (force required to dislodge boulders)		
	High (<49 N)	Medium (50–294 N)	Low (>294 N)
November 1975	1.7 (0.18)	3.7 (0.28)	2.5 (0.25)
May 1976	1.9 (0.19)	4.3 (0.34)	3.5 (0.26)
October 1976	1.9 (0.14)	3.4 (0.40)	2.3 (0.18)
May 1977	1.4 (0.16)	3.6 (0.20)	3.2 (0.21)

class tended to be smaller than boulders in the low disturbance class, yet they still had higher species richness (Sousa 1979).

Our discussion of factors that can contribute to the species diversity of a community has only scratched the surface. Historically, ecologists have proposed several hundred distinct hypotheses for factors contributing to species diversity. Thus our task is huge. Not only do we need to continue to test new hypotheses, we also need to identify when a particular hypothesis is likely to apply. For example, several researchers have shown that in their study system, species diversity is not greatest at intermediate levels of disturbance (Fox 2013). Ecologists need to develop a deeper understanding of when a particular factor will have an important impact on species diversity, and when and how multiple factors will interact to influence species diversity.

We have shown that biotic and abiotic factors interact to influence species diversity. Armed with this understanding, we can now turn to the question of what species diversity contributes to an ecological community. Does a diverse community have certain properties that distinguish it from a simple community?

17.3 WHAT IS THE RELATIONSHIP BETWEEN SPECIES DIVERSITY AND COMMUNITY STABILITY?

Intuitively, we might expect that a diverse community would be better in some ways than a simple community. That type of thinking begs the question of what we mean by "better." One possible answer is that a diverse community might be more productive, producing via photosynthesis more biomass over a defined period of time than does a simple community. A second possible answer is that a diverse community may be more stable – tending to retain its current structure and functioning even when disturbed by biotic or abiotic forces. Stability can have a number of different meanings. Resistance is a community's ability to avoid significant change in structure or function when disturbed, while resilience is the speed with which a community returns to its former structure and functioning following a shift caused by disturbance.

In 1973, Robert May published a landmark book entitled *Stability and Complexity in Model Ecosystems*. He reminded ecologists that stability can apply to the community as a whole, or to populations within the community. His major prediction that populations are less stable in communities of high species diversity was counterintuitive to the hypotheses of many ecologists.

Population versus community stability

We will not derive May's stability models in this text, but we will discuss the conclusions of two models, and the results of attempts to test these conclusions. The variables in the first model are the number of species (S), the directed connectance

of the food web (C), and the mean interaction strength (β). Recall from Chapter 17 that directed connectance is the number of links in a food web divided by the maximum possible number of links, and interaction strength is the strength of the link between two species in a food web.

May's model predicts that the community will tend to be stable if

$$\beta(SC)^{1/2} < 1$$

and will tend to be unstable if

$$\beta(SC)^{1/2} > 1$$

The qualitative prediction is that communities are destabilized by increasing interaction strength, increasing number of species, and increasing directed connectance.

This finding surprised many ecologists, because it seemed to imply that increased species diversity actually destabilizes communities. May clarified in the preface to the second edition of his book (released 28 years later) that the type of community stability considered by this model is *species stability*, which is the stability of population levels of individual species within the community, and not the stability of the community as a whole. The model predicts that adding species and increasing connectance and interaction strength will increase the fluctuations in the abundance of individual species within the community but does not address how the entire community responds to increased diversity (May 1973).

A second model presented later in his book predicts that the community as a whole will become more stable with increasing number of species. To summarize, there are two different predictions: (1) high species diversity should decrease the stability of individual populations of species within a community, and (2) high species diversity should increase the stability of communities as a whole.

 Thinking ecologically 17.3

Reconsider May's stability model. We presented evidence in Chapter 16 that increasing the number of species tends to decrease the webbiness of a food web. If this is true, how might it influence May's first prediction?

Rigorous testing of these predictions is difficult, because natural systems that vary only in species diversity do not exist in the real world. We can compare communities that vary in species diversity, and ask whether species in more diverse communities tend to fluctuate more in population size. The problem is that natural communities that differ in species diversity will always differ in biotic and abiotic factors that could influence population stability. A few studies have attempted to test the predictions of May's models while controlling these potentially confounding variables.

Small-scale grassland studies: high species diversity linked to reduced population stability

David Tilman and his colleagues (2006) tested the hypothesis that high diversity leads to low species stability (stability of population levels of individual species within the community). They established 168 grassland communities of 81 m² at Cedar Creek Natural History area in Minnesota, USA (Figure 17.16A). Each plot was planted with either 1, 2, 4, 8, or 16 species that were randomly chosen from a set of 18 perennial grassland plant species. Each plot received the same mass of seeds at planting time. The researchers harvested representative samples of each plot every year, so they could accurately estimate the biomass of each plant species over the 10 years of the study.

For each species, the researchers defined species stability (S) as equal to μ/σ, where μ is the mean biomass value for a time period and σ is the sample standard deviation over the same time period. A high value of S for a species indicates that the population biomass was relatively constant from year to year, while a low value of S indicates extreme fluctuation in biomass from year to year. The researchers found much higher stability for the monocultures (one species) than for each species in the multispecies communities (Figure 17.16B). This finding supports the prediction of May's first model that high species diversity leads to low species stability.

A second measure of stability is resilience, or how quickly a population returns to predisturbance state. Let's explore the relation between species diversity and resilience by shifting to an even smaller scale.

Microcosm studies of microorganisms: no effect of species diversity on resilience

Christopher Steiner and his colleagues (2006) tested the resilience of populations of communities of microorganisms within 200-ml microcosms. They established their communities with one of three species counts – one, two, or four species – from four trophic levels: producers (microalgae), bacteriovores (protozoa), algivores/bacteriovores (protozoa and rotifers), and omnivorous top predators (protozoa). Thus each community began with a total of 4, 8, or 16 species, but the realized species richness – the number of species actually present in the microcosm – changed over the course of the study as some species went extinct. The researchers also inoculated each microcosm with bacteria, which functioned as decomposers within the ecosystem but also served as food for many of the species. Steiner and his colleagues were also interested in how production influenced stability, so they varied the amount of food in their initial setup, creating a low-production and high-production treatment for each diversity level.

To measure resilience there must be a disturbance. Twenty-five days after setting up their microcosms, the researchers mixed up the contents of each microcosm, removed 90% of the contents, and added fresh medium to the remaining 10%. They set aside the

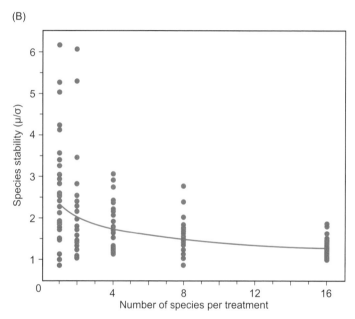

Figure 17.16 A. Experimental gardens at Cedar Creek Natural History Area. B. Species stability in relation to species richness.

90% they had removed as controls. They measured resilience as a function of the amount of time it took the disturbed (diluted) microcosms to approach the same biomass as found in the controls. Measuring resilience is not straightforward, so we will skip the statistical details of how this was actually done.

Overall there was no effect of species diversity on the mean resilience of species. In addition, there was no effect of productivity on species resilience (Figure 17.17). At least for microorganisms in mesocosms, the data do not support the prediction of May's first model, that high species diversity decreases the stability of individual populations of species within a community.

Many other studies have tested May's first prediction; like the two studies we've presented here, some support and others fail to support the model's prediction. Even with conflicting results, the model opens up new question to ask, and motivates new research

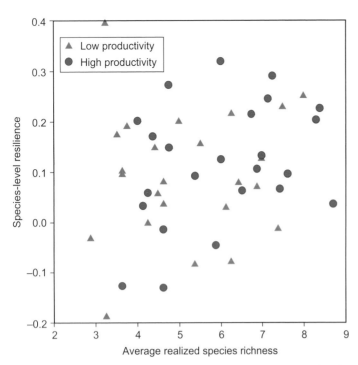

Figure 17.17 Species-level resilience in relation to average realized species richness at low and high levels of productivity. Note that realized species richness (x-axis) declined during the 25-day period before disturbance, while the researchers allowed the microcosms to stabilize from the initial species counts of 4, 8, and 16 species. $P > 0.15$ for both productivity levels.

directions. As one example, we can ask whether the studies that support the model's prediction have any factors in common that distinguish them from studies that do not support the model's prediction. If so, we can explore those factors to help us understand the relationship between diversity and the stability of populations within a community.

Let's now turn to May's second prediction – does high species diversity increase the stability of a community as a whole?

Grassland and microcosm studies: some support for May's second prediction

We can use data from the previous grassland and microcosm studies to test May's second prediction, that increased species diversity results in increased stability of the community as a whole. Tilman and his colleagues (2006) defined community stability (S) as μ/σ, where μ is the mean combined biomass of all the species in the grassland community over a time period and σ is the sample standard deviation. In this case, S is measuring the temporal stability of the entire community at different levels of diversity. In accordance with May's second prediction, community stability is correlated with species diversity (Figure 17.18A). Note, however, that there is a great deal of variation among plots for a given number of species.

The picture is a bit less clear in Steiner's microcosm experiment. Community-level resilience did increase somewhat at

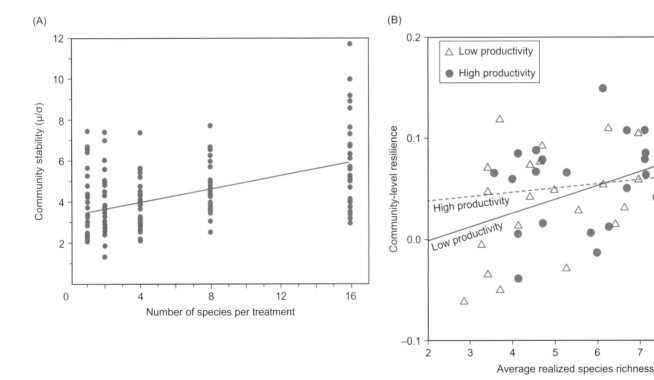

Figure 17.18 Testing the prediction that high diversity increases community stability. A. Community temporal stability in relation to species diversity within grassland plots. B. Community-level resilience in relation to average realized species richness at low and high levels of productivity, in microcosms of microorganisms. R^2 low productivity = 0.163, $P = 0.05$; R^2 high productivity = 0.022, $P = 0.487$.

higher species diversity for the low-productivity treatment, and increased insignificantly for the high-productivity treatment. In both cases, there was a great deal of scatter in the results, indicating that many other factors were also influencing community-level resilience (Figure 17.18B).

The results of these and other studies paint a complex picture, indicating that high species diversity may have a positive effect on overall community stability but that there are clearly other factors that influence stability. Given the number of different factors that affect species diversity in different communities, it is not surprising that a thorough understanding of the relationship between species diversity and community stability remains elusive. For now let's briefly consider three hypotheses for why we might expect that high species diversity would lead to increased community stability.

Hypotheses for greater community stability at higher species diversity

According to the *niche differentiation hypothesis*, communities with more species will partition resources more efficiently, and use a greater proportion of the available resources. Given this efficient use of resources, such a community will fluctuate less in biomass, and show greater resilience following a disturbance. The *facilitation hypothesis* proposes that a diverse community is more likely to have at least one species that enhances some aspect of ecosystem function. These facilitators can create conditions that support the presence of other species within the community.

A final hypothesis, the *sampling effect hypothesis* (also known as portfolio effect and selection effect) proposes that more diverse communities are more likely to contain a highly productive species that responds quickly to disturbance. A community with such a species will be more resilient simply because it happens to have this species. In partial support of the sampling effect hypothesis, Steiner and his colleagues (2006) found that

microcosm communities with the highest community-level resilience were dominated by one of two species of quick-growing green algae in the genera *Ankistrodesmus* and *Chlorella*. Communities with the lowest resilience tended to lack these species in their initial composition, and thus recovered more slowly from the disturbance.

Distinguishing between these three hypotheses is important for conservation ecologists. The first two hypotheses predict that high diversity should lead to overyielding – an increase in ecosystem function in response to high species diversity. One very important ecosystem function is production – thus these hypotheses argue that more diverse communities are more productive, and from a human-centered perspective, have the potential to feed more humans. In contrast, the sampling effect hypothesis does not predict overyielding, instead arguing that more diverse communities will become dominated by one highly productive species, which will competitively exclude other species. Ultimately, the sampling effect hypothesis predicts that a community with this productive species will approach the productivity of a community that is a monoculture of this productive species.

The relationship between diversity and stability will continue to be explored by community and ecosystem ecologists for the foreseeable future. As suggested by Steiner and his colleagues (2006), more productive communities may respond differently to increased species diversity than would less productive communities. In addition, different types of communities may respond differently to increased species diversity; for example, terrestrial communities with flowering plants as their primary producers may have different responses to increased diversity than marine communities that are based on phytoplankton and macroalgae.

Let's revisit our initial discussion of the relative diversity of terrestrial and marine communities to consider the question of why angiosperms became so diverse 110 million years ago.

REVIST: Keys to the success of apple trees and other flowering plants

Leaves are solar collectors; thus it makes sense that large leaves will collect more solar energy than small leaves. Large leaves need a large number of stomata for importing CO_2; these same stomata lose large quantities of water via transpiration. The maximum rate at which water exits a leaf is the hydraulic conductance, which can be very low if water must travel a long distance between the end of the minor veins and the stomata. Because water export and CO_2 import use the same pathways (Figure 17.19A), Tim Brodribb and his colleagues (2007) suggested that there should be a tight positive correlation between maximum CO_2 import rates (or maximum photosynthetic capacity) and hydraulic conductance. This is exactly what they discovered in a study of 43 different plant species (Figure 17.19B). Thus the researchers proposed that measuring the hydraulic conductance of a leaf allows accurate estimation of maximum photosynthetic rate.

In a follow-up study, Kevin Boyce and his colleagues (2009) tested the hypothesis that hydraulic conductance was limited by the density of minor veins that terminate near stomata.

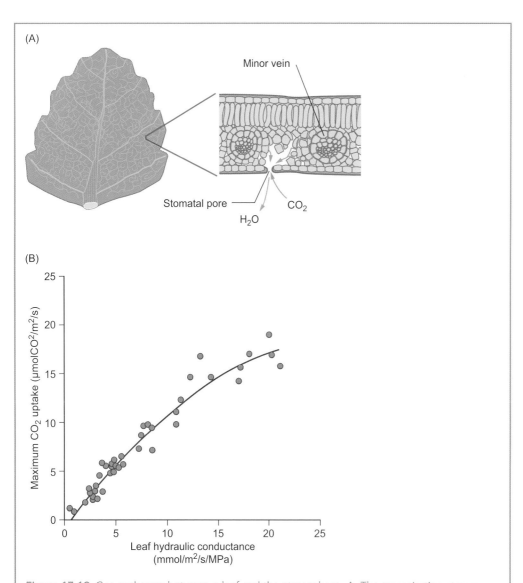

Figure 17.19 Gas exchange between a leaf and the atmosphere. A. The transpiration stream encounters major resistance as it travels from the end of the minor vein to the stomata (squiggly red line). A longer path leads to reduced hydraulic conductance, and slow diffusion of H_2O and CO_2 gases, ultimately limiting photosynthetic capacity. B. Relationship between hydraulic conductance and maximum photosynthetic rate in 43 plant species.

They reasoned that high vein density should reduce the distance that water needed to travel before reaching the nearest stoma, so high vein density should lead to high hydraulic conductance. Their research showed a very strong positive correlation between vein density and hydraulic conductance ($r = 0.93$).

Putting the two studies together, the researchers reasoned that high vein densities would increase photosynthetic rates in plants. Interestingly, flowering plants or angiosperms, such as Shel Silverstein's apple tree, have much higher vein densities than do most other types of plants, so we would expect them to have higher photosynthetic rates than do other plant species. I should point out that high vein density has its down side, because veins take up space but are not photosynthetic, their walls require a great deal of lignin, and they depend on an energetically costly network of cells that provide support services for them.

Brodribb and Taylor Feild (2010) suggest that the tremendous diversity of angiosperms arose from the evolution of high vein density in ancestral angiosperms, followed by an adaptive radiation, or a series of adaptive radiations in those lineages. Their reconstruction of vein density in various lineages of angiosperms supports this hypothesis. The vein densities of

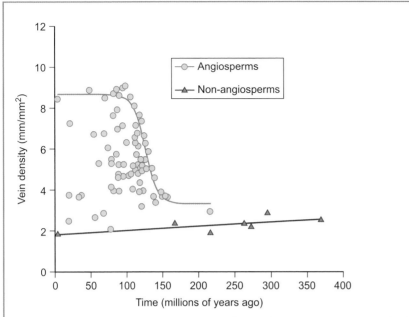

Figure 17.20 Vein density in angiosperm (circles) and non-Angiosperm (triangles) lineages in the past 400 million years, showing the sharp increase in angiosperm vein density beginning about 140 million years ago.

ancestral angiosperms were very similar to other plant species until about 120 MYA. At that time, in the middle of the Cretaceous period, there was a sudden increase in vein density (Figure 17.20). Interestingly, the explosion in angiosperm diversity began only a few million years after the increase in vein density, supporting the hypothesis that the increase in vein density was an important contributor to the dominance of angiosperms in terrestrial communities

High vein density may have caused important changes to the abiotic environment that would have benefited angiosperms. Boyce and his colleagues (2009) speculate that high vein density increased plant transpiration rates, causing increased vapor flux in tree canopies, and particularly in tropical biomes, led to substantially increased rainfall. Increased global rainfall could have potentially increased the amount of habitat highly suitable for angiosperms, ultimately facilitating the spectacular increase in angiosperm species diversity.

SUMMARY

The tremendous biological diversity of some plant communities may be a reflection of the variety of direct and indirect interactions that plants have with predators, competitors, and mutualists. Ecologists have several ways of measuring biological diversity; some diversity indices, such as the Shannon index, integrate species richness and evenness. Alpha diversity measures species richness within an area, beta diversity measures species turnover, while gamma diversity is the combined species richness of all communities under consideration.

Biotic and abiotic factors can influence community diversity directly and indirectly. For example, in southwest Finland, host plant abundance directly and positively influenced lepidopteran species abundance. In the African savanna, herbivorous mammals indirectly and negatively affected bird diversity by consuming trees and reducing the abundance of insects that served as food for the birds. Abiotic factors influencing community diversity include the type of habitat (e.g. land versus water), geological heterogeneity, nutrient levels, and the type and intensity of disturbance. Ecologists predict that diverse communities will be more stable than less diverse communities, but that the populations of species in diverse communities will be less stable. Ecologists are testing these predictions in a variety of different communities using controlled field and laboratory studies.

FURTHER READING

Borer, E. T., and 54 others. 2014. Herbivores and nutrients control grassland plant diversity via light limitation. *Nature* 508: 517–520.

    Controlled experiment conducted on 40 different grasslands (on six continents) – research addresses the question of how human alteration of nutrient cycles and herbivore communities influences biological diversity of plants. The researchers discovered that nutrient addition reduces plant diversity by increasing production and thereby limiting light penetration. This effect is counteracted by herbivores that eat the added production, and increase light availability.

Dornelas, M., and 6 others. 2014. Assemblage time series reveal biodiversity change but not systematic loss. *Science* 344: 296–299.

    Huge analysis of more than 35 000 species records that asks how global species diversity is changing over time. The researchers conclude that α-diversity is not changing significantly, but that species composition within communities is changing quite rapidly, leading to an increase in β-diversity.

Jousset, A., and 5 others. 2014. Biodiversity and species identity shape the antifungal activity of bacterial communities. *Ecology* 95: 1184–1190.

    Carefully designed laboratory study shows that more diverse communities of *Pseudomonas* bacteria produce greater concentrations of antifungal chemicals, which in turn suppressed the growth of certain fungal species. The researchers propose that the effects of microbial diversity need further study, so we can understand how bacteria may function to reduce the impact of plant pathogens on plants in soils.

Ogada, D. L., Gadd, M. E., Ostfeld, R. S., Young, T. P., and Keesing, F. 2008. Impacts of large herbivorous mammals on bird diversity and abundance in an African savanna. *Oecologia* 156: 387–397.

    Nicely designed experimental manipulation of habitat that demonstrates the indirect effects of large mammals on bird diversity in the African savanna. A good paper for discussing experimental design, and also for relating the experimental results to justifiable conclusions.

Pimm, S. L., and 8 others. 2014. The biodiversity of species and their rates of extinction, distribution, and protection. *Science* 344: 987.

    Excellent summary of the state of global eukaryotic biodiversity and the factors associated with probability of extinction. Nice discussion of past and current extinction rates, the biogeography of extinctions, and the patterns of global biodiversity we might expect to see in the future.

END-OF-CHAPTER QUESTIONS

Review questions

1. Contrast the relative diversity of terrestrial and marine communities. What factors may be responsible for these differences?
2. Identify and define each measure of diversity that we have described in this chapter.
3. From a global perspective, is species diversity always greatest in the tropics? Explain your answer.
4. How does the question of whether insects in tropical forests have greater host specificity than insects in temperate forests tie in to the question of why tropical forests are so diverse?

What is the controversy over the answer to this question, and what is the evidence supporting each side?

5. Make a list of each abiotic factor that influences species diversity. Give an example of a study that highlights the importance of each factor.

6. What were Robert May's predictions about the relationship between species diversity and community stability? How do the studies described in this text support or fail to support these predictions?

Synthesis and application questions

1. Stuart-Smith et al.'s (2013) study of fish diversity required a huge effort to systematically sample the world's reef fish communities. What would be an alternative approach to investigating the relationship between species richness and evenness? What are some of the challenges of using Stuart-Smith's approach and the approach you suggested? How might researchers deal with these challenges?

2. When introducing Ogada et al.'s (2008) study of bird and mammal interactions in Kenya, we discussed some ways in which birds and mammals might influence community structure. Can you come up with some other possible direct or indirect mechanisms?

3. Juha Pöyry and his colleagues (2009) found that species richness was correlated with plant height in the Lepidoptera? What mechanism might be responsible for this correlation?

4. Which property of air discussed by Vermeij and Grosberg (2010) do you think is most important for explaining high species diversity in terrestrial communities? Explain your reasoning, and how you would test your hypothesis.

5. One objection to Sousa's (1979) experiment in an intertidal boulder field is that rock size might be confounded with disturbance. In other words there might be some strange effect in which small and large boulders have low species richness, and intermediate boulders have high species richness, independent of the amount of disturbance. Sousa worried about this, and actually conducted an experiment to test whether this was a problem. What might he have done to separate out the effects of size and disturbance?

6. Refer to Table 17.5. How well do Sousa's (1979) data support the intermediate disturbance hypothesis? Explain your answer.

7. If Boyce et al. (2009) are correct that high vein density is an evolutionary breakthrough associated with angiosperm dominance, what question would you pose as a follow-up to their experiment? How might you approach answering this question?

8. Boris Worm and his colleagues (2005) report diversity peaks in the open oceans for tuna, billfish, and certain species of zooplankton at latitudes of between $15°$ and $30°$. Propose two hypotheses for why these groups have highest diversity at those latitudes. How would you test your hypotheses?

Analyse the data 1

Return to Table 17.3 and calculate the Sorensen's index (β) for each site in relation to Site 1. Then plot how β varies in relation to distance from Site 1 (recall that each site was separated by 25 m). Is there a statistically significant correlation?

Chapter 18
Dan Janzen and Winnie Hallwachs: community interactions and tropical restoration through biodiversity conservation

INTRODUCTION

In 1956, following his junior year of high school, Dan Janzen found himself alone in Mexico, with his primary means of transportation – a borrowed motor scooter – wrecked when it flew over a cliff where a bridge was supposed to be. His self-appointed mission was to wander throughout the country during his summer break and collect as many butterflies as he could. As he describes, "I lived in the backs of gas stations for nothing. I learned that you could go to a whorehouse and talk to the ladies, and I had no idea of what a whorehouse was, but I discovered I could get a free room because they were fascinated with this kid who collects butterflies. I learned to mooch food in the marketplace; I think I probably spent $20 on the entire summer."

His passion for butterflies began much earlier; when Janzen was 10, his family took a trip to the University of Minnesota Lake Itasca Biological Station, and Dan spent considerable time and effort there catching butterflies by hand. Seeing him in action, and sensing a budding biologist, Bill Marshall, a biology professor removed a butterfly net from the Itasca entomology laboratory and presented it to Dan, whose success rate and enthusiasm for butterfly catching soared to new heights. This larcenous liaison led to a career in collecting and inventorying, which still consumes many of Janzen's waking hours.

Much of Janzen's current butterfly and moth collection efforts occur at Area de Conservacion Guanacaste (ACG) in northwestern Costa Rica. In this chapter, we will discuss how Janzen began doing research there, and how he, and his wife and colleague Winnie Hallwachs, work with many government officials and private citizens from Costa Rica and elsewhere to conserve and help restore ACG back to a tropical forest. Early in his career, Janzen investigated numerous basic and important questions in evolutionary community ecology. Over time, he branched out into inventorying the diversity of Lepidoptera, their parasitoids and their host plants in ACG, so that their complex trophic relationships could be understood by researchers studying there and by students who were being educated in the ACG natural classroom.

Let's return to a much younger Janzen, so we can appreciate the series of events that led him to begin his research career in community ecology, and to develop the background, interest, and expertise to carry out his conservation efforts.

KEY QUESTIONS

18.1. What experiences and events launched Janzen's career?

18.2. How did viewing community interactions from the perspective of the organism and its evolutionary history shape Janzen's research program?

18.3. What factors motivated Janzen's conservation efforts in Costa Rica?

18.4. What factors supported the restoration of Santa Rosa, and ACG as a whole, and the continuation of scientific research there?

18.1 WHAT EXPERIENCES AND EVENTS LAUNCHED JANZEN'S CAREER?

Janzen credits both parents for his mixture of organizational and creative skills. He describes his father's background as "very stoic Russian–German farmer" with exceptional organizational skills, and his mother as "wild Irish and off-the-wall creative." In addition, both parents were very tolerant of his passion for all outdoor activities, and both were interested in nature, and they gave their son the freedom and emotional support to develop his interests.

Excursions into Mexico and beyond

One day Dan's father came home from work as the Regional Director of the U. S. Fish and Wildlife Service in Minneapolis, Minnesota, announcing that he had 2 months' vacation and that it was "use it or lose it." Dan, now in ninth grade, suggested that they go off to Mexico to collect butterflies. Back then there were no interstate highways, nor air conditioning; nonetheless the Janzens spent the entire 2 months traveling across Mexico, loving the wildlife, the culture, and the natural beauty of the region. The next summer they traveled even farther south, into Guatemala. (Note that these days collecting permits and export permits are required for all collections of this type – these requirements help to protect wildlife.)

Dan's butterfly collection continued to grow; the *Minneapolis Tribune* thought it was so marvelous that they displayed it as an advertisement in the window of a downtown bank building, and included his butterflies in some full-page ads that they ran in *Newsweek* magazine. Webb Disney, a relative of Walt Disney, saw the ad while Dan and his family were cruising through Mexico on their way to Guatemala after Dan's sophomore year in high school. Disney called the US Embassy in Mexico City, which somehow managed to find the car that was now full of Janzens and butterflies. Disney invited Dan to be his guest, all expenses paid, the following summer. This adventure marked Dan's introduction to airplanes, which he approved of, but also his introduction to people using him in ways that interfered with his developing interests. This did not meet with his approval.

After Dan was in Mexico for 3 days in 1956, it became clear that he was Disney's "prize," and that most of his time would be spent as a prop at business meetings and events associated with promoting the Disney corporate initiatives. It is an understatement to describe this as conflicting with Dan's summer plans, and after some intense discussion, Disney washed his hands of any responsibility for Dan's well-being, and loaned Dan the motor scooter that he could use to drive around Mexico collecting butterflies, and which he ultimately sailed over a cliff. In the process Dan learned Spanish, tolerance, and interest in other cultures, and began thinking about how he could somehow find a job that would take him to different places so he could continue indulging his need to explore. He settled on civil engineering because he envisioned himself collecting butterflies on the side while he was out building roads and bridges at interesting locations.

Figure 18.1 The natural world at ACG. The caterpillar, *Erinnyis ello*, is succumbing to an attack by *Microplitis Figuerasi* parasitoids that developed inside the caterpillar, and have punched out through its cuticle to spin cocoons and complete development.

Near the end of his first year at the University of Minnesota, a huge lightning storm hit, knocking off a chunk of the student union building and forcing Dan inside the zoology building next door. He stumbled into the zoology department, and as he describes it, "On the wall was a glass case with a pair of stuffed wood ducks. I turned around and asked a guy sitting behind a desk, 'What do you do?' Professor Magnus Olsen replied 'I teach zoology. Why not attend my lecture this afternoon.' I did, and I discovered these graduate students and faculty members all doing the same thing that I had been doing, but now in a formal academic setting. Boom, that was home."

In reality, Janzen is equally at home in two worlds. One is the world of urban academia, where he teaches an ecology course entitled "Humans and their environments" to University of Pennsylvania students, puts together grants to support his research programs, and writes papers describing his research findings. The second is the natural and political world in Costa Rica, which he shares with his wife and colleague, Dr Winnie Hallwachs. In Costa Rica, they investigate together the mysteries presented to them by tropical ecology and work with Costa Rican officials and scientists and neighbors, to conserve what remains of the natural world, and to restore ecosystems that have been severely damaged (Figure 18.1).

Costa Rica and academia

In Chapter 15, we discussed Janzen's research in Veracruz, Mexico, in which he describes the mutualism between ants and acacia trees as a "fully developed interdependency." Ants defend the acacias (now in the genus *Vachellia*) against herbivores and against other plants that might compete for space or nutrients, and acacias provide the ants with food and shelter (Janzen 1966). In 1963, while a graduate student at the University of California at Berkeley doing this field research in Mexico, Janzen received a phone call from a Professor Jay Savage inviting him to join a 2-month-long tropical ecology course in Costa

Rica that had previously been funded by the National Science Foundation for faculty only, but was now being expanded to include a group of 14 graduate students. Janzen was deep into his ant research but was coaxed by his graduate advisor to accept the invitation, and attended the first graduate course in tropical ecology sponsored by the Organization for Tropical Studies (OTS). In 1965 he taught this course, which has evolved into the intellectual birthplace of many of the world's most important tropical biologists.

While a student in that prototype OTS course, Janzen was dissecting an acacia tree in the reception area of a small hotel in the middle of San Jose, Costa Rica's capital city. An old man in a long overcoat, dripping from the rain, walked into the room and asked Janzen what he was doing. Janzen explained to him about his acacia research, and the man then asked what Janzen knew about the Universidad de Costa Rica. Janzen said that he knew nothing, except that he was being forced to do pointless coursework in their classrooms instead of exploring Costa Rica's wilds. Exit the old drippy man. A few weeks later, Janzen was digging for ants in a Costa Rican coffee plantation and he came across a mud-covered man excavating in the wall of a deep ditch with a pen knife, who explained to him that he was digging up a bee's nest. So they chatted for a few minutes, and then both resumed their respective digging operations.

A year later, Janzen was back in California; his major professor, Hal Daly at Berkeley, arranged a job interview for him at the University of Kansas. At the airport in Kansas, Janzen was met by the man from the bottom of the ditch, who it turns out was Charles Michener – one of the world's most famous entomologists and Chair of the entomology department at Kansas. Janzen's job seminar went well, and after the talk he was ushered in to meet with the dean, who turned out to be the old man in the drippy suit, who had been in Costa Rica to set up an exchange program between students from Costa Rica and Kansas. Despite his less-than-flattering portrayal of the University of Costa Rica coursework, Janzen was offered and accepted the job at Kansas, and worked there for 4 years, while continuing his research in Costa

Rica during breaks in his teaching schedule. He followed the Kansas position with 4-year stops at the University of Chicago and the University of Michigan, before finally settling at the University of Pennsylvania in 1977, where he remains to this day.

During the 1970s, Janzen's diverse research activities established him as one of the world's premier tropical ecologists. Part of Janzen's success was his unique perspective toward understanding community interactions.

18.2 HOW DID VIEWING COMMUNITY INTERACTIONS FROM THE PERSPECTIVE OF THE ORGANISM AND ITS EVOLUTIONARY HISTORY SHAPE JANZEN'S RESEARCH PROGRAM?

Janzen recognized that natural selection operates primarily on individual organisms, and that its mechanism is differential survival and reproduction of individuals with traits that are well adapted to their environment. Consequently, many of his studies ask questions that are explicitly framed in terms of what a particular species or group of species does in order to survive or reproduce successfully. Examples include: "How to be a fig" (Janzen 1979a), or the more pointed "How many babies do figs pay for babies?" (Janzen 1979b), "Why bamboos wait so long to flower" (Janzen 1976), "Mice, big mammals and seeds: it matters who defecates what where" (Janzen 1986), and "Why fruits rot, seeds mold, and meat spoils" (Janzen 1977). Central to all of these studies is that Janzen was able to imagine what was truly important for the specific species he was writing about; he tried to view the world from their perspective. He was particularly rankled when ecologists described plants as resources for animals, without also viewing them as unique species with their own strategies for success.

Plant reproductive success and life history strategies

In the mid-1960s, ecologists debated to what extent ecosystems are self-regulating. Some ecologists believed that ecosystems are not self-regulating – instead they are highly dynamic – and that they frequently change species composition and their types of community-level interactions. Other ecologists claimed that ecosystems are self-regulating – that even if one species goes extinct, another functionally similar species will fill its niche, and the same general types of community-level interactions will continue, albeit among different species. In support of the self-regulation hypothesis, Lawrence Slobodkin and his colleagues (1967) described how wholesale destruction of the American chestnut by the chestnut blight fungus, *Cryphonectria parasitica*, did not affect the functioning of eastern deciduous forest in the United States. Instead, other trees simply filled in the empty

space formed by chestnut deaths, and the forest retained its overall integrity.

In the process of making their argument, Slobodkin and his colleagues lumped nectar, pollen, and seeds into a category they called "plant products." Janzen did not take too kindly to this, and he wrote, "Seeds are clearly juvenile plants, not products. To place flower nectar and seeds in the same functional category is like placing human sweat and children in the same functional category." (Janzen 1969). His immersion in the forests of Costa Rica reminded him every day that plants are as much a part of the evolutionary theater as are animals, and that they fight similar battles. And many of their battles are fought against predacious insects trying to eat their seeds (children).

Janzen observed two basic approaches that plants could use to fight this battle. One approach, a low investment in each offspring strategy, is to produce huge numbers of very small seeds that barely have enough food to nourish the embryo during development, but that may escape predation through being so numerous that predators are satiated before they consume all available seeds. A second approach, a high investment in each offspring strategy, is to produce small numbers of large seeds that are well-provisioned with food to nourish the embryo, and with toxic chemicals to deter predators. Putting both approaches together, Janzen proposed a life history hypothesis that within a taxonomic group (such as a family), species that produce numerous small seeds will suffer high seed predation, while species that produce relatively few large seeds will suffer low or no seed predation.

Janzen chose Leguminosae as his family of plants, many of which are prey for pea weevils from the family Bruchidae (Figure 18.2) (today they are classified as the subfamily Bruchinae within the Chysomelidae). The beetles lay their eggs on the fruits (pods) or seeds; the larval bruchid drills into the seed, eats and develops there, and chews its way out when an adult. Janzen chose this system because he was familiar with the plants and their seed predators, and he had colleagues who could help identify the different species. In addition, there are many species of both legumes and bruchids in Central America, but not all species of legumes are attacked by bruchids. This allowed Janzen to compare seed size and number in legume species that were attacked by bruchids to those that were ignored by bruchids. Lastly, the seeds were large enough to be easily collected and counted.

Between 1965 and 1967, Janzen harvested the seed crops from 35 different species of legumes in lowland Central America between 9° and 21° latitude. He counted and weighed the seeds from each tree, and inspected the seeds to see if they were attacked by bruchids. In support of Janzen's life history hypothesis, legume species whose seeds were usually attacked tended to have seeds that were considerably smaller and much more numerous than species whose seeds were routinely ignored by bruchids (Table 18.1).

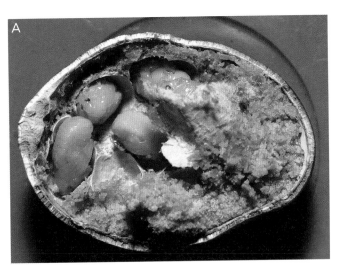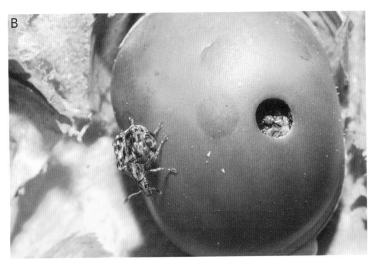

Figure 18.2 A. Several pea weevil pupae, *Caryedes brasiliensis* (Bruchidae), developing within a *Dioclea megacarpa* fruit. B. Adult pea weevil emerges.

Table 18.1 Mean values of dry weight per seed (g), total seed dry weight (g) per m³ of canopy, and seed number per m³ of canopy, of 35 species of Central American legumes, grouped according to whether they are attacked or ignored by bruchids (Janzen 1969).

Legumes attacked by bruchids				Legumes ignored by bruchids			
Species[a]	Dry weight per seed	Seed dry weight per m³	Seed number per m³	Species	Dry weight per seed	Seed dry weight per m³	Seed number per m³
Acacia farnesiana	0.0727	65.51	1157.6	*Mucuna mutisiana*	7.9559	116.97	14.7
A. cornigera	0.0747	61.27	1017.3	*Canavalia* sp. 1	0.4212	5.01	11.9
Mimosa pigra	0.0136	93.97	7925.0	*Entada polystachia*	24.0716	7.24	0.3
Cassia biflora	0.0034	19.40	5707.6	*Hymenaea courbaril*	3.5142	1.30	0.4
Mimosa sp. 1	0.0134	82.46	6153.8	*Canavalia* sp. 2	0.4061	15.90	39.2
Canavalia sp. 1	0.0222	2.19	98.6	*Gliricidia sepium*	0.0789	0.65	8.2
Pithecellobium saman	0.1256	19.55	155.6	*Leguminosae* sp. 1	0.0626	3.85	615.4
Dioclea reflexa	5.0000	5.36	1.4	*Erythrina standleyana*	0.2895	5.32	21.2
Rhynchosia pyramidalis	0.0596	17.20	288.6	*Erythrina macrophylla*	0.1534	0.53	3.4
Phaseolus lunatus	0.0601	97.89	1628.8	*Erythrina* sp. 1	0.2578	9.83	40.6
Leucaena shannonii	0.0614	1.89	30.8	*Schizolobium parahybum*	0.7936	1.05	1.3
Acacia sp. 1	0.0578	18.45	319.2	*Enterolobium cyclocarpum*	1.0764	2.57	2.4
A. collinsii	0.0335	57.43	1714.2				

Table 18.1 (cont.)

Legumes attacked by bruchids				Legumes ignored by bruchids			
Species[a]	Dry weight per seed	Seed dry weight per m³	Seed number per m³	Species	Dry weight per seed	Seed dry weight per m³	Seed number per m³
Acacia sp. 2	0.0860	73.71	857.1				
Lysiloma bahamensis	0.0189	26.81	141.8				
Pithecellobium albicans	0.0677	0.55	8.2				
Leucaena leucocephala	0.0492	62.62	1195.2				
A. gaumeri	0.0356	0.43	12.0				
A. gentlei	0.0565	29.72	530.2				
A. hindsii	0.0385	13.20	342.8				
A. cochliacantha	0.0242	7.78	321.7				
A. rigidula	0.0242	98.91	4300.1				
A. macracantha	0.0478	0.29	6.0				
Means	0.2629	37.24	1474.50		3.2568	14.19	63.25

[a] Some legumes could not be identified to species and are designated as species 1 or 2 of their taxon.

By viewing the battle between bruchid predators and legume seeds from the perspective of a predator attacking a group of juvenile prey, Janzen was able to support the hypothesis that natural selection has favored two distinct life history strategies in the legume family: predator saturation via huge number of offspring, and active defense with large body size and toxic deterrents. In his discussion of the research results (Janzen 1969), he points out that local human residents have used all 12 species ignored by bruchids as a poison against undesirable vertebrates. Janzen was subsequently drawn into studies of seed dispersal, seed chemistry, seed predation by mice and caterpillars, and finally the conservation of the plants and animals themselves.

Thinking ecologically 18.1

Seeds must be dispersed to suitable environments in order to germinate and mature. How might you evaluate the success of each of the two described life history strategies in relation to successful dispersal?

Viewing community interactions from the perspective of the organisms bore many fruits for Janzen. Much of his research and thinking involved considering how plants disperse their offspring to environments that are conducive to offspring survival. Many of Janzen's ideas germinated from a deep appreciation of plant and animal natural history, but have implications extending far beyond his initial observations.

The Janzen–Connell hypothesis and tropical forest species richness

One example of an observation with far-reaching implications began with Janzen standing in a rain forest in Corcovado in 1968 (Figure 18.3). He was collecting fruits underneath an *Aspidosperma megalocarpum* tree – a species of dogbane. The seeds of this tree are primarily wind dispersed into what he called a seed shadow – the distribution of seeds around the tree. Since these seeds are wind dispersed, the seed shadow is most dense near the parent tree, and becomes more diffuse with distance from the parent. Janzen broke open each seed that he gathered, and was stunned to discover that all of the seeds collected under the tree hosted a caterpillar, which meant they were effectively dead. However, the rare seeds he found farther from the tree often

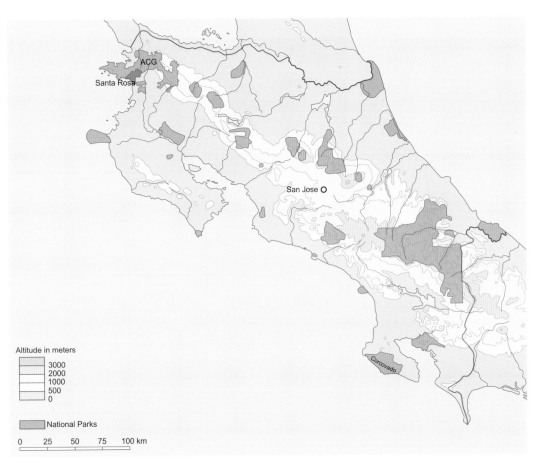

Figure 18.3 Costa Rica and some of its natural areas.

lacked caterpillars. He realized that seed predators could use the parent tree and the mass of seeds as an indicator of a good place to search for prey.

To visualize this process, Janzen (1970) constructed a simple graphical model that described what he had just observed. Based on measuring the seed shadow, he knew that seed density declines sharply with distance from the parent tree. Based on his observations of seed predation, he knew that seeds near their parent had little chance of surviving seed predation, but at a certain distance from the parent, their survival probability increased and presumably leveled off at some low value greater than zero (even seeds that were not eaten by caterpillars have a very low probability of reaching maturity). Depending on the shape of these curves, the product of the seed shadow curve and the survival curve generated a population recruitment curve that predicted the distance from the parent that this species would be most likely to produce surviving offspring (Figure 18.4).

Joe Connell (1971) developed a similar idea when watching seedling trees in Australia at about the same time, and together the published papers of these two researchers formed the basis of what is known today as the Janzen–Connell hypothesis. Because the hypothesis predicts that the peak of the population recruitment curve will be a considerable distance from the parent tree, the implication is that at least in tropical forests, it is unlikely that trees of the same species will be close to each other, thereby leaving space for other

species. This could be one important explanation for high species richness in tropical forests.

Testable predictions of the Janzen–Connell hypothesis

Janzen (1970) outlined several testable predictions generated by this hypothesis, including four to consider here. First, if seeds are placed or naturally found various distances from a parent tree, their survival should increase with distance from the parent. Second, the presence of other adult trees of the same species nearby should further depress seed survival, because seed predators that discover one of the trees will be more likely to also encounter the seed crops of other nearby trees. Third, if seeds are placed in patches of varying densities, individuals at low density will have a higher survival rate, because they are less likely to be discovered by predators. Lastly, if tree species are classified as having either regular, random, or clumped dispersion, the regularly dispersed species are more likely to conform to the predictions of the model.

Over the years this hypothesis and its predictions have served as the basis of many research programs that have explored the dynamic interactions between seeds and their predators. Janzen, however, left it to other researchers to determine whether this hypothesis applied more broadly to other systems or environments. One such study applied the Janzen–Connell hypothesis to interactions between trees and soil pathogens in a temperate forest.

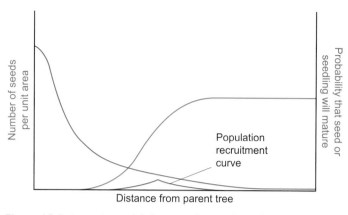

Figure 18.4 Janzen's model showing the number of seeds recruited into the population (the population recruitment curve) in relation to seed density (green curve) and the probability that a seed will survive (red curve).

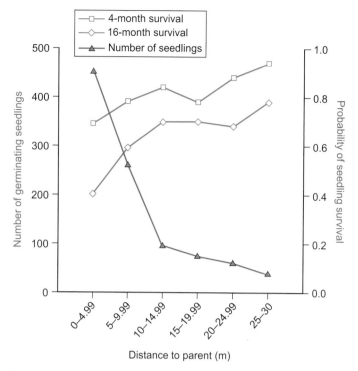

Figure 18.5 Relationship between distance from parent and number of germinating seedlings (green line – left y-axis), and probability of seedling survival (red lines – right y-axis) to 4 months (open squares) and 16 months (open diamonds).

Applying the Janzen–Connell hypothesis to soil pathogens in temperate forests

Alissa Packer and Keith Clay (2000) tested the Janzen–Connell hypotheses in a black cherry (*Prunus serotina*) temperate forest in Indiana, USA. In addition to working in a different biome, the researchers also investigated whether fungal pathogens in the soil could function like seed predators in the model, parasitizing the offspring more effectively when they are in close proximity to their parent tree.

In contrast to Janzen's findings with seed predators in the tropics, cherry seeds close to the parent tree did have high survival rates. Presumably these cherries were not plagued by seed predators like the bruchids that Janzen observed in the tropics. However, in support of Janzen's first prediction, seedling survival after 4 months and 16 months was much lower for trees that were close to their parent tree than it was for seedlings farther from the parent tree (Figure 18.5). The researchers also noted that if trees produced more than the average number of germinating seedlings, these seedlings tended to suffer higher mortality rates. Thus seedling mortality was density dependent.

Packer and Clay suspected soil pathogens at work, and hypothesized that these pathogens would be more prevalent near the parent tree, and like many pathogens, that they would be more successful if their host was living in high-density conditions. To test this line of reasoning, the researchers collected soil either 0–5 m (near) or 25–30 m (far) from the parent trees, and planted seedlings in pots of this soil in a greenhouse. Half of the seedlings from each distance were planted in sterilized soil, in which all pathogens were killed, while the remaining seedlings were planted in unsterilized soil. Half of the pots had three seedlings (high density) while the remaining pots had only one seedling (low density).

After 2 months, most of the seedlings survived in seven of the eight treatments. The one exception was that seedling survival was much reduced in high-density conditions in unsterilized soil that was collected near the parent (Figure 18.6). The researchers

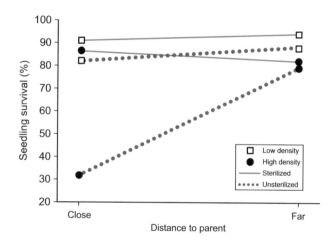

Figure 18.6 Effect of the distance between the parent tree and site of soil collection, plant density, and soil sterilization on seedling survival in a greenhouse experiment.

were able to isolate a water mold in the genus *Pythium* from the roots of the dying seedlings. When they inoculated healthy seedlings with this pathogen, mortality was more than three times greater in the inoculated seedlings in comparison to an uninoculated control.

Packer and Clay concluded that the Janzen–Connell mechanism operates within temperate forests, and discussed two important research directions. First they encouraged researchers to see how often and under what conditions the Janzen–Connell

(A)

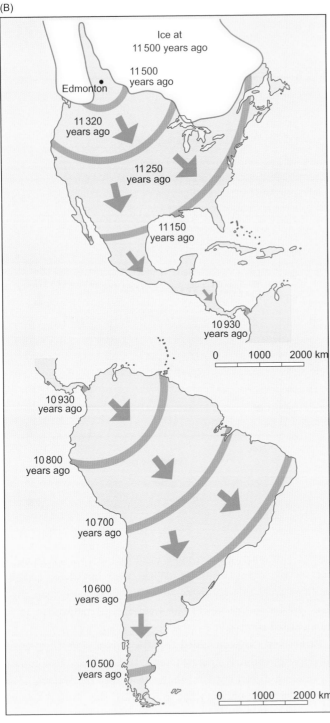

(B)

Ice at
11 500 years ago

Edmonton

11 500
years ago

11 320
years ago

11 250
years ago

11 150
years ago

10 930
years ago

0   1000   2000 km

10 930
years ago

10 800
years ago

10 700
years ago

10 600
years ago

10 500
years ago

0   1000   2000 km

hypothesis applies in other types of temperate forests. Second, they also encouraged researchers to recognize that soil pathogens may be equally likely as, or perhaps more likely than, many herbivores to kill seeds or seedlings in high-density populations near their parent trees.

Janzen recognized that plants are caught in a cruel bind, in that natural selection favors plants whose offspring are attractive to dispersers, but that these same dispersers may also function as predators on these offspring. Much of his early work evaluated the relative strength of positive and negative interactions within communities. This work is particularly challenging, because environments are dynamic, and natural selection operating in one environment may favor traits that are no longer adaptive after the environment has changed. Janzen was able to view the world from a perspective of an organism adapted to living in a world that no longer existed.

Ecological communities and historical selection pressures

The mammal community in Central America today is an impoverished remnant of what once roamed its lowland forests. Missing are giant ground sloths that tipped the scales at over 3000 kg, armadillo-like glyptodonts heavier than horses, 5000 kg mastodons, and their slightly smaller cousins, the gomphotheres (Figure 18.7A). Even rodents such as the giant capybara weighed in at over 100 kg. All of these dramatic mammals have one thing in common – they ate plants and dispersed seeds.

What happened to these giant mammals? Paul Martin (1966, 1973) proposed that about 11 000 years ago, a band of big-game hunters crossed the ice from Asia into North America, and spread southward explosively, killing most of the large mammals as they occupied most of the Americas (Figure 18.7B). Because these mammals had never experienced humans before, they did not fear them as predators and were easily dispatched by these big-game hunters with their advanced hunting technology. According to Martin's model, which assumes an annual human population growth rate of 3.4%, this entire wave of human conquest was accomplished in about 1000 years.

Janzen realized that today's tropical plants might have fruits adapted to being eaten by those extinct mammals, using them as their seed dispersal agents. He wrote to Martin, and together they built an argument that the rapid extinction of these large mammals had a profound influence on the life history of many tropical fruiting trees. They proposed that the fruits and seeds of many tropical trees were adapted to being eaten by these now-extinct megafauna, but that the surviving tree species have not yet responded to the absence of these animals. These surviving tree species have traits that seem puzzling, unless

Figure 18.7 A. Gomphotheres in North America before the invasion of big-game hunters. B. Sweep of hunters through North and then South America.

they are considered in light of these missing mammals (Janzen and Martin 1982).

Janzen and Martin introduce us to the large forest palm tree, *Scheelea rostrata* (now *Attalea rostrata*) and the animal community of 12 000 years ago associated with its egg-sized *drupes* –fruits with a fleshy outer part that surrounds a hardened shell or nut containing the seed or seeds (the peach is a familiar drupe). Each *Scheelea* tree drops thousands of drupes, which 12 000 years ago might have attracted a herd of gomphotheres, a group that would have made quick work of this massive drupe drop. As the gomphotheres moved on, the hard nut wall protected the soft fleshy seed from being crushed and digested, and the nuts were dispersed wherever the gomphotheres defecated. Thus *Scheelea* was well adapted to dispersing its offspring far away from the parent tree.

Of course, most of the *Scheelea* babies never germinated – instead falling victims to seed predators such as agoutis, which combed through the dung, gnawed open the nuts, ate their fill of seeds, and buried the rest of the nuts for future consumption. Some of these buried nuts did germinate, unless discovered by bruchids, squirrels, or other seed predators. Occasionally, a drupe was missed by the gomphotheres, and was then eaten by tapirs or peccaries, which chewed up the fruit and spat out the nuts. Despite this community of seed predators, some *Scheelea* seedlings did survive, in some cases many kilometers removed from their parents, and a few of them matured into adult trees and renewed the cycle.

Returning to the present, the gomphotheres are gone, and when the *Scheelea* drops its drupes, almost all remain under the parent tree and rot, or are killed by bruchids that are attracted by the odor cue released by the fruit. A few are eaten by agoutis and peccaries, and may be taken to new locations. An occasional drupe may roll far enough from its parent that it escapes its shadow and can germinate and survive. So *Scheelea* is still with us today, but it is probably much less common than previously. Ecologists may wander onto this scene and be astounded by this inefficient reproductive behavior, unless they consider *Scheelea* as a species that was well adapted to a community that no longer exists.

Janzen and Martin (1982) identified 39 plant species that followed this pattern – which they called the *megafaunal dispersal syndrome*. In general, the fruits are large and pulpy, nutrient rich, and dispersed biotically. The fruits are similar in appearance and structure to fruits eaten by large seed-dispersing mammals in Africa. The seeds are protected by a hard seed coat, which allows them to avoid being ground down by mammalian molars, and to survive the journey through a large mammal's digestive system. Many of these fruits fall off their parent trees before or immediately upon ripening – an adaptation that makes them available to their dispersers and removes them from arboreal predators such as monkeys, squirrels, and parrots. Lastly, these trees tend to be found on level ground, which coincides with the habitat favored by large mammalian herbivores.

This makes for a nice story, but is it true? Reconstructing history is challenging, but the megafaunal dispersal syndrome makes some testable predictions. For example, horses, which

had been present in the Americas during the Pleistocene, and only returned with the arrival of the European conquistadors in the 1500s, should function as dispersers of at least some of these fruits. In support of this prediction, horses readily feed on the fruits of many of the 39 species identified by Janzen and Martin in Costa Rica, effectively dispersing the seeds without killing all of them (Janzen 1981, 1982). When Janzen collected dung of horses that had recently consumed large amounts of fruit from the jicaro tree, *Crescentia alata*, in the dry forest of ACG, he discovered large numbers of jicaro seeds and determined that 97% of them were still viable. Furthermore, jicaro seedlings and saplings are very rare in areas without horses, even if those areas have adult jicaro trees producing large numbers of fruits. Lastly, seeds of the Guanacaste tree, *Enterolobium cyclocarpum*, can survive a 6-month journey through a horse's gut. Janzen concluded that the horse had been an important member of the megafaunal seed dispersal community during the Pleistocene.

Janzen and Martin concluded that many types of ecological interactions need to be considered in context of their history. This history may include a large number of interacting species, which may not be present today. They may also include species that, while present, are often overlooked.

Ecology of microorganisms that colonize fruits, seeds, and meat

As Janzen describes, "I love avocados, and I bought one for a very high price in an Ann Arbor [Michigan] supermarket. I took it home, sliced it in half and it was rotten inside. I was really upset, and I threw it across the kitchen at the garbage can, and as it was going through the air, this little piece of me inside said 'the microbes won.' Bang, of course that's why they [the microbes] make all those nasty things – it's to keep me from eating them." So Janzen mined the literature at the University of Michigan library for information on nasty chemicals made by fungi and bacteria, and turned it into a comprehensive review article, which he submitted to *American Naturalist*. The editor could not find any external reviewers, because there were so few people with expertise in the fields of microbiology, organic chemistry, evolutionary biology, and community ecology. But he liked the paper and published it anyhow (Janzen 1977).

Janzen's basic approach, which was novel at that time, was to view fruits, seeds, and meat as battlefields on which microorganisms compete with each other, and most importantly, with predators that may kill them if they swallow the fruit, seeds, or meat. His hypothesis was that microorganisms are under strong selection pressure to make the seeds, fruits, or carcasses objectionable or unusable by predators in the shortest possible time. He also pointed out that predators are under strong selection pressure to discriminate flavors or odors that offer false warnings of toxicity, and to evolve mechanisms to detoxify or bypass the chemical defenses of the microorganisms.

One recurring theme is that this evolutionary battlefield is complex. For example, upon ripening, fruits increase their sugar content and soften their flesh, which makes them more attractive to their dispersal agents. However, these same processes also make them more attractive and more vulnerable to colonization by microorganisms. These microorganisms may render the fruit inedible or undesirable to dispersal agents either by converting the nutrients to a less nutritious form, by producing toxins that can either kill the animal or dramatically reduce its fitness, or by producing antibiotics, which can kill animals by disrupting their gut flora (Figure 18.8). The primary selection pressure for the production of antibiotics is that they are part of the arsenal that microorganisms use to compete with each other for access to the resource, but a secondary effect is that antibiotics can also reduce predation by larger organisms.

Fungi will often victimize moist seeds. *Aspergillus* fungi produce aflatoxins, which can cause severe liver damage in all vertebrates, though humans are more tolerant than most other species. Aflatoxins are present in a wide variety of germinating and moldy seeds but don't have much impact on other

microorganisms, thus Janzen (1977) suggests they may be produced specifically as predator deterrents. Most animals, including many insect species, will avoid eating moldy foods, even if those animals are on the point of starvation.

Fresh meat is an incredibly rich source of nutrients, which, in very warm climates, becomes very quickly unacceptable to most carnivores within a few hours of death, due to high levels of bacterial colonization. Bacteria advertise their presence by producing odiferous amines that have been given suggestive names like cadaverine and putrescine. Janzen points out that the bacteria infecting dead organisms are under conflicting selection pressures. On one hand, they should produce toxins to keep away all vertebrates, but on the other hand, they may depend on vertebrates to open up the carcass so they can disperse to new locations. In dry climates, dead animals may mummify, effectively eliminating the long-term reproductive success of the infectious bacteria.

Dead meat cannot defend itself against invading microorganisms, and even if it could, there would presumably be no selective advantage for doing so. In contrast, fruits and seeds do benefit from defense. Janzen (1977) describes how ripe apples produce benzoic acid in response to attack by the *Nectria galligena* apple rot fungus, which greatly retards additional fungal growth. As a second example, many fruits will release tannins to wall off a boring insect and its associated microorganisms. There is a down side to this defense, as tannins give the fruit an astringent flavor that can repel dispersers.

Lastly, Janzen highlights that toxins produced by microorganisms can directly affect different trophic levels of a food web. For example, the anaerobic bacterium *Clostridium botulinum*, which is famous for causing botulism in humans, can infect and grow on soggy grain. Mice are highly sensitive to *Clostridium*, and are likely to die from eating infected grain. Fly larvae that colonize the mice carcasses become highly toxic and kill ducks and pheasants that eat them. All are victims of the botulinum neurotoxin, which causes muscle paralysis.

When Janzen's paper was published in 1977, the ecological world was unimpressed, in part because so few ecologists worked at multiple levels of the biological hierarchy. Most people were either plant ecologists or animal ecologists, and there were very few microbial ecologists. In contrast, those in the pharmaceutical world were extremely enthusiastic, because they recognized that they could screen rotting fruits, seeds, and meat for an entirely new source of pharmacologically active drugs. But there was another world that was important in Janzen's life that was undergoing some drastic changes at about this time. This was his world in Costa Rica.

18.3 WHAT FACTORS MOTIVATED JANZEN'S CONSERVATION EFFORTS IN COSTA RICA?

In the early 1970s, Janzen's field sites in Costa Rica were largely highway easements and private land. He was

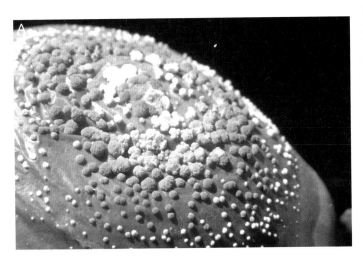

Figure 18.8 "The microbes won" – *Penicillium expansum* on apples. A. Micro view. B. Macro view.

conducting long-term studies of the natural history of the guapinol tree, *Hymenaea courbaril*, which involved keeping track of the seed and fruit production of about 500 trees over many years. Unfortunately for him, and for the trees, the Costa Rican government cut down about half of them when installing a power line along the Inter-American Highway in the late 1970s, and a rancher cut down many of the remaining trees when converting a piece of forest into a rice farm. Janzen had had enough of field site destruction, and established his research program in Santa Rosa National Park, newly established in 1971 (today Sector Santa Rosa of ACG) (Figure 18.3). More than 40 years later, he still does the vast majority of his research at this location.

Santa Rosa has changed dramatically over this 40-year period. Janzen played a major role in orchestrating some of these changes, but many other people also made major contributions. As the national park was establishing its identity as the beginning of today's ACG, Janzen's life and research were undergoing major transitions as well. Perhaps the most important influence on these transitions was Winnie Hallwachs, who in 1978 became Janzen's wife and lifelong partner.

Winnie Hallwachs

After graduating college, Winnie Hallwachs spent a year visiting Sweden, her mother's home country, and also visiting her sister in the Peace Corps in central Africa. At that point, she knew that she loved biology, but did not know what area to pursue, and having returned to Pennsylvania, she happened to sit in on Janzen's course entitled "Habitat and organisms." Inspired by what she learned, she went down to Costa Rica as a volunteer in Janzen's ecological research program. The rest is history.

Hallwachs (1986) established her research program on agoutis, *Dasyprocta punctata*, in part because she had a lifelong fascination with vertebrates, and agoutis are diurnal and relatively large. One of Hallwachs' observations was that agoutis were one of a few species that can get through the hard and thick wall protecting the fruit of the guapinol tree, and the only species that dispersed the seeds without necessarily killing them (Figure 18.9). She wondered how effective agoutis are as seed dispersers, because the megafaunal seed dispersers that coevolved with guapinol went extinct 11 000 years ago. Consequently, the long-term survival of the guapinol depends on agoutis being able to disperse guapinol seeds away from their parent's seed shadow. Unfortunately, Hallwachs's field studies indicated that guapinol survival from agouti dispersal was extraordinarily low – much less than 5%. Even those seeds that germinated were usually eaten by predators before they became established. The future for the guapinol was not bright.

While Hallwachs was doing her agouti research, Janzen was investigating the spiny pocket mouse, *Liomys salvini*, a seed

Figure 18.9 A. Guapinol tree in Costa Rican rain forest. B. Agouti forages on the forest floor.

predator on guapinol and many other species (Janzen 1986), and diving into a full-scale caterpillar inventory. Both researchers were simultaneously developing expertise and a deep understanding of what factors are important for successful tree reproduction. This understanding was critical for the next stage of their career together, in which they teamed up to help create ACG, which became a natural laboratory for the new field of restoration ecology.

Inspiration for restoration

Janzen used an analogy to describe Hallwachs's contributions to their efforts as follows, "I see it more as we're two people, like Rogers and Hammerstein, creating something; one of them thinks better, the other one plays the piano better. Or as I often put it, she thinks and I talk. We go to conferences, we always sit together, we talk, we discuss everything we see and hear, and very commonly Winnie notices stuff that I didn't notice, but I take what she noticed and verbalize it to the conference." Janzen is usually the spokesperson, in part because he is very quick, experienced, and charismatic. In some cases, Janzen's talk can be too direct, and may clash with people that it would be best not to clash with. For example, the Costa Rican culture puts a very heavy emphasis on avoiding direct conflict at all costs, and Janzen is not always aware when he is insulting someone or stepping on someone's toes. In those occasions Hallwachs may step in and defuse the situation before it can interfere with their common goal.

In his book entitled *Green Phoenix: Restoring the Tropical Forests of Guanacaste, Costa Rica*, William Allen (2001) describes three "peculiar accidents" that helped push Janzen in the direction of conservation ecology. The first was Janzen's trip in 1984 to Sweden to accept the Crafoord Prize, which is awarded annually to the world's top scientific researcher from a variety of fields, including ecology. There he met a Swedish journalist, Erika Bjerstrom, who asked him what he was doing to save tropical forests. Janzen explained that he was a researcher and that other people did conservation, but that if she raised money for conservation, he would help funnel it to the proper people.

The second "peculiar accident" was the assault on Corcovado National Park by Costa Rican nationals who were recently unemployed by banana plantation shutdowns, and had moved into Corcovado to mine for gold. The Director of the Costa Rican National Park Service, Alvaro Ugalde, asked Janzen in 1985 to put together a team of scientists to write up an environmental impact statement. After meeting with the miners, and studying the park and the impacts of mining, the team wrote an extensive report to the World Wildlife Fund and the Costa Rican government. The report explained that the miners had few other alternatives, but that their activities were destroying the parks. Former streams were reduced to mud pits, and subsistence hunting was wiping out the mammal populations. Janzen, of course, had a deep understanding of the role mammals play as seed dispersers, so wiping out the mammals would slow down the process of forest rejuvenation even after the miners left the park. The report concluded that "Corcovado National Park and the Costa Rican National Park System, and conservation biology in Costa Rica hang in the balance." Most important, Janzen realized that if a conserved wildland appeared to have no owners, and to be of no use to society, then people would feel morally comfortable using it for themselves.

The third "peculiar accident" was a trip Janzen and Hallwachs made to Australia that same year, at the Australian government's request, to examine what remained of the dry tropical biome, which had been decimated by fires – an event Janzen dubbed "the great Australian barbecue." Much of the biological diversity was gone, with most of the previous dry forest converted to unending grasslands. Gone too were the animals that interacted with those torched trees, shrubs, and lianas.

Janzen and Hallwachs now realized that Santa Rosa was as vulnerable as both northern Australia and Corcovado had been. The tropical dry forest biome in Australia was mostly gone, and Corcovado was saved but tenuous. They committed themselves to saving, restoring, and expanding Santa Rosa, focusing their energies on documenting and conserving its rich biological diversity.

18.4 WHAT FACTORS SUPPORTED THE RESTORATION OF SANTA ROSA, AND ACG AS A WHOLE, AND THE CONTINUATION OF SCIENTIFIC RESEARCH THERE?

To save Santa Rosa, several things needed to happen. First, someone needed to want to restore the tropical forests. Second, money was desperately needed for land purchases and maintenance. Third, the science needed to continue. Janzen and Hallwachs proposed to meet these needs using an approach they describe as biodiversity development – using the biodiversity of a conserved tropical wildland without harming it. Examples include ecotourism, biocultural education, and biodiversity prospecting for medicinal plants or biological control agents. For biodiversity development to succeed, the Costa Rican people needed to feel a sense of ownership for ACG.

The Costa Rican people and ACG

Many dedicated Costa Rican officials made tremendous sacrifices and exerted enormous energy toward the goal of conserving and restoring the natural habitat. But their efforts would have been largely futile without the participation of the Costa Rican general public, who are literate and curious, and in many ways open to new ideas. Janzen and Hallwachs' approach was to use Santa Rosa as a living classroom, so that people would become tied to it, expressing pride and a feeling of guardianship. Making the park relevant to the people who lived near it – what Janzen called

Figure 18.10 A parataxonomist discusses caterpillar genetics with a class of students from a neighboring school. Suspended from the ceiling are bags containing developing caterpillars and their host plants.

Figure 18.11 Parataxonomists at work in caterpillar rearing barn. Each hanging bag contains a developing butterfly or moth.

biocultural restoration – was implemented with a series of education initiatives, and by employing local residents to do the maintenance, restoration, education, and science.

For example, ACG teaches biology classes to all fourth, fifth, and sixth graders in the region (Figure 18.10). That comes to 2500 students per year, and the classes have been taught for 30 years. The teachers are trained biologists who don't have teaching degrees at all, and who were initially planning to do research or applied biology. Janzen and Hallwachs and the growing ACG persuaded them to change their career paths – instead devoting themselves to showing school children how to identify trees, catch insects, and ask questions that are raised by being immersed in tropical dry forest ecosystems (Janzen and Hallwachs 2011). But these efforts can cause conflict with existing belief systems; for example, one parent "knew" that fruit turns red because the moon enters a certain phase, and a second parent "knew" that baby rattlesnakes come from adult rattlesnake ribs. These teacher/researchers must be capable of teaching the science in the context of the culture.

Biocultural restoration can threaten the fabric of the Costa Rican culture, which is quite stratified. Janzen and Hallwachs related a story to me of a landowner who was negotiating a price for 4000 ha to be added to ACG. He was bragging to Janzen and Hallwachs about how he and his father successfully blocked the formation of the first local high school. He spoke with pride about his actions, because he assumed that Janzen and Hallwachs would be upper-class sympathizers with the idea of "keeping them dumb and down on the farms."

To provide opportunities for local residents, and to take advantage of local knowledge about the forest, ACG hires parataxonomists, who are taxonomists in all senses except they lack a degree. Currently there are 34 full-time parataxonomists, who have gone through a rigorous training program that teaches them many of the skills equivalent to a graduate degree (Figure 18.11). These outstanding scientists mostly started as farm kids with little formal scientific training, and no financial or educational future. The initial reaction by an administrator at the Universidad de Costa Rica was "How dare you offer those jobs to people with no university degree?" In his naïveté, Janzen did not realize at the time that the administrator was actually asking "How dare you move somebody from a lower social class up to a higher social class where they are a threat to compete with my sons and daughters?"

In addition to requiring sensitivity to existing cultural norms, these biocultural initiatives depended on major influxes of money, and beginning in the mid-1980s, much of Janzen and Hallwachs' time was spent raising funds. The previously mentioned Swedish journalist presented Janzen a check for US$24 000 that had been contributed by readers of her articles about Janzen's Crafoord Prize. This relatively small beginning was used to purchase a piece of land that in 1986 was added to Santa Rosa National Park, as the seed for ACG. Janzen and Hallwachs and the others working with them have raised over US$100 million dollars through an amazingly complex series of land swaps, grants, and matching funds arrangements that have kept ACG vital, and have expanded it to its present borders (Allen 2001 and personal communication). Meanwhile, Janzen and Hallwachs the scientists have continued to avidly pursue their research program whenever possible.

Scientific research in ACG

There are two components to the current and future Janzen and Hallwachs research initiative. One part is their inventory of juvenile and adult butterflies and moths of ACG and of the parasitoids that infect them. A second part is the science that helps to inform management and conservation decisions at ACG. Of course, the inventory plays a major role in the management and conservation process.

The inventory

In May 1978, Janzen took a fall, broke several ribs, and was confined to a chair next to the door of the house in Santa Rosa. Over the door hung one small light bulb, which attracted a huge number of moths, which Janzen, without leaving his chair, could grab and pin to a spreading board for preservation. After recovering, Janzen went out into the forest, which he described as "wall-to-wall caterpillars," that being the greatest year of caterpillar abundance in Santa Rosa's recorded caterpillar history. So he shifted from studying seed predators to studying caterpillar diversity, which was then almost completely unexplored. Most Lepidoptera are best known from the butterfly or moth adults, so he needed to raise each species to adulthood in order to have any chance of having them identified by taxonomists. This involved placing each caterpillar into its own plastic bag with its host plant, and allowing it to eat its fill, and metamorphose into an adult.

Since 1978, Janzen, Hallwachs, and a growing team of parataxonomists have collected, identified, preserved, described, and photographed the Lepidoptera of ACG at a frenzied pace. Based on these collections they have created a massive database, which is available to all interested people, including researchers, educators, managers, and private citizens (http://janzen.sas.upenn.edu). In addition, the research team is raising all of the parasitoids that attack the different species of caterpillars, primarily wasps and flies, but also nematodes and fungi. By describing each species of parasitoid, caterpillar host, and plant eaten by caterpillars, the researchers are identifying three trophic levels of the entire ACG food web, and the links connecting each species (Figure 18.12). As of 2015,

the researchers have reared and characterized approximately 9500 of the estimated 15 000 species of Lepidoptera in ACG and 3500 species of parasitoids (Janzen and Hallwachs, personal communication).

Though Figure 18.12 shows a huge number of interacting species, you should appreciate that each species has relatively few links with other species. Fewer than 10% of the caterpillars are generalists, and almost all of the parasitoids are highly specialized on one or two host species.

But why has Janzen decided to devote all of his energy to inventorying the Lepidoptera at ACG? From a personal standpoint, Janzen has studied butterflies and moths since his first clumsy efforts in northern Minnesota, and has developed extraordinary skill and knowledge in their collection, preparation, and identification. And from an ecological perspective, caterpillars are the last largely unknown species-rich group of animals in the tropics, and they are the most important group of consumers of tropical plants. Thus many roads to understanding tropical diversity and trophic interactions travel through caterpillars and the organisms with which they have direct and indirect interactions.

The researchers at ACG are expanding their research into the rain forest canopy, expecting to uncover an entirely new assemblage of species and species relationships. The basic information collected in this survey can then be used to address an array of more specific questions. For example, the climate at ACG is currently drying and warming. How are the species and species relationships shifting geographically in response to these climate changes? As a second example, ACG has dry, rain, and cloud forests that transition into each other. Are there differences in host specificity, parasitism rates, or patterns of ecological interactions among these three types of tropical forests? Third, why and how are the herbivores and their parasitoids so host specific and habitat specific? Lastly, some lepidopterans may live in one type of forest as caterpillars and then move to another ecosystem as adults. How common is this, and what benefits do they gain from this complex life history?

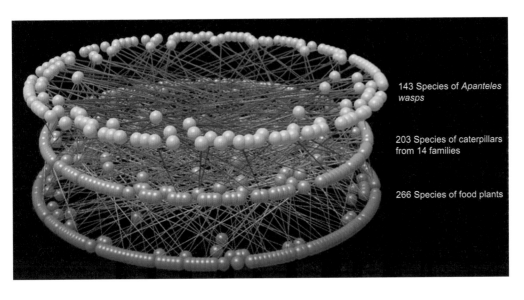

Figure 18.12 Food web showing three trophic levels of interactions, with parasitoid wasps parasitizing caterpillars that are eating plants. Based on Yoon *et al.* (2004).

143 Species of *Apanteles wasps*

203 Species of caterpillars from 14 families

266 Species of food plants

These are big questions, and they all share one important attribute: the researchers must be able to identify the species they are working on. This inventory is providing critical baseline data for investigations conducted by present and future generations of researchers interested in basic ecological questions. At the same time, it is also providing critical information for conservation and restoration ecologists working at ACG and elsewhere.

Science, conservation, management, and pragmatism at ACG

History plays an important role in understanding ecological relations at ACG. For the past 400 years, the central Santa Rosa section of ACG was a ranch, with most of the dry forest converted to pasture for cattle. Under court order, the last free-ranging cattle were removed from Santa Rosa National Park in 1978, triggering a dramatic effect. The introduced African jaragua grass on the site grew to its full height of 1–2 m and became fuel for fires set by ranchers nearby to clear their pastures of invading woody vegetation. Without cattle to consume the grass, fires burned out of control and inflicted heavy damage on the adjacent forests (Figure 18.13).

In 1986, the Guanacaste National Park Project was created, which in 1989 changed its name, as it was simultaneously

Figure 18.13 A wildfire consumes part of the forest at ACG.

expanding into the much larger ACG. The original mission of the new conservation area included the following three points. First, existing fragments of dry forests should provide seed and animal dispersers to restore the 700-km^2 area in hopes that all of the species known to originally inhabit the area would survive indefinitely. Second, this tropical wildland should be restored to offer a "menu of material goods [. . .] and basic biology data which in turn will be part of the cultural offering." And third, the tropical wildland should be the stimulus for a reawakening to the intellectual and cultural offerings of the natural world, and function as a user-friendly resource for a local, national, and international audience (Janzen 2000). The first and second points highlight the scientific goals, while the second and third points highlight the importance of cultural restoration and biodiversity development in the total restoration and conservation strategy.

Ecological restoration was a very controversial topic in the 1980s because it seemed to fly in the face of the argument being made by many conservation organizations that once a tropical forest disappeared it could never be resurrected. Janzen made the case instead that destroyed forests could be restored, if there was a seed source, but that compromises must be made. He recently wrote, "Stop trying to save all wild nature all the time, everywhere. Triage does have its place, especially when we are both the enemy and the partner" (Janzen 2010).

Janzen contends that there is no set scientific formula for ecological restoration, but that there are some guiding principles. The key to successful restoration is to build upon or enlarge existing habitat fragments – a process Janzen (1988) called "growing fragments." Fragments can grow and join together, forming even larger pieces of continuous habitat that can support an increasing number of species. There are several principles to apply when growing fragments. One principle is that management decisions will influence species richness, species composition, and habitat structure in a *place-specific* manner, so a management strategy at one location will have a very different effect than the same management strategy at another location. For example, fire suppression is essential for restoring the tropical dry forest at ACG, as demonstrated by the destruction caused by fires of the late 1970s and early 1980s. But fires are essential for efforts to restore many types of prairies by eliminating competing woody vegetation. A second principle is that the speed of recolonization will depend on how close the source areas are to the habitat that is being restored (see Chapter 22). Lastly, species will initially appear in a habitat as very small and non-viable populations. As we discussed in Chapter 11, small populations are very likely to go extinct. Thus observing that a species is present within a habitat does not mean that it will remain in that habitat.

Two initiatives highlight the importance of place-specific management practice for ecological restoration. In 1992, a large orange plantation owned by Del Oro Corporation was established just outside of ACG's northern border. When juiced, oranges

Figure 18.14 Huge mound of orange peels dumped by Del Oro Corporation.

generate very large amounts of useless orange peels that most plantations must truck to distant disposal sites. As an experiment, Janzen invited the orange producers to dump 100 truckloads of orange peels on a nearby, recently purchased low-grade jaragua pasture in ACG (Figure 18.14). As he expected, decomposers, including three fly species, colonized the peels and converted them into a nutrient-rich broth, which ultimately created a rich soil layer populated by a diverse community of broadleaf herbs, and eliminated the invasive jaragua. As an additional bonus, Del Oro agreed to donate large tracts of forested lands to ACG for this ecosystem service.

A second place-specific management practice began in 1998, following Janzen's recommendation to convert rain forest pastures recently purchased by ACG into plantations of the *Gmelina arborea* tree, which is grown for cardboard box fiber and low-grade lumber throughout Central America. In most cases, this import from Southeast Asia is a scourge for conservation biologists, because *Gmelina* plantations are often established by clear-cutting existing forests. But Janzen's vision was very different. As an alternative to clear-cutting rain forests for *Gmelina* plantations, Janzen recommended establishing *Gmelina* plantations in old pastures adjacent to rain forests. He pointed out that abandoned pastures regenerate back into forest very slowly, even if there is a forest nearby that provides seeds and animal dispersers. To help this process along, Janzen recommended purchasing old rain forest farms and going into business with experienced *Gmelina* planters. Planters would pay the expenses of planting and harvesting, and keep any profits. There were two key elements to this project. First, the plantation would not be weeded, so they would develop a dense shade-tolerant understory of rain forest shrubs, vines, and tree seedlings. Second, after one generation of

Gmelina, all trees would be harvested, and the former pasture could continue developing into restored rain forest (Janzen 2000).

 Thinking ecologically 18.2

As another example of a place-specific management practice, Janzen proposed that cows be reintroduced into ACG, so they could mow down the grass and reduce the number of jaragua-fueled fires. Many government officials opposed this proposal, because it was against the law to introduce livestock into a national park. If you had authority to resolve this controversy, how would you go about doing it?

Good management ideas come from good science, but they also come from improvements in technology. Janzen and Hallwachs are using technological advances to support both their science and their management recommendations at ACG.

Technology in science and management

In general, it is easier to distinguish between different species of closely related large animals than closely related small animals. But even large animals pose serious problems for species discrimination. For example, until recently, taxonomists had assumed that there was one species of killer whale, *Orcinus orca*. However, based on an analysis of complete sequences of mitochondrial DNA and considerations of differences in morphology and ecology, Philip Morin and his colleagues (2010) recently recommended reclassifying the killer whale into a complex of at least three distinct species. If taxonomists have trouble resolving species of an animal that large, you can imagine the problems they face with distinguishing among species of closely related caterpillars, and even more so among the tiny parasitoids they harbor.

One partial solution to this problem is DNA barcoding, a term coined by geneticist Paul Hebert, who was inspired by cans of food lining the aisles in the grocery store – each can with its own distinct identifying barcode. Hebert reasoned that each species of organism has its own distinct DNA barcode – its sequence of DNA base pairs. In theory, biologists could use a small portion of DNA extracted from organisms to identify them to species. For financial and engineering reasons, Hebert wanted a relatively short sequence, but it needed to be variable enough to differ in even closely related species. The solution for animals is a 650-base-pair segment of the mitochondrial cytochrome c oxidase subunit 1 (COI) gene. Each animal species' COI base pair sequence is its barcode. Individuals of the same species will usually have a very similar or identical COI base sequence. Because

mitochondrial DNA has a high spontaneous mutation rate, even very closely related species will usually have distinct COI base pair sequences. Thus the DNA barcode is one tool to distinguish among very closely related species (Janzen *et al.* 2009).

In 2003, Janzen, Hallwachs, and Hebert met and worked out an arrangement whereby the parataxonomists provide Hebert with a steady supply of lepidopteran legs from the ACG inventory. All information from each caterpillar, including its host plant, location, any relevant ecological data, and information about any parasitoids is stored by the parataxonomists in a massive database. After a caterpillar metamorphoses to adult, a leg is pulled off and shipped to Hebert's lab at the University of Guelph, in Ontario, Canada, where its barcode is determined and added to the database.

This partnership has allowed the researchers to address a large number of questions that were hampered by uncertainties

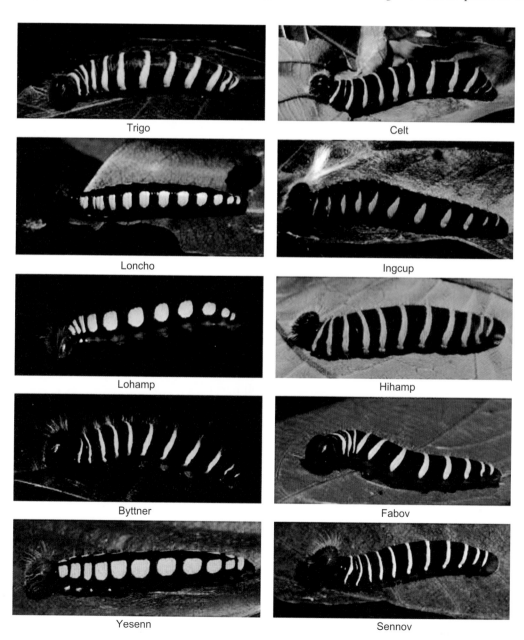

Figure 18.15 Final instar caterpillars of 10 species in the *A. fulgerator* complex. In most cases, the interim name below each species is shorthand for its primary food. All species were collected at ACG.

about species identification. As one example, the skipper butterfly, *Astraptes fulgerator*, was thought to be a very common, morphologically variable, and widely distributed species that ranged from the southern United States to northern Argentina. Janzen and his colleague John Burns suspected that there might be more than one species under the same name, because the caterpillar fed on so many different plant species, and individuals feeding on one host tended to look slightly different from individuals feeding on a different host at ACG. He shipped 484 legs to Hebert, who was able to barcode 479 of the legs. The analysis divided the butterflies into 10 genetically distinct groups of butterflies, each with its own DNA barcode, with the genetic divergence between groups averaging 2.97%, while there were almost no genetic differences between individuals of the same group (Hebert *et al.* 2004). In general, these distinctions were highly correlated with differences in food plants and with differences in caterpillar color patterns (Figure 18.15). The researchers concluded that there are at least 10 species in the *A. fulgerator* complex in ACG, and likely many more species that will be discovered over the entire geographical range. Indeed an 11th species was recently identified at ACG (Janzen and Hallwachs, personal communication).

This study raises several important ecological and evolutionary questions. For example, how did so many species form within this complex, given that there was no apparent geographic barrier that would prevent gene flow between populations? Second, why do the caterpillars differ in appearance, while the adults are almost morphologically indistinguishable? Third, why are there so many species with similar appearance in one place? These types of questions are likely to arise more often, as the inventory and the barcoding efforts continue to describe and identify more species of insects.

The researchers emphasize that barcoding is but one tool in the taxonomist's toolbox that can help resolve questions about species identity, ecology, and evolution. Like all tools, it does not always work, and like most new tools, its effectiveness will increase as the technology continues to develop. Janzen, Hallwachs, their students, parataxonomists, and many other researchers will continue to use whatever tools are available to uncover and understand the mysteries provided by ACG's biological diversity.

SUMMARY

For over 50 years, Dan Janzen has been studying the evolutionary and community ecology of tropical forests, primarily at ACG in northwest Costa Rica. His lifelong passion for butterflies and moths has evolved into a project that seeks to inventory all of the lepidopterans, their parasitoids and host plants in ACG. At the same time, his field research in the dry forest and rainforest, and the recognition that his field sites were in danger of disappearing, has encouraged Janzen and his wife and colleague, Winnie Hallwachs, to develop and implement a series of conservation and management plans to restore tropical forests in Costa Rica.

Part of Janzen's success as a scientist stems from spending enough time in the forest to have an intuition for when something interesting or unusual is happening. Janzen's evolutionary perspective, and viewing community interactions from the perspective of the organism, has allowed him to make important contributions to how ecologists understand community-level interactions, and tropical species distributions and patterns of species richness. His research stresses the importance of understanding ecological interactions in the context of their evolutionary history. Consequently, Janzen could understand that numerous tropical plant species that disperse seeds very ineffectively, evolved in, and were well adapted to, a different community of seed dispersal agents that no longer exists today.

FURTHER READING

Bagchi, R., and 7 others. 2014. Pathogens and insect herbivores drive rainforest plant diversity and composition. *Nature* 506: 85–88.
Recent test of the negative density-dependence prediction generated by the Janzen–Connell hypothesis. Isolates the effects of fungal pathogens and insect herbivores on the recruitment of seedlings into the plant population.

Janzen, D. H. 1970. Herbivores and the number of tree species in tropical forests. *American Naturalist* 104: 501–528.
Classic paper presenting Janzen's hypothesis (now know as the Janzen–Connell hypothesis) that accounts for high tree species richness in tropical forests. In addition to presenting the basic

hypothesis, Janzen graphically manipulates seed dispersal curves and seed survival curves in ways that do interesting things to the population recruitment curve.

Janzen, D. H. 2000. Costa Rica's Area de Conservacion Guanacaste: a long march to survival through non-damaging biodevelopment. *Biodiversity* 1: 7–20.
 Great concise summary of the history of ACG, and a few of the conservation and management initiatives employed there.

Janzen, D. H. 2010. Hope for tropical biodiversity through true bioliteracy. *Biotropica* 42: 540–542.
 Brief, optimistic and realistic essay that provides insight into Janzen and Hallwachs' approach to conservation at ACG.

Janzen, D. H., and Hallwachs, W. 2011. Joining inventory by parataxonomists with DNA barcoding of a large complex tropical conserved wildland in northwestern Costa Rica. *PLoS One* 6: e18123.
 Discusses benefits of both DNA barcoding technology, and the creation of a core group of parataxonomists, to fulfilling the mission of the ACG inventory.

END-OF-CHAPTER QUESTIONS

Review questions
1. What was Janzen's hypothesis for how plants allocate resources to their offspring in tropical forests? How did he support this hypothesis?
2. What is the Janzen–Connell hypothesis? Describe Packer and Clay's (2000) research that tested this hypothesis. What did they find, and what did they conclude?
3. How did research by Janzen and Martin (1982) demonstrate that past selection pressures influence present-day community interactions?
4. How did viewing the natural world from the perspective of the individual organisms help Janzen understand why fruits rot, seeds mold, and meat spoils?
5. What factors inspired Janzen and Hallwachs to pursue conservation and restoration at Guanacaste? What restoration and management activities have they done?
6. How has DNA barcoding helped Janzen and Hallwachs with the inventory project at ACG?

Synthesis and application questions
1. How do the two life history strategies used by legumes relate to the concept of *r*-selection and *K*-selection?
2. How does Janzen's analysis of seed number, size, and predation level relate to the study that we discussed in Chapter 8 in which Charnov and Ernest (2006) compared the relation between number of offspring and size of offspring in mammals?
3. In the discussion of his paper on seeds, predation, and dispersal, Janzen (1969) points out that seeds are made up of two parts: the seed coat, which is useless to the seedling, and the seed content, which is essential. Based on your understanding of surface area/volume ratios, how might this consideration influence how you would evaluate the costs and benefits of the two life history approaches selected for in legumes?
4. Some seed predators are highly host specific, preferring to eat only one or a few species of seeds, while other seed predators are generalists, and routinely feed on many different species. How might host specificity of seed predators influence the population recruitment curve (Figure 18.4)? How would that influence the predictions of the Janzen–Connell hypothesis?
5. Would you expect to observe the evolution of microorganisms that tasted or smelled bad without actually being toxic or non-nutritive?
6. Barcode analysis has increased the estimates of the number of species of parasitoids at ACG more than it has increased estimates of the number of species of lepidopterans. Why might this be expected? How would you test your hypothesis?

7. In some cases a lepidopteran species is sexually dimorphic. How could this influence a scientist's estimate of species diversity within a community? How could DNA barcoding help with this problem? How would behavioral observations help with this problem?

Analyse the data

Return to Table 18.1 and compare attacked versus ignored species of legumes to determine whether the differences between the means for each seed variable are significantly different. Are there any concerns you have with this data set?

PART V
Ecosystem and global ecology

Chapter 19

Ecosystem structure and energy flow

INTRODUCTION

About 50 years ago, Eugene Odum (1964) proposed that ecologists should "rally around" the ecosystem as the basic unit of ecology, just as cell and molecular biologists rally around the cell as their basic unit of interest. Ecosystems have important properties – such as production, energy flow, nutrient cycles, and stability – that need to be understood holistically, by considering the large-scale interactions between the organisms and their environment. Of course, other organisms are important parts of any individual's environment. So too are abiotic factors such as light, temperature, moisture, and nutrients.

Human activity has profoundly changed ecosystem properties since Odum's rally cry. To sustain food production, humans have converted many forests and savannas into cropland, and many mangrove swamps into shrimp or fish aquaculture facilities. Artificial fertilizers have altered basic nutrient cycles and patterns of energy flow. Economic development has altered these ecosystem properties in ways that threaten ecosystem structural and functional integrity.

We begin this chapter with a brief description of many types of ecosystem services – services that ecosystems provide that improve human well-being. To put these services into perspective, it is important to understand that some ecosystems are more productive than others, and that several different abiotic factors, particularly climate and nutrient availability, interact to influence production in different biomes. Producers are the basis of ecosystem production, while consumers, decomposers, and detritivores all play essential roles. Energy is lost as it passes from producer to consumer to detritivore and decomposer, effectively limiting most food chains to only a few trophic levels. Human well-being is tied to maintaining healthy and productive ecosystems: unfortunately many ecosystem services are being degraded by humans or used in an unsustainable manner.

KEY QUESTIONS

19.1. How do climate and nutrients influence ecosystem production?

19.2. What are the important components of ecosystem structure?

19.3. What is the relationship between food chain length and ecosystem structure and functioning?

CASE STUDY: Ecosystem services and ecosystem change

Applied ecologists use ecological principles to evaluate threats to ecosystem structure, and ultimately to ecosystem services. In an attempt to understand human behavior, some ecologists have connected with social and behavioral scientists in the fields of anthropology, sociology, and psychology. Other ecologists have built bridges with economists and business and government leaders. One such union resulted in the Millennium Ecosystem Assessment (MA), a report that was initially called for by United Nations Secretary General Kofi Annan in his address to the 2000 UN General Assembly. The MA's objective was "to assess the consequences of ecosystem change for human well-being and to establish the scientific basis for actions needed to enhance the conservation and sustainable use of ecosystems and their contributions to human well-being" (Millennium Ecosystem Assessment 2005). Over 2000 authors and reviewers contributed their knowledge to this massive undertaking.

The MA focuses on four types of ecosystem services – the benefits people receive from ecosystems. *Provisioning services* are essential material goods such as food, water, wood, timber (fuel), and fiber. *Regulating services* influence components of the environment, such as climate, floods, disease, waste, and water quality. *Cultural services* provide recreational, educational, spiritual, and aesthetic benefits. Underlying these three services are *supporting services*, which include functional properties of ecosystems such as soil formation, photosynthesis, and nutrient cycling. Together, these four types of services form important linkages to human well-being.

As one example of ecosystem services, consider the Ganges River and its services to humans along its course. At its source (Figure 19.1A), the river receives runoff from the Himalaya Mountains, which it sends downstream along with provisioning services such as food, water, and timber (Figure 19.1B and C). It provides regulating services by absorbing runoff during the monsoon seasons, and by supporting an army of decomposers that break down and detoxify wastes. It provides cultural services at many spiritually significant sites along its course (Figure 19.1A and C). And when the river does flood, it provides supporting services in the form of rich fertile soil that is deposited in the floodplain.

The MA recognizes that humans are important parts of ecosystems, and that there is important feedback between the human condition and ecosystem function. The MA identifies seven *drivers*

Figure 19.1 Ecosystem services and three faces of the Ganges Rivers. A. Near its source at Gaumukh. B. A mountain river passing through the forested and farmed highlands of Uttar Pradesh. C. People praying and bathing further downstream in Varanasi.

of ecosystem change – factors that directly or indirectly cause a change in ecosystem processes. These drivers are: (1) changes in local land use and cover; (2) species introduction and removal; (3) technology adaptation and use; (4) external inputs such as fertilizer, pest control, and irrigation; (5) harvest and resource consumption; (6) climate change; and (7) natural, physical, and biological drivers. Each of these drivers has many components; for example, climate change includes changes in global temperature, rainfall patterns, amount of solar radiation, and variation in the intensity of hurricanes. Notice that the first six drivers are primarily influenced by human activity, so the impact of humans on ecosystems is profound.

If ecosystems provide critical services for humans, and if humans have important influences on ecosystems, we need to understand how the balance of the human/ecosystem interaction plays out in the global economy. Before we do that, we need to identify the basic elements of ecosystem structure, and how energy flows through ecosystems.

19.1 HOW DO CLIMATE AND NUTRIENTS INFLUENCE ECOSYSTEM PRODUCTION?

In Chapter 2, we discussed how some biomes are more productive than others. When ecologists measure production, they are usually measuring the amount of carbon that is fixed by the autotrophs (via photosynthesis) that live within a particular area. In our discussions of ecosystems, gross primary production (GPP) describes the amount of carbon fixed within an ecosystem over some time period. But a portion of these carbon compounds is used to meet the autotrophs' metabolic needs. Consequently ecologists often use net primary production (NPP) to describe ecosystem production, which is equal to GPP minus the amount of carbon used by autotrophs for metabolism. NPP is most commonly expressed in terms of grams of carbon per meter square per year (g C/m^2/yr), but ecologists use many other measures as estimates of NPP, particularly when measuring carbon production is difficult or undesirable. A related concept, secondary production, is the amount of carbon that is passed on to the heterotrophs that live within an ecosystem.

There are several compelling reasons why ecologists must understand global patterns of NPP, and the factors that influence production. All models of human population size predict substantial increases over the next 100 years. Serious food shortages are already a problem, and will get worse with increased population. We can make better decisions about where to farm if we know the distribution of NPP on a global level. We can make better decisions about how to supplement our crops if we understand how climate and nutrients interact to influence NPP. Finally, sustained changes in NPP on a local or regional level can alert us that an ecosystem is being seriously stressed by changes in environmental factors.

Different patterns of NPP distribution in terrestrial versus marine ecosystems

Intuitively, you might predict that NPP would be greatest in areas that received the most sunlight and the greatest amount of precipitation. We all know that plants need light for photosynthesis and water for a variety of physiological processes, so terrestrial ecosystems that receive the greatest quantities of these two factors should have the highest NPP. For marine ecosystems, we might expect sunlight to still be important, but obviously water is no longer a limiting factor. Let's see how these factors actually play out to influence NPP distribution across the globe.

Global distribution of terrestrial NPP

Measuring NPP is no easy task. On a small scale, ecologists can use direct measures of change in biomass over time to estimate NPP. Researchers may also harvest a sample of plants to estimate the portion of production that different species allocate to roots over time. These approaches don't work very well over large spatial scales. Recently scientists have used satellite imagery to estimate regional and global NPP in both terrestrial and aquatic ecosystems. Recall that plant chlorophyll absorbs red and blue wavelengths and reflects green and infrared. Satellite images can capture these differences in reflectance to estimate how much chlorophyll is present in a certain region. These estimates of chlorophyll levels are used to estimate how much production is occurring at a particular time, with higher chlorophyll levels indicating greater production. Repeated measures allow researchers to estimate how much production is occurring over an extended period of time.

A global picture based on remote sensing is very revealing (Zhao and Running 2010). NPP is highest in the tropical latitudes and much lower at the poles. NPP is also very low in large deserts, which suggests that rainfall is also a critical factor influencing NPP. Even at similar latitudes, there are major differences in NPP, indicating that other factors besides latitude must be influencing NPP (Figure 19.2).

A clearer picture emerges if we look at terrestrial NPP for each biome (Saugier *et al.* 2001). Not surprisingly, tropical forests had the greatest NPP, followed by temperate forest and tropical savanna (Table 19.1). Notice how low the NPP is for deserts and tundra.

 Thinking ecologically 19.1

Given the total area of each biome in Table 19.1, calculate the total grams of carbon fixed per year for each biome. Then estimate the percentages of terrestrial carbon fixed per year for each biome. How do these percentages compare to what you know about where each biome is, and how much solar radiation it is likely to receive (refer back to Figure 2.2)?

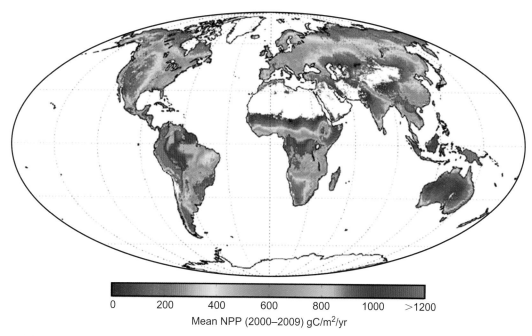

Figure 19.2 Mean global terrestrial NPP estimates for the period 2000–2009 based on satellite imagery using a moderate resolution imaging spectroradiometer. White terrestrial areas are non-vegetated.

Table 19.1 NPP of terrestrial biomes (Saugier *et al.* 2001).

Biome	NPP (g C/m²/yr)	Total area (× 10¹² m²)
Tropical forest	1250	17.5
Temperate forest	775	10.4
Tropical savanna	540	27.6
Mediterranean shrubland	500	2.8
Temperate grassland	375	15.0
Boreal forest	190	13.7
Desert	125	27.7
Tundra	90	5.6

We see a strikingly different distribution of NPP in marine systems.

Global distribution of marine NPP

Macrophytes (aquatic plants such as kelp and various large algae) and phytoplankton of all sizes need light in order to carry on photosynthesis. Thus we might expect NPP in marine systems to show a similar latitudinal effect to what we observed with terrestrial systems, with a peak of production near tropical latitudes. Satellite imagery reveals a very different story, with highest NPP associated with shallow shorelines, areas of upwelling (where oceanic currents bring nutrients up to the surface), and shorelines that receive extensive runoff from terrestrial sources (Figure 19.3).

Unlike the latitudinal pattern of NPP in terrestrial systems, marine NPP is relatively flat between 60° N and 30° S, but then rises to a peak at about 40° S before declining to very low levels in polar regions. This peak is associated with large areas of upwelling in the southern hemisphere near the southern tips of Africa, Australia, and South America.

These different regions with highest NPP have one feature in common – relatively good access to nutrients. Table 19.2 shows how global NPP is distributed in relation to marine biome or region.

Remember that NPP is calculated per unit area. This is important, because the marine biomes with the highest NPP are the smallest biomes represented in Table 19.2. Conversely, even though the open ocean has by far the lowest NPP overall, it is responsible for the majority of production by marine ecosystems.

We're now ready to take a closer look at these patterns of production on a regional and local scale.

Regional and local effects of climate and nutrients on NPP

Our understanding of the relationship between climate and NPP has undergone some drastic revisions over the past decade and is likely to change over the next few decades as ecologists develop

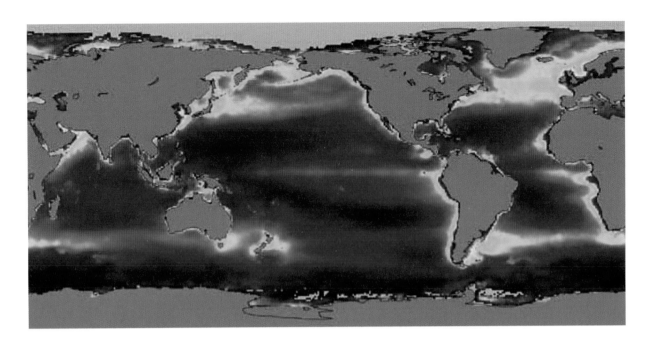

Net primary productivity (grams carbon per m² per year)

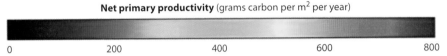

0   200   400   600   800

Figure 19.3 Annual marine NPP estimated by the Sea-viewing Wide Field-of-view Sensor that collected satellite data from 1997 to 2010.

Table 19.2 NPP of marine biomes and regions (Sumich and Morrissey 2004).

Biome or region	NPP (g C/m²/yr)	Total area (× 10¹² m²)
Open ocean (tropic/subtropics)	60	190
Open ocean (temperate/subpolar)[a]	220	100
Open ocean (polar)	30	52
Continental shelf (non-upwelling)	310	26.6
Continental shelf (upwelling)	970	0.4
Estuaries/salt marshes	1000	1.8
Coral reefs	1000	0.1
Seagrass	1000	0.02

[a]Includes the Antarctic upwelling.

improved methods for measuring NPP, and as the public becomes more aware of the importance of production to human existence. We begin by considering the relationship between abiotic factors and terrestrial NPP.

Effects of climate on local and regional terrestrial NPP

Early studies indicated reasonably strong positive correlations between terrestrial NPP and both mean annual precipitation and mean annual temperature (Lieth 1975). In subsequent years, researchers began collecting more data on tropical forests, a biome that had been grossly understudied in the mid-twentieth century. With more data collection on tropical forests, ecologists recognized that high rainfall leads to high NPP, but only up to a certain point.

As one example, Edward Schuur and Pamela Matson (2001) studied NPP in a tropical forest on the Hawaiian island of Maui, USA. They wanted to know how precipitation influenced production, but were aware that in many studies precipitation was confounded with other potentially important factors like temperature and elevation. They reasoned that taking advantage of the rain shadow created by a high volcano on the island would allow them to study a series of sites in close proximity to each other, which differed only in the amount of annual precipitation, which ranged from 2200 to 5050 mm at their six sites.

The proximity of the sites made it convenient for the researchers to make very precise measures of factors that are associated with NPP. To estimate aboveground NPP, they established a 25 × 25-m production plot in each site, where they measured changes in the diameter and height of all woody vegetation by repeated sampling. The investigators also harvested subsamples of the understory, including the litter layer, to measure rates of biomass production from shrubs, ferns, mosses, and herbaceous

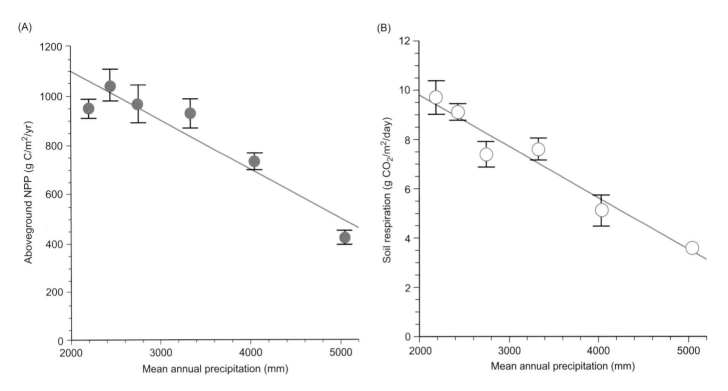

Figure 19.4 Data collected at six sites on the side of a volcano on the island of Maui showing A. Aboveground NPP in relation to annual precipitation. (R^2 = 0.89. P = 0.005). B. Average daily soil CO_2 production in relation to annual precipitation (R^2 = 0.95, P = 0.001)

vegetation. Lastly, they estimated belowground NPP by collecting samples of roots from soil cores that penetrated 51 cm into the ground, and by measuring changes in the *soil respiration rate* – the rate of CO_2 production by roots and heterotrophs that live in the soil.

At precipitation levels above 2500 cm, aboveground NPP at these tropical forest sites decreased in a nearly linear manner with increased mean annual precipitation (Figure 19.4A). In addition, soil respiration rates showed a very similar pattern, indicating that belowground NPP also decreased with increasing mean annual precipitation (Figure 19.4B). In a related study, Schuur (2001) provides evidence that high soil water levels slow down the process of decomposition, which could reduce the amount of nutrients available to producers, and reduce NPP.

Schuur (2003) then applied the results of the Maui study to more recent findings by other researchers to provide a more comprehensive analysis of how precipitation and temperature influence NPP. He found that NPP increases with mean annual precipitation up to about 2000 mm of precipitation, but levels off and begins to decline above 2500 mm. In contrast, NPP increases with mean annual temperature up to 30 degrees C, which was the highest mean annual temperature of studies that Schuur analysed.

Stephen del Grosso and his colleagues (2008) followed up on Schuur's study to reconsider the relationship between precipitation and temperature on NPP, taking advantage of the vast increase in available data that they analysed at the National Center for Ecological Analysis and Synthesis (NCEAS). With a

larger data set, the NCEAS model predicted lower NPP than Schuur's model at precipitation levels below 3000 mm, but somewhat higher NPP at mean annual precipitation above 3500 mm. In agreement with Schuur, the NCEAS model also showed a decrease in NPP at high mean annual precipitation, but the decrease was substantially more gradual than that of Schuur's model (Figure 19.5A). Most notably the NCEAS model also makes an important distinction between forested and nonforested ecosystems. In general, at most precipitation levels, forested ecosystems fix 100–150 g $C/m^2/yr$ more than nonforested ecosystems (Figure 19.5B). Interestingly, when these additional data are considered, temperature becomes a much poorer predictor of NPP than previously estimated.

To summarize, both the Schuur and NCEAS models show nonlinear effects of mean annual precipitation on NPP, and some effects of temperature as well. The NCEAS model shows that forested ecosystems are more productive than nonforested ecosystems at all precipitation levels. But we also know that adding nutrients to crops in the form of fertilizers increases plant growth, thus we can imagine that the distribution of nutrients within ecosystems can influence NPP. We will now discuss briefly how nutrients influence terrestrial ecosystems, and explore this relationship more fully in Chapter 20.

Effects of nutrients on local and regional terrestrial NPP

At low air temperatures, ectotherms such as decomposers have reduced metabolic rates. One consequence is that decomposition rates of litter on the forest floor are much lower in polar and subpolar regions than they are in tropical and subtropical biomes. David LeBauer and Kathleen Treseder (2008) hypothesized that nutrients such as nitrogen should be less available to plants at colder latitudes, and would be likely to limit production under those circumstances.

A veritable army of researchers had already tested the hypothesis that nitrogen limits NPP at numerous locations around the world. One approach used by these researchers was supplementing a defined region with nitrogen, and observing whether that region showed higher NPP than an equivalent unsupplemented control. If nitrogen is limiting, nitrogen supplements should enhance NPP. LeBauer and Treseder combined 126 previously conducted nitrogen supplementation experiments in a meta-analysis that tested their hypothesis that nitrogen would be more likely to limit production in cold ecosystems.

Most of the studies estimated aboveground NPP. To combine data from each study into a meaningful statistic, the researchers calculated the response ratio, which is the NPP of the supplemented site(s) divided by the NPP of the control (unsupplemented) site(s). A response ratio of greater than 1.0 indicates that nitrogen supplementation was associated with increased NPP. The researchers found that in all biomes, except for desert, there was a positive response to supplementation (Figure 19.6). In contrast to their

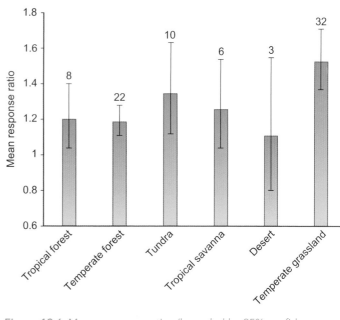

Figure 19.6 Mean response ratios (bounded by 95% confidence intervals) in response to nitrogen supplementation in six terrestrial biomes. Sample sizes are above each bar.

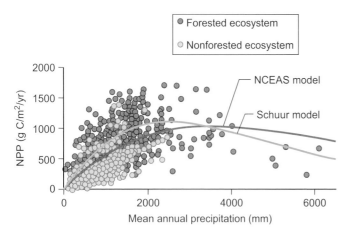

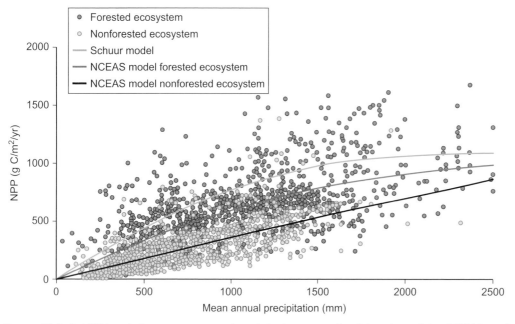

Figure 19.5 A. NPP in relation to mean annual precipitation comparing Schuur's model to NCEAS model over entire range of mean annual precipitation values. B. NPP in relation to mean annual precipitation below 2500 mm.

expectation, there was little evidence of an overall effect of latitude, though there was some effect of latitude when the analyses were restricted to specific biomes. Of interest to us for our present discussion is the central finding that nitrogen availability does appear to limit production across most terrestrial biomes.

Clearly, both climate and resources influence NPP in terrestrial ecosystems. What patterns do we observe in marine and freshwater ecosystems?

Effects of nutrient availability on freshwater and marine NPP

Given our discussion thus far, you will not be astonished to learn that some climate factors are much less important in influencing freshwater and marine NPP than they are in terrestrial ecosystems. Clearly water is not a limiting factor for plants that live in water. However, two major challenges for macrophytes and phytoplankton are getting enough light for photosynthesis, which is always at a maximum near the surface, and getting enough nutrients to build complex molecules needed for growth and reproduction. This problem is particularly acute in the *pelagic zones* (away from the shoreline) of the open ocean or deep lakes, both of which are relatively unproductive.

As with terrestrial ecosystems, there have been numerous studies on the effects of adding fertilizers to freshwater and marine ecosystems. Overall, the prevailing doctrine among ecologists is that phosphorus limits NPP in freshwater, and nitrogen limits NPP in marine ecosystems. Phosphorus limitation in lakes is supported by Schindler and Fee's (1974) research that we described in Chapter 2, in which the half of an experimental lake that received phosphorus supplements showed a tremendous increase in production in comparison to the other half that received no phosphorus supplement. However, several subsequent studies have shown that nitrogen can also limit production in some freshwater systems.

In marine ecosystems, numerous studies have shown increases in production in response to nitrogen supplements. But also recall the 1988 study by Martin and Fitzwater (see Chapter 4) that demonstrated iron limitation in some portions of the open ocean. And most recently, several studies have shown that phosphorus can limit production in the open ocean as well.

While it appears that nutrients limit freshwater and marine NPP, it is still not clear whether the prevailing doctrine is still valid that phosphorus limits freshwater production and nitrogen limits marine production. James Elser and his colleagues (2007) attempted to resolve this issue with a meta-analysis of 653 freshwater, 243 marine, and 173 terrestrial experiments. One of the challenges with this meta-analysis is that different researchers measured different correlates of NPP; for example, one experiment might look at changes in aboveground biomass, a second study might measure soil respiration rates, and a study of marine production might measure the concentration of chlorophyll to estimate phytoplankton abundance. To combine all of these variables

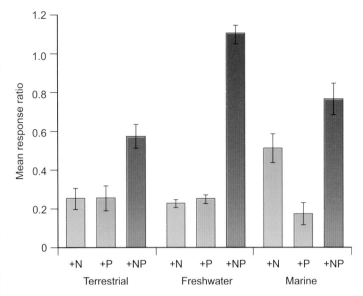

Figure 19.7 Mean responses ratios (±1 SE) of autotrophs to nitrogen (+N), phosphorus (+P) and nitrogen + phosphorus (+NP) enrichment in terrestrial, freshwater, and marine ecosystems.

into a common unit, the researchers calculated the response ratio $(RR) = \ln (E/C)$, where ln is the natural logarithm, E is the measured value of the response to the supplement received by the experimental group, and C is the associated value in the unsupplemented control system. A value of RR less than zero indicates the control has higher production, and an RR value greater than zero indicates that the supplemented ecosystem has higher production. Because RR is a logarithmic function, $RR = 1$ indicates that the supplemented ecosystem has 2.718 (base of the natural logarithm) times the production of the control.

The analysis has several intriguing findings. Overall, NPP is limited by both nitrogen and phosphorus in terrestrial, freshwater, and marine systems. Supplementing with both nitrogen and phosphorus increased production considerably in all three systems. However, in contrast to the prevailing doctrine, phosphorus limits production to the same extent in all three types of ecosystems. In support of the prevailing doctrine, nitrogen does limit production to a greater extent in marine than in freshwater systems (Figure 19.7), but nitrogen limitation clearly is important in many freshwater systems as well.

Addition of nutrients to ecosystems is a major source of water pollution across the globe. Increased production resulting from adding nutrients causes eutrophication of lakes and extreme overproduction of phytoplankton in many estuaries, leading to immense dead zones. The meta-analysis compiled by Elser and colleagues also highlights that local effects of adding nutrients may vary in different lakes; for example, phosphorus inputs from detergents may have little impact on one lake and a substantial impact on a second lake. Second, phosphorus and nitrogen together have a much stronger effect than either nutrient alone, thus ecosystem managers should take a balanced approach to restricting excessive inputs of both nutrients throughout the world.

Now that we have a handle on the distribution of production across the world and on the factors that influence NPP, we are ready to explore how the energy synthesized by primary producers moves through an ecosystem. To do that, we first introduce the cast of characters that make up an ecosystem.

19.2 WHAT ARE THE IMPORTANT COMPONENTS OF ECOSYSTEM STRUCTURE?

Our discussion of food webs in Chapter 16 introduced some of the biotic players in ecosystems. Producers, mostly plants, occupy the first trophic level, and can manufacture food from simple organic substances. The remainder of the biotic ecosystem comprises heterotrophs – organisms that get their nourishment from external sources. Primary consumers ingest producers and occupy the second trophic level, secondary consumers ingest primary consumers and occupy the third trophic level, and tertiary consumers ingest secondary consumers and occupy the fourth trophic level (Figure 19.8). In theory there could be fourth-, fifth-, and sixth-level consumers, and so on. Later in this chapter we will discuss why higher-level consumers are relatively rare.

This linear model, while useful, is an unrealistic distortion of ecosystem structure. A more realistic description recognizes two important features. First, there are other types of organisms that play essential roles in ecosystems. Second, much of the energy content in food is lost as it passes through an ecosystem.

Detritivores and decomposers

There are millions of species of herbivores in the world, and yet most plants never get eaten, or at most are only partially eaten. As we've discussed previously, many plants defend themselves effectively against herbivores, or they may live in inconvenient locations, or have low nutritive value. In addition, herbivores are limited by competition and by predators and parasites. The upshot is that many plants die without feeding a primary consumer. In many cases, the nutritional value of dead plants is not lost to the ecosystem, however, instead passing to organisms that specialize on dead organic matter. These are the detritivores that eat dead organic matter, and the decomposers, primarily bacteria and fungi, that break down dead organic matter, use the energy released to fuel their metabolic processes, and absorb the released nutrients to build complex molecules.

Consumers at all levels also produce dead organic matter while they are alive, in the form of metabolic waste such as feces and urine, and when they die, if they are uneaten or only partially

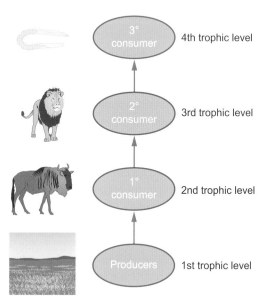

Figure 19.8 Linear food chain with trophic levels. Arrows indicate the direction of energy flow. In this and subsequent figures, 1° = primary, 2° = secondary, and 3° = tertiary.

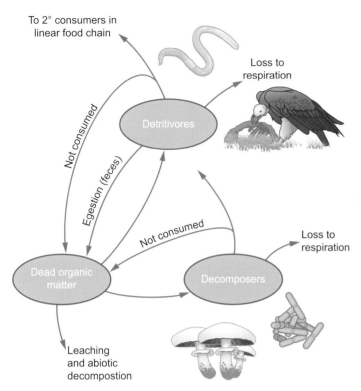

Figure 19.9 Detrital food chain. Arrows indicate the direction of energy flow.

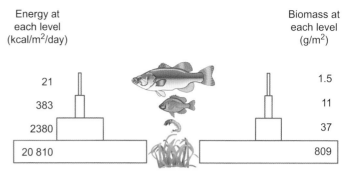

Figure 19.10 Left – Howard Odum's energy pyramid. Energy (kcal/m²/day) is lost at each trophic level. Right – Odum's biomass pyramid. Higher trophic levels have much less biomass (g/m²). The width of each level is not drawn to scale.

eaten. Dead organisms and their products then enter the detrital food chain (Figure 19.9).

There are several important features of the detrital chain. First, detritivores are both competing with and preying upon decomposers when they consume detritus. Second, both detritivores and decomposers are mortal, and thus add to the pool of dead organic matter if they die without being eaten. Lastly, detritivores that are eaten by consumers form an important linkage with the linear food chain described in Figure 19.8.

One feature shared by both food chains is that energy is lost on several paths as it moves through the ecosystem.

Energy loss with each transfer

A great deal of energy is lost to cellular respiration at all levels, with producers generally losing less than consumers. Energy is also lost in the linear food chain because much of the production is simply not consumed, and thus passes to the detrital chain. This can happen at any level; for example, not all of the herbivores (primary consumers) are eaten by secondary consumers, and their dead tissues pass to the detrital food chain. In addition, not all ingested organic matter can be used by consumers and is egested as feces, also passing on to the detrital chain (see Figure 19.9). As a consequence of all of these losses, there is much less energy available to each higher trophic level within an ecosystem.

Energy and biomass pyramids

The amount of energy available to one trophic level is a small fraction of what is available to the trophic level below it. This

relationship, the trophic efficiency, is equal to $(P_N/P_{N-1}) \times 100$, where P_N is equal to the production of one trophic level, and P_{N-1} is equal to the production of the trophic level below it. In a very early study, Howard Odum (1957), Eugene Odum's brother and colleague, explored the freshwater ecosystem at the headwaters of Silver Springs, Florida. This was an outstanding choice of research system because the waters there are a constant 23°C and very clear, and deep enough for humans to dive into but shallow enough that the vast majority of the organisms could be sampled and quantified. Odum measured production directly, by observing growth rates of individuals over time, and indirectly, primarily by measuring rates of photosynthesis and respiration.

When graphed, Odum's measures of production at each trophic level formed an energy pyramid, with decreasing amounts of production for each trophic level. The trophic efficiency ranged from 5 to 20% for each energy transfer. The Silver Springs community also formed a biomass pyramid, with higher trophic levels having the lowest biomass. Because energy is lost at each trophic level, all ecosystems form an upright energy pyramid (Figure 19.10).

However, some freshwater and marine ecosystems may form an inverted biomass pyramid, where there is a greater biomass of consumers than there is of their phytoplankton prey. On the surface, this seems to violate the idea that substantial energy is lost with each energy transfer. How could a community of consumers have a larger biomass than the community of producers on which it subsists? The solution to this conundrum is that nutrient-poor waters tend to be dominated by tiny phytoplankton (picoplankton) that have very short lifespans and very high reproductive rates. During its brief life, a picoplankton individual will probably get eaten by a consumer. Before this happens, it may give rise to numerous picoplankton generations during the lifespan of the average zooplankton consumer. If most of the picoplankton are consumed, they are producing more than enough energy to support a relatively large number of consumers. Whether eaten or dying of natural causes, the picoplankton has a fleeting existence, and thus the biomass of living picoplankton at any particular time is relatively small. In contrast, nutrient-rich waters, such as those near the coast or in areas of upwelling, are dominated by much larger phytoplankton (microplankton), which have much longer lifespans and thus contribute to a much larger biomass of producers.

We would expect that increased phytoplankton biomass should lead to increased consumer biomass, and it does. When Carlos Duarte and his colleagues (2000) systematically varied nutrient levels in a mesocosm experiment on a Mediterranean Sea ecosystem off the coast of Spain, they discovered that high nutrient levels increased phytoplankton biomass, in particular that of the large

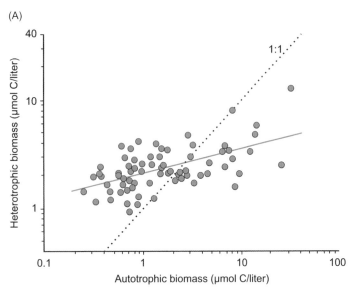

(A)

microplankton, by a factor of 50. The zooplankton responded to the added nutrients, but their biomass only increased by a factor of 10 (Figure 19.11A). Thus at high nutrient levels the ratio of heterotrophs to autotrophs declined dramatically, yielding a more familiar upright pyramid. In contrast, low nutrient levels were associated with a higher ratio of heterotrophs to autotrophs, and an inverted pyramid. This research helps to explain why open ocean and deep lake ecosystems, with their low nutrient levels, tend to have inverted pyramids, while coastal waters and areas of upwelling are associated with upright pyramids (Figure 19.11B)

These energy and biomass pyramids are very useful models for visualizing how energy is moving through ecosystems, and how ecosystems are structured. But their apparent simplicity can obscure important differences between ecosystems that become apparent when we dig deeper. We will focus our attention on how energy moves through ecosystems, keeping in mind that each step in energy transfer involves some energy being lost from the initial input.

Efficiency of energy transfer

Let's begin with a linear food chain that produces a quantity of energy (P_{N-1}). If we look at a portion of the food chain, we can easily visualize that energy is lost from three processes. First, some of the initial energy is consumed (I_N), but the rest is not consumed and is lost to this food chain. Ecologists define the consumption efficiency as the percentage of available energy

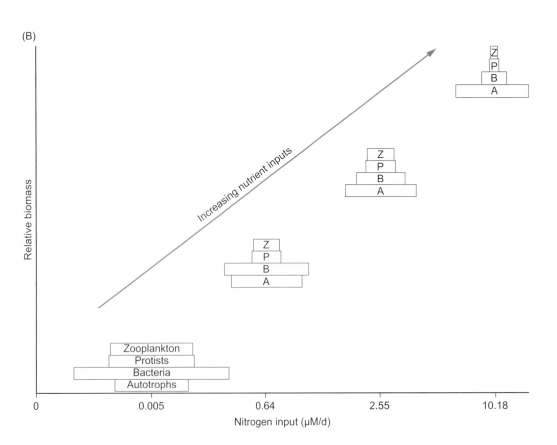

(B)

Figure 19.11 A. Heterotroph biomass increases more slowly than autotroph biomass in response to nutrient enrichment in mesocosms. B. The result is a shift from a somewhat inverted biomass pyramid at low nutrient levels to an upright biomass pyramid at high nutrient levels. Autotroph abundance is normalized to the same width in each pyramid, to allow comparison of relative abundance of autotrophs and heterotrophs.

that is actually ingested or consumed (=$(I_N/P_{N-1})100$ in Figure 19.12). Second, some of the consumed energy is egested in feces and urine and not assimilated. Ecologists define the assimilation efficiency as the percentage of energy in food that is assimilated by the organism for metabolic processes (= (A_N/I_N) 100 in Figure 19.12). Because consumed energy must be either assimilated or egested, the assimilation efficiency is also equal to (100% − % egested). Lastly, a substantial portion of the assimilated energy is lost in cellular respiration. Ecologists define the production efficiency as the percentage of assimilated energy that is allocated to growth or production of offspring (=(P_N/A_N) 100 in Figure 19.12). The product of these three efficiencies = (consumption efficiency)(assimilation efficiency)(production efficiency), which defines the trophic efficiency, which is equal to $(P_N/P_{N-1})100$.

Ecologists want to know whether there are any patterns of energy flow that are common to all ecosystems. If we know how much energy enters an ecosystem from solar radiation, can we predict how much energy will be available to different trophic levels? Or do different types of ecosystems have different patterns of energy flow?

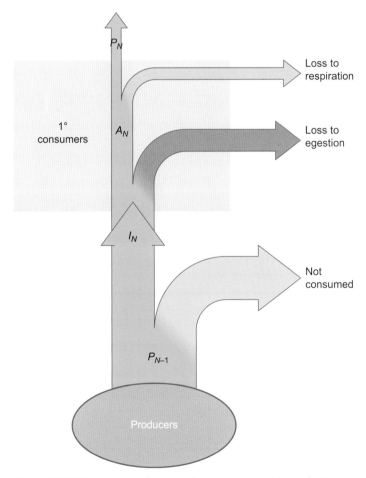

P_N

Loss to respiration

1° consumers

A_N

Loss to egestion

I_N

Not consumed

P_{N-1}

Producers

Figure 19.12 Movement of energy through one trophic level. P_{N-1} is the amount of energy produced by the producers. I_N is the amount of energy that enters the consumers, A_N is the amount of energy assimilated by the consumers, P_N is the amount of energy produced by the consumers.

Differences in energy transfer across ecosystems

While it is interesting to know that energy is lost as it passes through ecosystems, it would be much more informative to know what that pattern of energy transfer actually is. Nelson Hairston Jr. and Nelson Hairston Sr. (1993) estimated patterns of energy flow from 20 studies of terrestrial ecosystems and 12 studies of lakes or ponds. First we will discuss some of the highlights of their findings for temperate forest ecosystems (Figure 19.13). Then we will compare these results with their findings for temperate grasslands and pelagic lake ecosystems.

It may seem odd to have an ecosystem diagram that has dead organic matter at its heart, but as the authors point out, about 62% of the organic matter in temperate forests is dead organic matter, mostly plants. The abundance of dead organic matter arises because the consumption efficiency of the primary consumers (the herbivores) is only 3.7% (Figure 19.13), ultimately diverting over 96% of production to the pool of dead organic matter. Furthermore, the assimilation efficiency of the primary consumers is only 33%, so about two-thirds of the production is egested by consumers, returning to the dead organic matter pool. In sharp contrast, almost all of the primary consumers ultimately get eaten by secondary consumers (carnivores) in temperate forests; the consumption efficiency is 89.9%. Furthermore, carnivores have a higher assimilation efficiency than herbivores (56% compared to 33%), which reflects the generally higher nutrient content of meat.

Most of the dead organic matter passes to decomposers (bacteria and fungi) and detritivores, but some is decomposed abiotically. The decomposers themselves are consumed by detritivores, with a very high assimilation efficiency (78%). Like herbivores, almost 90% of detritivores are consumed by secondary consumers. The authors estimate that the detritivore biomass is about four times greater than the herbivore biomass in temperate forest ecosystems.

There are three important messages in this analysis. First, dead organic matter has a central role in temperate forest ecosystems. It receives input from every other trophic level and at any point in time contains the bulk of the production. Second, most ecosystem production is actually passed onto decomposers, who process the bulk of the energy stored in the dead organic matter. Lastly, consumption efficiencies are highly variable among trophic levels, ranging from 3.7% of producers consumed by primary consumers, to almost 90% of primary consumers and detritivores consumed by secondary consumers.

The authors' comparison of temperate forest to temperate grassland and pelagic lake ecosystems revealed notable differences. The structure of the pelagic lake ecosystem is very complex, in part because that includes more levels of higher-level consumers. Piscivorous (fish-eating) fish eat planktivorous (plankton-eating) fish, which eat predatory zooplankton, which eat grazing zooplankton and planktonic bacteria that get their energy from dead organic matter. Thus there are four major levels of consumers in pelagic lake ecosystems, in contrast to two major levels found in temperate forests and temperate grasslands. As a result, pelagic lake food chains tend to be longer than most terrestrial food chains.

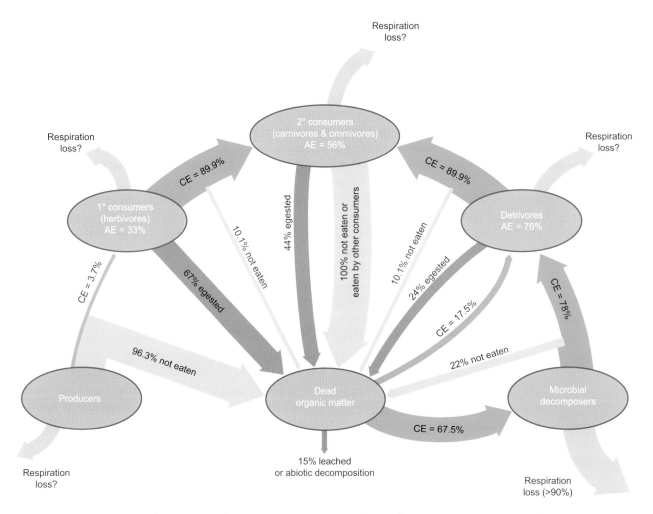

Figure 19.13 Energy flow diagram for temperate forest ecosystem. AE, assimilation efficiency; CE, consumption efficiency.

Decomposers are situated primarily at the bottom of the lake, forming a part of the benthic food chain. Because water is more viscous than air, most of the dead phytoplankton never make it to the lake bottom but are instead eaten by grazers or colonized by bacteria while still in the pelagic zone. Overall, only about 15% of pelagic production enters the benthic food chain, in contrast to the 95% of production from forested systems that reaches the forest floor, where it is processed by detritivores and decomposers. However, the consumption efficiency by detritivores and decomposers in the benthic chain is only about 25%, with the rest of the dead organisms building up as sediment.

Table 19.3 summarizes some of the differences among the three ecosystems studied by Hairston and Hairston.

 Thinking ecologically 19.2

Contrast the physical structure of the three ecosystems in Table 19.3, and the types of foods each ecosystem provides for primary consumers. How might these differences account for differences in consumption efficiencies and assimilation efficiencies among the three ecosystems?

Table 19.3 Some distinctions between the three ecosystems considered by Hairston and Hairston (1993).

	Temperate forest	Temperate grassland	Pelagic lake
Consumption efficiency of primary consumers	3.7	9.3	32.8
Assimilation efficiency of primary consumers	33	16.1	75–80
Consumption efficiency of secondary consumers[a]	89.9	77	98
Number of major levels of consumers	2	2	4
Number of pools of dead organic matter	1	1	2

[a] For a pelagic lake ecosystem, this represents the percentage of grazing zooplankton eaten by predatory zooplankton and planktivorous fish.

Hairston and Hairston do not detail losses to cellular respiration, because they were more interested in trophic interactions. The only exception is microbial decomposers, which lost over 90% of their energy to respiration. However, even within trophic groups, there are substantial differences in the production efficiencies of different groups of animals. For example, production efficiency of most endotherms is a remarkably low 1–3%, while many ectotherms average between 20 and 50% (Chapin *et al.* 2002). As we discussed in Chapter 5, endotherms use much of their energy intake to regulate their body temperatures. In general, smaller endotherms, with their relatively greater surface area, tend to lose heat more easily than larger endotherms. By diverting more energy to temperature regulation, small homeotherms tend to have exceedingly low production efficiencies.

Lastly, Hairston and Hairston appear to take some liberties by simplifying ecosystem structure. For example, their temperate forest and grassland ecosystems only have one level of carnivore. This seems strange, since we all know that carnivores do eat other carnivores. In fact, the first data presented by the authors demonstrate that in some salamander populations, more than 50% of prey are other carnivores. The authors do not deny that carnivores eat other carnivores, but rather argue that most carnivores don't specialize on other carnivores but are equally likely, or usually more likely, to eat herbivores in addition to carnivores, and that other carnivores are a relatively small part of the average carnivore's diet. Of course, there are a few exceptions, such as peregrine falcons that eat only birds, but they are rare and make only minor contributions to the flow of energy through food chains in terrestrial systems.

Let's explore in more detail the question of how long food chains really are, and what regulates their length.

19.3 WHAT IS THE RELATIONSHIP BETWEEN FOOD CHAIN LENGTH AND ECOSYSTEM STRUCTURE AND FUNCTIONING?

Our understanding of ecosystem structure and energy flow allows us to address the central question of food chain length (FCL), which measures the mean number of transfers from the base to the top of a food web. To understand how energy flow and trophic structure interact to determine FCL, consider the following simple ecosystems. In Figure 19.14A, consumer species C_1 and C_2 both get all their energy from the producer (P), and FCL = 1. In Figure 19.14B, consumer C_2 eats only consumer C_1, so FCL = 2. But Figure 19.14C describes a more complex three-species ecosystem, in which consumer C_2 gets 80% of its energy from C_1 (FCL = 2) and 20% from the producer (FCL = 1). Because some energy is flowing directly from P to C_2, this shortens the FCL somewhat. In Figure 19.14D, consumer C_2 gets most of its energy from the producer (FCL = 1), which has the effect of

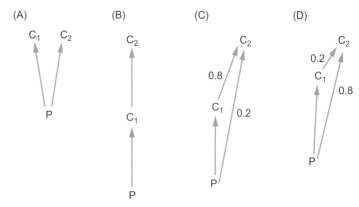

Figure 19.14 Food chain length varies because there are differences in energy flow among the three species in the food chain.

shortening the food chain much more, making food chain D more similar in structure, and length, to food chain A.

Ecological processes and food chain length

FCL can influence primary and secondary production, and the flow of nutrients through ecosystems. It can also influence other related ecological processes such as biomagnification – the tendency for substances, particularly toxins, to have higher concentrations at higher trophic levels.

Influence of food chain length on primary production and nutrient cycles

One of our greatest environmental concerns is the global increase in atmospheric CO_2 levels. A lake can add to the problem by exporting CO_2 into the atmosphere. This occurs when carbon inputs from terrestrial sources are so high that lakes become supersaturated with CO_2 and release CO_2 into the atmosphere. However, a nutrient-rich lake can reduce atmospheric CO_2 if it harbors large populations of phytoplankton, which fix sufficient carbon in the process of photosynthesis to reduce lake CO_2 levels low enough so that CO_2 diffuses from the atmosphere into the lake.

Daniel Schindler and his colleagues (1997) were interested in two related questions. First, they wanted to know whether adding nutrients to a lake could increase phytoplankton production sufficiently to convert a lake from a CO_2 exporter to a CO_2 importer. Second, they wanted to determine whether lakes with longer food chains would have a greater tendency to act as CO_2 exporters. Lakes with short food chains only had minnows to feed on zooplankton. The researchers reasoned that these minnows would keep zooplankton populations in check, so that phytoplankton would be able to increase. In contrast, lakes with longer food chains also hosted bass populations, which fed on minnows and reduced the minnow population to levels where large zooplankton became abundant. These large zooplankton would be sufficiently abundant to reduce phytoplankton population and reduce NPP of the lake. If true, this would be a good example of top-down control of these lake ecosystems (see Chapter 16).

(A)

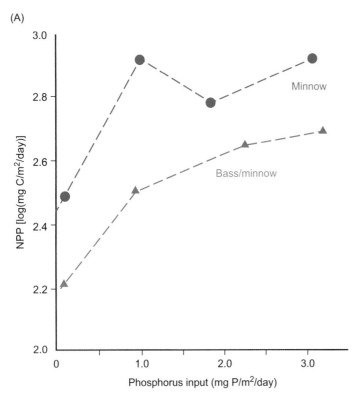

(B)

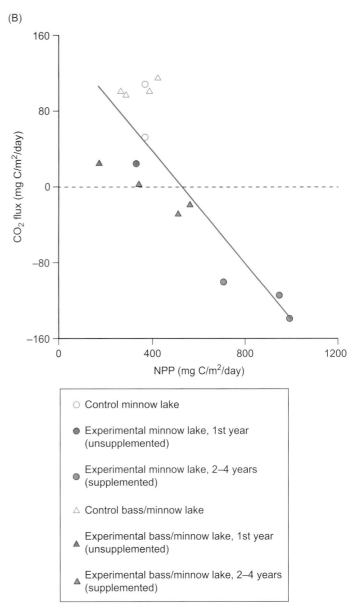

○ Control minnow lake

● Experimental minnow lake, 1st year (unsupplemented)

◔ Experimental minnow lake, 2–4 years (supplemented)

△ Control bass/minnow lake

▲ Experimental bass/minnow lake, 1st year (unsupplemented)

▲ Experimental bass/minnow lake, 2–4 years (supplemented)

Figure 19.15 A. Increases in primary production in response to phosphorus supplementation in a minnow-only lake (circles) and a bass/minnow lake (triangles). B. Increases in CO_2 import in response to higher production resulting from nutrient supplementation. Negative flux values indicate that the lake is importing CO_2 from the atmosphere. The researchers measured flux in all four years, with the exception of the control minnow-only lake, which was measured only in the third and fourth year of the study.

To address these two questions, the researchers found four similar-sized lakes, two of which had bass and minnows, and two with only minnows. In the first year of the study, all four lakes were monitored for nutrient levels and production levels. In addition, the researchers measured CO_2 flux – the rate at which CO_2 is exchanged with the atmosphere. For the second, third, and fourth years of the study, one bass/minnow lake and one minnow-only lake received varying amounts of supplemental nitrogen and phosphorus. The other bass/minnow lake and minnow-only lake remained as unsupplemented controls. Schindler and his colleagues continuously monitored nutrient and production levels and CO_2 flux throughout the study.

Nutrient enrichment increased primary production in both the bass/minnow and minnow-only lakes (Figure 19.15A). This increase in NPP was sufficient to convert the minnow-only lake from a carbon exporter to a carbon importer for each treatment year (Figure 19.15B). The increase in NPP was also sufficient to convert the bass/minnow lake to a carbon importer in two of three years, and to an equilibrium state for the third year. Finally, the bass/minnow lakes, with their longer food chains, had substantially lower production than the minnow-only lakes. In this case,

longer food chains were associated with lower production, and a greater tendency to export carbon into the atmosphere.

 Thinking ecologically 19.3

In which lake was flux more affected by nutrient enrichment? Can you think of a mechanism that might be responsible for this difference in responsiveness?

The relationship between FCL and nutrient export will vary with different ecosystems. However, the relationship between FCL and biomagnification is much more predictable

Food chain length and biomagnification

Some substances are lipophilic and not easily broken down in the body. These include novel organic substances, such as DDT, that have been introduced by humans into ecosystems. They are novel in the sense that consumers are not adapted to them, and thus may not have effective ways of detoxifying them. Many heavy metals, such as mercury, are also difficult for consumers to degrade, and can cause serious health problems for top-trophic-level consumers. Both of these types of substances can build up to dangerously high levels as they move up long food chains.

Rachel Carson (1961) made the general public aware of some of the problems associated with biological magnification with her eloquent portrayal of a toxic world lacking the insect and frog choruses that we associate with the abundance of springtime life. Her analysis of the scientific research on environmental toxins and her popular writing inspired many ecologists to pursue research into ecotoxicology. For example, George Woodwell and his colleagues (1967) demonstrated biomagnification of DDT at higher trophic levels (Figure 19.16). Later studies showed that birds of prey with high DDT levels tended to have thinner eggshells, which broke when the eggs were incubated by the parents. As a result of this partnership of popular activism and scientific research, DDT was banned in the United States in 1972.

FCL is a fundamental property of ecosystems, and it has important impacts on how substances (nutrients and poisons) move through ecosystems. Unfortunately, it is difficult to measure, because it requires knowledge of how energy is transferred from producers to consumers, and from consumers to other consumers. Fortunately, the application of stable isotope analysis has allowed researchers to more easily estimate FCL within an ecosystem.

Estimating food chain length with stable isotope analysis

Estimating FCL can be very tedious when it involves analysing the diet of each species in an ecosystem to estimate what

	DDT residues (ppm)
Water	0.0005
Mixed plankton	0.040
Shrimp	0.16
Herbivorous fish	0.3 → 1.3
Carnivorous fish	1.3 → 2.07
Shore birds (carnivorous)	3 → 75.5

Figure 19.16 Biomagnification of DDT (in parts per million) at higher trophic levels.

proportion of that species' energy comes from each item in its diet. Researchers now routinely use stable isotope analysis to estimate the trophic position of each species in an ecosystem. Recall from Chapter 3 that the $\delta^{15}N$ value is typically elevated in consumers by 3.4‰ in relation to the value in their prey. Thus a consumer's trophic position is calculated with the following formula:

$$\text{Trophic position} = [(\delta^{15}N_{consumer} - \delta^{15}N_{producer})/3.4] + 1$$

FCL is simply equal to the trophic position of the top consumer.

David Post (2002) reminds us that there are several issues associated with using stable isotope analysis to calculate trophic position and FCL. One important concern is that there is considerable variation in the $\delta^{15}N$ value at the base of any ecosystem, so a baseline value must be established for each ecosystem. Also, the $\delta^{15}N$ value does change somewhat over the lifespan of long-lived organisms, and this change needs to be accounted for, especially in small samples. Despite these problems, stable isotope analysis has opened up many avenues of research into trophic structure. Using this technique, we are now ready to address the question of what factors influence FCL.

Factors influencing food chain length in ecosystems

There are three major hypotheses to explain what factors control FCL in ecosystems. The *ecosystem size hypothesis* proposes that larger ecosystems should have longer food chains because their greater habitat heterogeneity provides more ecological niches for a diversity of species. The *environmental stability hypothesis* proposes that food chains will be shorter in highly variable environments, because long food chains are less stable and more easily disturbed or disrupted. The *resource availability or energy supply hypothesis* proposes that FCL is limited by an ecosystem's production. Because trophic efficiencies are generally quite low, less productive ecosystems will tend to have shorter FCLs.

Let's consider two studies that attempt to identify the factors influencing FCL in aquatic and terrestrial ecosystems.

Factors influencing food chain length in rivers

John Sabo and his colleagues (2010) used stable isotopes to estimate FCL in a study of 36 rivers in North America. Ecosystem size was the area of the river drainage system, which ranged from less than 1 to over 10^6 km^2. Environmental stability was based on the variation in annual maximum flow rate over a 16–20-year period, which differed among streams by more than a factor of a thousand. GPP was the rate of oxygen evolution, which also varied by more than a factor of a thousand when comparing different river systems. This extraordinary amount of variation provided a great opportunity for the researchers to test the predictions of each of the three hypotheses.

FCL varied from a low of just under three to a high of almost five. Larger ecosystems tended to have greater FCL, but there was substantial variation in this trend (Figure 19.17A). Rivers with greatest environmental stability tended to have longer food chains, but there was also substantial variation in this trend (Figure 19.17B). Finally, there was no relation between GPP and FCL (Figure 19.17C).

One issue with this analysis is that the largest ecosystems tend to have the lowest variation in flow rate. Thus it is difficult to evaluate whether the data support the ecosystem size or the environmental stability hypothesis. The authors argue that

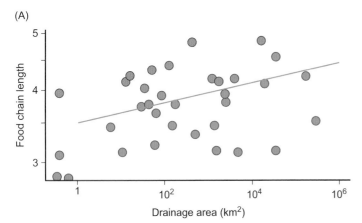

(A)

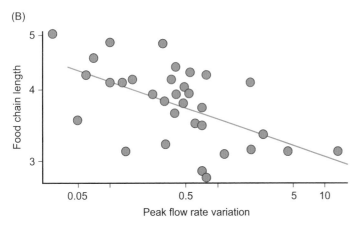

(B)

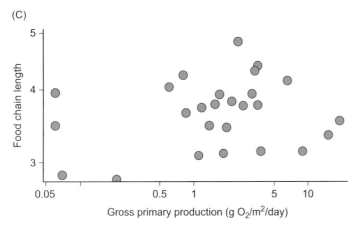

(C)

Figure 19.17. Studies of food chain length in 36 river systems that vary in ecosystem size (A), variation in peak flow rate (B), and gross primary production (C). The actual calculation of peak flow rate variation is based on the standard deviation of year-to-year differences in peak flow.

ecosystem size effects are somewhat independent of variation in river flow rates. Two pieces of evidence support this. First, rivers that flooded frequently tended to have shorter food chains than rivers that rarely flooded. This effect was independent of ecosystem size. Second, on average, intermittent rivers (rivers that occasionally dried up in stretches) had shorter food chains than perennial rivers of similar drainage area. Perennial rivers tended to have piscivorous fish as their top predators, while intermittent rivers more commonly had invertebrates or small-bodied fish. The researchers caution us that human changes to water flow patterns (for example, those caused by building dams or diverting rivers for irrigation) can have profound effects on food webs in river ecosystems.

Let's now compare these results with a similar study conducted on terrestrial systems.

Factors influencing food chain length in terrestrial ecosystems

Hillary Young and her colleagues (2013) also used stable isotope analysis to estimate FCL in 23 islands that make up the tropical Palmyra Atoll in the central Pacific Ocean (Figure 19.18A). They were able to test the hypothesis that larger ecosystems have greater FCL, because island area varied from about 500 m^2 to 2.6×10^6 m^2. They were able to test the hypothesis that more productive ecosystems have greater FCL, because the islands varied enormously in their production as a result of fecal nitrogen contributed by nesting and roosting seabirds that frequented some islands, but not others. Because island choice by seabirds was independent of island size, the fecal fertilizer effect was also independent of island size. For technical reasons, the researchers measured production based on nitrogen levels (rather than g C/m^2/yr) but were able to show in a series of experiments that the two measures of production were highly correlated.

In contrast to the findings of the Sabo team for rivers, production on these islands emerged as a very important factor influencing FCL, while island size had no effect (Figure 19.18B). Young and her colleagues proposed two mechanisms to explain why production has such a strong effect on FCL. First, the researchers discovered that more productive islands had a greater abundance and greater species richness of insects that could feed the top predators, which were two species of spiders and two species of geckos. Some of these insects were predators on other insects, so increased species richness introduced extra trophic levels to the food chain. Second, some species of carnivores grew to larger body size on the more productive islands. This would enable the carnivores to eat larger (and presumably higher trophic level) prey, including prey from their own trophic level, thereby effectively lengthening FCL.

At this time, we don't know whether more studies will show that ecosystem size is an important factor influencing FCL in aquatic systems and production a more important factor influencing FCL in terrestrial systems. The general consensus among researchers is that different types of ecosystems will be

(B)

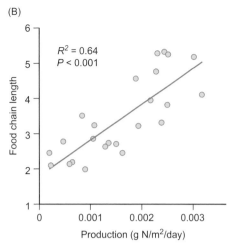

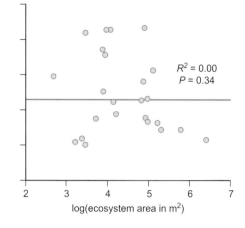

Figure 19.18 A. The Palmyra atoll. B. The relationship between production and food chain length (left) and log(ecosystem area in m^2) and food chain length (right).

influenced by varying combinations of ecosystem size, environmental variation, and production. The challenge now is to identify patterns in which factors are more likely to be important for particular types of ecosystems. Applied ecologists can then use that knowledge to make good decisions about resource use that balance the dual challenges of maintaining ecosystem integrity, and achieving a sustainable quality of life for the growing human population.

REVISIT: The Millennium Ecosystem Assessment

If ecosystems provide critical services for humans, and if humans have important influences on ecosystems, we need to understand how the balance of the human/ecosystem interaction plays out in the global economy. The MA report is mixed; overall, 15 of 24 ecosystem services are being degraded or used unsustainably. The good news is that global food production has increased 160% since 1960, while human population has only doubled. This increase results from improvements in agricultural technology, and from increased conversion of natural biomes to agriculture. In theory, this translates to more food per person, but recent trends are mixed. Globally, the number of undernourished people has declined steadily from more than 1 billion in 1990–2 to 842.5 million in 2011–13. However, the number of undernourished people has actually increased over the same time period in some of the economically poorest countries or regions, for example increasing 28.7% in sub-Saharan Africa and 69.4% in the Near East and North Africa (FAO, IFAD, and WFP 2013).

The conversion of natural biomes to agriculture is very problematic; once 100% of a biome has been converted to agriculture, there is no further conversion possibility. Already, large fractions of many biomes have been converted to agriculture, so there is little remaining potential for conversion (see Figure 11.10). In addition, the remaining unconverted areas are, in general, not conducive to agriculture for a variety of reasons, such as the areas being too steep or too rocky (Millennium Ecosystem Assessment 2005).

Though the situation is mixed in some aspects of provisioning services, the picture is much bleaker in regulating services. Most regulating services have degraded since 1960, including regulation of air quality, water purification systems, soil erosion, and agricultural pests. Ecosystems are also providing inferior pollination services and showing reduced protection against storms and tides as a result of loss of wetlands and mangrove buffers. Cultural services have also declined, providing fewer spiritual and aesthetic values to people.

The coalition of ecologists and economists working together on the MA estimated the value of these ecosystem processes on local, regional, and global economies. These scientists recognized that the true value of an ecosystem is a composite of the value of any resource it generates and any other service it renders. If a mangrove swamp is converted to a shrimp aquaculture facility, the value of that converted ecosystem is the value of the shrimp in terms of increased production (provisioning service) minus the loss of natural hazard regulation, water purification regulation, and aesthetic values.

The authors of the MA do not recommend that all ecosystems be allowed to revert to their natural state (even if it could be done). They recognize that humans need food and other resources, and that difficult tradeoffs must be accepted. For example, in many cases, increasing ecosystem production will be costly in terms of loss of ecosystem services. The authors do, however, argue that we remain aware of these tradeoffs, so that we can make rational decisions after accounting for the benefits and costs of converting an intact ecosystem.

SUMMARY

Ecosystem structure and functioning is the focus of much ecological research because many ecosystem properties such as production, energy flow, nutrient cycles, and stability lie at the core of understanding ecological processes. A related reason for this focus is that applied ecologists must understand ecosystem structure and functioning to assess how human changes to ecosystems are influencing critical ecosystem services, so they can make rational management recommendations.

Net Primary Production (NPP) is primarily influenced by climate and nutrients. On a global scale, NPP in terrestrial biomes tends to be greatest near the tropics, where the combination of constant and moderately high temperature and adequate rainfall promote plant growth. NPP in marine biomes peaks at about 40° S latitude, which is associated with large areas of upwelling and high nutrient availability. On a regional and local scale, the availability of nutrients such as nitrogen and phosphorus influence terrestrial, marine, and freshwater production.

Ecosystem structure is based on the interactions between producers, consumers, detritivores, and decomposers. A substantial but variable amount of energy is lost with each transfer from one trophic level to the level above, which has the effect of limiting food chain length (FCL). In some aquatic systems, longer food chains are associated with CO_2 export from the water into the atmosphere, and with the biomagnification of toxic substances. Researchers are studying the effects of ecosystem size, ecosystem production, and ecosystem variability on FCL in terrestrial and aquatic ecosystems.

FURTHER READING

Dirzo, R., and 5 others. 2014. Defaunation in the Anthropocene. *Science* 345: 401–406.

Very nice review paper that discusses global patterns of human-caused extinctions, describes how these extinctions will influence ecosystem functioning, and then ties the changes in ecosystem functioning to the loss of ecosystem services.

Elser, J. J., and 9 others. 2007. Global analysis of nitrogen and phosphorus limitation of primary producers in freshwater, marine and terrestrial ecosystems. *Ecology Letters* 10: 1135–1142.

Large-scale meta-analysis of how nitrogen and phosphorus availability limit production in terrestrial and aquatic ecosystems. Good opportunity to practice analysing meta-analyses in this clearly laid-out study that shows global prevalence of nitrogen and phosphorus limitation and also some synergistic effects.

Del Grosso, S., and 7 others. 2008. Global potential net primary production predicted from vegetation class, precipitation and temperature. *Ecology* 89: 2117–2226.

Good synthetic study showing how land cover, precipitation and temperature influence terrestrial NPP. The logistic regression equations may be a bit daunting, but can be interpreted with some persistence.

Jackrel, S. L., and Wootton, J. T. 2014. Local adaptation of stream communities to intraspecific variation in a terrestrial ecosystem subsidy. *Ecology* 95: 37–43.

Great study demonstrating that red alder leaves taken from trees that overgrow a stream decompose substantially more rapidly in their native stream than do red alder leaves taken from trees that overgrow nearby streams. The researchers conclude that selection pressure favoring decomposition efficiency has led to extraordinarily fine-scale adaptations in the community of decomposers inhabiting the streams.

END-OF-CHAPTER QUESTIONS

Review questions

1. What are the different types of services provided by ecosystems and what are the drivers of these services?
2. On a global level, how do terrestrial and marine ecosystems differ in their distribution of NPP?
3. What are the basic trophic levels of ecosystems and how do these different trophic levels interact?
4. What are energy and biomass pyramids and what causes them to have a pyramidal shape. Why are some energy pyramids inverted?
5. How does food chain length influence biomagnification and CO_2 export from lakes? What factors have been proposed to influence food chain length in ecosystems?

Synthesis and application questions

1. Imagine a situation in which a corporate executive used the NPP data from Table 19.2 to argue that we don't need to worry about polluting the open ocean because it is so unproductive. How would you use the total area (column 2) data to respond to this argument? Use actual numbers for the total amount of fixed carbon to support your response.
2. Schuur (2001) provides evidence that high soil water levels slow down the process of decomposition, which could reduce the amount of nutrients available to producers, and reduce NPP. Can you think of other factors that might reduce NPP in areas with high mean annual precipitation?
3. Combining studies for a meta-analysis can be challenging. What obstacles can you think of that LeBauer and Treseder (2008) faced in their nitrogen supplementation meta-analysis, and how might they have resolved those obstacles?
4. Chapin *et al.* (2002) estimate a consumption efficiency of 5–15% for herbivores consuming herbaceous old fields abandoned for 1–7 years, but only 1.1% for herbivores consuming old fields abandoned for 30 years. Why might there be such a big difference in consumption efficiencies?
5. Fish have production efficiencies that are much lower than most other ectotherms. Why might this be the case? How might you test your explanation?
6. Some people argue that while it's nice that we care about bioaccumulation of toxic substances in ecosystems, it is more important that we concern ourselves with human health, and that we should go back to using DDT, because it kills mosquitoes that carry malaria. How would you respond to such a proposal?

Analyse the data 1

The research by Schindler and colleagues (1997) used two controls: (1) the experimental lake the year before supplementation and (2) a similar lake that was not supplemented. Why did they need two controls and was this a good idea? Would it have been a better study to double their sample size by supplementing two minnow-only lakes and two bass/minnow lakes, using the year before supplementation as the only control? Explain your reasoning?

Analyse the data 2

Knowing that most phosphorus becomes available to ecosystems by the weathering of rocks, Elser and his colleagues (2007) predicted more nitrogen limitation than phosphorus limitation in the tundra. Explain the logic behind this prediction. Does Figure 19.19 support this prediction? What type of statistical analysis would help you answer this question?

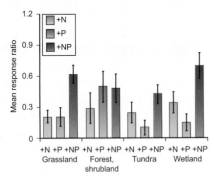

Figure 19.19 Mean response ratios (±1 SE) of autotrophs to nutrient enrichment in four biomes.

Chapter 20

Nutrient cycles: global, regional, and local

INTRODUCTION

In the 1970s and 1980s, the Black Sea could have been renamed the Greenish Brown Sea in testimony to a series of related dramatic changes in its constitution. Each spring and summer, beginning in 1973, a large area of the sea at the mouths of the Danube and Dnipro Rivers hosted dense blooms of phytoplankton near the surface, in some years expanding to over 40 000 km^2. Increases in marine production can be beneficial, but in this case, there was too much of a good thing. The phytoplankton blooms reduced water transparency so that sufficient light no longer reached the red algae, *Phyllophora nervosa*, a keystone species for the benthic offshore ecosystem. *Mytilus* mussels and other bivalves had thrived in these algal beds, filtering the waters of plankton, including algae and bacteria. Before the phytoplankton blooms, 118 invertebrate and 47 fish species were identified in these waters, and many others were undoubtedly present but not identified (Mee *et al.* 2005). In association with these phytoplankton blooms, most of these species either died off or migrated away.

There are many other examples of human activities causing significant changes to nutrient cycles on a local, regional, and global level. It is difficult to evaluate all of the ways human activities affect nutrient cycles, because the cycles are so complex with many interconnected linkages. In this chapter we will learn how the nitrogen cycle depends heavily on the actions of microorganisms to convert nitrogen from one molecular form to another, and to move nitrogen from the atmosphere into terrestrial and aquatic ecosystems. Ecologists have recently learned a great deal about how the nitrogen cycle works, with an increased understanding of anaerobic ammonium oxidation and the discovery that some nitrogen enters the cycle from weathered bedrock. In contrast to the nitrogen cycle, the atmosphere plays a relatively minor role in the phosphorus cycle, which is dominated by fluxes through oceans, freshwater systems, and soils. On a regional or local scale, decomposition and nutrient uptake retain and recycle nutrients within ecosystems. As the Black Sea case study illustrates, human actions may change the abundance of nutrients moving within and between ecosystems, leading to dramatic changes in species composition, distribution, and abundance.

KEY QUESTIONS

20.1. How do microorganisms move nitrogen compounds within and between ecosystems?

20.2. How do biological, geological, chemical, and physical processes interact in the global nitrogen cycle?

20.3. How do oceans, freshwater systems, and soils interact in the global phosphorus cycle?

20.4. What processes cycle nutrients within ecosystems?

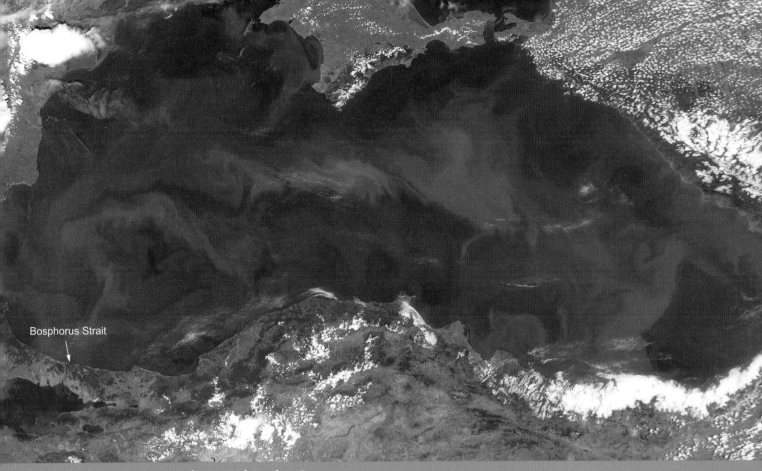

Bosphorus Strait

CASE STUDY: The Black Sea ecosystem

In the Black Sea, **eutrophication** – the addition of excessive nutrients such as nitrogen and phosphorus – appears to be the primary cause of massive phytoplankton blooms. Between 1950 and 1980, levels of both nitrogen and phosphorous compounds more than doubled in the Danube and Dnipro river systems as a consequence of agricultural runoff and emissions from fossil fuel use. Huge pig farms produced large quantities of nitrogenous waste, and growing urban populations of humans generated increased nitrogen-rich sewage and phosphate-rich detergents, all of which ran off into the river systems, and ultimately into the shallow shelf in the northwest corner of the Black Sea.

Elimination of the red algae fields was an early step in the formation of a dead zone – a region of *hypoxia* or low oxygen, where dissolved oxygen levels drop below 2.0 ml/l for an extended period of time, causing the death of most aerobic benthic organisms. When the massive assemblage of phytoplankton die off, their bodies are decomposed by large communities of bacteria and eaten by detritivores, both of which use the oxygen in the water for cellular respiration. Under these conditions the water may become *anoxic*, losing virtually all of its dissolved oxygen. When this happens, the water is unsuitable for aerobic organisms for several months during the spring and summer.

Overfishing by humans may also have played a supporting role in forming the Black Sea dead zone. Analysis of fish catch data by Georgi Daskalov (2002) shows a strong negative correlation between the abundance of predatory fish such as bonito, mackerel, and blue fish, and the abundance of planktivorous (zooplankton-eating) fish such as sprat and horse mackerel. Daskalov hypothesizes that increased harvesting of predatory fish in the 1970s allowed planktivorous fish to increase, which reduced zooplankton abundance to levels where they were no longer controlling phytoplankton abundance. This trophic cascade operated in conjunction with eutrophication to favor the development of the Black Sea dead zone (Figure 20.1).

To make a bad situation even worse, in 1982 the comb jelly *Mnemiopsis leidyi* was introduced into the Black Sea, presumably in the ballast water of a ship. *Mnemiopsis* is native to the western Atlantic Ocean, and had already been identified as a voracious and effective predator on many

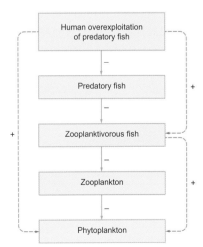

Figure 20.1 Hypothesized Black Sea trophic cascade culminating in phytoplankton blooms. Direct effects are solid arrows and indirect effects are dashed arrows. Not all effects are shown.

zooplankton species in its native habitat. It has an unusually high reproductive rate, producing up to 8000 eggs at 13 days after its own birth. It did not reach its highest population in the Black Sea until 1989, which coincided with the collapse of the anchovy fishing industry there. Anchovies had constituted about 60% of the total Black Sea fish catch in the second half of the twentieth century. Ahmet Kideys and his colleagues (2005) propose that *Mnemiopsis* outcompeted the anchovies for zooplankton, leading to the precipitous decline in this critical fishery, and at the same time, increasing phytoplankton abundance.

Looking closely at the photograph of the Black Sea, you will notice the narrow Bosphorus Strait, which provides the only outflow from the Black Sea into the much larger Mediterranean Sea. Because this connection is so small, averaging 1.6 km wide and 36 m deep, water in the Black Sea has an unusually long turnover time, taking an average of 300–400 years to renew itself. This contributes to small tides and weak currents; consequently, nutrients are slow to disperse from their origins, enhancing the impact of eutrophication.

While some dead zones occur naturally, most have been created by increased nutrient loads arising from human activity. There are now over 400 eutrophication-associated dead zones, and the number is doubling approximately every decade (Diaz and Rosenberg 2008). The emergence of dead zones as an environmental issue of deep concern reminds us that both abiotic and biotic factors can interact in ways that are difficult to foresee or anticipate. These interactions can occur on a global, regional, or local scale. In this chapter, we will follow the movements of two important nutrients, nitrogen and phosphorus, both between and within ecosystems.

20.1 HOW DO MICROORGANISMS MOVE NITROGEN COMPOUNDS WITHIN AND BETWEEN ECOSYSTEMS?

In the previous chapter, we saw how energy is dissipated as it passes through ecosystems, creating, in most cases, an energy pyramid. In contrast, nutrients are used or recycled over and over again, a process ecologists call nutrient cycling. We can't predict precisely how a nitrogen atom will pass through an ecosystem; for example, we don't know whether a zooplankton, with its numerous nitrogen atoms, will get eaten by a fish, or will die uneaten, with its tissues subsequently broken down and its nitrogen atoms liberated by a host of detritivores and decomposers. But we can quantitatively describe the possible pathways a nitrogen atom may take.

Perhaps more than any other nutrient cycle, the nitrogen cycle is characterized by the *transformation* or conversion of molecules from one chemical form to another (Figure 20.2). These transformations may occur with or without the help of organisms. As one example, nitrogen gas (N_2) makes up 78% of our

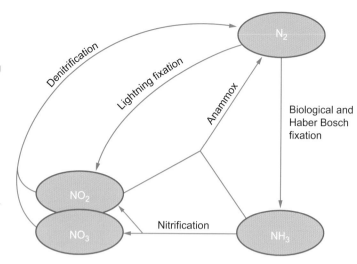

Figure 20.2 Relationships among some of the important transformation reactions in the nitrogen cycle.

atmosphere, but the N_2 molecule is held together by a powerful triple bond, which makes it inaccessible to most organisms. Nitrogen fixation is the reaction that transforms N_2 into a form that is biologically accessible to most organisms. Lightning does this process abiotically, by oxidizing N_2 to nitrogen oxides and ultimately forming nitrates (NO_3^-). More important is biotic fixation, in which several types of microorganisms have adaptations that allow them to break those triple bonds and convert N_2 to ammonia (NH_3). Ammonia can be further processed to construct biologically important molecules such as proteins and nucleic acids. Nitrogen transformations benefit the microorganisms, but also have far-reaching consequences to ecosystems that ultimately support the diversity of life on Earth. Let's explore some of these transformations in more detail.

Nitrogen fixation

There is a rapidly growing list of microorganisms that can fix nitrogen, including the *Rhizobium* bacteria, which can live symbiotically with legumes (see Chapter 15), and the *Frankia* bacterium, which can live symbiotically in the root systems of many vascular plants that consequently can establish themselves in nutrient-poor soils. In addition, there are many free-living bacteria and archaeans that can fix nitrogen. All of them use some form of the nitrogenase enzyme complex to catalyse the following reaction:

$$N_2 + 8H^+ + 8e^- \rightarrow 2NH_3 + H_2$$

This is an energy-demanding process, which requires the conversion of at least 16 ATP to ADP (ATP $\rightarrow$ ADP). Oxygen (O_2) inactivates nitrogenase, thus aerobic nitrogen fixers have evolved a variety of mechanisms to protect nitrogenase from O_2 inactivation.

Humans have made important contributions to the process of nitrogen fixation in three ways. Combustion by transportation vehicles and power plants transforms N_2 to nitric oxide (NO) and

Table 20.1 Amount of nitrogen (in teragrams: 1 Tg = 10^{12} g) fixed by natural and human sources.

Natural source of fixation	Amount (Tg)	Human source of fixation	Amount (Tg)
Lightning	5	Combustion	30
Natural biological N-fixation (e. g. plant–bacteria mutualisms)	58	Agricultural biological N-fixation (e. g. legume/bacteria mutualisms)	60
Marine biological fixation (e. g. cyanobacteria)	140	Haber–Bosch process (fertilizers)	120
Total	203		210

nitrogen dioxide (NO_2). Second, we have increased the rate of biological nitrogen fixation by cultivating plant crops, such as soybeans, that carry symbiotic nitrogen-fixing bacteria. Most important, the Haber–Bosch process – an industrial application – converts N_2 to ammonia (NH_3), and allows large-scale production of inorganic fertilizers. Nitrogen fertilizer application increased by 800% between 1960 and 2000, with wheat, rice, and corn accounting for 50% of global nitrogen use (Canfield *et al.* 2010).

Two important studies summarized the effect of humans on the global abundance of *reactive nitrogen* – inorganic nitrogen other than N_2 (Galloway *et al.* 2004; Fowler *et al.* 2013). The authors show that human demand is driving the increase in reactive nitrogen abundance. In fact, we can see that humans are now adding more reactive nitrogen to the global nitrogen cycle than is added by natural processes of nitrogen fixation (Table 20.1). Much of this reactive nitrogen is converted into the highly mobile nitrate ion (NO_3^-), where it can wash into watersheds, leading to eutrophication and the production of dead zones, as we described in the introductory case study.

A second source of ammonia in the nitrogen cycle arises from the decomposition of complex nitrogen-rich molecules – such as proteins and nucleotides – that are released from dead organisms. The ammonia generated by this process of ammonification is readily assimilated by plants and microorganisms, and is often kept within the ecosystem. However, like the ammonia produced by nitrogen fixation, it may also be transformed into highly mobile nitrate.

Nitrification

Much movement of terrestrial nitrogen results from nitrification, the conversion of ammonia to nitrate and nitrite (NO_2^-). At a neutral or nearly neutral pH, ammonia is primarily present as ammonium (NH_4^+). Because ammonium has a positive charge, it tends to bond to the negatively charged clay particles in many soils. However, nitrification oxidizes ammonium to negatively charged nitrites or nitrates. These forms are much more mobile, and may run off into watersheds to continue through the global nitrogen cycle.

Recall from Chapter 4 that chemosynthetic bacteria use inorganic electron donors as energy sources. *Nitrosifying bacteria* oxidize ammonia into nitrite using the summed reaction

$$NH_3 + 1\frac{1}{2}O_2 \rightarrow NO_2^- + H_2O (\Delta G = -275\,kJ)$$

Nitrifying bacteria oxidize nitrite into nitrate with the reaction

$$NO_2^- + \frac{1}{2}O_2 \rightarrow NO_3^- (\Delta G = -74\,kJ)$$

Notice that both oxidation reactions have negative values for *Gibbs energy* (ΔG). This indicates that these reactions tend to occur spontaneously, and that when these reactions occur, free energy is available to do work for these bacteria. In the case of these two oxidation reactions, this work is the phosphorylation of ADP into ATP, which fuels the bacterium's metabolism. From the standpoint of an individual bacterium, natural selection has favored the evolution of pathways that use energy-rich ammonia to drive its metabolic machinery. A byproduct of natural selection leading to the evolution of these metabolic pathways is that much of the ammonium within ecosystems is ultimately converted to nitrate by nitrosifying and nitrifying bacteria working in tandem. Not surprisingly, these bacteria are commonly found in areas that are rich in ammonia, such as sewage treatment plants and eutrophic rivers, lakes, and estuaries.

Nitrogen cycles within and between ecosystems, so nitrates cannot be a dead end for nitrogen atoms. There are two major processes by which nitrogen makes its way back to stable nitrogen gas.

Conversion of reactive nitrogen back into nitrogen gas

Ecologists have known for over a century that denitrification – the reduction of nitrates and nitrites to nitrogen gas – plays an important role in biogeochemical processes. They also noticed that NH_4^+ was disappearing from various environments, and assumed that the NH_4^+ had first been oxidized to nitrates and then undergone denitrification. But it was not until the end of the twentieth century that ecologists described the first direct evidence of anaerobic oxidation of NH_4^+, a process named anammox (anaerobic **ammon**ium **ox**idation).

Denitrification

Many species of bacteria, archaeans, and even some eukaryotes are denitrifiers. Denitrification is an anaerobic process in which nitrate or nitrite is the final electron acceptor in the respiratory process, which is coupled with the phosphorylation of ADP to ATP. Depending on the species and environmental conditions, denitrification can produce a diverse assemblage of nitrogenous products as described below:

$$NO_3^- \rightarrow NO_2- \rightarrow NO(gas) + N_2O(gas) \rightarrow N_2(gas)$$

Some denitrifier species tend to produce exclusively N_2, while others may produce considerable quantities of NO (nitric oxide) and N_2O (nitrous oxide). The complete denitrification process is expressed as:

$$2\,NO_3^- + 10e^- + 12H^+ \rightarrow N_2 + 6\,H_2O$$

Nitrification, the production of nitrate, requires oxygen. Denitrification requires the absence of oxygen. So how are these two processes linked in natural systems, given these conflicting environmental requirements? Sybil Seitzinger and her colleagues (2006) describe three broad categories of environmental conditions in which denitrification is favored. In diffusion-dominated systems, there is a very sharp gradient in oxygen concentrations over a relatively small spatial scale. For example, even in highly oxygenated soils there are microsites, such as earthworm castings, which have an oxygen-rich exterior (where nitrification can occur) and an oxygen-poor interior (where denitrification can occur) (Figure 20.3A). Similarly, as we have discussed, in aquatic and marine systems there often are well-oxygenated zones that transition sharply into hypoxic or anoxic zones.

A second category of conditions favoring denitrification is advection-dominated systems, which have relatively stable low-oxygen conditions and receive their nitrate and nitrite inputs from distant sources. For example, immense marine oxygen minimum zones are widespread in the East Pacific Ocean near the coasts of the Americas, and in the east Atlantic Ocean near the west African coast at depths of 100–1000 m. Despite being oxygen poor, these regions routinely receive high nitrate levels courtesy of ocean currents, providing ideal conditions for denitrification reactions (Figure 20.3B). The last category of conditions favoring denitrification is systems whose oxygen concentration oscillates over time. Such systems can build up high nitrate concentrations during times when oxygen levels are relatively high, and have high denitrification rates when oxygen is depleted episodically, by events such as algal blooms in marine or aquatic systems, and by heavy rainfall that can saturate soils of terrestrial systems, creating anoxic conditions.

While denitrification has long been recognized as an integral part of the global nitrogen cycle, researchers are becoming increasingly convinced that anammox also plays an important role in generating atmospheric forms of nitrogen.

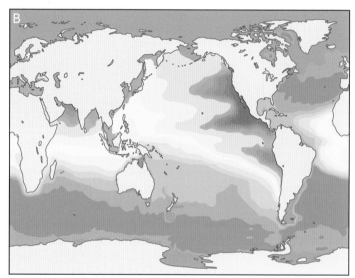

Figure 20.3 Denitrification sites. A. Earthworm castings (worm manure) are generally released below the surface and provide microsites for denitrification. B. Marine oxygen minimum zones. Deep red indicates lowest oxygen levels (Paulmier and Ruiz-Pino 2009).

Anammox

Some bacteria in the order Plantomycetales can use anammox to fuel their metabolic processes. Anammox is a complex process, involving the coordinated efforts of more than 200 genes (Strous *et al*. 2006). We don't have the space here to go into the details of this reaction, which is outlined as:

$$NH_4^+ + NO_2^- \rightarrow N_2H_4(hydrozine) \rightarrow N_2 + H_2O$$

Hydrozine is an energy-rich compound, and in fact is an important component of rocket fuel. Its four high-energy electrons are used to phosphorylate ADP into ATP, which provides energy to the bacterium. Hydrozine is toxic and is encased in a compartment – an anammoxosome – making this group of bacteria that produce it unique in having membrane-bound organelles. Research on these bacteria is very difficult, because they reproduce so slowly (Kuenen 2008).

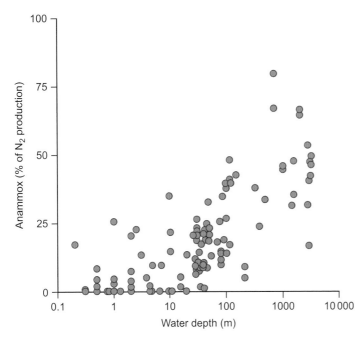

Figure 20.4 Percentage of nitrogen gas production in marine and freshwater sediments that results from anammox. A value of 30 means that 30% of the nitrogen gas production is from anammox, and 70% is from denitrification.

We still don't know how important anammox is in the global nitrogen cycle. Bo Thamdrup (2012) summarized data from a series of papers that showed that anammox tended to become more important in comparison to standard denitrification in deep-water sediments (Figure 20.4). Researchers such as Tage Dalsgaard and colleagues (2005) have argued that given that 87% of the ocean is more than 1000 m deep, and that two-thirds of marine N_2 production occurs in sediments rather than in the water column, then we can extrapolate that about half of marine N_2 production occurs from anammox. In addition, recent investigations have expanded the known distribution of anammox globally, with high anammox activity found in the Black Sea, in major upwelling zones off the coasts of Africa and South America, and in association with hydrothermal vents at temperatures above 65°C.

Thinking ecologically 20.1

We've discussed in some detail four general types of nitrogen transformation reactions – nitrogen fixation, nitrification, denitrification, and anammox – and briefly mentioned a fifth type, ammonification. Keeping in mind that natural selection favors organisms that are efficient at survival and reproduction, which of these processes do you think will be represented by microorganisms with the greatest species diversity?

Having established the important role of microorganism metabolism in nitrogen transformations, let's explore more quantitative estimates of the global nitrogen cycle.

20.2 HOW DO BIOLOGICAL, GEOLOGICAL, CHEMICAL, AND PHYSICAL PROCESSES INTERACT IN THE GLOBAL NITROGEN CYCLE?

In addition to biochemical interactions such as those we've already described, geological and physical processes interact to move nutrients around the world. Black-box or input–output types of models can be used to describe nutrient storage in compartments we call pools or reservoirs. Usually pools are smaller and tend to cycle more quickly than reservoirs, but researchers may use the two terms interchangeably. Nutrient flux describes the rate of nutrient movement from one pool or reservoir to another. Influx describes movement into a pool, and efflux is movement out of a pool. A pool in which influx is greater than efflux is a nutrient sink, while a pool in which efflux is greater than influx is a nutrient source. The actual calculation of nutrient flux is challenging, and estimates change as ecologists develop better measurement technology and learn about new processes.

Extrapolations of global nitrogen flux

There is no sensor yet developed that allows a scientist to scan the entire globe and measure nutrient flux. Ecologists must instead synthesize information from observations and experiments conducted on a much smaller spatial scale to estimate flux. We have greater confidence in estimates that are derived from large numbers of studies conducted over relatively large spatial scales, but sometimes those data are not available. For example, one important flux in the nitrogen cycle is fixation of N_2 by cyanobacteria in the genus *Trichodesmium*. These organisms are extraordinarily abundant and have been identified as making substantial contributions to the amount of fixed nitrogen in the oceans. One way to estimate this contribution is to incubate ocean water with *Trichodesmium* in an airtight container that has a known amount of the stable isotope of nitrogen gas ($^{15}N_2$) added to its atmosphere. If *Trichodesmium* is fixing nitrogen, its stable isotope ratio ($\delta^{15}N$) should increase, and the amount of that increase can be translated into a N_2 fixation rate (Montoya *et al.* 1996).

James Galloway and his colleagues (2004) compiled data collected by nine groups of researchers who studied nitrogen fixation rates in *Trichodesmium* in seas and oceans around the world. Based on stable isotope ratios and other techniques, the researchers estimated a mean fixation rate for *Trichodesmium* of 1.82 mg N/m^2/day. From this estimate, they extrapolated the global fixation rate of *Trichodesmium*. They knew that *Trichodesmium* was globally abundant in ocean waters above 25°C, so they assumed that all such waters had *Trichodesmium* fixing nitrogen at that rate. Multiplying 1.82 mg N/m^2/day by the surface area of marine water over 25°C yielded an estimate of 85 Tg N/yr fixed by *Trichodesmium* (1 Tg = 10^{12} g).

The researchers point out several reasons that this is actually an underestimate of global marine nitrogen fixation. First, these estimates are based on average concentrations of *Trichodesmium*, and don't account for the extraordinarily dense populations associated with relatively common *Trichodesmium* blooms. Second, these estimates are only for open oceans, and there is considerable nitrogen fixation in shallow waters as well. Third, Galloway and colleagues predicted (correctly) that subsequent studies would uncover numerous other species of marine microorganisms that carry out nitrogen fixation (for example, Moisander *et al.* 2010). Lastly, the researchers assumed that no fixation occurs below 25°C, even though *Trichodesmium* is found in waters as cool as 20°C.

To varying degrees, all estimates of global cycles make assumptions, and Galloway and his colleagues suggest we assume a margin of error of at least 20% for many of the estimates, and up to 50% for those with the greatest degree of uncertainty. Fortunately, the error in these estimates is narrowing as we learn more about global cycles. With these degrees of uncertainty in mind, let's take a closer look at the details of the nitrogen cycle.

The global nitrogen cycle: a changing reality

One feature of the present-day nitrogen cycle is that most of the flux stays within the ecosystem, where nutrients are recycled internally. For example, organisms (mostly plants) transfer 1200 Tg of reactive nitrogen per year to the soil, from leaf fall, death, and leaching, and at the same time, organisms take up 1200 Tg of nitrogen per year, mostly through plant roots and mycorrhizae (Figure 20.5). The 1200 Tg of recycled nitrogen is much greater than the total flux of nitrogen between the atmosphere and the soils. As we will soon discuss, the process of decomposition plays a critical role in this process.

Human activities are transforming the nitrogen cycle in several ways. In addition, our understanding of some of the processes is improving, which allows ecologists to make more accurate predictions about how the nitrogen cycle will operate in the future.

Present impacts of human activities

One significant feature of the present-day nitrogen cycle is that terrestrial inputs from humans have become substantially more important than terrestrial inputs from natural fixation and from natural physical processes, such as lightning. Most of this increased human input is from the Haber–Bosch process used to make fertilizers, from increased cultivation of nitrogen-fixing crops, and from fossil fuel combustion, which creates nitric oxide as a byproduct. If you add up the numbers shown in Figure 20.5, you will note that for terrestrial systems, the amount of reactive nitrogen created by various forms of nitrogen fixation is much greater than the amount of reactive

nitrogen lost from denitrification. Marine systems may be more in balance, but our estimates of efflux from denitrification and anammox are very variable at this time. Ecologists still need to come up with reliable estimates for the amount of N_2 generated by anammox before we can claim a reasonable understanding of the global marine nitrogen cycle.

When farmers apply fertilizers to soil, on average about half of the reactive nitrogen is lost to the atmosphere and to runoff. A substantial portion of nitrogen-rich runoff makes its way to large bodies of water, and may cause hypoxia and anoxia, as we discussed in the introduction to the chapter. But there are also important indirect effects that result from the greater abundance of reactive nitrogen. As one example, as nitrates increase, denitrification rates are also increasing in both terrestrial and marine systems. In addition to producing N_2, denitrification also produces a substantial quantity of nitrous oxide (N_2O), which makes its way to the atmospheric pool, which is increasing at a rate of 0.3% per year (Schlesinger and Bernhardt 2013). In the atmosphere, N_2O has a mean residence time of 120 years, during which time it acts as a powerful greenhouse gas, and contributes substantially to global warming. In addition, N_2O is now the most potent destroyer of ozone molecules in the stratosphere. We will discuss both of these topics at length in Chapter 23.

Given the central role of the nitrogen cycle in ecosystem functioning, it is fair to ask, "What is the future of nitrogen in the environment?" The answer is complicated because we are still discovering new elements of the nitrogen cycle, and because we are very poor at predicting human behavior.

Newly discovered geological sources of reactive nitrogen

Our understanding of the nitrogen cycle is still changing in important ways. Until recently, researchers had assumed that new reactive nitrogen entered ecosystems solely from the atmosphere, courtesy of the transformation reactions we have already discussed. Recently, Scott Morford and his colleagues (2011) proposed that weathering bedrock (the rock located just below the soil) could be an important source of reactive nitrogen for some ecosystems. They argue that as the nitrogen-rich bedrock breaks down, it releases important nutrients into the soils that are taken up by soil organisms, including the plants. Their basic approach compares forests that experience similar environmental conditions, such as temperature, rainfall, soil texture, and slope, but that differ in the nitrogen content of the underlying bedrock. If underlying bedrock is a source of nitrogen to these ecosystems, then trees growing above nitrogen-rich bedrock should show evidence of nitrogen enrichment.

The researchers compared two adjacent coniferous forests in northern California, USA. The bedrock below South Fork Mountain had more than ten times the nitrogen content than did the

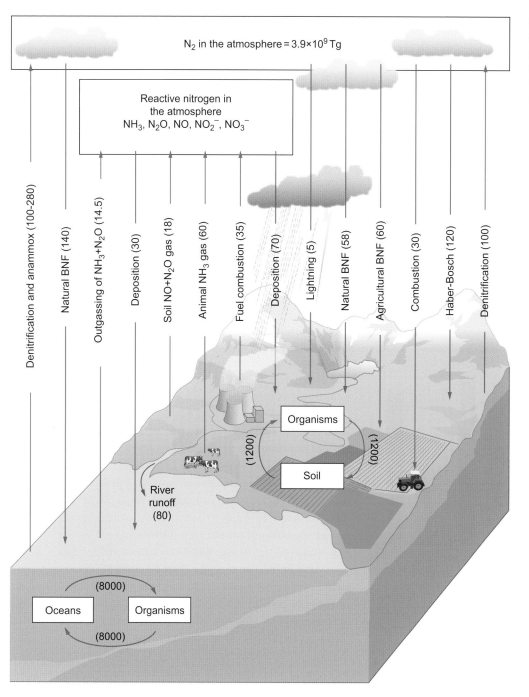

Figure 20.5 The global nitrogen cycle. Fluxes (in Tg/yr) are in parentheses. BNF, Biological nitrogen fixation. (Based on Galloway *et al.* 2004 and Fowler *et al.* 2013).

bedrock underlying the adjacent Bear Wallow Diorite Complex. If underlying bedrock is a source of nitrogen to these ecosystems, the researchers expected to find more nitrogen in the surface soils, and in the trees at South Fork Mountain than at Bear Wallow. This is exactly what they found; soil nitrogen was more than twice as high, and leaf nitrogen of the four major tree species was substantially higher at South Fork Mountain (Figure 20.6). In addition, the trees had greater biomass at South Fork Mountain, indicating that higher nutrient availability translated to improved growth.

Even with this evidence, we could still propose that there was an unidentified source of nitrogen to the soils underlying South Fork Mountain, that was then being passed on to the foliage. To explore this possibility, the researchers measured the stable isotope ratio ($^{15}N/^{14}N$) of the bedrock, soil and foliage at both sites in relation to the $^{15}N/^{14}N$ ratio of nitrogen in the air. In support of the hypothesis that much of the nitrogen at South Fork Mountain was coming from the rocks, the researchers discovered almost identical stable isotope ratios for bedrock, soil, and foliage at South Fork Mountain, all of which were

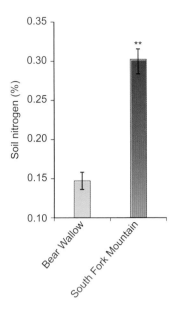

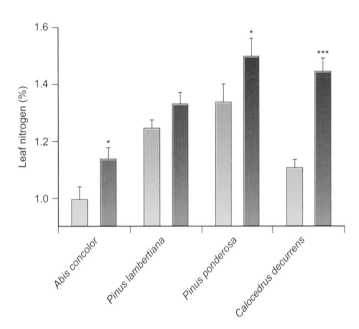

Figure 20.6 Soil nitrogen concentration (top) and leaf nitrogen concentration (bottom) at Bear Wallow (green bars) and South Fork Mountain (red bars). * $P < 0.05$, ** $P < 0.01$, *** $P < 0.001$.

the potential contribution to global nitrogen cycles could be substantial.

Our understanding of the nitrogen cycle is changing as a result of new discoveries. But at the same time, our ability to forecast the future is muddied by our inability to predict human behavior.

The future of the nitrogen cycle and human behavior

As we discussed in Chapter 10, the United Nations has six different projections of human population growth, each of which makes its own set of assumptions about future fertility rates, death rates from infectious disease, and the influence of migration patterns on regional population growth rates. Given the challenge of projecting human population growth, the problem of estimating how much reactive nitrogen this population will use becomes even more daunting. Much of the added reactive nitrogen comes from the Haber–Bosch process, thus it makes sense that more people will translate to more nitrogen fertilizer. Other important sources of extra reactive nitrogen are industrial combustion and crops that have symbioses with nitrogen fixers. These too would, on the surface, be expected to increase with population size.

Wilfried Winiwarter and his colleagues (2013) have begun modeling how changes in human behavior could influence the amount of nitrogen fixation in the twenty-first century. For their models, the researchers consider five potential behavioral drivers of nitrogen fixation: population growth, *efficiency increase*, *food equity*, *diet optimization* and production of *biofuels*. Efficiency increase refers to improvements in agricultural processes that reduce the demand on nitrogen fertilizers. Food equity assumes that much of the world will have substantial increases in the amount of protein in their diet. Diet optimization occurs if human switch their meat intake to animals that more efficiently convert nitrogen to protein. Lastly, increased biofuel production for the purposes of reducing use of coal and oil will require additional nitrogen fertilizers. Most of their projections show a small increase in nitrogen fixation early in this century, but that nitrogen fixation levels off as world population growth slows, and agriculture becomes more efficient.

 Thinking ecologically 20.2

Winiwarter and his colleagues proposed five important drivers of nitrogen fixation over the next 100 years. What other possible drivers can you identify, and how might they influence future nitrogen fixation rates?

considerably greater than the stable isotope ratio for the air. In contrast, the foliage at Bear Wallow had a stable isotope ratio that was indistinguishable from that of the air above, and much lower than that found in the bedrock, indicating that those trees were getting much of their fixed nitrogen from atmospheric sources.

While this study provides evidence that nitrogen enters ecosystems from weathering bedrock, we still don't know how general this phenomenon is. Is bedrock an important nitrogen source to ecosystems around the globe? Given that sedimentary rock such as found beneath South Fork Mountain contains 99% of global fixed nitrogen and covers 75% of the Earth's surface,

It is somewhat ironic that increased biofuel production for the purposes of reducing carbon emissions will lead to increased use of nitrogen fertilizers, which has the potential to cause numerous undesirable environmental effects, including the release of extra

N₂O – an important greenhouse gas. Let's turn our attention to the phosphorus cycle, which has very different dynamics from the nitrogen cycle, but also is under stress from human activity.

20.3 HOW DO OCEANS, FRESHWATER SYSTEMS, AND SOILS INTERACT IN THE GLOBAL PHOSPHORUS CYCLE?

In common with nitrogen, there is a great deal of phosphorus in the world, and most of it is inaccessible to organisms. The problem with phosphorus is physical rather than chemical, with the vast majority stored in marine sediments or in deep ocean waters. Phosphorus may remain locked in marine sediments for hundreds of millions of years before tectonic uplift returns it to the Earth's surface, where weathering of rocks can make it available to organisms. Phosphorus in the form of inorganic phosphate is readily taken up by autotrophs, which use it to make nucleic acids, ATP, and other essential molecules. There is also mounting evidence that dissolved organic phosphates may play an important role in some regions that have low levels of inorganic phosphates (Lomas *et al.* 2010). Fortunately, from the standpoint of life on Earth, phosphorus tends to be recycled internally within ecosystems, passing repeatedly through various trophic levels before being lost to ocean sediments (Figure 20.7).

Atmospheric sources of phosphorus include wind erosion of soil, pollen, burning of plant material and fossil fuels, and volcanic eruptions. In addition to being unpredictable and variable, the sum total of these sources yields, on average, a seemingly paltry quantity of phosphorus to the atmosphere. Gravity, wind currents, and rain or snow can return these relatively small quantities of phosphorus to the Earth's surface. Though its load of nutrients is relatively small, the atmosphere delivers important nutrients to ecosystems across the globe.

As one example, consider the Bodélé Depression in the southern Sahara Desert in northern Chad. This depression is the dried-up lake bottom of the former Lake Megachad, and is dominated by the remains of diatoms that thrived in the lake before it dried up over a thousand years ago. Strong winds scour the lake

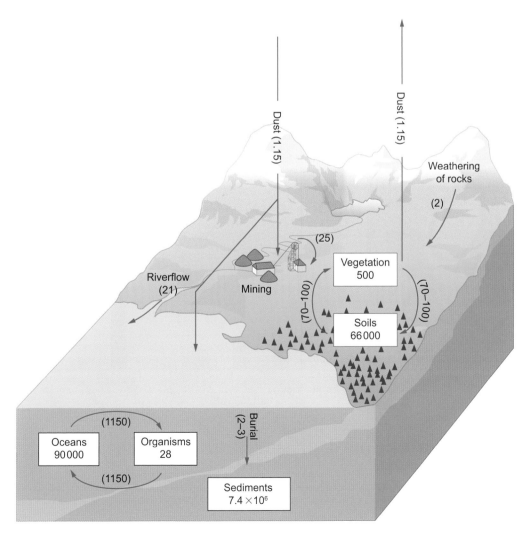

Figure 20.7 The global phosphorus cycle. Fluxes (in Tg/yr) are in parentheses. Boxes indicate amount stored in reservoirs (Tg). Based on data in Mahowald *et al.* (2008) and Schlesinger and Bernhardt (2013).

bottom, forcing the light fluffy sediment into the air, where it forms plumes that drift westward releasing nutrients as dry deposition or as wet deposition in rainstorms along the way. Researchers estimate that about 0.12 Tg of phosphorus are exported from the Bodélé Depression annually, which is about 10% of the global total of phosphorus that cycles in the atmosphere (Bristow *et al.* 2010).

A significant portion of Bodélé Depression dust makes a 10–12 day journey over the Atlantic Ocean to the Amazon, providing much-needed nutrients to the tropical forests. Plants take up most of the phosphorus, but some is lost as runoff into the Amazon River and its tributaries, which carry large nutrient loads into the ocean. However, these nutrients tend to accumulate in estuaries and do not make it very far into the ocean. Consequently, some open oceans that are far from any landmass can be severely phosphorus limited.

One example of oceanic phosphorus limitation is the North Pacific Subtropical Gyre, which comprises a huge portion of the North Pacific Ocean. This gyre's currents form a large clockwise circulation, encompassing an area of about 20 million km^2. For reasons still poorly understood, since the 1970s this gyre has shown enhanced photosynthesis, high rates of nitrogen fixation, increased levels of organic carbon, and decreased levels of inorganic phosphorus. Concurrently, the phytoplankton community has changed from a high diversity of eukaryotic algae to a lower diversity but super-abundance of cyanobacteria – primarily in the genus *Prochlorococcus* (Figure 20.8A).

Benjamin Van Mooy and his colleagues (2006) were interested in why this shift had happened. They reasoned that *Prochlorococcus* might have been a better competitor for a limiting nutrient. Given the unusually low levels of phosphorus in many portions of this gyre, phosphorus was a strong candidate. There are two ways for an organism to compete successfully for a nutrient that is in short supply. One way is to be better at taking it up, and the second way is to use much less of it than its competitors. Van Mooy and his colleagues were able to identify two ways that *Prochlorococcus* uses much less phosphorus than other phytoplankton. First, it has an unusually small genome, which reduces the number of inorganic phosphates needed for its DNA molecule. Second, it produces much fewer phospholipids than most phytoplankton, so it requires less phosphorus as a building block.

The researchers measured phospholipid formation by capturing a portion of the phytoplankton community and incubating the organisms with radioactive phosphate ($^{33}PO_4^3$). They then measured the percentage of this phosphorus tracer that was devoted to creating phospholipids for cell membrane synthesis. Cultures of *Prochlorococcus* showed almost no phosphorus uptake for phospholipid synthesis, while other bacteria species allocated 12–22% of their phosphate uptake to

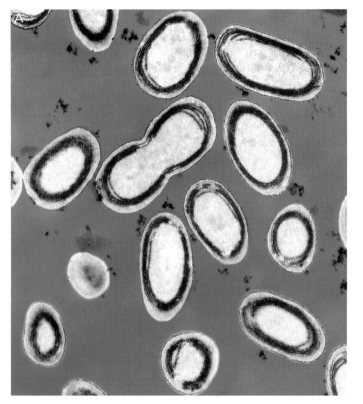
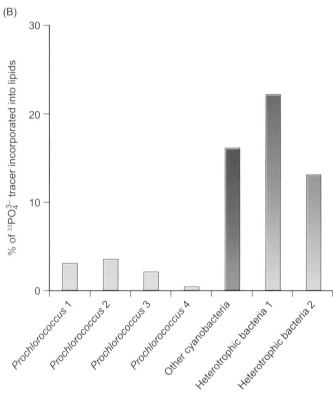

Figure 20.8 A. Transmission electron micrograph (23 000×) of *Prochlorococcus* cyanobacteria highlighting the chlorophyll pigment positioned just inside the sulfate-rich cell membrane. B. Percentage of $^{33}PO_4^3$ that was recovered as phospholipids in incubations of four strains of *Prochlorococcus* (green bars), other photosynthetic bacteria (red bar), and heterotrophic bacteria (blue bars).

phospholipid production (Figure 20.8B). The researchers' chemical analysis of *Prochlorococcus* demonstrated that the cell membrane consisted primarily of lipids with polar head groups of sulfate or sugar-based molecules, instead of the usual phosphates found in almost all other species. Thus *Prochlorococcus* is, in effect, winning the war for phosphates by refusing to participate.

Some researchers contend that the North Pacific Subtropical Gyre is the largest ecosystem in the world. Organisms such as *Prochlorococcus* play an important role in cycling nutrients within ecosystems. We will reduce our spatial scale from the global to ecosystem level to explore how nutrients cycle within ecosystems, focusing on the roles played by organisms in these regional and local processes.

20.4 WHAT PROCESSES CYCLE NUTRIENTS WITHIN ECOSYSTEMS?

Organisms have a great deal of free energy in their carbon-based macromolecules. When an organism dies, saprotrophs, which comprise decomposers (bacteria and fungi) and detritivores (animals), use the free energy locked in these macromolecules to drive their metabolic processes. The process of mineralization breaks down organic molecules into inorganic molecules, and allows these molecules to be taken up by cells. The reverse process, immobilization, occurs when inorganic molecules (what we have been calling nutrients) are taken in by an organism and converted to organic molecules. In this sense, nutrient cycling within ecosystems simply consists of repeated rounds of mineralization and immobilization. Let's begin our discussion with very simple box models of nutrient cycling in terrestrial and aquatic systems.

Simple box models of nutrient cycles

Looking carefully at Figures 20.5 and 20.7, you will notice that the majority of nutrient flux is between organisms and the environment and then back into organisms. A relatively small quantity of nutrients passes from organisms into a reservoir. This is very fortunate, because, particularly in the phosphorus cycle, fluxes from reservoirs into ecosystems are so low that without this efficient recycling process there simply would not be enough phosphorus to support construction of sufficient DNA, phospholipid bilayers, and ATP.

Thus a very general and oversimplified model of the recycling process has soluble nutrients being taken up either by plants or by microorganisms. These organisms can either pass the nutrients on to consumers, or else the plants and microorganisms can die without being consumed, becoming detritus. There may be several trophic levels of consumers, all of whom suffer the ultimate fate of turning into detritus. Some detritus is directly eaten by consumers, such as vultures, but most detritus is

decomposed and its nutrients are released to begin the cycle anew (Figure 20.9A).

There is one twist (literally) in stream ecosystems, because nutrients cycle through the ecosystem while drifting downstream. The sequential process of nutrient uptake, nutrients passing through a food web, nutrient release, and decomposition repeats itself as nutrients drift downstream, forming nutrient spirals. Spiraling length is simply the average length of stream it takes for a nutrient to complete one spiral, which tends to be longer in faster streams (Figure 20.9B).

Organisms don't need to die to decompose. Insects shed exoskeletons, all animals egest fecal material, and perennial plants

(A)

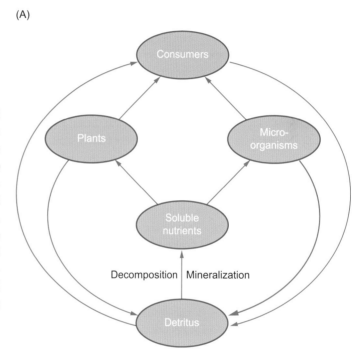

(B)

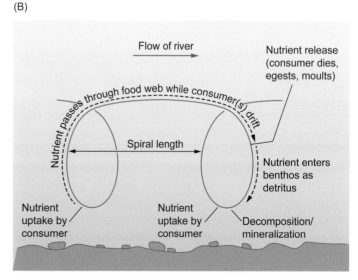

Figure 20.9. A. Generalized nutrient cycle that operates in many ecosystems. B. Nutrient spiral in a stream.

shed leaves, branches, and roots every year. All of these substances provide substantial decomposable materials. Given the vast amount of plant matter that enters both terrestrial and aquatic ecosystems every year, we will focus our discussion of decomposition on a few of the interactions between nitrogen and carbon when leaves fall on to the forest floor.

Litter decomposition in terrestrial systems

William Parton and his colleagues (2007) wanted to know how climate, litter chemistry, and soil characteristics influenced decomposition rates. The researchers had the benefit of learning from many previous studies, but their attempts to come up with general patterns were frustrated by the short time periods and local scales of many of those studies. With funding from the U. S. National Science Foundation, they began the Long-Term Intersite Decomposition Experiment, which, over a 10-year period, investigated how various factors influenced decomposition and immobilization. They chose 21 sites in seven different biomes, at which they systematically and repeatedly measured climate and soil conditions. They used six species of leaf litter and three species of root litter at each site; species were chosen that varied substantially in their chemical makeup – particularly in the amount of nitrogen, lignin, and cellulose present. The researchers constructed litter bags with a large enough mesh to admit decomposers, and filled the bags with an equal mass of leaves, with only one species per bag. Bags were placed on the surface, remaining there for up to 10 years. The researchers periodically collected a subsample of the litter bags, and weighed and chemically analysed the contents.

Recall that decomposers in forested systems use the chemical bond energy in carbon-based macromolecules such as lignin and cellulose to convert ADP to ATP and fuel their metabolism. In this process CO_2 is released, and there is an overall decrease in litter mass. One clear trend that emerged from this study is that decomposition was much more rapid, and much more complete, in the tropics than in any other biome. After 2 years, only 20% of the original leaf mass remained in the tropics, a level not reached until 6 years by leaves from deciduous forest, the biome with the second greatest rate of decomposition. Root decomposition rates yielded qualitatively similar findings (Figure 20.10).

A second important trend in this long-term research effort gives us an appreciation for the process of litter decomposition. The decomposing organisms must have sufficient nitrogen (among many other factors) to survive and reproduce. Thus we might expect that decomposition rates would be greater in the leaf litter with the highest nitrogen content, and they are. But, to meet their nutrient needs, decomposers can import nitrogen from the environment. Consequently, the percentage of nitrogen in litter actually increased as the decomposer population increased and the leaf litter mass decreased from the breakdown of the carbon-based macromolecules (Figure 20.11). The time of net nitrogen release (TNET) is the number of years that pass until

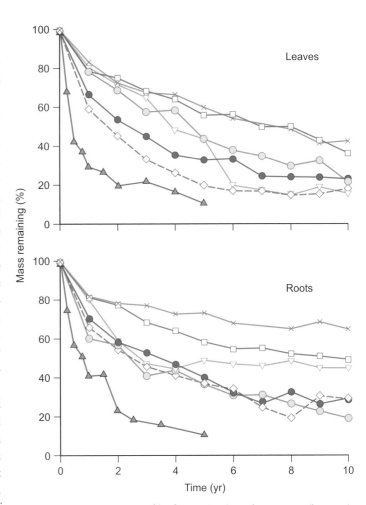

Figure 20.10 Percentage of leaf mass (top), and root mass (bottom), remaining in relation to time (in years) for seven biomes.

the leaf litter mass reaches its peak percentage of nitrogen, after which the percentage begins to decline as the population of decomposers decreases, decomposition slows down, and the nutrient needs of the remaining decomposers are lower.

Table 20.2 shows the TNET values for the six species of leaf litter for the six biomes in which TNET was measured. There are two important trends to note. First, the TNET values are considerably lower for the species with the highest initial nitrogen levels, presumably because the decomposers did not need to import very much nitrogen from the surrounding environment. Second, tropical and deciduous forests, which had the highest decomposition rates (see Figure 20.10) had the lowest TNET values, indicating that their decomposers were able to import all the nitrogen they needed relatively rapidly.

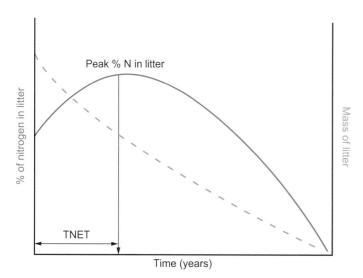

Figure 20.11 Changes in litter mass and percentage of nitrogen in the remaining mass in relation to time. TNET is the amount of time that elapses until the percent nitrogen in the litter reaches its peak.

Based on these findings, Parton and his colleagues conclude that initial nitrogen content in leaves plays a critical role in net nitrogen immobilization and release during long-term litter decomposition. Because nitrogen levels also influence NPP, we need to consider nutrient uptake and release properties of decomposing microorganisms to understand patterns of production.

In aquatic systems, nutrient availability plays an important role in microbial processes and nutrient cycles. Let's revisit the process of denitrification to see how nitrogen availability influences the rate at which denitrifiers convert nitrates into nitrogen gases.

Efficiency of denitrification in aquatic systems

In general, only about 20–25% of the nitrogen added to streams actually makes it into oceans, which indicates that streams and rivers are important sinks for nitrogen. Aquatic organisms may take up the nitrogen directly, storing it in organic form either in live tissues or in particles of detritus that lie in the streambed, or else denitrifiers may convert the nitrates to nitrogen gas as part of their metabolic cycle. Patrick Mulholland and his colleagues (2008) wanted to know how the efficiency of these processes varied in response to nutrient loading, the amount of nutrients – in this case nitrates – that are present in streams. To do so, they added small quantities of tracer nitrate ($^{15}NO_3$) molecules to 72 different small streams in the United States and Puerto Rico. They then used mass spectrometry to detect small decreases in nitrates and increases in the products of the denitrification process – nitrous oxide (N_2O) and N_2. The dependent variable used by the researchers in these analyses was nitrate uptake velocity, which is the speed (in cm/s) at which nitrate moves toward the benthos, the community of organisms on the bottom of the stream, where it is taken up by an organism. ^{15}N lost from the stream is used to measure total nitrate uptake velocity (due to organism nitrate uptake and storage and to denitrification by bacteria). The quantity of ^{15}N present in N_2O and N_2 gases is used to measure nitrate uptake velocity due to denitrification.

An important finding is that the total nitrate uptake velocity decreases in streams carrying greater nitrate concentrations (Figure 20.12A). In addition, the nitrate uptake velocity

Table 20.2 Climatic variables for each biome, and time of net nitrogen release (TNET in years) in relation to initial percentage of nitrogen (value in parentheses) in each species.

Biome	Mean annual precipitation (mm)	Mean annual temperature (°C)	*Triticum aestivum* (0.38%)	*Pinus resinosa* (0.58%)	*Thuja plicata* (0.62%)	*Acer saccharum* (0.80%)	*Quercus prinus* (1.02%)	*Drypetes glauca* (1.98%)
Humid grassland	807	9.8	3.5	3	3.9	1.75	1.6	0.25
Tundra	788	−6.0	9.5	7.6	6.4	2.6	1.6	0.5
Boreal forest	750	0.0	6.4	4	4	2.6	1.3	0.5
Coniferous forest	1840	11.2	3.8	2.6	2.2	1.4	0.7	0.25
Deciduous forest	1485	8.8	1.85	1.1	0.9	0.8	0.33	0.2
Tropical forest	3210	23.6	0.4	0.6	0.35	0.5	0.2	0.1

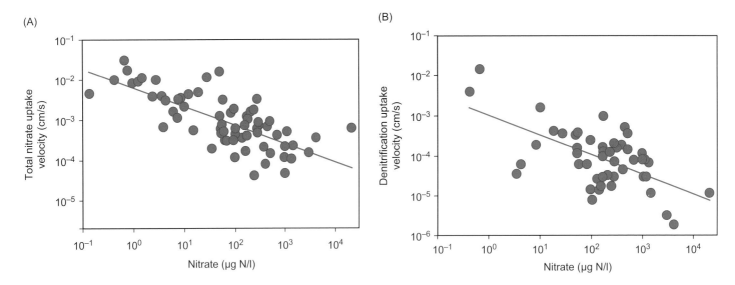

Figure 20.12 A. Total nitrate uptake velocity in relation to nitrate loading in 72 streams. $R^2 = 0.53$, $P < 0.0001$. B. Denitrification uptake velocity in relation to nitrate loading in streams. $R^2 = 0.355$, $P < 0.0001$.

due to denitrification also decreased with higher nitrate loading (Figure 20.12B). Note that both axes in Figure 20.12 are graphed on logarithmic scales. If you were to convert these to a linear scale, you would see that the range in nitrate concentration in the different streams was enormous, ranging from less than 1 to greater than 10 000 μg/l. Finding that total nitrate uptake velocity and that nitrate uptake velocity due to denitrification decrease sharply with nutrient loading, indicates that streams are able to be efficient nitrate sinks up to a point, but that additional nutrient loads are handled very inefficiently.

The studies on litter decomposition and denitrification discussed here both demonstrate the importance of initial nitrogen concentration on reaction rate. The first study provides a mechanism for why initial nitrogen levels are important, in that the decomposers must have sufficient nitrogen to meet their own nutrient needs. Both studies were unusual, because they repeated the same experiment in many different local ecosystems, which enabled researchers to make general conclusions about processes that occur over a relatively small spatial scale.

Let's consider how organisms influence the movement of nutrients in terrestrial ecosystems on a relatively small spatial scale.

Roles of organisms in cycling nutrients in terrestrial ecosystems

One of the most famous experiments in the history of ecology was conducted by a research team headed by Frank Bormann (1968) and Gene Likens (1970) at Hubbard Brook Experimental Forest in New Hampshire, USA, which has subsequently become one of 26 Long Term Ecological Research Sites (LTER) funded by the National Science Foundation. Hubbard Brook is a deciduous forest, primarily of beech, maple, and birch trees, with cold winters, cool summers,

and annual precipitation of about 125 cm, which is spread evenly over the year. The researchers wanted to know what role the forest trees played in recycling nutrients within this ecosystem.

The Hubbard Brook Experiment

To address this question, the researchers collected baseline data on nutrient input and export on several small watersheds within the forest. They measured nutrient input by collecting rainwater and chemically assaying the nutrient levels, and export by collecting water from the stream that drained each watershed. Then, in the fall of 1965, the researchers cut down all of the trees in one watershed, leaving all the harvested vegetation in place, and continued monitoring a second, uncut watershed as a reference (Figure 20.13A). They prevented regrowth of the vegetation over the next 2 years by spraying herbicide in June 1966 and during the summer of 1967. After that time, plants were allowed to begin recolonizing the forest. For the next 45 years, the researchers sampled water from both streams – from the clear-cut and the undisturbed watershed – to see how nutrient flow changed over time.

One important early finding was that undisturbed watersheds were highly conservative of nitrogen. Both watersheds, when undisturbed, showed substantially more ammonium and nitrate in the rainwater than was exported in the stream. Presumably, organisms within the ecosystem were taking up most of these nutrients, and were also recycling nutrients already present within the watershed, rather than exporting them.

Cumulative runoff in the clear-cut watershed in 1966 was about 40% higher than it would have been if the watershed had been undisturbed, presumably because transpiration was no longer taking up water from the soil. Comparing the nutrient exports in the clear-cut ecosystem with the undisturbed ecosystem yielded some dramatic differences. Nitrate export increased from 1.3 kg/ha in 1965 to 58.1 kg/ha in 1966 in the clear-cut ecosystem, contrasting with a tiny increase from 1.0 kg/ha to 1.5 kg/ha in the undisturbed ecosystem. In contrast, ammonium export increased only slightly in the clear-cut watershed, and was still below rainwater input levels. The researchers explain this lack of ammonium export as arising from rapid nitrification of ammonium into nitrate, which then ran off into the stream.

Nitrification also generates an excess of hydrogen ions; if these ran off, they could be responsible in part for the decrease in pH observed in the stream in the clear-cut watershed. Recall that pH is measured on a log10 scale, so the observed decrease of pH from 5.3 to 4.3 indicates a tenfold increase in the acidity of the stream (Figure 20.13B).

The researchers followed the fate of all measurable ions over the course of the study. The first year after treatment, the clear-cut watershed showed very substantial increases in calcium (Ca^{2+}), magnesium (Mg^{2+}), potassium (K^+), sodium (Na^+), and aluminum (Al^{3+}) export, but almost no change in sulfate (SO_4^{2-}) export. The cations are usually bound to negatively charged binding sites in the organic soil horizon. The researchers interpret increase in cation export as indicating that under conditions of low pH, the abundant H^+ ions may be outcompeting some of the larger cations for binding sites on the clay or humus in the organic horizon. The outcompeted cations are water soluble and can easily run off into the stream and are exported from the ecosystem. In contrast, SO_4^{2-} with its negative charge is rarely bound to soil, so there was very little sulfate to run off into the stream of the deforested ecosystem.

After 3 years, the forest began recovering (Likens *et al.* 1978). The changes were rather rapid and complete for some processes, and were still incomplete for other processes. Stream flow was still somewhat elevated and evapotranspiration was somewhat depressed after 6 years. Calcium and potassium export were reduced but still elevated from pretreatment levels 6 years into the recovery process, while nitrate took only 2 years to return to pretreatment export levels. NPP was still down 4 years into recovery, but increasing rapidly (Figure 20.14).

These results highlight some of the ways that trees are central players in maintaining an ecosystem's nutrient budget. They also demonstrate that this ecosystem is remarkably resilient. But the research team warns us that the ecosystem is drawing heavily on stored reserves in the soil that will take many years to replenish. Though many of the processes were approaching "normal" rates, the nutrient losses after 10 years were still greater than those experienced by the undisturbed forest. Some soils have much lower nutrient levels than those found at Hubbard Brook and will take much longer to recover from disturbance.

Thinking ecologically 20.3

Look again at Figure 20.14. Notice that many of the processes had already shown a decrease toward the end of the experimental period in the clear-cut ecosystem. Can you think of why this might have occurred? How would you test your hypothesis?

We might wonder what other biotic influences are important for nutrient cycles in other terrestrial ecosystems. Let's take a

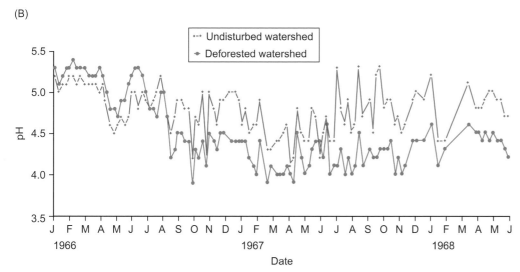

Figure 20.13 A. Deforested watershed at Hubbard Brook. B. pH levels of streams draining the deforested and undisturbed watershed at Hubbard Brook.

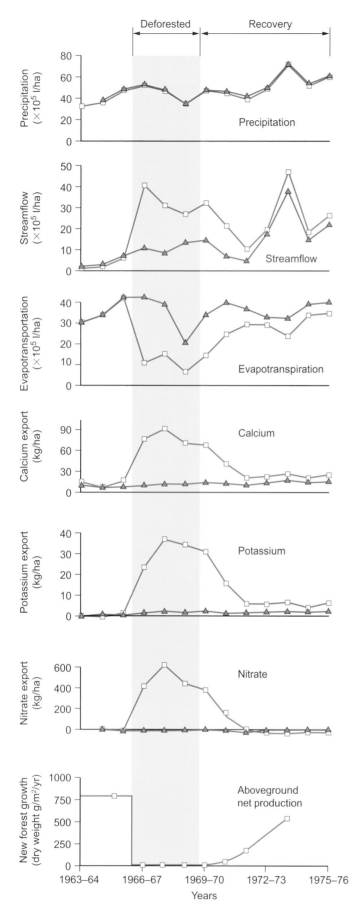

quick look at grasslands for a different perspective on how organisms influence nutrient cycles.

Grass and mycorrhizae

As the Hubbard Brook experiment showed us, disturbance can cause nutrient loss via runoff. But nutrients can also be lost when they move vertically through the water column – a process known as leaching. Marcel van der Heijden (2010) wanted to know whether arbuscular mycorrhizae helped retain nutrients within a grassland ecosystem. Recall from Chapter 15 that arbuscular mycorrhizae form belowground symbiotic relationships with most land plants, and may substantially enhance the rate of nutrient uptake, particularly phosphorus. Van der Heijden reasoned that increasing uptake rates would decrease losses due to leaching, particularly in sandy soils that drain very quickly.

Van der Heijden collected sandy soil from a grassland in Denmark. He sterilized and transferred the soil to 60, 2-l pots in a greenhouse. He inoculated each pot with the mycorrhiza *Glomus intradices*, but in some treatments the mycorrhiza was killed before inoculation. He fertilized each pot with a nutrient solution that contained (1) high nutrient levels that were comparable to what farmers use to fertilize their grasslands, or (2) low nutrient levels that were 20% of that amount. He planted one of three grass species in each pot, and allowed the grasses to grow for 16 weeks (Figure 20.15A). At the end of that period, he gave each pot 45 min of controlled simulated rain, and then collected and analysed the water or leachate that ran out the bottom of each pot for PO_4^-, NH_4^+, and NO_3^- concentrations. His dependent variable for all comparisons was the percent mycorrhizal reduction of nutrient leaching, which he defined for each nutrient as the [nutrient concentration in nonmycorrhizal leachate – nutrient concentration in mycorrhizal leachate]/[nutrient concentration in nonmycorrhizal leachate].

Under nutrient-poor conditions, for all three species, pots with living mycorrhizae lost about 60% less phosphate than pots without living mycorrhizae (Figure 20.15B). In nutrient-rich soils there was a much weaker reduction in phosphate loss due to mycorrhizae, and, in fact, the reduction was not statistically significant. The effects on ammonium and nitrate loss were much smaller, and the only statistically significant reduction in nutrient leaching caused by mycorrhizae was a 10% reduction in ammonium loss for *Festuca* (Figure 20.15C, D). The presence of mycorrhizae had no effect on plant growth, but nutrients did affect plant growth, with the total biomass of high-nutrient treatments about double that of low-nutrient treatments.

Figure 20.14 Values of hydrological, ecological, and geochemical variables in the stream that drained the clear-cut ecosystem (red), and in the undisturbed reference stream (green). Shaded time begins at deforestation and ends when herbicide spraying ceased.

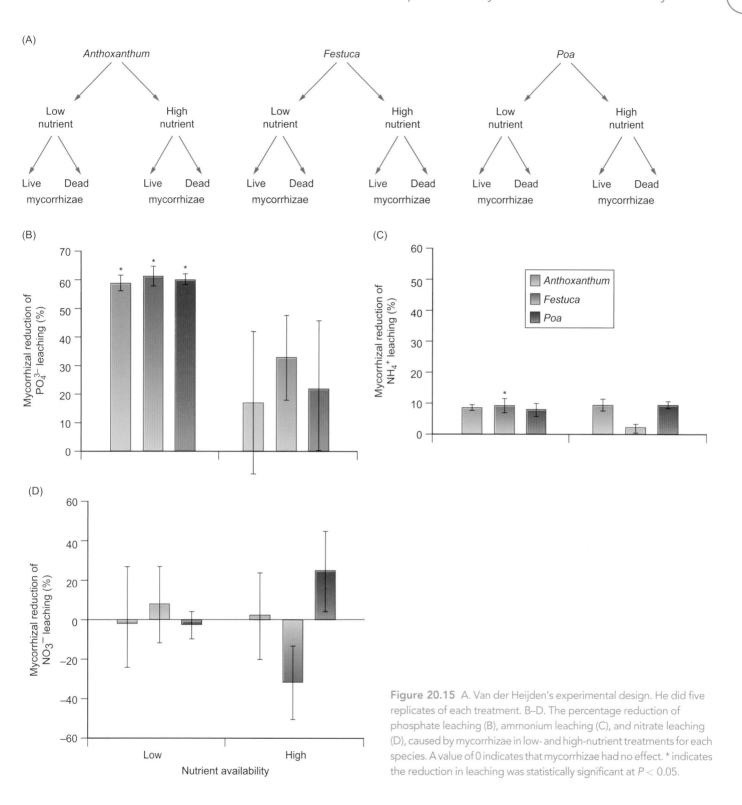

Figure 20.15 A. Van der Heijden's experimental design. He did five replicates of each treatment. B–D. The percentage reduction of phosphate leaching (B), ammonium leaching (C), and nitrate leaching (D), caused by mycorrhizae in low- and high-nutrient treatments for each species. A value of 0 indicates that mycorrhizae had no effect. * indicates the reduction in leaching was statistically significant at $P < 0.05$.

Van der Heijden concluded that mycorrhizal fungi substantially reduce the loss of phosphate in this greenhouse experiment and, more importantly, would have the same effect in nutrient-poor grassland ecosystems, such as are typically found in sandy grasslands. Preventing losses from leaching retains phosphorus in the local nutrient cycle, so that it remains available to other organisms in the food web.

Leached nutrients may find their way to streams, rivers, and ultimately the oceans. Let's return to the Black Sea to explore how recent changes in nutrient inputs have influenced that ecosystem.

REVIST: The Black Sea ecosystem

The fall of communism in 1989–91 ended central economic planning in countries bordering the Black Sea, and in other countries harboring rivers that run into the Black Sea. Almost overnight, farmers had little money to buy fertilizers, and huge animal processing plants closed. The upshot of this economic collapse was a sudden dramatic decrease in nutrient export into the Black Sea (Figure 20.16A), and a decline in ecosystem production (Figure 20.16B). A major environmental shift coincided with this change in the regional nutrient cycles: the dead zone in the Black Sea failed to develop (Figure 20.16C).

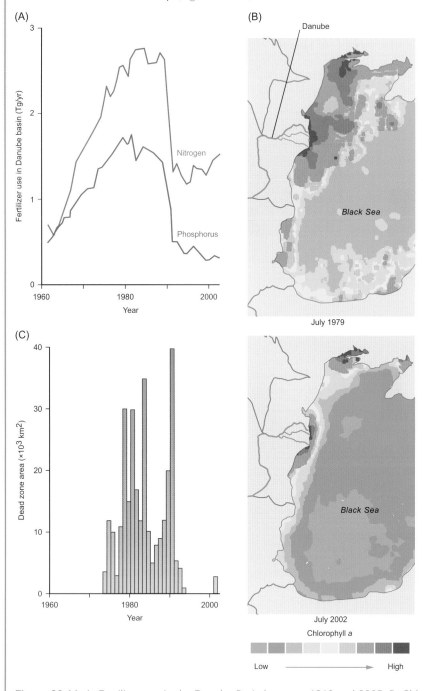

Figure 20.16 A. Fertilizer use in the Danube Basin between 1960 and 2005. B. Chlorophyll *a* concentration associated with high production in the northwest basin of the Black Sea in 1979 (top right) and 2002 (bottom right). C. Size of dead zone in the Danube Basin between 1960 and 2005.

While the dead zone was eliminated very rapidly, ecosystem recovery in the Black Sea has been much more gradual. Benthic algae are back, though previously abundant species are still rare in many locations. Mussel beds took more than 10 years to become reestablished and their presence is still tenuous. Part of the challenge for restoration has to do with ecosystem resilience. For years, the Black Sea was able to withstand increasing levels of environmental disturbance from increased nutrient input and overfishing. Once a disturbance threshold was crossed, the ecosystem was changed in ways we've discussed at length. The problem is that the new ecosystem may also be resilient, meaning that returning nutrient inputs to pre-1970s levels will not bring back the diverse biotic community. To restore the undisturbed Black Sea ecosystem, it may be necessary to reduce nutrient input and fishing to much lower levels than those of the early 1970s when the first dead zones appeared (Mee 2006).

The restoration of the Black Sea has been very costly in terms of human suffering over the loss of the fishing industry and general socioeconomic upheaval. Ecologists hope that ecological restoration can occur without economic collapse and overthrow of governments. Mee and his colleagues (2005) argue that planners should adopt an adaptive management process, in which government organizations focus on two goals: first, establishing the facts about how the ecosystem is operating, and what factors may still be impeding recovery; and second, clearly identifying what important information is lacking, so that researchers know what investigations need to be carried out. Having identified what is known and needs to be known, involved parties – including the general public – can agree upon and establish ecosystem quality objectives, a vision of what the environment should be like in 10 years. The current ecosystem quality objective reads, "to take measures to reduce the loads of nutrients and hazardous substances discharged to such levels necessary to permit Black Sea ecosystems to recover to conditions similar to those observed in the 1960s" (Mee et al. 2005). While this objective sounds nice, it will only succeed if the general public understands its value – and perhaps, if central governments are strong enough to enforce its implementation.

SUMMARY

Humans have profoundly changed nutrient cycles on a global, regional, and local level. Agricultural runoff carrying heavy loads of nitrogen and phosphorus compounds caused eutrophication of the Black Sea. This led to a series of events that culminated in the annual formation of a dead zone within the Black Sea, and the consequent loss of biological diversity of several trophic levels. Subsequent reduction of nutrient application has partially reversed this process, though species composition has not fully recovered.

The nitrogen cycle depends heavily on the activities of microorganisms to fix nitrogen, and to transform nitrogen in the processes of nitrification, ammonification, denitrification, and anammox. Technological advances such as the Haber–Bosch process have vastly increased the amount of reactive nitrogen entering ecosystems, leading to increases in agricultural production, but also polluting many aquatic systems. The phosphorus cycle is similar to the nitrogen cycle, in that globally there are vast stores of phosphorus compounds, but most of it is inaccessible to organisms. In contrast to the nitrogen cycle, there is only a small atmospheric component to the phosphorus cycle; most phosphorus becomes available through weathering of rocks. Both nutrient cycles are similar in one very important way; nitrogen and phosphorus are recycled many times between organisms and the environment before exiting an ecosystem.

FURTHER READING

Canfield, D. E., Glazer, A. N., and Falkowski, P. G. 2010. The evolution and future of Earth's nitrogen cycle. *Science* 330: 192–196.

Nice overview of most features of the nitrogen cycle, including its evolutionary history and a discussion of present and future human impacts.

Chen, X., and 32 others. 2014. Producing more grain with lower environmental costs. *Nature* 514: 486–489.

The greatest change to the nitrogen cycle results from adding nitrogen fertilizers to crops in an effort to enhance production, and then losing much of that nitrogen to runoff. This paper shows how an integrated crop and soil management system tested in China was able to increase the yield of rice, wheat, and maize, while reducing nitrogen export.

Handa, I. T., and 17 others. 2014. Consequences of biodiversity loss for litter decomposition across biomes. *Nature* 509: 218–221.

Demonstrates that decomposition rates are slower when biological diversity is reduced in either the community of decomposers, or in the leaf litter that is being decomposed. Researchers conducted these experiments in five different biomes in both aquatic and terrestrial ecosystems, so their findings have general application.

Houle, D., Moore, J. D., Ouimet, R., and Marty, C. 2014. Tree species partition N uptake by soil depth in boreal forests. *Ecology* 95: 1127–1133.

Provides evidence that nitrogen partitioning by soil depth promotes the coexistence of black spruce and jack pine in the Canadian taiga. This study uses three different approaches to build the case that each species is specializing on a different pool of available nitrogen.

Kaspari, M., Clay, N. A., Donoso, D. A., and Yanoviak, S. P. 2014. Sodium fertilization increases termites and enhances decomposition in an Amazonian forest. *Ecology* 95: 795–800.

Termites and fungi are two important groups of decomposers in Ecuadorian rainforests. Adding modest amounts of sodium salt to small rainforest plots caused dramatic increases in termite abundance and had almost no effect on fungi. The net effect was a sharp increase in decomposition rates, mediated by the increase in sodium availability, which limits termite abundance in this ecosystem. This is a very well-designed (and conceptually straightforward) experimental study.

END-OF-CHAPTER QUESTIONS

Review questions

1. What factors led to the loss of biological diversity in the Black Sea ecosystem? What factors led to its partial recovery?
2. What are the roles played by microorganisms in the nitrogen cycle?
3. In what ways are the nitrogen cycle and phosphorus cycles similar, and how do they differ?
4. How have human activities changed the nitrogen cycle? How might human activities change the nitrogen cycle in the twenty-first century?
5. What is the role of carbon and nitrogen in decomposition? How did Parton *et al.* (2007) evaluate how abiotic factors influenced the rates of litter decomposition across biomes? What did they discover?

Synthesis and application questions

1. In the last century, nitrogen fixation has increased dramatically, yet the flux of nitrogen into estuaries has increased much more modestly. What has happened to all of that extra fixed nitrogen?
2. Figures 20.12 A and B show the relationship between uptake velocity and nitrate loading on a log–log scale. What would be the effect of untransforming these data – in other words converting the data to a linear scale? How does that influence your understanding of these relationships?
3. The Hubbard Brook research team hypothesized that the lack of ammonium exported from the clear-cut ecosystem arose because most of the ammonium was quickly converted to nitrate, which ran off into the stream. How might you test this hypothesis?
4. Based on the Hubbard Brook findings, what recommendations would you make to foresters who want to use clear-cutting to harvest wood from forests?

5. The researchers used the term "reference" rather than "control" to describe the undisturbed watershed at Hubbard Brook. Why did they use that term?
6. It is easy to imagine that restoring the Black Sea runs into some of the problems we identified in our discussion of the tragedy of the commons in Chapter 6. What are these problems, and what might be some viable solutions?
7. Referring to the Black Sea ecosystem, what would be the predicted effect if humans increased their harvest of fish that feed primarily on zooplankton?

Analyse the data 1

Return to Figure 20.15 (Van der Heijden 2010) and note that even in nutrient-rich pots, there was a substantial mycorrhizal reduction of phosphate leaching. The author states that this reduction was not statistically significant, but then argues that the 10% reduction in ammonium leaching for *Festuca* was statistically significant. How can this be?

Analyse the data 2

Table 20.2 provides climatic data and data on time to the initiation of net nitrogen release for six biomes.
1. For each biome, what is the relationship between percent nitrogen and the time until net nitrogen release? Which relationships are statistically significant based on this table?
2. What is the relationship between mean annual precipitation in each biome and time until net nitrogen release? Is this relationship statistically significant?

Chapter 21
Disturbance and succession

INTRODUCTION

On May 18, 1980, I was nailing some siding to my house located in the Willamette Valley in the state of Oregon in the Pacific Northwest, USA. A winter of rain had overpowered the previous siding, threatening to require my son to swim for his life from the confines of his crib. Suddenly I heard a loud boom, but looking skyward, I could see no offending plane overhead. I went inside and turned on the radio to discover that Mount St. Helens, over 200 km to the north and east, had finally erupted! People in the area, and across the world, were riveted by tales of the mountain collapsing in an unimaginable avalanche of dust and debris, of the raging Touttle River overpowering all in its path, of mudflows and pyroclastic flows of rock fragments and hot gases that buried everything, and of human triumph and human tragedy. Only later did people consider the ecological and environmental consequences of such devastation.

Some disturbances, such as the explosion of Mount St. Helens, disrupt ecosystem structure to such an extent that most organisms are killed. Primary succession, as described by ecologists for some habitats at Mount St. Helens and Glacier Bay in Alaska, is the recovery process that follows disturbances of such intensity. Secondary succession describes the pattern of recolonization following a disturbance in which only some of the organisms are killed, such as occurred in the 1988 fires at Yellowstone National Park. In this chapter, we apply three different models of recovery – facilitation, tolerance, and inhibition – to describe how the plant community recovers following disturbance. Though many succession studies have focused on interactions between plants and abiotic factors, ecologists are now appreciating the roles played by animals and the importance of legacy effects, in succession. Complicating our understanding of succession is that an ecosystem may change abruptly from one stable state to another in response to changes in the biotic or abiotic environment, and once changed, it may be difficult for an ecosystem to revert back to its predisturbance state.

KEY QUESTIONS

21.1. What is primary succession, and when does it occur?

21.2. What is secondary succession, and when does it occur?

21.3. What mechanisms underlie the process of succession?

21.4. How do animal communities respond to disturbance?

21.5. Can an ecosystem shift abruptly from one stable state to another?

CASE STUDY: Extreme disturbances

Eight years after Mount St. Helens erupted, and a few hundred kilometers to the east, the natural world provided another flamboyant demonstration of its might, when Yellowstone National Park endured 3 months of fires, which burned about 36% (321 0000 ha) of the park. About 80% of the park had been covered with forests of lodgepole pine, *Pinus contorta*, which under very dry conditions can fuel large and devastating crown fires. Historical evidence from pollen cores and tree rings reveal that these intense conflagrations have occurred, on average, every 100 to 500 years over the past 10 000 years. The most intense fires burned during the end of August through the middle of September, during periods of high wind and very low humidity. These fires were so intense that they were able to jump natural firebreaks such as the Grand Canyon of the Yellowstone River (Turner *et al.* 2003). Though described as a devastated moonscape by some sensationalist reporters, the reality is that the fires of 1988 created a complex mosaic of patches: some blackened by crown fires, others experiencing less severe ground fires, and others completely unaffected (Figure 21.1).

The eruption of Mount St. Helens and the fires at Yellowstone were extreme examples of a **disturbance** – a relatively discrete event that disrupts the structure of an ecosystem, community, or population by changing either the physical environment or the resources it contains. Ecologists want to understand what happens following a disturbance. Does the ecosystem revert to its predisturbance condition, or might it change into something entirely different? If so, how long does this process take, and what mechanisms underlie the ecosystem response? Also, do different types of ecosystems show different responses to the same type or intensity of disturbance? Lastly, is there a clear distinction in how an ecosystem responds to a completely devastating disturbance, such as the blast zone just north of Mount St. Helens, in comparison to a more benign disturbance, such as occurred in areas farther from the blast zone?

These are central questions in ecology, because the science of ecology attempts to describe and predict the distribution and abundance of organisms. In addition, applied ecologists need to understand the role of disturbance in ecosystem structure, so they can know whether action to prevent disturbance is warranted. Should forest ecologists try to prevent forest fires, or should they

Figure 21.1 A. Fire approaches the Old Faithful complex at Yellowstone National Park. B. Mosaic of burned and unburned areas resulting from the 1988 fires at Yellowstone.

encourage them to burn, in the belief that fire plays an important role in maintaining the structural integrity of ecosystems? Let's begin our discussion with some descriptive studies of what happens to ecosystems following disturbances. We will periodically return to the eruption of Mount St. Helens and the fires of Yellowstone, when they can help answer some of the questions we have posed.

21.1 WHAT IS PRIMARY SUCCESSION, AND WHEN DOES IT OCCUR?

Primary succession describes the changes that occur to a habitat following a severe disturbance that kills virtually all organisms, including plant seeds and microorganisms, in that habitat. Many insights about primary succession come from studying chronosequences – habitats that are similar to each other in all respects except for the time since they were colonized. William S. Cooper (1923a and b) initiated one of the most important chronosequence studies of primary succession caused by glacial action in Glacier Bay, Alaska. Since Cooper's pioneering work, numerous ecologists have journeyed to Glacier Bay to describe, analyse, and interpret the plant communities and their interactions with soils and associated soil organisms.

Stages of succession at Glacier Bay

Glaciers are dynamic, advancing and retreating in response to changes in climate. In 1794, George Vancouver described a huge glacier that almost completely filled the entire expanse of Glacier Bay. But when Cooper first visited Glacier Bay, he found an environment that was vastly different, with the glacier having retreated well over a hundred kilometers. By consulting historical records, talking to others who had traveled to the region, and making some assumptions based on the vegetation, he could reconstruct the history of the glacial retreat. Thus Glacier Bay

provided Cooper with a chronosequence for studying primary succession.

When the glaciers retreated from Glacier Bay they left behind a complex surface mosaic of rock, gravel (over 100 m thick in some places), and a thin layer of finely crushed glacial till (sediment). One feature common to all of these surfaces is that they provide a very inhospitable habitat for any plant life, having almost no nutrients associated with them. To study succession in plant communities of different ages, Cooper established nine different stations in various locations along the bay.

Cooper described three stages of development in the plant community. First was the pioneer stage, which began immediately after glacial retreat and consisted primarily of lichens, mosses, three perennial herbs, several species of willows, and small cottonwood trees. Second was the willow–alder stage, dominated by the alder, *Alnus tenuifolia*, and several species of willows. These trees are about 5–7 m tall and support a relatively rich community of undergrowth made up of several species of moss, ferns, and various forbs. Last was the conifer forest stage, which was dominated by Sitka spruce, *Picea sitchensis*, but also hosted cottonwoods and hemlocks. This forest stage also hosted an increasingly diverse array of mosses, ferns, and various forbs, as well as some old alders and willows – relics from the previous stage that Cooper described as "obviously having a hard time of it" (Figure 21.2A).

Cooper, and many researchers who followed him, viewed the conifer forest stage (often called the spruce stage) as the climax community – a stable biotic community that is in equilibrium with existing environmental conditions and represents the last stage of succession. Figure 21.2B shows a very simplified conceptual model of primary succession based on Cooper's research at Glacier Bay. Based on the chronosequences he observed, Cooper predicted that most of the communities would follow the same trajectory of pioneer → willow-alder → spruce. But using chronosequences forces the ecologist to make the important assumption that all of the communities have experienced the same abiotic and biotic influences since they were established.

(A)

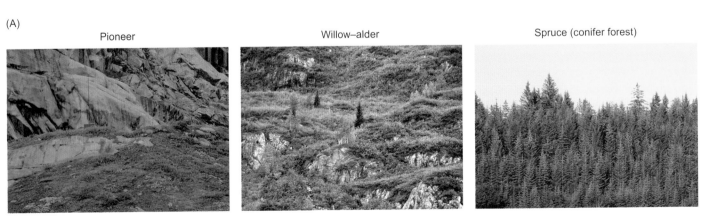

Figure 21.2. A. Successional stages at Glacier Bay. Pioneer stage (left). Willow–alder stage (middle). Spruce stage (right). B. Simple conceptual model of primary succession showing catastrophic disturbance leading to a predictable series of developmental stages.

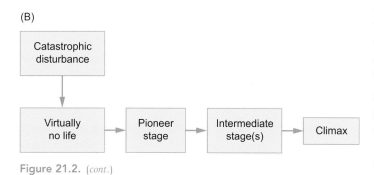

Figure 21.2. (*cont.*)

Multiple pathways for succession at Glacier Bay

Christopher Fastie (1995) suspected that not all communities at Glacier Bay were passing through the same successional stages. He established 10 chronosequence sites at Glacier Bay, which differed in date of establishment; the oldest site was deglaciated about 1768, while the more recent sites were from the mid-twentieth century (Figure 21.3A). Fastie collected increment cores from all live Sitka spruce, western hemlock, and black cottonwood trees, which were the major tree species in each plot. This allowed him to calculate the age of each tree, and to distinguish periods of rapid growth (wide rings) from periods of slow growth (narrow rings). He also collected cores from many of the dead trees, and estimated time since death based on appearance.

Fastie's most important finding was that there are substantial differences in species composition, in community development, and in the mechanisms influencing development across the chronosequence. In particular there are important differences when we compare the older sites (8–10) to the more recently established sites (1–7).

Differences in species composition

Based on tree coring data, Fastie was able to estimate the number of overstory, subcanopy, understory, and seedling trees that were alive at each site at any time. Some differences stand out.

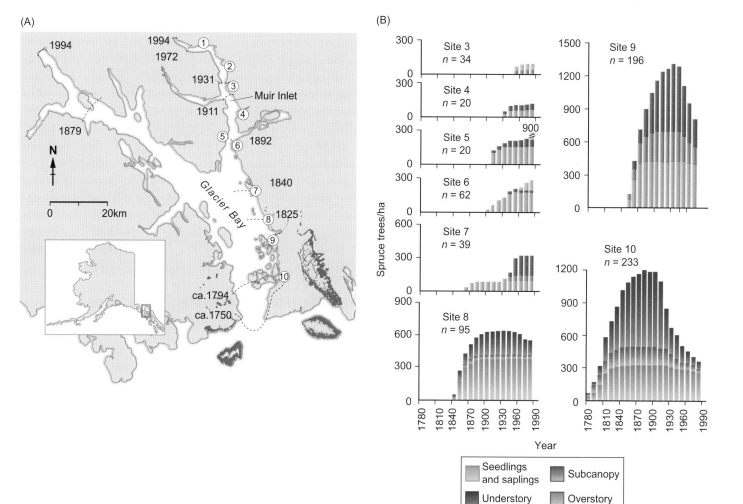

Figure 21.3 A. Fastie's research sites at Glacier Bay. Year followed by dotted isoline indicates estimated year of deglaciation. Dark green shading indicates spruce and hemlock refugia. B. Density of Sitka spruce trees (per ha) at different stages of development reconstructed for eight chronosequence sites at Glacier Bay.

The older stands reached a much higher spruce density, and did so relatively quickly after deglaciation (Figure 21.3B). In addition, only the older stands had western hemlock, and they too were relatively abundant. In contrast, black cottonwoods were present only in the younger stands, where they were relatively abundant.

Older sites reached high hemlock densities rather quickly; for example, site 10 exceeded 200 hemlocks a mere 70 years after deglaciation. In contrast, the younger sites, even those that were deglaciated for 150 years, were never colonized by hemlock. Similarly, the older sites were never colonized by black cottonwood, even though the younger sites reached densities of 200 black cottonwood trees/ha within 40 years of deglaciation.

Differences in community development

Returning to Figure 21.3B, you can visualize the different patterns of spruce colonization. In the older sites (8–10), spruce increased very rapidly in abundance following their initial colonization. In contrast, the spruce colonization rate for sites 3–7 has been much slower.

 Thinking ecologically 21.1

Looking at Figure 21.3B, you can see that spruce abundance appears to increase more quickly over the first 50 years in the older sites than in the younger sites. Try to convert the figures to a data file in which the first column is years since first colonization by Sitka spruce and the second column is spruce abundance after 50 years (you will need to estimate your values). Then calculate the slope from year 0 to year 50 for each site. Finally, analyse the data to see if the slopes are statistically significantly different when comparing sites 8–10 with sites 3–7. What can you conclude based on your analysis?

Given the differences in patterns of succession between the older and younger sites, Fastie's next task was to determine the underlying mechanisms for these differences in succession. He was able to identify two factors that appeared to play a role.

Factors underlying differences in succession

Fastie proposed two hypotheses for why the older and younger sites were so different in species composition and the pattern of succession. First, he reasoned that the older sites might be nearer to spruce and western hemlock seed sources. To test this hypothesis, he consulted aerial photographs of the surrounding region and discovered that the three oldest sites were within 4 km of mature forests that had 400+-year-old Sitka spruce and also supported extensive stands of mature western hemlock. These upland forests refugia had managed to avoid the last glacial advance. The younger sites were more distant from these old-growth refugia, and thus did not receive a comparable "rain" of potential colonizing seeds.

Fastie's second hypothesis was that spruce seedlings may have faced more competition in the younger sites than in the older sites. Given the abundance of alder at 30–50 years post colonization, he reasoned that perhaps alders slowed down the growth of spruce in the younger sites by shading out spruce saplings to a much greater extent than they did in the older sites. He reasoned that if alder suppressed spruce growth in the younger sites, spruce should show a long period of very slow growth (narrow rings), which should suddenly accelerate (much wider rings) once they grew taller than alder height. In support of this hypothesis, Fastie found this process of tree-ring release much more frequently at the younger sites than in the older sites. To test the hypothesis that tree-ring release occurred as a result of the loss of light competition once the spruce trees grew taller than the alders, Fastie tied back alder branches that were shading out spruce trees, and took core samples on an annual basis following this manipulation. He found that spruce tree rings got substantially wider when the alder branches were tied back so they were no longer shading out the spruce trees.

Fastie concluded that the Glacier Bay chronosequence actually has three distinct successional patterns. The older sites went very quickly to a spruce/hemlock climax, and probably did not have an alder stage that interfered with spruce development. The middle-aged sites had an extensive alder stage, but were not colonized by cottonwood, while the youngest sites will probably go through the following sequence: pioneer → alder → cottonwood → spruce.

Based on Fastie's research, it is clear that the assumption underlying chronosequences, that all of the communities have experienced the same abiotic and biotic influences since they were established, does not hold in the case of Glacier Bay succession. Thus the explosion of Mount St. Helens interested ecologists specifically because it provided them with the opportunity to study primary succession of a large ecosystem in real time from the very beginning of the succession process.

Mount St. Helens: a theater for the study of succession

Several factors influenced the intensity of disturbance experienced by the communities near Mount St. Helens. A moderate-intensity earthquake caused much of the northern flank of the mountain to collapse, which gave rise to the largest *debris avalanche* in recorded history, burying about 60 km^2 of the valley in rock, sand, and gravel to a mean depth of 45 m. When the north flank collapsed, the magma that had been compressed within the volcano escaped in a devastating blast that produced a hot cloud filled with rock debris that removed, toppled, or scorched most aboveground vegetation over a 570 km^2 area north of the mountain. This region was divided into four progressively less affected zones (Figure 21.4A–F). Closest to the blast was the *tree-removal*

Figure 21.4 Disturbances resulting from eruption of Mount St. Helens. A. Debris-avalanche deposit. B. Tree removal zone. C. Blowdown zone. D. Scorch zone. E. Tephra fall zone. F. Mudflow. G. Map of disturbance zones.

Table 21.1 Volcanic events associated with eruption of Mount St. Helens.

Event/zone	Area (km^2)	Deposit thickness (m)	Temperature (°C)	Organic matter
Pyroclastic flow/pumice plain	15	0.25–40	300–850	None
Debris avalanche	60	10–195	70–100	Rare
Blast/tree-removal	90	Variable	Variable	Common
Blast/blowdown	370	0.01–1.0	100–300	Common
Blast/scorch	110	0.01–0.1	50–250	Common
Blast/mudflow	50	0.1–10	30	Common
Tephra fall	1000	>0.05[a]	<50	Common

[a] There was a much greater area covered by a thin layer of tephra (<0.05 cm thick).

zone in which almost all trees were removed and blown away. Farther from the blast was the *blowdown zone* where the trees were toppled in place by the force of the blast. Farther away still was the *scorch zone*, a layer of standing dead forest – killed by the heat cloud, but far enough from the blast that the trees remained standing. Farthest away, and mostly to the northeast was the *tephra fall zone*, an expansive layer of tephra – uncompacted rock and sand – deposited over the terrain (Swanson and Major 2005).

We will be discussing the ecological effects of two other processes. After the blast, large quantities of very hot pumice spewed from the volcano for about 5 hours, forming *pyroclastic flows* that covered about 15 km^2 of the surface north of the volcano in a layer up to 40 m thick. When these pyroclastic flows cooled and solidified, they formed a large *pumice plain*. Lastly, *mudflows* filled the river channels with thick layers of hot mud that flowed mostly northwest down the North and South Touttle River Valleys, but also southeast down the Muddy River Valley and several smaller creeks as well (Figure 21.4G).

The Mount St. Helens eruption, with its many component events, provides us with a continuum of disturbances of varying intensity. The pumice plain and debris avalanche were most disturbed, with virtually no surviving organisms, and relatively little organic matter to sustain any recolonizing organisms. The tree-removal and mudflow zones were also devastated, but did have a relatively rich supply of organic matter. Communities in the blowdown zone, scorch zone, and tephra fall zones were progressively less disturbed (Table 21.1).

Of course, ecologists have long recognized that disturbances are of varying intensities. Thus they distinguish between primary succession of habitat that is completely devoid of life as a result of an intense disturbance, such as was experienced on the

pumice plain or debris avalanche, and secondary succession of a habitat in response to less intense disturbance.

21.2 WHAT IS SECONDARY SUCCESSION, AND WHEN DOES IT OCCUR?

Secondary succession describes the pattern of recolonization following a disturbance in which only some of the organisms are killed, such as occurred in the scorch or tephra fall zones at Mount St. Helens, or in the habitat burned by the fires at Yellowstone Park. In both cases, there was considerable variation in the intensity of disturbance, which influenced the pattern of recolonization.

Variation in disturbance intensity

In reality, the distinction between primary and secondary succession is somewhat arbitrary, because there is no clear boundary between the two processes. Though we've discussed the pumice plain and debris-avalanche areas as experiencing total destruction, and thus undergoing primary succession, in reality some species did survive. These include organisms that were buried beneath the snow, or sheltered from the blast by geological formations, or – in the case of migrating species – had not yet returned from their winter homes. But particularly in the pumice plain and the avalanche debris area, the process is closer to the primary succession end of the continuum. At Yellowstone Park, the most devastated areas were similarly mostly devoid of life following the fires, but, in general, there were considerable *residuals*, or

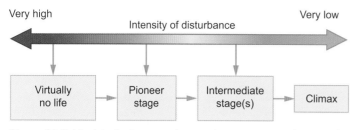

Figure 21.5 Model of primary and secondary succession showing the two processes as responses to a continuum of disturbance intensity.

remaining organisms there, so we can view the recovery process at Yellowstone as primarily a process of secondary succession. These examples highlight that the disturbances triggering succession fall along a continuum of intensity (Figure 21.5).

We might expect that habitats that experienced a more intense disturbance would take longer to recover. Monica Turner and her colleagues (1999, 2003) have spent the past 25 years

monitoring the recovery process at Yellowstone to test this, and other hypotheses about secondary succession.

How disturbance intensity influences the rate of secondary succession

Turner and her colleagues describe three tree-fire burn severity classes at Yellowstone Park. *Crown fires* are most severe: All canopy trees are killed, their needles are completely consumed, and the soil organic layer is nearly completely consumed, leaving only bare ground and no soil litter. Less intense is a *severe surface burn*, with extensive canopy tree mortality but many needles unconsumed, and some surface litter present. Least intense is a *light surface burn*, in which trees have scorched stems but little mortality, and most of the soil organic layer is intact. The researchers established three 1 km² sites, each with a mosaic of these three burn severity classes and some unburned areas as well. Their analysis showed that recovery was substantially quicker in the sites with lowest burn intensity (Figure 21.6A).

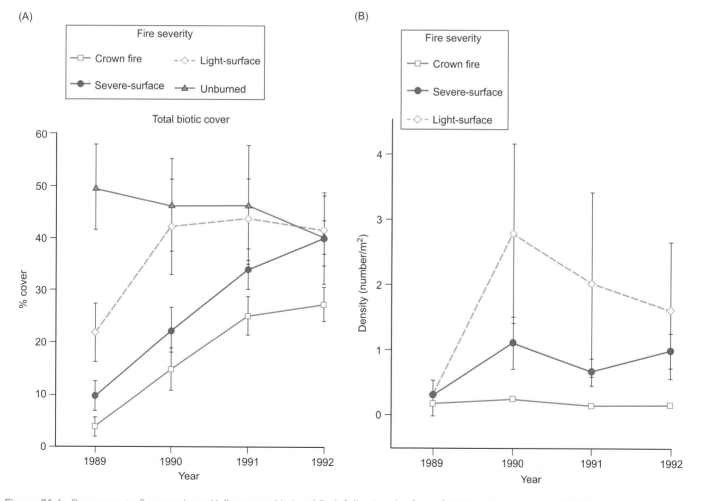

Figure 21.6. Responses to fire severity at Yellowstone National Park following the fires of 1988. A. Percent cover (±SE) in relation to fire severity. Note that by 1990, sites with light surface burn had already achieved similar percent cover as unburned sites. B. Mean density of lodgepole pines in relation to fire severity.

When Turner and her colleagues studied the effect of fire intensity on subsequent recruitment of lodgepole pine seedlings, they discovered that sites with the most intense fires had by far the lowest seedling recruitment. In sites with crown fires, seedling density in 1989 averaged 0.19 seedlings/m^2, and there was no evidence of any subsequent increase over time. In contrast, sites with light surface fires showed a sharp increase in seedling density between 1989 and 1990, but actually decreased slightly in 1991 and 1992, indicating that some seedlings were dying out and were not being replaced at a significant rate by new colonizers (Figure 21.6B).

Numerous other factors can influence succession. Life history traits of potential colonizers may play an important role.

How life history traits influence the process of succession

Certain life history traits are favored early in succession. Pioneer and mid-successional species tend to have high reproductive rates, are usually able to disperse long distances, germinate and grow quickly in poor soils, and usually are most successful in the absence of competitors for nutrients or sun. But other life history traits play an important role in succession at Yellowstone.

We have already mentioned that lodgepole pine is the dominant tree at Yellowstone. This species can, under certain conditions, produce serotinous cones, which remain closed for many decades, are retained by the tree, and only open when heated by intense fire, releasing seeds onto the mineral soil that becomes available after fires (Figure 21.7A). It is important to realize that serotiny is only adaptive under certain fire regimes. If the time span between fires is longer than a tree's expected lifespan, then natural selection should favor trees that produce open cones, because seeds in serotinous cones that are still in a tree when it dies will fail to germinate. If the time span between fires is shorter than a tree's expected lifespan, then selection favors serotiny. But if the time span between fires is very short, fires will be relatively low intensity due to lack of fuel, and may not be hot enough to open serotinous cones. Thus serotiny may be favored when fire intervals are long enough to burn hot enough to open the cones, but are shorter than a tree's expected lifespan.

Tania Schoennagel and her colleagues (2003) asked whether lodgepole pines switch adaptively from open cones to serotiny in relation to expected fire interval. The researchers measured fire interval using historical records, and also by counting tree rings in cores taken from the largest trees in a given plot. In one study, the researchers established 50 plots, which comprised 25 matched pairs of pine stands. One plot of each pair was young, with a short fire interval (mean = 38 years), while the second plot of the same pair was much older, with a much longer fire interval (mean = 211 years). Serotiny occurrence was 4% in the young plots and 14% in the old plots. But these differences were

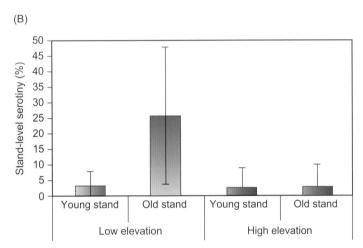

Figure 21.7 A. Serotinous cones (left) and open cones (right) on a lodgepole pine tree at Yellowstone National Park. B. Mean serotiny of young and old stands of lodgepole pine at Yellowstone in relation to elevation and age of stand.

strongly influenced by elevation. Old plots had much higher serotiny at low-elevation sites, where serotiny averaged over 25%, than they did at high-elevation sites, where serotiny averaged below 5% (Figure 21.7B).

Why should elevation play such an important role? As it turns out, fires are much more common at low elevations at Yellowstone than they are at high elevations. Low elevations tend to be drier and warmer, leading to relatively short fire intervals of 135–185 years. In contrast, the estimated mean interval between fires at high-elevation sites was 280–310 years. The researchers propose that lodgepole pines are phenotypically plastic at low-elevation sites: younger trees produce primarily open cones, which can colonize sites if a fire occurs after a short time interval, but the pines switch to serotiny as they get older and are more likely to experience a very hot fire. In contrast, at high elevations, fires are rare enough that serotiny is not adaptive and thus not common.

But does serotiny lead to higher reproductive success? Monica Turner and her colleagues (2003) measured serotiny at three locations in Yellowstone Park before the 1988 fires, and found

65% serotiny at Cougar Creek, 10% serotiny at Fern Cascades, and <1% serotiny at Yellowstone Lake. Five years after the fire, they measured lodgepole pine seedling density at these three locations. The Cougar Creek site had a mean of 211 000 stems/ha, Fern Cascades had a mean of 2300 stems/ha, and Yellowstone Lake had a mean of only 600 stems/ha. The results of these two studies supported the hypothesis that serotiny is a viable strategy for increasing reproductive success in regions prone to fire intervals of intermediate length.

 Thinking ecologically 21.2

One variable we have not discussed is how the size of a burned patch might influence the rate of succession. What hypothesis would you make about the relationship between patch size and recolonization rate? Second, some plants reproduce from seeds, while others do so vegetatively, usually from surviving rootstock. How might that affect your hypothesis?

We have discussed primary and secondary succession in some detail, and concluded that primary succession and secondary succession are ecosystem responses to a continuum of disturbance intensity. We have discussed the relationship between intensity of disturbance and subsequent recolonization, and addressed how life history traits influence succession. However, we still need to explore the mechanisms underlying succession in enough detail so we can have a deeper understanding of why succession happens, and also consider whether it is realistic to expect succession to proceed in a predictable manner.

21.3 WHAT MECHANISMS UNDERLIE THE PROCESS OF SUCCESSION?

Arguments about the mechanisms underlying succession have dominated ecological discussions for over a century, and promise to do so for the foreseeable future. Joe Connell and Ralph Slatyer (1977) summarized the state of the field 40 years ago with an important paper that continues to be influential to this present day.

Three conceptual models of succession

Connell and Slatyer (1977) proposed three distinct models of succession in terrestrial ecosystems. The facilitation model proposes that the first colonizers are shade-intolerant pioneer species, with life history traits such as high fecundity, good dispersal powers, efficient germination in open spaces, and rapid growth, that modify the environment in such a way that it becomes less suitable for future pioneer species but more suitable for later successional species. This can be accomplished by a variety of mechanisms including providing shade, increasing moisture levels, or adding nutrients or organic matter to the soil.

Late successional species that move in tend to be more shade tolerant and compete well for limiting resources.

The tolerance model proposes that both early- and late-succession species colonize a habitat following disturbance, but that the majority of early colonizers are pioneer species, simply because they are excellent dispersers and colonizers. Early species modify the environment so it becomes less suitable for early-succession species but do not make it more or less suitable for late-succession species. Some late-succession species do arrive earlier, and survive long periods of slow growth under conditions of limiting resources, such as lack of sunlight or nutrients. Over time, late-succession species, which can tolerate conditions of limiting resources, tend to dominate the community.

The inhibition model also proposes that early- and late-succession species colonize a habitat following disturbance, with the majority of early colonizers having the characteristics of pioneer species. Once an individual secures a space or other limiting resource, it inhibits the invasion of all other plant species. When a resident dies, individuals of any species may move into the space and secure the limiting resource. Thus an early-succession or late-succession species may move in. The key is that late-succession species tend to live longer, so over time, late-succession species increase in abundance in the community.

Connell and Slatyer predicted that the facilitation model should often apply to primary succession. They reasoned that after very severe disturbances associated with primary succession, soils are usually so nutrient poor that only a few species can initially get established. Once these pioneers are in place, they can improve the soil when they either shed leaves or die, and decomposers break down their bodies and body parts. This allows other species to move into the improving habitat. Let's turn our attention back to primary succession at Glacier Bay to see how this prediction plays out.

Facilitation and inhibition at Glacier Bay

Stuart Chapin and his colleagues (1994) combined systematic observations with a series of experiments to investigate the mechanisms underlying succession at Glacier Bay. In contrast to Cooper's original description, they considered four stages of succession, by dividing Cooper's willow-alder stage into an early *Dryas* stage, which develops at about 30 years and is characterized by a high abundance of the nitrogen-fixing forb *Dryas drummondii*, and a later alder stage (already discussed), which generally becomes dominant at about 50 years. The researchers established one 2 km^2 study area at locations representing each stage, and then chose 10 smaller sites within each study area.

Facilitation of early stages
Soil collections from the four successional stages revealed some interesting trends. Organic content and moisture levels of the soils were lowest in the first three stages, but significantly higher in soil collected from the spruce stage. Nitrogen content climbed steadily from the *Dryas* to alder stage, and more so from the alder

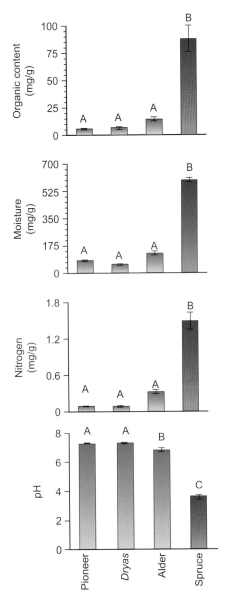

Figure 21.8 Soil properties at four stages of succession at Glacier Bay. Different letters above the bars indicate that the means are statistically significantly different from each other at $P < 0.05$.

to spruce stage, while pH decreased somewhat in the alder and sharply in the spruce stage (Figure 21.8).

It would seem like these changes (with the exception of reduced pH) should be beneficial to most species, in support of the facilitation model. To test this in the laboratory, the researchers collected soils from each stage at Glacier Bay, and grew alder and spruce seedlings in the greenhouse under controlled conditions. Spruce seedling biomass was greater in soils from later stages than the earlier stages. Alder showed an interesting trend, with much greater biomass in *Dryas* than in pioneer soil (as expected), but then a sharp decline in alder soil followed by a sharp increase in spruce soil (Table 21.2). Thus, with the exception of alder not growing well in soil from the alder stage, there appears to be the expected facilitation of growth in soils from later successional stages.

Table 21.2 Plant biomass (mg) of alder and spruce trees ($\pm$SE) grown in soil from four successional stages at Glacier Bay.

Species	Successional stage			
	Pioneer	*Dryas*	Alder	Spruce
Spruce	27 (2)	41 (3)	43 (5)	147 (29)
Alder	13 (2)	1567 (317)	298 (181)	3225 (583)

Inhibition

Having shown that facilitation was important in primary succession (Figure 21.8 and Table 21.2), the researchers wanted to know whether other processes also played an important role. Greenhouse experiments allow researchers to control independent variables; however, plants at Glacier Bay must endure some challenging and variable environmental conditions. Consequently, Chapin and his colleagues combined observations of plants growing under natural conditions, and transplants of seeds and seedlings into plots at different stages of succession, in an effort to understand the mechanisms of succession at Glacier Bay.

Let's consider the establishment and growth of spruce seedlings – the climax community at Glacier Bay. The researchers were able to show from their series of experiments and observations that germination of spruce was inhibited in the pioneer, *Dryas*, and alder stages. In addition, spruce survival was inhibited in the *Dryas*, alder, and spruce stages, as was spruce growth. Inhibition was caused by increased seed predation, and competition for light and nutrients. Given all this inhibition, the researchers make the point that succession to the spruce stage requires that either a very large number of spruce seeds are sown, or that at least a few spruce seeds get established during the first stages of succession, when inhibition is weaker.

Based on this study, we can see the influence of both facilitation and inhibition on succession at Glacier Bay. Let's turn our attention next to the pattern of succession following the eruption at Mount St. Helens, in search of evidence supporting the facilitation, tolerance or inhibition models.

Succession following the Mount St. Helen's eruption

Mount St. Helens produced the most powerful debris avalanche in recorded history, so the recovery process was expected to be quite slow. To study primary succession in this area, Virginia Dale and her colleagues (2005b) established a large series of 250-m^2 plots down the length of the debris-avalanche zone, which they surveyed seven times in the 20 years following the eruption. They discovered that the recovery process was actually proceeding quite rapidly, with mean vegetation cover approaching 70% at 20 years after the eruption. In addition, mean plant species

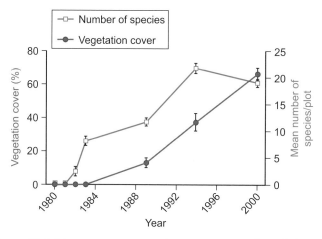

Figure 21.9 Mean percent vegetation cover and number of plant species per plot in 1980–2000.

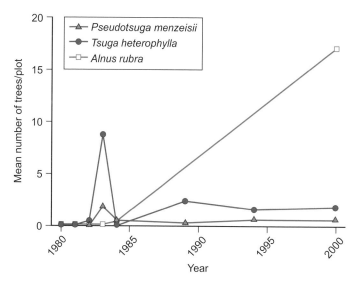

Figure 21.10 Abundance of western hemlock, Douglas fir and red alder in Mount St. Helens debris-avalanche zone in 1980–2000.

richness per plot increased rapidly until 1994 but actually declined somewhat by 2000, as some early succession plants were replaced by late succession species (Figure 21.9).

In contrast to succession at Glacier Bay, some tree species recolonized almost immediately after the disturbance. Before the eruption, conifers dominated the debris-avalanche area, so the researchers were interested to discover relatively large populations of two conifer species, western hemlock (*Tsuga heterophylla*) and Douglas fir (*Pseudotsuga menzeisii*), only 2 years after the eruption. However, a drought the following year killed most of the seedlings, but new seedlings became established in subsequent years. In 2004, a few Douglas fir trees produced cones. But the most abundant tree species 20 years after the eruption was red alder, *Alnus rubra*, which has an important mutualism with nitrogen-fixing *Frankia* bacteria. The mutualism allows red alder to survive in the nitrogen-deficient soils characteristic of the debris-avalanche zone. In addition, the mutualism imports nitrogen into the ecosystem, which can then be used by other plants as a nitrogen source when alder trees die or drop their leaves, or as a nitrogen source by animals that eat alder plants (Figure 21.10).

One of the important take-home lessons of the Mount St. Helens eruption is the importance of nitrogen fixation for the recovery process. Both red alder and the legume, broadleaf lupine (*Lupinus latifolius*), were abundant during primary succession. Both of these plants have mutualisms with nitrogen-fixing bacteria, and thus are likely to increase nitrogen availability in nutrient-poor soils. In addition, these plants created shade, conserved moisture, and added organic material to the soil. From this perspective, succession at Mount St. Helens tends to support the facilitation model. However, some late successional plants, such as western hemlock, also established themselves on these plots, though in low numbers, only a few years after succession, as predicted by both the tolerance and inhibition models. Dale and her colleagues predict that succession will be a three-step process, with early-succession plants having dominated for the

first 15 years, mid-succession deciduous trees such as red alder dominating during years 15–80, and finally transitioning to coniferous forest during years 80–115. However, periodic disturbances such as drought, erosion and mudflows, herbivory, fire and perhaps more volcanic activity, may reset the successional clock, producing a patchwork of successional stages.

A second important lesson of the Mount St. Helens eruption is the importance of biological legacies – organisms, the remains of organisms, and organically derived environmental patterns that persist through a disturbance, and are incorporated into the recovering ecosystem (Franklin 2000). All of the surviving plants in the debris-avalanche zone were *cryptophytes* – species with dormant buds located below the surface (Dale *et al.* 2005b). Some western hemlocks (*Tsuga heterophylla*) survived in protected snow banks outside the debris-avalanche zone and may have contributed seeds to the abundant hemlock seedlings in 1983. The snags and downed logs provided nutrients and substrates for fungi and plants, and cover for small animals that moved in following the eruption.

Until now we have discussed primary succession in the context of plants that colonize an area following disturbance. However, animals are important members of ecological communities, and can go through their own successional processes, as well as having profound influences on the pattern of plant succession.

21.4 HOW DO ANIMAL COMMUNITIES RESPOND TO DISTURBANCE?

Animals experience similar responses as plants do to disturbance. Intense disturbance can extirpate animals from an ecosystem, while mild disturbance can have a variety of effects on their populations. In addition, animal populations can have an important influence on how plant communities respond to disturbance. Animals eat plants, and can wipe out species or groups of species. Animals can also facilitate plant recolonization. They add limiting nutrients to the soil and can act as pollinators and seed dispersers, and carnivorous animals eat grazers and browsers that could potentially devastate recovering plant populations.

Let's return to Mount St. Helens to study the early stages of animal recovery after intense disturbance.

Return of arthropod communities to Mount St. Helens

John Edwards and Patrick Sugg (2005) studied the arthropod response to the Mount St. Helens eruption and the subsequent recolonization. They argued that given the total barrenness of the pumice plain, it would not be surprising if arthropods were the first colonizers following the eruption. Arthropods could potentially eke out an existence by scavenging nutrients from materials swept into the area by the wind. Arthropods are important for primary succession in two ways. First, many are excellent dispersers who can move into a denuded area in large numbers, restoring diversity and abundance of species to the area. In addition, even if they die following dispersal, insect bodies are loaded with nutrients, and can help replenish nutrient levels in a nutrient-poor habitat.

To study arthropod survival, the researchers set up a series of pitfall traps in numerous sites in the pumice plain and the blowdown zone. Pitfall traps are steep-sided cups that are buried at regular intervals, with the top of each cup level with the surface of the ground. Terrestrial insects that have the misfortune of stumbling into these traps slide down the wall of the cup into a toxic vat of ethylene glycol solution. The researchers collected insects and spiders from these traps approximately every 2 weeks between 1981 and 1987.

Almost all of the species present in the first samples collected from the pumice plain were aerial dispersers. With two minor exceptions, no ground species were found, indicating that none survived the blast. In contrast, in the blowdown zone, many species of ants, ground beetles, spiders, millipedes, centipedes, and crickets survived the explosion. These biological legacies survived beneath the snowpack, or were protected by trees from the force of the blast. The number of species surviving was inversely correlated to the intensity of the disturbance as measured by tephra depth in the blowdown zone (Figure 21.11).

Dispersal of immigrants to the pumice plain was rapid and impressive. The researchers estimate that at least 1500 different arthropod species dispersed into the pumice plain during the 2 years after the explosion. This dwarfs the number of plant species that colonized during the same time period. Though most of these arthropods died soon after arrival, presumably from lack of food or water, approximately 20 species from four families of beetles and two families of true bugs were able to establish

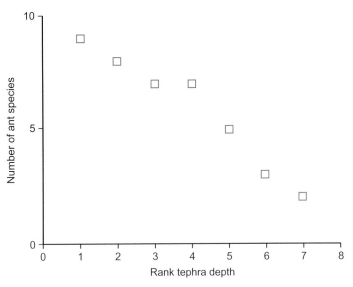

Figure 21.11 Number of ant species found in relation to tephra depth in the Mount St. Helens blowdown zone. Tephra depth ranged from 12 to 52 cm (rank 1 = 12 cm).

breeding populations soon after the eruption. The most abundant species were from the genus *Bembidion*, which comprised ground beetles that consumed primarily organic fallout blown in by the wind, including dead insects that were unable to survive the barren conditions. These beetles were very successful, producing large numbers of offspring, and increasing by a factor of ten between 1983 and 1985.

Edwards and Sugg acknowledge that plant colonization is a critical process following intense disturbance. However, they argue that some species of arthropods are the true pioneer species following intense disturbance, surviving on the organic fallout that rains down upon the desolate landscape, and adding nutrients to the soil that help facilitate colonization by early succession plants.

Of course, aquatic habitats also suffered from the eruption of Mount St. Helens. Spirit Lake was the site of Mount St. Helens Lodge, made famous by the refusal of its owner, Harry Truman, to vacate it despite pleas from various officials. Truman, and his 16 cats, were buried beneath approximately 50 m of volcanic debris. Life in the lake was also obliterated.

Chemical and biotic changes at Spirit Lake

The debris avalanche blocked the lake's only outlet, which formed the North Fork of the Toutle River. Consequently, one effect of the eruption was raising the level of the lake. Far more profound were chemical changes to the lake; a few are summarized in Table 21.3. The obliteration of all life in the lake from high temperatures and chemical pollution set the stage for a remarkable series of successional stages (Dahm *et al*. 2005).

Bacteria were the first colonizers of this incredibly concentrated nutrient broth, reaching levels greater than those

Table 21.3 Water chemistry at Spirit Lake shortly before and shortly after the eruption. After Dahm *et al.* (2005).

	pH	Carbon (mg/l dissolved)	Ca (mg/l)	Al (mg/l)	PO_4 (mg/l)	NO_3 (mg/l)	NH_3 (mg/l)	SO_4 (mg/l)
Before	7.35	0.83	2.2	9.98	2.85	3.72	1.19	0.80
After	6.21	39.9	66.9	301.0	707.0	9.0	16.0	124.9

measured in any other lake in history, including the world's most polluted lakes. The lake was cleansed of all photosynthetic organisms, including phytoplankton, and most energy transformations were either heterotrophic or chemosynthetic. Chemosynthetic bacteria oxidized the vast supply of reduced metals and gases in the lake. This proliferation of bacterial activity – and the absence of photosynthetic organisms – made the lake anoxic, except at the surface, where a small amount of oxygen dissolved in from the air above. Also note from Table 21.3 that nitrogen levels did not rise as much as carbon and other nutrients. Consequently, nitrogen was a limiting resource for these bacteria, and this selected for very high levels of nitrogen-fixing bacteria in the lake.

Within a year, conditions started changing dramatically. Heavy winter snow and rain diluted the nutrient broth, and bacterial populations, though still very high, diminished considerably from maximum levels. The lake became reoxygenated from the autumnal lake turnover and vertical mixing of the lake during the winter, and from the diminished microbial population. This oxygen revolution was short-lived; increasing summer temperatures and increasing microbial activity transitioned the lake back into anoxic conditions during the summer. Ultimately, however, the oxygen levels increased in 1982, 2 years after the eruption, setting the stage for the return of aerobic organisms.

Phytoplankton returned very slowly in 1981 and 1982, but began increasing dramatically in 1983 when the water started clearing. Accordingly, primary production shifted from exclusively chemosynthetic in 1980 to mostly photosynthetic by 1986 (Figure 21.12). Phytoplankton diversity increased substantially between 1983 and 1986, with over 135 species identified. However, the phytoplankton species composition was more characteristic of what might be expected in a eutrophic lake, which is not surprising, given that nutrient levels in Spirit Lake were still elevated from the eruption. Zooplankton were first observed in 1982, and gradually increased in abundance and species richness, so that by 1986 the zooplankton community resembled what might be expected of a subalpine lake in the western USA Cascade Mountains.

Fish were much slower to return. Before the eruption, Spirit Lake was home to four species of trout courtesy of repeated stocking events by the Washington Department of Fish and Wildlife. In addition, three species of anadromous fish visited

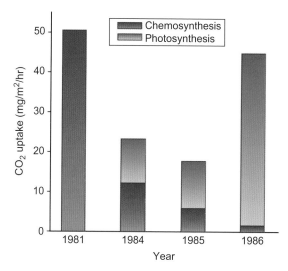

Figure 21.12 Production by chemosynthesis and photosynthesis in Spirit Lake in years following the eruption of Mount St. Helens.

the lake from the ocean via the Touttle River. All fish were killed by the eruption. The first rainbow trout was captured in a gillnet in 1993. How it got there is a complete mystery, though several people called in to claim responsibility for illegal stocking. Between 2000 and 2002, officials captured 151 rainbow trout with an unusual age distribution of age 1+ = 3.4%, age 2+ = 13.8%, and age 3+ = 82.8%. These rainbow trout are huge by most standards, weighing, on average, more than 2 kg in 2002. Peter Bisson and his colleagues (2005) expect that trout will remain abundant for a few years, but will begin to decline in size and abundance as nutrient levels decrease back to pre-eruption conditions.

 Thinking ecologically 21.3

Compare succession in Spirit Lake with succession of terrestrial organisms on the pumice plain in relation to similarities and differences in geochemical factors. How might differences in geochemical factors influence the process of succession?

We don't really know why rainbow trout are thriving in Spirit Lake. Their primary food, aquatic insects, are abundant, presumably because of high nutrient levels. There are also few food competitors; at present, there are no other species of salmonids in a lake that formerly was host to seven species. Even though the population and body-size response is impressive, fish species richness is still very low, more than 30 years after the eruption. One problem for fish is that recolonization of the lake is blocked by the debris dam at the Toutle River. Let's look at other ecosystems to see how quickly animals tend to recolonize following disturbance.

Animal recovery rates following disturbances in other ecosystems

We can apply our understanding of plant succession to address the question of how quickly animals will recolonize an area following disturbance. One hypothesis is that in analogy to the facilitation model of plant succession, early arriving animals may modify the habitat in such a way that it becomes less suitable for themselves but more suitable for later successional species. If these facilitators are already present in the community after the disturbance, then recovery can be rapid. A second hypothesis is that animals with effective dispersal capabilities will recolonize relatively quickly following disturbance. Lastly, in analogy to the findings of Turner and her colleagues (1999) at Yellowstone, we might expect that animal communities will recolonize more quickly following mild disturbance than intense disturbance.

Recovery of an isolated coral reef system

The Scott system of reefs is positioned on the edge of Western Australia's continental shelf. It is more than 250 km from other reefs, and more than 1000 km from a major human urban center. Because it is so isolated, researchers had predicted that coral would recover relatively slowly from a major disturbance, as there would be no recruits from other reefs. In 1998, increases in SSTs caused a mass bleaching of this and other reef systems worldwide (see Chapter 15). James Gilmour and his colleagues (2013) followed the response of the coral reef ecosystem for 15 years to evaluate the rate and process of recovery.

The effects of this mass coral bleaching were quite severe, killing most of the coral, and reducing percent coral cover from

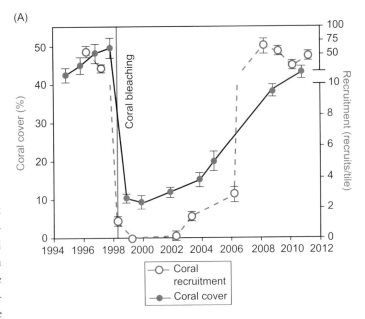

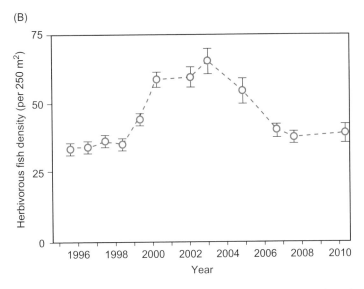

Figure 21.13 A. Percentage (±SE) of coral cover and mean number of recruits (±SE) per tile between 1995 and 2011 at Scott Reef. Note the break in the right y-axis. B. Mean density (±SE) of herbivorous fish per 250 m² between 1995 and 2011 at Scott Reef.

about 50% to 10% (Figure 21.13A). Coral diversity was seriously affected, with the number of coral genera declining by 50%. There are two possible avenues for coral ecosystem recovery. One avenue is for propagules from other reefs to disperse into the ecosystem, but we have already indicated that this was highly unlikely in this system, due to its isolation. The second avenue is to produce a large number of broods locally.

Coral *recruitment* is the process in which tiny planulae (larvae) attach to a substrate and establish themselves in the reef. To measure recruitment, the researchers set down a series of ceramic tiles and counted the number of planulae that attached to the tiles. After the bleaching, there was almost no coral recruitment until 2003, when recruitment began to climb. By 2008, recruitment was back to pre-1998 levels and by 2011, coral cover and generic diversity had also recovered (Figure 21.13A)

The researchers were surprised by this relatively quick recovery. They expected that once the corals were gone, that either macroalgae or sponges, both of which effectively compete with coral for substrate, would dominate the system. Instead, the system was colonized by smaller algae, which were kept in check by herbivorous fish, which doubled in abundance after the bleaching event (Figure 21.13B). These abundant herbivorous fish probably prevented the macroalgae from becoming dominant. Thus substrate was available for the coral, and almost 90% of the recruits survived, in comparison with less than 50% survival in other comparable coral systems. High coral survival and rapid growth resulted in rapid maturation, high reproductive rates of the newly established coral, and relatively rapid recovery of ecosystem structure and functioning. Once coral abundance recovered, herbivorous fish density returned to pre-bleaching levels.

In this coral ecosystem, following disturbance, abundant small algae were eaten by herbivorous fish. By limiting algal expansion, the herbivorous fish facilitated the return of coral, allowing the coral communities to recover rapidly. Once the coral returned, the habitat could no longer support as many herbivorous fish, which declined to their predisturbance densities.

Let's turn to some terrestrial ecosystems to test the hypothesis that animals with effective dispersal capabilities will recolonize relatively quickly following disturbance.

Ant colonization following disturbances in tropical forests

As old-growth forests become increasingly rare, conservation biologists are concerned that secondary forests may not provide suitable habitat for many forest-adapted species. Jochen Bihn and his colleagues (2008) wanted to know how quickly animal species richness recovers during the process of forest succession. They chose ants as their taxonomic group, reasoning that ants are a keystone group in tropical forests and have strong effects on forest ecosystems. Ants are also very diverse and are easily captured in large numbers.

The researchers distinguished between two functional groups of ants. *Epigeic* ants live in the litter layer, while *hypogeic* ants live in nests buried in the soil. Bihn and his colleagues proposed that epigeic ants, because they tend to travel longer distances, and move their nests quite frequently, might recover more quickly following disturbance than hypogeic ants, which tend to travel shorter distances, and only rarely shift nest locations. In addition, many hypogeic ants are often dietary specialists, so they might take more time to recolonize an area that was severely disturbed, simply because their dietary needs are not met in a secondary forest.

The researchers restricted their study to habitats that had been converted to pastures and subsequently abandoned. They

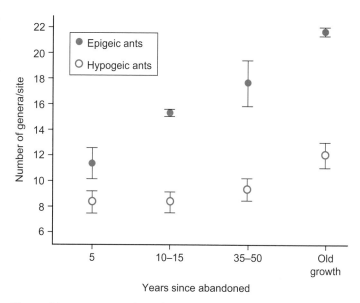

Figure 21.14 Mean number of ant genera in Brazilian forest sites in relation to successional stage.

collected litter samples (for epigeic ants) and soil samples (for hypogeic ants) at Rio Cachoeira Nature Reserve in southern Brazil at sites representing four successional chronosequences. Very young forests were pastures that were abandoned for about 5 years, young forests were abandoned for 10–15 years, old secondary forests were abandoned for 35–50 years, and the final stage was old-growth forests. The researchers' dependent variable was the number of different ant genera at each forest successional stage.

Even forests that were abandoned for 35–50 years had not recovered to richness levels of old-growth forests (Figure 21.14). This was particularly true for hypogeic ants, in which genus richness of old secondary forests was almost the same as found in very young secondary forests. Even for epigeic ants, genus richness was considerably lower in old secondary forests in comparison to old-growth forests, though the recovery was considerably more advanced.

The finding that epigeic species had recovered more than hypogeic species supports the hypothesis that effective dispersers will recolonize more rapidly following disturbance. However, it does not address the question of how disturbance intensity influences secondary succession, as all of these sites were very disturbed. Bihn and his colleagues make the important point that conservation biologists and planners must recognize that even well-established secondary forests do not support the full complement of species richness found in undisturbed primary forest.

Some other studies describe a somewhat different picture of animal recovery following disturbance.

Species richness and species composition

One approach to evaluating the speed of recovery of animal communities involves consolidating a series of studies into a meta-analysis to see if any patterns emerge. Robert Dunn (2004) compiled 33 papers that represented research on 22 tropical rainforest sites in the Americas and West Africa. To qualify, sites had to be clear-cut, and subsequently abandoned for known periods of time. Dunn had two questions: First, how long did it take for animal species richness to return to levels present in adjacent undisturbed forests? Second, how long did it take for animal species composition to resemble that in nearby undisturbed forests?

Most of the recovery studies were on ants and birds, but several other taxonomic groups were included in the compilation. In contrast to the ant study we just described (Bihn *et al.* 2008), Dunn's meta-analysis shows species richness only taking, on average, about 20–40 years to reach the levels present in the nearby undisturbed areas (Figure 21.15A). There were no differences in his findings when he carried out separate analyses on ant species and on bird species. Again, both achieved mature forest species richness in 20–40 years.

However, Dunn did discover profound differences in species composition. For the analysis of species composition he calculated the proportion of identical species shared by the disturbed and undisturbed sites. He then compared that proportion to the proportion of identical species shared by two nearby undisturbed forest plots of the same size, to calculate the Morisita–Horn index of similarity (a value of 0 indicates no overlap in species composition and a value of 1 means that species composition is identical in both sites). He found that species composition in the disturbed sites was very different from species composition in undisturbed sites, and that, by extrapolation, species composition in disturbed forest would not become similar to that of undisturbed forest for several hundred years (Figure 21.15B). Even though his findings on species richness differed from those of Bihn and his colleagues, Dunn too makes the argument that we must preserve old-growth forests, because species composition takes so long to recover from disturbance.

Why did Dunn's comparative study, and Gilmour and colleagues' coral study show quick recovery of species or generic richness, while Bihn and colleagues' ant research showed much slower recovery? There are two important factors to consider when comparing these two studies. First, the intensity of disturbance varies in each study. For example, Dunn included many sites where forests were cut down and immediately allowed to regenerate. In these cases, it is likely that recovery would be much faster than in Bihn and colleagues' ant study, where forests were converted to pasture for an extended period of time, and thus the disturbance was much more severe. Second, each ecosystem is unique, and recovery rates will vary depending on biotic and abiotic factors that directly and indirectly influence recovery for that particular ecosystem.

To summarize thus far, animal succession following disturbance can be very rapid, depending on the biotic and abiotic factors that can be unique to a particular ecosystem, and the

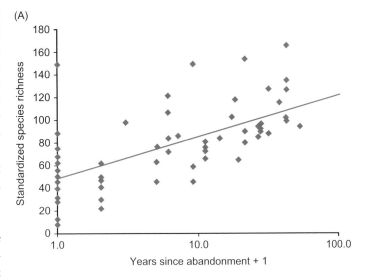

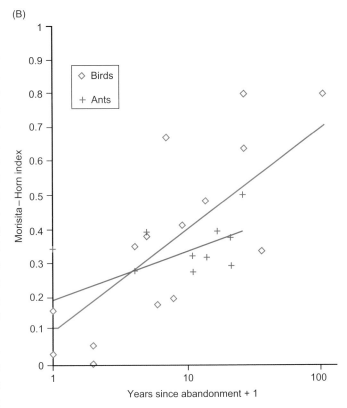

Figure 21.15 A. Standardized animal species richness at rainforest sites in relation to year since abandonment for 39 taxa. Standardized species richness = (species richness of disturbed area/species richness of nearby undisturbed area) × 100. A value of 100 indicates a return to predisturbance species richness. Note that 1 is added to the x-axis, as the value "0" is not defined on a log scale. B. Morisita–Horn index in relation to years since abandonment.

dispersal abilities of the species considered. But species composition often takes a very long time to approach predisturbance conditions, and may never actually get there. Finally conservation ecologists need to consider how intensity of disturbance influences animal recolonization patterns.

Interactions between animals and plants are a two-way process, with plants influencing animal recovery, and animals influencing plant recovery. Let's see how animals can influence plant recovery following disturbance.

How animals influence plant secondary succession

We are accustomed to thinking about animals having a positive or negative effect on plant diversity and abundance in a variety of ecosystems. For example, we discussed in Chapter 15 how chipmunks are important agents of seed dispersal for Jeffrey pine trees, in effect planting seeds at depths that increase seedling emergence. In contrast, we discussed in Chapter 14 how deer grazing may lead to ginseng extinction in West Virginia, USA. But do animals influence the process of plant recovery following disturbance?

Soil invertebrates

When disturbed, for example by fire, grasslands may go through a process of succession similar to that we described at length for forested ecosystems. Gerlinde De Deyn and her colleagues (2003) wanted to know whether soil invertebrates sped up the rate of plant secondary succession. They sterilized soil from experimental grasslands, which they seeded with a diverse assortment of plants characteristic of early, middle, and late stages of grassland succession. They then added native soil fauna, consisting of nematodes, beetle larvae, and mites, to some of the plots, while reserving other plots as unsupplemented controls. They then compared the biomass of early-, mid-, and late-succession plants in the supplemented versus unsupplemented groups to see if soil invertebrates preferentially enhanced plant growth in the late-successional stages.

If soil invertebrates increase the rate of secondary succession, the researchers predicted that relative shoot biomass would increase in late-succession plants in the supplemented plots in comparison to the unsupplemented plots. After 4 months of growth, a trend began to emerge, and after 6 months, there was a clear enhancement in the biomass of late-succession plants and a clear inhibition of early- and mid-succession plants (Figure 21.16). In the unsupplemented plots, the communities actually became dominated by mid-succession plants and never achieved the species composition of the late-successional community. Soil organisms facilitated plant succession by eating the roots of the early- and mid-successional species, allowing the later successional species to achieve dominance more quickly.

But animals can also slow down the rate of plant secondary succession.

Seed predation by mice after fire

If you were a mighty Ponderosa pine or Douglas fir tree, you might be quaking in your roots when approached by a tiny deer mouse, because deer mice are voracious pine and fir seed predators (Figure 21.17A). Rafal Zwolak and his colleagues (2010) wanted to know whether deer mouse seed predation influenced

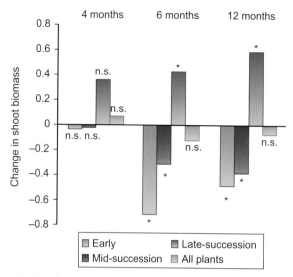

Figure 21.16 Changes in shoot biomass in early-, mid-, and late-successional stages in a grassland community with soil fauna added in comparison to a community without soil fauna. The y-axis is $(B_f - B_{nf})/B_{nf}$, where B_f and B_{nf} = mean shoot biomass with and without soil fauna. Thus a value greater than 0 indicates enhanced growth and a value less than 0 indicates suppressed growth in the presence of soil fauna.

the rate of secondary succession after fire in the Rocky Mountains in the western United States.

Their preliminary research indicated that deer mice were almost twice as abundant in burned regions in comparison to unburned regions. They also knew that seed removal at night, the time when deer mice are active, was also higher in burned versus unburned areas. Deer mouse feces was more common in association with Ponderosa pine and Douglas fir seeds that were placed in Petri plates in burned versus unburned areas. All of these findings indicate that deer mice are important seed predators of both tree species.

To investigate the ecosystem-level importance of the deer mouse effect, the researchers sowed seeds in soil protected by wire mesh cages that excluded deer mice. The control cages had a hole cut in each side to allow deer mice free access. Emergence success was the proportion of seedlings present the following June, while survival was the proportion of emerged seedlings that were alive the following September.

The results were quite dramatic. Burned areas had much higher emergence success, but only when deer mice were excluded from eating the seeds. When rodents had access to seeds, they completely devastated almost the entire population of both tree species (Figure 21.17B). Lastly, once trees did emerge, survival was much higher in burned areas than unburned areas (Figure 21.17C). The researchers concluded that wildfires do provide highly favorable conditions for seedling recruitment, but that seed predation by deer mice nullifies that effect, slowing down the process of secondary succession following fires.

Based on our discussion thus far, patterns of succession vary in relation to the intensity of disturbance and in relation to the

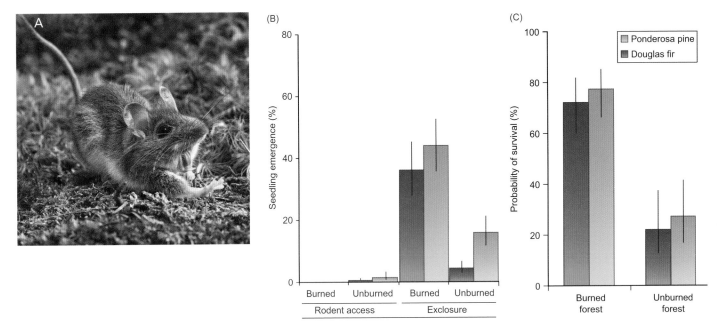

Figure 21.17 A. Ferocious deer mouse. B. Seedling emergence in cages with rodent access vs. cages from which rodents were exclosed in burned and unburned areas in the Rocky Mountains. C. Probability of survival of seedlings that emerged in Figure 21.17B. Survival data are from exclosure cages only, because no seedlings emerged in burned areas with rodent access.

particular ecosystem. But as research continues it is becoming clearer that the idea of an endpoint or climax to succession is often not a useful perspective. Let's conclude our discussion of succession by highlighting this uncertainty.

21.5 CAN AN ECOSYSTEM SHIFT ABRUPTLY FROM ONE STABLE STATE TO ANOTHER?

After reading this chapter, I expect that you have an easy time appreciating that ecosystems can change in species composition or even in the identity of dominant or keystone species in response to disturbance. Related to our discussion of disturbance and climax, does a disturbed ecosystem always return to its predisturbance state once environmental conditions are restored?

Regime shifts

As Marten Scheffer and Stephen Carpenter (2003) point out, there are many intuitive examples of small environmental changes causing large shifts to a system. A small overloaded boat may tip and capsize if only one person shifts their position, or an earthquake may result from a small increase of tension along a crustal plate. In ecology, there are several examples of similar phenomena; for example, in Chapter 20 we discussed how the Black Sea developed very large dead zones over a very short time period in response to an increase in nutrients. This abrupt change of state is an example of a regime shift – a relatively rapid change in ecosystem structure and functioning.

Scheffer and Carpenter describe rapid changes that occurred in Caribbean coral reefs, which converted them from coral

dominated to algae dominated. Increased nutrient loading favored algal growth, but this was kept in check by herbivorous fish. Over time, fishing reduced herbivore abundance, but sea urchins increased in abundance and kept algae at bay. However, when most urchins were killed by pathogens, algae were released from herbivore control and overgrew the reefs.

It is difficult to isolate factors leading to regime shifts in natural ecosystems, because some obvious experiments, such as intentionally killing off herbivores, are unethical and/or illegal. In addition, correlation studies are suspect because of small sample size and uncontrolled variables. But some experimental manipulations can and have been done to help us understand regime shifts in coral reefs. As one example, Terence Hughes and his colleagues (2007) set up an experiment on Australia's Great Barrier Reef to study the resilience of coral reef ecosystems, which was of particular concern because of recent incidents of coral bleaching. Recall from Chapter 17 that resilience is the speed with which a community returns to its former structure and functioning following a shift caused by disturbance. In some cases, coral bleaching was followed by the establishment of macroalgae-dominated ecosystems. The researchers knew that bleaching was primarily caused by ocean warming, and they hypothesized that bleaching was reducing ecosystem resilience, and that subsequent overfishing was leading to loss of herbivores and the release of macroalgae from herbivore control in the weakened ecosystem.

To test this hypothesis, the researchers built 5 × 5 m cages that excluded all large- and medium-sized fish. As controls, they set up equal numbers of similar cages with open sides, and also 5 × 5 m plots with no exclusion cages (Figure 21.18A). They did this experiment in 1998 immediately following a mass bleaching event. They then studied ecosystem response to fish exclusion,

(A)

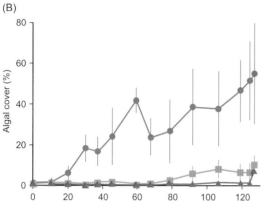

(B)

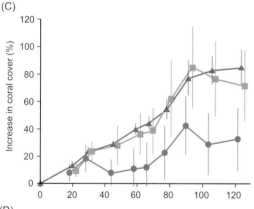

(C)

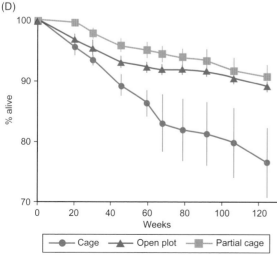

(D)

Figure 21.18 Great Barrier Reef coral experiments. A. Cages set up by researchers. B. Algae cover (%). C. Increase in coral cover (%). D. Survival of original coral (%), in relation to number of weeks since the beginning of the experiment.

predicting that the exclusion of large grazing herbivores would have two important effects. First, herbivore exclusion should lead to increases in macroalgae, and possibly to an experimentally induced regime shift. Second, the bleached corals would recover more quickly, in other words show higher resilience, in the presence of herbivorous fish.

Both predictions were supported. Algae cover over 28 months increased sharply in the completely caged plots in comparison to both controls (Figure 21.18B). In addition, the unprotected plots showed much higher resilience following the bleaching event; with coral cover increasing much more rapidly in comparison to the caged treatment (Figure 21.18C). Lastly, the survival rate of the original corals was substantially higher in the two controls in comparison to the completely caged treatment (Figure 21.18D). The researchers conclude that regime shifts can and are occurring in coral reefs as a result of fishing pressure and reduced resilience to bleaching events.

This study raises the question of whether regime shifts are temporary or long-term. In the case of the coral reef experiment, when the researchers removed the cages, herbivores came in and had a lovely meal of succulent macroalgae. But under natural conditions, does disturbance lead to stable long-term regime change? The answer is that sometimes it does, and sometimes it doesn't, but ecologists are only beginning to understand how novel and stable regimes might develop in natural systems.

Disturbance and alternative stable states

An ecosystem can exist in alternative stable states if it can have more than one set of distinct biotic and abiotic conditions that is relatively stable over time. Each discrete state is separated by some type of ecological or environmental threshold, so if the state is perturbed (for example by a disturbance) it should revert back to its previous state when the threshold is not crossed, but should switch to a different state if the threshold is crossed. Let's explore the theory of alternative stable states very briefly, and then consider one study that investigates this process in a natural ecosystem.

Alternative stable states: theory
Alternative stable state theory proposes that a change in environmental conditions can cause an abrupt transition in the state of an ecosystem. Once this change occurs, the system will then tend to remain in its new state. This has important implications for the process of succession, because the implication is that once a disturbance occurs, there may be multiple pathways for succession, and thus it may be naïve to believe that a final outcome (climax) can be predicted.

We will first discuss ecosystems with an ecological state that varies linearly with environmental conditions (Figure 21.19A). Environmental conditions (biotic and abiotic) are on the x-axis and ecosystem state is on the y-axis. The thick line through the graph is ecosystem equilibrium, so that if environmental conditions change, the equilibrium state will also change. Just like the

(A)

(B)

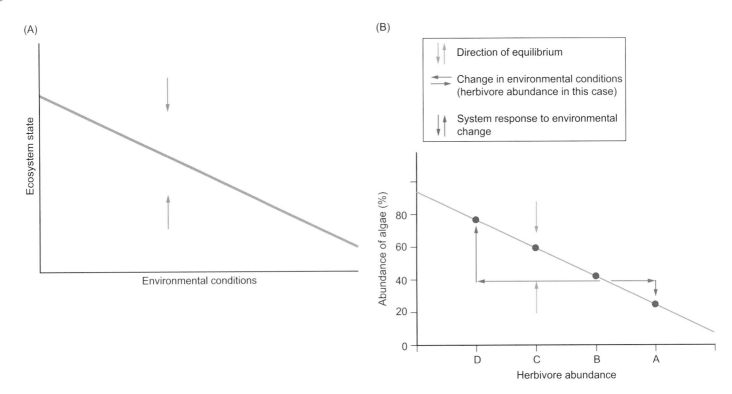

Figure 21.19 A. Environmental conditions influence ecosystem state in a continuous linear fashion. B. Coral reef ecosystem in which herbivore abundance has linear effect on percent macroalgae.

Lotka–Volterra competition models that we discussed in Chapter 13, we can draw arrows that demonstrate that any deviation in ecosystem state from equilibrium (for example changes caused by disturbance) will tend to be forced back to the equilibrium state.

Let's use the coral reef example to illustrate the predictions of this graph, with the x-axis being herbivore abundance and the y-axis being the amount of algae present (Figure 21.19B). We have four points representing different abundances of herbivores. If we start out with herbivore abundance at point B, our equilibrium value of algae abundance will be 40%; if we increase herbivore levels to abundance A, our percent algae drops to 20%, and if we decrease to point D, algae abundance climbs to 80% (the red arrow shows the disturbance, and the green arrow shows the response)

Figure 21.20 shows a very different pattern. There are two equilibrium curves that are joined by a dashed line. As in the first graph, the blue arrows are forcing the ecosystem back to equilibrium – the dashed line is the boundary separating the upward and downward arrows between the two equilibrium curves. At value A, there are many herbivores, and consequently low algae abundance (point 1 on the y-axis). At values B and C, herbivore abundance decreases, and algae abundance increases slightly (points 2 and 3). But moving to point D, herbivore abundance continues to decrease, and suddenly algae abundance increases dramatically, because the system is controlled by the attraction of the upper equilibrium curve (point 4). Most important, if the system moves to herbivore abundance C and even B, algae abundance remains much higher than when the system was

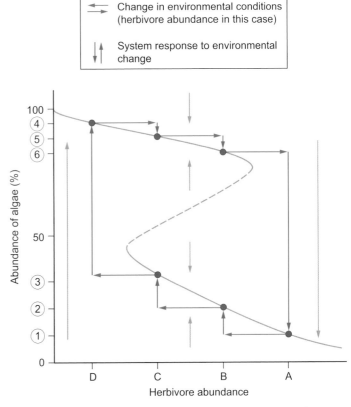

Figure 21.20 Coral reef ecosystem showing the presence of two alternative stable states with hysteresis.

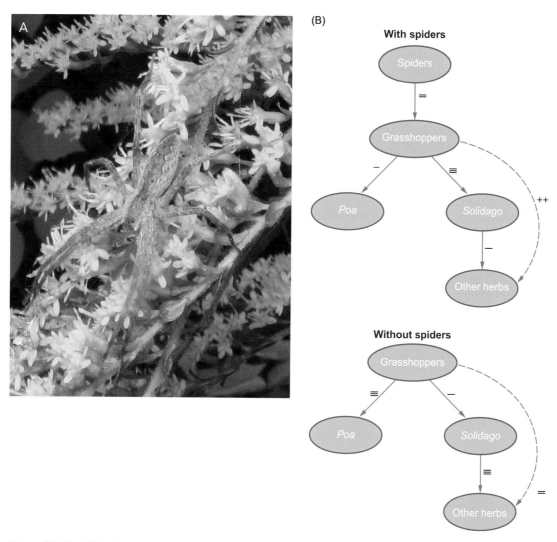

Figure 21.21 A. The hunting spider, *Pisaurina mira* on goldenrod. B. Meadow ecosystem interactions webs. Solid arrows indicate direct effects, and dashed arrows indicate indirect effects. With spiders (top). Without spiders (bottom).

controlled by the lower equilibrium curve (points 5 and 6). This describes *hysteresis*, where multiple (in this case two) ecosystem states are possible under identical ecosystem conditions. This point of sudden transition from one state to another is known as an ecological threshold (E on the graph).

Do alternative stable states occur in the real world, and if so, can we predict when they are most likely to arise? I will describe one example of alternative stable states under natural conditions, and reserve the second question to more advanced texts.

Alternative stable states: a field study

Oswald Schmitz (2004) studied a small meadow ecosystem composed of hunting spiders, *Pisaurina mira*, their grasshopper prey, *Melanoplus femurrubrum*, the grass *Poa pratensis*, and a competitively dominant goldenrod, *Solidago rugosa*, which if present in high numbers, competitively suppressed the abundance of other herbs (Figure 21.21A). Under natural conditions, in the presence of spiders, grasshoppers hide on and eat large quantities of *Solidago*, decreasing *Solidago*'s abundance and reducing its

suppression of other herbaceous plants. The result is that plant species richness is relatively high when spiders are present, and the overall productivity of the system is relatively low (presumably the other herbaceous species don't grow as rapidly as the *Solidago* would have). But when Schmitz removed spiders from the ecosystem, by setting up spider exclosures, grasshoppers switched to eating primarily their preferred grass, which allowed *Solidago* abundance to increase, and increased its suppression of other herbaceous plants (Figure 21.21B). This resulted in reduced species richness, and increased productivity as a result of high *Solidago* growth rates.

To see whether the differences in richness and production showed hysteresis, Schmitz removed the exclosures after they had been set up for either 1, 2, or 3 years. This procedure restored spiders to the ecosystem. After 1 year of spider exclusion, the restored ecosystem quickly returned to its previous high-diversity/low-production state. But after 2 or more years of spider exclusion, the restored ecosystem stayed in its new low-diversity/high-production state for several more years. Schmitz concluded that this meadow has two alternative stable states that are determined by the presence or absence of spiders, and by their effect on herbivory by grasshoppers.

From the perspective of understanding succession, the existence of alternative stable states suggests that disturbance can lead to multiple outcomes, and that return to climax communities might not be a reasonable expectation in many circumstances. Let's return to Mount St. Helens, and to Yellowstone to see how some of the ecosystems have responded a few decades after disturbance.

REVISIT: Mount St. Helens and Yellowstone updates

The Mount St. Helens eruption changed Spirit Lake in many ways. The lake is substantially shallower, but covers a much larger area, and receives runoff from a relatively treeless watershed, which contrasts dramatically with the forested slopes that surrounded it before the eruption. These differences have changed its chemical and thermal profiles. Despite these and several other differences, the lake has returned to its previous oligotrophic condition, though some of the nutrient levels are still somewhat elevated. The most recent survey (Charlie Crisafulli, personal communication) shows a much more normal size and age distribution of rainbow trout in the lake, with juveniles now represented, and reproduction occurring in small feeder streams and springs. In addition, the mean weight of sampled trout has declined to 1.10 kg in 2010, down from 1.96 kg in 2000. The pH has returned to its pre-eruption range of 7.3 to 7.7. However, because of the debris dam, the lake still lacks the other trout and salmon that were former residents in the lake.

In the aftermath of the Yellowstone fires, both scientists and citizens wanted to know whether the ecosystem needed human intervention to restore it to its previous status. William Romme and Monica Turner (2004) identified six questions that need to be addressed before determining whether active restoration is needed (Table 21.4).

Their follow-up investigations indicated, that for the most part, the answer to the first three questions is "no" and to questions 4 and 5 is "yes." For question 6, the major unexpected result was that the quaking aspens (*Populus tremuloides*) at Yellowstone were stimulated by the fires into a rare episode of sexual reproduction, resulting in the production of large numbers of aspen offspring. Presumably sexual reproduction increased genetic variation

Table 21.4 Criteria for evaluating the need for active restoration. From Romme and Turner (2004).

1. Is the current disturbance regime within the historical or natural range of variability?
2. Are current stand structure and landscape structure within the historical or natural range of variability?
3. Are any species or communities extinct or threatened with extinction because of changes in the disturbance regime?
4. Have recent disturbances been accompanied by normal return of community structure and composition?
5. Have recent disturbances been accompanied by normal return of ecosystem functioning, for example, energy flow and material cycling?
6. Have recent disturbances been associated with any novel or unexpected effects that are regarded as undesirable?

among the aspen populations, and enhanced the viability of this important species. Given these results, the researchers concluded that there is no need for any active restoration process, and that Yellowstone is an excellent example of a naturally functioning and resilient ecosystem.

SUMMARY

Disturbances come in many forms, and have variable impacts on ecosystem structure and functioning. Sometimes ecosystems experience an extreme disturbance, such as a glacier or volcanic eruption that kills virtually all life. Following an extreme disturbance, the ecosystem may go through the process of primary succession, which is characterized by a predictable series of developmental stages that culminate in a climax community – a stable biotic community that represents the final stage of succession. As evidenced by research on communities at Glacier Bay and Mount St. Helens, there can be different patterns of succession that are caused by differences in how abiotic and biotic factors influence the recovery process.

In many cases a disturbance will only kill some of the organisms within the ecosystem. In these cases, the ecosystem may go through a process of secondary succession, in which many factors, including the intensity of the disturbance, the life history traits of colonizing species, and the presence of biological legacies influence the recovery process. Ecologists have described three conceptual models of succession: facilitation, tolerance, and inhibition, that apply under different conditions in different ecosystems.

Animals play an important role in the recovery process. Many animal species are excellent dispersers and can quickly return to a disturbed ecosystem. Even if they are unable to establish a breeding population, animals can import seeds or nutrients into a disturbed habitat. Alternatively animals can inhibit the recovery process by eating seeds or young plants before they get established. In some cases, disturbance can causes ecosystems to experience a regime shift – a very rapid change of ecosystem structure and functioning from one stable state to a second stable state.

FURTHER READING

Dale, V. H., Swanson, F. J., and Crisafulli, C. M. 2005a. *Ecological Responses to the 1980 Eruption of Mount St. Helens*. New York: Springer Science.

This book is the source of a vast amount of information on how life recovered following the St. Helens explosion. The chapters are well-written, and the information is great, but it is time for an update.

Fletcher, M. S., Wood, S. W., and Haberle, S. G. 2014. A fire-driven shift from forest to non-forest: evidence for alternative stable states? *Ecology* 95: 2504–2513.

Historical study based on analyses of sediment cores in Tasmania, Australia. The researchers provide evidence that a catastrophic fire caused a regime shift from a wet temperate forest to a nonforested bog, which represent two alternative stable states in this ecosystem.

Freschet, G. T., Östlund, L., Kichenin, E., and Wardle, D. A. 2013. Aboveground and below-ground legacies of native Sami land use on boreal forest in northern Sweden 100 years after abandonment. *Ecology* 95: 963–977.

This is a great paper! The authors describe legacy effects at two remote former settlements that were abandoned by reindeer-herding Sami settlers in northern Sweden about 100 years ago. They explore a very large number of ecosystem variables in relation to the intensity of land use, and demonstrate a fascinating series of interacting legacy effects.

Seddon, P. J., Griffiths, C. J., Soorae, P. S., and Armstrong, D. P. 2014. Reversing defaunation: restoring species in a changing world. *Science* 345: 406–412.

Nice summary of options for restoring animal species using the full spectrum of conservation tools. Argues that restoration efforts must consider the roles that had been played by now-extinct species in stabilizing ecosystem functioning.

Turner, M. G., Romme, W. H., and Tinker, D. B. 2003. Surprises and lessons from the 1988 Yellowstone fires. *Frontiers in Ecology and the Environment* 1: 351–358.

Clear summary of what we have learned from secondary succession of Yellowstone plant communities after the fires. Fires create a spatially complex mosaic of burned and unburned areas, which had a major impact on the recovery process. Even within the burned areas, there were profound differences in disturbance intensity, which influenced the recovery process in important ways.

END-OF-CHAPTER QUESTIONS

Review questions

1. Distinguish between primary and secondary succession. Did the Mount St. Helens ecosystem go through primary or secondary succession. Explain your reasoning?
2. Outline the stages of succession at Glacier Bay as described by Cooper. Then discuss how Fastie's research changed our understanding of primary succession at Glacier Bay.
3. What factors influence the rate of recovery during the process of secondary succession. Illustrate with examples.
4. What are three conceptual models of succession in terrestrial ecosystems? Describe studies in which each model may help explain succession dynamics in a particular ecosystem.
5. Use examples to highlight how abiotic and biotic factors influence the rate of animal succession following a disturbance.

Synthesis and application questions

1. Suppose you joined the Virginia Dale research team, and were given funding for 30 years, and the task of determining the relative importance of facilitation, tolerance, and inhibition on succession at Mount St. Helens. How would you set up your observations and experiments?
2. Edwards and Sugg (2005) conclude that arthropods are the true pioneer species following primary succession. What life history traits would you expect to find in pioneer arthropods? How do these compare to life history traits of pioneer plant species?
3. The age distribution of rainbow trout in Spirit Lake was age 1+ = 3.4%, age 2+ = 13.8%, and age 3+ = 82.8%. If this were the true population age distribution, what would you predict about the viability of this population, based on the principles we discussed in Chapter 11? If these data are not correct, what factors could be responsible for the error?
4. Recall the intermediate disturbance hypothesis that we introduced in Chapter 17. Apply the prediction of the intermediate disturbance hypothesis to extrapolate the regression line in Dunn's research (Figure 21.15A) to 500 years. Explain your reasoning.
5. Both Terence Hughes and his colleagues (2007) and James Gilmour and his colleagues (2013) conducted research on the response of coral reef ecosystems to disturbances. Compare and contrast their research methods and the response of their study reefs to disturbance.
6. Return to Figure 21.20. Can you draw a graph that shows hysteresis leading to three distinct alternative stable states?
7. Human activity has converted many tropical forest ecosystems into grassland. One of the goals of restoration ecology is to undo this process, and to reconvert the grasslands back into tropical forest (see Chapter 18). How can the concept of alternative stable states be used by restoration ecologists to help guide them in this process?
8. Apply the concept of alternative stable states to eutrophication and changes in species composition in the Black Sea ecosystem that we discussed in Chapter 20.

Analyse the data 1

Table 21.5 shows the percentage of Sitka spruce trees that showed ring-width release in each site, and the mean ring width before and after release. Calculate the ring-width release factor (ring-width after/ring-width before) and state whether there is any statistically significant difference when comparing ring-width release factor in old versus young sites. Then calculate the mean percentage of Sitka spruce trees showing ring-width release at old versus young sites and test whether these means are statistically significantly different. Interpret your findings.

Table 21.5 Ring-width releases at eight study sites (3–7 are young, and 8–10 are old).

Site #	% of trees	Ring width before (mm)	Ring width after (mm)
3	100	3.8	7.9
4	84.6	1.6	5.5
5	78.3	2.3	6.0
6	91.1	2.2	5.8
7	72.2	2.1	5.1
8	22.8	1.5	3.4
9	17.2	1.4	3.4
10	22.7	1.1	3.1

Chapter 22

Geographic and landscape ecology

INTRODUCTION

In his autobiography, E. O. Wilson writes, "Without a trace of irony I can say I have been blessed with brilliant enemies. They made me suffer, but I owe them a great debt, because they redoubled my energies and drove me in new directions [. . .] James Dewey Watson, the co-discoverer of the structure of DNA, served as one such adverse hero for me. I found him the most unpleasant human being I had ever met." (Wilson 1994). Wilson and Watson began their careers as Harvard assistant professors in the same year, both while in their 20s. Wilson describes Watson as ill-mannered, and overtly contemptuous of anybody who did not believe, as Watson did, that all important questions in biology would ultimately be answered by molecular biologists, and that traditional biologists were merely stamp collectors, who collected natural history observations but did not know what to do with them.

Trapped in this very contentious departmental atmosphere, Wilson had a few choices. One was to attempt to split the department into two faculty groups that could work separately on different types of questions, with molecular biologists looking at small-scale processes and ecologists and evolutionary biologists investigating biological processes that operated on a larger scale. A second was to go back to the tropics to continue his fieldwork on a variety of tropical ants that lived on islands off the coast of Venezuela, and on mainland South America. A third option, which Wilson chose, was to fight back and create a modern vision for "mainstream biology" by teaming up with like-minded faculty members at other universities who were facing similar conflicts. This chapter begins with a discussion of how Wilson and Robert MacArthur formed a fruitful partnership that stimulated the development of the field of geographic ecology. We then apply island biogeography theory to make predictions of species richness and species turnover within ecosystems. Given that Earth's geography is ever-changing, we then consider how historical events can help explain the current distribution of a species or taxonomic group. Finally, in recognition that geographic features are spatially variable, we explore how landscape structure, including landscape scale, fragmentation, and edge effects, influences the distribution and abundance of species.

KEY QUESTIONS

22.1. What are the major components of island biogeography theory?

22.2. How do historical events help explain the current distribution of a species or taxonomic group?

22.3. How does landscape structure influence the distribution and abundance of species?

CASE STUDY: Wilson meets MacArthur

To create his modern vision for mainstream biology, E. O. Wilson needed to address his own deficiencies in mathematics, as he believed the future principles of evolutionary ecology would be written in equations and described in mathematical models. While working on all three approaches, he met up with Robert MacArthur, a young ecologist from the University of Pennsylvania. MacArthur and Wilson shared the perspective that evolutionary ecology could be invigorated by developing a foundation in population biology. Wilson had spent several years collecting ants on the Maluku and Melanesia islands of Southeast Asia. One of his findings was that there was a power function relationship between ant species richness and the area of the island, as described by the following general formula:

$$S = cA^z$$

where S = the number of species, c and z are constants that are fitted to the actual data, and A is the area of each island (Figure 22.1). Ecologists refer to this relationship as a species–area relationship. The exponent in this formula (z) describes how quickly species richness increases in relation to island area. In almost all cases the z-value is much less than 1, indicating, for example, that if one island has 10 species, a second island that is twice as large might be predicted to have 12 or 13 species, rather than 20 species. (See Dealing with data 22.1 for further discussion of the species–area relationship.) For his ant data, Wilson found that z was approximately equal to 0.30 (Wilson 1961, updated in MacArthur and Wilson 1967).

Many other researchers had previously discovered a power function relationship between island area and species richness. Comparing studies showed substantial variation in the z-values. Why is there so much variation in z-values among different studies, and why are species–area relationships oftentimes best described as a power function? When shown these data, MacArthur got very excited by the challenge of interpreting them.

Exploring the relationship between island area and species richness initiated a relationship between MacArthur and Wilson that led to numerous co-authored papers and one of the most influential books in the history of ecology, *The Theory of Island Biogeography* (MacArthur and

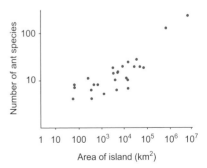

Figure 22.1 Number of ponerine and cerapachyine ant species on Maluku and Melanesia islands in relation to the area of each island.

Wilson 1967). This book encouraged researchers to use a more conceptual approach to understand biogeographic patterns, and to develop predictive models of biodiversity based on variables that could be measured and manipulated in the field. Several of the principles discussed in MacArthur and Wilson's book form important cornerstones for conservation biology and the developing field of landscape ecology. Unfortunately, MacArthur did not live long enough to fully appreciate the impact of his contributions, dying of renal cancer at the age of 42. But island biogeography theory continues to stimulate researchers interested in understanding basic questions about patterns of biological diversity and landscape structure.

22.1 WHAT ARE THE MAJOR COMPONENTS OF ISLAND BIOGEOGRAPHY THEORY?

As recounted in his book *Naturalist*, Wilson presented MacArthur with two important pieces of information based on his own research, or research conducted by other biogeographers (Wilson 1994). First, as we've already described, the relationship between species richness and island area can be modeled with a simple power function. Second, when one ant species spreads out from Asia or Australia to an adjacent island, it often eliminates another species that settled there previously. MacArthur quickly developed a conceptual and mathematical model to explain these two pieces of information.

Island species richness: equilibrium of extinction and immigration rates

MacArthur explained Wilson's species richness patterns by considering species diversity from the perspective of extinction and immigration. He reasoned that as more species establish themselves on an island, the probability that one or more species will go extinct increases. At the same time, as more species are established on an island, the immigration rate of *new* species will decline, simply because many immigrants will represent species that are already present. As the island fills up, extinction rates increase and immigration (of new species) rates decrease until the two processes are happening at the same rate. At that equilibrium, species richness should be constant, but there should be species turnover due to continuous extinctions and arrival of new immigrant species (Figure 22.2A).

MacArthur reasoned that extinction rates would mostly be influenced by island size, because mean population sizes of each species will be lower on small islands and more prone to extinction. In contrast, immigration rates would mostly be influenced by how far each island was from a source of colonists (e.g. the mainland), with distant islands having lower immigration rates. Putting these two factors together predicts that small distant islands should have the fewest species, and large islands near a source of colonists should have the most species (Figure 22.2B).

Dealing with data 22.1

It is not surprising that larger islands tend to have more species. Geographical ecologists have observed that the relationship between island size and species richness can often be fitted to a simple power function of the form $S = cA^z$, where S = the number of species, and c is a constant that depends on a variety of factors including the units of area being considered, the taxonomic group being considered, and several other factors whose importance is debated by geographical ecologists. A is the area of the island and z is the constant that helps describe the shape of the curve. For almost all species–area relationships, the value of z is much less than 1, which means that, for example, doubling the area will increase species richness by much less than a factor of two. To appreciate the power function relationship, let's consider a data set of plant species richness on the Galápagos Islands (Johnson and Raven 1973).

Table D22.1.1 Plant species richness on the Galápagos Islands (Johnson and Raven 1973).

Island	Area (km^2)	Number of species
Darwin	2.33	6
Espanola	58.27	9
Fernandina	634.49	15
Genovesa	17.35	8
Isabela	4669.32	20
Marchena	129.49	14
Pinta	59.56	16
Pinzon	17.95	14
Rabida	4.89	12
San Cristobal	551.62	16
San Salvador	572.33	20
Santa Cruz	903.82	19
Santa Fe	24.08	12
Santa Maria	170.92	16
Wolf	2.85	7

If we plot these data on a graph with linear axes, we get a graph that has some problems. All of the data are squashed onto the left side of the graph, and it is difficult to see any pattern or relationship between island area and species richness (Figure D22.1.1 (top)).

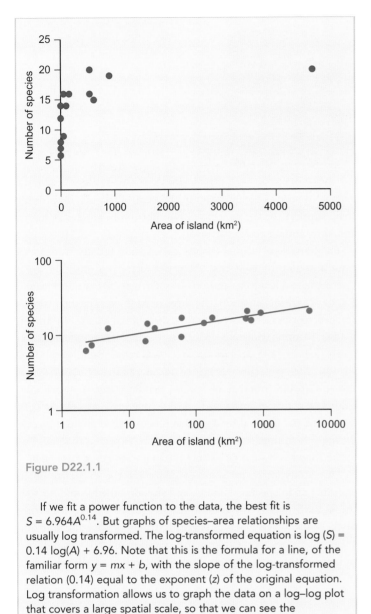

Figure D22.1.1

If we fit a power function to the data, the best fit is $S = 6.964A^{0.14}$. But graphs of species–area relationships are usually log transformed. The log-transformed equation is $\log(S) = 0.14 \log(A) + 6.96$. Note that this is the formula for a line, of the familiar form $y = mx + b$, with the slope of the log-transformed relation (0.14) equal to the exponent (z) of the original equation. Log transformation allows us to graph the data on a log–log plot that covers a large spatial scale, so that we can see the relationship between area and species richness more clearly as in the graph above (Figure D22.1.1 (bottom)).

Both MacArthur and Wilson were delighted with this model, but it needed to be tested in the natural world. Wilson had his predictions, but he needed a laboratory. Ultimately, he settled on islands formed by red mangrove trees, *Rhizophora mangle*, in the Florida Keys. The smallest islands were formed by only one mangrove, while the largest islands comprised a small mangrove forest. These islands also varied quite conveniently in their distance from sources of immigrant species. In addition to identifying a natural laboratory, Wilson also found himself a collaborator in Dan Simberloff, who was beginning his PhD research with Wilson as his advisor.

Testing the theory: defaunation without defloration

Wilson and Simberloff were determined to test the predictions of island biogeography theory by doing actual experiments in their

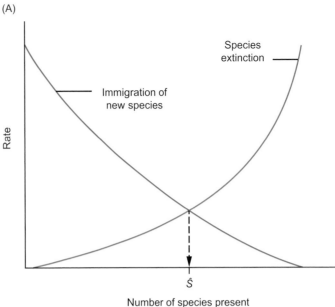

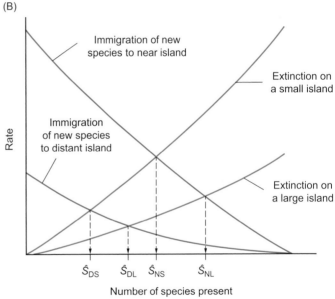

Figure 22.2 A. The basic graphical model of island biogeography. $\hat{S}$ represents the equilibrium number of species on an island. B. The model of island biogeography accounting for the effect of distance from a colonist source on immigration rates. $\hat{S}_{DS}$, $\hat{S}_{DL}$, $\hat{S}_{NS}$, $\hat{S}_{NL}$, represent the equilibrium number of species on distant small, distant large, near small, and near large islands, respectively. Note that $\hat{S}_{DL}$ may be greater than or less than $\hat{S}_{NS}$, depending on the specific immigration and extinction curves used for the graphs.

natural laboratory. One approach involved killing off all of the animals (defaunation) while minimizing damage to the mangroves. After unsuccessfully trying a mixture of insecticides, which killed most but not all of the arthropod inhabitants, the researchers, with the help of a local exterminator, were able to fumigate entire islands by constructing a tent to completely enclose each island, and spraying with methyl bromide gas (Figure 22.3). This

Figure 22.3 Erection of a fumigation tent in preparation for defaunation.

approach killed virtually all of the arthropods, and caused only minor damage to the mangroves (Wilson and Simberloff 1969).

Testing the theory: distance effects, equilibrium, and species turnover

Simberloff and Wilson (1969, 1970) elected to test the hypothesis that islands farther from sources of colonists would take longer to recover and achieve lower equilibrium species richness values than islands close to a source of colonists. Island size was controlled by choosing islands that were 11–15 m in diameter. The basic experimental protocol began with a pre-defaunation survey of all arthropods on four experimental islands, and two islands that received no treatment. This was followed by periodic surveys, at approximately 3-week intervals, of arthropods on the four experimental islands after defaunation, and a final survey on all six islands.

The two control islands showed very little change in species richness; one island increased from 20 to 23 species, while the other decreased from 30 to 28 species. Species composition changed considerably over the time of the experiment, but the rate of change could not be measured in the controls because there were only two control surveys separated by approximately a year. The small change in species richness, and the substantial change in species composition in the control islands, do provide some support for the concept of equilibrium in species richness, and substantial species turnover.

The four experimental islands varied substantially in distance from a source of colonists. E2 was nearest, E3 and ST2 were intermediate distances, and E1 was farthest. By approximately 150 days post-defaunation, the three nearest experimental islands reached species richness values near their initial levels. In contrast, it took E1 approximately 280 days to approach pre-defaunation levels (Figure 22.4). Species turnover was rather rapid for all four islands. The researchers estimated turnover rates of about 1% of the equilibrium number of species per day, but they acknowledged that this estimate had considerable error associated with it because the interval between samples averaged about 3 weeks, during which time some species would be expected to immigrate to the island and then go extinct without ever being surveyed.

This study provides some evidence for the existence of an equilibrium in species richness, and supports the hypothesis that the equilibrium is lower and reached more slowly following disturbance, when distance is greater from the pool of colonizing species. It does not explore the question of how island size influences this equilibrium. Dan Simberloff directly addressed this question a few years later.

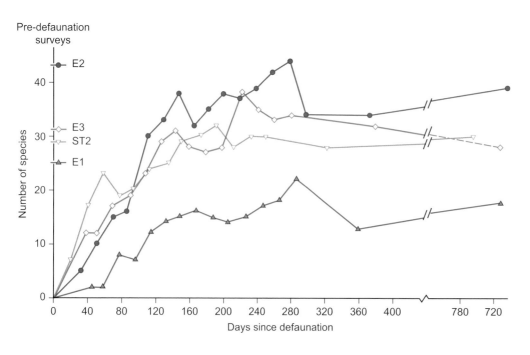

Figure 22.4 Colonization curves of experimental islands after defaunation. The distances separating each experimental island from a major source of colonists was as follows: E2 = 2 m, ST2 = 154 m, E3 = 172 m, and E1 = 533 m.

Testing the theory: size effects

Simberloff (1976) considered two hypotheses for why large islands might support more species than small islands. One hypothesis was that large islands have a greater diversity of microhabitats than small islands, providing more ecological niches for a variety of species. The second hypothesis was that larger islands support larger populations of each species, reducing the probability of extinction. To distinguish between the two hypotheses, Simberloff went back to the Florida Keys, but worked on islands that were much larger than the ones he studied with Wilson. Mangrove islands contain a very homogeneous habitat, so he reasoned that the diversity of microhabitats would be reasonably similar on islands of different sizes. To test the effect of island size on species richness, Simberloff experimentally reduced the size of eight islands by removing some mangrove trees, including the root system, on each island. For four of the islands he repeated this procedure for a second year. Most important, he only surveyed arthropods on the portion of each island that he was not planning to remove, even in his initial survey. That way, reduction in arthropod diversity would not be attributed to reduction in local microhabitat diversity, but could only be attributed to increases in extinction rate or (less likely) decreases in colonization rates of islands that were reduced in size.

Each island showed a reduction in species richness in association with area reduction. Over the same time period, the control island (IN1) showed no change between 1969 and 1970, and an increase of five species between 1970 and 1971 (Table 22.1).

Simberloff argued that the decrease in species richness on experimental islands was due primarily to an increase in extinction rates as a result of the decrease in island size. One piece of evidence supporting this contention was that the species that went extinct following reduction were those that tended to be present on only one or two islands; Simberloff described these species as being poorly adapted to mangrove islands and thus prone to extinction. In contrast, species that were present on all of the islands tended to persist following island reduction.

 Thinking ecologically 22.1

You might argue that the time period from when islands were reduced in size to when they were surveyed was not long enough to allow for species richness to decline to equilibrium. How might you use the data in Table 22.1 to address this concern?

Let's explore the relationship between area and species richness in more detail.

Species–area relationships

Biogeographers have devoted considerable attention to two important questions related to species–area relationships. First, they want to know whether the same power function can be applied across a large spatial scale. In other words, would the same formula accurately describe species richness on a huge

Table 22.1 Changes in species richness in association with reduction of island size.

	Area (m^2)			Number of species		
Island name	1969	1970	1971	1969	1970	1971
IN1 (control)	264	264	264	63	63	68
J1	1263	779		75	71	
R1	721	478		103	85	
SQ1	1082	731		88	82	
MUD2[a]	942	327	327	79	62	61
CR1	343	104	54	74	65	62
G1	519	327	169	86	77	69
MUD1	990	565	320	79	76	71
WH1	320	261	123	86	73	72

[a]Note that MUD2 was surveyed in 1971, despite being the same size as in 1970.

island like New Guinea or Madagascar, and on an island that is only a millionth as large? Second, they want to know whether a power function is an appropriate description of species–area relationships for most or all species. The most direct approach to answering these two questions is to look at a variety of studies that operate over large spatial scales and to consider a diversity of organisms.

Macroorganisms and species–area relationships

Mark Williamson and his colleagues (2001) were interested in whether the species–area relationship flattened out at very large spatial scales. Some researchers had argued for such a leveling out, because they reasoned that the number of species of a particular taxonomic group is finite. If an island were very large, it might have all of those species already present. If so, getting larger would not increase the island's species richness (Figure 22.5A).

Williamson's team responded in two ways. First, they argued that not only is the pool of species finite, so too is the area of an island. So to them, it did not make sense to consider an island so big that all of the species of a particular taxonomic group were already present, and then to imagine a second island that was

even larger. But their second approach was to present some data that considered species richness over very large spatial scales to see if there was any tendency to asymptote at large areas.

Islands in tropical and subtropical oceans have an enormous range of sizes. Because birds are so popular among bird watchers and scientists, we have relatively complete data sets on bird species richness on a large number of islands. Williamson and his colleagues compiled data from several studies, which indicated that log (number of bird species) increases with log (island area) with a slope of 0.33. Though there is considerable scatter in the data, the slope shows no evidence of leveling off, even on the largest islands (New Guinea and Madagascar). The graph shows a second trend, which is that more isolated islands (more than 300 km away) tended to have lower species richness than islands closer to a source of potential immigrants (Figure 22.5B). But the slopes of the species–area relationships are identical for near and far islands.

It is important to recognize that species–area relationships need not be restricted to islands. They can also be applied to mainland populations, by comparing surveys that measure species richness across different-sized pieces of mainland. One such study by Michael Judas (1988) compiled data from 90 papers on the earthworm (lumbricid species) across Europe (that would be one big earthworm). In this case the largest area surveyed was France, while the smallest area was 0.0001 km^2. Over this huge spatial scale the relationship was best described by a power function of $S = 7.9A^{0.09}$. Again, there is no evidence of an asymptote at a large spatial scale, but note that the z-value (the exponent of the species–area relationship) is much lower than what we've observed up to this point.

Some geological formations might not, on the surface, seem like islands, but from an ecological perspective they function as such. For example, forested mountains surrounded by scrubby desert function as islands to the forest-adapted species that live there, and lakes function as islands to their assemblage of aquatic species. On a much smaller spatial scale, water-filled tree holes can function as islands to aquatic organisms in forested areas. Not only are these tree hole islands small, but they also host some very small organisms that until recently had not been studied as extensively by biogeographers.

Microorganisms and species–area relationships

Thomas Bell and his colleagues (2005) wanted to know whether bacterial diversity increased with tree hole size in European beech trees (*Fagus sylvatica*). Tree hole size was identified by the volume of water present, and bacterial diversity was measured using a type of gel electrophoresis. As with many studies of much larger organisms, the researchers found that diversity increased as a power function of island (tree hole) size, with an exponent of 0.26 (Figure 22.6).

On a slightly larger scale, Kabir Peay and his colleagues (2007) explored the species–area relationship among a group of ectomycorrhizal fungi associated with the bishop pine, *Pinus*

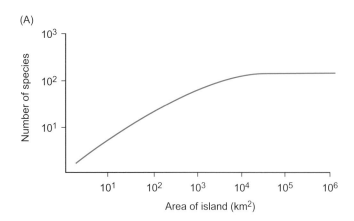

(A)

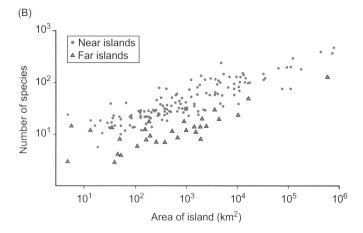

(B)

Figure 22.5 A. Hypothetical species–area curve with the slope leveling out at large island size. B. Species–area relationship for birds over a tremendous range of island sizes. Green triangles represent islands located more than 300 km from a source of colonists.

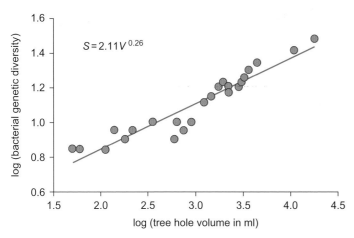

Figure 22.6 Bacterial genetic diversity in relation to tree hole volume. $R^2 = 0.91$, $P < 0.001$.

muricata, in California, USA. This one tree species dominates the ecosystem. These trees can grow as individuals, or in large clumps, so that the area encompassed by the root system varied from 3 m² for small individuals to 70 000 m² for the largest clump. The researchers specifically chose a group of even-aged trees that had sprouted after a fire, so that tree age would not be a confounding variable. They used visual inspection or PCR amplification to identify ectomycorrhizae to species.

As found with birds, lumbricid worms, and bacteria, species richness of ectomycorrhizae increased with island area. The data were described by a power function, with the exponent (z-value) equal to 0.20. Like Simberloff before them, the researchers specifically chose a relatively uniform type of habitat, so they could more confidently attribute the species–area relationship to the effects of area per se, rather than the relationship between area and habitat heterogeneity. But they were interested in distance effects as well. Specifically, the researchers were interested in whether increasing tree-island isolation would decrease species richness by reducing dispersal rates.

To control for size effects, Peay and his research team (2010) limited their study to tree islands formed by a single tree, which resulted in island size only varying by a factor of four. Distance to the mainland was the distance to any patch of bishop pine. The researchers found that there was a modest distance effect, with close islands having somewhat greater species richness than distant islands (Figure 22.7).

To summarize thus far, distance effects appear to influence species richness, as does island size over a broad range of spatial and organism size scales. Island size effects are well described by a power function, but the exponent (z-value) of that relationship varies substantially from system to system. Remember that a high z-value indicates that species richness increases rapidly with area, while a low z-value indicates a more gradual increase. In the studies we've discussed, the z-value is between 0.09 and 0.30, and in some other studies it approaches 1.0. Are there any trends in species–area relationships across a wide range of species and habitats?

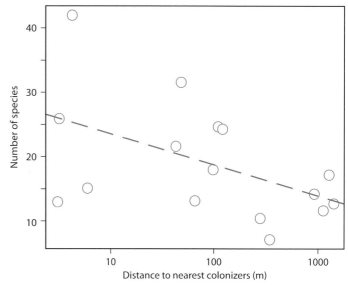

Figure 22.7 Number of ectomycorrhizal species in relation to distance from a source of colonists. $R^2 = 0.25$, $P = 0.048$.

Patterns of species–area relationships

One approach to looking for broad trends in species–area relationships is to conduct a meta-analysis of existing data. Stina Drakare and her colleagues (2006) identified 794 studies that considered species–area relationships across a variety of habitats, species and locations. The researchers discovered that forests tended to have a greater z-value than nonforested areas, and that marine ecosystems tended to have a greater z-value than lakes (Figure 22.8). They speculate that two contrasts could account for these differences. First, oceans have a much longer evolutionary history than lakes, and may host a greater number of species that

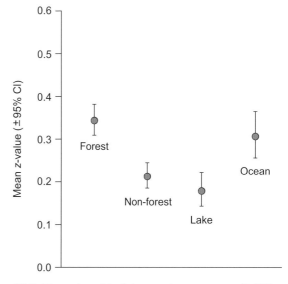

Figure 22.8 Mean slope (z) of the species–area curve ($\pm95\%$ confidence interval) in relation to type of habitat that is being surveyed for species richness.

Figure 22.9 A. The common puffball (*Lycoperdon* perlatum) disperses a thick cloud of spores (left). A crab spider (*Mecaphesa sp.*) balloons to a new home in Austin, Texas (right). B. Slope of species–area relationship (*z*) in relation to body weight. C. Slope of species–area relationship (*z*) in relation to latitude.

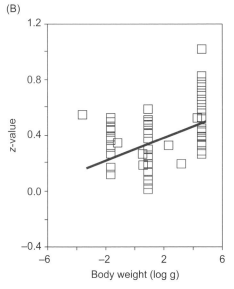 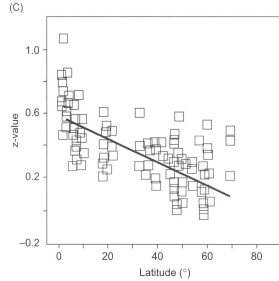

could disperse into a particular area. Second, there are important differences in vertical dimension, with forests being much taller than non-forests, and oceans being much deeper than lakes. This difference in vertical dimension could provide an opportunity for stacking more species into a given area, which would tend to increase species richness for larger areas.

Drakare and her colleagues also wanted to know whether small-bodied species would have a lower *z*-value than large-bodied species. They predicted this relationship because many small-bodied species have two related features: They are very numerous, and they can disperse long distances through the air (Figure 22.9A). Thus for many small-bodied species there is a high probability of finding a few individuals in any sized habitat, so increasing area should not substantially increase the chance of them being present in a sample, resulting in a relatively shallow species–area relationship. The meta-analysis confirmed this prediction, with a significant positive correlation between body size and *z*-value (Figure 22.9B).

In contrast, the researchers had no clear prediction about how latitude should influence the slope of the species–area curve. We know that alpha diversity is high in the tropics, but there are conflicting findings on beta diversity, our measure of species turnover (see Chapter 17). The meta-analysis revealed that the steepness of the species–area curve tends to decrease with latitude, with higher latitudes having the shallowest species–area relationships and tropical habitats showing the steepest species–area relationships (Figure 22.9C). These high tropical *z*-values indicate that species ranges tend to be smaller in tropical than higher latitudes.

A clear understanding of geographical influences on species richness remains elusive. Drakare's meta-analysis, and other large-scale studies, are helping to identify broad trends. But biogeographers recognize that historical events, including changes to the physical environment, can also influence both species richness and species composition. Let's briefly explore some of these relationships.

22.2 HOW DO HISTORICAL EVENTS HELP EXPLAIN THE CURRENT DISTRIBUTION OF A SPECIES OR TAXONOMIC GROUP?

Perhaps more than any ecological subdiscipline, historical biogeography has benefited from some important improvements in our understanding of natural processes, and in our ability to use technology. Observations that were initially puzzling to us no longer are, but, with greater knowledge, biogeographers find themselves addressing more interesting and more difficult puzzles. Let's go back to the nineteenth century to introduce some of these observations and to discuss their resolution.

Wallace's observations of global biogeographical patterns

Alfred Wallace is best known as the co-discoverer of the theory of natural selection. Given natural selection's status as the unifying theory of biology, and that Darwin's name always comes up in association with the theory, historians have agonized over whether Wallace deserves more credit than he gets for his astonishing insights about evolution. But there is no disagreement over Wallace's position as the greatest figure in the history of zoogeographical explorations. He spent over 20 years exploring the world, mapping out the distribution of animals worldwide. One puzzling observation generated by his journey to Malaysia was that mammals in the Philippines were more similar to mammals in far-away Africa than they were to mammals in nearby New Guinea (Wallace 1876).

A related observation was that, from a global perspective, there appeared to be very sharp geographical boundaries separating one group of mammal families from another group. (The same was true for birds, though the boundaries were more diffuse.) For example, the mammals and birds of Australia and New Guinea and several smaller islands are very distinct from those in Borneo, Sumatra, and the mainland of Southeast Asia. This demarcation became known as Wallace's line (Figure 22.10A). Based on his explorations and his correspondences with other naturalists, Wallace divided up the world into six biogeographic regions, each containing relatively distinct mammal fauna (Figure 22.10B). As ecologists learn more about the interaction of geography and species distribution, they are developing a more refined understanding of biogeographical regions. An analysis by Ben Holt and his colleagues (2013) used information from the phylogenetic analysis of over 20 000 terrestrial vertebrates to further subdivide Wallace's six biogeographic regions into 20 such regions.

Puzzles of biogeographic regions

Remember that Wallace's biogeographic findings were being presented in concert with evolutionary theory, which most of the scientific (and non-scientific) community had already accepted. Consequently, naturalists were interested in understanding patterns of animal distribution as described by Wallace and others in the context of the new evolutionary theory. So it was not surprising that Nearctic mammals should be different from Palearctic mammals, as there are some very unswimmable oceans between them, and people now understood that the lineages could have evolved independently. But why were Palearctic mammals so different from the adjacent Oriental mammals? Why would these two regions have such different evolutionary outcomes?

The opposite question was raised by disjunctions, which are distributions of closely related species or species groups over widely separated regions of the world. A familiar disjunction is the distribution of marsupial mammals: over 200 marsupial species in Australia, about 100 species in South and Central America, and one species in North America. Biogeographers were puzzled about how marsupial ancestors could have hopped from one continent to another, and why they were entirely absent from some continents (Figure 22.11A). A less familiar but perhaps more compelling disjunction is the worldwide distribution of trees in the genus *Glossopteris*. These unusual conifers, common during the Permian Period, had tongue-shaped leaves, which they shed during the winter. But most unusual in light of our current discussion is their distribution, with over 70 species in India and a widespread distribution in Australia, southeast Africa, South America and northern Antarctica (Figure 22.11B). These trees had heavy seeds that sunk in ocean water, so it should have been impossible for them to float from one continent to another.

All of these puzzles have the same solution, which is that geological conditions in the past were very different from current geological conditions.

Solution – continental drift and plate tectonics

Geologists have for several centuries noted how the continents seem to fit together like pieces in a jigsaw puzzle. Early in the twentieth century, the meteorologist Alfred Wegener proposed his theory of continental drift, arguing that the landmasses were shaped like matching pieces of a jigsaw puzzle because they had once formed a single landmass, Pangaea, which broke apart into smaller plates that we know as today's continents. These continents float atop a dense viscous mantle, and are still in motion today. The breakup of Pangaea began with the formation of a rift valley, a linear lowland created when a geological rift or fault caused the Earth's crust to spread apart. As spreading continued, these lowlands were filled in with ocean waters, and the floating crustal plates developed into the continents, which Wegener contended were still moving at a rapid rate (Wegener 1929). Interestingly, Wegener even used biogeographical evidence to support his claims, arguing that closely related earthworm species are found on either side of the Atlantic, and challenging biogeographers to explain how the worms managed to make the crossing.

(A)

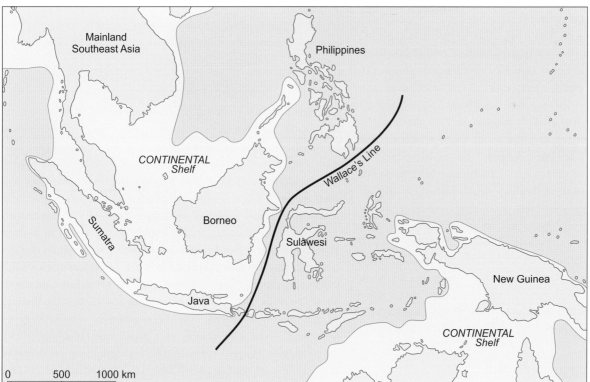

Figure 22.10 A. Wallace's line separates Asian and Australian fauna. Drawings of fauna from Borneo (left) and New Guinea (right) are by Wallace (1876).

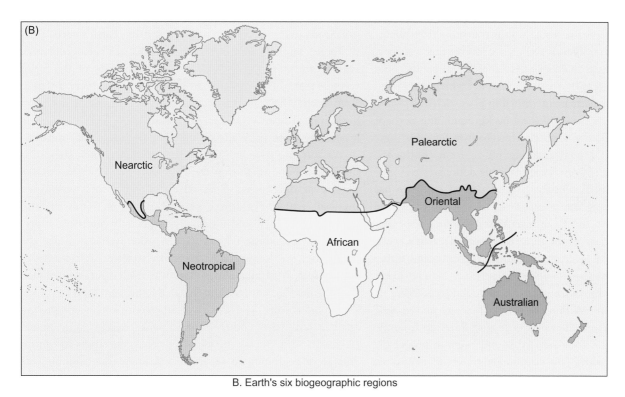

B. Earth's six biogeographic regions

Figure 22.10 (*cont.*) B. Earth's six biogeographic regions.

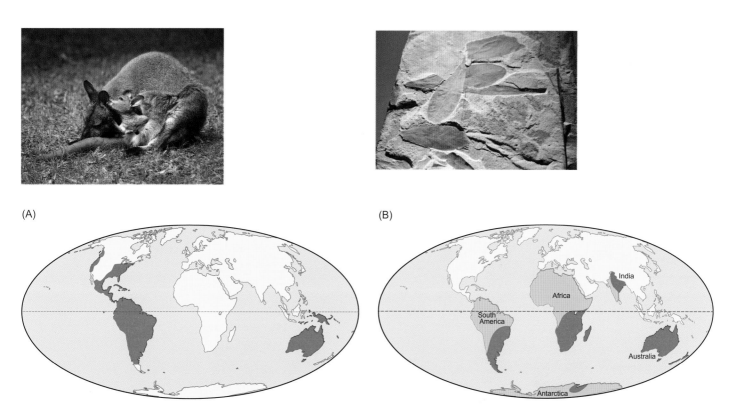

Figure 22.11 A. Marsupials, named for the marsupium or pouch in which most species carry their young, form a disjunct distribution. (Pictured: Rock-wallaby *Petrogale assimillis*.) B. *Glossopteris* species formed a disjunct distribution (as based on the fossil record).

For a variety of reasons, most biologists rejected Wegener's ideas. There is an interesting history of the transition from almost universal rejection to universal acceptance of Wegener's theory. Scientists found it easier to believe that animals and plants could move around more easily than continents, which they had always assumed were permanently fixed in place. Scientists were skeptical of Wegener's meteorological background, and perhaps most importantly, there was no clear mechanism for why these pieces of crust would go wandering off to form continents, though Wegener did suggest some plausible options.

By the mid-twentieth century, a great deal of evidence has accumulated supporting continental drift. Alas, we cannot delve into this history. But the basic mechanism of plate tectonics describes the movement of the thin rigid layer of Earth's crust, which rides the upper layer of the mantle. The crust and upper mantle together form the *lithosphere* (Figure 22.12). The lower layer of the mantle, the *asthenosphere*, is more fluid, receiving its heat from Earth's core. Within the ocean, there are mid-oceanic ridges that form where molten rock or magma flows up from the mantle, which cools to form new crust and helps to propel the older crust to either side. This continuous process of new crust forming and older crust being pushed away from the oceanic ridge is called seafloor spreading. The solid lithosphere rides, very slowly, on top of the asthenosphere, gradually making its way to a plate boundary. Finally, at a plate boundary, one crust may sink deep into the asthenosphere, forming a *subduction zone*. As gravity pulls the leading edge of the dense subducting plate into the mantle, the rest of the plate is moved along in the direction of the subduction. Thus heat from the mantle and gravity at subduction zones both provide the force to move continental plates. Subduction zones often have violent seismic activity, and contribute to the uplifting of mountain ranges around the world.

Working out the details of plate tectonics theory will continue to keep geologists busy for the foreseeable future. For biogeographers attempting to understand the distribution of organisms living atop these plates, a drifting Earth's surface has profound implications. For example, Wallace's biogeographic regions are mostly congruent to the boundaries of the major tectonic plates. Thus the animals in each region are very different from each other simply because they have a very different geographical history, and in some cases have come together in space only recently. Wallace's line defines one such boundary.

We can also use continental drift theory to understand disjunct distributions. During the Permian Period, which was the heyday of *Glossopteris* dominance, the continents were fused together, with the southern supercontinent of Gondwana still attached to what would ultimately become the northern supercontinent of Laurasia (Figure 22.13A). Gondwana included the landmasses we know today as South America, Africa, India, Madagascar, Australia, and Antarctica, while Laurasia included North America, Europe, and Asia. By the late Jurassic, about 150 million years ago, the two supercontinents were pulling apart, but the other connections were still intact (Figure 22.13B). But by the mid-Cretaceous, the southern continents were beginning to separate, and India was beginning its rapid northern journey, which culminated with a geologically violent fusion with Asia about 50 million years ago (Figure 22.13C). By 45–50 million years ago, Australia and Antarctica had broken apart, and Australia too was beginning to migrate northward. The final break between Australia and Antarctica occurred about 35 million years ago.

The glossopterids had a widespread distribution in temperate latitudes of Permian Gondwana. Their southern distribution was limited by the glaciers of the future Antarctica, and their northern distribution reached about 30° S (McLoughlin 2001). But they went extinct in Gondwana during the late Permian, leaving a rich and disjunct fossil history that puzzled biogeographers until continental drift theory rose into preeminence. We now

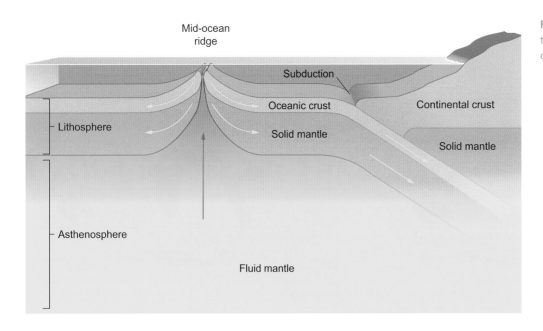

Figure 22.12 Model of plate tectonics. Both continents and oceans ride atop these plates.

(A)

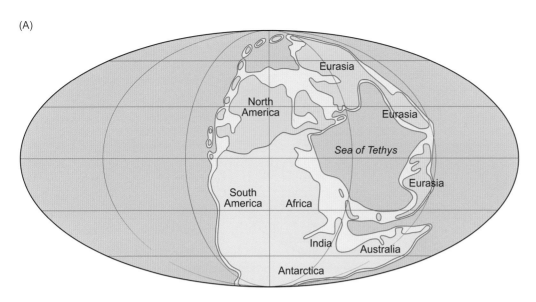

(B)

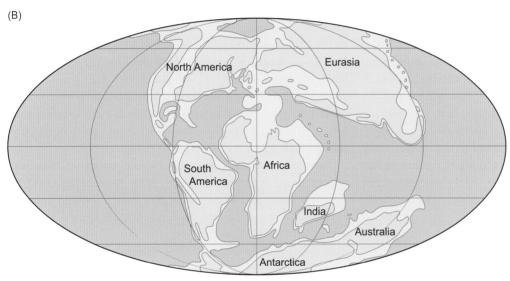

(C)

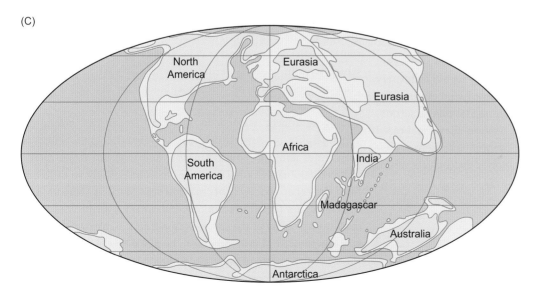

Figure 22.13 A. Earth in the late Permian – 255 million years ago (MYA). B. Earth in the mid-late Cretaceous – 94 MYA. C. Earth in the middle Eocene – 50 MYA.

(D)

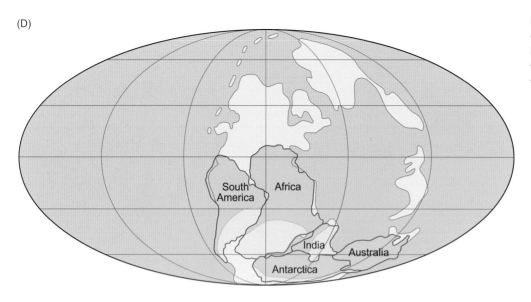

Figure 22.13 (*cont.*) D. Distribution of glossopterids (dark green) on late Permian Earth. Compare this to today's distribution of glossopterid fossils as shown in Figure 22.11B.

understand their disjunct distribution as the remains of what was a continuous distribution during the Permian (Figure 22.13D).

The question then arises whether we can use continental drift theory to understand all puzzling problems associated with species distributions. In reality there are two other ways for related species to form disjunct distributions. First, the ancestral species may have formed a continuous distribution, but populations of species in the middle of the distribution may have gone extinct, leaving disjunct species along the edges of the former continuous distribution. Second, one group may have undergone long-distance dispersal (see Chapter 9 for a discussion of long-distance dispersal). Fortunately, biogeographers have a growing arsenal of tools to help them infer patterns of species distributions.

Using molecular analyses to infer mechanisms underlying current species distributions

We began this chapter with a discussion of how James D. Watson believed that all important questions in biology would ultimately be answered by molecular biologists. Subsequent discoveries, such as plate tectonics, proved him wrong in that assertion. However, ecologists, including E. O. Wilson, have embraced the use of molecular techniques to resolve questions about species distribution. In fact, the application of molecular analyses of evolutionary relationships in the context of geographical distributions has led to the formation of the new field of phylogeography.

Drift, dispersal, and extinctions in marsupials

Let's consider the marsupials to see whether continental drift can help explain their disjunct distribution in the context of what we now know about their phylogeny. There has been considerable debate about marsupial phylogeny for a number of reasons. One

problem is the relatively spotty fossil record, which tells us that the first known marsupial evolved in China about 125 MYA. Marsupials spread westward into North America and then into South America, which was connected to North America until about 65 MYA. All northern hemisphere marsupials went extinct during the great global extinction at the end of the Cretaceous, but the southern marsupials spread throughout much of the large southern continent.

There are two marsupial superorders, based on both taxonomy and current distribution. The Ameridelphia, which comprise the orders Didelphimorphia (opossums) and Paucituberculata (shrew opossums), live in South and Central America. The Australidelphia, which comprise the orders Dasyuromorphia (many species including the Tasmanian devil), Peramelemorphia (bandicoots), Notoryctemorphia (marsupial moles), and Diprotodontia (koalas, wombats, possums, and kangaroos) live in Australia. One problematic order outside of these groups is Microbiotheria, consisting of only one extant (current) species, the monito del monte, *Dromiciops gliroides*, which lives in South America but is similar to the Australidelphia species, based on the morphology of its tarsal bones (Figure 22.14A).

Biogeographers wanted to know if *Dromiciops* belong to Ameridelphia or Australidelphia. If Australidelphia, what is *Dromiciops* doing in South America, and how did it get there? If Ameridelphia, why is *Dromiciops* morphologically so different from the other species in its superorder? Maria A. Nilsson and her colleagues (2004) used mitochondrial DNA analysis to work out the phylogeny of the seven extant orders of marsupials. Their analysis showed Microbiotheria to be more closely related to the three smaller orders of Australidelphia than they were to Diprotodontia. Most important, the Microbiotheria were only distantly related to the Ameridelphia (Figure 22.14B). This suggested that Microbiotheria must have somehow traveled back to South America after its evolution in Australia. This could have

(A)

occurred, because Australia and Antarctica were connected until about 45 MYA.

The researchers admitted that there could be considerable error associated with this particular analysis; so later Nilsson and a different research team (2010) used another molecular approach to resolve the phylogeny. Occasionally pieces of DNA are transcribed to RNA, then reverse transcribed to DNA, and reinserted back into the genome at a random location. These *retroposons* are wonderful genetic markers, because it is very unlikely that two independent events would retropose identical sequences back into identical, randomly selected locations. So two species sharing the same retroposon are likely to be related to each other. The researchers did an exhaustive genomic analysis of species from each order, and found 53 informative retroposons that could be used to resolve evolutionary relationships. Their findings help us to understand marsupial evolution and distribution.

The retroposon analysis shows that *Dromiciops* is only distantly related to the Australian marsupials, and not particularly closely related to the Diprotodontia. Thus there was no need to have an ancestor to Microbiotheria traveling back to South America. Instead there was probably a single Gondwanan migration of marsupials to Australia, presumably via Antarctica. Let's

(B)

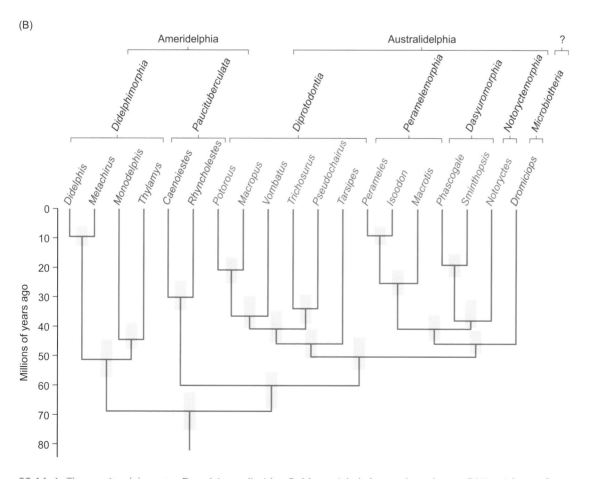

22.14 A. The monito del monte, *Dromiciops gliroides*. B. Marsupial phylogeny based on mtDNA evidence. Green genera are South American and blue are Australian. Note the relatively distant relationship between Microbiotheria and the Ameridelphia. Shaded regions indicate 1 SD for estimate of divergence time.

work through Figure 22.15 to understand what we think happened. The sequence of evolution in the figure is moving from left to right.

In South America, about 75 MYA there was a species ancestral to all extant marsupials. The 10 adjacent circles over the continent of South America indicate that all marsupials share those ten retroposons. Didelphimorphia was the first to branch off and has five unique retroposons. The remaining marsupials share two unique retroposons, which indicates that all remaining marsupials share a common ancestor after the divergence of Didelphimorphia. Next to branch off was Paucituberculata, as represented by three unique retroposons. All of the remaining extant

marsupials (superorder Australidelphia) share 13 more retroposons. The first Australidelphian to branch off was an ancestor to *Dromiciops* (with no further retroposon evolution in its lineage). The rest of the Australidelphians share four retroposons that evolved after their ancestor branched off from the lineage leading to Microbiotheria. These four retroposons evolved in their shared lineage at about the time that their ancestor, the ancestor to all of the Australian marsupials, migrated to Australia – about 50 MYA, just before Australia broke apart from Antarctica (Figure 22.15).

The beauty of this hypothesis is that it conforms to the data, and it is simple – requiring only one migration event, which happened just before Australia split from Antarctica. But is it

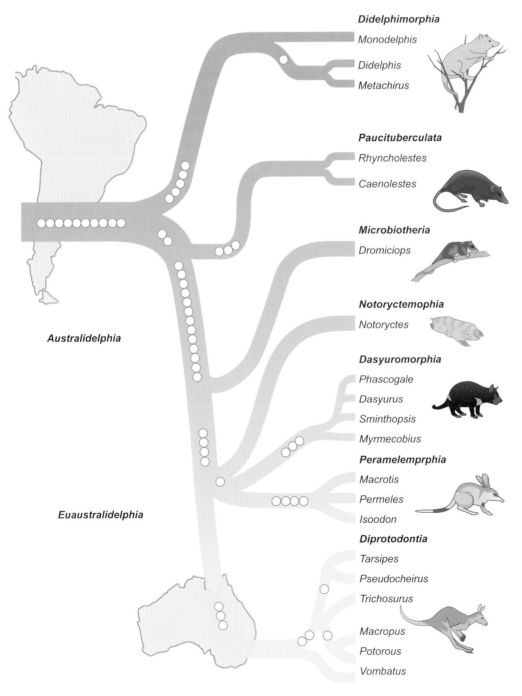

Figure 22.15 Marsupial phylogeny based on retroposon evidence.

correct? We eagerly await more evidence from fossil finds, morphological analyses, and new molecular approaches.

Of course, many evolutionary events occur on a smaller scale in space and/or time, so in those cases continental drift is not a viable candidate for explaining the distribution of different groups. The Hawaiian *Drosophila* provides one such example.

Dispersal, colonization, and speciation of Hawaiian *Drosophila*

If you have spent any time on the Hawaiian archipelago, you would not be surprised that ecologists love to work there. It's not the sandy beaches, awesome waves, nor perfect climate that ecologists find attractive. Rather it is the known geological history, environmental heterogeneity, and spectacular adaptive radiations that lure researchers. Well, maybe some of both. The islands sit on a geological hotspot, where molten rock is forced to the surface, constantly creating new islands and shifting older islands to the north and west courtesy of seafloor spreading. As they move, they subside and eventually sink into the ocean. Thus the youngest island is the big island of Hawaii and the oldest is Nihau, but still older islands form a continuous submarine chain that reaches almost to Asia. All of these islands have been dated using potassium–argon techniques (Figure 22.16A).

One of the most-studied adaptive radiations is the Hawaiian drosophilids, comprising about 1000 species. The age of this clade is estimated at about 26 million years, making them the oldest known radiation of plants or animals on the islands. Based on geographical reconstruction of the time period, James Bonacum and his colleagues (2005) knew that the initial colonization was through long-distance dispersal. They decided to limit their analysis to 17 related rainforest species within the planitibia group, which comprises four subgroups. The researchers wanted

to unravel the pattern of colonization and speciation. We will discuss only a portion of their findings.

To work out the phylogeny, the researchers used DNA base sequences from two mitochondrial and four nuclear gene loci to estimate the times when two lineages diverged from each other. To conduct this analysis, the researchers used the *molecular clock hypothesis*, which assumes that DNA base substitutions occur as a linear function of time. By counting the number of different bases in the six chosen genes for each species, Bonacum and his colleagues were able to estimate how long ago each species shared a common ancestor. Their analysis revealed that the *picticornis* subgroup was the first to break off from the other species, about 6.1 MYA. If we confine our analysis to the *planitibia* subgroup, we see two important findings. First, closely related species tend to be on the same or on adjacent islands. Second, the sequence of

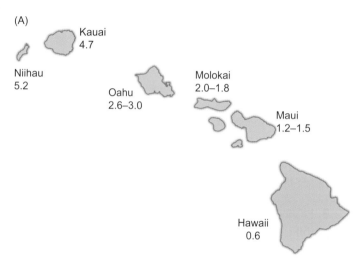

(A)

Kauai 4.7

Niihau 5.2

Oahu 2.6–3.0

Molokai 2.0–1.8

Maui 1.2–1.5

Hawaii 0.6

Figure 22.16 A. The Hawaiian Island chain with dates (millions of years ago) of island emergence B. Phylogeny of *planitibia* and *picticornis* subgroups; numbers at each node represent estimated time of divergence (in MYA).

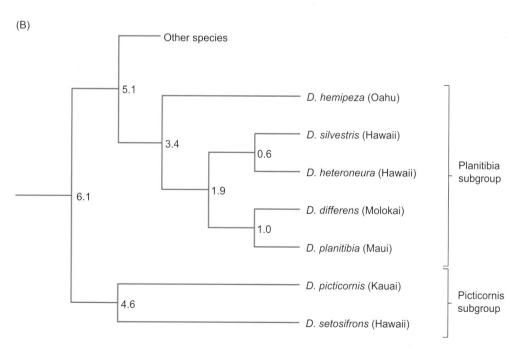

(B)

Other species

5.1

3.4

0.6 D. silvestris (Hawaii)

D. hemipeza (Oahu)

D. heteroneura (Hawaii)

1.9

6.1

1.0 D. differens (Molokai)

D. planitibia (Maui)

Planitibia subgroup

4.6 D. picticornis (Kauai)

D. setosifrons (Hawaii)

Picticornis subgroup

speciation is consistent with the sequence of island formation. Thus *D. silvestris* and *D. heteroneura* diverged at about the time that Hawaii emerged. And *D. differens* and *D. planitibia* diverged shortly after the emergence of Maui. And the most ancient lineage is found on the relatively old island of Oahu (Figure 22.16B).

 Thinking ecologically 22.2

The phylogenetic analysis shows the *D. picticornis* and *D. setosifrons* lineages diverging about 4.6 MYA. *D. setosifrons* is endemic to the big island of Hawaii, which rose out of the ocean about 0.6 MYA. What might the *D. setosifrons* lineage have done for 4 million years without its island? As a hint, you should keep in mind that this phylogeny only considers extant species; many species in the *picticornis* subgroup could have evolved and gone extinct since the 4.6 MYA divergence of the ancestor to *D. picticornis* and *D. setosifrons*.

Our discussions of geographical ecology have looked at patterns of species richness based on the size of the habitat, the size of the organism, and how size or scale influences dispersal patterns. In addition, biogeographical studies have helped us to appreciate that landforms are dynamic and have a history that can influence dispersal and species composition. Considerations of scale and of variation in landforms lie at the heart of the developing field of landscape ecology.

22.3 HOW DOES LANDSCAPE STRUCTURE INFLUENCE THE DISTRIBUTION AND ABUNDANCE OF SPECIES?

A landscape can be defined in numerous ways. We will simply define a landscape as an area that is spatially heterogeneous in one factor of interest (Turner *et al.* 2001). We have been talking about landscape ecology throughout this book without calling it by name. For example, our discussion of metapopulation ecology in Chapter 10 introduced habitat patches, which are relatively discrete areas of a landscape that differ in some important way from their surroundings. The patches in Chapter 10 were meadows that provided food plants for the Glanville fritillary butterfly. Later, in Chapter 11, we asked how the distance between patches and the connectivity of patches influences the probability of colonization. And our discussions in Chapter 21 of disturbances in Yellowstone National Park and Mount St. Helens addressed how spatial heterogeneity influenced ecological processes. Now we will more explicitly see what landscape ecologists do, so we can ask some questions about what they have discovered.

Variation in spatial configuration

Spatial configuration is the spatial arrangement of the component or components of interest. As ecologists, we might ask how the

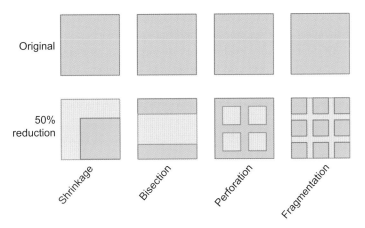

Figure 22.17 Types of changes in spatial configuration that can accompany habitat reduction.

spatial configuration of a particular habitat attribute influences species distribution or species richness. Humans are drastically influencing spatial distributions and landscape configuration. Consequently, landscape ecologists are attempting to understand how changes in configuration interact with changes in habitat attributes to influence ecological processes and structure.

As one example, let's consider types of change in spatial configuration that can accompany reduction in the size of a habitat. *Shrinkage* is reduction in habitat size without fragmentation or subdivision. *Bisection* divides the habitat into two smaller areas. *Perforation* introduces holes of unsuitable habitat into the native habitat, and fragmentation divides a habitat into a larger group of smaller unconnected pieces. (Figure 22.17).

The area surrounding suitable habitat is the matrix. Depending on the type of study, attributes of the matrix may be ignored, or they may be a crucial feature of the study. The quality of the matrix will influence the ability of individuals to move between fragments in a bisected or fragmented landscape. Matrix quality is a function of the particular species; a wet boggy matrix may be inviting to a salamander or newt, but form an effective barrier to the passage of a sun-loving lizard.

One important feature of landscape ecology is that the processes it investigates rely heavily on spatial scale.

Variation in spatial scale

Much of landscape ecology is viewed from the perspective of large brainy mammals (that's us!), so the scale is generally quite large. For example, a landscape ecologist studying the distribution of farms in relation to soil nutrient content might consider nutrient heterogeneity over a very large expanse of hundreds of square kilometers. However, landscape ecologists emphasize choosing scales that best characterize the relationship between spatial heterogeneity and the ecological process that is being studied. So a landscape ecologist studying the process of nutrient decomposition by dung beetles would need to consider nutrient heterogeneity on a much smaller scale, perhaps on the scale of the distance a dung beetle travels to lay her eggs.

Scale-related terminology

Scale is described by grain and extent. Grain is the finest resolution used for viewing a phenomenon. It could be the pixel size of a remote camera, or the cell size of a GIS map. Extent is the overall size of the study area (Figure 22.18). When ecologists refer to fine-scale studies, they are describing studies that have small extent, high resolution, and more detail than coarse-scale studies.

The problem of scale is not unique to landscape ecology, but applies to all types of scientific inquiry. Just because chemists can make a reaction happen in a test tube does not guarantee that it will happen under large-scale production. Ecologists routinely use microcosms or mesocosms as models of ecosystem function, but that does not mean that the same processes occur on the scale of ecosystems. Because landscape ecologists tend to ask questions about large-scale phenomena, they often need to address the problem of scaling up. But some important ecological processes occur across relatively small landscapes as well.

Effects of scale on hawthorn tree fitness

Daniel Garcia and Natacha Chacoff (2007) wanted to know how the scale with which they viewed forest fragmentation in the Cantabrian Range in northern Spain influenced their interpretation of how spatial heterogeneity affected the fitness of hawthorn trees, *Crataegus monogyna*. Measures of hawthorn fitness were pollination rate, fruit set, and the frequency of fruit and seeds being eaten by predators. The forest was fragmented by humans for grazing their cattle, but also had some undisturbed areas of rocky outcrops that were not suitable for tree growth.

The researchers evaluated fragmentation at three spatial scales, by measuring the percent of tree cover surrounding 60 focal hawthorn trees, in three concentric circles of 10 m, 20–50 m, and 50–100 m radius. Circles with greater tree cover were defined as less fragmented than were circles with reduced tree cover. (Figure 22.19).

The pollinators of these hawthorn trees are primarily honeybees, *Apis mellifera*, and flies (several generalist families). The fruits are consumed by several species of birds, including thrushes (genus *Turdus*), which then defecate the seeds that may then be scavenged and eaten by woodmice (genus *Apodemus*). Thus the animals that influence plant fitness represent a wide gradient of foraging range, with birds foraging farther than insects, which forage farther than the woodmice.

Garcia and Chacoff discovered that the mean number of pollinator visits to flowers increased with percent cover on large and medium spatial scales, but not on small spatial scales (Figure 22.20A). Fruit set showed a very similar pattern. In contrast, frugivory increased most significantly with percent cover on the small and medium spatial scales, but not on the largest spatial scale (Figure 22.20B).

The researchers examined seed predation rates, by gluing 10 hawthorn and 10 holly, *Ilex aquifolium*, seeds onto strips of plastic (so they would not blow away), and then resampling these seed depots after 2 and 4 weeks. They used holly seeds because previous work showed that rodents like to eat them, and the researchers wanted to see if similar scale effects operated in more than one species. They discovered that seed predation was greatest at low tree cover for all spatial scales, and that the reduction in predation at high tree cover was greatest at the intermediate spatial scale for both holly and hawthorn (Figure 22.20C).

Overall, trees in highly fragmented landscapes (with low percent tree cover) had lower fitness resulting from decreased pollination and increased seed predation. The pollination effects were greatest at the largest spatial scale, which may be related to the wide foraging range and mobility of the insect pollinators. The increased seed predation in fragmented areas is strongest at intermediate scales, which correspond in area to the home range

(A)

Increasing grain size

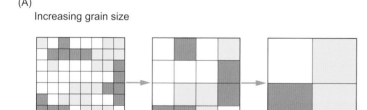

(B)

Increasing extent

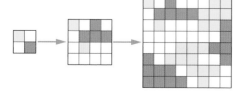

Figure 22.18 Spatial scale with increasing grain size at the same extent (A), and increasing extent at the same grain size (B).

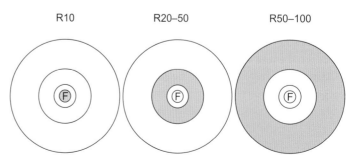

Figure 22.19 Garcia and Chacoff's experimental design. R10 fragments were within 10 m of the focal tree, R20–50 was a donut shape of 20–50 m radius, R50–100 was a donut shape of 50–100 m radius. F indicates position of focal tree.

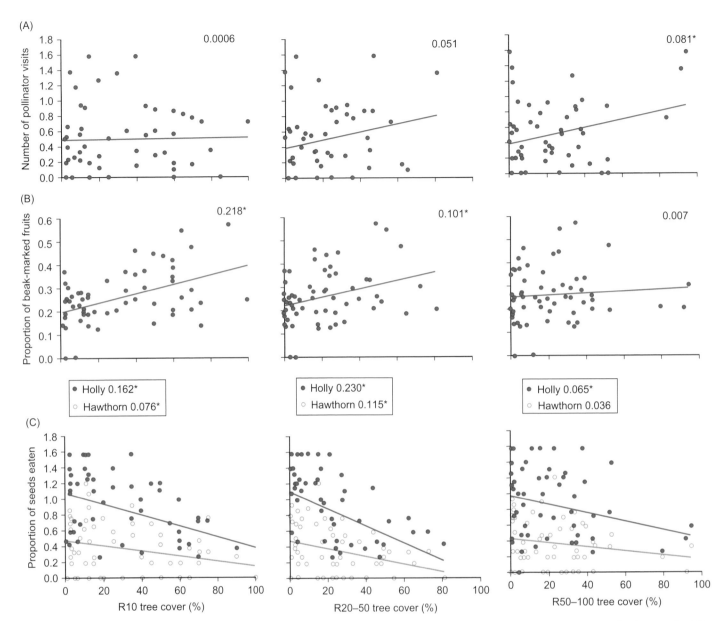

Figure 22.20 Responses of focal tree in relation to percent tree cover at each spatial scale. Numbers above each graph = R^2 value. * = $P<0.05$. A. Number of visits to flowers by pollinators. B. Frugivory as measured by the proportion of fruits with beak marks. C. Proportion of seeds eaten by predators. All dependent variables were arc sine square root transformed for the analysis and graphs.

of individual woodmice. The authors suggest that woodmice know the landscape features of their home range, and those living in highly fragmented landscapes are forced to concentrate their feeding under the few isolated trees that are producing seeds for them to eat.

Frugivory decreased with fragmentation. It was somewhat surprising that the frugivory effect was most significant on a fine scale, given a thrush's wide foraging range. The researchers suggest that the birds are able to make coarse-scale evaluations of habitat quality during exploratory flights, and then fly to patches that appear to have good attributes, such as high fruit density and protection from predators, on a finer scale. It doesn't matter to the birds if there are trees 50 m away; what is important is the fruit density and tree cover in the immediate area.

Garcia and Chacoff's study highlights how landscape elements can be important on one scale, but not another. Their study also introduces us to a discussion of landscape fragmentation, a core area of research in landscape ecology. Landscapes are naturally heterogeneous, and many are naturally fragmented. However, human activity is causing a tremendous increase in fragmentation, which can have important impacts on landscape processes. Let's consider fragmentation in more detail.

Wide-ranging effects of fragmentation on species diversity and ecological processes

The popular press, both digital and analog, abounds with tales of rainforest destruction, usually focusing on the loss of charismatic

(A)

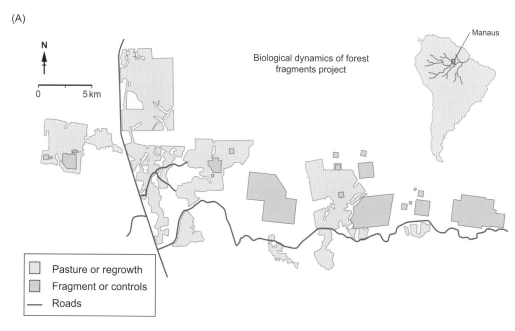

Figure 22.21 **Figure 22.21** A. Map of the BDFFP site. B. Aerial view of 10 ha and 1 ha fragments

(B)

animals such as mountain gorillas and chimpanzees, and on its long-term effects on global carbon cycles and the implications for global warming. Removal of rainforest is only part of the problem, because the remaining forest fragments also suffer serious ecological consequences. But fragmentation is a problem across other landscapes as well. We'll begin with a discussion of fragmentation in tropical Brazilian forests, and then briefly turn to some other landscapes.

In 1979, a group of scientists began the Biological Dynamics of Forest Fragments Project (BDFFP) to assess the effects of fragmentation on the Amazonian landscape (Lovejoy and Oren 1981). The project's original goals were to assess the effect of fragment size on species richness and ecosystem processes, to determine a minimum size for effective rainforest reserves, and to resolve a related debate over whether it was best to have a small number of very large reserves or a large number of smaller reserves. The

project has continued for about 35 years, and continues to yield insights that reflect its original goals, but also that have moved into some new directions.

The BDFFP research area occupies about 1000 km² of relatively flat terrain just north of Manaus, Brazil. The study area has three large cattle ranges, and some very large expanses of continuous forests that serve as controls; in addition, the researchers created 11 artificial fragments by burning and clearing the surrounding vegetation. Five fragments were 1 ha, four were 10 ha, and two were 100 ha (Figure 22.21). Before creating the fragments, the researchers surveyed all the macroscopic life within the soon-to-be-created fragments for a baseline. The forests are very diverse, with tree species richness often exceeding 280 species per hectare, and comparably high levels of diversity in other plant and animal groups. The basic experimental protocol involves systematic surveys of the organisms and the biological processes within the fragments, and comparing these results to similar measures for the unmanipulated controls. We will discuss some of the findings in the context of fragment size and also the impact on ecological processes near a fragment's edge.

Effects of fragment size at the BDFFP

In most cases, BDFFP fragments had much lower species richness than the unmanipulated control regions, and small fragments had much lower diversity than large fragments. For example, Jeffrey Stratford and Philip Stouffer (1999) compared the species richness of understory insectivorous birds in 1, 10, and 100 ha BDFFP fragments with equal-sized plots of unfragmented forests. Their results show that fragments supported fewer species than unfragmented areas of the same size, and that smaller fragments supported fewer species than did larger fragments (Figure 22.22).

The researchers were able to use prefragmentation data to compare species presence before and after fragmentation as a

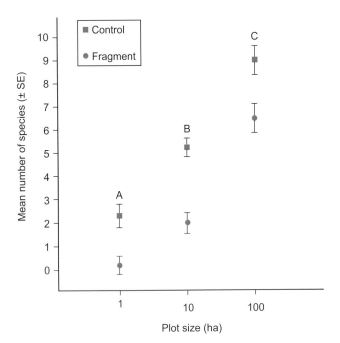

Figure 22.22 Mean species richness of insectivorous birds in experimental fragments and control plots of the same size.

measure of extinction rate. As Table 22.2 shows, extinctions were most common in the smallest fragments and least common in the largest fragments. Recolonization after extinction was rare in all fragments.

We can view these findings from the perspective of MacArthur and Wilson's theory of island biogeography. Smaller fragments (islands) should have higher extinction rates, because they support smaller populations. Fragments of all sizes should reach equilibrium, where immigration and extinction rates are approximately equal. The turnover rate may be higher in small fragments due to the high extinction rates. Using the same fragments, but considering all understory birds (not only insectivores), Stouffer and two other colleagues (2009) investigated whether these fragments had achieved equilibrium by 1992, by comparing species richness and species composition between 1992 and 2000.

The researchers found that, as predicted, extinction rates were much higher in the smallest fragments (Figure 22.23A). Second,

mean species richness did not change appreciably between 1992 and 2000 for small and large fragments, and actually increased slightly for intermediate fragments (Figure 22.23B). But landscape features play an important role in species extinctions, with generally higher extinction rates in borders that were kept relatively clear of plant growth, and lower extinction rates associated with borders in which secondary forest was allowed to return (Figure 22.23C). Thus the matrix buffers a fragment against extinctions, by providing habitat that is suitable for some of the bird species.

The BDFFP showed that small fragments hosted fewer species of bryophytes, tree seedlings, palms, several insect groups, primates, and large herbivorous mammals. But some species were unaffected, particularly those that specialized in more edge-type habitat, and species richness of some groups, including small mammals and frogs, actually increased (Laurance *et al.* 2011). Though the fragment-size effect applies to most species, clearly we need to make predictions on responses to fragmentation based on the life history traits of each species, and how each species interacts with other species and with the environment.

Geometrically, small fragments will tend to have a substantial portion of their habitat located near the fragment's edge. This edge effect can have an important influence on species richness and many ecosystem processes.

Edge effects at the BDFFP

William Laurance and his colleagues (2002) published a paper that summarized 33 edge effects that had been identified by the BDFFP. Some of these effects, such as increased wind disturbance and elevated tree mortality, penetrated at least 300 m into the forest, while others, such as reduced density of fungal fruiting bodies and increased invasion of disturbance-adapted plants, were only observed to penetrate a few meters in from the edge of the fragment. Some important microclimatic effects, such as reduced humidity, increased light, and greater temperature variability, made significant inroads into the forest, and had a negative impact on species adapted to dark, moist conditions characteristic of forest interiors.

Table 22.2 Number of understory insectivorous birds present before fragmentation, and number of species extinctions and colonizations after fragmentation.

Fragment size (ha)	1	1	1	1	1	10	10	10	10	100	100
Number species present before fragmentation	8	7	6	6	6	7	6	7	5	9	7
Number of extinctions after fragmentation	8	7	6	6	6	6	5	3	3	3	2
Number of colonizations after fragmentation	0	0	0	0	1	0	0	0	0	0	2

(A)

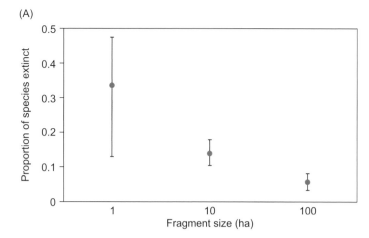

(B)

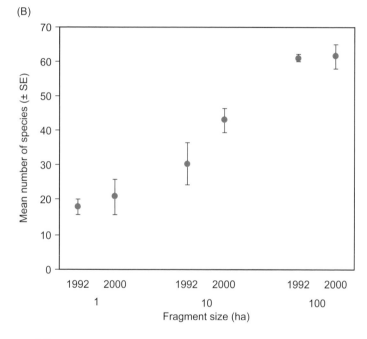

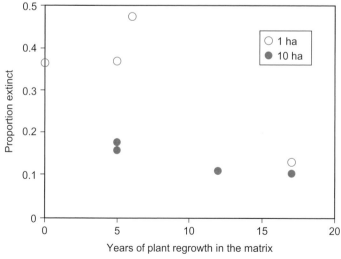

Figure 22.23 A. Mean proportion of bird extinctions between 1992 and 2000 for each sized fragment (error bars represent the minimum and maximum values for each size). B. Mean (±SE) species richness of birds in relation to fragment size in 1992 and 2000. C. Proportion of bird species extinct in relation to years of plant regrowth in the matrix.

One of the most important edge effects is an increase in tree mortality. This comes from several sources. When an edge is created, some trees simply drop their leaves and die, presumably a result of changes to the microclimate. Other trees are snapped by high winds, which develop over cleared areas and may penetrate several hundred meters into the forest interior. Still others are killed more slowly, by the increase in lianas – woody vines – that parasitize the trees, reducing their survival, growth, and reproductive rates. Laurance and his colleagues (2000) were able to show that increased mortality from edges primarily affected the largest trees. When the researchers divided plots into edge (located less than 300 m from the nearest edge) and interior (located more than 300 m from the edge, mortality rates were about double for small and medium-sized edge trees and almost triple for very large edge trees. (Figure 22.24A). Large trees are relatively inflexible, so they may be more prone to being snapped by high winds near edges. They also reach intense sunlight above the canopy, and thus may be more susceptible to periodic droughts, which are amplified near the forest edge.

Proximity to edges also has important influences on species composition. From our discussion of ecological succession you might expect that habitats near edges might have a greater abundance and richness of pioneer tree species. Laurance and his colleagues (2006) approached this question in an interesting way, by looking at pioneer species density and abundance in relation to the number of edges that were within 100 m of each tree. They predicted that more edges would be associated with greater density and richness of pioneer tree species. They discovered that the density of pioneer species increased dramatically with the number of nearby edges, and that pioneer species richness increased with the number of nearby edges as well (Figure 22.24B and C).

In the world outside the BDFFP some edges are created by humans as a byproduct of agricultural, industrial, or urban development, while other edges form naturally, often in association with changes in abiotic conditions. For example, forests may naturally grade into grasslands as rainfall diminishes across a landscape, or as soil type changes. Grasslands or forests may grade into wetlands with increases in moisture or decreases in soil permeability. Broad transition zones, or *ecotones*, often have a rich community of plants and animals that are adapted to the diverse niches found in the ecotone.

This BDFFP is immense in its scope and its duration. Unfortunately, we have only scratched the surface of a very complex suite of interacting processes that shape landscapes. Let's briefly turn to some other studies that introduce us to some different landscape-level processes.

Effects of other landscape-level processes on species distribution and abundance

Edge-adapted species are likely to benefit from any changes to the landscape that increase the amount of available edge. For

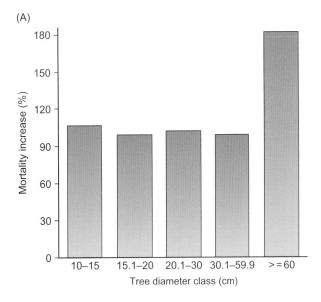

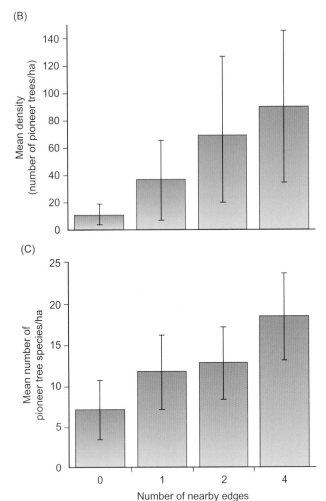

Figure 22.24 A. Mortality increase of trees located within 300 m of an edge in comparison to interior trees. Note that a mortality increase of 100% indicates that edge trees had twice the mortality rate than did interior trees. B. Mean density (± SD) of pioneer tree species in relation to number of nearby edges. C. Mean pioneer tree species richness in relation to number of nearby edges.

example, I live in a small town with extensive forest fragments that come right up to traditional suburban homes. Almost any night I can go outside my home, and easily find a few deer browsing on vegetation that emerges from this edge-type habitat. But even edge-adapted species may not always benefit from recent human-caused increases in edge habitat.

Ecological traps

Species of birds that nest along forest edges require disturbed habitats that historically have occurred in forest openings, or on a larger scale, in fire-maintained successional habitats. Natural selection has favored a preference for forest edge for a variety of species-specific reasons, which could include lower predation rates, better access to food, or less competition with other species. The availability of edge habitat is increasing worldwide as a result of human activity. Aimee Weldon and Nick Haddad (2005) wanted to know whether this landscape feature was acting as an ecological trap for indigo buntings, *Passerina cyanea*, in deciduous forest edge habitat in South Carolina, USA. An ecological trap has two important characteristics. First, individuals must actively select habitat based on some attribute. Second, their selected habitat must result in lower fitness than the alternatives. The researchers hypothesized that indigo buntings are attracted to complex forest edges for nesting, but that the edges with more complex shapes have a negative effect on reproductive success.

To test their ecological trap hypothesis, Weldon and Haddad established experimental perforations by clearing and burning pine trees to form patches in one of two shapes. One was a simple rectangle, while the other was a smaller square with two long wings on opposite sides. Both patches had equal area, but the winged patch had much greater edge (Figure 22.25A). In many species of birds, the older males generally get the best territories, so the researchers used male age as their measure of habitat preference. They found that older males controlled 83% of the territories in winged patches but only 53% of the territories in rectangular patches. Thus, at least to the males, winged patches were more desirable.

But how did this preference for complex shape play out in terms of reproductive success? In this population, over 80% of nest failures resulted from predation by a variety of vertebrate predators, including mammals, snakes, and birds. Mean reproductive success – measured by number of fledglings, or young birds leaving the nest – was over 50% higher in the rectangular patches, and this difference resulted from higher predation-induced nest failures in the complex landscape (Figure 22.25B). The authors conclude that habitat complexity is an ecological trap for these edge-adapted birds, and given that bunting abundance is declining, conservation managers need to incorporate the landscape-level behavioral responses of birds into their management plans.

By their nature, landscape studies involve multiple ecosystems. In some cases there may be a disconnect between the

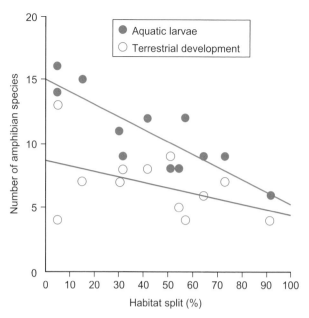

Figure 22.26 Species richness in relation to habitat split for amphibians with aquatic larvae ($R^2 = 0.74$, $P < 0.001$), and species with terrestrial development (correlation not statistically significant).

Figure 22.25 A. Indigo bunting (top) and experimental design (bottom) with rectangular and winged perforations or patches. B. Mean ($\pm$SE) number of offspring fledged per female in rectangular and winged patches ($F = 7.98$, $P = 0.03$).

habitats used by organisms at different stages of their life. This is particularly true of amphibians.

Habitat split and the decline of amphibians

Amphibian populations are declining at an alarming rate worldwide (Stuart *et al.* 2004). There are numerous reasons underlying this decline, including habitat loss and habitat fragmentation, as well as a host of non-landscape-related causes such as the virulent fungus, *Batrachochytrium dendrobatidis*, UVB radiation, introduction of exotic species, and pollutants. Amphibian species with aquatic larvae experience a major niche shift during development, switching from aquatic to terrestrial habitats as they mature, and returning to aquatic habitats to breed. In undisturbed habitats, suitable aquatic and terrestrial habitats are often near each other. However, in disturbed habitats, there may be a large spatial separation between these two types of habitats. Carlos Becker and his research team (2007) describe this division as habitat split, and argue that it may underlie a substantial portion of the decline in amphibian populations.

The researchers conducted their study in Brazil's Atlantic forest, where about 93% of the forest has been removed by human development. This forest hosts about 480 known amphibian species, of which approximately 80% have an aquatic stage of development. Becker and his team defined habitat split as the percentage of the total stream length that does not overlap with forest cover. If their habitat split hypothesis is true, the researchers predicted that amphibian species with terrestrial development should not be affected by habitat split, since they spend their entire lives within the confines of the forest, and don't suffer from spatial separation caused by habitat split. In contrast, species with aquatic larvae should suffer from habitat split, and the researchers predicted that local species richness should decline with increases in habitat split.

The results support the habitat split hypothesis. There was a very strong negative correlation between habitat split and species richness for species with aquatic larvae, and only a weak, and not statistically significant, negative correlation for species with terrestrial development (Figure 22.26).

The habitat split hypothesis was compared to two competing hypotheses, habitat loss and habitat fragmentation, and was shown to have a much stronger influence on the species richness of amphibians with aquatic larvae.

Most of these landscape studies have important management implications. Understanding how landscape heterogeneity influences ecological structure and process will allow conservation ecologists to make good decisions about how to design conservation areas that optimize species richness and ecological function.

REVISIT: Management implications of island biogeography

MacArthur and Wilson (1967) were fully aware of the management implications of their theory of island biogeography. On the second page of their text they wrote "The same principles apply, and will apply to an accelerating extent in the future, to formerly continuous natural habitats now being broken up by the encroachment of civilization." They then went on to graphically show the process of reduction and fragmentation of a woodland. Other researchers have extended these findings and made the analogy between islands and fragments even more explicit.

One important application is designing nature reserves that have high immigration rates from other species sources, which should lead to higher species richness equilibria within the reserve. Conservation ecologists have built immigration corridors connecting nature reserves so they form a continuous landscape (Figure 22.27). One problem with this approach is that corridors are structurally very edgy, so losses due to edge effects could possibly eliminate the gains from higher immigration rates.

Lynne Gilbert-Norton and her colleagues (2010) did a meta-analysis that investigated whether immigration corridors actually increased the movements of organisms between fragments. Based on 35 studies, they conclude that the average corridor increased immigration by about 50%. The increases were greatest for plants and non-avian vertebrates, and smallest for birds and invertebrates. Overall, natural corridors were slightly more effective than artificially constructed corridors. The researchers emphasize that not all corridors work, and that about 25% had negative effects on their target species. Thus managers will need to construct or modify existing corridors in a way that optimizes immigration opportunities for their target species.

Figure 22.27 Group of pronghorns, *Antilocapra americana*, cross a migration corridor in Wyoming, USA.

Of course, increasing immigration rates does not necessarily lead to increased species richness within a reserve, but it is a good start. Insights arising from landscape ecology research can help management professionals decide which factors to consider when designing reserves that maximize species richness and enhance important ecosystem processes.

SUMMARY

E. O. Wilson and Eugene MacArthur teamed together to write a landmark book, *The Theory of Island Biogeography* (1967), which helped invigorate the field of geographic ecology. Wilson was originally inspired by his ant studies in Southeast Asia that indicated that there was a power function relationship between ant species richness and the area of an island. Island biogeography theory views island species richness as an equilibrium of extinction rates and the immigration rates of novel species to an island. At equilibrium, MacArthur and Wilson's model predicts that species composition will change over time, but species richness will remain relatively stable. In addition, large islands with low extinction rates and high immigration rates will tend to support more species than will small islands.

Geographic ecologists also want to understand why particular species or groups of species have a particular geographic distribution. One puzzle was Alfred Wallace's nineteenth-century description of sharp demarcations in the mammalian faunal distributions, which Wallace ultimately used to define six distinct biogeographic regions. A second puzzle was raised by the existence of disjunctions, distributions of closely related species distributed over widely separated regions of the world. The theories of continental drift and plate tectonics have helped to solve these puzzles. More recently, developments in molecular technology have allowed biogeographers to answer numerous questions about species distributions.

Landscape ecology explores how variation in landscape structure, such as configuration or scale, influences the distribution and abundance of species. Conservation ecologists are particularly concerned that industrial, agricultural, and urban development have led to increased fragmentation of habitat that is suitable for sustainable wildlife populations. Even nature preserves are highly fragmented islands of relatively intact habitat surrounded by an ocean of unsuitable habitat. Unfortunately edge effects and low immigration rates can reduce species richness within nature preserves. Applying the lessons of island biogeography, ecologists recommend erecting immigration corridors to increase immigration rates of novel species into nature preserves, thereby increasing species richness.

FURTHER READING

Hadley, A. S., Frey, S. J., Robinson, W. D., Kress, W. J., and Betts, M. G. 2014. Tropical forest fragmentation limits pollination of a keystone understory herb. *Ecology* 95: 2202–2212.

This study investigates the effect of forest patch size on the fitness of a tropical forest herb, *Heliconia tortuosa*. Larger patches had higher seed set, possibly because there were more hummingbirds present to pollinate the flowers.

Jousimo, J., and 6 others. 2014. Ecological and evolutionary effects of fragmentation on infectious disease dynamics. *Science* 344: 1289–1293.

Great long-term study of the effects of connectivity and patch size on the ecological and evolutionary dynamics between populations of a host plant and its fungal pathogen. Highly connected host plant populations had lower fungal colonization rates than expected, because high connectivity promoted the spread of genes that made the plants resistant to fungal infection.

Nilsson, M. A., and 6 others. 2010. Tracking marsupial evolution using archaic genomic retroposon insertions. *PLoS Biology* 8: e1000436.

Very readable paper that demonstrates how ecologists use molecular approaches to reconstruct phylogeny in the context of geographic change.

Ramalho, C. E., Laliberté, E., Poot, P., and Hobbs, R. J. 2014. Complex effects of fragmentation on remnant woodland plant communities of a rapidly urbanizing biodiversity hotspot. *Ecology* 95: 2466–2478.

Woodlands near Perth, Australia have become fragmented as a result of increased urbanization. This study considers a wide range of abiotic and biotic factors that can influence species richness on different spatial scales.

Wilson, E. O., and Simberloff, D. S. 1969. Experimental zoogeography of islands: defaunation and monitoring techniques. *Ecology* 50: 267–278.

Simberloff, D. S., and Wilson, E. O. 1969. Experimental zoogeography of islands: the colonization of empty islands. *Ecology* 50: 278–296.

These two classic papers appeared back-to-back in *Ecology*. They are important because of the lessons learned about island biogeography and its applications to modern conservation ecology. They are important for two other reasons. First, these papers show how much difficult field work can be needed to generate one graph in a paper. Second, these papers open up the question of what types of manipulations are ethical, and also how today's cultural values may differ from the values from a half-century ago.

END-OF-CHAPTER QUESTIONS

Review questions

1. What do we mean when we say there is a power function relationship between species richness and the area of an island. What do all the symbols in the relationship represent? What does a high *z*-value indicate?
2. Outline the theory of island biogeography, and the role played by island size and distance to a source of colonists.
3. What is Wallace's Line? Use the theory of continental drift to explain why Wallace's Line exists.
4. What are four different types of reduction in spatial configuration? In which cases of reduction is the surrounding matrix most likely to be an important consideration?
5. What do the results of the Biological Dynamics of Forest Fragments Project tell us about the importance of fragmentation and edge effects for understanding landscape processes. Explain your answer by discussing studies in the text.

Synthesis and application questions

1. If you look at Table 22.1, you will notice that there is little evidence for an island size effect on species richness, if you just consider the data for 1 year. For example, in 1969, a medium-sized island (R1) had the most species, while the largest island (J1) had relatively few species. Why might we fail to see a strong relationship between unmanipulated island size and species richness in this data set?
2. One can argue that larger islands might have higher immigration rates than smaller islands, because they form a larger target for immigrants. How would this affect the basic graphical model and the predictions generated by the theory of island biogeography? Draw a new graphical model that highlights this effect.
3. Looking at Figure 22.4, you will note that species richness for the three nearest islands actually climbed slightly above the initial survey values. The researchers suggest that population densities on all four islands were still very low at this point, and that the excess species could result from the absence of competition at low population densities. Can you

think of other factors that could cause this increase above equilibrium? How would you test your idea?

4. Figure 22.15 is drawn in a way that highlights the admittedly complicated analysis of retroposons. Redraw the figure representing the seven orders, using the more familiar format that we used in Figure 22.14.

5. In the text I mention that the drosophilids are the oldest Hawaiian clade, dating back to 26 million years. How can this be, if the oldest island dates to 5.2 MYA?

6. In Chapter 1, we defined a landscape as two or more interacting ecosystems. Compare that definition to our definition in this chapter. Use the concept of scale to show how they can be saying the same thing.

7. Suppose you were to do a meta-analysis that showed that corridors have no effect on species richness. Does that mean that corridors have no value? Explain your answer?

8. Figure 22.20 shows three different measures of hawthorn tree fitness at three spatial scales. Which relationship are you most confident about? What can you conclude based on the variation you observe in the data?

Analyse the data 1

Use Table 22.1 to investigate whether islands that received greater reduction in area also showed greater reduction in species richness.

Analyse the data 2

Use Table 22.2 to analyse whether extinction rates between each sized fragment were significantly different from each other.

Chapter 23

The carbon cycle and climate change ecology

INTRODUCTION

The May 7, 2010, issue of the journal *Science* included a letter entitled "Climate Change and the Integrity of Science" (Gleick *et al.* 2010). The letter was signed by 255 members of the United States National Academy of Sciences, including 11 Nobel Prize winners, and many scientists whose research is discussed in this book. Because it makes its point so cogently, I am presenting the letter's first paragraph as written:

> We are deeply disturbed by the recent escalation of political assaults on scientists in general and on climate scientists in particular. All citizens should understand some basic scientific facts. There is always some uncertainty associated with scientific conclusions; science never absolutely proves anything. When someone says that society should wait until scientists are absolutely certain before taking any action, it is the same as saying society should never take action.

This editorial was, in part, a response to several incidents that were covered by the popular press, including one involving the Intergovernmental Panel on Climate Change (IPCC). We begin this chapter with a discussion of the IPCC's mission, and how the language used by scientists (including the IPCC) is different from the language used by people in everyday conversations. Changes in the atmospheric concentration of carbon compounds are the foundation of climate change, so we explore the carbon cycle in considerable detail. We discover that global climate is influenced by direct effects, indirect effects, and many different types of feedback interactions. Some of this complexity is handled by climate models that make projections of future climate based on our understanding of physical processes and our measures of current global conditions. We conclude this chapter with a discussion of global change ecology, and how natural ecosystems are already changing in ways predicted by climate change models.

KEY QUESTIONS

23.1. How do biogeochemical processes move carbon through the global system?

23.2. How do greenhouse gases directly influence Earth's temperature?

23.3. How do indirect effects and feedback interactions influence global climate?

23.4. What are climate models and what do they tell us?

23.5. Are natural ecosystems changing in response to climate change?

CASE STUDY: The IPCC report, and public perception of climate change

The United Nations Environment Programme and the World Meteorological Organization established the IPCC to provide a clear scientific view on the current state of knowledge in climate change and its environmental and socioeconomic impacts. Its comprehensive 2007 report earned it the Nobel Peace Prize, but the report also had one factual mistake – a prediction that all Himalayan glaciers could melt by 2035. The IPCC subsequently retracted this prediction, but climate change deniers used this error to arouse suspicions about all climate change science.

The IPCC report also used language that seems strange to most nonscientists. Its core finding was that recent global warming was very likely caused by human activity (Pachauri and Reisinger 2008). What does "very likely" mean? And does this core finding mean that it is somewhat possible that global warming is not caused by human activity? The letter by Gleick and his colleagues in *Science* magazine attempts to explain that "very likely" is about as close to certainty as most scientists get about difficult questions. But using terms like "very likely" makes the public believe that the evidence for human-caused global warming is weak, and that there is serious debate in the scientific community about the IPCC's core finding.

Like other members of the general public, scientists use jargon and shorthand in electronic messages. Thus when a climate scientist had about 1000 email messages hacked from his computer server, it was not surprising that a few messages had imprecise language and statistical jargon that when taken out of context, suggested he was trying to hide findings that temperatures had decreased over a time period (Adam 2010). Members of the climate change denial community dubbed this incident "Climategate," and argued that it proved that scientists fudge their data and can't be trusted to dispassionately present their findings. And if their data are no good, then we can't trust their conclusions and recommendations.

An editorial by the editors of *Nature* entitled "Closing the Climategate" argues that "attitude changes are needed for science to ensure it holds the public's trust." The editorial describes that several investigations had exonerated the researcher whose files were hacked from any wrongdoing, but too late to forestall the damage in the court of public opinion. Using uncharacteristically strong language, the editorial continued, "climate scientists have to accept that they are in a street fight. They should expect a few low blows. The key is to learn which punches to roll with and which to counter."

So what is the science underlying climate change? How will climate change affect organisms, populations, communities, ecosystems, and landscapes? Has climate change already started to influence these different levels of the biological hierarchy? To address these questions, we must begin by exploring the carbon cycle.

23.1 HOW DO BIOGEOCHEMICAL PROCESSES MOVE CARBON THROUGH THE GLOBAL SYSTEM?

The carbon cycle describes how carbon moves between the atmosphere, the oceans and the land. Most of the world's carbon is stored in two reservoirs: carbonate rocks and deep ocean waters. Carbon in both of these reservoirs cycles very slowly, though upwelling of ocean waters contributes considerable quantities of carbon to the carbon cycle. To understand the link between the carbon cycle and climate change, we need to understand how carbon compounds build up in our atmosphere based on molecules that cycle relatively quickly as part of biogeochemical processes between the atmosphere and the oceans, and the atmosphere and the land (IPCC 2013).

The oceans: a critical CO$_2$ sink

Through a combination of chemical reactions and uptake and release by organisms, carbon compounds move back-and-forth between the oceans and the atmosphere. The oceans can hold a considerable amount of carbon dioxide because CO$_2$ is a weak acid, and ocean water is somewhat alkaline. CO$_2$ tends to diffuse rapidly into cold water, such as found in high latitudes, and out from warm tropical water into the atmosphere, particularly in areas of upwelling. Dissolved inorganic carbon is used by photosynthetic marine organisms (primarily phytoplankton) at the rate of about 50 Pg/yr (Pg = petagram = 10^{15} g). When these organisms die, much of this organic carbon (37 Pg/yr) is metabolized by decomposers and returned to the ocean surface carbon pool. But a smaller portion (11 Pg/yr) sinks deeper in the ocean to join the intermediate and deep ocean reservoir (Figure 23.1).

Most of the CO$_2$ uptake from the atmosphere reacts with ocean water to form H$_2$CO$_3$ (carbonic acid), which is converted into HCO$_3^-$ (bicarbonate) and CO$_3^{2-}$ (carbonate), as described by the following two equations:

$$CO_2 + H_2O \rightleftharpoons H_2CO_3 (\text{carbonic acid})$$
$$H_2CO_3 \rightleftharpoons H^+ + HCO_3^- (\text{bicarbonate})$$
$$H^+ + HCO_3^- \rightleftharpoons 2H^+ + CO_3^{2-} (\text{carbonate})$$

and

$$CO_2 + H_2O + CO_3^{2-} \rightleftharpoons HCO_3^- + H^+ + CO_3^{2-} \rightleftharpoons 2HCO_3^-$$

The approximate ratio of CO$_2$, HCO$_3^-$, and CO$_3^{2-}$ (collectively known as dissolved inorganic carbon) in the ocean is 1:100:10. Because CO$_2$ makes up such a small fraction of dissolved inorganic carbon, there is usually a favorable concentration gradient for diffusion of additional CO$_2$ into the ocean. Since industrialization, and the consequent increases in atmospheric CO$_2$ levels, the ocean has become an important CO$_2$ sink, currently importing 80 Pg/yr and exporting 78.4 Pg/yr of CO$_2$, which yields a net influx of 1.6 Pg/yr. While this feedback mechanism helps to regulate CO$_2$ levels in the atmosphere, it can cause some serious issues within marine ecosystems. Carbonate is alkaline, and serves as a buffer in the ocean. Adding CO$_2$ to the ocean reduces the amount

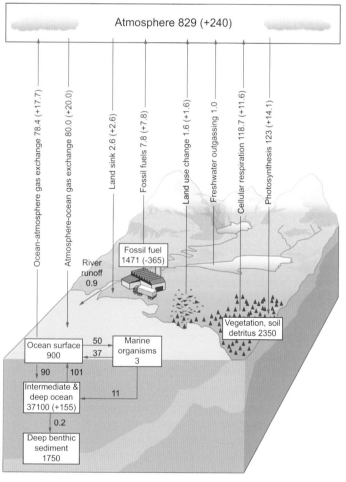

Figure 23.1 Major elements of the carbon cycle. Fluxes (in Pg/yr) are within arrows, while pools (in Pg) are enclosed within boxes (1 Pg = 10^{15} g). When pool ranges were given, I used the midpoint of the range. Numbers in parentheses represent the change in flux or pool since 1750, when humans began influencing carbon flow in a substantial way. Data from IPCC (2013).

of carbonate (as a result of the last equation above), which reduces the buffering capacity of the ocean and causes more (acidic) H^+ ions to remain in solution. The upshot is that high atmospheric CO_2 levels make the ocean more acidic. We introduced this problem in Chapter 2, and will discuss some of the ecological implications later in this chapter. We will now turn to CO_2 exchange between the atmosphere and terrestrial systems.

The terrestrial environment: a critical CO₂ source

From a quantitative standpoint, most of the CO_2 exchanges between the atmosphere and terrestrial systems are a result of the interplay of photosynthesis, which takes up CO_2 in the process of carbon fixation, and cellular respiration, which generates CO_2. Plants and decomposers in the soil do almost all of the cellular respiration; animals contribute minimally to global respiration. At present, these two processes are roughly in balance with each other, based on the estimates from direct measurements that are then extrapolated to the global system. Before industrialization, there were no other significant fluxes of CO_2 into the atmosphere from terrestrial sources.

By the middle of the twentieth century, scientists were fairly certain that industrialization and the burning of fossil fuels were increasing atmospheric CO_2 levels. Hans Suess (1955) tested the relationship between burning fossil fuels and CO_2 in the atmosphere indirectly. He knew that the radioactive form of carbon, ^{14}C, is present in minute quantities (about 1 part per trillion) in the atmosphere, and that it decays with a half-life of just over 5000 years. He also knew that fossil fuels formed from plants that lived in the Carboniferous period, about 300 MYA. These plants, like plants today, fixed CO_2 with similar small quantities of ^{14}C. When these plants died, the fixed $^{14}CO_2$ began to decay, and modern-day fossil fuel has virtually no ^{14}C. Consequently, when we burn fossil fuels, the CO_2 we add to the atmosphere has virtually no ^{14}C, thereby diluting the $^{14}CO_2$ content of the atmosphere. Following this logic, Suess predicted that atmospheric $^{14}CO_2$ concentration should be decreasing over time as more fossil fuel is burned.

Unfortunately, Suess was not a time traveler, but he recognized that trees kept a record of atmospheric CO_2 concentrations by fixing CO_2 during photosynthesis using the available atmospheric CO_2. When Suess extracted CO_2 from tree rings of various ages, he discovered, as he predicted, a much higher level of ^{14}C in the oldest rings – from about 1700 – than in the rings formed in the middle of the twentieth century. He concluded that the burning of fossil fuels was reducing the relative ^{14}C concentration in the atmosphere and reducing ^{14}C levels in CO_2 taken up during the carbon fixation stage of photosynthesis (Figure 23.2).

Since industrialization, humans have increased the global CO_2 flux, by burning fossil fuels and by manufacturing cement, which together add about 7.8 Pg/yr of CO_2, and by land-use changes, which add another 1.6 Pg/yr. The land-use changes include converting forests to less productive ecosystems, and using agricultural practices that reduce soil production and may lead to soil erosion and desertification, and further reduction in production. The fossil fuel flux is still increasing steadily, from an average of 5.4 Pg/yr during the 1980s to 6.4 Pg/yr during the 1990s, to 7.8 Pg/yr in the 2000–9 decade. The value for 2013 was up to 9.9 Pg/yr (Le Quéré et al. 2014).

Estimating net CO₂ flux: a work in progress

Understanding the net change in atmospheric CO_2 should be a simple accounting exercise, with influx balancing efflux. Researchers are confident about some fluxes, which have been measured accurately for a considerable period of time. Researchers are less confident about the mechanisms underlying some of the others fluxes, and consequently there is more error associated with these estimates. We will begin our analysis of net CO_2 flux by looking at some very reliable data on atmospheric CO_2 levels recorded at the Mauna Loa Meteorological Observatory in Hawaii.

Keeling's recordings of CO₂ levels at Mauna Loa
After getting his graduate degree in 1954, C. David Keeling landed a position as a postdoctoral associate at California Institute

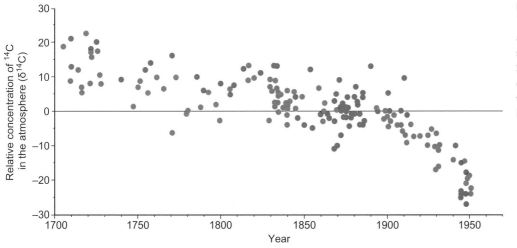

Figure 23.2 Decline in ^{14}C concentration in the atmosphere since 1700. Green symbols indicate data collected by Suess; red symbols are data collected by later researchers. Data were normalized to a value of 0 in 1880.

of Technology. Keeling's advisor was interested in the amount of carbonate in groundwater. Keeling was looking for a research project, but he knew two things about himself: he loved being outdoors, and he enjoyed making machines. So Keeling proposed that he would collect the groundwater measurements, in a variety of different (usually aesthetically beautiful) locations, and build a machine capable of taking the measurements. As it turns out, his protocol also required that he take accurate measurements of CO_2 in the air. With his high-quality instrumentation, Keeling found that all locations he measured had similar atmospheric CO_2 concentrations, which in 1956 were approximately 310 parts per million (ppm). This result was in direct conflict with the established dogma, which, based on inaccurate instrumentation, measured CO_2 levels ranging from 150 ppm in the Arctic to 350 ppm in some tropical regions. Once Keeling showed that regions in a variety of locations had similar values, he proposed to study variation over the course of a year, and later across many years, to identify any patterns of global atmospheric CO_2 levels (Keeling 1998).

Keeling found more permanent employment at Scripps Institution of Oceanography in California, where he helped establish two long-term CO_2 monitoring stations. One station was in Antarctica, while the second was on the slopes of Mauna Loa, Hawaii's tallest volcano, and home to the Mauna Loa Meteorological Observatory, which was established just as Keeling began his research career (Figure 23.3A). Keeling had incomplete data from Mauna Loa in 1958, which showed peak CO_2 levels in March–May and lowest CO_2 levels during October and November. The 1959 data were complete and showed the same trend more clearly (Figure 23.3B). Keeling reasoned that total global photosynthetic activity declined in fall and winter, allowing CO_2 to build up gradually in the atmosphere, peaking in late winter and early spring. By May and June, photosynthesis rates were increasing, and atmospheric CO_2 levels began to decline.

Intrigued, Keeling continued monitoring, and he observed that CO_2 levels were increasing from year to year. He recognized that the rate of CO_2 increase was variable, and appeared to have periods of rapid increase, followed by periods of slower increase. Fifty-five years of continuous data collection at the Mauna Loa observatory clearly show this increase (Figure 23.3C). Since 2000, the annual increase in atmospheric CO_2 levels has averaged about 4.2 Pg/year, or about 2.0 ppm/year.

 Thinking ecologically 23.1

Looking carefully at the Mauna Loa data, you can appreciate that the slope of the increase in CO_2 levels is somewhat variable from year to year. Use Figure 23.1 to propose hypotheses for this variation. Then propose ways of testing your hypotheses.

Direct data and using models

Keeling's data are extremely useful, but at the same time they are somewhat limited because they don't tell the complete story on

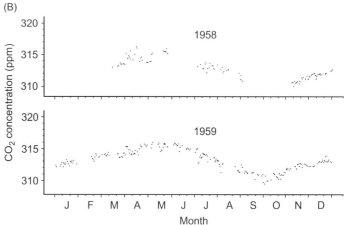

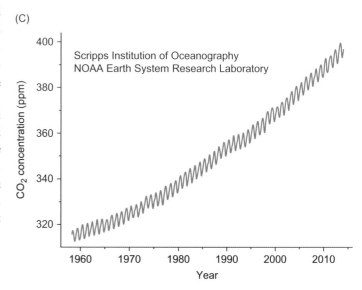

Figure 23.3 A. Mauna Loa observatory. B. Mean daily atmospheric CO_2 concentration (in parts per million) at Mauna Loa Observatory in 1958 (top) and 1959 (bottom). C. Mean atmospheric CO_2 concentration (ppm) at Mauna Loa Observatory.

CO_2 levels. Keeling chose the Mauna Loa observatory because it measures CO_2 levels in a well-mixed troposphere far removed from significant terrestrial inputs. For this reason, and because it has the oldest reliable data, scientists often cite the Mauna Loa data to show how atmospheric CO_2 levels are changing over time. However, there still are some differences between locations. Some stations record more often than others, and some use more accurate instruments. There are more recording stations in north temperate latitudes than in tropical and south temperate latitudes. Some locations are susceptible to local or regional influences on CO_2 levels from power plants or other industrial sources. Thus when scientists are interested in local and regional CO_2 levels, they use models to account for some of these variables.

I point this out because climate change science has been criticized for its extensive use of forecasting models, but the reality is that even direct data measurement must, in many cases, be modeled before it has real value. As we discussed in Chapter 9, models must be validated before they are accepted for general use, and all models should be viewed as works in progress that are refined as we learn more about the processes that influence the data they represent and the projections they make. Models have been particularly helpful in estimating CO_2 transfers between the atmosphere and oceans and between the atmosphere and land.

The ocean CO_2 flux

The flux of CO_2 between the atmosphere and the ocean is determined by a variety of environmental factors, which differ at each location. The reaction is expressed as follows:

$$\text{Flux} = -K_g K_0 (pCO_2{}^{sur} - pCO_2{}^{at})$$

Where

K_g = the air–sea gas transfer coefficient, which varies in relation to temperature, wind speed, and salinity,

K_0 = the solubility of CO_2 at a particular temperature and salinity

$pCO_2{}^{sur}$ = partial pressure of CO_2 in the surface water

$pCO_2{}^{at}$ = partial pressure of CO_2 in the atmosphere

Remember that *partial pressure* is the pressure that a gas would exert if it were the only gas occupying a particular volume. So all things being equal, increasing CO_2 concentration will increase CO_2 partial pressure. The flux equation tells us that CO_2 diffusion into the ocean (influx) happens more quickly when the difference between the concentration of CO_2 in the atmosphere above the ocean and the surface water $[pCO_2{}^{sur} - pCO_2{}^{at}]$ is at a maximum. As humans increase the amount of CO_2 in the atmosphere, $[pCO_2{}^{sur} - pCO_2{}^{at}]$ becomes more negative, making total flux more positive, and increasing the influx of CO_2. In contrast, there are regions of the ocean with unusually high $pCO_2{}^{sur}$, which results in a positive value for $[pCO_2{}^{sur} - pCO_2{}^{at}]$ and a negative flux value, causing an outgassing of CO_2.

K_g and K_0 are influenced by several environmental factors. Physicists can relatively easily describe the relationship between solubility, temperature, and salinity under controlled laboratory conditions. But the influence of wind speed on K_g is much more difficult to estimate because it depends on several related

phenomena that are very difficult to predict, such as the amount of ocean turbulence caused by the wind on the ocean surface and the size and number of bubbles the wind creates in the ocean. But in general, diffusion occurs more rapidly at high wind speed, as more ocean surface is exposed to the atmosphere

Oceanographers have embarked on research cruises to collect data on atmospheric and ocean CO_2 concentrations, ocean salinity, and wind speed, which they then enter into the ocean flux equation. Of course, this approach has problems. For example, wind speed is not constant, and the effect of wind speed on solubility is not linear and has error associated with it. Second, one location in the ocean is very different from another, so it is critical to measure from a very large number of locations. Third, different locations have different CO_2 concentrations in relation to time of day and year. Mathematical models allow us to extrapolate flux estimates for unmeasured locations and times, but these extrapolations, of course, introduce some error or uncertainty into our measures of flux.

There are several other methods oceanographers use to estimate flux. Two important points must be made in regard to these additional estimates of ocean flux. The first is that by and large these approaches use independent methods of making their estimates, so they are unlikely to be biased by using the same assumptions. Second, the different approaches to estimating flux all have generated very similar values. Consequently, despite all the challenges associated with estimating ocean flux, scientists are quite confident that the estimate is reasonably accurate.

Some ocean waters are CO_2 sinks, while other regions are sources. We can understand a portion of these differences based on physical properties. For example, SSTs are much colder at high latitudes than they are in the tropics. The solubility of CO_2 in water is greatest at cold temperatures (which is why some of our favorite carbonated beverages tend to go flat if we leave them out on a summer day).

Focusing on the north Atlantic Ocean, surface winds and the Coriolis effect bring water from middle latitudes to higher latitudes, where they rapidly cool. CO_2 diffuses into these cold surface waters very quickly. Cold water is denser than warm water, so these CO_2-rich waters sink into the ocean depths, beginning a very slow southern journey that is part of the meridional overturning circulation, also known as the *great ocean conveyer* (Broecker 2010). This entire path travels south through the Atlantic, and then eastward into tropical oceans, where these waters move into shallower waters as they approach continental landmasses (Figure 23.4A). Nearing the surface, these waters warm, which reduces CO_2 solubility. Consequently, tropical waters in the Indian and Pacific Oceans tend to release CO_2 into the atmosphere, serving as CO_2 sources (Figure 23.4B).

Scientists are very confident about how much CO_2 is being emitted from human activity over land, and reasonably confident about how much CO_2 the ocean is taking up. They also have a solid understanding of why certain parts of the ocean are sinks and others are sources, and how that varies over time. However,

(A)

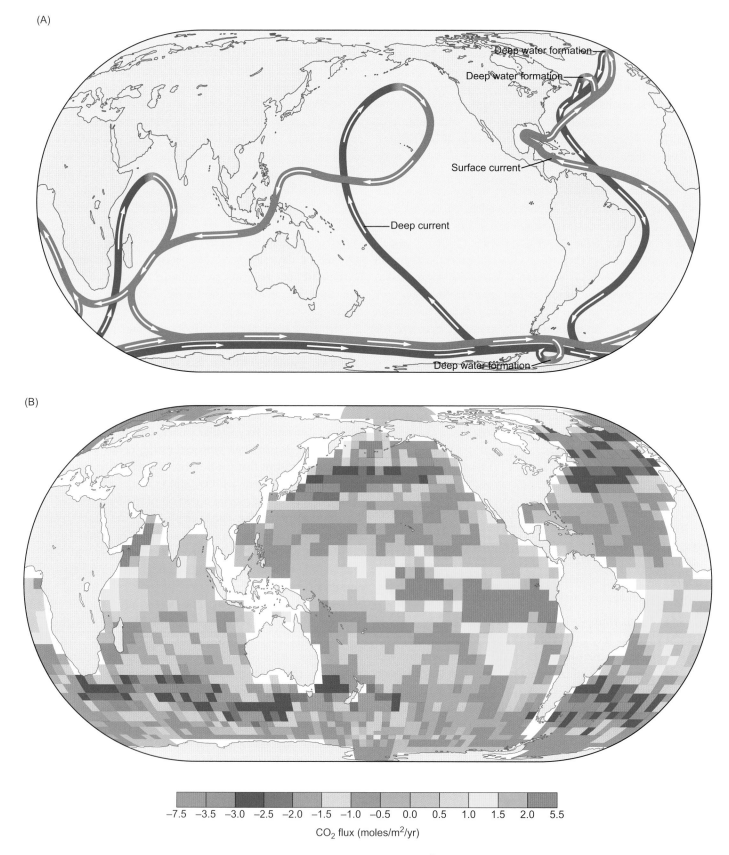

(B)

−7.5 −3.5 −3.0 −2.5 −2.0 −1.5 −1.0 −0.5 0.0 0.5 1.0 1.5 2.0 5.5

CO_2 flux (moles/m^2/yr)

Figure 23.4 A. The meridional overturning circulation. B. Global CO_2 flux (moles/m^2/yr).

scientists are still attempting to understand processes underlying CO_2 uptake over land.

The land CO_2 flux

The amount of CO_2 cycling through the global system is conserved, so that CO_2 sources must equal CO_2 sinks. Since industrialization, anthropogenic, or human-caused, sources are fossil fuel burning and cement production, and land-use changes that reduce global photosynthesis. Additional sinks are uptake by the atmosphere, uptake by water, and uptake by land (Table 23.1). These numbers change from year to year based on variations in emissions and land use and also environmental changes that influence the amount of CO_2 uptake by land and water. The land flux is the most poorly understood process in Table 23.1 and has considerable uncertainty associated with it.

Recently researchers have identified several important components of the land sink. Some regions of the world are recovering from previous deforestation or other types of land use, and thus are taking up more carbon from increased photosynthetic rates. Unfortunately, tropical forests are being logged or burned at very high rates, and logging-related activities are increasing decomposition rates, converting previously forested areas from carbon sinks into sources. But on balance, forests across the world are a net carbon sink. Yude Pan and her colleagues (2011) estimated changes in forest carbon pools in aboveground and belowground biomass, forest soils, deadwood, and leaf litter. When possible, they used a *stock change* approach, which measured carbon stocks (based on plant biomass) at two different times, and then used the difference between stock estimates as the rate of change over that time period. If these data were lacking, the researchers used carbon stock measures taken at one point in time, and assumed default rates of carbon increase from forest growth, and carbon loss from harvest, fires, and decomposition, for each country in the world (Figure 23.5). Based on this complex bookkeeping operation, the researchers estimated that the world's forests were a carbon sink of about 2.30 (±0.49) Pg/yr, which accounts for almost all of the land sink identified in Table 23.1.

In 2011, the estimated land carbon sink was a record 4.03–4.10 Pg/yr. This increase was driven by record high rainfall in the

Table 23.1 Estimated mean anthropogenic CO_2 sources and resulting sinks for 2000–9 (based on IPCC 2013).

Global sources/amounts (Pg/yr)		Global sinks/amounts (Pg/yr)	
Fossil fuels and cement	7.8	Atmospheric uptake	4.0
Land use	1.1	Ocean sink	2.3
		Additional land sink	2.6
Total	8.9	Total	8.9

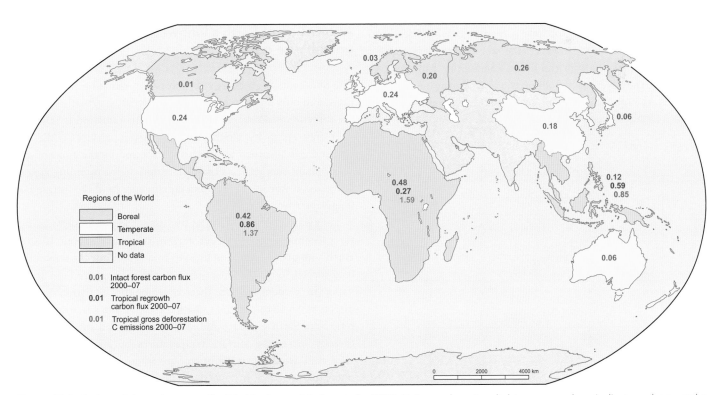

Figure 23.5 Carbon sinks and sources (Pg C/yr) in the world's forests for 2000–7. For each region, light green numbers indicate carbon uptake by intact forests. Tropical forests also contain (dark green) numbers that represent carbon uptake by regrowth forests and (in red) numbers that represent carbon emissions in recently deforested areas. Thus for each tropical region, green represents a carbon sink, and red a carbon source.

southern hemisphere caused by a several year stretch of La Niña conditions. La Niña episodes are caused by a buildup of unusually cool subsurface waters in the tropical Pacific, which are brought to the surface by strong currents in areas of upwelling. This brings regional cooling and increased rainfall to many locations. The primary effect of greater southern hemisphere rainfall was a substantial expansion of vegetation cover in semi-arid biomes in the southern hemisphere, which increased photosynthetic uptake of CO_2 from the atmosphere. A secondary effect of this rainfall was a reduction in fire emissions, which decreased CO_2 export from these biomes (Poulter *et al.* 2014).

A less intuitive carbon sink has emerged from studies of inorganic absorption of CO_2 by alkaline soils in deserts. Previously, researchers had assumed that deserts played a limited role in the carbon cycle, with the sparse vegetation in this arid biome supporting relatively little photosynthetic CO_2 uptake. Similarly, CO_2 efflux would be limited by the low cellular respiration rates of the impoverished group of plants, animals, and soil microorganisms.

Surprisingly, several recent studies indicate that alkaline desert soils may be an important sink in the global carbon cycle. Using an automated soil flux system, Jingxia Xie and his colleagues (2009) compared soil CO_2 flux on the floor of the Gubantonggut Desert in southern China to CO_2 flux from a location that had previously been very similar desert, but was converted to cropland in 1990. When they placed their sensors in bare soil in both locations, the resulting time series showed a much higher rate of cellular respiration in the cropland soil. This was expected, because the cropland soil should contain a much more abundant community of soil microorganisms, which leads to higher CO_2 production. However, the researchers were startled to find that the soils from the desert community had negative flux – that they were functioning as a sink during most of the day, even though there were no photosynthetic organisms in the soil to fix CO_2 (Figure 23.6).

Xie and colleagues suggested that highly saline and alkaline desert soils are behaving very much like the ocean waters we discussed previously. CO_2 in the atmosphere is converted by the alkaline soil water to carbonic acid, which neutralizes the alkalinity of the soil by forming bicarbonate and carbonate. In a sense, these desert soils are acting like giant ocean sinks. Researchers need to find out how common this process is before they can estimate its contribution to the global carbon cycle.

There is still much to learn about the global carbon cycle, but there is no doubt that industrial and agricultural development is increasing atmospheric CO_2 to historically high levels. We can use physical principles, historical data, and modeling to predict, with some precision, the long-term consequences of these changes. To do so, we must consider numerous other factors that influence Earth's temperature.

23.2 HOW DO GREENHOUSE GASES DIRECTLY INFLUENCE EARTH'S TEMPERATURE?

Carbon dioxide is only one type of greenhouse gas – a molecule in the atmosphere that warms Earth's surface because it absorbs some of the longwave radiation emitted from Earth's surface. Other greenhouse gases include methane, chlorofluorocarbons, nitrous oxide, water vapor, and ozone. Based on its molecular structure, each greenhouse gas has its particular effect on the radiation that is emitted from Earth.

Absorption of Earth's longwave radiation by greenhouse gases

The Sun emits high-energy or shortwave radiation in all directions, and only a paltry portion of this radiation actually heads in the direction of Earth. Almost 30% of this radiation is reflected back into space by clouds and by Earth's surface, while the remainder is absorbed by Earth's surface and by the atmosphere. This radiant energy heats the surface and the surrounding atmosphere. As it warms, Earth emits radiant energy in the infrared or longwave spectrum.

Longwave radiation emitted by Earth is key to understanding the greenhouse effect. Under current conditions, about 10% of Earth's longwave radiation escapes into space, but the remainder is absorbed by greenhouse gases in the atmosphere (and water vapor in clouds), which then re-emits this radiation in any direction. Some re-emitted radiation escapes into space, and some returns to Earth's surface and may be reabsorbed and re-emitted, thereby increasing surface temperatures by about 21°C above what they would be without any greenhouse gases in the atmosphere (Figure 23.7). Increased greenhouse gas emissions are reducing the amount of escaping longwave radiation, increasing the amount of longwave radiation returning to Earth's surface, thereby increasing Earth's surface temperature.

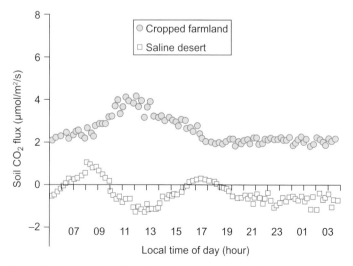

Figure 23.6 Soil CO_2 efflux (positive values) and influx (negative values) over the course of a day.

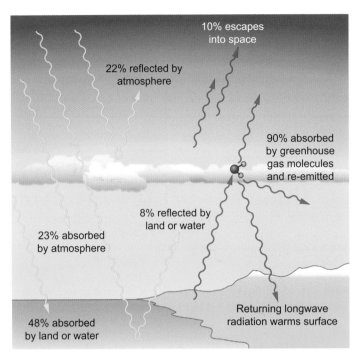

Figure 23.7 Fate of solar radiation (yellow) as it enters Earth's atmosphere, and of longwave radiation (red) as it is emitted from Earth's surface.

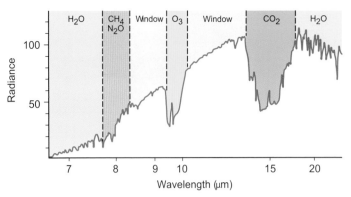

Figure 23.8 Amount of thermal radiation transmitted to a satellite orbiting above the Mediterranean Sea. Note that much of the emitted energy got through to the satellite at 8–14 μm (the atmospheric window). However, you can see the depression in the amount transmitted at 9–10 μm, corresponding to absorption by ozone. Also note the reduction in transmission at 13.5–17 μm corresponding to absorption by CO_2.

Neither N_2 nor O_2, the atmosphere's two most abundant gases, absorb significant longwave radiation. Rather, water vapor and CO_2 are the world's most important greenhouse gases, because they are relatively abundant in the atmosphere and they absorb longwave radiation at several different wavelengths. Later we will discuss water vapor's complex impact on climate. For now, we will introduce some of the other greenhouse gases.

Radiative forcing of greenhouse gases

Methane (CH_4), nitrous oxide (N_2O), and ozone (O_3) are trace gases naturally found in Earth's atmosphere that have a substantial effect on the amount of infrared radiation that is absorbed by the atmosphere. Each molecule's structure makes it effective at absorbing longwave radiation of one or more specific wavelengths, and ineffective at absorbing most other wavelengths. For example, ozone is most effective at absorbing longwave radiation with wavelengths between 9.5 and 10 μm. Thus an atmosphere with more ozone will absorb more of Earth's thermal radiation in that wavelength range (Figure 23.8). With the exception of ozone's activity, no natural greenhouse gases absorb in the wavelength range of 8.2–14 μm, which because of its transparency to radiation is known as the *atmospheric window* of the spectrum.

As a result of human activities, the abundances of methane and nitrous oxide have climbed quite sharply since 1850, but the factors leading to these increases are not as well understood as those affecting CO_2, and the path of increase has been less steady than we've seen for CO_2. According to the IPCC report (IPCC 2013), the two major sources of human-generated methane emissions are the emissions from coal, oil, and natural gas mining activities, and the slightly more natural fermentation processes that release natural gas from the digestive systems of livestock. Methane concentrations actually leveled out in the early 2000s but have resumed their climb in recent years.

Halocarbons such as $CFCl_3$ and CF_2Cl_2 (popularly known as CFCs, which stands for chlorofluorocarbons) are present in the atmosphere almost entirely as a result of humans using them to pressurize spray cans and as refrigerants. Because they destroy ozone in the stratosphere, several types of halocarbons were banned by the Montreal Protocol, which required them to be phased out and completely eliminated by 1996 in industrialized countries and by 2006 in developing countries. Other types of halocarbons that don't destroy ozone, but do act as greenhouse gases, are still being manufactured, so halocarbons remain an important problem.

Radiative forcing (measured in W/m^2) describes the influence that a change in a climatic factor, such as a greenhouse gas, has on altering the balance of incoming and outgoing radiation, as measured at the top of the troposphere. Essentially, radiative forcing measures the global warming effect of a particular factor. By reducing the amount of longwave radiation that escapes Earth's atmosphere, CO_2 has a positive radiative forcing. In most cases, a radiative forcing of 1.0 W/m_2 should increase global surface temperatures by about 0.5°C.

The global warming potential measures the radiative forcing by a known mass of a greenhouse gas in comparison to the radiative forcing of the same amount of CO_2 over a certain time period (usually 100 years). Because it is a relative measure that compares gases, the global warming potential is expressed numerically without units. All things being equal, a molecule that has a quick turnover time in the atmosphere would have a lower global warming potential than a molecule with a longer

Table 23.2 Direct radiative forcing since 1750, and global warming potential over a 100-year timeframe, of some greenhouse gases.

Molecule	Abundance in atmosphere[a]	Mean turnover time (years)	Global warming potential	Direct radiative forcing (W/m^2)
Carbon dioxide	400 ppm	not easily measured[b]	1	1.82
Methane	1.8 ppm	12.4	28	0.48
Nitrous oxide	324.2 ppb	121	265	0.17
CFC-12 (CCl_2F_2)	0.528 ppb	100	10 720	0.17

[a]ppm = parts per million, ppb = parts per billion.

[b]While more than half of the CO_2 emitted is currently removed from the atmosphere within a century, some fraction (about 20%) of emitted CO_2 remains in the atmosphere for thousands of years. Data from IPCC (2013).

turnover time. Also, greenhouse gases that operate in the atmospheric window of the spectrum and are present in very low quantities, including most halocarbons, have a higher global warming potential than more common gases such as CO_2. You can make an analogy to a cup of black coffee. Adding one teaspoon of milk makes your coffee much whiter, but adding a second teaspoon causes a much smaller change, and a third teaspoon a still smaller change. Halocarbons are very long-lived, operate within the atmospheric window of the spectrum, and are relatively rare in the atmosphere. Consequently one halocarbon molecule may have more than 10 000 times the global warming potential of one molecule of the already-abundant CO_2.

Greenhouse gases comprise a complex cast of molecular characters, each with their own direct effects on the atmosphere (Table 23.2). Carbon dioxide is the most important greenhouse gas, as measured by radiative forcing, because it is by far the most abundant. Methane is also very important, because it is relatively abundant and has a much greater global warming potential than CO_2, but its impact is somewhat reduced because it is relatively short-lived in the atmosphere. Several other gases are also quite important, and there are many others that we have not mentioned, but that have – at this point – relatively small effects on radiative forcing.

But we have only considered the direct effects of greenhouse gases. We need to recognize that greenhouse gases can also have indirect effects on radiative forcing. In addition, there are numerous other climatic factors that influence global climate directly and indirectly.

23.3 HOW DO INDIRECT EFFECTS AND FEEDBACK INTERACTIONS INFLUENCE GLOBAL CLIMATE?

In addition to direct radiative forcing by greenhouse gases, numerous other factors influence climate change in a variety of different ways. These factors are both physical and biological, and in many cases they have complex interactions with each other. One example of this complexity is the interaction between stratospheric ozone and halocarbons.

Indirect effects of halocarbons on global temperature

We've already discussed how halocarbons have increased global temperature by absorbing longwave radiation emitted by Earth in the atmospheric window of the spectrum. But halocarbons can influence global temperatures through a second, less direct effect. In 1974, Mario Molina and F. S. Rowland published a paper that alerted the scientific community to the threat of CF_2Cl_2 and $CFCl_3$, two halocarbons that were commonly used as refrigerants and as propellants for foam and aerosol products (including asthma sprays). The researchers argued that when these two commonly used halocarbons made their way to the upper stratosphere, they could be broken down by ultraviolet radiation, which liberated highly reactive chlorine atoms. They proposed that in the stratosphere, the following two reactions would occur:

$$Cl + O_3 \rightarrow ClO \text{ (chlorine monoxide)} + O_2$$
$$ClO + O \text{ (atomic oxygen)} \rightarrow Cl + O_2$$

Overall, no chlorine is lost in this proposed process of ozone destruction. Consequently, a relatively small quantity of atmospheric chlorine could cause extensive ozone destruction in the stratosphere.

Ozone's peak absorption is at 9.6 μm, which is in the atmospheric window of the spectrum. Because halocarbons destroy ozone, they also indirectly reduce global temperatures, with a radiative forcing estimated at -0.15 ± 0.10 W/m2. Thus our efforts to eliminate halocarbons and conserve ozone, which we will discuss in more detail at the end of this chapter, actually increase global warming by increasing absorption of longwave radiation in the atmospheric window spectrum. Unfortunately, most of the refrigerant and propellant molecules that have replaced halocarbons are also powerful greenhouse gases with very high global warming potentials.

Many interactions associated with climate change involve scenarios in which one factor influences a second factor, which then influences the first factor. These feedback interactions can involve both physical and biological processes.

Complex feedback interactions between surface temperature and water vapor

Warmer temperatures cause higher water evaporation rates, leading to increased atmospheric water vapor. Water vapor is a powerful greenhouse gas, so the direct effect of higher surface temperature is to increase the absorption of Earth's longwave radiation, thereby increasing Earth's surface temperature. This positive feedback interaction is a very important source of global warming.

But water vapor condenses to form clouds, which would be expected to increase in abundance with the increase in the amount of water in the atmosphere. Clouds reflect a substantial portion of the Sun's solar radiation back into space, and thus tend to cool the Earth's surface. This is an example of negative feedback. The amount of radiation reflected by a surface is its albedo, which is very important for understanding global temperature change. Shiny surfaces with lighter color, such as clouds, have higher albedo than do dark surfaces.

Finally, clouds also absorb longwave thermal radiation emitted from Earth's surface, and behave like greenhouse gases by emitting some of this radiation back toward Earth's surface, thereby increasing Earth's surface temperature. Those of you who periodically sleep outdoors have probably noticed that cloudy nights tend to be warmer than clear nights. This is another example of positive feedback.

Thinking ecologically 23.2

When glaciers and snowfields melt, they expose Earth's surface underneath. Will this tend to increase or decrease surface temperatures? Is this an example of positive or negative feedback? Explain your answer in terms of the albedo of the surfaces involved in this process.

Many climate feedback interactions involve biotic factors. One interaction that we've already introduced is the effect of increased atmospheric CO_2 on plant primary production.

Increases in CO_2 uptake by plants at elevated atmospheric CO_2 concentrations

Greater atmospheric CO_2 concentrations should, in theory, make more CO_2 available for plant primary production. If true, plant primary production would have a negative feedback effect on atmospheric CO_2 levels. It is very difficult to isolate CO_2 effects on a global level, however, because there are so many other

Figure 23.9 FACE experiments set up at the Harshaw Experimental Forest in northern Wisconsin, USA, where researchers can manipulate CO_2 and ozone concentrations by controlling gas outflow from the pipes surrounding each study site.

factors that influence plant production, and these factors differ throughout the globe. One approach is free-air CO_2 enrichment (FACE) experiments, in which researchers release a constant flow of CO_2 into the environment above open-air vegetation plots, thereby increasing the atmospheric CO_2 concentration within a prescribed area (Figure 23.9). This approach allows ecologists to test for CO_2 effects under controlled conditions.

Elizabeth Ainsworth and Stephen Long (2005) wanted to know what these FACE experiments were actually telling us about CO_2 effects on plants. They compiled data from 120 studies in which CO_2 levels were increased to 475–600 ppm, approximately double the preindustrial concentration. The ecosystems studied included forests, grasslands, deserts, and crops of various types. In all of their meta-analyses, the researchers compared the high-CO_2-concentration plots to unmanipulated control plots.

Effects on photosynthesis and water use

Exposure to high CO_2 concentrations increased photosynthetic rates by 31% and carbon uptake during the day by 28%. In addition, *transpiration efficiency* – a measure of the amount of biomass produced in relation to the amount of water lost from transpiration – increased by 50% in comparison to controls. Presumably, at higher CO_2 levels, plants were able to keep their stomata closed or partially closed for extended periods of time, which reduced their transpiration rates. Trees showed the greatest enhancement in photosynthetic rates, followed by crops and grasses (Figure 23.10A)

Recall from Chapter 4 that rubisco, the enzyme that binds CO_2 in C_3 plants, is less efficient than PEP carboxylase, the corresponding enzyme in C_4 plants. Consequently C_3 plants must keep their stomata open for longer periods, and thus tend to lose more

(A)

(B)

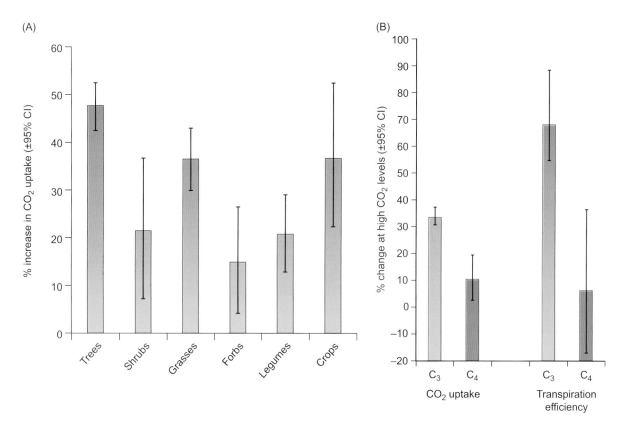

Figure 23.10 A. Mean percent increase in CO_2 uptake (±95% confidence intervals) for different plant groups at high CO_2 levels, based on a meta-analysis of 120 CO_2 supplementation studies. B. Mean % change (±95% confidence intervals) in CO_2 uptake and transpiration efficiency at high CO_2 levels in C_3 and C_4 plants.

water through transpiration, so we would expect C_3 plants to benefit more than C_4 plants from CO_2 enrichment. Specifically, we would expect a greater increase in CO_2 uptake and transpiration efficiency, as C_3 plants would have the luxury of keeping their stomata partially closed during the day, and thus suffer less water loss than they would under ambient CO_2 conditions. Ainsworth and Long's meta-analysis showed substantially greater benefits for C_3 plants in both of these variables (Figure 23.10B). However, these findings were based on only five species of C_4 plants, so they should be viewed with caution.

Plant growth and crop yield

One important question for understanding the negative feedback between CO_2 in the atmosphere and CO_2 uptake by plants is whether the increased carbon uptake is actually stored by plants. A second important question, from the perspective of feeding future generations of humans, is whether crop yields increase with higher photosynthetic rates. Based on Ainsworth and Long's meta-analysis, the answer to both questions is a provisional "yes." Aboveground biomass increased about 20% in C_3 plants, but did not increase in C_4 plants. Tree biomass increased 28% in this analysis, while grasses and legumes showed much smaller increases from CO_2 enrichment. Lastly, crop yield increased 17% overall, but only significantly in one crop (cotton) and not significantly in the three food crops (rice, wheat, and sorghum).

This value is much lower than was found in CO_2 enhancement studies conducted in isolated chambers, and is disappointing from the standpoint of projecting yields under the high atmospheric CO_2 conditions that we expect to experience in future years.

Overall, this meta-analysis indicates that there is an important negative feedback involving atmospheric CO_2 and photosynthesis. Increased atmospheric CO_2 levels cause greater CO_2 uptake by plants, which should reduce the rate of increase of atmospheric CO_2 levels. Unfortunately, this increase in CO_2 uptake is not sufficient to offset the greater CO_2 emissions by the increasing human population. Also, unfortunately, there are many indirect effects that offset the benefits of this negative feedback.

Indirect effects of CO_2 uptake by plants

Recall from the previous meta-analysis that high atmospheric CO_2 levels cause plants to use water more efficiently, losing less water to transpiration. One possible consequence is that soil moisture should increase, which in some ecosystems could reduce the oxygen content of the soil. Anoxic soils are hospitable environments for anaerobic organisms, including methanogens that generate methane as a byproduct of their metabolic processes, and anaerobic denitrifying bacteria that produce nitrous oxide. The upshot is that high CO_2 levels could indirectly generate more greenhouse gases (Figure 23.11)

Figure 23.11 Anaerobic conditions created in rice paddies, such as these in China, release CH_4 and N_2O, partially offsetting gains from increased CO_2 uptake.

Kees Jan van Groenigen and his colleagues (2011) tested this hypothesis with a meta-analysis of 49 studies in which CO_2 levels were artificially enhanced. They used some of the FACE studies that we described in the previous section as well as greenhouse and growth chamber studies. In support of the indirect effects hypothesis, nitrous oxide increased by 18.8% and methane shot up by 43.4% in rice paddy soils and 13.2% in wetland soils. The researchers extrapolated these results to the global abundance of rice paddy and wetland soils, and converted these methane and nitrous oxide increases to CO_2 equivalents using the global warming potentials of each gas (see Table 23.2). These extrapolations indicated that about 17% of the increased CO_2 uptake from enhanced photosynthesis is negated by this indirect process that creates greater methane and nitrous oxide emissions.

At this point, the world is beginning to sound like a very complicated place, with an almost unmanageable complexity of gas input and output, considerations of albedo, and interactions with organisms. The reality is that we have only scratched the surface of this complexity. There are numerous known effects that we have not discussed, and doubtless other effects that have not been discovered. Fortunately, researchers have developed models that can handle some of this complexity, and scientists have developed computer technology that can run these models.

23.4 WHAT ARE CLIMATE MODELS AND WHAT DO THEY TELL US?

General circulation models, and their most recent extensions, *coupled atmosphere–ocean general circulation models* (AOGCMs) and *coupled earth system models* (CESMs), use the movement of the atmosphere around Earth, and the interaction of the atmosphere with the oceans and with biological processes, to project future climate. These models have certain structural and computational features in common with each other, but they also differ in some important ways. These differences lead to quantitatively different projections about future global climate change. Impressively, most of the projections are qualitatively very similar.

Climate models: a set of equations that operate on a three-dimensional grid of Earth's surface, the oceans, and the atmosphere

Climate models simulate planetary heat exchange using equations that were derived and tested by physicists more than a century ago, coupled with more recent data and equations that describe the movement of the atmosphere, such as Hadley and Ferrell cells and the Coriolis effect (as discussed in Chapter 2), SST, humidity, friction, and gravity. Later models incorporated the carbon cycle. There is no doubt that more complex models will develop, particularly as our technology allows for more precise measurement, and for processing increasingly larger data sets in more powerful computers. Presently more than 20 coupled AOGCMs have been developed, and are constantly being improved. The most recent models have taken advantage of greater data availability and greater computer power to consider inputs on an increasingly finer spatial scale.

Climate model projections: a much warmer and wetter global climate

What are the projections of global change for the next century? The answer depends on human behavior. Richard Moss and his colleagues (2010) prepared a report for the 2013 IPCC that outlined four representative concentration pathways (RCPs) that integrate what scientists know about radiative forcing with an analysis on how socioeconomic conditions influence human behavior regarding adaptation and mitigation. These RCPs are named after the amount of radiative forcing caused by CO_2 levels assumed for the year 2100 (Table 23.3).

It is impossible for us to know which of these RCPs will most closely represent reality; they were established as a way of projecting what is most likely to occur if we make certain assumptions about emission levels. Different global models use these RCPs as a basis for projecting climate change into the future using a broad range of emissions scenarios. Let's explore how global models project climate change over the next century and beyond.

All of the AOGCMs project substantial temperature increases over the next century (Figure 23.12A). Not surprisingly, the temperature increase is smallest for *RCP2.6* and greatest for *RCP8.5*. Until 2035, there is very little difference in response of the four RCPs, but thereafter, the differences become much more pronounced. Temperatures continue to increase slowly for *RC4.5* through 2300, even though emissions are stabilized by 2100. This indicates that there is a time lag between the time when emissions

Table 23.3 The four RCPs used by the 2013 IPCC.

Name	Radiative forcing	Concentration of CO_2 (ppm)	Pathway
RCP8.5	>8.5 W/m$_2$ by 2100	>1370	CO_2 still rising rapidly at 2100
RCP6.0	6 W/m$_2$ by 2100	850	CO_2 rising slowly at 2100
RCP4.5	4.5 W/m$_2$ by 2100	650	CO_2 stable at 2100
RCP2.6	Peaks at 3 W/m$_2$ at 2035, then declines to 2.6 W/m$_2$ by 2100	Peak at 490, then declines	CO_2 declining at 2100

(A)

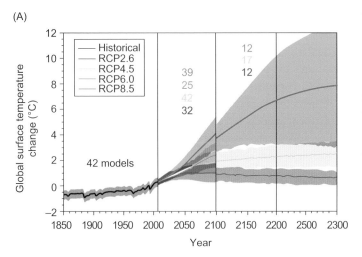

(B)

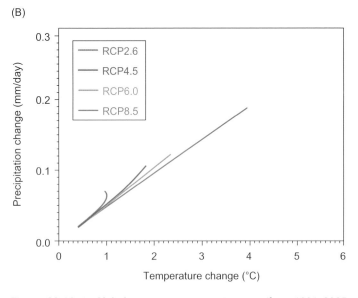

Figure 23.12 A. Global mean temperature increase (from 1986–2005 average) projected by AOGCMs under the four RCPs. Column of numbers represents the number of different models used to calculate the mean curve until 2100 (left column) and 2300 (right column). *RC6* was only run until 2100. B. Global mean precipitation increase (from 1986–2005 average) in relation to temperature increase projected by AOGCMs under the four RCPs.

are stabilized, and when warming is stabilized. The models also indicate that air temperatures warm more over land than over oceans, and more in the northern hemisphere than the southern hemisphere. As we discussed in Chapter 2, water has a greater specific heat than land, so we would expect the air over land to heat relatively rapidly, and the air in the northern hemisphere, with its relatively greater land surface area, to also heat more rapidly.

All of the models agree that higher temperatures will lead to greater evaporation rates of the ocean waters, resulting in a higher atmospheric water content. This water will condense to form clouds, which will increase the amount of rain globally by about 1–3% per 1°C temperature increase. This corresponds to about 0.5 mm/day increase for each 1°C temperature increase (Figure 23.12B). Most of the models also project an increase in the intensity of storms, because cloud formation releases thermal energy into the atmosphere, resulting in a more intense hydrological cycle and in more severe storms. Somewhat counterintuitively, most of the models project an increase in droughts, partially from high rates of water loss from evapotranspiration, and partially from most precipitation falling during relatively few very intense storms that are scattered throughout the year. With these intense storms, much of the water will run off without being absorbed by the soils.

Are these models correct in their projections? Though the specific numbers vary, all of the available evidence supports these projections, but there are still some areas of uncertainty.

Testing climate models

Long-term projections are notoriously difficult to test. This situation is exacerbated by the problem of having only one Earth. There are no parallel worlds (that we know of) to allow us to increase our sample size above one. Fortunately, there are approaches researchers can use to increase their confidence in the projections. These approaches include increasing our understanding of the underlying mechanisms, testing the projections of early models against current conditions, and using models to explain climatic variation in the distant past.

Increased understanding of mechanisms

One reason for having confidence in current climate models is that they are based on well-established physical principles such as conservation of energy, mass, and angular momentum. Early models were somewhat more flawed, because some of the feedbacks were not very well understood, or not even considered. But scientists now have more data on these physical, chemical, and biological feedbacks that are entered into the AOGCMs and CESMs as parameters.

Testing the projections against current conditions

We now have about 30 years of data to use in evaluating some of the earliest global circulation models. How has climate changed in the past 30 years, and how well have the climate models predicted these changes?

GLOBAL TEMPERATURE. AOGCMs are used to simulate past climatic events, as well as to make projections into the future. The 2013 IPCC report presented the results of combining 14 models that simulated global temperature from 1900 until 2006. The simulations were run under two conditions. The first simulation assumed only natural forcings, that temperature variation was only influenced by variation in natural events (Figure 23.13A). The second simulation assumed both natural and anthropogenic forcings, that temperature variation was influenced by natural factors and human activity (Figure 23.13B).

The natural forcing simulation worked pretty well until 1965, but failed to predict the relatively sharp temperature increase thereafter. The natural and anthropogenic forcing simulation did an excellent job at predicting the sharp increase in global surface temperatures after 1965. A second finding, captured by both models, was that volcanic eruptions tended to lead to brief periods of global cooling. This happens because volcanoes belch large quantities of aerosols – small solid or liquid droplets – that tend to cool Earth by reflecting incoming solar radiation back into space. Human activities are also important in aerosol emission. The cooling effect of aerosols that are emitted into the atmosphere by humans is considerable, and, in fact, global warming would be substantially greater were it not for the cooling effect of aerosol-bearing sulfur dioxide emissions that accompany the burning of fossil fuels.

Since 1998, the rate of global warming has slowed down considerably. The IPCC report attributes some of this slowdown to a decrease in the amount of solar radiation entering Earth's atmosphere, which tends to go through 11-year cycles. In 2000 the solar radiation cycle was at a maximum, and it declined about 0.1% to its 2009 minimum. A study by Yu Kosaka and Siang-Ping Xe (2013) attributes much of the slowdown to an unusual number of La Niña episodes during the same time period. When Kosaka and Xe added the effects of La Niña cooling to their climate model, the correlation coefficient (r) of their model predictions to the actual mean global surface temperature

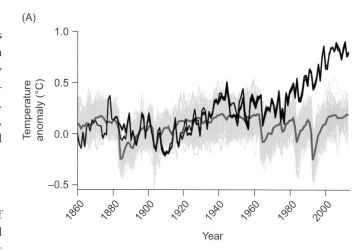

(A)

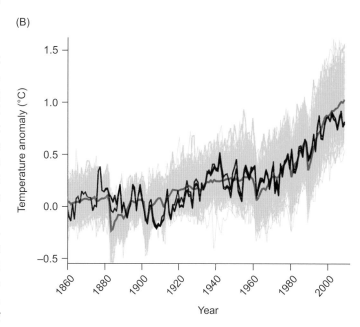

(B)

Figure 23.13 Global mean surface temperature anomalies relative to the period 1880–1919 – the 0 value on the y-axis. The black line represents the actual observations, while the red line represents the predictions of a combined group of AOGCMs and CESMs. The range of model predictions are outlined in yellow. A. Models run with natural forcing only. B. Models run with natural and anthropogenic forcing.

between 1970 and 2012 was an astoundingly strong 0.97. Unfortunately, La Niña episodes are still very difficult to predict. One important lesson is that such short-term (decadal) fluctuation will always be part of the global temperature profile.

Finally, as predicted by all AOGCMs, global warming is much greater over the northern hemisphere. Part of this difference arises because downwelling ocean currents in the southern oceans take substantial heat into much deeper waters. Another factor determining this difference is the much greater land surface area existing in the northern hemisphere. As we discussed previously, land heats up more quickly than water, and, in fact, global warming is about twice as rapid over land than over water.

GLOBAL HYDROLOGICAL TRENDS. The amount of water vapor in the air since 1988 has increased about 1.2% per decade, but this effect is highly variable. Over the same time period, there have been more unusually wet years than unusually dry years, but again there is substantial year-to-year variation in global precipitation patterns. Global ocean heat content has increased steadily since 1985, following a time period which showed a general upward trend, although with a great deal of variability. We have much better data on global mean sea level, which has risen steadily since 1880 (Figure 23.14A). At the same time, Arctic sea ice has declined sharply, while much of Greenland and a portion of Antarctica have experienced substantial ice loss. (Figure 23.14B). These trends were predicted by most AOGCMs.

Most (84%) of the total heating of Earth's system goes into the oceans, so Tim Barnett and his colleagues (2005) figured that the oceans were a good place to look for global warming. Researchers had just published a comprehensive data set that recorded temperatures at a variety of depths in the world's major oceans. Since the AOGCMs took into account ocean circulation patterns and atmospheric influences on ocean temperatures, Barnett and his colleagues reasoned that this new data set was a great opportunity to test the validity of existing global climate models.

The researchers used *signal strength*, the average difference in ocean temperature over a time period (1960–2000 in this study), as the dependent variable in their analysis. One set of simulations of the Parallel Climate Model (an AOGCM) considered only natural forcing from the cooling effects of volcanoes and differences in incoming solar radiation to see if changes in signal strength could be explained from natural forcings. A second set of simulations of the Parallel Climate model included anthropogenic forcing in addition to natural forcings (Figure 23.15). Clearly, the models with anthropogenic forcing did a much better job at predicting variation in ocean temperatures at different depths. It is critical to appreciate that these models were created before the ocean depth data became available, so these findings can reassure us that AOGCMs are capturing the important influences of human activities on ocean temperatures.

One last approach to validating climate models is to see if some of the factors influencing global temperature in the present-day models are also correlated with variation in climate regimes that occurred in the distant past.

Understanding past climatic variation from climate models

Climate scientists have a toolbox of indirect methods for learning about past climates. The fossil record provides one rich source of information about how climate has changed over time, though interpretation becomes more difficult the farther back in time we go. Perhaps the most analogous situation to the climate change we are experiencing today occurred about 55 MYA, a time period known as the *Paleocene–Eocene Thermal Maximum*. Based on several independent lines of evidence, across the world, mean

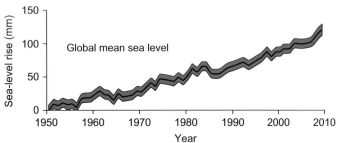
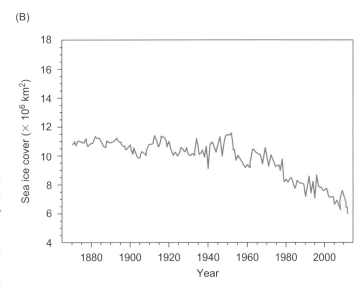

Figure 23.14 A. (Top) Sea-level rise is creating climate change refugees in Bangladesh and other low-lying regions of the world. (Bottom) Mean global sea-level rise (mm) since 1950. Shaded area represents 90% confidence intervals. B. Summer Arctic sea ice cover (× 10⁶ km²) since 1870.

surface temperatures increased by about 5°C over a time period of 1000–10 000 years (Denman *et al*. 2007).

Was this sudden rise in temperature associated with an increase in greenhouse gases? One compelling piece of evidence is a tremendous decrease in δ¹³C in foraminifera fossilized from that time period (Figure 23.16A). The most plausible explanation is that there was a sudden release of carbon that had a low

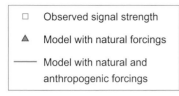

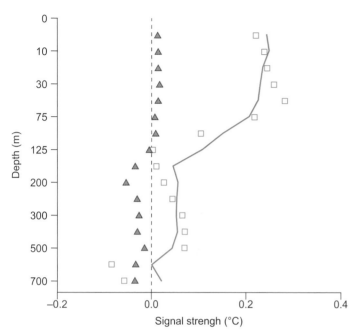

Figure 23.15 Signal strength (°C) in relation to ocean depth (m) for the north Indian Ocean. Signal strength for this study is the average change in temperature between 1960 and 2000. Squares are the observed data, triangles are predictions of the Parallel Climate Model based on natural forcings only, and the red line is the prediction of the Parallel Climate Model that accounts for natural and anthropogenic forcings combined. For all of the world's oceans, adding anthropogenic forcing to the Parallel Climate Model produced a much better fit to the observed signal strength.

concentration of ^{13}C (probably in the form of CO_2 or CH_4 or both) into the atmosphere that caused this universal decline (see Dealing with data 3.3 for a review of stable isotope ratios). There are several possible sources for this sudden release, which would have been sufficient to force global warming of this magnitude.

If high atmospheric carbon levels were responsible for these elevated temperatures, we would expect to observe evidence of substantial ocean acidification during the Paleocene–Eocene Thermal Maximum. As we discussed in the beginning of this chapter, ocean acidification is associated with reduced carbonate levels in the oceans. Marine organisms use carbonate (in the form of calcium carbonate – $CaCO_3$) to make their shells. The ocean acidification hypothesis predicts reduced levels of $CaCO_3$ in sediment core samples taken from the Paleocene–Eocene Thermal Maximum time period. Sediment cores taken from two different sites during this time period support this prediction, with a precipitous drop in $CaCO_3$, followed by a very quick recovery (less than 100 000 years) (Figure 23.16B). The IPCC report (Denman et al. 2007) concludes

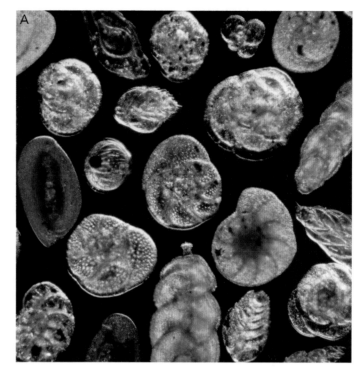

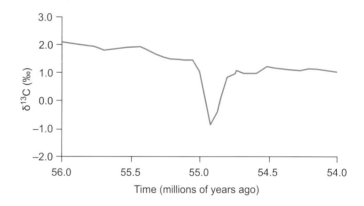

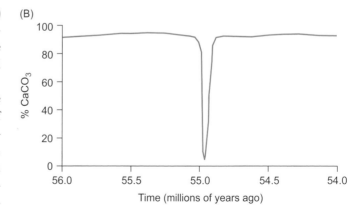

Figure 23.16 A. Fossilized remains of foraminifera (top) can be used to assess the ocean's chemical makeup. (Bottom) Abrupt reduction in $\delta^{13}C$ indicates increase in atmospheric carbon. B. Sharp and relatively short-lived decrease in calcium carbonate in sediment cores collected from the Paleocene–Eocene Thermal Maximum.

that these findings support the hypothesis that high atmospheric carbon levels caused this relatively rapid increase in surface temperature. They also point out that the changes we are currently experiencing are much more rapid than those that occurred during the Paleocene–Eocene Thermal Maximum timeframe.

Based on improved understanding of the mechanisms responsible for climate variability, observations or inferences from the recent and distant past, and validation of models from multiple sources, climate scientists are very confident that we are experiencing rapid climate change. We will now explore some of the ecological impacts of these changes.

23.5 ARE NATURAL ECOSYSTEMS CHANGING IN RESPONSE TO CLIMATE CHANGE?

Scientists have conducted a large number of experiments to evaluate what impacts climate change is expected to have on global ecosystems. Here we will limit our discussion to considering whether species, communities, ecosystems, and biomes are already showing evidence of responding to the effects of the global warming of the past few decades.

Effects of climate change on coral growth

As we just discussed, $CaCO_3$ is the primary raw material in the shells of marine organisms such as foraminifera, and it is equally essential for coral skeletons. At high ocean temperatures, the photosynthetic capability of zooxanthellae, the symbiotic algae that live with corals, declines. This can result in corals having insufficient organic carbon to fuel the calcification reaction, which can slow – or eventually stop – coral growth. Ultimately this can kill coral colonies.

When they lose their zooxanthellae, corals appear bleached. Without observing coral bleaching, it is difficult to tell if a coral is growing well or poorly in a particular year. Neal Cantin and his colleagues (2010) extracted cores from six coral *Diploastrea heliopora* skeletons with a submersible electric drill. They visualized these cores with a computerized tomography (CT) scanner, which enabled them to measure annual growth rings that they could then correlate with mean annual SSTs experienced by corals over the 80 years sampled in the core (Figure 23.17A). Coral growth was relatively stable when mean summer SSTs were below 30.2°C. At higher temperatures, growth rates declined sharply, about 0.12 mm for every 0.2°C increase in SST (Figure 23.17B).

Since 1998, the SSTs at the investigators' Red Sea study site has remained 0.9–2.2°C above historical means. At the same time, calcification rates have been at historical lows. There is still sufficient $CaCO_3$ there for corals to build shells at high rates, so the researchers propose that, at least in the Red Sea, the limiting factor is the impact of thermal stress on physiological processes involved in calcification. Using the data they have, and the mean SST projected by 17 climate models, the researchers extrapolate

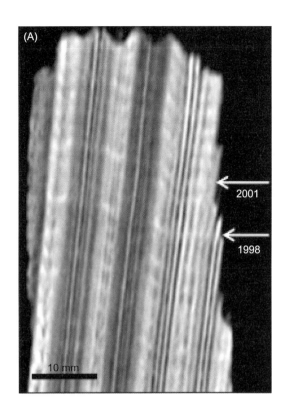

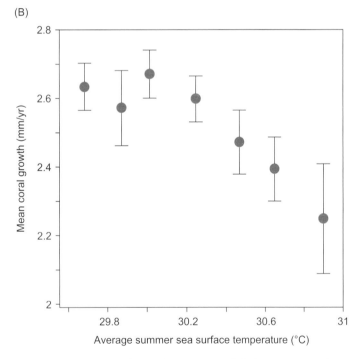

Figure 23.17 A. Annual growth rings in coral skeletons. B. Coral growth rate in relation to mean summer sea surface temperature ($F_{5,300} = 2.73$, $P < 0.05$).

that corals in the Red Sea will stop growing by the year 2070. But they caution that most corals will die before that time, from bleaching and subsequent depletion of energy reserves.

Another important concern for the future of corals is the acidification of the ocean that is resulting from high atmospheric CO_2 levels, and the associated depletion of carbonate from the oceans. Carbonate depletion is greatest in cold waters, and, particularly in the southern ocean, is expected to reduce carbonate levels to a point where there is insufficient carbonate for organisms to efficiently construct their shells. James Orr and his colleagues (2005) project that carbonate depletion will become critical when atmospheric CO_2 levels climb above 550 ppm, which will occur by the year 2100, according to most IPCC scenarios.

Climate change will of course have important effects on the distribution of marine organisms as well. For example, many species of marine fish have moved into higher latitudes that presently harbor waters that were formerly too cold for them. But we will shift our focus to terrestrial systems to continue exploring the range of responses exhibited by organisms to changing climates.

Effects of climate change on food availability

Polar bears, *Ursus maritimus*, depend on sea ice for feeding and breeding, but as we discussed earlier (Figure 23.14B), the extent of sea ice has been declining rapidly in recent decades. Bears prefer to live on sea ice, because seals, a highly preferred food, are very abundant on or near sea ice, but not on land (Figure 23.18A). Eric Regehr and his colleagues (2010) wanted to know how polar bear demography was influenced by the number of days the environment lacked sea ice, which melts during the summer months on the southern Beaufort Sea. Over a 5-year period they used helicopters to find 627 different bears, which they anesthetized and fitted with radio collars. They then followed these bears to measure survival and reproductive success.

Long periods without ice were very damaging to bear populations. In good years (2001–2003), the number of ice-free days averaged 101, while in bad years (2004, 2005), the number of ice-free days averaged 134.5. Good years were associated with relatively high survival of bears at all stages and sexes – for example, breeding females had a 0.99 survival probability in good years and only a 0.79 survival probability in bad years. The breeding probability in good years averaged 0.58, while in bad years it averaged 0.19. Lastly, cub survival averaged 0.54 in good years and 0.40 in bad years. During 2004 and 2005, the researchers saw examples of cannibalism, and recovered remains of bears that had starved to death. Taken together, these data indicate that bear populations have better survival and reproduction when sea ice is present for long periods of time during the year.

Christine Hunter and her colleagues (2010) used these polar bear data to construct a life table, which allowed them to calculate the population growth rate (λ) for each year, as well as the confidence intervals for each value of λ. Remember that populations with $\lambda < 1.0$ will tend to decrease, while populations with $\lambda > 1.0$ will tend to increase. The mean value of λ was 1.05 in good years, and 0.78 in bad years (Hunter *et al.* 2010).

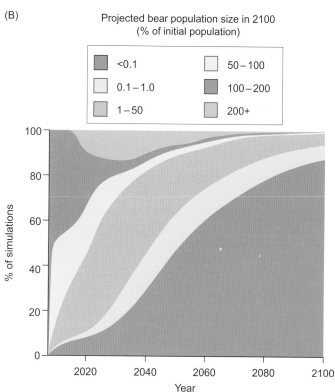

Figure 23.18 A. Two polar bears adrift. B. Percentage of simulations in which polar bear population size changes to specified percentages of their initial population size.

The researchers then asked whether climatic conditions predicted by the climate models in the 2007 IPCC report could be used to make projections of polar bear population viability. They ran repeated simulations using their data and the climate projections, and found that over 80% of the simulations resulted in the bear population declining to less than 0.1% of its original size (of 627) by 2100. Fewer than 1% of the simulations resulted in the population actually increasing over that same time period (Figure 23.18B). The researchers estimated that the probability of this population going extinct before 2100 is 80–94%. This study resulted in polar bears being listed as a threatened species under the US Endangered Species Act in 2008.

 Thinking ecologically 23.3

Imagine you were given the task of protecting polar bears from extinction. What questions would you need to ask about polar bears so you could better protect them? How would you find out the answers to those questions (what research would you do)?

Not all species are suffering from the effects of global warming. For example, yellow-bellied marmots in the Rocky Mountains of Colorado have tripled in abundance since 2000 (Ozgul *et al.* 2010). The researchers studying the marmots attribute this increase to higher temperatures allowing earlier emergence from hibernation, which results in a longer feeding season and much higher growth rates. Heavier marmots survive better, leading to sharp population increases.

Overall, however, climate change is more of a problem for populations than it is a boon. One major response of species to global warming is migrating to appropriate climates, which could be either higher latitudes or altitudes.

Effects of climate change on geographic range

In Chapter 5, we discussed a study by Parmesan and Yohe (2003) that demonstrated that species ranges had moved to higher latitudes at a mean rate of 6.1 km per decade, and to higher elevations at a mean rate of 6.1 m per decade. I-Ching Chen and her colleagues (2011) wanted to know whether species were changing habitats at the same rate as their habitats were warming, either by changing latitude or elevation or both. For example, if regional temperature increased one degree per decade, and if temperature decreased 1°C with each 60 km of latitude in a poleward direction, then the predicted distance of migration would be 60 km over that decade. But if regional temperature increased 2°C per decade, then a species would be predicted to migrate 120 km over that decade to keep pace with regional warming. So range shift would be most rapid in locations experiencing the greatest rate of global warming.

The researchers studied 764 species responses for latitude shifts and 1367 species responses for elevational changes. They then grouped each species by taxonomic group and location – for example, amphibians in Chile or plants in Switzerland. They discovered that across species, organisms are changing latitudes about as quickly as temperatures are increasing – about 16.9 km per decade (Figure 23.19A). However, organisms are increasing elevation at an average of 11.0 m per decade, which is considerably slower than climate is warming (Figure 23.19B). Importantly, for both latitude and elevation, range shift is greatest where warming is most rapid. This supports the hypothesis that range shift is a response to global warming.

Chen and her colleagues discovered a much greater change in latitude and elevation than was found by Parmesan and Yohe. These differences may indicate that species are moving more quickly in this later study, or the differences may simply result

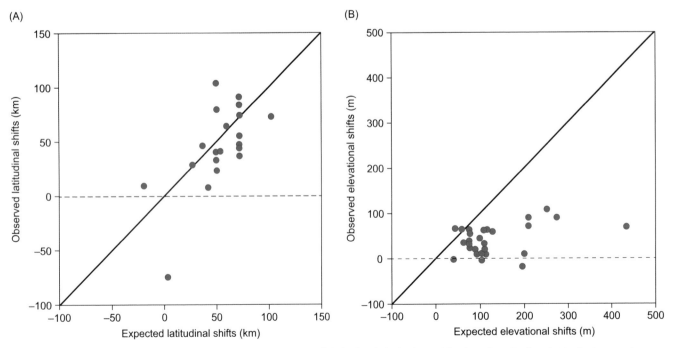

Figure 23.19 Observed population shifts in comparison to expected shifts for A. latitude and B. elevation. Each point indicates one taxonomic group at one location. Taxonomic groups were birds, mammals, arthropods, herptiles, fish, mollusks, and plants. Points on the diagonal line are migrating according to expectation.

from Chen and her colleagues focusing primarily on northern latitudes, where climate is changing more rapidly.

Migration is not a viable option for many species. Immobile species or habitat specialists are least likely to shift their range. Species are unlikely to shift their range if their food species do not shift. Migration to colder climates may not be an option for cold-adapted alpine species, which may find themselves marooned on a warming mountain island, with no place to go, but up.

Effects of range contraction on genetic diversity of populations

The studies above show that as climates warm, populations of alpine-adapted species will tend to migrate to higher elevation, though more slowly than expected. One possible outcome of upward migration is that formerly continuous populations may get broken into small fragments. With fragmentation, we expect population size to decrease, which should reduce genetic

diversity within the small subpopulations, and could increase the probability that a population will go extinct (see Chapter 11).

Yosemite National Park in California, USA, is home to four chipmunk species, including the alpine chipmunk, *Tamias alpinus*, which is adapted to life above the treeline, and the lodgepole chipmunk, *Tamias speciosus*, which is adapted to life in lower-elevation lodgepole pine forests (Figure 23.20A). A research team led by Joseph Grinnell (1924) studied the animal community in Yosemite in 1915 and 1916, and collected careful distribution data along with a large sample of mammal skins from that time period. Since then, the mean temperature in Yosemite has increased by 3°C. Associated with this warming, a previous study had shown that the lower-elevation distribution limit of *T. alpinus* shifted upwards by more than 500 m, while the lower elevation limit of *T. speciosus* was relatively stable. Emily Rubidge and her colleagues (2012) first needed to determine whether this increase in the lower elevation limit caused population fragmentation in *T. alpinus*. If so, they could then test the hypothesis that fragmentation had reduced the genetic diversity

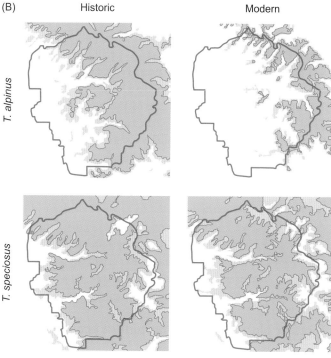

Figure 23.20 A. The alpine chipmunk, *Tamias alpinus* (top) and the lodgepole chipmunk, *Tamias speciosus* (bottom). B. Historic and modern distributions of *T. alpinus* and *T. speciosus* in and near Yosemite National Park.

of *T. alpinus*. In contrast, they hypothesized that *T. speciosus* should maintain the same genetic diversity, because its distribution was relatively unaffected

The researchers set up traplines throughout Yosemite National Park, and were able to show that the population of *T. alpinus* had become much more fragmented, breaking into smaller subpopulations as they moved up the mountains (Figure 23.20B). In contrast, there was little evidence of any fragmentation in *T. speciosus*. Thus the stage was set for testing the hypothesis that *T. alpinus* had lost genetic diversity over the 90 years that elapsed between their modern samples (from 2003 to 2008) and Grinnell's historical samples (from 1915 and 1916). The researchers expected no loss of genetic diversity in *T. speciosus* over that time period. They were able to test these hypotheses by extracting, analysing, and comparing the DNA from historic and modern skins.

Rubidge and her colleagues used seven different microsatellite loci for their comparisons. (Recall from Chapter 3 that microsatellites are genes that have a variable number of repeating sequences of DNA: for example a microsatellite with five repeating TA DNA base sequences is considered a different allele from the same microsatellite with six repeating TA DNA base sequences.) The chipmunk populations were polymorphic (had multiple alleles) for each of these loci. As predicted, mean allelic diversity, the number of alleles present at each microsatellite locus, was about 25% lower in modern *T. alpinus* than it was in historic *T. alpinus* ($P = 0.04$). Six of the seven microsatellite loci lost at least one allele between the two surveys. In contrast, there was no difference in mean allelic diversity between modern and historic *T. speciosus*. The researchers concluded that one possible outcome of climate change is fragmentation of populations which can lead to a loss of genetic diversity and an increase in the probability of extinctions.

On a larger scale, the movement of species across landscapes can have important effects on the ecosystem or biome level. Given that northern latitudes have experienced the greatest temperature increases, we might expect to see some important changes to species distribution and abundance in the tundra.

Effects of global warming on tundra ecosystems

We have already discussed in detail some of the physical changes that have occurred as a result of global warming of extreme northern latitudes. From an ecosystem standpoint, Arctic tundra is relatively unproductive because of low temperatures, brief growing seasons, and low nutrient levels. James Hudson and Greg Henry (2009) wanted to know whether a heath community in northern Canada dominated by a relatively small group of vascular plants and mosses was responding to climatic changes between 1981 and 2007.

The researchers actually did two separate analyses. The first experiment involved comparing the biomass of plants collected on two dates separated by 27 years: 21 July, 1981 (reported in Nams and Freedman, 1987), and 20 July, 2007 (this study). As Nams and Freedman had done in 1981, the researchers visually estimated percent cover, and harvested plants from 25 × 25 cm quadrats spaced every 10 m across the study site. Aboveground biomass was the dry weight of (formerly) living tissue, while standing crop was the dry weight of living and dead tissue. Aboveground biomass increased sharply from 33.49 to 86.91 g/m^2, while standing crop increased from 347.88 to 489.28 g/m^2. The total percent cover was 65–75% for both years.

In a second experiment, the researchers established 18 1 × 1 m quadrats in 1995, for which they measured species composition and abundance in 1995, 2000, and 2007. They also set up weather stations so they could measure changes in local temperature and length of growing season over the course of the study. Mean annual temperature increased by 2.86°C and the growing season length expanded by about 16 days (there were no data for the 12th year because polar bears destroyed the weather station). Production increased in these permanent plots over the course of this study, primarily from a 60% increase in evergreen shrubs and a 74% increase in bryophyte aboveground biomass (Figure 23.21). Species diversity (measured with the Shannon index) did not change. Mean canopy height was only measured in 2000 and 2007, and it doubled during that 7-year time span.

These two experiments show an increase in production without a change in species composition. In fact, there was considerable bare ground throughout the study, which suggests that

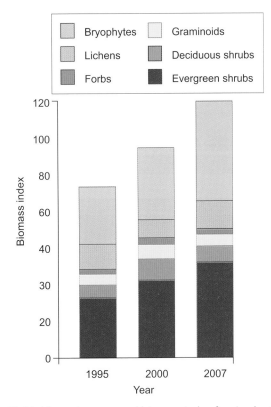

Figure 23.21 Mean aboveground biomass index for six plant functional groups at three time periods during the 12-year study.

competition between species might not have been important. The researchers propose that climate warming directly enhanced growth by lengthening the growing season, increasing soil temperature, deepening the layer of unfrozen soil, and thus increasing nutrient availability.

Of course, changes to plant communities can have profound influences on species that interact with plants. One important impact of climate change is that it can lead to a disruption of the synchrony between a species' life cycle and the availability of an essential resource. An example of synchrony is that herbivores should be active at a time when food items (plants) are easily accessible. A second example from the perspective of a plant and its pollinators is that both benefit when flower and pollinator abundance co-occur. *Mismatches* may occur if two interacting species show different responses to climate change; for example, if regional or global warming causes a plant to flower earlier in the year, and its pollinator's emergence depends on photoperiod, that could reduce pollination and seed set in the plant, and presumably lead to a population of starving pollinators. Ecologists use the term trophic mismatch when mismatches are between species from two different trophic levels.

Historically, in west Greenland, there has been a very close match between the time when caribou, *Rangifer tarandus*, give birth and when food plants emerge from the tundra soil (Figure 23.22A). Jeffrey Kerby and Eric Post (2013) knew that caribou calf production was quite variable from year to year, and hypothesized that this variation was tied to the extent that calf production was matched to the emergence of plants that served as their food source. Simplifying their analysis, this population had no predators. For this study, they defined a trophic mismatch index based on the proportion of plant species that have emerged on the date that 50% of the caribou have been born into the population. Based on 11 years of data, the researchers discovered that caribou did not adjust the birth dates of their young to the emergence of food plants (Figure 23.22B). This caused considerable trophic mismatch in some years.

But what are the consequences of trophic mismatch to caribou? Kerby and Post measured calf production as the ratio of calves to total individuals in the population several weeks after the conclusion of the calving season, to account for early calf mortality due to starvation. The researchers discovered that years with greatest trophic mismatch tended to have much lower calf production (Figure 23.22C).

As climates continue to warm, we expect trophic mismatch to increase in some trophic interactions. There are several possible ecological and evolutionary responses. First, assuming there is genetic variation in the timing of traits such as calving dates, natural selection would favor individuals who gave birth earlier in the season. Over generations, as climates continued to warm, there would be a shift to earlier calving dates. A second possibility is that many species have phenotypic plasticity (see Chapter 8), and can shift the timing of their trophic interaction in a way that maximizes their success. In this case, caribou would

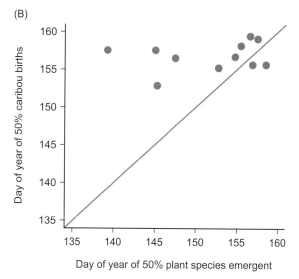

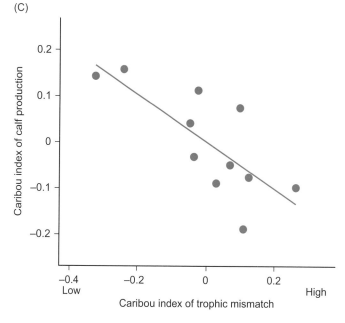

Figure 23.22 A. A caribou mother and calf forage in the tundra. B. Day of year when 50% of caribou have been born in relation to day of year when 50% of plant species have emerged. Diagonal line indicates a perfect trophic match. C. Caribou index of calf production in relation to caribou index of trophic mismatch.

use some environmental cue to produce calves earlier in years with early plant emergence and later in years with late plant emergence. Unfortunately for caribou, there is good evidence that they lack this flexibility. In a species lacking phenotypic plasticity, if the response to natural selection is not fast enough to reduce the costs of warming-induced trophic mismatch, a possible outcome is extinction of the population.

This study and the others discussed in this section support the hypothesis that climate change is influencing life on Earth in profound ways. Some of these effects may actually be beneficial, but many are clearly destructive. To reduce these destructive effects, we need to change our behavior, and reduce the emissions of greenhouse gases. But many people (and nations) are resistant to these changes, and I'd like to briefly explore this resistance.

REVISIT: Public perception of climate change

The world community was swift to ban ozone-destroying substances, and we now have good evidence that the abundance of ozone-destroying substances in the stratosphere is either stabilizing or actually beginning to decrease. In contrast, the world community has been slow to respond to the threat of climate change. Why has the response been so different?

Part of the problem has to do with the complexity of the mechanisms. There are a huge number of variables associated with climate change (which we've really only touched on in this chapter). In contrast, the science of the halocarbon/ozone/human health interaction is comparatively simple. Most atmospheric ozone resides in the stratosphere between 12 and 30 km above Earth's surface. This ozone layer is critical to life on Earth, because ozone absorbs UV-B radiation from the Sun, protecting organisms from UV's harmful carcinogenic and mutagenic effects. We have wonderful data and illustrations in biology medical textbooks showing how UV-B causes mutations and cancer.

As a result of the Molina and Roland (1974) study, subsequent researchers began doing more systematic measurements of atmospheric ozone. Between 1975 and 1995, the amount of ozone in the stratosphere declined about 5% globally. But this decline was not uniform; the polar regions were particularly hard hit, with an overall reduction of over 60% over the Antarctic, and with most of the loss between 12 and 20 km altitude, which unfortunately is the altitude of greatest ozone abundance (Figure 23.23). Researchers were puzzled by the Antarctic ozone hole, both because it was seasonal, and also because studies showed that atomic oxygen was relatively rare at the altitude of greatest ozone depletion. Susan Solomon and her colleagues (1986) proposed a different chemical reaction for halogen-catalysed ozone

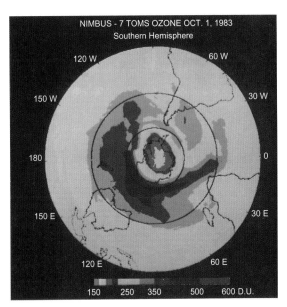

Figure 23.23 Antarctic ozone hole pictured from Nimbus 7 satellite on October 1, 1983. D.U. are Dobson Units, the measure of ozone concentration.

destruction. Importantly, this reaction only worked efficiently on surfaces such as those provided by clouds in the stratosphere, which were only abundant over the Antarctic during the months of greatest ozone depletion.

The ozone hole is highly photogenic, and was seen by millions of people with the assistance of some graphics that had only recently become available. Though most people did not really understand the link between the ozone hole, stratospheric clouds, and the chemical reaction, they believed that the hole was real because they could see it, and they knew that ozone protected them from cancer. Given an improved understanding of the link between CFCs and ozone depletion, and an appreciation of the medical risks caused by ozone depletion, scientists had a relatively easy case to make, and representatives of the global community convened the Montreal Protocol in 1987. The resulting accord to phase out the production of many halocarbons (including CF_2Cl_2 and $CFCl_3$) was signed by all members of the United Nations.

One more important issue in this comparison is that it is much easier and much less costly to get rid of ozone-destroying substances than it is to reduce CO_2 emissions. Many corporations have a very strong financial incentive to convince the public that global climate change is a myth. Though a few corporations spent millions of dollars attempting to discredit the science underlying ozone destruction, that is a pittance compared to the financial investment that is now being made by multinational corporations to discredit the science of climate change.

As we learn more about climate change, it will be easier to make an increasingly compelling argument for changing our behavior to reduce emissions. However, early action will be much less painful than waiting before acting, particularly given the time lag between increased emissions of greenhouse gases and their effects on global climate. While committing to responsible action, we must understand that our current scientific understanding of climate change is incomplete. But science, in its messy way, is getting closer, and will continue to get closer, to an accurate understanding of climate change.

SUMMARY

The IPCC was established to provide a clear up-to-date presentation of the science underlying climate change. Using data from direct observations, experimental mesocosms, field experiments, and complex computer models, the IPCC has made a very strong case supporting the hypothesis that human behavior is leading to rapid and substantial climate change. One important anthropogenic effect is changes to the carbon cycle, primarily greater CO_2 export into the atmosphere from industrial activity. In recent years, both oceans and terrestrial sources have taken up some of this excess CO_2, but ocean uptake is particularly problematic, because it leads to acidification.

There are many other important greenhouse gases that influence Earth's surface temperatures, including methane, nitrous oxide, ozone, and a diverse group of halocarbons. Though less abundant, these gases have a much greater global warming potential than CO_2, on a per molecule basis. Many effects of greenhouse gases on global climate are complex; for example, some halocarbons absorb longwave radiation and thus directly raise surface temperatures. These same halocarbons also destroy ozone, itself a greenhouse gas, so by reducing ozone, halocarbons can indirectly decrease surface temperatures. Increased atmospheric CO_2 can increase photosynthetic rates and reduce water lost to transpiration, which is particularly beneficial to C_3 plants. Unfortunately, crop yields in mesocosm experiments improved only marginally, and were partially offset by indirect effects of increased CO_2 uptake by plants.

There are many different types of climate models that use the movement of the atmosphere around Earth, and the interaction of the atmosphere with the oceans and with biological processes, to project future climate. Though there are quantitative differences between the projections of each model, these models all project a much warmer and wetter global climate over the next century, with northern latitudes experiencing the greatest impact of climate change. Many regions have already experienced an ecological impact. For example, coral growth has slowed in response to warming in the Red Sea, and polar bear reproductive rates have declined in response to loss of sea ice. Climate change has caused many species to shift their geographic range and/or altitudinal range. Unfortunately, population fragmentation from altitudinal range shift can reduce genetic diversity, thereby reducing population viability.

FURTHER READING

Garcia, R. A., Cabeza, M., Rahbek, C., and Araújo, M. B. 2014. Multiple dimensions of climate change and their implications for biodiversity. *Science* 344: 1247579.

This exhaustive synthesis projects the future effects of climate change in relation to biodiversity for Earth's different biomes. The researchers integrate multiple measures of climate change into a predictive framework to make their projections.

Kerby, J., and Post, E. 2013. Capital and income breeding traits differentiate trophic match–mismatch dynamics in large herbivores. *Philosophical Transactions of the Royal Society B* 368: 20120484.

Explores the impact of climate-induced trophic mismatch (the lack of synchrony between resource need and resource availability) for populations of caribou and muskoxen in western Greenland. The researchers ascribe differences between the response of caribou and muskoxen to differences in their reproductive life histories.

Palumbi, S. R., Barshis, D. J., Traylor-Knowles, N., and Bay, R. A. 2014. Mechanisms of reef coral resistance to future climate change. *Science* 344: 895–898.

A study with some hopeful findings on the impact of climate change on ecosystems. Reciprocal transplants of coral between unusually hot and more moderate-temperature environments showed evidence for short-term acclimatization and long-term adaptation to high temperatures.

Poulter, B., and 12 others. 2014. Contribution of semi-arid ecosystems to interannual variability of the global carbon cycle. *Nature* 509: 600–603.

Nice study of factors that influence the land sink, and how researchers use models and direct measurements to help quantify the contributions of a variety of (sometimes competing) factors to the carbon cycle.

Rubidge, E. M., and 5 others. 2012. Climate-induced range contraction drives genetic erosion in an alpine mammal. *Nature Climate Change* 2: 285–288.

Comparative historical study on two related species of alpine chipmunks. The study shows a substantial loss of genetic diversity in a species that has shown extreme climate-change-driven range contraction and fragmentation in the past 100 years, in comparison to a related species whose range has stayed relatively constant and continuous over the same time period.

END-OF-CHAPTER QUESTIONS

Review questions

1. Why did people embrace legislation to protect stratospheric ozone, but resist legislation to mitigate climate change?
2. What are the biggest sources and biggest sinks of atmospheric carbon? Which sources and sinks have experienced the greatest percent change since 1750?
3. Distinguish between radiative forcing and global warming potential? Which greenhouse gas has greatest radiative forcing, and which has greatest global warming potential?

4. What are the projections that are agreed upon by climate models? How have these projections been tested and validated?
5. In what two ways does increased atmospheric CO_2 threaten coral populations? How is increased atmospheric CO_2 threatening cold-adapted species?

Synthesis and application questions

1. There are regions of the tropics and subtropics with highly acidic soils. Reflecting on the research of Xie *et al.* (2009), how might these soils influence the global carbon cycle?
2. Burning coal releases aerosols and soot into the atmosphere (in addition to CO_2). What are the positive and negative feedbacks associated with reducing our use of coal as a fuel source?
3. For each of the following, what ecological effect would you expect to result from global climate change: relative abundance of C_4 plants, relative abundance of CAM plants, geographic range of the tropical forest biome, geographical range of the world's deserts?
4. The "tragedy of the commons" (Chapter 6) made the point that people were unlikely to act for the general good, because as individuals they may get a very small benefit and incur high costs from doing so. How does this argument tie into the question of climate change, and the role played by individuals, corporations, and nations in addressing the problem?
5. The study by Rubidge and her colleagues (2012) demonstrated that populations can become fragmented as they move up mountains in response to climate change. What are the possible implications for the genetic structure of populations and for future speciation events in response to climate change? How might you check to see if there have been any changes to the genetic structure of populations?
6. We discussed that phenotypic plasticity can help some species deal with the problem of climate change-induced trophic mismatch. Can you suggest plausible scenarios (including life history evolution, habitat, and other environmental considerations) that would increase the probability that a species is sufficiently phenotypically plastic to avoid major trophic mismatch?

Analyse the data 1

Here are the data for 24 years of carbon flux into oceans, land, and the atmosphere (Le Quéré *et al.* 2014). How have these fluxes changed and which changes are statistically significant? Which trends seem most concerning to you? Which data show the greatest amount of variation from year to year? Use what you've learned about the carbon cycle to suggest plausible hypotheses for this variation.

Year	Fossil fuel + cement	Land-use change	Atmospheric increase	Ocean sink	Land sink
1990	6.13	1.44	2.48	2.03	3.06
1991	6.22	1.64	1.67	2.15	4.03
1992	6.16	1.68	1.42	2.45	3.98
1993	6.16	1.55	2.59	2.43	2.69
1994	6.27	1.50	3.58	2.20	1.98
1995	6.40	1.49	4.11	2.08	1.69

(*cont.*)

Year	Fossil fuel + cement	Land-use change	Atmospheric increase	Ocean sink	Land sink
1996	6.54	1.47	2.29	2.06	3.66
1997	6.65	2.29	4.18	2.18	2.59
1998	6.64	1.59	6.02	2.29	-0.07
1999	6.61	1.34	2.84	2.14	2.97
2000	6.77	1.23	2.65	2.12	3.23
2001	6.93	0.97	3.82	1.97	2.12
2002	7.00	1.05	5.05	2.34	0.66
2003	7.42	0.90	4.75	2.46	1.11
2004	7.81	1.03	3.41	2.31	3.12
2005	8.09	1.03	5.15	2.36	1.61
2006	8.37	1.01	3.69	2.51	3.18
2007	8.57	0.94	4.43	2.51	2.56
2008	8.78	0.66	3.75	2.43	3.25
2009	8.74	0.71	3.58	2.56	3.31
2010	9.17	0.83	5.11	2.55	2.34
2011	9.46	0.88	3.63	2.69	4.03
2012	9.67	0.93	5.09	2.85	2.66
2013	9.86	0.89	5.36	2.88	2.51

Analyse the data 2

Here are data analysed by Hudson and Henry (2009) on changes to the High Arctic heath community between 1995 and 2007. The equation is the least square regression, where y is the aboveground biomass (g/m^2), and x is the year (for 1995, $x = 0$, for 1996, $x = 1$, etc., increasing until 2007 where $x = 12$). Based on these equations, which variable shows the greatest % increase over the course of the study?

Variable	Equation	F-value	P-value
Live vegetation	$y = 80.15 + 3.33x$	64.67	<0.01*
Vascular	$y = 43.61 + 1.53x$	18.38	<0.01*
Woody	$y = 35.15 + 1.50x$	18.68	<0.01*
Nonwoody	$y = 7.87 + 0.16x$	0.16	0.69
Non-vascular	$y = 37.64 + 1.77x$	33.16	<0.01*
Evergreen shrubs	$y = 28.15 + 1.39x$	25.21	<0.01*
Deciduous shrubs	$y = 6.86 + 0.11x$	0.65	0.42
Graminoids	$y = 5.39 + 0.04x$	0.12	0.73
Forbs	$y = 2.50 + 0.01x$	0.05	0.82
Lichens	$y = 10.90 + 0.11x$	0.44	0.51
Bryophytes	$y = 26.71 + 1.66x$	54.03	<0.01*
Litter	$y = 51.13 - 0.96x$	3.11	0.08

Chapter 24

Jane Lubchenco:
community, ecosystem,
and global ecology

INTRODUCTION

In 2003, Jane Lubchenco and several colleagues published a paper in the journal *Ecological Applications* entitled "Plugging a hole in the ocean: the emerging science of marine reserves." This metaphorical hole, a result of radical degradation of marine ecosystems, was threatening to consume biological diversity and upset ecosystem structure and functioning. The authors argued that establishing a network of marine reserves would help plug this hole, promoting recovery of depleted oceans and helping to maintain healthy ocean ecosystems (Lubchenco *et al.* 2003).

How ironic that seven years later, Lubchenco's attentions would be consumed by plugging a literal hole in the ocean, a result of the explosion of the *Deepwater Horizon* rig, and the failure by British Petroleum Corporation (BP) to contain the ensuing torrent of oil that gushed forth from the devastated Macondo well (Figure 24.1). Equally ironic, when the well exploded on April 20, 2010, Lubchenco was en route from Michigan to the Florida Keys to be part of a celebration of Earth Week's 40th anniversary. Michigan and Florida were two of 22 states that had received funding from the National Oceanic and Atmospheric Administration (NOAA) to do lakeside and marine habitat restoration projects. Lubchenco, as Director of NOAA, and other top officials from NOAA and the United States Department of Commerce, were visiting funded sites to help promote the benefits of habitat restoration. Earth Week was established in part as a response to the Santa Barbara oil spill of 1969, which also resulted from the explosion of an offshore oil platform, and which devastated portions of the California, USA, coastline.

We begin this chapter with a discussion of NOAA's response to the *Deepwater Horizon* explosion, and an introduction to the scientific and political issues it raised. We then step back to briefly consider Jane Lubchenco's years growing up and then going to college in Colorado, before moving to Washington to begin her graduate career. As a marine ecologist, Lubchenco began with small-scale studies of animal and algal communities, but expanded over time to investigate ecosystem and regional level patterns of structure and functioning. Ultimately, she entered the world of politics and scientific communication, which led to her accepting the position of NOAA director. In that position, she applied her understanding of ecosystem and global ecology and her expertise in communication to keep a concerned public informed about the impact of the oil spill, and to develop new programs that used ecosystem services in a sustainable manner.

KEY QUESTIONS

24.1. How did NOAA respond to the explosion of the *Deepwater Horizon*?

24.2. How did Lubchenco get started in marine ecology?

24.3. How did moving to New England affect Lubchenco's research direction?

24.4. What factors influence the structure and functioning of temperate and tropical ecosystems?

24.5. How did Lubchenco expand her scope into politics and scientific communication?

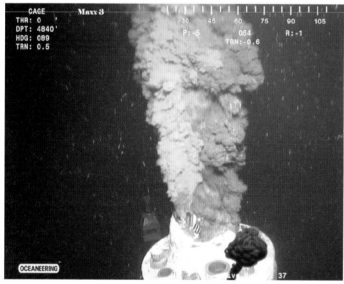

Figure 24.1 Oil and natural gas gush out from the damaged wellhead as pictured by a British Petroleum video.

Let's return our attentions to the exploding *Deepwater Horizon* oil rig, and see how the US government responded to the many challenges it presented.

24.1 HOW DID NOAA RESPOND TO THE EXPLOSION OF THE *DEEPWATER HORIZON*?

As Director of NOAA, Lubchenco was one of a small team of people that coordinated the response to the oil spill on behalf of the President. The numerous aspects to be coordinated included how to kill the well, how to estimate oil and methane flow rates from the well, how and where to use chemical dispersants, how to prevent the public from eating contaminated seafood, how to protect vulnerable marine organisms, and how to assess and mitigate the damage.

When Lubchenco was first informed about the explosion, she focused on NOAA's role in providing weather and ocean forecasts to guide the US Coast Guard's search and rescue mission. As information became available about the sad loss of life, and the rescue mission ended, her attention shifted to the fate of the diesel fuel that had been on board the rig. At the time, few were concerned about a massive oil spill because BP had assured various federal agencies that automatic shut-off valves would close the well in case it was damaged. Unfortunately, those mechanisms failed, and oil began gushing out in an uncontrolled torrent from about 1500 m below the ocean's surface.

One of the earliest challenges was estimating how much oil was flowing from the well. Initially BP denied that any oil was leaking, but that position became untenable as the surface oil slick continued to grow. NOAA scientists challenged BP's first assertion that oil was flowing at approximately 1000 barrels (790 m^3) per day (bpd), arguing that based on the size and nature of the accumulating oil slick, the flow rate had to be at least 5000 bpd and likely much more. The government responded by establishing a team of scientists to determine the flow rate. While the team was developing the technology to estimate flow rate, the public was becoming increasingly frustrated and concerned about the environmental and economic impact, particularly after attempts to cap the well failed. A common story line was that the government was hiding information, or underestimating the severity of the spill. Neither was true.

Lubchenco's philosophy was to share information with the public as soon as it was deemed reliable. She understood that preliminary results are sometimes inaccurate, and so she directed her team to share information once it had been confirmed. As she describes, "we made a commitment to share information as soon as we were able to verify its accuracy; we thought that was the right balance in a crisis. We didn't want to just pop everything out before we knew it was correct. But we didn't want to sit on it and hold it forever." As one example, NOAA had partnered with researchers at the University of New Hampshire to develop a tool for oil-spill purposes. This tool – the Emergency Response Management Application (ERMA) – uses GIS technology (see Dealing with data 9.1) to visualize in real time the position of all factors that may play an important role in any crisis. Using ERMA, it was easy to create maps of where the oil slick was, where the booms were positioned, where various responders were, and from a biological perspective to locate sensitive groups of marine mammals, pelican rookeries, and turtle nesting sites. Consistent with Lubchenco's commitment to transparency, NOAA took the lead in creating a much more robust online platform that could withstand a huge volume of public traffic, and posted ERMA in the public domain. On its first day, ERMA received over 3 million hits.

Despite major communication efforts, several additional issues aroused public distrust. There were reports of a mysterious vast plume of subsurface oil that was expanding at an alarming rate. In this case, a few scientists fueled media and public hysteria with findings that were never confirmed or published. And there was the famous pie chart that provided initial estimates about the fate of oil from the well (Figure 24.2). The bottom-line conclusion was inadvertently misrepresented initially by the director of the White House Office of Energy and Climate Change Policy as showing that 75% of the oil was gone from the Gulf of Mexico. Lubchenco immediately corrected the director's statement to a much lower 50% value, pointing out that chemically dispersed (8% of total) and naturally dispersed (16% of total) oil still exists in the environment, but that researchers expect it to be broken down by microorganisms. However, the initial announcement of the higher figure was panned as wishful thinking and bad science. Scientists were also concerned that a graph format was used without error bars or some indication of confidence levels. That concern fueled public skepticism, especially since few people could believe that even 50% of the oil could be gone. Though the 50% figure was later validated, the storyline that "the government is hiding something" persisted.

It is quite ironic that communication was so difficult throughout the crisis. As we shall see later in this chapter, Lubchenco has been a leader in training scientists to explain their science to the public. But, she explains, communication during an immense, complex, and lengthy crisis with multiple players and an

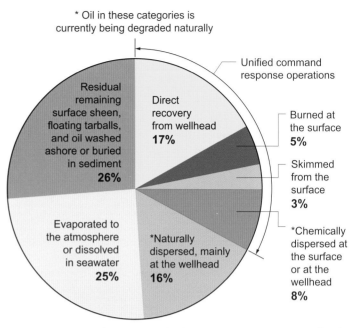

Figure 24.2 Fate of oil released by the *Deepwater Horizon*, as of early August 2010.

insatiable media is quite unlike normal scientific communications. Much new information flowed in that related to the ecological effects of the oil spill, and more will be forthcoming over the next years and decades. At the end of this chapter, we will learn about some of the most recent scientific findings. For now, let's go back in time to explore how Lubchenco's scientific interests have evolved over the course of her career.

24.2 HOW DID LUBCHENCO GET STARTED IN MARINE ECOLOGY?

Lubchenco is the oldest of six daughters born to two physicians. Her dad was a surgeon and her mom a pediatrician. She describes her mother, LaMeta, as "a pioneer, who found a way to balance family and career. She had polio when she was 18 months old, and so she grew up limping, with one leg shorter than another. But she never felt that she had a disability." She encouraged her children to be active in school and in diverse extracurricular activities, including sports; all of the Lubchenco girls were Junior Olympic champions multiple times. Lubchenco also journeyed far and wide.

Early marine experiences

Girl Scouts formed an important presence in the Lubchenco household. Jane was president of the Mile High Girl Scouts of Denver, and her troop was a mariner troop, which specialized in canoeing and sailing. Canoe opportunities were widespread in Colorado, but there had been no conveniently situated oceans for sailing in the Denver area for about 70 million years. So, using funds from bake sales and from recycling newspapers, Lubchenco's troop traveled twice to the West Coast to spend time on the Pacific Ocean. On one trip, Lubchenco's aunt arranged for the Denver Mariner Scout troop to tour the USS *Kitty Hawk* aircraft carrier, which was being serviced in San Diego. The officer in charge was absolutely panicked when these female mariners arrived; he was expecting boys and was concerned about inappropriate male behavior and forbade them from boarding; but Lubchenco's aunt persevered and they toured the *Kitty Hawk* without incident. Lubchenco also traveled to Mexico, Canada, the Eastern United States, and the Philippines for various scouting adventures, relishing learning about other cultures and places.

A well-traveled Lubchenco entered Colorado College. She had spent some time shadowing her father in a Denver emergency room and loved the experience. But in college, she shied away from medicine because she enjoyed the company of philosophers and liberal arts majors more than that of the pre-med students. She enrolled in a unique independent study major in which students designed their own curriculum. There were no grades, no mandatory classes, and no tests – only a general exam after 2 years, and a more specialized exam at the end of 4 years. Lubchenco loved this freedom, and explored many different disciplines. Her transformational experience was a summer program at the Marine Biological Laboratory in Woods Hole, Massachusetts, which provided her first significant exposure to the world of marine biology and invertebrates. As she describes, "I just couldn't get enough of it. It was intoxicating." In addition to the invertebrates, many famous humans came through to lecture, including Nobel laureates such as Albert Szent-Györgyi, the discoverer of vitamin C and parts of the Krebs cycle. Immersed in this oceanographic and intellectual sea, Lubchenco found her calling, and she knew that graduate school was the next step.

Graduate school with the stars

Lubchenco began her graduate career at the University of Washington in 1969. She chose Washington because the department was very diverse, and she did not know precisely what she wanted to do. Her initial research advisor was a physiologist who believed that graduate students were his slaves, and women graduate students existed primarily to cater to the men. Needless to say, this relationship did not last very long. Lubchenco was attracted to the dynamic group of experimental marine ecologists and evolutionary biologists in the department, both the faculty and the students. She explains, "It was an amazing time for ecology, and a wonderful, just spectacular group of graduate students. They were so excited, so loved reading papers and arguing about them, and teaching each other things."

Early in her stint at the University of Washington, Lubchenco met fellow graduate student Bruce Menge, who later became her husband. She accompanied Menge on his field research near Friday Harbor, which is situated in the San Juan Islands, along the Washington State coastline. Menge was studying the sea star,

Leptasterias hexactis, and how it competed with the much larger sea star, *Pisaster ochraceus*. He showed that the agile *Leptasterias* was more efficient at capturing and processing food than the much larger *Pisaster*. He argued that *Leptasterias'* small size resulted in a reproductive strategy of brooding its young – producing relatively few, large offspring, which the female keeps for about 2 months in a brood chamber formed by wrapping its arms in toward the main axis of its body. If *Leptasterias* were to broadcast its young into the environment, like *Pisaster* does, then its reproductive success would be vanishingly low. *Pisaster*, because it is about 100 times as large at maturity than is *Leptasterias*, can produce a massive number of eggs, which it can broadcast into the environment. Because these eggs are small and parents don't care for the young, survival is very low. Menge argued that broadcast spawning has evolved in *Pisaster* to increase dispersal to new environments that may have excellent opportunities for survival and subsequent reproductive success. Even though almost all of the young die before maturity, the few survivors may be highly successful (Menge 1975). In addition to being much larger, *Pisaster* also has nasty pedicellariae, pincer-like structures on its upper surface, which it used to beat up on *Leptasterias* whenever they encountered one another. Lubchenco wondered what happened when individuals of the two species were both the same size, so she began studying a natural *Pisaster* nursery at Point Caution, near Friday Harbor Laboratories. In contrast to the size distribution usually found in nature, at Point Caution the young *Pisaster* were about the same size as mature *Leptasterias* (Figure 24.3).

Lubchenco started collecting a variety of data on interactions between the two species, and also began thinking of observations she could make and experiments that she could conduct. After analysing a substantial amount of data, she built up the nerve to approach Robert Paine at the University of Washington to show him her findings. Paine had already established himself as one of the icons of ecology; now, 50 years after his research on *Pisaster* as a keystone species was first published, his work is still presented in most ecology and introductory biology textbooks (see Chapter 16 of this text). So Lubchenco was relieved when Paine's response was "This is fantastic. I love it. Are you asking me to be your advisor?" Lubchenco describes Paine as a spectacular advisor, very challenging with very high expectations, but extremely supportive. Paine's philosophy, which Lubchenco also favors, is that graduate students should carve out their own research niche, and that a student's research should not simply be an extension of her advisor's research. So Paine was delighted when approached by a student who had taken the initiative to do so much of the research on her own.

Lubchenco suspected that the two sea star species were competing for food at Point Caution, in part because the primary food available to the sea stars were the barnacle, *Balanus glandula*, and the snail, *Lacuna* spp., which Menge (1972) had already shown were relatively low-quality food items from the standpoint of energy. So she studied both sea star species to see whether there

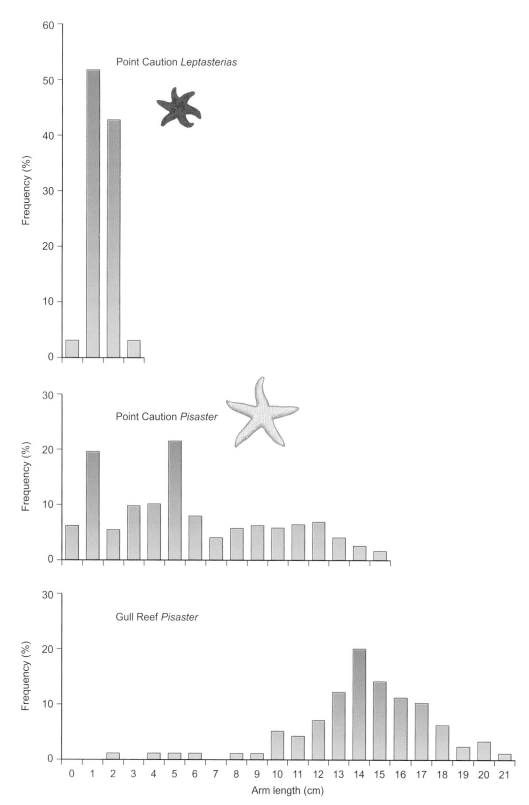

Figure 24.3 Size distributions (based on arm length) of *Leptasterias* at Point Caution, *Pisaster* at Point Caution, and *Pisaster* at Gull Reef. The size distribution at Gull Reef was typical of most habitats that supported *Pisaster*.

were differences in how they partitioned their resources. Lubchenco considered and rejected several resource partitioning hypotheses. For example, she observed that both species have their peak foraging activities during the late spring and summer months. Both species forage during high tide, though *Leptasterias* has a shorter foraging period. Both species forage equally on the two available substrates – solid rock and cobble (loose, rounded rock). Both species migrate vertically on a seasonal basis, venturing into deeper waters in the winter and favoring shallower waters in summer. And both species take similar size prey. Taken together, these observations provided no evidence for resource partitioning.

However, Lubchenco's studies showed that the average *Balanus* eaten by *Leptasterias* was about 30% wider than the average *Balanus* eaten by small *Pisaster*. Given that *Balanus* constituted about 70% of the prey at Point Caution, these differences could, in theory, have important implications for the energy budget of the two species. A second dietary distinction between the two species was that the more agile *Leptasterias* could subdue a substantial amount (about 30%) of prey that have a much higher calorie content than does *Balanus*, while small *Pisaster* were completely dependent on low-calorie prey. Once *Pisaster* grew larger, it too could subdue higher-quality prey. But while small, *Pisaster* suffered a competitive foraging disadvantage against *Leptasterias* (Menge and Menge 1974).

Despite this foraging disadvantage, *Pisaster* grew much larger than *Leptasterias*. Lubchenco's research indicates that *Pisaster* aggression depressed *Leptasterias*' feeding rate. When *Pisaster* was placed in a tank with *Leptasterias*, *Pisaster* extended its pedicellariae and clamped them onto *Leptasterias*; this pinching resulted in *Leptasterias* contorting into unusual positions. To explore whether this event depressed *Leptasterias*' feeding rate, Lubchenco placed five *Leptasterias* in a container with at least 20 prey items (the snail *Littorina scutulata*) and monitored the feeding for 2 days. She then added one *Pisaster* for 30 min, long enough for pinching to occur, and monitored the feeding for 2 more days. Based on two replicates, the number of snails consumed by *Leptasterias* decreased from 251 to 172 after 30 min of *Pisaster* exposure. In contrast, snail consumption actually increased in the unexposed controls, from 249 to 281. At least in the laboratory, contact with *Pisaster* substantially decreased *Leptasterias*' feeding rate.

To test the effect of *Pisaster* on *Leptasterias* foraging under field conditions, Lubchenco and Menge teamed up. They selected a site that was divided into two channels, both containing only *Leptasterias*. They then added 125 *Pisaster* to the north channel, and compared the foraging behavior of *Leptasterias* in both channels. Once *Pisaster* recovered from handling and began foraging normally, *Leptasterias* foraging declined substantially. Interestingly, after 2 weeks of foraging inhibition, *Leptasterias* was able to resume normal foraging activities; apparently it acclimatized to *Pisaster* presence (Figure 24.4). Further observations suggested *Leptasterias* individuals simply avoided getting too near to *Pisaster*. One observation supporting this hypothesis is that *Pisaster* began foraging when submerged by the incoming tide, while *Leptasterias* did not begin foraging until much later, about 2 h before high tide, presumably to avoid *Pisaster*. The researchers could not determine when the sea stars stopped foraging, but observed that *Leptasterias* rested deeper in crevices during low tide, and thus could avoid contact with *Pisaster*. Thus foraging time for *Leptasterias* was a compromise between selection favoring maximum time exposed to prey, and selection favoring avoidance of *Pisaster*.

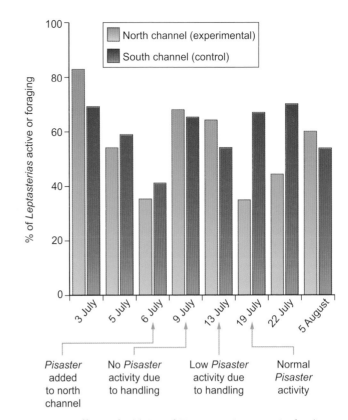

Figure 24.4 Effects of addition of *Pisaster* on *Leptasterias* feeding activity at Cattle Point, San Juan Island. *Pisaster* was added to the north channel on July 6, 1971, while the south channel was the unmanipulated control. The researchers first observed five *Pisaster* active on July 13, and the first normal level of *Pisaster* activity on July 19, when several *Leptasterias* were observed with pedicellariae attached.

 Thinking ecologically 24.1

Menge demonstrated that *Leptasterias* had a substantially longer foraging period in geographical areas with no *Pisaster*. As we described above, pedicellarial pinching depressed *Leptasterias* foraging, and presumably shortened the foraging period. Do you think that each individual *Leptasterias* needed to be pinched before it shortened its foraging period, or was there some type of sensory system for detecting *Pisaster* activity? How might you evaluate this in the field and/or laboratory?

Lubchenco and Menge married in May 1971, and soon thereafter journeyed to Boston, Massachusetts, where Menge began a faculty position at the University of Massachusetts.

24.3 HOW DID MOVING TO NEW ENGLAND AFFECT LUBCHENCO'S RESEARCH DIRECTION?

In moving to Boston, Lubchenco had to forgo her PhD ambitions at Washington, and get herself into a new graduate program in the Boston area. Fortunately, there were many options for her in the area, but she was particularly impressed with the program at Harvard, which boasted some of the top young ecologists in the country.

The atmosphere at Harvard was different from conditions Lubchenco had encountered elsewhere in several ways. Graduate students were mostly on their own; if they knew what they wanted and went after it, they did well, but if not, they simply dropped out. Lubchenco was happy with that arrangement, because she had National Science Foundation funding and was accustomed to and comfortable with working on her own. Her advisors were supportive, and always helpful in discussing ideas and concepts, despite not knowing much about the rocky shores, seaweeds, and invertebrates she wanted to study.

But there were a few daunting moments for Lubchenco. One occurred early on when a faculty advisor sat all of the first-year graduate students into a room and began,

> I want you to look around the room. You will notice that half of this class is women. This is a major departure for us. We're very excited about it. We have discovered from our past experience that if classes are almost exclusively men, then the men are unhappy. And so we've made a conscious decision to have a more balanced approach. While we fully suspect that most of you [women] won't make it, we urge you to do your best . . .

Despite this inauspicious start, Lubchenco loved her time at Harvard, and made it through the program very quickly. She had already begun some research in the rocky intertidal zone in New England, and was using her research to address some basic questions. In particular, she wanted to understand the causes of patterns of species diversity and community stability in marine communities, with a focus on seaweeds and herbivores. She was particularly interested in making comparisons between East Pacific and West Atlantic shoreline communities.

Community development and persistence in the low rocky intertidal zone

Ocean shorelines are dynamic, with extreme wave action and alternating sequences of submersion in water followed by exposure to hot sun, driving winds, violent storms, and occasional pounding by rounded rocks or cobbles. Some organisms are adapted to these conditions, and Lubchenco and Menge wanted to understand what factors determined community patterns such as who lives where. They decided to divvy up the rocky shore community, with Lubchenco focusing on herbivores and seaweeds and Menge focusing on predators and their prey. Let's start our examination of this effort by considering their research on the lower intertidal zone, which is the ocean zone that is exposed only during very low tides. Species richness can be very high in the lower intertidal zone, because it is relatively benign from a physical standpoint but still has plenty of sunlight

penetrating through the shallow water, so photosynthetic organisms and their consumers can thrive there (Figure 24.5). Lubchenco and Menge began their research with careful surveys of organism distribution and abundance. Most of their survey results are based on counting the numbers of organisms or the percent cover in a large series of 0.25-m^2 quadrats along the lower intertidal zone of the New England coastline.

Observations of distribution and abundance

Ocean wave action can have profound impacts on the structure of intertidal communities, so Lubchenco and Menge (1978) decided to study communities that varied in their exposure to strong wave action (Figure 24.6A). One pattern that emerged from these studies was that exposed sites were dominated by the mussel *Mytilus edulis*, sometimes accompanied by the barnacle *Balanus balanoides*. Sites with intermediate exposure had abundant populations of *Mytilus* and the red alga *Chondrus crispus*, while protected sites were dominated by *Chondrus*, with very low populations of *Mytilus* and *Balanus*. The researchers wanted to understand why these patterns existed (Figure 24.6B).

Lubchenco and Menge discovered that the distribution and abundance of consumers varied in relation to exposure to wave action. Sea stars, voracious predators on both mussels and barnacles, were rare in the most exposed site. Whelks (*Thais lapillus*), which feed almost exclusively on mussels in the intertidal, were relatively rare in the most protected site. The most common herbivores, including the periwinkle, *Littorina littorea*, were rare in the most exposed site and common in the most protected site. Mussels and barnacles were patchily abundant in the intermediate sites, but uniformly rare at the most protected sites (Table 24.1).

One other group of organisms found in the intertidal zone is ephemeral algae, a group of algae that are abundant for relatively short durations. The researchers identified 20 ephemeral algae species, which increased in abundance during the spring and became relatively rare in the winter. These algae differed from *Chondrus* in that they had high growth and reproductive rates. Previous work by Lubchenco indicated that *Littorina*, the primary herbivore, had a strong preference for these ephemeral algae but would also eat *Chondrus*, the abundant red alga, though very reluctantly.

Lubchenco and Menge wanted to uncover the mechanisms underlying these patterns of diversity, distribution, and abundance. Observations were essential for documenting the patterns, but experimental manipulations would be the key to uncovering the underlying mechanisms.

Experimental manipulations – secondary succession

To study the effect of predation on community structure, the researchers cleared small patches of organisms from the rocks at Little Brewster Cove, an area of intermediate wave action that hosts most of the species we've been discussing, and then

Mytilus edulis

Balanus (Semibalanus) balanoides

Chondrus crispus

Littorina littorea

Figure 24.5 Important intertidal species that Lubchenco and Menge studied along the New England coast.

Table 24.1 Mean density of animals (per m^2) at five sites ordered from most exposed (Pemaquid Point) to least exposed (Canoe Beach Cove) along the lower rocky intertidal zone.

Taxa	Pemaquid Point (very high exposure)	Chamberlain (moderately high exposure)	Little Brewster Cove (intermediate exposure)	Grindstone Neck (moderately low exposure)	Canoe Beach Cove (very low exposure)
Sea stars (predators)	0.1	57.6	15.8	56.4	16.0
Whelks (predators)	41.9	42.6	37.3	712	6.2
Snails and limpets (herbivores)	4.4	1.8	385	154	812
Barnacles (*Balanus*) (filter feeders)	1970	59	21	8.4	0
Mussels (*Mytilus*) (filter feeders)	18 775	10 493	150	6376	16

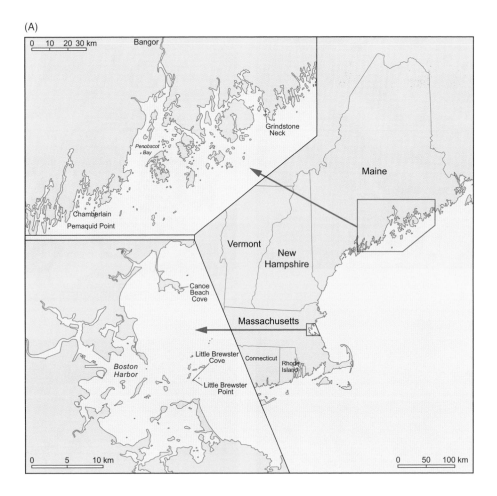

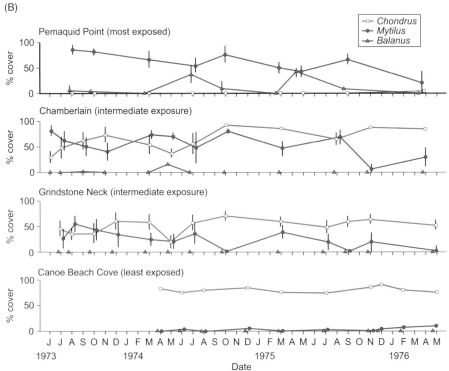

Figure 24.6 A. Lubchenco and Menge's study site along the New England coast. B. Mean (±SE) percent cover of dominant species at four sites that vary in exposure to wave action, based on data from 10 0.5-m² quadrats at each site. Total percent cover is greater than 100% in cases where the organisms are layered (such as a snail grazing a patch of algae, or algae that have settled atop a mussel bed, or grown above a second algal species).

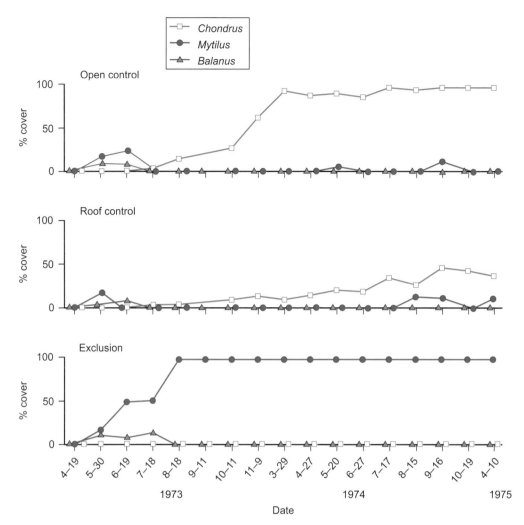

Figure 24.7 Effect of predation by sea stars and whelks on competitive interactions and subsequent secondary succession in the low intertidal zone at Little Brewster Cove, a site with intermediate wave action.

allowed secondary succession to occur under three different conditions. For the experimental treatment, Lubchenco and Menge covered the cleared area with stainless steel mesh cages that excluded predators such as sea stars and whelks. One control was a similar-sized area with no cage (open), while a second control was a roof that provided the same amount of shade as a cage, but allowed free passage of predators.

This series of experiments showed the importance of predators in this system. When predators were allowed, *Chondrus* slowly increased in abundance, while *Mytilus* and *Balanus* cover remained relatively low. Both open controls and roof controls showed this same pattern. However, when predators were excluded from the system, *Balanus* and *Mytilus* both increased, but *Mytilus* then quickly outcompeted *Balanus*, ultimately increasing to 100% cover (Figure 24.7). This is an example of how indirect interactions influence community structure. Under normal conditions at Little Brewster Cove, sea stars and whelks eat *Mytilus* and *Balanus*, and in so doing eliminate *Chondrus*'

major competitors for space. Thus, indirectly, these predators are facilitating *Chondrus*.

One exception to this pattern was experiments that were set up on rocks that initially had no substantial *Chondrus* thalli, or upright branches, but instead were covered with *Chondrus*' encrusting holdfast, the structure that attaches the alga to the rock, and the source of thallus growth. Structurally, these sites had large numbers of cobbles, or rounded stones, strewn among the more established rocks that supported the *Chondrus* holdfasts. Over time, the control quadrats remained encrusted, but never grew any thalli, while experimental cages eventually filled with *Mytilus*. Lubchenco and Menge explained that the cobbles were often tossed about during winter storms, scouring the substrate of thalli that were beginning to grow from the holdfasts. These regular disturbances returned the rocks to a very early successional stage. Taken together, these findings show how biotic factors such as predators and abiotic factors such as substrate structure are important in structuring the community.

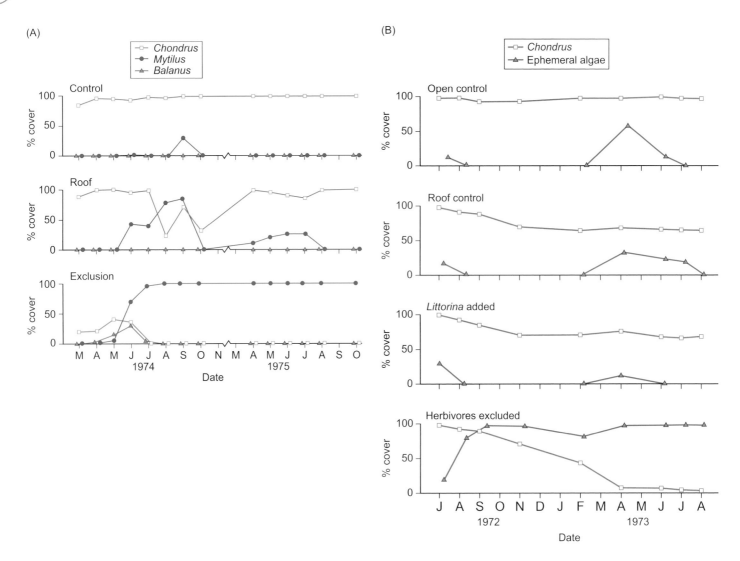

Figure 24.8 A. Role of predators in maintaining already established communities. *Chondrus* maintains its dominance in open (top) and roof (middle) controls, but is outcompeted by *Mytilus* when predators are excluded (bottom). B. Role of herbivores in maintaining already established communities. *Chondrus* maintains its dominance in open and roof controls (top two graphs) and in exclosure that had the major herbivore (*Littorina*) added to the cage. However, when herbivores are excluded, ephemeral algae outcompete *Chondrus* (bottom graph)

Next, Lubchenco and Menge wanted to understand the role played by herbivores in structuring the community. Working at Canoe Beach Cove, the most protected site, they cleared all animal species and algal thalli from rocks. As before, they set up cages that could exclude various species, but this time the focus was on herbivores. In one cage they excluded all herbivores, while in another cage they placed five *Littorina littorea*, the most abundant herbivore in the low intertidal zone. Exclusion of all herbivores resulted in extensive growth of the ephemeral green alga *Ulva lactuca*, and reduction in *Chondrus* regeneration. In contrast, both controls (which allowed herbivores) and the *Littorina*-added treatment had steady regeneration of *Chondrus* thalli. Replication of this experiment at a second site showed very similar results. When herbivores were present, they ate the ephemeral algae, and *Chondrus* grew well, but when herbivores were absent, ephemeral algae were much more successful, and *Chondrus* growth was depressed.

The researchers concluded that both predators and herbivores exert important top-down effects on community development during secondary succession. Their next series of experiments explored the role of predators and herbivores in maintaining already established communities.

Experimental manipulations – established communities

To explore the direct role of predators in maintaining existing community structure, the researchers excluded predators from some quadrats at two sites – Little Brewster Cove and Grindstone Neck – while also monitoring open and roof controls. All controls (predators allowed) maintained high *Chondrus* levels throughout 2 years of observation, although *Mytilus* did colonize some of the quadrats before being discovered by predators. However, in the predator-free quadrats, *Mytilus* increased to 100% cover, and competitively excluded the *Chondrus* (Figure 24.8A).

To study the direct role of herbivores in maintaining *Chondrus* domination, the researchers excluded *Littorina* from some quadrats, and also monitored open and roof control quadrats. As expected, in the control quadrats, *Chondrus* maintained its overall dominance, though ephemerals did increase during the spring and early summer. In the herbivore exclusion quadrats, the ephemerals increased to very high numbers, while *Chondrus* abundance dropped sharply (Figure 24.8B).

To summarize some of these findings, the development and maintenance of communities in the low rocky intertidal depend on three factors. First, predators remove *Mytilus*, which allows *Chondrus* to colonize and persist. Second, herbivores remove ephemeral algae, releasing *Chondrus* from competition with these algae. Lastly, heavy wave action can remove predators, which allows exposed habitat to become dominated by the competitively superior *Mytilus*.

Thinking ecologically 24.2

We've discussed two important questions. First, what is the role of predators and herbivores in establishing community structure after disturbance (secondary succession)? Second what is the role of predators and herbivores in maintaining the structure of established communities? Consider a third question: What is the role of predators and herbivores in establishing community structure after a major disturbance that destroys almost all life (primary succession)? How would you set up such an experiment, and what types of predictions would you make? Explain your reasoning.

This research program brings to light an important dimension common in ecological research. Lubchenco and Menge demonstrated the importance of biotic interactions in determining species composition and abundance. Recall that in Chapter 1 we introduce an ecological community as an assemblage of interacting populations. As such we could argue that Lubchenco and Menge are community ecologists. However, Lubchenco and Menge also showed that wave action and storms – abiotic factors – are the basis for different biotic interactions in different habitats. We've defined an ecosystem as a community of organisms plus the abiotic factors that influence the community, so since Lubchenco and Menge investigate both biotic and abiotic factors, we can also argue that they are ecosystem ecologists. Are they community or ecosystem ecologists? The answer is both. As we emphasized in Chapter 1, ecologists, more than most other biologists, often ask questions that link different levels of the biological hierarchy. Later, we will see how Lubchenco's more recent work considered even higher levels of the biological hierarchy.

Creeping up and down the shoreline

As you might imagine, Lubchenco and Menge were not confining their studies to the low intertidal, but also conducted research

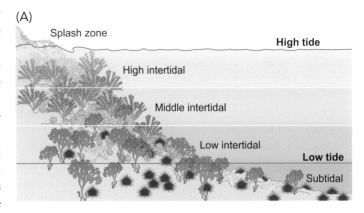

Figure 24.9 A. Five zones along the New England coast. B. New England rocky intertidal with conspicuous zonation. At this site, the wave-protected shore at Little Brewster Cove, the high zone is dominated by barnacles, the middle zone by *Fucus* and *Ascophyllum* algae, and the low zone by *Chondrus*.

throughout the various shoreline zones (Figure 24.9A). Previous researchers had described the striking zonation patterns along the New England coast, with the low intertidal dominated by *Chondrus*, as we've just described, the middle intertidal dominated by the brown algae *Fucus* or *Ascophyllum*, the high intertidal dominated by *Balanus*, and the splash zone immediately above the intertidal dominated by cyanobacteria and lichens (Figure 24.9B). Most researchers attributed this pattern solely to differences between species in their timing of emergence, their use of light, or their sensitivity to desiccation. Lubchenco and Menge suspected that biotic interactions might also play a role in determining these patterns.

When Lubchenco (1980) removed *Chondrus* thalli from experimental quadrats, the *Chondrus* quickly regenerated its thalli from the residual crust. However, when Lubchenco also removed the *Chondrus* crust, *Fucus* quickly moved in and colonized the exposed rock. These brown algae were very healthy and also developed reproductive structures. They maintained a presence for about 3 years, during which time

Chondrus was gradually able to recolonize and eventually reassert its dominance. Lubchenco concluded that the absence of *Fucus* from the low zone is a consequence of being outcompeted by *Chondrus*. Competition could set a lower limit to seaweeds much like Connell had shown in his study of barnacle distribution (see Chapter 13). Herbivores also play a role in determining *Fucus* success. In locations such as Canoe Beach Cove, where *Littorina* was abundant, *Fucus* was much slower to colonize areas where *Chondrus* was removed, because *Littorina* eats young *Fucus*. But at Chamberlain, where *Littorina* was rare or absent, *Fucus* colonized areas where *Chondrus* was removed much more rapidly

The lower limits of *Chondrus* distribution are also influenced by herbivore distribution, in this case the sea urchin, *Strongylocentrotus droebachiensis*. Sea urchins are common up to the boundary between the subtidal and the low intertidal zone at Grindstone Neck, and the *Chondrus* zone also ends right at that boundary. One winter, sea urchins mysteriously disappeared from the Grindstone Neck subtidal zone, and *Chondrus* extended its range substantially deeper. Furthermore, at Canoe Beach Cove, which has very few sea urchins in the subtidal zone, *Chondrus* continues into the subtidal to a depth of 10–20 m. Both of these observations support the hypothesis that sea urchins are responsible for the absence of *Chondrus* in the subtidal zone.

In summary, zonation is a prominent feature of the rocky intertidal zone, and interspecific interactions help to establish and maintain these zones. At this stage of their research, Lubchenco and Menge were ready to expand their research to other ecosystems, to see whether interspecific interactions contribute to the structure of rocky intertidal ecosystems in other biomes.

24.4 WHAT FACTORS INFLUENCE THE STRUCTURE AND FUNCTIONING OF TEMPERATE AND TROPICAL ECOSYSTEMS?

Having completed her graduate program at Harvard in 1975, Lubchenco was offered a tenure-track position there, which rarely happened to a woman at Harvard at that time. This timing coincided with the great civil war at Harvard (see Chapter 22), which resulted in the Organismal and Evolutionary Biology portion of the Biology Department splitting from Cell and Molecular Biology. There were intense politics afoot, but numerous opportunities and great colleagues to interact with. Lubchenco was hesitant to accept the position, in part because she knew that Menge wanted to move back west, but he agreed they should stay in the Boston area, because it was such an excellent opportunity for her. So they stayed for 2 more years, continued their New England studies, and expanded their research into the tropics.

Lubchenco and Menge chose the Pacific Coast of Panama because it suited their research needs ideally. They rejected the Caribbean coast, because it generally has a small intertidal zone, with the difference between high and low tides only about 1 m in most areas. In contrast, the Pacific shore has an intertidal of about 6 m, which is comparable to, though slightly greater than, the tide differential in Washington State and New England. In addition, Panama's Smithsonian Tropical Research Institute Laboratory was an excellent base of operations for the two researchers.

Like today's ecologists, researchers in the 1970s wanted to understand global patterns of biological diversity. They knew that biological diversity was greatest in the tropics, but they did not understand why, though there were no shortages of hypotheses. Lubchenco and Menge wanted to understand the relative importance of physical and biological factors in structuring the communities in tropical biomes, just as they had with temperate biomes. One obvious difference was the nearly complete absence of obvious zonation in the tropical intertidal, which is much more barren than its temperate counterpart. The prevailing hypothesis at the time was that strong sun and high temperatures created conditions on the intertidal that were too harsh for substantial growth of algae and other sessile organisms. Based on their experiences in temperate biomes, Lubchenco and Menge suspected that interspecific interactions might be important as well. They also believed that understanding the interaction between physical factors and ecological interactions in the tropics might help explain differences in global biodiversity between temperate and global biomes.

Observations by Lubchenco and her colleagues (1984) summarized many differences between the tropical and temperate intertidal zones. At their tropical site on Taboguilla Island in the Bay of Panama, primary space is dominated by algal crusts in the mid and low intertidal, and by bare rock in the high intertidal. Sessile animals are rare, in stark contrast to the huge densities of mussels and barnacles that are found in exposed sites in the temperate intertidal zones. Animal species diversity is much greater in the tropical intertidal, but population densities of individual species are relatively low. The cover of erect algae is very low, again contrasting with the abundant red and brown algae and ephemeral green algae in the temperate intertidal zone. Lastly, the few erect algae are very short in stature.

Why do these differences exist? To determine if consumers were maintaining low levels of upright algae and sessile invertebrates in the tropics, the researchers used cages to exclude all consumers from the high, middle, and low intertidal zones, as described previously. At the same time they also established control plots, and compared changes in the density of sessile invertebrates, as well as percent cover of algae and bare ground over a period of 1–2 years. They found sharp increases in barnacle density in the high intertidal and in bivalve density in the middle and low intertidal. In addition, the percent cover of upright algae increased substantially, particularly in the lower intertidal (Table 24.2). The researchers concluded that the

Table 24.2 Effects of consumers on prey abundance over a 1- to 2-year time period. Densities of barnacles in the upper intertidal are no./0.04 m². all other densities are no./0.25 m². Algal abundance is percent cover (SE in parentheses). *N* = number of plots.

Zone	Category	Control		Consumers excluded	
		Initial	Final	Initial	Final
High	Barnacles	203.1 (81.9)	434.4 (227.8)	59.4 (9.4)	1253.1 (128.1)
	Bare rock	97.2 (0.7)	91.7 (0.9)	95.5 (1.5)	78.5 (8.5)
	N	10	10	2	2
Middle	Barnacles	4.1 (1.8)	14.8 (6.7)	2.0 (2.0)	10.5 (7.5)
	Bivalves	1.6 (0.4)	5.4 (1.4)	1.0 (0)	95.5 (55.5)
	Algal crusts	80.9 (8.0)	84.2 (4.0)	95.9 (0.6)	85.5 (4.9)
	Upright algae	0	0	0	8.3 (3.2)
	N	15	15	2	2
Low	Barnacles	0.7 (0.3)	0.3 (0.1)	0	26.0 (3.6)
	Bivalves	13.6 (2.2)	34.5 (5.2)	4.0 (0)	247.5 (128.5)
	Algal crusts	91.1 (0.9)	88.9 (0.8)	86.2 (3.3)	12.0 (0)
	Upright algae	0.1 (0.1)	1.0 (0.2)	0.5 (0.7)	30.8 (7.0)
	N	15	15	2	2

diverse assemblage of consumers in the tropical intertidal was keeping the algae cropped, so that only the crusts were present, except in a few protected refuges. When consumers were excluded, the algae were released from consumption, and were able to grow a more upright form. Sessile invertebrates were also being eaten by predators, and were released from consumption by the exclosures.

Menge and Lubchenco (1981) further explored the importance of four different types or functional groups of consumers in structuring the community. Slow-moving predators included predaceous gastropods and sea stars. Slow-moving herbivores were most snails, limpets, and chitons. Large fishes were defined as those that could not fit under the 5 cm roofs of the steel cages that only had two sides. Lastly, small fishes and crabs could fit under the 5 cm roof overhang. Using a combination of manual removal of slow-moving organisms and exclusion using either complete cages or roofs, Menge and Lubchenco were able to create quadrats that excluded either one, two, three, or all four

functional groups, and to compare the effect of such removal on colonization by upright algae and sessile invertebrates.

These experiments showed that removal of increasing numbers of functional groups had a cumulative effect on eliminating free space in the experimental quadrats (Figure 24.10). Removal of all four functional groups caused a rapid and dramatic increase in upright algae, barnacles, oysters, mussels, and several other groups. Menge and Lubchenco concluded that the barrenness of the intertidal zone at Taboguilla Island results from intense herbivory and predation by many different species and types of species. No individual species, or even functional group made up of many species, is responsible for the observed pattern. This is a strong contrast to Robert Paine's findings for the mid- and low-intertidal zones along the Washington coast, where removal of the keystone species, the sea star *Pisaster ochraceus*, caused a sequence of dramatic changes to community structure (Paine 1966).

If this large diversity of predators is so effective at consuming algae and sessile invertebrates, how then do these prey species

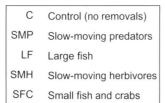

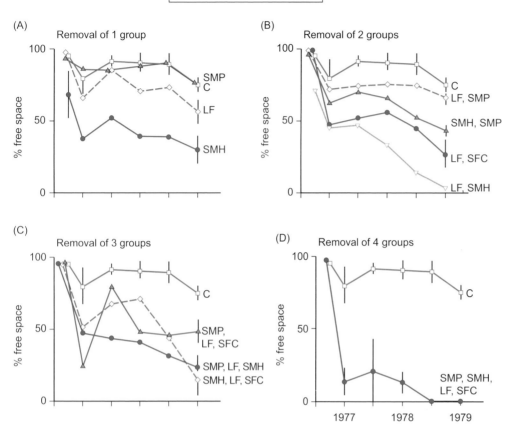

Figure 24.10 Effect of removal of functional groups on amount of free space present in the low intertidal zone off Taboguilla Island. A. One group removed. B. Two groups removed. C. Three groups removed. D. All four groups removed. The letters next to the curves represent the functional groups that were removed. Where no error bars appear on the graph, the SE was smaller than the symbol.

persist in the tropical intertidal community? The distributions of solitary invertebrates provide an important clue. Over 30% of barnacles, and more than 90% of bivalves, limpets, chitons, and predacious snails, were found in protected surfaces such as cracks, depressions, crevices, and holes at Taboguilla Island. Presumably these refuges allowed individuals to escape predation by the larger visually oriented predators common in the tropical intertidal zone.

 Thinking ecologically 24.3

There are at least two alternative hypotheses for why intertidal organisms occupy holes or crevices. First, holes and crevices may provide protection against excessive heat or desiccation. Second, wave shock may be too severe to allow attachment to open substrata. What predictions do these alternative hypotheses make, and how might you collect data to test the predictions?

Menge and Lubchenco concluded that the barren appearance and absence of obvious zonation in their tropical site was a result of the diverse community of active visually oriented predators and herbivores that occupy almost all parts of the intertidal zone, consuming most of the vulnerable prey species. In temperate biomes, prey species can often find habitats that are predator-free, or that have greatly reduced predator abundance. In addition, many temperate predators and herbivores lack eyes, and thus will often miss prey within their range. Finally, most temperate consumers have extended periods of seasonal inactivity during late autumn and winter, providing a temporal refuge from consumption for many prey species. None of these refuges are available for tropical prey species, so their distribution is mostly restricted to protected surfaces.

As Lubchenco points out, zonation is not an isolated phenomenon limited to rocky intertidal zones but is also a feature of many terrestrial ecosystems. For example, if you climb a mountain, you will experience different life zones during your ascent, which may begin in lush forest and terminate in alpine tundra. Lubchenco argues that on a very large scale, our understanding of altitudinal and latitudinal patterns of biological diversity will benefit from a two-pronged approach that includes considering differences in the physical environment characteristic of each biome, and differences in the biotic environment from the direct and indirect effects of species interactions.

Shortly after beginning their research in Panama, Lubchenco and Menge began to think about having children. Both of them wanted to continue doing research and teaching, but also wanted family time. The convention at that time for academic couples had the husband working as a full-time faculty member while the wife stayed home with the children, possibly employed in a part-time position with little future. Rejecting that approach, Lubchenco and Menge sought universities willing to do something non-traditional: split a single faculty position into two separate tenure-track positions. Oregon State University was intrigued and in 1977 offered each of them a tenure-track, half-time faculty position. Both were delighted with the prospect of returning to the West Coast and having access to the relatively unstudied local intertidal ecosystem. They moved to Oregon, started research there while maintaining their Panama research, and started a family. At the same time, Lubchenco's professional interests were expanding into new areas as well.

24.5 HOW DID LUBCHENCO EXPAND HER SCOPE INTO POLITICS AND SCIENTIFIC COMMUNICATION?

Following the move to Oregon State, Lubchenco and Menge's scientific approaches expanded in a number of important ways. First, they were curious about whether similar processes influenced ecosystem structure and functioning globally, so while continuing their Panama work and starting their Oregon research, they began laying the groundwork for a series of collaborations with colleagues in Chile, New Zealand, and South Africa. Because their field work on temperate sea shores was concentrated in the spring and summer, they could work in Oregon during those seasons, and shift to the southern hemisphere for spring and summer seasons there. Later they began looking at patterns at a much larger scale by studying Large Marine Ecosystems (LMEs), which are regions of the world's oceans bounded on the land side by the shoreline, including estuaries and river basins, and on the ocean side by the continental shelf and the major ocean currents. NOAA has identified 64 distinct LMEs, each of which has characteristic physical and biological attributes, such as ocean currents, sea floor topography, and patterns of productivity and biogeography.

Exploring larger scales requires working with a large number of researchers who possess a variety of skills and expertise. Many of their studies used the *comparative-experimental approach*, in which researchers conduct identically designed experiments at different sites (for example different LMEs) to test specific hypotheses. In many cases, there may be similar environmental gradients operating at these different LMEs; for example, wave exposure may vary in different locations within each LME that is being studied. Other environmental gradients within an LME might be differences in salinity, or in the strength and pattern of upwelling (Menge *et al.* 2002).

The comparative-experimental approach enables researchers to understand how natural ecosystems operate. But this approach also allowed Lubchenco and Menge to see how natural ecosystems were changing and ultimately, how these changes might influence people, as well as how the activities of people were influencing natural ecosystems.

As her approach to doing research was evolving, Lubchenco was becoming more involved in politics on a variety of levels. Locally, she served as the Zoology Department Chair at Oregon State from 1989 to 1992. On a national level, she was very active in the Ecological Society of America (ESA) and the American Association for the Advancement of Science, serving as president of both organizations during the 1990s. Perhaps her most lasting achievement will prove to be her role in supporting the development of a landmark in the history of ecological research, the Sustainable Biosphere Initiative (SBI).

The Sustainable Biosphere Initiative

When I was in graduate school in the 1980s, applied research, research that has the potential of benefiting society, was almost considered a four-letter word. It was okay if our studies had the potential to benefit society, but only so long as that was not the primary focus of our research. We were trained to pursue basic research, which advances our understanding of ecological processes and patterns but does not necessarily carry with it any societal benefit beyond improving our understanding of how the world works. This tension between applied and basic research has a long history in ecology and many other scientific fields. For example, in 1951 the Nature Conservancy, one of the world's leading conservation organizations, broke away from the ESA because the society believed that relevant applied research was inappropriate for a professional scientific society (Lubchenco 2012). But by the late 1980s, the ecological community was beginning to recognize that basic and applied investigations were not mutually exclusive.

Part of the motivation for the shift in attitude was money: funding for ecological research was meager. In hopes of addressing the funding problem and influencing the kind of research being funded, the ESA formed an *ad hoc* Research Agenda Committee, chaired by Lubchenco, which was charged with articulating ecological research priorities. The final product of this committee was the SBI, which highlighted that much of the significant research that is needed now is cutting edge, pushing the boundaries of our understanding while simultaneously helping society. After wrangling through an extraordinarily difficult process that demanded great cooperation, the committee created a document they could embrace and presented it to the 1990 ESA meeting in Snowbird, Utah, where Lubchenco's presentation received a standing ovation. From that time onward, the SBI has provided impetus for research that is relevant to society, and emphasizes the importance of sharing knowledge and using scientific information for management and policymaking.

Published in 1991, the SBI established a framework for the acquisition, dissemination, and utilization of ecological knowledge based on three processes: (1) basic research for the acquisition of ecological knowledge, (2) communication of that knowledge to citizens, and (3) incorporation of that knowledge into policy and management decisions. It established three research priorities. One priority, *global change*, included "the ecological causes and consequences of changes in climate; in atmospheric, soil and water chemistry (including pollutants); and in land and water use patterns." The second priority, *biological diversity*, included "natural and anthropogenic changes in patterns of genetic, species and habitat diversity; ecological determinants and consequences of diversity; the conservation of rare and declining species; and the effects of global and regional change on biological diversity." The third priority, *sustainable ecological systems*, included "the definition and detection of stress in natural and managed systems; the restoration of damaged systems; the management of sustainable ecological systems; the roles of pests, pathogens and disease; and the interface between ecological processes and human social systems" (Lubchenco *et al.* 1991). The SBI emphasizes that these three areas are tightly linked. For example, deforestation may alter regional climate by influencing the hydrological cycle, it may reduce biological diversity by altering the habitat, and it may threaten sustainability of fisheries by increasing sedimentation and temperature regimes within streams. These linkages are highlighted by the SBI logo, a Borromean knot; if any single ring of this knot is broken, the entire structure falls apart (Figure 24.11).

The SBI framework is broad – not surprising when you consider that its purpose is to outline the direction for a major scientific discipline. Despite its breadth, it articulated a group of 10 key research topics that are central to the field of ecology. Some of them were traditional topics that had been part of the focus of ecological inquiry for many years. For example, one research topic was to inventory the patterns of biological diversity. Ecologists should accelerate research on factors that determine biological diversity on the genetic, species, habitat, and ecosystem level. Included in their inventory, ecologists should determine the rates of change of biological diversity, and how those changes affected community structure and ecosystem processes. Under this topic was a list of related questions, all of which are discussed in this textbook. While it was critical to highlight this research topic, no ecologist was surprised by its inclusion. In contrast, the final topic in the list was relatively new to the ecological community. Ecologists should determine the principles that govern disease outbreak, and the patterns underlying the spread of disease and pest organisms. The list of

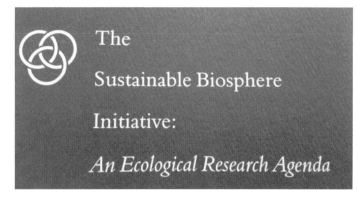

Figure 24.11 SBI logo showing Borromean knot.

questions related to this topic outlined research programs that were novel to many ecologists. These included evaluating the effects of climate change on the distribution of disease-causing organisms, and determining how environmental changes such as deforestation and drought might alter the transmission of infectious diseases in human populations. By extension, the public at large had absolutely no idea that understanding the dynamics and spread of disease was an ecological problem. By including this topic, ecological research suddenly became much more mainstream.

The SBI would be of little value if it sat on a shelf, or in a computer file; it needed to go public. Lubchenco and several other ESA members made the rounds in Washington DC, to share the SBI, and to make its proposals understandable not only to policymakers, but also to journalists and to industry. At that time, most scientists did not want to talk to the media, which often took the science out of context, did not understand important nuances, and made the science appear irrelevant and frivolous. So Lubchenco and her colleagues designed a program to train scientists to talk to a variety of groups about ecological issues.

Leopold Leadership Program and COMPASS

Named after the famed conservationist, gifted scientist and communicator, and author of the iconic *A Sand County Almanac*, the Aldo Leopold Leadership Program trains 20 mid-career ecologists per year to be better communicators about environmental science, and to be ambassadors to the public about what environmental scientists do and why it is important. Lubchenco and her colleagues decided on mid-career ecologists because "going public" was considered risky from a career perspective to untenured faculty members who still did not have their research programs established. The advisory group consisted of ecologists from around the world, and of media people and policymakers – some of the people that formed the target audience for the Leopold Fellows. Many of the Leopold Fellows have gone on to design their own science communication courses at their own institutions, teaching the principles of scientific communication they learned during their fellowship.

The basic program gathers 20 Leopold Fellows for two 1-week interactive training workshops in which they develop a vision for how they can use their backgrounds to influence environmental change, and they develop the skills for doing so. Over the course of the next year they practice those skills at their home institution to see what works and what doesn't. Lastly, they gather together the following year to develop a more sophisticated vision and action plan for promoting environmental change.

While the Aldo Fellowship Program was getting started, Lubchenco and a few other colleagues were developing a second initiative, the Communication Partnership for Science and the Sea (COMPASS). Its mission is to train marine scientists (including the Leopold Fellows) to become more effective communicators with policymakers and the media. In the process, these trained marine scientists also develop connections with each other so they are aware of each other's expertise and can use this collective expertise to solve specific problems. COMPASS also serves as a mechanism for policymakers to receive reliable scientific information that is not tainted by special interests. For example, COMPASS recently organized a briefing for leaders of three branches of the federal government – the Office of Management and Budget, the Council on Environmental Quality, and the Office of Technology Policies – to help them understand the major principles of landscape-level ecology, so they could apply these principles to specific issues that were under their jurisdiction. COMPASS is now in the process of expanding its mission from marine ecology to include terrestrial and atmospheric ecology.

When asked if there was any one experience that convinced her to expand her work into the realm of environmental policy, Lubchenco described her return to Discovery Bay, Jamaica, two decades after she had taught a course there for the Organization for Tropical Studies (see Chapter 18). During her original visit she was impressed by the incredibly diverse and rich coral, fish, and invertebrate communities. When she returned there many years later she discovered a wasteland that had been completely overfished – especially the grazers. Nutrients were pouring off the land, a product of bad land-use practices, which in turn resulted in weedy seaweeds taking over the whole ecosystem, and a tremendous loss of species richness. "It was just such a dramatic difference and so incredibly depressing. I remember crying underwater into my mask." The change was particularly apparent because of the time interval between her two visits, but she emphasizes that at many of her other field sites, she could observe environmental degradation from one year to the next.

Obama calls

Given Lubchenco's joint commitments to her research program with her husband and to her initiatives to improve scientific communication, she was initially not inclined to accept the offer to head up NOAA when Obama's transition team called her in December of 2008, while she was vacationing with her family in Tasmania. Nonetheless, she agreed to meet with the President-Elect and discuss the offer. She described Obama as "very inspiring, very curious, and full of questions." He was also "not at all hesitant to ask about something he doesn't know about, and committed to restoring scientific integrity and basing decisions on science." Because Obama was so inspiring to her and had a compelling vision of change, Lubchenco decided that she could not pass up the opportunity to make meaningful contributions to environmental science policy in the United States.

One of Lubchenco's most important contributions has been to establish programs that reduce overfishing and focus more on sustainable fisheries. NOAA, in partnership with local fishermen and state and regional agencies, has established *catch-share* programs that provide conservation benefits to ensure sustainable fisheries, while allowing fishermen to make wise business decisions that maximize profitability, safety, and the long-term health of the fishery. The basic premise is that the total allowable catch in the fishery is divided into shares that are controlled by the fishermen. Fishermen cannot exceed their quota over the course of the season, but they are guaranteed access to that quota, freeing them to fish when and where the yield is likely to be high, when the market demand is at a peak, and when the weather is good. Moreover, they retain their share from year to year unless they choose to sell it. Catch shares thus give fishermen both a stake in the present and in the future, ending the infamous "race to fish" that typically results in overfishing and depleted fisheries.

In 2011, the western US states adopted a comprehensive catch-share program for their groundfish fishery, which includes over 90 different species, with Pacific whiting, *Merluccius productus*, the most abundant species. In its first year, the program, which limits individual fishermen to an annual harvest quota, showed evidence of being highly successful. Despite skepticism from local fishermen, annual revenues reached record highs in 2011. As one trawl vessel operator described, "I had some pretty big reservations about the catch-share program prior to implementation. However, after the first year, I'm happy to say I was wrong. Now my discards are almost nonexistent, and I can plan my groundfish landings when it's convenient to my operation" (NOAA 2012).

One reason discards were so low in 2011 is that under a catch-share program, fisherman are given incentives to avoid non-target species. Reduced competition between fishermen makes it more likely that they share with each other the location of fishing hotspots and, equally important, locations where non-target species are abundant, and thus should be avoided. Each fisherman has a quota for the year. In addition, fishermen who exceed their quota for a particular species can trade or buy quota pounds from other fishermen. The upshot is that by carefully managing their harvest and fishing effort, fishermen can efficiently harvest when weather and market conditions are optimal, with much less waste. In the catch-share program's first year

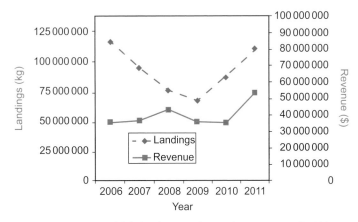

Figure 24.12 Annual fish landings (in kg) and revenues in the West Coast Groundfish Individual Fishing Quota Fishery in 2011 compared to the previous 5 years.

(2011), the number of fish brought to market increased sharply. More important, the total revenue also increased sharply (Figure 24.12), despite a slight reduction in the number of boats in the fleet (Matson 2012).

During her tenure as NOAA chief, Lubchenco oversaw many other successful ventures, as well as others that were less successful. She expects that the many lessons learned during those years will allow NOAA to move forward as she returns to academia.

Let's go back to the *Deepwater Horizon* episode to explore some of those lessons.

Lessons learned from the *Deepwater Horizon* spill

Lubchenco and her colleagues (2012) remind us that the response to the *Deepwater Horizon* oil spill required a team effort involving thousands of scientists, technicians, response workers, and managers from a variety of government and non-government agencies and academia. Many interagency teams of experts were created to deal with the spill, which differed from previous spills because it was so huge, and because it occurred approximately 1500 m below the surface. The spill generated a large number of far-reaching questions, whose answers will allow a more effective response to future oil spills.

Important questions for all oil spills include the following: (1) Where is the oil and where will it go? (2) What remains of the oil in the environment? (3) Should the oil be dispersed chemically? (4) Is seafood contaminated with hydrocarbons? (5) What is the environmental impact of the spill? (6) What is the best path to restoration? As Lubchenco and her colleagues point out, based on the response to the *Deepwater Horizon* blowout, we are developing answers to some of these questions. In addition, the spill encouraged researchers to develop new technology that

allows them to more effectively track oil once it is released into the environment.

As an example, let's consider the question of whether to use chemical dispersants at the source of the blowout, 1500 m below the surface. Several factors favored doing so. First, direct injection of dispersant into the source would maximize oil exposure to the dispersant before the oil could weather and emulsify, resulting in far less dispersant used per volume of oil treated. Second, breaking up the oil into smaller droplets would increase surface area to volume ratio, resulting in faster decomposition by microorganisms. Third, the dispersant had been shown to be non-toxic in several test species, in contrast to numerous oil derivatives that are known carcinogens. Fourth, the dispersant breaks down rapidly. Lastly, workers in boats on the surface would be spared the health risks associated with exposure to the dispersant. However, several factors supported not using dispersant deep below the surface. Perhaps most important, it had not been tested at such depths and no one knew what the consequences would be. Second, if the dispersant worked, and it did increase the rate of microbial action, a spike in the number of microbes could deplete the treated area of oxygen, causing hypoxia or anoxia, and ultimately creating dead zones. And finally, although the testing that had been done indicated that the dispersant was not toxic, the chemicals had not been tested on the full range of Gulf of Mexico species. In addition, impacts of the combination of dispersant plus oil had not been tested.

The US Environmental Protection Agency (EPA), which by law has the relevant authority, made the decision to use the dispersant contingent on strict monitoring of dissolved oxygen levels. Research conducted by Terry Hazen and his colleagues (2010) discovered a complex oil plume 1000–1300 m below the surface that had upward of 200 times the hydrocarbon concentration of a nearby area that was outside of the plume. Within the plume, bacterial cell density was substantially elevated, but surprisingly, dissolved oxygen levels were only slightly depressed from those outside the plume (Figure 24.13A and B). Interestingly, though bacterial abundance was increased, ribosomal RNA analysis indicated a substantial decline in bacterial richness, based on the number of subfamilies present. All of the bacterial subfamilies within the plume that had elevated cell densities were gamma-proteobacteria, most of which were structurally and taxonomically similar to known hydrocarbon degraders.

The scientific community was surprised with how quickly the bacteria metabolized the tiny oil droplets. Part of the explanation revolves around ocean currents, which cause waters to mix in ways that are difficult to predict or even describe. David Valentine and his colleagues (2012) were able to show how these complex mixing processes caused the bacteria community in the 1000–1300 m plume to metabolize the hydrocarbons so rapidly, and to do so without causing a major hypoxic event. Part of

(A)

(B)

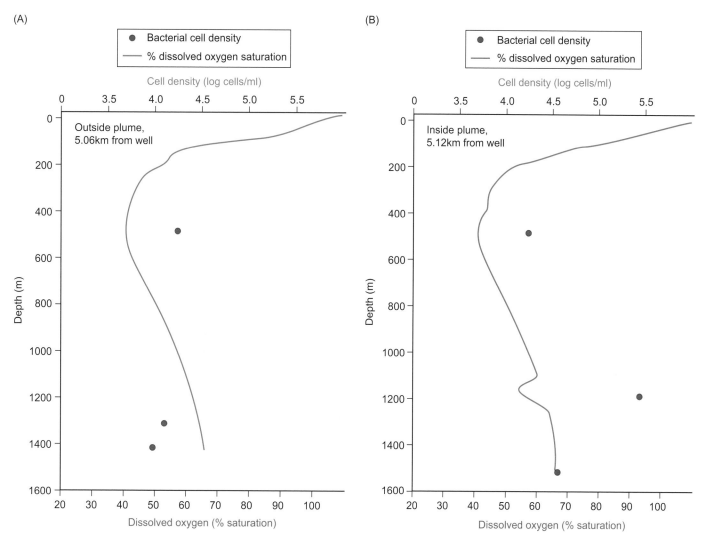

Figure 24.13 Percent oxygen saturation and bacterial cell density in relation to depth in the Gulf of Mexico. Both sites were approximately 5 km from the *Deepwater Horizon* well. A. Control site outside the oil plume. B. Site within the oil plume.

their model is based on the process of *autoinoculation*, in which water masses which had previously experienced hydrocarbon blooms, and still have high levels of hydrocarbon-degrading bacteria, return back to a location where fresh oil has recently entered the environment. This large hydrocarbon-degrading bacterial population is thus primed to rapidly dispatch the new hydrocarbons they encounter.

Unfortunately, most marine life forms do not use oil and natural gas as an energy source, and are instead poisoned by even low levels of these substances in their environment. However, in many cases it is difficult to pinpoint the causes of mortality, though a wide range of organisms, including sea-turtles, dolphins, corals, and fishes, have experienced high mortality since the spill (White *et al*. 2012; Whitehead *et al* 2012). Some of the scientific findings regarding the environmental impact of the blowout are known but have not yet been

released, because they are part of the court proceedings in the US Government versus BP trial; other findings will not be known for years, particularly for long-lived species.

Lubchenco has devoted her research career to understanding linkages between organisms and their environment. Two important take-home messages from her research are that indirect ecological interactions are common, and that the effects of environmental factors on ecosystems are often surprising and not straightforward. She cautions that the long-term environmental consequences of the *Deepwater Horizon* blowout will not be known for many decades, and urges the ecological community to learn as much as possible about all of the processes and effects, so that the environmental impacts of future oil spills can be anticipated and mitigated, and that future disasters of this magnitude can be avoided.

SUMMARY

Assuming directorship of the National Oceanic and Atmospheric Administration (NOAA) was one step in Jane Lubchenco's career that demonstrated her commitment to both basic and applied ecology. During her time at NOAA she applied her understanding of marine ecosystems and her communication skills to establish a catch-share program that has been popular with fishermen and successful in reducing pressure on threatened fisheries. In her role as NOAA director, she helped coordinate the efforts of thousands of responders to the *Deepwater Horizon* spill, and helped evaluate the short- and long-term effects of the spill on marine ecosystems.

Lubchenco's research career began with an investigation into how two species of seastars coexist in intertidal communities. This experience led to a series of comparative studies of intertidal communities off the eastern and western US coastline, and a collaborative study off the Panama coastline. Her research highlighted that ecosystems are structured from the interactions of biotic factors such as herbivory and predation, and abiotic factors such as wave intensity and the presence of refuges to escape predation. A common thread running through her research is that indirect biotic interactions are important and easy to overlook.

Field experiences and interactions with many colleagues motivated Lubchenco to get involved in a variety of initiatives that defined the future of ecological research and developed a core of researchers who were effective communicators of ecological applications. She helped establish two programs aimed at communicating environmental science: the Leopold Leadership Program and the Communication Partnership for Science and the Sea. As part of her work with the ESA, Lubchenco helped write the SBI, which articulates a framework for ecological research based on the processes of acquiring basic ecological knowledge, communicating that information to citizens, and incorporating that knowledge into effective policy.

FURTHER READING

Cornell, S. and 13 others. 2013. Opening up knowledge systems for better responses to global environmental change. *Environmental Science & Policy* 28: 60-70.

This report addresses the question of how society can most effectively deal with meeting the needs of the world's inhabitants in a sustainable manner. The authors argue that concerned citizens should collaborate with scientists to help set the research agenda, and that scientists must be encouraged to frame the problems and their solutions in a transparent manner that clearly identifies uncertainty.

Edgar, G. J. and 24 others. 2014. Global conservation outcomes depend on marine protected areas with five key features. *Nature* 506: 216–220.

An analysis of 964 sites in 87 marine protected areas identifies five key features – no harvesting of fish, good enforcement, old, large size, and spatial isolation – that maximize conservation benefits.

Lubchenco, J. 1978. Plant species diversity in a marine intertidal community: importance of herbivore food preference and algal competitive abilities. *American Naturalist* 112: 23–39.

This is one of Lubchenco's most important papers that did not make it into this chapter. She showed that the snail *Littorina littorea* influences algal diversity by preying preferentially on algal species that are highly successful competitors against other algal species. By reducing the abundance of the best competitors, space becomes available for a diverse group of less effective competitors.

Lubchenco, J. and 15 others. 1991. The Sustainable Biosphere Initiative: an ecological research agenda: a report from the Ecological Society of America. *Ecology* 72: 371–412.

This is essential reading for anyone interested in environmental policy, or who is considering a career in ecology. It lays out a foundation for ecological research priorities and approaches that is still highly relevant 25 years after its conception.

END-OF-CHAPTER QUESTIONS

Review questions

1. How might seastars partition resources? Did *Pisaster* and *Leptasterias* partition resources? Explain.
2. What distribution patterns did Lubchenco and Menge observe in intertidal communities off the New England coast?
3. What experiments did Lubchenco and Menge conduct to study (a) secondary succession and (b) maintenance of existing intertidal communities? What did they discover?
4. What is the evidence that the interactions of functional groups of marine organisms are responsible for the absence of zonation in tropical ecosystems?
5. How does the Sustainable Biosphere Initiative address the perceived dichotomy between basic and applied research in ecology?

Synthesis and application questions

1. Given the relatively high survival rate for *Leptasterias* young, why hasn't *Pisaster* also evolved a strategy of brooding its young? What are some of the physiological costs associated with brooding a large clutch of offspring?
2. It seems that natural selection would favor herbivores that could feast on *Chondrus* in temperate rocky intertidal communities, yet there do not seem to be any present. Propose hypotheses for why there are no herbivores in New England that control this alga. How might you test these hypotheses?
3. Given the high predation pressure in the tropical intertidal zone, what types of adaptations might you expect to observe in some of the common prey species? Would there be life history consequences for these adaptations?
4. Lubchenco made two observations that support the role played by sea urchins in excluding *Chondrus* from the subtidal zone. How might you experimentally test the hypothesis that sea urchins exclude *Chondrus*? What problems might you encounter while carrying out your experiment?
5. Return to Table 24.2 and note the small sample size for each exclosure treatment. Why do you think the researchers only had two exclosures per treatment, and does this influence how you interpret their findings?
6. One take-home lesson of Lubchenco's research is the importance of indirect effects. Go through the text and find three different cases of indirect facilitation described in Lubchenco's research.
7. One reason for why there is so little information available on the impact of the *Deepwater Horizon* oil spill on the surrounding ecosystem is that most of the findings are tied up in litigation. Lubchenco argues that the most important goal is to hold the responsible party accountable, and to build the strongest possible court case to get the amount of money that is actually needed for restoration. That goal is best served by not revealing the government's case prematurely. Do you agree with her argument, or do you think that more information on the ecosystem impact should be released? Explain your answer.

Analyse the data 1

As we discussed earlier, to test whether *Pisaster* depressed the *Leptasterias* feeding rate, Lubchenco placed 5 *Leptasterias* in a container with at least 20 *Littorina scutulata* and monitored the feeding for 2 days. She then added 1 *Pisaster* for 30 min, long enough for pinching to occur, and monitored the feeding for 2 more days. She found that the total number of snails consumed by *Leptasterias* decreased from 251 to 172 after 30 min of *Pisaster* exposure. In contrast, snail consumption actually increased in the unexposed controls, from 249 to 281. Based on these data, how certain are you that *Pisaster* actually depresses the feeding rate of *Leptasterias*?

Analyse the data 2

Figure 24.14 shows the relationship between *Littorina* density and percent cover of ephemeral algae at 10 different areas that vary in exposure to wave action (Lubchenco and Menge 1978). Explain what relationship you see, and provide a plausible explanation for the outlier to the trend.

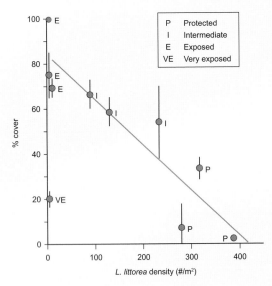

Figure 24.14 *Littorina* density and percent cover of ephemeral algae at 10 different areas that vary in exposure to wave action.

Epilogue

My true introduction to research programs came after college, when I was hired by the late George Streisinger as a research assistant. During my job interview he explained that by studying the zebrafish, *Danio rerio*, he was developing a vertebrate model that had many of the attributes of his beloved bacteriophage (small size, easy to keep, short generation time), but that was endowed with the same basic circuitry and underlying genetics of more complex vertebrates. I did not understand enough about how science worked to realized how audacious it was for someone to decide to establish a new vertebrate model, nor was I prescient enough to foresee how outstandingly successful his effort would be. But joining his team of researchers taught me that science is much more than just hypothesize, experiment, analyse, and conclude. And later exposure to other research programs convinced me that there are innumerable different paths to being a successful scientist. Let's consider an analogy between the structure of ecological research programs and the structure of the biological hierarchy.

ECOLOGICAL RESEARCH PROGRAMS AND HIERARCHICAL STRUCTURE

In analogy to the biological hierarchy, the anatomy of research programs can be investigated at a variety of levels. We can consider researchers as *individuals*, to understand how they developed their passion for ecology, their skill set, and approach to asking questions and conducting research. In many cases these interests develop at an early age, but in some cases these interests don't arise until later in life. Bernd Heinrich was a 5-year-old entomologist, and his home served as a surrogate zoo. Jane Goodall was fascinated by chimpanzees, and determined to get to Africa from an early age. In contrast, Anne Pusey wanted to study human development, and though she was always interested in chimpanzees, her initial application to graduate school was to work on humans. Dan Janzen was hooked on butterflies at a very early age, and has never flitted away from his passion, while Winnie Hallwachs happened to sit in on one of Janzen's lectures. And from an early age, Jane Lubchenco spent as much time as she could on, or in, water. All of these scientists were fortunate to have the opportunity to study organisms and systems that were innately fascinating to them. And their fascination was both intellectual and visceral.

Very few researchers work by themselves. Collaboration begins in graduate school when each student is given an academic advisor. On one level, the academic advisor plays a similar role as the student's parents, nourishing each student with ideas, and with financial and logistical support. Often graduate students work in a laboratory or in the field with several (or many) other researchers, all of whom are asking related questions using the same basic model or system. Success in graduate school and later is strongly influenced by how well each student functions intellectually and socially within the *population* of researchers within the research program. Maturing students become increasingly self-reliant, gradually developing their own ideas and means of support as they become postdoctoral associates and ultimately develop their own research programs. Escaping the shadow of their mentor may be seamless, or may be fraught with tension reminiscent of parent–offspring conflict.

Continuing the analogy, research programs interact with each other in numerous ways as part of a research *community*. Different research groups get together at meetings to share ideas, findings, and methodologies. On a very large scale, groups may be formed to address a common problem – one example is the IPCC, which meets every few years to assess the status of the climate change problem, and to project the future global and regional repercussions of climate change. In many cases, like the IPCC, these interactions are mutualistic, but in other cases, there is a strong element of competition and even exploitative interactions, including parasitism. On a smaller scale, different research groups also get together informally to work on common problems. Lastly, some research groups develop technical expertise that they share with others. For example, Dan Janzen and Winnie Hallwachs used Paul Hebert's expertise in DNA barcoding to discriminate between closely related lepidopteran species. And Jane Lubchenco collaborated with numerous scientists with diverse skills, including many chemists, to study large marine ecosystems throughout the world.

Because resources play such an important role in any research activity, we can also consider ecological research from an *ecosystem* perspective. Research programs compete with each other for resources directly when they submit proposals to funding agencies, and indirectly when they publish papers or give talks at meetings, which have the effect of increasing their probability of getting funded. Research programs also have highly mutualistic interactions in relation to resource use. For example, the research station at Gombe Stream National Park provides facilities for researchers working on a diverse array of research questions, which may have nothing to do with chimpanzees. The LTER sites and numerous field stations around the world provide housing, laboratory space, and equipment to a diverse array of researchers asking very different questions. I'd like to conclude by discussing how researchers come up with these questions.

ASKING ECOLOGICAL QUESTIONS

One essential feature of being a successful ecologist (or scientist of any type) is being able to ask good research questions. For ecologists, good questions come from observations of ecological processes in the field, and from observations of interesting findings in books, the web, videos, lectures, and discussions. These observations must then be put together to create questions that are important to the researcher, and to other ecologists. Identifying and articulating those good questions is not easy, and there is no magic formula. Researchers use many different approaches. Heinrich was required by his advisor to come up with a description of six different research topics and associated questions for his doctoral work. He was able to do so relatively easily because he had such a strong background as a naturalist, and because he was intensely curious. In contrast, Louis Leakey sent Jane Goodall out into the field with food, a tent, and binoculars and instructions to watch what was going on. He encouraged her to let the chimpanzees provide her with interesting questions, which they were happy to do.

Most ecologists are also evolutionary biologists. Many ecological studies are initiated by observations of evolutionary puzzles; organisms bearing traits that on the surface seem to be maladaptive. Heinrich could not understand why ravens called out to other ravens, when such behavior required them to share their food. And Janzen could not understand why many tropical trees produced fruits that had no efficient dispersal mechanism. Their efforts to resolve these and other evolutionary puzzles led them on fascinating and unanticipated journeys.

Most ecologists are fascinated by and revel in complexity. Given their intellectual prowess and complex social systems, chimpanzees have provided population biologists with an array of complex challenges. As we move up the biological hierarchy into the realm of communities, ecosystems, and global ecology, there are more and more factors that need to be considered when asking any question. Lubchenco needed to consider both biotic and abiotic factors in her studies of community structure and ecosystem functioning, and then had to dig deeper by exploring whether indirect effects are also important. By extending her studies to other ecosystems, she could then ask whether similar processes were operating across the world.

Lastly, over the course of their careers, some researchers become more interested in questions in applied ecology. Goodall abandoned her science to devote herself full-time to conservation efforts, while Janzen, Hallwachs, and Lubchenco continued doing basic research while addressing applied questions. One reason for this shift in emphasis may be that it is easier for an ecologist to get funding for applied research once they have established themselves within the scientific community and have achieved some level of fame within the public at large. A second reason for this shift in emphasis may be that some scientists want to leave a legacy. I suspect that the reasons underlying a shift from basic to applied research questions are as individualistic as the scientists themselves, and demonstrate that ecological questions are a complex function of each ecologist's experiences, skill set, and passions.

As Earth's ecosystems change from species loss, climate change, land-use practices, and many other factors, new discoveries and new questions arise and become more pressing. At the same time, developing technology allows researchers to answer questions with more precision. "Ecology in Action" is not just the name of this textbook; it is also a goal that can be realized only when people develop a sufficiently deep understanding of ecology to identify and articulate important questions, and to take the necessary actions to answer them.

GLOSSARY

Abundance (population) (1) – The number of individuals in a population

Acclimation (acclimatization) (5) – A physiological or morphological change in an organism in response to exposure to a change in its environment. Some researchers use acclimation to describe responses in the laboratory and acclimatization to describe responses under field conditions.

Accumulation curve (10) – For estimating population size, a curve formed by plotting the number of individuals sampled on the x-axis versus the number of unique individuals in the sample on the y-axis. The estimated population size is the value at which the number of unique individuals is projected to reach an asymptote.

Adaptation (3) – A naturally selected trait that increases an individual's reproductive success in a particular environment in comparison to the reproductive success of individuals without that trait.

Adaptive radiation (3) – Divergence of an ancestral species into numerous descendant species adapted to different habitats available in the new environment.

Aerosols (23) – Small solid or liquid droplets in a gas.

Age structure (population) (10) – The proportion of individuals in different age groups.

Albedo (23) – The amount of radiation reflected by a surface.

Allee effect (11) – Reduced per capita growth rate at very low population density.

Allele (3) – Variant form of a gene, or variant form of nucleotides at a particular DNA locus.

Allopatric speciation (3) – The origin of new species that occurs when at least two populations become geographically isolated from each other and diverge because natural selection (and possibly genetic drift) operates independently on each population.

Alpha diversity (17) – Species richness within a defined area.

Alternative stable state (21) – One of two or more sets of distinct ecosystem biotic and abiotic conditions that are relatively stable over time.

Ammonification (20) – Production of ammonia arising from the decomposition of complex nitrogen-rich molecules that are released from dead organisms.

Anadromous (fish) (9) – A fish that spends much of its life at sea, and moves into freshwater to breed.

Analysis of variance (ANOVA) (6) – Statistical analysis (or collection of statistical models) that analyses the differences among group means.

Anammox (20) – Anaerobic oxidation of NH_4^+ into nitrogen gas, using hydrozine as an intermediate.

Anthropogenic (23) – Caused by humans.

Anthropogenic biomes (2) – Biomes whose nature and character are determined by humans.

Anthropogenic forcings (23) – Regarding global temperature, assumption that temperature variation is influenced by human activity.

Apparent competition (13) – Interaction in which each of two (or more) prey species is indirectly having a negative effect on the other(s), because they have a positive effect on the predator.

Applied research (24) – Scientific research that has a goal of providing tangible benefits to society or some individuals within society.

Arbuscular mycorrhizal association (endomycorrhizal associations) (15) – Association between plant roots and fungi of the phylum Glomeromycota. The fungal hyphae grow between the root cells, ultimately entering some of them and forming finely branched arbuscles.

Assimilation efficiency (19) – The percentage of energy in food that is assimilated by the organism for metabolic processes.

Autotroph (2) – An organism that converts energy from the Sun in the process of photosynthesis, or from inorganic chemical compounds in the process of chemosynthesis, to chemical bond energy.

Basic research (24) – Scientific research that has a goal of increasing understanding of patterns or processes, but that does not necessarily provide tangible benefits to society or individuals within society.

Benthic zone (2) – The region at the bottom of a body of water.

Beta diversity (17) – A measure of the change in species composition, usually along an environmental gradient.

Biocultural restoration (18) – Restoring a habitat by making a natural area relevant to the people who live near it, and involving them directly and indirectly in sustaining the natural area.

Biodiversity development (18) – Using the biodiversity of a conserved tropical wildland without harming it.

Biological diversity (11) – The diversity of life considered from at least three different levels: genes, species, and ecosystems.

Biological legacy (21) – Organisms, the remains of organisms, and organically derived environmental patterns that persist through a disturbance, and are incorporated into the recovering ecosystem.

Biological pump (4) – Slow process of carbon circulation in the oceans, in which some carbon sinks deep into the water column, but returns to the surface as a result of upwelling.

Biomagnification (19) – The tendency for substances to have progressively higher tissue concentrations in organisms from higher trophic levels.

Biomass pyramid (19) – a graph of the total biomass of each trophic level, forming a pyramid because there is usually decreasing biomass with increasing trophic level.

Biome (2) – A large geographical area with characteristic groups of organisms adapted to that particular environment.

Biosphere (1) – The part of the world that supports life.

Bottom-up control (16) – Hypothesis that producers are primarily responsible for community structure.

C_3 photosynthesis (4) – Form of carbon fixation used by most plants in which rubisco catalyses a reaction that fixes CO_2, and ultimately yields two 3-carbon PGA molecules.

C_4 photosynthesis (4) – Form of carbon fixation used by about 5% of plant species, mostly from hot and dry environments, in which PEPCase catalyses a reaction that fixes CO_2, and ultimately yields a 4-carbon acid.

CAM photosynthesis (4) – Form of carbon fixation in which CO_2 is taken up during the night, thereby minimizing water loss through the stomata.

Canopy cover (1) – The percentage of the ground shaded by the tree canopy.

Carbon cycle (23) − The cycling of carbon between the atmosphere, the oceans, and the land.

Carbon fixation reactions (4) − In plants, a series of chemical reactions in which the chemical bond energy in ATP and NADPH is used to create complex carbohydrates from CO_2.

Carnivore (4) − An animal that eats other animals.

Carrying capacity (K) (10) − The maximum population size that can be supported or sustained by the environment.

Chemosynthesis (2) − Conversion of energy from chemical compounds in the environment into chemical bond energy stored in carbohydrates.

Chi-square test (χ^2 test) (5) − Statistical analysis that tests for a relationship between two categorical variables.

Chlorophyll (4) − Pigment molecules that trap light energy.

Chronosequences (21) − Habitats that are similar to each other in all respects except for the time since they were colonized.

Clade (15) − A set of species descended from a particular common ancestor.

Climax community (21) − A stable biotic community that is in equilibrium with existing environmental conditions and represents the last stage of succession.

Clumped dispersion (9) − Pattern in which individuals are more likely to be situated near other individuals within the habitat.

CO_2 flux (19) − The rate at which CO_2 is exchanged, usually in reference to the atmosphere.

Coefficient of relatedness (r) (6) − The probability that two individuals share genes that are identical by descent.

Coefficient of determination (R^2) (1) − The proportion of the variance of the data explained by the line of best fit in a regression analysis.

Coevolution (14) − A process in which interactions between species lead to reciprocal adaptation.

Colony collapse disorder (15) − Honeybee disease characterized by the sudden disappearance of most workers, leaving only the queen, her brood, and a few workers.

Commensalism (15) − Interaction that benefits one species and has no effect on the second species.

Community (1) − A group of populations of different species that interact with each other.

Community structure (16) − Attributes of an ecological community including species composition and diversity, and the distribution and abundance of each species.

Competition (13) − Interaction between individuals (or species) that has a negative effect on the fitness of the interacting individuals (or species). Competition may be intraspecific (within species) or interspecific (between species).

Competition coefficient (α) (13) − The per capita competitive effect of one species on the other.

Competitive exclusion (13) − Interspecific competition that results in one species outcompeting a second species for a limiting resource, and excluding it from a habitat.

Conductive heat transfer (5) − Transfer of heat energy arising from molecular collisions within or between substances.

Confidence interval (3) − A measure of confidence that an interval calculated based on collected data includes the true population mean.

Connectivity (11) − A measure of the distance between a focal habitat patch and nearby populations, weighted by the sizes of the nearby populations and their distances to the focal patch.

Constitutive defense (14) − A defense mechanism that is expressed continuously.

Constraint (16) − A factor that limits one or more other factors.

Consumption efficiency (19) − The percentage of available energy that is actually ingested or consumed.

Convective heat transfer (5) − Transfer of heat energy arising from the movement of fluids.

Coral bleaching (15) − A breakdown in the coral–algal mutualism that occurs when the coral rejects its algal symbiont thereby losing its algal pigmentation.

Coriolis effect (2) − Deflection of air currents to the right in the northern hemisphere and to the left in the southern hemisphere.

Correlation (1) − A measure of the association between two numeric or continuous variables.

Correlation coefficient (r) (1) − A measure of the strength of the association between two numeric or continuous variables.

Dead zone (2) − A region of hypoxia in an aquatic system where dissolved oxygen levels drop below 2.0 ml/l for an extended period of time, causing the death of most aerobic benthic organisms.

Decomposer (4) − An organism that physically and/or chemically breaks down the tissue of dead organisms, releasing chemicals and nutrients that may be taken up by the decomposer or by other organisms.

Demography (8) − The quantitative study of the size and structure of populations, and of how populations change over time.

Denitrification (20) − The reduction of nitrates and nitrites to nitrogen gas.

Density-dependent factor (10) − A factor that reduces a population's growth rate as population density increases, by decreasing the birth rate and/or increasing the death rate at high population density.

Density-independent factor (10) − A factor that influences a population's growth rate in the same manner regardless of population density.

Dependent variable (1) − The variable whose value is influenced by the independent variable.

Desert (2) − Hot and dry biome commonly found near 30° latitude.

Detritivore (4) − An organism that feeds on the remains of dead organisms.

Detritus (4) − The remains of dead organisms.

Direct fitness (6) − A measure of the genetic contribution an individual makes to the next generation by producing surviving offspring.

Directional selection (3) − Natural selection that favors either a higher or lower value of the mean value of a trait, resulting in the mean value of that trait shifting in the favored direction over time.

Disjunctions (22) − Distributions of closely related species or species groups over widely separated regions of the world.

Dispersal (9) − The movement of individuals from one location to another location.

Dispersion (9) − Spatial pattern of distribution of individuals within a population.

Distribution (1) − The geographic location in which individuals of a population or species are located.

Disturbance (21) − A relatively discrete event that disrupts the structure of an ecosystem, community, or population, by changing either the physical environment or the resources it contains.

Diversity − See biological diversity.

Dominant species (foundation species) (16) − A species that has a major influence on community structure primarily by virtue of its great abundance or biomass.

Dormancy (5) – An extended period with no growth or development.

Ecological trap (22) – Landscape process in which individuals actively select habitat based on some attribute, which ultimately results in lower fitness than the alternatives.

Ecological stoichiometry (4) – The study of the balance of multiple chemical substances in ecological processes.

Ecosystem (1) – A community of organisms plus the abiotic (nonliving) factors that interact with the community.

Ecosystem diversity (11) – The number of different ecosystems within a larger area – either a landscape or a region.

Ecosystem engineer (14) – An organism that substantially influences the availability of essential resources within an ecosystem by causing important changes to the biotic or abiotic environment.

Ecosystem services (18) – The benefits people receive from ecosystems.

Ectomycorrhizal associations (15) – Associations between plant roots and fungi, in which the fungal hyphae form a sheath around the exterior of roots, and also penetrate into the spaces between individual root cells.

Ectotherms (5) – Organisms whose body temperature is influenced primarily by the external environment.

Edge effect (11) – Reduction of biological diversity arising from differences in environmental conditions and species interactions that occur along the edge of a particular habitat.

Efflux (nutrient) (20) – The movement of a nutrient out of a pool.

El Niño Southern Oscillation (ENSO) (1) – A large-scale atmospheric system that affects global climate. An El Niño is associated with high sea temperature and low barometric pressure in the eastern Pacific Ocean region.

Emerging infectious disease (14) – An infectious disease that is projected to become much more prevalent in the near future.

Endangered species (8) – Any species that is in danger of extinction through all or a significant portion of its range.

Endemic species (2) – Species found in one geographic region and nowhere else.

Energy pyramid (19) – A graph of the amount of energy at each trophic level, forming a pyramid because there is decreasing production with increasing trophic level.

Endotherms (5) – Organisms that maintain a relatively constant body temperature by conserving metabolic heat.

Environmental gradient (17) – Change in an abiotic factor through space or time.

Epiphyte (2) – Plants that grow upon other plants and depend on their hosts for support.

Essential nutrient (4) – Nutrient required by an organism for metabolism, growth or reproduction.

Eusocial (6) – A social system characterized by cooperative care of young, and the presence of sterile castes of non-reproductive workers.

Eutrophication (2) – The addition of excessive nutrients such as nitrogen and phosphorus to an aquatic ecosystem.

Evenness (community) (17) – The amount of variation in the distribution of abundances of each species within a community. Low variation implies high evenness.

Evolution (3) – Changes in the characteristics of a population over time; more specifically, changes in allele frequencies within a population over time.

Evolutionary puzzle (7) – Possession of a trait that, on the surface, appears to be maladaptive.

Exclosure (1) – Experimental plot that is barricaded in a way that prevents the intrusion of unwanted factors.

Exponential growth model (10) – For a population with a continuous reproductive life history, the change in population size by a constant proportion over a vanishingly small time interval.

Exploitative interaction (14) – Interaction in which an organism of one species benefits by consuming all or part of an organism of another species, or by diverting resources that would otherwise nourish the organism.

Extent (22) – The overall size of the study area.

Extrapolation (2) – Inferring data points that lie outside the range of known or existing data points.

Facilitative interaction (15) – Interaction between two organisms, or two species, that benefits at least one of them and harms neither.

Facilitation model of succession (21) – A model that proposes that the first colonizers are shade-intolerant pioneer species that modify the environment in such a way that it becomes less suitable for future pioneer species but more suitable for later successional species.

Fecundity (11) – The number of offspring produced by females of a particular age class.

Feedback interaction (23) – Interaction in which one factor influences a second factor, which then influences the first factor.

Ferrel cell (2) – Pattern of atmospheric and surface circulation located between the Hadley cell and the Polar cell.

Fission–fusion social system (12) – Social system in which the main social group breaks up into a series of much smaller unstable social groups which persist for a relatively small time period.

Fitness (3) – Measure of an individual's genetic contribution to future generations.

Food chain length (19) – A measure of the mean number of transfers from the base to the top of a food web.

Food web (2) – A summary of the feeding relationships within an ecological community.

Forbs (2) – Nonwoody plants other than grasses.

Flux (2) – Movement or flow.

Fragmentation (11) – Breaking up a large, continuous habitat into small isolated patches.

Functional group (16) – A group of species that perform the same function within the ecological community.

Functional response (14) – The relationship between prey density and predator per capita consumption rate.

Fundamental niche (9) – The potential set of environmental conditions and resources in which a species can survive, grow, and reproduce.

Gamma diversity (17) – The combined species richness of all the communities that are being evaluated.

Gene flow (3) – The movement of alleles from one population to another.

Genetically effective population size (N_e) (11) – The average number of individuals in a population that effectively contribute genes to succeeding generations.

Genetic diversity (11) – A measure of differences in nucleic acid composition among individuals within a population or species.

Genetic drift (3) – Change in allele frequency in a population resulting from chance events.

Geographic information system (GIS) (9) – A system designed to store, display, and analyse the locations and characteristics of Earth's spatial features.

Geometric growth model (10) – For a population with a discrete reproductive life history, the change in abundance by a constant proportion from one time period to the next.

Global warming potential (23) – The radiative forcing by a known mass of a greenhouse gas in comparison to the radiative forcing of the same amount of CO_2 over a certain time period.

Grain (22) – In reference to scale, the finest resolution used for viewing a phenomenon.

Grazer (14) – An exploiter that eats a portion of its prey, usually without killing it, and does not live in or on its prey.

Greenhouse gas (23) – A molecule in the atmosphere that warms Earth's surface because it absorbs some of the longwave radiation emitted from Earth's surface.

Gross primary production (GPP) (19) – The amount of carbon fixed within an ecosystem over some time period.

Guild (16) – A group of species that depend on the same resource for survival and reproduction.

Gyre (2) – Large-scale surface circulation of ocean water.

Haber–Bosch process (20) – An industrial application that converts N_2 to ammonia (NH_3), and allows large-scale production of inorganic fertilizers.

Habitat degradation (destruction) (11) – Reduction of habitat quality for a particular species, so that its population is no longer viable in that habitat.

Habitat heterogeneity (17) – A measure of how different a habitat is over space or time.

Habitat split (22) – Landscape process characterized by a species requiring two distinct spatially separated habitats over the course of its life history.

Hadley cell (2) – Pattern of atmospheric and surface circulation, in which a hot air mass rises near the equator, flows away from the equator at an elevation of 10–15 km, descends in the subtropics, and flows back toward the equator above the surface.

Haplodiploid (6) – A genetic system which produces haploid offspring that develop into males, and diploid offspring that develop into females.

Herbivore (4) – An animal that eats plant tissues or fluids.

Heritability (3) – Proportion of variation in a trait that is due to genetic variation.

Heterotroph (4) – An organism that gets its energy by consuming carbon compounds produced by other organisms.

Hibernation (5) – A state of torpor lasting for several months during the cold season, in which body temperature approaches environmental temperature.

Homeostasis (4) – The maintenance of a constant internal environment within an organism's cells, tissues, organs, and organ systems.

Home range (6) – Geographic area that an animal traverses over the course of its lifetime.

Host specificity (17) – A measure of the number of species used as a resource by a consumer. Greater host specificity indicates that a consumer uses fewer species.

Human footprint (9) – A global map that shows human influence on the land surface.

Hyperosmotic regulators (5) – Organisms that maintain internal solute levels that are greater than those found in their environment.

Hypoosmotic regulators (5) – Organisms that maintain internal solute levels that are less than those found in their environment.

Hypothesis (1) – Possible answer to a question or explanation for an observation.

Immobilization (20) – Process in which inorganic molecules (nutrients) are taken in by an organism and converted to organic molecules.

Inbreeding depression (11) – A reduction in survival or viability of offspring produced when two relatives mate with each other.

Inclusive fitness (6) – The sum of direct fitness and indirect fitness.

Independent variable (1) – The variable that is proposed to influence the value of the dependent variable.

Indirect fitness (6) – A measure of the genetic contribution an individual makes to the next generation by enhancing the production or survival of genetic relatives.

Indirect interaction (13) – Interaction between two species in which the outcome is influenced by the interaction of one or both species with other species.

Induced defense (4) – A defense mechanism that is expressed in response to an injury or threat.

Influx (nutrient) (20) – The movement of a nutrient into a pool.

Inhibition model of succession (21) – Model that proposes that early- and late-succession species colonize a habitat following disturbance, with the majority of early colonizers having the characteristics of pioneer species. Late-succession species tend to live longer, so over time, late-succession species increase in abundance in the community.

Interaction strength (16) – The strength of the link between two species in a food web.

Interactions web (16) – A web which includes not only feeding relations among species from different trophic levels, but also competitive and mutualistic interactions that may occur among species that occupy the same trophic level.

Interbirth interval (12) – The amount of time between births.

Interference competition (13) – Competitive interaction in which individuals (or species) engage in direct aggressive acts.

Intersexual selection (3) – Differential mating success among individuals of one sex arising from interactions with individuals of the opposite sex.

Intertidal zone (13) – The region between the highest and lowest tides along the ocean's shore.

Intertropical Convergence Zone (ITCZ) (2) – Region of Earth receiving maximum solar radiation, characterized by upwelling of hot air mass, formation of low-pressure systems, and heavy rainfall.

Intrasexual selection (3) – Differential mating success among individuals of one sex arising from competition among individuals of the same sex.

Intrinsic growth rate (10) – The per individual birth rate minus the per individual death rate when there is no competition or predation.

Invasive species (9) – A non-native species that is introduced into a new habitat, and that often adversely affects numerous species in the new habitat.

Iteroparous (8) – Having multiple reproductive efforts over the course of a lifetime.

Janzen–Connell hypothesis (18) – A hypothesis that reasons that seed recruitment is the product of seed density, which is greatest near the parent tree, and seed survival, which is greatest far from the parent tree. Consequently, peak population recruitment will be a considerable distance from the parent tree.

Keystone species (16) – A species that has a much greater impact on the community than would be expected by measuring its abundance or biomass.

K-selected species (8) – Species whose individuals compete effectively for resources in predictable and stable environments, and whose populations tend to be at or near their carrying capacity.

Landscape (1) – Used in two contexts: (1) a group of interacting, and usually spatially connected, ecosystems, or (2) an area that is spatially heterogeneous in one factor of interest.

La Niña (23) – A stage of El Niño Southern Oscillation caused by a buildup of unusually cool subsurface waters in the tropical Pacific, which are brought to the surface by strong currents in areas of upwelling.

Large Marine Ecosystems (24) – Regions of the world's oceans bounded on the land side by the shoreline, including estuaries and river basins, and on the ocean side, by the continental shelf and the major ocean currents.

Leaching (20) – Loss of nutrients as they move through a water column.

Legacy – See biological legacy.

Liana (2) – A woody vine that climbs trees.

Life history traits (8) – Adaptations that influence, growth, survivorship, and a variety of reproductive parameters for individuals of a particular species.

Life table (11) – A summary of survival and reproductive rates for different age classes of a population.

Light reactions (4) – In plants, a series of chemical reactions in which the radiant energy from sunlight is converted into chemical bond energy in the form of ATP and NADPH.

Limiting nutrient (resource) (4) – Nutrient or resource that is in short supply and limits the growth and/or reproductive success of an organism.

Logistic model of population growth (10) – A model of population growth that is nearly exponential at low abundance but that levels out as population abundance approaches the carrying capacity.

Lotka–Volterra competition model (13) – An extension of the logistic model that predicts the outcome of interspecific competition.

Lotka–Volterra predation model (14) – An extension of the logistic model that predicts the outcome of interactions between a predator and prey population, or a parasite and host population.

Marginal value theorem (6) – Optimization model that predicts how long an animal should persist at an activity.

Mark–recapture study (10) – A type of population analysis in which individuals are captured, marked, and released into the environment, allowed to mix back into the source population, and then recaptured a second time.

Mating system (6) – The general pattern of copulatory partners for each sex over the course of a breeding season.

Matrix (22) – Area surrounding a fragment of suitable habitat.

Meridional overturning circulation (great ocean conveyer) (23) – Global pathway of ocean circulation.

Meta-analysis (4) – Systematic analysis of data collected by many other researchers.

Metapopulation (10) – A group of local populations inhabiting networks of somewhat discrete habitat patches.

Microsatellite (3) – Sections of DNA that are made up of repetitive DNA sequences of variable length surrounded by unique regions of constant length.

Migration (9) – The back-and-forth intentional movement of individuals or populations between two locations.

Millennium Ecosystem Assessment (MA) (19) – A report completed in 2005 that assessed the consequences of ecosystem change for human well-being and established the scientific basis for actions needed to enhance the conservation and sustainable use of ecosystems.

Mineralization (20) – Breaking down of organic molecules into inorganic molecules.

Modifier gene (3) – A gene that modifies the expression of a second gene.

Modular organisms (10) – Species that develop an undetermined number of repeated copies of similar structures, so it can be difficult to distinguish individual organisms.

Monogamy (6) – A mating system in which one male mates with one female over the course of the breeding season.

Morphs (3) – Variant forms of a trait within a population of organisms.

Mycorrhizal associations (15) – Symbiotic and usually mutualistic associations between plant roots and various types of basidiomycete or ascomycete fungi.

Mutation (3) – A random change in the base sequence of the genetic material.

Mutualism (2) – Interaction in which both species benefit.

Natural forcings (23) – Regarding global temperature, assumption that temperature variation is only influenced by variation in natural events.

Negative feedback (23) – Interaction in which an increase in one factor causes an increase in a second factor, which then causes a decrease in the first factor. Also, an interaction in which a decrease in one factor causes a decrease in a second factor, which then causes an increase in the first factor.

Net primary production (NPP) (19) – A measure of ecosystem production, equal to gross primary production minus the amount of carbon used by autotrophs for metabolism, most commonly expressed in terms of grams of carbon per square meter per year (g C/m^2/yr).

Net reproductive rate (R_0) (11) – The mean number of offspring produced by an individual over the course of its lifetime.

Niche (9) – The set of environmental conditions and patterns of resource availability, in which a species can survive, grow, and reproduce.

Niche partitioning (13) – A process in which ecologically similar species coexist by using different resources or specializing on different factors in the environment.

Niche shift (13) – An adaptive change in a species in one or more of its niche dimensions.

Nitrification (20) – The conversion of ammonia to nitrate and nitrite.

Nitrogen fixation (4) – A biological process in which bacteria convert nitrogen gas (N_2) to ammonia (NH_3), and a physical process in which lightning powers the conversion of nitrogen gas to nitrogen oxides. Most recently, humans have developed chemical means (the Haber–Bosch process) for converting vast quantities of N_2 to NH_3.

Nitrogen mineralization (14) – The process that converts organic nitrogen from decaying organisms into inorganic nitrogen that can be taken up by plants.

Nonconsumptive effects (14) – Changes in the behavior, growth, development, or reproduction of prey or hosts in response to the presence or absence of predators or parasites.

Nutrient (4) – Chemical that supports growth, development, and reproductive success of an organism.

Nutrient cycle (20) – The cyclical movement of nutrients within and between ecosystems.

Nutrient flux (20) – The rate of nutrient movement from one pool or reservoir to another.

Nutrient sink (20) – A pool in which influx is greater than efflux.

Nutrient source (20) – A pool in which efflux is greater than influx.

Nutrient spiral (20) – Sequential process of nutrient uptake, nutrients passing through a food web, nutrient release, and decomposition as nutrients drift downstream.

Oligotrophic lake (2) – A cold, unproductive, and nutrient-poor lake.

Omnivore (4) – Formally, an organism that eats from more than one trophic level. More generally, an animal that eats plants and animals.

Organism (1) – An individual life form.

Osmoconformer (5) – An organism whose tissues have the same osmolarity as its environment.

Osmolarity (5) – Total solute concentration in terms of moles of solutes per liter of solution.

Osmoregulator (5) – An organism that maintains an internal osmolarity that is different from its external environment.

Overexploitation (11) – Using a species as a resource beyond its ability to replenish itself numerically.

Overyielding (17) – An increase in ecosystem function in response to high species diversity.

P-value (1) – Probability of obtaining the results that are observed in a particular data set (or more extreme results), under the assumption that the null hypothesis is correct.

Pangaea (22) – Single global landmass that broke apart beginning about 200 MYA into smaller plates that we know as today's continents.

Parasite (14) – An exploiter that lives in or on its host, feeding on its tissues, and usually does not kill it.

Parasitoid (13) – An exploiter that lives in or on its host, feeding on its tissues, and usually kills it.

Parataxonomist (18) – A citizen who is trained to conduct specific systematic research, but has no formal degree.

PEPCase (4) – The enzyme phosphoenol pyruvate carboxylase, which catalyses carbon fixation in C_4 plants.

Percent cover (10) – The percent of a substrate covered by organisms when viewed from above.

Per capita growth rate (10) – Per individual birth rate minus the per individual death rate.

Phenology (5) – The relationship between climate and the timing of periodic ecological events, such as flowering, breeding, and migration.

Phenotypic plasticity (8) – An ability to change phenotypes (in developmental, physiological or behavioral traits) in response to different environments.

Photic (epipelagic) zone (2) – Depth in a body of water to which sufficient sunlight penetrates to support photosynthesis (generally about 200 m under ideal conditions).

Photorespiration (4) – A sequence of chemical reactions in plants that begins when rubisco catalyses the uptake of oxygen, rather than CO_2, resulting in a loss of carbon and energy to the plant.

Photosynthesis (4) – Biochemical process in most plants, algae, and some bacteria, that uses light energy to synthesize energy-rich organic molecules.

Photosystem (4) – Light-capturing apparatus within the chloroplast that converts light energy into chemical bond energy in the light reaction of photosynthesis.

Phylogeny (3) – The evolutionary history of ancestry and descent of a group of related organisms.

Phylogeography (22) – The study of evolutionary relationships in the context of geographical distributions.

Phytoplankton (4) – Photosynthetic free-floating, freshwater or marine organisms.

Pioneer species (5) – Species that are adapted to colonizing newly available habitats.

Plate tectonics (22) – A mechanism for the movement of the thin rigid layer of Earth's crust that rides the upper layer of the mantle.

Polar cell (2) – Pattern of atmospheric and surface circulation in which cold air descends near each pole, flows along the surface away from the pole, rises into the troposphere near 60° latitude, and then flows back toward the pole.

Pollination (15) – Transfer of pollen from the anther of a plant to the stigma of a plant.

Polyandry (6) – A mating system in which one female mates with more than one male over the course of the breeding season.

Polygynandry (6) – A mating system in which males and females have multiple mates over the course of the breeding season.

Polygyny (6) – A mating system in which one male mates with more than one female over the course of the breeding season.

Pool (20) – Representation of compartment where nutrients are stored.

Population (1) – A group of interacting individuals of the same species.

Population density (10) – The number of individuals within a population per unit area.

Population viability analysis (PVA) (11) – A quantitative analysis of extinction risk that allows ecologists to recommend management options to improve the prognosis for continued survival of a population.

Polymorphic (3) – A population (or gene) for which there is more than one form of a trait (or allele).

Positive feedback (23) – Interaction in which an increase in one factor causes an increase in a second factor, which then causes an increase in the first factor. Also, an interaction in which a decrease in one factor causes a decrease in a second factor, which then causes a decrease in the first factor.

Predation (14) – Exploitative interaction in which one organism kills and eats another organism.

Predator (1) – An exploiter that kills and eats its prey, but does not live in or on its prey.

Predator saturation (8) – An antipredator adaptation in which prey populations occur at high densities, thereby reducing the probability of a particular individual being eaten by predators.

Prediction (1) – A logical outcome of a hypothesis that is used to test a hypothesis. A correct prediction provides some support for a hypothesis, while an incorrect prediction is used to reject a hypothesis.

Primary consumer (16) – A consumer that eats producers, occupying the second trophic level of a food web.

Primary production (2) – The chemical energy generated by autotrophs from photosynthesis and chemosynthesis.

Primary succession (21) – The changes that occur to a habitat following a severe disturbance that kills virtually all organisms in that habitat.

Prisoner's Dilemma (6) – A game theory model that identifies the constraints against the evolution of cooperation.

Producer (16) – An organism that produces its own energy by photosynthesis or chemosynthesis and occupies the first trophic level of a food web.

Production efficiency (19) – The percentage of assimilated energy that is allocated to growth or production of offspring.

Quadrat (10) – A small (often rectangular) sampling plot.

r-selected species (8) – Species with potentially high reproductive rates that are most successful in unpredictable environments including those that have been recently disturbed by catastrophic events such as fires or floods.

Radiative forcing (23) – The influence that a change in a climatic factor, such as a greenhouse gas, has on altering the balance of incoming and outgoing radiation.

Random dispersion (9) – Pattern in which each individual within a population has an equal probability of occupying any given space in the habitat.

Realized niche (9) – The set of environmental conditions and resources in which a species can survive, grow, and reproduce in the presence of competitors and predators.

Recruitment (9) – The addition of new individuals to a population.

Regime shift (21) – A relatively rapid change in ecosystem structure and functioning.

Region (1) – A large geographical area that experiences a common set of environmental and evolutionary influences.

Regression analysis (1) – Statistical analysis for estimating the relationship between a dependent variable and one or more independent variables.

Representative concentration pathways (23) – Alternative amounts of radiative forcing from CO_2 emissions predicted by the year 2100, based on an analysis of how socioeconomic conditions influence human behavior regarding adaptation and mitigation.

Reproductive effort (8) – The amount of resources an organism allocates to a reproductive event.

Reproductive success (3) – The number of genetic offspring an individual produces that survive until reproductive age.

Reservoir (2) – Representation of compartment where nutrients are stored.

Resilience (17) – The speed with which a community returns to its former structure and function following a shift caused by disturbance.

Resistance (17) – A community's ability to avoid significant change in structure or function when disturbed.

Resource allocation (8) – The quantity of key resources, such as energy and nutrients, that a parent can devote to reproduction.

Resource competition (13) – Competitive interaction in which individuals (or species) compete for a limiting resource such as food, space, or nest sites.

Resting metabolic rate (8) – The amount of energy used by an organism over a given time period while at rest in a thermally neutral environment.

Root/shoot ratio (5) – A measure of plant relative investment in roots, equal to the weight of the root system divided by the weight of the shoots.

Rubisco (4) – The enzyme ribulose-1,5-bisphosphate carboxylase oxygenase, which catalyses carbon fixation in the Calvin cycle, and also oxygen uptake in the process of photorespiration.

Saprotrophs (4) – Decomposers and detritivores.

Savanna (2) – Biome at about 15–20° latitude characterized by a prolonged dry season, abundant grasses, and occasional trees.

Secondary compound (14) – Molecule that is not essential for growth, development, or reproduction, but that functions as a constitutive chemical defense.

Secondary consumer (16) – An organism that eats primary consumers, and thereby occupies the third trophic level of a food web.

Secondary production (19) – The amount of carbon that is passed on to the heterotrophs that live within an ecosystem.

Secondary succession (21) – The pattern of recolonization following a disturbance in which only some of the organisms are killed within a habitat.

Seed bank (10) – The collection of viable seeds stored in the soil.

Seed shadow (18) – A pattern of dispersion in which seed density is greatest near the parent tree, and declines with distance from the parent.

Semelparous (8) – Having only one reproductive effort over the course of a lifetime.

Sensitivity (of a model) (9) – The number of correct positive predictions divided by the total number of positive occurrences.

Serotiny (21) – A life history trait of some plants in which they retain their seeds in the canopy for many years after the seeds mature, and release the seeds after exposure to the heat of a fire.

Sexual dimorphism (3) – Difference in form between females and males of the same species.

Sexual selection (3) – A form of natural selection that favors individuals who have traits that improve their mating success.

Shannon index (17) – A measure of species diversity that considers species richness and species evenness.

Sink – See nutrient sink.

Sorensen's index (17) – A measure of how similar two sites are in species composition.

Source – See nutrient source.

Speciation (3) – The origin of new species.

Species–area relationship (22) – Power function relationship between species richness and the area of the island.

Species composition (4) – The identity of species present in a community.

Species diversity (11) – A measure of the number of species, and the relative abundance of these species, within a particular area.

Species richness (1) – The number of different species within a particular area (often a community).

Species turnover (17) – The change in species composition over space or time.

Specificity (of a model) (9) – The number of correct negative predictions divided by the total number of negative occurrences.

Stability (17) – A community's tendency to retain its current structure and functioning even when it is disturbed by biotic or abiotic forces.

Stable isotope ratio (3) – A measure of the ratio of a rare (and nonradioactive) isotope of an element to its standard common isotope.

Standard error (of the mean) (2) – A measure of sample variation used most commonly in this text. It is equal to the standard deviation divided by the square root of the sample size

Stomata (2) – Pores most commonly found on a leaf's surface through which gas exchange occurs.

Sustainable Biological Initiative (24) – An initiative adopted by the Ecological Society of America that established a framework for the acquisition, dissemination, and utilization of ecological knowledge based on basic research, communication of knowledge derived from basic research to citizens, and incorporation of that knowledge into policy and management decisions.

Symbiotic interaction (2) – An interaction in which two species live together.

Sympatric speciation (3) – The evolution of new species within a single geographical area.

T-test (6) – Statistical analysis most commonly used to test for statistically significant differences between two group means.

Taiga (boreal forest) (2) – Terrestrial biome characterized by cold temperature, moderate precipitation, and continuous forest.

Temperate forest (2) – Terrestrial biome characterized by moderate temperature, substantial precipitation, and a moderately diverse assemblage of plant and animal species.

Temperate grassland (2) – Biome in interior Eurasia and North America, with cold winters, warm summers, and much grass.

Temperate shrubland (2) – Biome at 30–40° latitude on the southwest side of large landmasses, with mild, wet winters, hot dry summers, occasional fires, and a diverse community of drought-adapted shrubby plants.

Territory (6) – Any defended area.

Tertiary consumer (16) – A consumer that eats secondary consumers, and thereby occupies the fourth trophic level of a food web.

Thermal neutral zone (5) – Environmental temperature range within which an animal's metabolic rate is constant.

Thermal stratification (2) – In aquatic systems, the formation of layers of water with distinct temperature profiles.

Threshold population size (S_T) (14) – The threshold number of susceptible hosts needed for an infectious disease to spread in a population of hosts.

TIT for TAT (6) – Solution to the Prisoner's Dilemma that involves cooperating for the first interaction, and then doing whatever the other player does in future interactions.

Tolerance (5) – The ability to survive and function under stressful or extreme environmental conditions.

Tolerance model of succession (21) – Model that proposes that both early- and late-succession species colonize a habitat following disturbance, but that the majority of early colonizers are pioneer species, simply because they are excellent dispersers and colonizers. Over time, late-succession species, which can tolerate conditions of limiting resources, tend to dominate the community.

Top-down control (16) – Hypothesis that predators are primarily responsible for community structure.

Torpor (5) – A state of dormancy with greatly reduced body temperature.

Tradeoff (3) – Evolutionary compromise between one function and another that prevents adaptations from being optimal in all environments.

Transpiration (2) – The evaporation of water into the atmosphere through a plant's stems and leaves.

Treefall gaps (17) – Openings in the canopy formed when an overstory tree falls down.

Trophic cascade (16) – A representation of the feeding relations within a food web, in which the effects of consumption cascade down the food web from higher to lower trophic levels.

Trophic efficiency (19) – The production of one trophic level divided by the production of the trophic level below it.

Trophic level (16) – The feeding level of a food web.

Trophic mismatch (23) – Disruption of synchrony between a species' life cycle and an essential resource (from a different trophic level).

Tropical dry forest (2) – Terrestrial biome characterized by high temperature and seasonal rainfall in a broadleaf deciduous forest that loses its leaves during the dry season. Species diversity is very high.

Tropical rain forest (2) – Terrestrial biome characterized by constant high temperature and rainfall in a broadleaf evergreen forest with extraordinary species diversity.

Tundra (2) – Terrestrial biome characterized by very cold and dry conditions, permafrost, and low species diversity.

Turnover (lake) (2) – The mixing of layers of the water column in a thermally stratified lake.

Turnover (species) – See species turnover.

Turnover time (2) – The average amount of time a substance remains in its pool or reservoir.

Uniform (regular) dispersion (9) – Pattern in which individuals tend to be evenly spread out within a habitat.

Unitary organisms (10) – Species in which each individual is an easily distinguished unit.

Upwelling (2) – The rising of deep ocean or lake water to the surface.

Wallace's line (22) – Geographic boundary that separates New Guinea and Australian mammals and birds from a very distinct fauna in Borneo, Sumatra, and the southeast Asian mainland.

Water potential (5) – The sum of potential energy components within an aqueous system.

Zero growth isocline (13) – Equilibrium line that defines a stable population size.

Zonation (intertidal zone) (24) – A pattern of distribution of different species groups into distinct layers or zones within the intertidal zone.

REFERENCES

Adam, D. 2010. The hottest year. *Nature* 468: 362–364.

Aerts, R., and Chapin, F. S. 2000. The mineral nutrition of wild plants revisited: a reevaluation of processes and patterns. *Advances in Ecological Research* 30: 1–67.

Ainsworth, E. A., and Long, S. P. 2005. What have we learned from 15 years of free-air CO_2 enrichment (FACE)? A meta-analytic review of the responses of photosynthesis, canopy properties and plant production to rising CO_2. *New Phytologist* 165: 351–372.

Alcaraz, C., Bisazza, A., and Garcia-Berthou, E. 2008. Salinity mediates the competitive interactions between invasive mosquitofish and an endangered fish. *Oecologia* 155: 205–213.

Allee, W. C. 1931. *Animal Aggregations: A Study in General Sociobiology*. Chicago, IL: University of Chicago Press.

Allen, W. 2001. *Green Phoenix: Restoring the Tropical Forests of Guanacaste, Costa Rica*. New York: Oxford University Press.

Allison, F. R. 1979. Life cycle of *Curtuteria australis* n.sp. (Digenea: Echinostomatidae: Himasthlinae), intestinal parasite of the South Island pied oystercatcher. *New Zealand Journal of Zoology* 6: 13–20.

Altevogt, B.M., Pankevich, D. E., Shelton-Davenport, M. K., and Kahn, J. P. 2011. *Chimpanzees in Biomedical and Behavioral Research: Assessing the Necessity*. Washington DC: Board on Health Sciences Policy, Institute of Medicine.

Anderson, R. M., and May, R. M. 1979. Population biology of infectious diseases. Part I. *Nature*, 280: 361–367.

Anderson, R. M., and May, R. M. 1982. Directly transmitted infectious diseases: control by vaccination. *Science*, 215: 1053–1060.

Andrade, M. C. B. 1996. Sexual selection for male sacrifice in the Australian redback spider. *Science* 271: 70–72.

Andrade, M. C. B. 2003. Risky mate search and male self-sacrifice in redback spiders. *Behavioral Ecology* 14: 531–538.

Angilletta Jr, M. 2009. *Thermal Adaptation: A Theoretical and Empirical Synthesis*. New York: Oxford University Press.

Arcese, P., and Smith, J. N. M. 1988. Effects of population density and supplemental food on reproduction in song sparrows. *Journal of Animal Ecology* 57: 119–136.

Athens, J. S., Tuggle, H. D., Ward, J. V., and Welch, D. J. 2002. Avifaunal extinctions, vegetation change, and Polynesian impacts in prehistoric Hawaii. *Archaeology Oceana* 37: 57–78.

Axelrod, R., and Hamilton, W. D. 1981. The evolution of cooperation. *Science* 211: 1390–1396.

Azevedo, L. B., van Zelm, R., Hendriks, A. J., Bobbink, R., and Huijbregts, M. A. 2013. Global assessment of the effects of terrestrial acidification on plant species richness. *Environmental Pollution*, 174: 10–15.

Bacastow, R. B., Keeling, C. D., Woodwell, G. M., and Pecan, E. V. 1973. Atmospheric carbon dioxide and radiocarbon in the natural carbon cycle. II. Changes from AD 1700 to 2070 as deduced from a geochemical model (No. CONF-720510–). University of California, San Diego, La Jolla; Brookhaven National Lab., Upton, NY.

Baez, S., Collins, S., Lightfoot, D., and Koontz, T. L. 2006. Bottom-up regulation of plant community structure in an aridland ecosystem. *Ecology* 87: 2746–2754.

Bagchi, R., and 7 others. 2014. Pathogens and insect herbivores drive rainforest plant diversity and composition. *Nature* 506: 85–88.

Baker, C. S., and 5 others. 1990. Influence of seasonal migration on geographic distribution of mitochondrial DNA haplotypes in humpback whales. *Nature* 344: 238–240.

Bakker, R. 1986. *The Dinosaur Heresies: New Theories Unlocking the Mystery of the Dinosaurs and their Extinction*. New York: William Morrow and Company.

Balda, R. P., and Kamil, A. C. 1992. Long-term spatial memory in Clark's nutcracker, *Nucifraga columbiana*. *Animal Behaviour* 44: 761–769.

Baldwin, I. T. 1998. Jasmonate-induced responses are costly but benefit plants under attack in native populations. *Proceedings of the National Academy of Sciences* 95: 8113–8118.

Baldwin, J. 1971. Adaptation of enzymes to temperature: acetylcholinesterases in the central nervous system of fishes. *Comparative Biochemistry and Physiology* 40: 181–187.

Barbeito, I., Dawes, M. A., Rixen, C., Senn, J., and Bebi, P. 2012. Factors driving mortality and growth at treeline: a 30-year experiment of 92 000 conifers. *Ecology* 93: 389–401.

Bardgett, R. D., and 6 others. 2006. Parasitic plants indirectly regulate below-ground properties in grassland ecosystems. *Nature* 439: 969–972.

Barlow, J., and 18 others. 2011. Humpback whale abundance in the north Pacific estimated by photographic capture-recapture with bias correction from simulation studies. *Marine Mammal Science* 27: 793–818.

Barnett, T. P., and 6 others. 2005. Penetration of human-induced warming into the world's oceans. *Science* 309: 284–287.

Barrett, R. D. H., Rogers, S. M., and Schluter, D. 2009. Environment specific pleiotropy facilitates divergence at the *ectodysplasin* locus in threespine stickleback. *Evolution* 63: 2831–2837.

Barthlott, W., and 5 others. 1999. Biodiversity: the uneven distribution of a treasure. In *Forests in Focus: Proceedings Forum, Biodiversity-Treasures in the World's Forests*, 3–7 July 1998. Hamburg, Germany: Alfred Toepfer Foundation, pp. 18–28

Barthlott, W., Lauer, W., and Placke, A. 1996. Global distribution of species diversity in vascular plants: towards a world map of phytodiversity. *Erdkunde* 50: 317–328.

Barthlott, W., Mutke, J., Rafiqpoor, D., Kier, G., and Kreft, H. 2005. Global centers of vascular plant diversity. *Nova Acta Leopoldina* 92: 61–83.

Bartlett, R. M., Matthes-Sears, U., and Larson, D. W. 1990. Organization of the Niagara Escarpment cliff community. II. Characterization of the physical environment. *Canadian Journal of Botany* 68: 1931–1941.

Barton, B. T., and Ives, A. R. 2014. Species interactions and a chain of indirect effects driven by reduced precipitation. *Ecology* 95: 486–494.

Bauer, S., and Hoye, B. J. 2014. Migratory animals couple biodiversity and ecosystem functioning worldwide. *Science* 344: 54–62.

Bearhop, S., and 8 others. 2005. Assortative mating as a mechanism for rapid evolution of a migratory divide. *Science* 310: 502–504.

Becker, C. G., Fonseca, C. R., Haddad, C. F. B., Batista, R. F., and Prado, P. I. 2007. Habitat split and the global decline of amphibians. *Science* 318: 1775–1777.

Bednaršek, N., Feely, R. A., Reum, J. C. P., *et al.* 2014. *Limacina helicina* shell dissolution as an indicator of declining habitat suitability due to ocean acidification in the California Current Ecosystem. *Proceedings of the Royal Society B* 251 (1785) 20140123.

Bell, M. A., Aguirre, W. E., and Buck, N. J. 2004. Twelve years of contemporary armor evolution in a threespine stickleback population. *Evolution* 58: 814–824.

Bell, T., and 6 others. 2005. Larger islands house more bacterial taxa. *Science* 308: 1884.

Benton, M. J. 2009. The Red Queen and the Court Jester: species diversity and the role of biotic and abiotic factors through time. *Science* 323: 728–732.

Bergstrom, C. A. 2002. Fast-start swimming performance and reduction in lateral plate number in threespine stickleback. *Canadian Journal of Zoology* 80: 207–213.

Berlin, L. 2008. Tuberculosis: resurgent disease renewed liability. *American Journal of Roentgenology* 190: 1438–1444.

Berthold, P., Helbig, A. J., Mohr, G., and Querner, U. 1992. Rapid microevolution of migratory behaviour in a wild bird species. *Nature* 360: 668–670.

Beschta, R. L., and Ripple, W. J. 2009. Large predators and trophic cascades in terrestrial ecosystems of the western United States. *Biological Conservation* 142: 2401–2414.

Beschta, R., and Ripple, W. 2013. Are wolves saving Yellowstone's aspen? A landscape-level test of a behaviorally mediated trophic cascade. Comment. *Ecology* 94: 1420–1425.

Beschta, R. L., Eisenberg, C., Laundré, J. W., Ripple, W. J., and Rooney, T. P. 2014. Predation risk, elk, and aspen: comment. *Ecology* 95: 2669–2671.

Best, P. B., and Kishino, H. 1998. Estimating natural mortality rate in reproductively active female southern right whales, *Eubalaena australis*. *Marine Mammal Science* 14: 738–749.

Best, P. B., Brandão, A., and Butterworth, D. S. 2001. Demographic parameters of southern right whales off South Africa. *Journal of Cetacean Research and Management* 2: 161–169.

Bihn, J. H., Verhaagh, M., Brandle, M., and Brandl, R. 2008. Do secondary forests act as refuges for old growth forest animals? Recovery of ant diversity in the Atlantic forest of Brazil. *Biological Conservation* 141: 733–743.

Bisson, P. A., and 5 others. 2005. Responses of fish to the 1980 eruption of Mount St. Helens. In V. H. Dale, F. J. Swanson, and C. M. Crisafulli (eds) *Ecological Responses to the 1980 Eruption of Mount St. Helens*. New York: Springer Science, pp. 163–181.

Bjork K. E., Averbeck G. A., and Stromberg B. E. 2000. Parasites and parasite stages of free-ranging wild lions (*Panthera leo*) of Northern Tanzania. *Journal of Zoo and Wildlife Medicine* 31: 56–61.

Blockstein, D. E., and Tordoff, H. B. 1985. The passenger pigeon (*Ectopistes migratorius*): a new look at an extinct bird. *American Birds* 39: 845–851.

Boag, P. T. 1983. The heritability of external morphology in Darwin's ground finches (*Geospiza*) on Isla Daphne Major, Galapagos. *Evolution* 877–894.

Boag, P. T., and Grant, P. R. 1984. The classical case of character release: Darwin's finches (*Geospiza*) on Isla Daphne Major, Galapagos. *Biological Journal of the Linnean Society* 22: 243–287.

Bonacum, J., O'Grady, P. M., Kambysellis, M., and DeSalle, R. 2005. Phylogeny and age of diversification of the planitibia species group of Hawaiian *Drosophila*. *Molecular Phylogenetics and Evolution* 37: 73–82.

Boonstra, R., Hik, D., Singleton, G. R., and Tinnikov, A. 1998. The impact of predator-induced stress on the snowshoe hare cycle. *Ecological Monographs* 79: 371–394.

Borer, E. T., and 54 others. 2014. Herbivores and nutrients control grassland plant diversity via light limitation. *Nature* 508: 517–520.

Bormann, F. B., and Likens, G. E. 1968. Nutrient loss accelerated by clear-cutting of a forest ecosystem. *Science* 159: 882–884.

Boyce, C. K., Brodribb, T. J., Feild, T. S., and Zwieniecki, M. A. 2009. Angiosperm leaf vein evolution was physiologically and environmentally transformative. *Proceedings of the Royal Society B* 276: 1771–1776.

Braude, S. 2000. Dispersal and new colony formation in wild naked mole-rats: evidence against inbreeding as the system of mating. *Behavioral Ecology* 11: 7–12.

Brey, T., Müller-Wiegmann, C., Zittier, Z. M. C., and Hagen, W. 2010. Body composition in aquatic organisms: a global data bank of relationships between mass, elemental composition and energy content. *Journal of Sea Research* 64: 334–340.

Briggs, J. S., Vander Wall, S. B., and Jenkins, S. H. 2009. Forest rodents provide directed dispersal of Jeffrey pine seeds. *Ecology* 90: 675–687.

Bristow, C. S., Hudson-Edwards, K. A., and Chappell, A. 2010. Fertilizing the Amazon and equatorial Atlantic with West African dust. *Geophysical Research Letters*, 37(14).

Brittain, C., Williams, N., Kremen, C., and Klein, A. 2013. Synergistic effects of non-*Apis* bees and honey bees for pollination services. *Proceedings of the Royal Society B* 280: 20122767.

Brodribb, T. J., and Feild, T. S. 2010. Leaf hydraulic evolution led a surge in leaf photosynthetic capacity during early angiosperm diversification. *Ecology Letters* 13: 175–183.

Brodribb, T. J., Feild, T. S., and Jordan, G. J. 2007. Leaf maximum photosynthetic rate and venation are linked by hydraulics. *Plant Physiology* 144: 1890–1898.

Broecker, W. S. 2010. *The Great Ocean Conveyor*. Princeton, NJ: Princeton University Press.

Bro-Jørgensen, J. 2013. Evolution of sprint speed in African savannah herbivores in relation to predation. *Evolution* 67: 3371–3376.

Brower, L. P., and 7 others. 2004. Catastrophic winter storm mortality of monarch butterflies in Mexico during January 2002. In K. Oberhauser and M. J. Solensky (eds), *Monarch Butterfly Biology and Conservation*. Ithaca, NY: Cornell University Press, pp. 151–166.

Brown, J. L. 1964. The evolution of diversity in avian territorial systems. *Wilson Bulletin* 76: 160–169.

Buchanan, B.B., Gruissem, W., and Jones, R. L. 2000. *Biochemistry and Molecular Biology of Plants*. Rockville, MD: American Society of Plant Physiologists.

Buffo, J., Fritschen, L. J., and Murphy, J. L. 1972. Direct solar radiation on various slopes from 0 to 60 degrees north latitude. USDA Forest Service Research Paper PNW-142.

Bugnyar, T., and Heinrich, B. 2005. Ravens, *Corvus corax*, differentiate between knowledgeable and ignorant competitors. *Proceedings of the Royal Society B* 272: 1641–1646.

Büyükgüzel, K., Tunaz, H., Putnam, M., and Stanley, D. 2002. Prostaglandin synthesis by midgut tissue isolated from the tobacco hornworm, *Manduca sexta*. *Insect Biochemistry and Molecular Biology* 32: 435–443.

Byrne, P. G., and Keogh, J. S. 2009. Extreme sequential polyandry insures against nest failure in a frog. *Proceedings of the Royal Society B* 276: 115–120.

Calambokidis, J., and Barlow, J. 2004. Abundance of blue and humpback whales in the eastern north Pacific estimated by capture–recapture and line-transect methods. *Marine Mammal Science* 20: 63–85.

Callaway, R. M., and 12 others. 2002. Positive interactions among alpine plants increase with stress. *Nature* 417: 844–848.

Cambefort, Y., and Hanski, I. 1991. Dung beetle population biology. In I. Hanski and Y. Cambefort (eds), *Dung Beetle Ecology*. Princeton, NJ: Princeton University Press, pp. 36–50.

Campagna, C., and Le Boeuf, B. J. 1988. Reproductive behaviour of Southern sea lions. *Behaviour* 104: 233–261.

Campbell, A. H., Vergés, A., and Steinberg, P. D. 2014. Demographic consequences of disease in a habitat-forming seaweed and impacts on interactions between natural enemies. *Ecology* 95: 142–152.

Canestrari, D., and 5 others. 2014. From parasitism to mutualism: unexpected interactions between a cuckoo and its host. *Science* 343: 1350–1352.

Canfield, D. E., Glazer, A. N., and Falkowski, P. G. 2010. The evolution and future of Earth's nitrogen cycle. *Science* 330: 192–196.

Cantin, N. E., Cohen, A. L., Karnauskas, K. B., Tarrant, A. M., and McCorkle, D. C. 2010. Ocean warming slows coral growth in the central Red Sea. *Science* 329: 322–325.

Carroll, L. 1871. *Through the Looking Glass and What Alice Found There*. London: Macmillan and Co.

Carroll, S. P., and 8 others. 2014. Applying evolutionary biology to address global challenges. *Science* 346: 313.

Carson, R. 1961. *Silent Spring*. Boston, MA: Houghton Mifflin Pub.

Cary, J. R., and Keith, L. B., 1979. Reproductive change in the 10-year cycle of snowshoe hares. *Canadian Journal of Zoology* 57: 375–390.

Chan, Y. F., and 15 others. 2010. Adaptive evolution of pelvic reduction in sticklebacks by recurrent deletion of a *Pitx1* enhancer. *Science* 327: 302–305.

Chandler, J. 2014. Ghosts of the rainforests. *New Scientist* 233: 42–45.

Chapin, F. S. III, Matson, P. A., and Mooney, H. A. 2002. *Principles of Terrestrial Ecosystem Ecology*. New York, NY: Springer-Verlag Publishing.

Chapin, F. S., Walker, L. R., Fastie, C. L., and Sharman, L. C. 1994. Mechanisms of primary succession following deglaciation at Glacier Bay, Alaska. *Ecological Monographs* 64: 149–175.

Charmantier, A., and 5 others. 2008. Adaptive phenotypic plasticity in response to climate change in a wild bird population. *Science* 320: 800–803.

Charnov, E. L. 1976. Optimal foraging: the marginal value theorem. *Theoretical Population Biology* 9: 129–136.

Charnov, E. L., and Ernest, S. K. M. 2006. The offspring-size/clutch-size trade-off in mammals. *American Naturalist* 167: 578–582.

Chen, I., Hill, J. K., Ohlemuller, R., Roy, D. B., and Thomas, C. D. 2011. Rapid range shifts of species associated with high levels of global warming. *Science* 333: 1024–1026.

Chen, X., and 32 others. 2014. Producing more grain with lower environmental costs. *Nature* 514: 486–489.

Christner, B. C., and 37 others. 2014. A microbial ecosystem beneath the West Antarctic ice sheet. *Nature* 512: 310–313.

Ciszek, D. 2000. New colony formation in the "highly inbred" eusocial naked mole-rat: outbreeding is preferred. *Behavioral Ecology* 11: 1–6.

Clark, A., and Gaston, K. J. 2006. Climate, energy and diversity. *Proceedings of the Royal Society B* 273: 2257–2266.

Cleveland, C. C., and Liptzin, D. 2007. C:N:P stoichiometry in soil: is there a "Redfield ratio" for the microbial biomass? *Biogeochemistry* 85: 235–252.

Coale, K. H., and 18 others. 1996. A massive phytoplankton bloom induced by an ecosystem-scale iron fertilization experiment in the equatorial Pacific Ocean. *Nature* 383: 495–501.

Collinge, S. K. 1998. Spatial arrangement of habitat patches and corridors: clues from ecological field experiments. *Landscape and Urban Planning* 42(2): 157–168.

Colosimo, P. F., and 6 others. 2004. The genetic architecture of parallel armor plate reduction in threespine sticklebacks. *PLoS Biology* 2: 635–641.

Colosimo, P. F., and 8 others. 2005. Widespread parallel evolution in sticklebacks by repeated fixation of ectodysplasin alleles. *Science* 307: 1928–1933.

Colwell, M. A., and Oring, L. W. 1989. Extra-pair mating in the spotted sandpiper: a female mate-acquisition tactic. *Animal Behaviour* 38: 675–684.

Comas, I., and 9 others. 2012. Whole genome sequencing of rifampicin-resistant *Mycobacterium tuberculosis* strains identifies compensatory mutations in RNA polymerase genes. *Nature Genetics* 44: 106–110.

Connell, J. 1961. The influence of interspecific competition and other factors on the distribution of the barnacle *Chthamalus stellatus*. *Ecology* 42: 710–723.

Connell, J. 1978. Diversity in tropical rain forests and coral reefs. *Science* 199: 1302–1310.

Connell, J. 1980. Diversity and the coevolution of competitors, or the ghost of competition past. *Oikos* 35: 131–138.

Connell, J. H. 1971. On the role of natural enemies in preventing competitive exclusion in some marine animals and in rain forest trees. In P. J. Den Boer and G. Gradwell (eds), *Dynamics of Populations*. Wageningen, the Netherlands: PUDOC, pp. 298–312.

Connell, J. H., and Slatyer, R. O. 1977. Mechanisms of succession in natural communities and their role in community stability and organization. *American Naturalist* 111: 1119–1144.

Connolly, B. M., Pearson, D. E., and Mack, R. N. 2014. Granivory of invasive, naturalized, and native plants in communities differentially susceptible to invasion. *Ecology* 95: 1759–1769.

Cook, S. E., Conway, K. W., and Burd, B. 2008. Status of the glass sponge reefs in the Georgia Basin. *Marine Environmental Research* 66: S80–S86.

Cooper, W. E. Jr., Perez-Mellado, V., and Hawlena, D. 2006. Magnitude of food reward affects escape behavior and acceptable risk in Balearic lizards, *Podarcis lilfordi*. *Behavioral Ecology* 17: 554–559.

Cooper, W. S. 1923a. The recent ecological history of Glacier Bay, Alaska: I. The interglacial forests of Glacier Bay. *Ecology* 4: 93–128.

Cooper, W. S. 1923b. The recent ecological history of Glacier Bay, Alaska: II. The present vegetation cycle. *Ecology* 4: 223–246.

Corcoran, A. J., and Conner, W. E. 2014. Bats jamming bats: food competition through sonar interference. *Science* 346: 745–747.

Cornell, S. and 13 others. 2013. Opening up knowledge systems for better responses to global environmental change. *Environmental Science & Policy* 28: 60–70.

Côté, I. M., and 6 others. 2014. What doesn't kill you makes you wary? Effect of repeated culling on the behaviour of an invasive predator. *PLoS One* 9(4): e94248.

Courchamp, F., Berec, L., and Gascoigne, J. 2008. *Allee Effects in Ecology and Conservation*. New York: Oxford University Press.

Cox-Foster, D. L., and 23 others. 2007. A metagenomic study of microbes in honeybee colony collapse disorder. *Science* 318: 283–287.

Creel, S., Christianson, D., Liley, S., and Winnie Jr., J. A. 2007. Predation risk affects reproductive physiology and demography of elk. *Science* 315: 960.

Da Silva, P. E. A., and Palomino, J. C. 2011. Molecular basis and mechanisms of drug resistance in *Mycobacterium tuberculosis*: classical and new drugs. *Journal of Antimicrobial Chemotherapy* 66: 1417–1430.

Dahm, D. N., Larson, D. W., Petersen, R. R., and Wissmar, R. C. 2005. Response and recovery of lakes. In V. H. Dale, F. J. Swanson, and C. M. Crisafulli (eds), *Ecological Responses to the 1980 Eruption of Mount St. Helens*. New York: Springer Science, pp. 255–274.

Dainat, B., vanEngelsdorp, D., and Neumann, P. 2012. Colony collapse disorder in Europe. *Environmental Microbiology Reports* 4: 123–125.

Dale, V. H., Swanson, F. J., and Crisafulli, C. M. 2005a. *Ecological Responses to the 1980 Eruption of Mount St. Helens*. New York: Springer Science.

Dale, V. H., and 6 others. 2005b. Plant succession on the Mount St. Helens debris-avalanche deposit. In V. H. Dale, F. J. Swanson, and C. M. Crisafulli (eds), *Ecological Responses to the 1980 Eruption of Mount St. Helens*. New York: Springer Science, pp. 59–73.

Dalsgaard, T., Thamdrup, B., and Canfield, D. E. 2005. Anaerobic ammonium oxidation (anammox) in the marine environment. *Research in Microbiology* 156: 457–464.

Darwin, C. 1839. *The Voyage of the Beagle*. London: Henry Colburn.

Darwin, C. 1859. *On the Origin of Species by Means of Natural Selection*. London: John Murray.

Daskalov, G. M. 2002. Overfishing drives a trophic cascade in the Black Sea. *Marine Ecology Progress Series* 225: 53–62.

Davidson, J. 1938. On the growth of the sheep population in Tasmania. *Transactions of the Royal Society of South Australia* 62: 342–346.

Davies, N. B. 1992. *Dunnock Behavior and Social Evolution*. Oxford: Oxford University Press.

Davies, N. B., and Lundberg, A. 1984. Food distribution and a variable mating system in the dunnock, *Prunella modularis*. *Journal of Animal Ecology* 53: 895–912.

Dawkins, R., and Krebs, J. R. 1979. Arms races between and within species. *Proceedings of the Royal Society B* 205: 489–511.

Decaestecker, E., and 6 others. 2007. Host-parasite "Red Queen" dynamics archived in pond sediment. *Nature* 450: 870–873.

De Deyn, G. B., and 7 others. 2003. Soil invertebrate fauna enhances grassland succession and diversity. *Nature* 422: 711–713.

Del Grosso, S., and 7 others. 2008. Global potential net primary production predicted from vegetation class, precipitation and temperature. *Ecology* 89: 2117–2226.

Denman, K.L., and 14 others. 2007. Couplings between changes in the climate system and biogeochemistry. In S. Solomon, and 7 others. (eds), *Climate Change 2007: The Physical Science Basis. Contribution of Working Group I to the Fourth Assessment Report of the Intergovernmental Panel on Climate Change*. Cambridge, UK: Cambridge University Press, pp. 499–587.

DeWalt, S. J. 2006. Population dynamics and potential for biological control of an exotic invasive shrub in Hawaiian rainforests. *Biological Invasions* 8: 1145–1158.

Diamond, J. 1975. Assembly of species communities. In M. L. Cody and J. M. Diamond (eds), *Ecology and Evolution of Communities*. Cambridge, MA: Harvard University Press, pp. 342–444.

Diamond, J. 2005. *Collapse: How Societies Choose to Fail or Succeed*. New York: Viking Press.

Diaz, R. J., and Rosenberg, R. 2008. Spreading dead zones and consequences for marine ecosystems. *Science* 321: 926–929.

Dirzo, R., and 5 others. 2014. Defaunation in the Anthropocene. *Science* 345: 401–406.

Dixson, D. L., Abrego, D., and Hay, M. E. 2014. Chemically mediated behavior of recruiting corals and fishes: a tipping point that may limit reef recovery. *Science* 345: 892–897.

Dornelas, M., and 6 others. 2014. Assemblage time series reveal biodiversity change but not systematic loss. *Science* 344: 296–299.

Doroff, A. M., Estes, J. A., Tinker, M. T., Burn, D. M., and Evans, T. J. 2003. Sea otter population declines in the Aleutian archipelago. *Journal of Mammalogy* 84: 55–64.

Drakare, S., Lennon, J. J., and Hillebrand, H. 2006. The imprint of the geographical, evolutionary and ecological context on species–area relationships. *Ecology Letters* 9: 215–227.

Drezner, T. D., and Garrity, C. M. 2003. Saguaro distribution under nurse plants in Arizona's Sonoran Desert: directional and microclimate influences. *The Professional Geographer* 55: 505–512.

Duarte, C. M., Agustí, S., Gasol, J. M., Vaqué, D., and Vasquez-Dominguez, E. 2000. Effect of nutrient supply on the biomass structure of planktonic communities: an experimental test on a Mediterranean coastal community. *Marine Ecology Progress Series* 206: 87–95.

Dublin, H. T. 1995. Vegetation dynamics in the Serengeti-Mara ecosystem: the role of elephants, fire and other factors. In A. R. E. Sinclair and P. Arcese (eds), *Serengeti II: Dynamics, Management and Conservation of an Ecosystem*. Chicago, IL: University of Chicago Press, pp. 71–90.

Dublin, H. T., Sinclair, A. R. E., and McGlade, J. 1990. Elephants and fire as causes of multiple stable states for Serengeti-Mara woodlands. *Journal of Animal Ecology* 59: 1157–1164.

Duggins, D. O. 1980. Kelp beds and sea otters: an experimental approach. *Ecology* 61: 447–453.

Duncan, D. J. 2001. *My Story as Told by Water*. San Francisco, CA: Sierra Club Books.

Dunn, R. 2011. *The Wild Life of Our Bodies: Predators, Parasites and Partners That Shape Who We Are Today*. New York: HarperCollins Publishers.

Dunn, R. R. 2004. Recovery of faunal communities during tropical forest regeneration. *Conservation Biology* 18: 302–309.

Dunn, R. R., Harris, N. C., Colwell, R. K., Koh, L. P., and Sodhi, N. S. 2009. The sixth mass coextinction: are most endangered species parasites and mutualists? *Proceedings of the Royal Society B* 276: 3037–3045.

Dybas, C. L. 2008. Deep sea lost and found. *Bioscience* 58: 288–294.

Dyer, L. A., and 12 others. 2007. Host specificity of Lepidoptera in tropical and temperate forests. *Nature* 448: 696–699.

Eberhardt, L. L., White, P. J., Garrott, R. A., and Houston, D. B. 2007. A seventy-year history of trends in Yellowstone's northern elk herd. *Journal of Wildlife Management* 71: 594–602.

Edgar, G. J., and 24 others. 2014. Global conservation outcomes depend on marine protected areas with five key features. *Nature* 506: 216–220.

Edwards, J. S., and Sugg, P. M. 2005. Arthropods as pioneers in the regeneration of life on the pyroclastic-flow deposits of Mount St. Helens. In V. H. Dale, F. J. Swanson, and C. M. Crisafulli (eds), *Ecological Responses to the 1980 Eruption of Mount St. Helens*. New York: Springer Science, pp. 127–138.

Ehrlich, P. R., and Hanski, I. 2004. *On the Wings of Checkerspots: A Model System for Population Biology*. New York: Oxford University Press.

Ehrlinger, J. 1985. Annuals and perennials of warm deserts. In B. F. Chabot and H. A. Mooney (eds), *Physiological Ecology of North American Plant Communities*. New York: Chapman and Hall, pp. 162–180.

Eliassen, S., and Jørgensen, C. 2014. Extra-pair mating and evolution of cooperative neighbourhoods. *PloS One* 9(7): e99878.

Ellis, E. C., and Ramankutty, N. 2008. Putting people in the map: anthropogenic biomes of the world. *Frontiers in Ecology* 6: 439–447.

Ellison, A. M., and Farnsworth, E. J. 1990. The ecology of Belizean mangrove-root fouling communities: I. Epibenthic fauna are barriers to isopod attack of red mangrove roots. *Journal of Experimental Marine Biology and Ecology* 142: 91–104.

Ellison, A. M., Farnsworth, E. J., and Twilley, R. R. 1996. Facultative mutualism between red mangroves and root-fouling sponges in Belizean mangal. *Ecology* 77: 2431–2444.

Elser, J. J., and 9 others. 2007. Global analysis of nitrogen and phosphorus limitation of primary producers in freshwater, marine and terrestrial ecosystems. *Ecology Letters* 10: 1135–1142.

Elser, J. J., Dobberfuhl, D. R., MacKay, N. A., and Schampel, J. H. 1996. Organism size, life history, and N:P stoichiometry:

Towards a unified view of cellular and ecosystem processes. *BioScience* 46: 674–684.

Elton, C., and Nicholson, M. 1942. The ten-year cycle in numbers of the lynx in Canada. *Journal of Animal Ecology*, 215–244.

Ernest, S. K. M. 2003. Life history characteristics of placental nonvolant mammals. *Ecology* 84: 3402.

Erwin, T. L. 1982. Tropical forests: their richness in Coleoptera and other arthropod species. *The Coleopterists Bulletin* 36: 74–75.

Estes, J. A., and Duggins, D. O. 1995. Sea otters and kelp forests in Alaska: generality and variation in a community ecological paradigm. *Ecological Monographs* 65: 75–100.

Estes, J. A., Smith, N. S., and Palmisano, J. F. 1978. Sea otter predation and community organization in the western Aleutian Islands, Alaska. *Ecology* 59: 822–833.

Estes, J. A., Tinker, M. T., Williams, T. M., and Doak, D. F. 1998. Killer whale predation on sea otters linking oceanic and nearshore ecosystems. *Science* 282: 473–476.

Estes, J. A., Tinker, M., and Bodkin, J. L. 2010. Using ecological function to develop recovery criteria for depleted species: sea otters and kelp forests in the Aleutian archipelago. *Conservation Biology* 24: 852–860.

Ewbanks, M. D., Styrsky, J. D., and Denno, R. F. 2003. The evolution of omnivory in heteropteran insects. *Ecology* 84: 2549–2555.

Falkowski, P. G. 2002. The ocean's invisible forest. *Scientific American* 287: 54–62.

FAO, IFAD, and WFP. 2013. *The State of Food Insecurity in the World 2013. The Multiple Dimensions of Food Security*. Rome: FAO.

Fastie, C. L. 1995. Causes and ecosystem consequences of multiple pathways of primary succession at Glacier Bay, Alaska. *Ecology* 76: 1899–1916.

Faulkes, C. G., and Bennett, N. C. 2009. Reproductive skew in African mole-rats: behavioural and physiological mechanisms to maintain high skew. In R. Hager and C. B. Jones (eds), *Reproductive Skew in Vertebrates: Proximate and Ultimate Causes*. Cambridge: Cambridge University Press, pp. 369–396.

Fitzpatrick, J. W. and 17 others. 2005. Ivory-billed woodpecker (*Campephilus principalis*) persists in continental North America. *Science* 308: 1460–1462.

Fletcher, M. S., Wood, S. W., and Haberle, S. G. 2014. A fire-driven shift from forest to non-forest: evidence for alternative stable states? *Ecology* 95: 2504–2513.

Ford, A. T., and 8 others. 2014. Large carnivores make savanna tree communities less thorny. *Science* 346: 346–349.

Fourrier, M., Sussman, R. W., Kippen, R., and Childs, G. 2008. Demographic modeling of a predator-prey system and its implication for the Gombe population of *Procolobus rufomitratus tephrosceles*. *International Journal of Primatology* 29: 497–508.

Fowler, D., and 17 others. 2013. The global nitrogen cycle in the twenty-first century. *Philosophical Transactions of the Royal Society B: Biological Sciences* 368 (1621) Discussion Meeting Issue.

Fox, D. 2014. Antarctica's secret garden. *Nature* 512: 244–246.

Fox, J. W. 2013. The intermediate disturbance hypothesis should be abandoned. *Trends in Ecology & Evolution* 28: 86–92.

Franklin, J. F., and 7 others. 2000. Threads of continuity. *Conservation in Practice* 1: 8–17.

Frederickson, M. E., Greene, M. J., and Gordon, D. M. 2005. "Devil's gardens" bedeviled by ants. *Nature* 437: 495–496.

Freeman, A. S., and Byers, J. E. 2006. Divergent induced responses to an invasive predator in marine mussel populations. *Science* 313: 831–833.

Freschet, G. T., Östlund, L., Kichenin, E., and Wardle, D. A. 2013. Aboveground and belowground legacies of native Sami land use on boreal forest in northern Sweden 100 years after abandonment. *Ecology* 95: 963–977.

Froeschke, G., Harf, R., Sommer, S., and Matthee, S. 2010. Effects of precipitation on parasite burden along a natural climatic gradient in southern Africa: implications for possible shifts in infestation patterns due to global changes. *Oikos* 119: 1029–1039.

Fujita, Y. and 11 others. 2014. Low investment in sexual reproduction threatens plants adapted to phosphorus limitation. *Nature* 505: 82–86.

Gagneux, S., and 5 others. 2006. The competitive cost of antibiotic resistance in *Mycobacterium tuberculosis*. *Science* 312: 1944–1946.

Gallagher, T. 2005. *The Grail Bird*. New York: Houghton Mifflin Company.

Gallai, N., Salles, J., Settele, J., and Vaissière, B. E. 2009. Economic valuation of the vulnerability of world agriculture confronted with pollinator decline. *Ecological Economics* 68: 810–821.

Galloway, J. N., and 14 others. 2004. Nitrogen cycles: past present and future. *Biogeochemistry* 70: 153–226.

Garcia, D., and Chacoff, N. P. 2007. Scale-dependent effects of habitat fragmentation on hawthorn pollination, frugivory, and seed predation. *Conservation Biology* 21: 400–411.

Garcia, R. A., Cabeza, M., Rahbek, C., and Araújo, M. B. 2014. Multiple dimensions of climate change and their implications for biodiversity. *Science* 344: 1247579.

Gardner, T. A., Coté, I. M., Gill, J. A., Grant, A., and Watkinson, A. R. 2003. Long-term region-wide declines in Caribbean corals. *Science* 301: 958–960.

Gause, G. F. 1934. *The Struggle for Existence*. Baltimore, MD: Williams and Wilkins.

Gilbert, B., Srivastava, D. S., and Kirby, K. R. 2008. Niche partitioning at multiple scales facilitates coexistence among mosquito larvae. *Oikos* 117: 944–950.

Gilbert-Norton, L., Wilson, R., Stevens, J. R., and Beard, K. H. 2010. A meta-analytic review of corridor effectiveness. *Conservation Biology* 24: 660–668.

Gilbey, I. C. 2006. Meat sharing among the Gombe chimpanzees: harassment and reciprocal exchange. *Animal Behaviour* 71: 953–963.

Gilmour, J. P., Smith, L. D., Heyward, A. J., Baird, A. H., and Pratchett, M. S. 2013. Recovery of an isolated coral reef system following severe disturbance. *Science* 340: 69–71.

Gleick, P. H., and 254 others. 2010. Climate change and the integrity of science. *Science* 328: 689–690.

Gonzalez, A., Lawton, J. H., Gilbert, F. S., Blackburn, T. M., and Evans-Freke, I. 1998. Metapopulation dynamics, abundance, and distribution in a microecosystem. *Science* 281: 2045–2047.

González-Bergonzoni, I., and 5 others. 2012. Meta-analysis shows a consistent and strong latitudinal pattern in fish omnivory across ecosystems. *Ecosystems* 15: 492–503.

Goodall, J. 1964. Tool-using and aimed throwing in a community of free-living chimpanzees. *Nature* 201: 1264–1266.

Goodall, J. 1967. *My Friends: The Wild Chimpanzees*. Washington DC: National Geographic Society,.

Goodall, J. 1971. *In the Shadow of Man*. Boston, MA: Houghton Mifflin Co.

Goodall, J. 1986. *The Chimpanzees of Gombe: Patterns of Behavior*. Cambridge, MA: Belknap Press.

Goodall, J. 1987. A plea for the chimpanzee. *New York Times Magazine* (17 May).

Gotelli, N. J. 2008. *A Primer of Ecology*. 4th edn. Sunderland, MA: Sinauer Associates Inc.

Gotelli, N. J., and Elllison, A. M. 2004. *A Primer of Ecological Statistics*. Sunderland, MA: Sinauer Associates Inc.

Goudswaard, K., Witte, F., and Katunzi, E. F. B. 2008. The invasion of an introduced predator, Nile perch (*Lates niloticus*, L.) in Lake Victoria (East Africa): chronology and causes. *Environmental Biology and Fishes* 81: 127–139.

Goulson, D. 2014. Pesticides linked to bird declines. *Nature* 511: 295–296.

Grant, B. R., and Grant, P. R. 2003. What Darwin's finches can tell us about the evolutionary origin and regulation of biodiversity. *Bioscience* 53: 965–975.

Grant, P. R., and Grant, B. R. 2002. Unpredictable evolution in a 30-year study of Darwin's finches. *Science* 296: 707–711.

Grant, P. R., and Grant, B. R. 2008. *How and Why Species Multiply: the Radiation of Darwin's Finches*. Princeton, NJ: Princeton University Press.

Graves, J. E., and Somero, G. N. 1982. Electrophoretic and functional enzyme evolution in four species of Eastern Pacific barracudas from different thermal environments. *Evolution* 36: 97–106.

Gray, S. T., Betancourt, J. L., Jackson, S. T., and Eddy, R. G. 2006. Role of multidecadal climate variability in a range extension of pinyon pine. *Ecology* 87: 1124–1130.

Griffith, S. C., Owens, I. P. F., and Thuman, K. A. 2002. Extra pair paternity in birds: a review of interspecific variation and adaptive function. *Molecular Ecology* 11: 2195–2212.

Grime, J. P. 1977. Evidence for the existence of three primary strategies in plants and its relevance to ecological and evolutionary theory. *American Naturalist* 111: 1169–1194.

Grinnell, J., and Storer, T. I. 1924. *Animal Life in the Yosemite*. Berkley, CA: California University Press.

Groom, M. J., Meffe, G. K., and Carroll, C. R. 2006. *Principles of Conservation Biology*. Sunderland, MA: Sinauer Associates.

Hadley, A. S., Frey, S. J., Robinson, W. D., Kress, W. J., and Betts, M. G. 2014. Tropical forest fragmentation limits pollination of a keystone understory herb. *Ecology* 95: 2202–2212.

Hairston, N. G., Jr., and Hairston, N. G., Sr. 1993. Cause-effect relationships in energy flow, trophic structure, and interspecific interactions. *American Naturalist* 142: 379–411.

Hall, A. R., and Kelson, K. R. 1959. *The Mammals of North America*. New York: Ronald Press.

Halliwell, D. B., Whittier, T. R., and Ringler, N. H. 2001. Distributions of lake fishes of the northeast USA: III. Salmonidae and associated coldwater species. *Northeastern Naturalist* 8: 189–206.

Hallmann, C. A., Foppen, R. P., van Turnhout, C. A., de Kroon, H., and Jongejans, E. 2014. Declines in insectivorous birds are associated with high neonicotinoid concentrations. *Nature* 511: 341–344.

Hallwachs, W. 1986. Agouti (*Dasyprocta punctata*): the inheritors of guapinol (*Hymenaea courbaril*: Leguminosae). In A. Estrada and T. H. Fleming (eds), *Frugivores and Seed Dispersal*. Dordrecht, the Netherlands: Dr W. Junk Publishers, pp. 285–304.

Halofsky, J., and Ripple, W. 2008. Linkages between wolf presence and aspen recruitment in the Gallatin elk winter range of southwestern Montana, USA. *Forest Ecology and Management* 256: 1004–1008.

Hamilton, W. D. 1964. The genetical evolution of social behaviour. *Journal of Theoretical Biology* 7: 1–52.

Hanby, J. P., and Bygott, J. D. 1979. Population changes in lions and other predators. In A. R. E. Sinclair and M. Norton-Griffiths (eds), *Serengeti: Dynamics of an Ecosystem*. Chicago, IL: University of Chicago Press, pp. 249–262.

Handa, I. T., and 17 others. 2014. Consequences of biodiversity loss for litter decomposition across biomes. *Nature* 509: 218–221.

Hanel, R. A., and 4 others. 1971. Spectrum taken with the infrared interferometer spectrometer flown on the satellite Nimbus 4 in 1971. *Applied Optics* 10: 1376–1382.

Hanna, C., Foote, D., and Kremen, C. 2014. Competitive impacts of an invasive nectar thief on plant–pollinator mutualisms. *Ecology* 95: 1622–1632.

Hanski, I. 2011. Eco-evolutionary spatial dynamics in the Glanville fritillary butterfly. *Proceedings of the National Academy of Sciences* 108: 14397–14404.

Hanski, I., Kuussaari, M., and Nieminen, M. 1994. Metapopulation structure and migration in the butterfly *Melitaea cinxia*. *Ecology* 75: 747–762.

Hanski, I., Pakkala, T., Kuussaati, M., and Lei, G. 1995. Metapopulation persistence of an endangered butterfly in a fragmented landscape. *Oikos* 72: 21–28.

Hardin, G. 1968. The tragedy of the commons. *Science* 162: 1243–1248.

Harpole, W. S., and Tilman, D. 2007. Grassland species loss resulting from reduced niche dimension. *Nature* 446: 791–793.

Harrington, L. A., and 6 others. 2009. The impact of native competitors on an alien invasive: temporal niche shifts to avoid interspecific aggression? *Ecology* 90: 1207–1216.

Hatfield, T., and Schluter, D. 1999. Ecological speciation in sticklebacks: environment-dependent hybrid fitness. *Evolution* 53: 866–873.

Hatton, H. 1938. Essais de bionornie explicative sur quelques especes intercotidales d'algues et d'animaux. *Annales de l'Institut Oceanographique* 17: 241–348.

Hauser, M. D., Chen, M. K., Chern, F., and Chuang, E. 2003. Give unto others: genetically unrelated cotton-top tamarin monkeys preferentially give food to those who altruistically give food back. *Proceedings of the Royal Society B* 270: 2363–2370.

Hautier, Y., Niklaus, P. A., and Hector, A. 2009. Competition for light causes plant biodiversity loss after eutrophication. *Science* 324: 636–638.

Hazen, T. C., and 31 others. 2010. Deep-sea oil plume enriches indigenous oil-degrading bacteria. *Science* 330: 204–208.

Heath, J. E., and Adams, P. A. 1965. Temperature regulation in the sphinx moth during flight. *Nature* 205: 309–310.

Heath, J. E., and Adams, P. A. 1967. Regulation of heat production by large moths. *Journal of Experimental Biology* 47: 21–33.

Hebert, P. D. N., Penton, E. H., Burns, J. M., Janzen, D. H., and Hallwachs, W. 2004. Ten species in one: DNA barcoding reveals cryptic species in the neotropical skipper butterfly *Astraptes fulgerator. Proceedings of the National Academy of Sciences* 101: 14812–14817.

Heinrich, B. 1971a. Temperature regulation of the sphinx moth, *Manduca sexta*. I. Flight energetics and body temperature during free and tethered flight. *Journal of Experimental Biology* 54: 141–152.

Heinrich, B. 1971b. Temperature regulation of the sphinx moth, *Manduca sexta*. II. Regulation of heat loss by control of blood circulation. *Journal of Experimental Biology* 54: 153–166.

Heinrich, B. 1979. Foraging strategies of caterpillars: leaf damage and possible predator avoidance strategies. *Oecologia* 42: 325–337.

Heinrich, B. 1988. Winter foraging at carcasses by three sympatric corvids, with emphasis on recruitment by the raven, *Corvus corax. Behavioral Ecology and Sociobiology* 23: 141–156.

Heinrich, B. 1989. *Ravens in Winter*. New York: Summit Books.

Heinrich, B. 1993. *The Hot-Blooded Insects: Strategies and Mechanisms of Thermoregulation*. Cambridge, MA: Harvard University Press.

Heinrich, B. 1995. An experimental investigation of insight in common ravens (*Corvus corax*). *The Auk* 112: 994–1003.

Heinrich, B., and Bugnyar, T. 2005. Testing understanding in ravens: string-pulling to reach food. *Ethology* 111: 962–976.

Heinrich, B., and Collins, S. 1983. Caterpillar leaf damage, and the game of hide-and-seek with birds. *Ecology* 64: 592–602.

Heinrich, B., and Heinrich, M. J. 1984. The pit-trapping foraging strategy of the ant lion, *Myrmeleon immaculatus* DeGeer (Neuroptera: Myrmeleontidae). *Behavioral Ecology and Sociobiology* 14(2): 151–160.

Heinrich, B., and Mommsen, T. P. 1985. Flight of winter moths near 0°C. *Science* 228: 177–179.

Henry, M., and 8 others. 2012. A common pesticide decreases foraging success and survival in honey bees. *Science* 336: 348–350.

Hess, J. 2004. *A population genetic study of the eusocial naked mole-rat (Heterocephalus glaber)*. Unpublished PhD thesis. Washington University, USA.

Hillborn, R., and Sinclair, A. R. E. 1979. A simulation of the wildebeest population, other ungulates and their predators. In A. R. E. Sinclair and M. Norton-Griffiths (eds), *Serengeti: Dynamics of an Ecosystem*. Chicago, IL: University of Chicago Press, pp. 287–309.

Hinz, H. L., and 5 others. 2012. Biogeographical comparison of the invasive *Lepidium draba* in its native, expanded and introduced ranges. *Biological Invasions* 14: 1999–2016.

Hobson, K. A., Bowen, G. J., Wasswnaar, L. I., Ferrand, Y., and Lormee, H. 2004. Using stable hydrogen and oxygen isotope measurements of feathers to infer geographical origins of migrating European birds. *Oecologia* 141: 477–488.

Holland Jones, J., Wilson, M. L., Murray, C., and Pusey, A. 2010. Phenotypic quality influences fertility in Gombe chimpanzees. *Journal of Animal Ecology* 79: 1262–1269.

Holland, R. A., Wikelski, M., and Wilcove, D. S. 2006. How and why do insects migrate? *Science* 313: 794–796.

Holling, C. S. 1959a. Some characteristics of simple types of predation and parasitism. *Canadian Entomologist* 91: 385–398.

Holling, C. S. 1959b. The components of predation as revealed by a study of small-mammal predation of the European pine sawfly. *The Canadian Entomologist* 91: 293–320.

Hollowed, A. B., and Sundby, S. 2014. Change is coming to the northern oceans. *Science* 344(6188): 1084–1085.

Holmes, J. A. 2008. How the Sahara became dry. *Science* 320: 752–753.

Holt, B. G., and 14 others. 2013. An update of Wallace's zoogeographic regions of the world. *Science* 339: 74–78.

Hoppe, P. P., Qvortrup, S. A., and Woodford, M. H. 1977. Rumen fermentation and food selection in East African sheep, goats, Thomson's gazelle, Grant's gazelle and impala. *Journal of Agricultural Science* 89: 129–135.

Houle, D., Moore, J. D., Ouimet, R., and Marty, C. 2014. Tree species partition N uptake by soil depth in boreal forests. *Ecology* 95: 1127–1133.

Howard, E., and Davis, A. K. 2004. Documenting the spring movements of monarch butterflies with Journey North, a citizen science program. In K. Oberhauser and M. J. Solensky (eds), *Monarch Butterfly Biology and Conservation*. Ithaca, NY: Cornell University Press, pp. 105–116.

Howard, E., and Davis, A. K. 2011. A simple numerical index for assessing the spring migration of monarch butterflies using data from Journey North, a citizen-science program. *Journal of the Lepidopterists' Society* 65: 267–270.

Hudson, J. M. G., and Henry, G. H. R. 2009. Increased plant biomass in a High Arctic heath community from 1981 to 2008. *Ecology* 90: 2657–2663.

Hughes, T. P., and 9 others. 2007. Phase shifts, herbivory, and the resilience of coral reefs to climate change. *Current Biology* 17: 1–6.

Hunt, T. L. 2006. Rethinking the fall of Easter Island: new evidence points to an alternative explanation for a civilization's collapse. *American Scientist* 94: 412–419.

Hunt, T. L, and Lipo, C. P. 2006. Late colonization of Easter Island. *Science* 311: 1603–1606.

Hunt, T. L., and Lipo, C. 2011. *The Statues that Walked: Unraveling the Mystery of Easter Island*. New York: Tree Press.

Hunter, C. M., and 5 others. 2010. Climate change threatens polar bear populations: a stochastic demographic analysis. *Ecology* 91: 2883–2997.

Hutchinson, G. E. 1957. Concluding remarks. *Cold Spring Harbor Symposium on Quantitative Biology* 22: 415–427.

Inagaki, F., and 13 others. 2006. Biogeographical distribution and diversity of microbes in methane hydrate-bearing deep marine sediments on the Pacific Ocean margin. *Proceedings of the National Academy of Science* 103: 2815–2820.

IPCC. 2007. Climate change 2007: the physical science basis. Contribution of Working Group I to the fourth assessment report of the Intergovernmental Panel on Climate Change. S. Solomon and 7 others (eds). Cambridge: Cambridge University Press.

IPCC. 2013. Climate change 2013: the physical science basis. Contribution of Working Group I to the fifth assessment report of the Intergovernmental Panel on Climate Change. T. F. Stocker, and 9 others (eds). Cambridge: Cambridge University Press.

Jackrel, S. L., and Wootton, J. T. 2014. Local adaptation of stream communities to intraspecific variation in a terrestrial ecosystem subsidy. *Ecology* 95: 37–43.

Jackson, J. A. 2006. Ivory-billed woodpecker (*Campephilus principalis*): hope, and the interfaces of science, conservation, and politics. *The Auk* 123: 1–15.

Janzen, D. H. 1966. Coevolution of mutualism between ants and acacias in Central America. *Evolution* 20: 249–275.

Janzen, D. H. 1969. Seed-eaters versus seed size, number, toxicity and dispersal. *Evolution* 23: 1–27.

Janzen, D. H. 1970. Herbivores and the number of tree species in tropical forests. *American Naturalist* 104: 501–528.

Janzen, D. H. 1976. Why bamboos wait so long to flower. *Annual Review of Ecology and Systematics* 7: 347–391.

Janzen, D. H. 1977. Why fruits rot, seeds mold, and meat spoils. *American Naturalist* 111: 691–713.

Janzen, D. H. 1979a. How to be a fig. *Annual Review of Ecology and Systematics* 10: 13–51.

Janzen, D. H. 1979b. How many babies do figs pay for babies? *Biotropica* 11: 48–50.

Janzen, D. H. 1981. *Enterolobium cyclocarpum* seed passage rate and survival in horses, Costa Rican Pleistocene seed dispersal agents. *Ecology* 62: 593–601.

Janzen, D. H. 1982. Natural history of guacimo fruits (Sterculiaceae: *Guazuma ulmifolia*) with respect to consumption by large mammals. *American Journal of Botany* 69: 1240–1250.

Janzen, D. H. 1986. Mice, big mammals, and seeds: it matters who defecates what where. In A. Estrada and T. H. Fleming (eds), *Frugivores and Seed Dispersal*. Dordrecht, the Netherlands: Dr W. Junk Publishers, pp. 251–271.

Janzen, D. H. 1988. Management of habitat fragments in a tropical dry forest: growth. *Annals of the Missouri Botanical Garden* 75: 105–116.

Janzen, D. H. 2000. Costa Rica's Area de Conservacion Guanacaste: a long march to survival through non-damaging biodevelopment. *Biodiversity* 1: 7–20.

Janzen, D. H. 2010. Hope for tropical biodiversity through true bioliteracy. *Biotropica* 42: 540–542.

Janzen, D. H., and Hallwachs, W. 2011. Joining inventory by parataxonomists with DNA barcoding of a large complex tropical conserved wildland in northwestern Costa Rica. *PloS One* 6: e18123.

Janzen, D., and Martin, P. S. 1982. Neotropical anachronisms: the fruit the gomphotheres ate. *Science* 215: 19–27.

Janzen, D. H., and 45 others. 2009. Integration of DNA barcoding into an ongoing inventory of complex tropical biodiversity. *Molecular Ecology Resources* 9 (Suppl. 1): 1–26.

Jarvis, J. U. M. 1981. Eusociality in a mammal: cooperative breeding in naked mole-rat colonies. *Science* 212: 571–573.

Johnson, H. E., and 5 others. 2013. Evaluating apparent competition in limiting the recovery of an endangered ungulate. *Oecologia* 171: 295–307.

Johnson, M. P., and Raven, P. H. 1973. Species number and endemism: the Galápagos Archipelago revisited. *Science* 179: 893–895.

Johnson, S. D., and Steiner, K. E. 2000. Generalization versus specialization in plant pollination system. *Trends in Ecology and Evolution* 15: 140–143.

Johnson, S. J., Neal, P. R., Peter, C. I., and Edwards, T. J. 2004. Fruiting failure and limited recruitment in remnant populations of the hawkmoth-pollinated tree *Oxyanthus pyriformis* subsp. *pyriformis* (Rubiaceae). *Biological Conservation* 120: 31–39.

Jonasson, K. A., and Willis, C. K. R. 2012. Hibernation energetics of free-ranging little brown bats. *Journal of Experimental Biology* 215: 2141–2149.

Jones, A. M., Berkelmans, R., van Oppen, M. J. H., Mieog, J. C., and Sinclair, W. 2008. A community change in the algal endosymbionts of a scleractinian coral following a natural bleaching event: field evidence of acclimatization. *Proceedings of the Royal Society B* 275: 1359–1365.

Jordan, D. B., and Ogren, W. L. 1984. The CO_2/O_2 specificity of ribulose 1,5-bisphosphate carboxylase/oxygenase. *Planta* 161: 308–313.

Journey North (2015). *Journey North: A Global Study of Wildlife Migration and Seasonal Change*. Available at: http://www.learner.org/jnorth/.

Jousimo, J., and 6 others. 2014. Ecological and evolutionary effects of fragmentation on infectious disease dynamics. *Science* 344: 1289–1293.

Jousset, A., and 5 others. 2014. Biodiversity and species identity shape the antifungal activity of bacterial communities. *Ecology* 95: 1184–1190.

Judas, M. 1988. The species–area relationship of European Lumbricidae (Annelida, Oligochaeta). *Oecologia* 76: 579–587.

Kacelnik, A. 1984. Central place foraging in starlings (*Sturnus vulgaris*). I. Patch residence time. *Journal of Animal Ecology* 53: 283–299.

Kalka, M. B., Smith, A. R., and Kalko, E. K. V. 2008. Bats limit arthropods and herbivory in a tropical forest. *Science* 320: 71.

Kandori, I., Hirao, T., Matsunaga, S., and Kurosaki, T. 2009. An invasive dandelion reduces the reproduction of a native congener through competition for pollination. *Oecologia* 159: 559–569.

Karban, R. 1982. Increased reproductive success at high densities and predator satiation for periodical cicadas. *Ecology* 63: 321–328.

Karban, R. 2014. Transient habitats limit the development time for periodical cicadas. *Ecology* 95: 3–8.

Kaspari, M., Clay, N. A., Donoso, D. A., and Yanoviak, S. P. 2014. Sodium fertilization increases termites and enhances decomposition in an Amazonian forest. *Ecology* 95: 795–800.

Kauffman, M. J., and 5 others. 2007. Landscape heterogeneity shapes predation in a newly restored predator–prey system. *Ecology Letters* 10: 690–700.

Kauffman, M. J., Brodie, J. F., and Jules, E. S. 2010. Are wolves saving Yellowstone's aspen? A landscape-level test of a behaviorally mediated trophic cascade. *Ecology* 91: 2742–2755.

Kauffman, M. J., Brodie, J. F., and Jules, E. S. 2013. Are wolves saving Yellowstone's aspen? A landscape-level test of a behaviorally mediated trophic cascade: reply. *Ecology* 94: 1425–1431.

Kays, R. W., and Wilson, D. E. 2002. *Mammals of North America*. Princeton, NJ: Princeton University Press.

Keele, B. F., and 24 others. 2009. Increased mortality and AIDS-like immunopathology in wild chimpanzees infected with SIVcpz. *Nature* 460: 515–519.

Keeley, J., E., and Fotheringham, C. J. 2001. Historic fire regime in southern California shrublands. *Conservation Biology* 15: 1536–1548.

Keeling, C. D. 1998. Rewards and penalties of monitoring the Earth. *Annual Review of Energy and the Environment* 23: 25–82.

Keller, L. F., Grant, P. R., Grant, B. R., and Petren, K. 2001. Heritability of morphological traits in Darwin's finches: misidentified paternity and maternal effects. *Heredity* 87: 325–336.

Kennedy, D. 2003. Sustainability and the commons. *Science* 302: 1861.

Kenward, R. E. 1978. Hawks and doves: factors affecting success and selection in goshawk attacks on woodpigeons. *Journal of Animal Ecology* 47: 449–460.

Kerby, J., and Post, E. 2013. Capital and income breeding traits differentiate trophic match–mismatch dynamics in large herbivores. *Philosophical Transactions of the Royal Society B* 368: 20120484.

Kessler, A., and Baldwin, I. T. 2001. Defensive function of herbivore-induced plant volatile emissions in nature. *Science* 291: 2141–2144.

Kideys, A. E., Roohi, A., Bagheri, S., Finenko, G., and Kamburska, L. 2005. Impact of invasive ctenophores on the fisheries of the Black Sea and Caspian Sea. *Oceanography* 18: 77–85.

Kiers, E. T., Rousseau, R. A., West, S. A., and Denison, R. F. 2003. Host sanctions and the legume–rhizobium mutualism. *Nature* 425: 78–81.

Kingsland, S. E. 1985. *Modeling Nature: Episodes in the History of Population Ecology*. Chicago, IL: University of Chicago Press.

Kinnear, J. E., Onus, M. L., and Sumner, N. R. 1998. Fox control and rock-wallaby population dynamics II. An update. *Wildlife Research* 25: 81–88.

Klak, C., Reeves, G., and Hedderson, T. 2004. Unmatched tempo of evolution in southern African semi-desert ice plants. *Nature* 427: 63–65.

Kleiber, M. 1932. Body size and metabolism. *Hilgardia* 6: 315–353.

Kleypas, J. A., and 5 others. 2006. *Impacts of Ocean Acidification on Coral Reefs and Other Marine Calcifiers: A Guide for Future Research*. Report of a workshop held 18–20 April 2005, St. Petersburg, FL, sponsored by the National Science Foundation, the National Oceanic and Atmospheric Administration, and the US Geological Survey.

Knight, T. M, McCoy, M. W., Chase, J. M., McCoy, K. A., and Holt, R. D. 2005. Trophic cascades across ecosystems. *Nature* 437: 880–883.

Knowlton, A., and Kraus, S. D. 2001. Mortality and serious injury of northern right whales (*Eubalaena glacialis*) in the western North Atlantic Ocean. 2001. *Journal of Cetacean Research and Management (Special Issue)* 2: 193–208.

Knowlton, A. R., Hamilton, P. K., Marx, M. K., Pettis, H. M., and Kraus, S. D. 2012. Monitoring North Atlantic right whale *Eubalaena glacialis* entanglement rates: a 30 year retrospective. *Marine Ecology Progress Series* 466: 293–302.

Koenig, W. D., and Liebhold, A. 2005. Effects of periodical cicada emergences on abundance and synchrony of avian populations. *Ecology* 86: 1873–1882.

Kohn, M. H., and 5 others. 1999. Estimating population size by genotyping faeces. *Proceedings of the Royal Society B* 266: 657–663.

Kosaka, Y., and Xie, S. P. 2013. Recent global-warming hiatus tied to equatorial Pacific surface cooling. *Nature* 501: 403–407.

Kraus, S. D., and 15 others. 2005. North Atlantic right whales in crisis. *Science* 309: 561–562.

Kröpelin, S., and 14 others. 2008. Climate-driven ecosystem succession in the Sahara: the past 6000 years. *Science* 320: 765–768.

Kruuk, H. 1972. *The Spotted Hyena*. Chicago, IL: University of Chicago Press.

Kuenen, J. G. 2008. Anammox bacteria: from discovery to application. *Nature Reviews: Microbiology* 6: 320–326.

Kunte, K. 2008. Competition and species diversity: removal of dominant species increases diversity in Costa Rican butterfly communities. *Oikos* 117: 69–76.

Kuussaari, M., Saccheri, I., Camara, M., and Hanski, I. 1998. Allee effect and population dynamics in the Glanville fritillary butterfly. *Oikos* 82: 384–392.

Lack, D. 1947. *Darwin's Finches*. Cambridge: Cambridge University Press.

LaDeau, S. L., Kilpatrick, A. M., and Marra, P. P. 2007. West Nile virus emergence and large-scale declines of North American bird populations. *Nature* 447: 710–714.

Lafferty, K. D., Dobson, A. P., and Kuris, A. M. 2006a. Parasites dominate food web links. *Proceedings of the National Academy of Sciences* 103: 11211–11216.

Lafferty, K. D., Hechinger, R. F., Shaw, J. C., Whitney, K. L., and Kuris, A. M. 2006b. Food webs and parasites in a salt marsh ecosystem. In S. Collinge and C. Ray (eds), *Disease Ecology: Community Structure and Pathogen Dynamics*. New York: Oxford University Press, pp. 119–134.

Laliberté, A. S., and Ripple, W. J. 2004. Range contractions of North American carnivores and ungulates. *Bioscience* 54: 123–138.

Landrum, J. V. 2001. Wide-band tracheids in leaves in Aizoaceae: the systematic occurrence of a novel cell type and its implications for the monophyly of the subfamily Rushioideae. *Plant Systematics and Evolution* 227: 49–61.

Landry, S. 2012. Right whale entanglement response. *Right Whale News* 20(2): 6–10.

Laundré, J. W., Hernandez, L., and Altendorf, K. B. 2001. Wolves, elk, and bison: reestablishing the "landscape of fear" in Yellowstone National Park USA. *Canadian Journal of Zoology* 79: 1401–1409.

Laundré, J. W., and 9 others. 2014. The landscape of fear: the missing link to understand top-down and bottom-up controls of prey abundance? *Ecology* 95: 1141–1152.

Laurance, W. F., Delmonica, P., Laurance, S. G., Vasconcelos, H. L., and Lovejoy, T. E. 2000. Rainforest fragmentation kills big trees. *Nature* 404: 836.

Laurance, W. F., and 9 others. 2002. Ecosystem decay of Amazonian forest fragments: a 22-year investigation. *Conservation Biology* 16: 605–618.

Laurance, W. F., and 6 others. 2006. Rain forest fragmentation and the proliferation of successional trees. *Ecology* 87: 469–482.

Laurance, W. F., and 16 others. 2011. The fate of Amazonian forest fragments: a 32-year investigation. *Biological Conservation* 144: 56–67.

Lawler, A. 2014. In search of green Arabia. *Science* 345: 994–997.

Laxminarayan, R. 2014 Antibiotic effectiveness: Balancing conservation against innovation *Science* 345: 1299–1301.

LeBauer, D. S., and Treseder, K. K. 2008. Nitrogen limitation of net primary productivity in terrestrial ecosystems is globally distributed. *Ecology* 89: 371–379.

LeBrun, E. G., Plowes, R. M., and Gilbert, L. E. 2009. Indirect competition facilitates widespread displacement of one naturalized parasitoid of imported fire ants by another. *Ecology* 90: 1184–1194.

LeBrun, E. G., Jones, N. T., and Gilbert, L. E. 2014. Chemical warfare among invaders: a detoxification interaction facilitates an ant invasion. *Science* 343: 1014–1017.

Le Corre, M., and 9 others. 2012. Tracking seabirds to identify potential Marine Protected Areas in the tropical western Indian Ocean. *Biological Conservation* 156: 83–93.

Le Quéré, C., and 59 others. 2014. Global Carbon Budget 2014. *Earth System Science Data Discussions*, doi: 10.5194/essdd-7-521-2014.

Lesica, P., and Young, T. 2005. A demographic model explains life-history variation in *Arabis fecunda*. *Functional Ecology* 19: 471–477.

Leung, T. L. F., and Poulin, R. 2007. Interactions between parasites of the cockle *Austrovenus stutchburyi*: hitch-hikers, resident-cleaners, and habitat-facilitators. *Parasitology* 134: 247–255.

Li, Z., Wang, W., and Zhang, Y. 2014. Recruitment and herbivory affect spread of invasive *Spartina alterniflora* in China. *Ecology* 95: 1972–1980.

Lieth, H. 1975. Modeling the primary productivity of the world. In H. Lieth and R. Whittaker (eds), *Primary Productivity of the Biosphere*. Berlin: Springer-Verlag, pp. 237–263.

Likens, G. E., Bormann, F. H., Johnson, N. M., Fisher, D. W., and Pierce, R. S. 1970. Effects of forest cutting and herbicide treatment on nutrient budgets in the Hubbard Brook watershed-ecosystem. *Ecological Monographs* 40: 23–47.

Likens, G. E., Bormann, F. H., Pierce, R. S., and Reiners, W. A. 1978. Recovery of a deforested ecosystem. *Science* 199: 492–496.

Lin, L.-H., and 13 others. 2006. Long-term sustainability of a high-energy, low-diversity crustal biome. *Science* 314: 479–482.

Link, J. 2002. Does food web theory work for marine ecosystems? *Marine Ecology Progress Series* 230: 1–9.

Lloyd, J. E. 1965. Aggressive mimicry in *Photuris*: firefly femmes fatales. *Science* 149: 653–654.

Lomas, M. W., and 6 others. 2010. Sargasso Sea phosphorus biogeochemistry: an important role for dissolved organic phosphorus (DOP). *Biogeosciences* 7: 695–710.

Lonsdale, P. 1977. Clustering of suspension-feeding macrobenthos near abyssal hydrothermal vents at oceanic spreading centers. *Deep-Sea Research* 24: 857–863.

Lotka, A. J. 1925. *Elements of Physical Biology*. Baltimore, MD: Williams & Wilkins.

Lovejoy, T. E., and Oren, D. C. 1981. Minimum critical size of ecosystems. In R. L. Burgess and D. M. Sharp (eds), *Forest Island Dynamics in Man-dominated Landscapes*. Springer-Verlag, New York, NY, pp. 7–12.

Lubchenco, J. 1980. Algal zonation in the New England rocky intertidal community: an experimental analysis. *Ecology* 61: 333–344.

Lubchenco, J. 2012. Reflections on the Sustainable Biosphere Initiative. *Bulletin of the Ecological Society of America* 93(4): 260–267.

Lubchenco, J., and Menge, B. 1978. Community development and persistence in a low rocky intertidal zone. *Ecological Monographs* 59: 67–94.

Lubchenco, J., and 8 others. 1984. Structure, persistence and role of consumers in a tropical rocky intertidal community (Taboguilla Island, Bay of Panama). *Journal of Experimental Marine Biology and Ecology* 78: 23–73.

Lubchenco, J., and 15 others. 1991. The Sustainable Biosphere Initiative: an ecological research agenda: a report from the Ecological Society of America. *Ecology* 72: 371–412.

Lubchenco, J., Palumbi, S. R., Gaines, S. D., and Andelman, S. 2003. Plugging a hole in the ocean: the emerging science of marine reserves. *Ecological Applications* 13(1) Supplement: S3–S7.

Lubchenco, J., and 7 others. 2012. Science in support of the *Deepwater Horizon* response. *Proceedings of the National Academy of Sciences* 109: 20212–20221.

Ludwig, F., De Kroon, H., and Prins, H. H. T. 2008. Impacts of savanna trees on forage quality for a large African herbivore. *Oecologia* 155: 487–496.

Luttge, U. 2004. Ecophysiology and crassulacean acid metabolism (CAM). *Annals of Botany* 93: 629–652.

Lyons, T. W., Reinhard, C. T., and Planavsky, N. J. 2014. The rise of oxygen in Earth's early ocean and atmosphere. *Nature* 506: 307–315.

MacArthur, R. H., and Wilson, E. O. 1967. *The Theory of Island Biogeography*. Princeton, NJ: Princeton University Press.

Mace, G. M. 2014. Whose conservation? *Science* 345: 1558–1560.

MacLulich, D. A. 1937. Fluctuations in the numbers of the varying hare (*Lepus americanus*). *University of Toronto Studies; Biological Series* 43: 1–136.

Mahowald, N., and 17 others. 2008. Global distribution of atmospheric phosphorus sources, concentrations and deposition rates, and anthropogenic impacts. *Global Biogeochemical Cycles* 22(4).

Main, T., Dobberfuhl, D. R., and Elser, J. J. 1997. N:P stoichiometry and ontogeny of crustacean zooplankton: a test of the growth rate hypothesis. *Limnology and Oceanography* 42: 1474–1478.

Markham, A. C., Lonsdorf, E. V., Pusey, A. E., and Murray, C. M. 2015. Maternal rank influences the outcome of aggressive interactions between immature chimpanzees. *Animal Behaviour* 100: 192–198.

Marquez, L. M., Redman, R. S., Rodriguez, R. J., and Roossinck, M. J. 2007. A virus in a fungus in a plant: three-way symbiosis required for thermal tolerance. *Science* 315: 513–515.

Martin, J. H., and 43 others. 1994. Testing the iron hypothesis in ecosystems of the equatorial Pacific Ocean. *Nature* 371: 123129.

Martin, J. H., and Fitzwater, S. E. 1988. Iron deficiency limits phytoplankton growth in the north-east Pacific subarctic. *Nature* 331: 341–343.

Martin, P. S. 1966. Africa and Pleistocene overkill. *Nature* 212: 339–342.

Martin, P. S. 1973. The discovery of America. *Science* 179: 969–974.

Mascarelli, A. 2014. Designer reefs. *Nature* 508: 444–446.

Matson, S. E. 2012. West coast groundfish IFQ fishery catch summary for 2011: first look. *Supplemental NMFS Report March* 2012.

Matthes, U., and Larson, D. W., 2006. Microsite and climatic controls of tree population dynamics: an 18-year study on cliffs. *Journal of Ecology* 94: 402–414.

May, R. M. 1973. *Stability and Complexity in Model Ecosystems*, Vol. 6. Princeton University Press.

May, R. M. 1983. Parasitic infections as regulators of animal populations: the dynamic relationship between parasites and their host populations offers clues to the etiology and control of infectious disease. *American Scientist* 71: 36–45.

May, R. M., and Godfrey, J. 1994. Biological diversity: differences between land and sea [and discussion]. *Philosophical Transactions of the Royal Society of London Series B: Biological Sciences* 343: 105–111.

Maynard, J. A., Anthony, K. R. N., Marshall, P. A., and Masiri, I. 2008. Major bleaching events can lead to increased thermal tolerance in corals. *Marine Biology* 155: 173–182.

Mayr, E. 1942. *Systematics and the Origin of Species*. New York: Columbia University Press.

McClanahan, T. T., Ateweberhan, M., Muhando, C. A., Maina, J., and Mohammed, M. S. 2007. Effects of climate and seawater temperature variation on coral bleaching and mortality. *Ecological Monographs* 77: 503–525.

McGraw, J. B., and Furedi, M. A. 2005. Deer browsing and population viability of a forest understory plant. *Science* 307: 920–922.

McGrew, W. C. 1974. Tool use by wild chimpanzees in feeding upon driver ants. *Journal of Human Evolution* 3: 501–508.

McGrew, W. C. 1992. *Chimpanzee Material Culture: Implications for Human Evolution*. Cambridge: Cambridge University Press.

McGroddy, M., Daufresne, T., and Hedin, L. 2004. Scaling of C:N:P stoichiometry in forests worldwide: implications of terrestrial Redfield-type ratios. *Ecology* 85: 2390–2401.

McLoughlin, S. 2001. The breakup history of Gondwana and its impact on pre-Cenozoic floristic provincialism. *Australian Journal of Botany* 49: 271–300.

McNatty, A., Abbott, K. L., and Lester, P. J. 2009. Invasive ants compete with and modify the trophic ecology of hermit crabs on tropical islands. *Oecologia* 160: 187–194.

McNaughton, S. J. 1976. Serengeti migratory wildebeest: facilitation of energy flow by grazing. *Science* 191: 92–94.

McNickle, G. G., and Cahill, J. F., Jr. 2009. Plant root growth and the marginal value theorem. *Proceedings of the National Academy of Sciences* 106(12): 4747–4751.

McNutt, M. K., and 7 others. 2012. Review of flow rate estimates of the *Deepwater Horizon* oil spill. *Proceedings of the National Academy of Sciences* 109(50): 20260–20267.

Mee, L. 2006. Reviving dead zones. *Scientific American* 295: 78–85.

Mee, L. D., Friedrich, J., and Gomoiu, M. T. 2005. Restoring the Black Sea in times of uncertainty. *Oceanography* 18: 100–111.

Menge, B. 1972. Competition for food between two intertidal starfish species and its effect on body size and feeding. *Ecology* 53: 635–644.

Menge, B. 1975. Brood or broadcast? The adaptive significance of different reproductive strategies in the two intertidal sea stars *Leptasterias hexactis* and *Pisaster ochraceus*. *Marine Biology* 31: 87–100.

Menge, B. A. 1976. Organization of the New England rocky intertidal community: role of predation, competition, and environmental heterogeneity. *Ecological Monographs* 46(4): 355–393.

Menge, B. A., and Lubchenco, J. 1981. Community organization in temperate and tropical rocky intertidal habitats: prey refuges in relation to consumer pressure gradients. *Ecological Monographs* 51: 429–450.

Menge, J. L., and Menge, B. A. 1974. Role of resource allocation, aggression and spatial heterogeneity in coexistence of two competing intertidal starfish. *Ecological Monographs* 44: 189–209.

Menge, B. A., Berlow, E. L., Blanchette, C. A., Navarrete, S. A., and Yamada, S. B. 1994. The keystone species concept: variation in interaction strength in a rocky intertidal habitat. *Ecological Monographs* 64: 249–286.

Menge, B., and 5 others. 2002. Inter-hemispheric comparison of bottom-up effects on community structure: Insights revealed using the comparative-experimental approach. *Ecological Research* 17: 1–16.

Mercado-Silva, N., Olden, J. D., Maxted, J. T., Hrabik, T. R., and Zanden, M. J. V. 2006. Forecasting the spread of invasive rainbow smelt in the Laurentian Great Lakes region of North America. *Conservation Biology* 20: 1740–1749.

Michaelidis, C. I., Demary, K. C., and Lewis, S. M. 2006. Male courtship signals and female signal assessment in *Photinus greeni* fireflies. *Behavioral Ecology* 17: 329–335.

Mieth, A., and Bork, H. R. 2010. Humans, climate or introduced rats: which is to blame for the woodland destruction on prehistoric Rapa Nui (Easter Island)? *Journal of Archaeological Science* 37: 417–426.

Migge, A., Kahmann, U., Fock, H. P., and Becker, T. W. 1999. Prolonged exposure of tobacco to a low oxygen atmosphere to suppress photorespiration decreases net photosynthesis and results in changes in plant morphology and chloroplast structure. *Photosynthetica* 36: 107–116.

Millennium Ecosystem Assessment. 2005. *Ecosystems and Human Well-being: Biodiversity Synthesis*. Washington DC: World Resources Institute.

Mills, L. S. 2007. *Conservation of Wildlife Populations*. Malden, MA: Blackwell Publishing.

Mitchell, C. E., and Power, A. G. 2003. Release of invasive plants from fungal and viral pathogens. *Nature* 421: 625–627.

Moisander, P. H., and 7 others. 2010. Unicellular cyanobacterial distributions broaden the oceanic N_2 fixation domain. *Science* 327: 1512–1514.

Molina, M. J., and Rowland, F. S. 1974. Stratospheric sink for chlorofluoromethanes: chlorine atom catalysed destruction of ozone. *Nature* 249: 810–812.

Monarch Watch (2015) http://www.monarchwatch.org/.

Mondal, M. S., and Wasimi, S. A. 2006. Generating and forecasting monthly flows of the Ganges River with OPAR model. *Journal of Hydrology* 323: 41–56.

Montoya, J. P., Voss, M., Kahler, P., and Capone, D. G. 1996. A simple, high-precision, high-sensitivity tracer assay for N_2 fixation. *Applied and Environmental Microbiology* 62: 986–993.

Morell, V. 2014. Science behind plan to ease wolf protection is flawed, panel says. *Science* 343: 719.

Morford, S. L., Houlton, B. Z., and Dahlgren, R. A. 2011. Increased forest ecosystem carbon and nitrogen storage from nitrogen rich bedrock. *Nature* 477: 78–81.

Morin, P. A., and 15 others. 2010. Complete mitochondrial genome phylogeographic analysis of killer whales (*Orcinus orca*) indicates multiple species. *Genome Research* 20: 908–916.

Moss, R. H., and 18 others. 2010. The next generation of scenarios for climate change research and assessment. *Nature* 463: 747–756.

Mouritsen, K. N. 2002. The parasite-induced surfacing behaviour in the cockle *Austrovenus stutchburyi*: a test of an alternative hypothesis and identification of potential mechanisms. *Parasitology* 124: 521–528.

Mouritsen, K. N., and Poulin, R. 2003. Parasite-induced trophic facilitation exploited by a non-host predator: a manipulator's nightmare. *International Journal for Parasitology* 33: 1043–1050.

Mulder, C., and Elser, J. J. 2009. Soil acidity, ecological stoichiometry and allometric scaling in grassland food webs. *Global Change Biology* 15: 2730–2738.

Mulholland, P. J., and 30 others. 2008. Stream denitrification across biomes and its response to anthropogenic nitrate loading. *Nature* 452: 202–205.

Mumme, R. L. 1992. Do helpers increase reproductive success? An experimental analysis in the Florida scrub jay. *Behavioral Ecology and Sociobiology* 31: 319–328.

Murray, C. M., Eberly, L. E., and Pusey, A. E. 2006. Foraging strategies as a function of season and rank among wild chimpanzees (*Pan troglodytes*). *Behavioral Ecology* 17: 1020–1028.

Murray, C. M., Gilby, I. C., Mane, S. V., and Pusey, A. E. 2008. Adult male chimpanzees inherit maternal ranging patterns. *Current Biology* 18: 20–24.

Murray, C. M., and 6 others. 2014. Early social exposure in wild chimpanzees: mothers with sons are more gregarious than mothers with daughters. *Proceedings of the National Academy of Sciences* 111(51), 18189–18194.

Nams, M. L., and Freedman, B. 1987. Ecology of heath communities dominated by *Cassiope tetragona* at Alexandra Fiord, Ellesmere Island, Canada. *Ecography* 10: 22–32.

Newman, D., and Pilson, D. 1997. Increased probability of extinction due to decreased genetic effective population size: experimental populations of *Clarkia pulchella*. *Evolution* 51: 354–362.

Nieminen, M., Siljander, M., and Hanski, I. 2004. Structure and dynamics of *Melitaea cinxia* metapopulations. In P. R. Ehrlich and I. Hanski (eds), *On the Wings of Checkerspots: A Model System for Population Biology*. New York: Oxford University Press, pp. 63–91.

Nilsson, M. A., Arnason, U., Spencer, P. B. S., and Janke, A. 2004. Marsupial relationships and a timeline for marsupial radiation in South Gondwana. *Gene* 340: 189–196.

Nilsson, M. A., and 6 others. 2010. Tracking marsupial evolution using archaic genomic retroposon insertions. *PLoS Biology* 8: e1000436.

Normile, D. 2008. Driven to extinction. *Science* 219: 1606–1609.

Norton-Griffiths, M. 1979. The influence of grazing, browsing, and fire on the vegetation dynamics of the Serengeti. In A. R. E. Sinclair and M. Norton-Griffiths (eds), *Serengeti: Dynamics of an Ecosystem*. Chicago, IL: University of Chicago Press, pp. 310–352.

Novotny, V., and 7 others. 2005. An altitudinal comparison of caterpillar (Lepidoptera) assemblages on *Ficus* trees in Papua New Guinea. *Journal of Biogeography* 32: 1303–1314.

Novotny, V., and 6 others. 2006. Why are there so many species of herbivorous insects in tropical rain forests? *Science* 313: 1115–1118.

Novotny, V., and 15 others. 2007. Low beta diversity of herbivorous insects in tropical forests. *Nature* 448: 692–695.

Nowak, M. A., and Sigmund, K. 2005. Evolution of indirect reciprocity. *Nature* 437: 1291–1296.

Nunes, A. L., Orizaola, G., Laurila, A., and Rebelo, R. 2014. Rapid evolution of constitutive and inducible defenses against an invasive predator. *Ecology* 95: 1520–1530.

Nuñez, M. A., Horton, T. R., and Simberloff, D. 2009. Lack of belowground mutualisms hinders Pinacae invasions. *Ecology* 90: 2352–2359.

Nunn, C. L., and Altizer, S. M. 2005. The global mammal parasite database: an online resource for infectious disease records in wild primates. *Evolutionary Anthropology* 14: 1–2.

O'Riain, M. J., Jarvis, J. U., and Faulkes, C. G. (1996). A dispersive morph in the naked mole-rat. *Nature* 380: 619–621.

Oberhauser, K., and Peterson, A. T. 2003. Modeling current and future potential wintering distributions of eastern North American monarch butterflies. *Proceedings of the National Academy of Sciences* 100: 14063–14068.

Odum, E. P. 1964. The new ecology. *Bioscience* 14: 14–16.

Odum, H. T. 1957. Trophic structure and productivity of Silver Springs, Florida. *Ecological Monographs* 27: 55–112.

Ogada, D. L., Gadd, M. E., Ostfeld, R. S., Young, T. P., and Keesing, F. 2008. Impacts of large herbivorous mammals on bird diversity and abundance in an African savanna. *Oecologia* 156: 387–397.

Oki, T. and Kanae, S. 2006. Global hydrological cycles and world water resources. *Science* 313: 1068–1072.

Oring, L. W., Fleischer, R. C., Reed, J. M., and Marsden, K. E. 1992. Cuckoldry through stored sperm in the sequentially polyandrous spotted sandpiper. *Nature* 359: 631–633.

Orliac, C. 2000. The woody vegetation of Easter Island between the early 14th and the mid-17th centuries AD. In C. Stevenson and W. Ayres (eds), *Easter Island Archaeology: Research on Early Rapanui Culture*. Los Osos, California: Easter Island Foundation, pp. 211–220.

Orr, J., and 26 others. 2005. Anthropogenic ocean acidification over the twenty-first century and its impact on calcifying organisms. *Nature* 437: 681–686.

Orrock, J. L., Witter, M. S., and Reichman, O. J. 2008. Apparent competition with an exotic plant reduces native plant establishment. *Ecology* 89: 1168–1174

Ozgul, A., and 7 others. 2010. Coupled dynamics of body mass and population growth in response to environmental change. *Nature* 466: 482–486.

Pachauri, R. K., and Reisinger, A. 2008. *Climate change 2007. Synthesis report. Contribution of working groups I, II and III to the fourth assessment report*. Geneva, Switzerland: IPCC.

Packer, A. P., and Clay, K. 2000. Soil pathogens and spatial patterns of seedling mortality in a temperate tree. *Nature* 404: 278–281.

Paine, R. T. 1966. Food web complexity and species diversity. *American Naturalist* 100: 65–75.

Paine, R. T. 1969. A note on trophic complexity and community stability. *American Naturalist* 103: 91–93.

Paine, R. T. 1992. Food-web analysis through field measurement of per capita interaction strength. *Nature* 355: 73–75.

Palumbi, S. R., Barshis, D. J., Traylor-Knowles, N., and Bay, R. A. 2014. Mechanisms of reef coral resistance to future climate change. *Science* 344: 895–898.

Pan, Y., and 17 others. 2011. A large and persistent carbon sink in the world's forests. *Science* 333: 988–993.

Pardonnet, S., Beck, H., Milberg, P., and Bergman, K. O. 2013. Effect of tree-fall gaps on fruit-feeding nymphalid butterfly assemblages in a Peruvian rain forest. *Biotropica* 45: 612–619.

Parker, G. A. 1970. Sperm competition and its evolutionary effect on copula duration in the fly *Scatophaga stercoraria*. *Journal of Insect Physiology* 16: 1301–1328.

Parker, G. A. 2001. Golden flies, sunlit meadows: a tribute to the yellow dungfly. In L. A. Dugatkin (ed.), *Model Systems in Behavioral Ecology: Integrating Conceptual, Theoretical and Empirical Approaches*. Princeton, NJ: Princeton University Press, pp. 3–26.

Parker, G. A., and Stuart, R. A. 1976. Animal behavior as a strategy optimizer: evolution of resource assessment strategies and optimal emigration thresholds. *American Naturalist* 110: 1055–1076.

Parks, S. E., Warren, J. D., Stamieszkin, K., Mayo, C. A., and Wiley, D. 2012. Dangerous dining: surface foraging of North Atlantic right whales increases risk of vessel collisions. *Biology Letters* 8: 57–60.

Parmesan, C., and Yohe, G. 2003. A globally coherent fingerprint of climate change impacts across natural systems. *Nature* 421: 37–42.

Parolin, P. 2006. Ombrohydrochory: rain-operated seed dispersal in plants: with special regard to jet action dispersal in Aizoaceae. *Flora* 201: 511–518.

Parton, W., and 10 others. 2007. Global-scale similarities in nitrogen release patterns during long-term decomposition. *Science* 315: 361–364.

Paulmier, A., and Ruiz-Pino, D. 2009. Oxygen minimum zones (OMZs) in the modern ocean. *Progress in Oceanography* 80: 113–128.

Pearson, O. P. 1954. Habits of the lizard *Liolaemus multiformis* at high altitudes in southern Peru. *Copeia* 1954: 111–116.

Peay, K. G., Bruns, T. D., Kennedy, P. G., Bergemann, S. E., and Garbelotto, M. 2007. A strong species–area relationship for eukaryotic soil microbes: Island size matters for ectomycorrhizal fungi. *Ecology Letters* 10: 470–480.

Peay, K. G., Garbelotto, M., and Bruns, T. D. 2010. Evidence of dispersal limitation in soil microorganisms: isolation reduces species richness on mycorrhizal tree islands. *Ecology* 91: 3631–3640.

Peckarsky, B. L., and 10 others. 2008. Revisiting the classics: considering nonconsumptive effects in textbook examples of predator–prey interactions. *Ecology* 89: 2416–2425.

Pennisi, E. 2014. Baboon watch. *Science* 346: 292–295.

Perry, A. L., Low, P. J., Ellis, J. R., and Reynolds, J. D. 2005. Climate change and distribution shifts in marine fishes. *Science* 308: 1912–1915.

Peterson, D. 2006. *Jane Goodall: The Woman who Redefined Man*. Boston, MA: Houghton Mifflin Co.

Peterson, R. O., Vucetich, J. A., Bump, J. M., and Smith, D. W. 2014. Trophic cascades in a multicausal world: Isle Royale and Yellowstone. *Annual Review of Ecology, Evolution, and Systematics* 45: 325–345.

Petit, S., Moilanen, A., Hanski, I., and Baguette, M. 2001. Metapopulation dynamics of the bog fritillary butterfly: movements between habitat patches. *Oikos* 92: 491–500.

Petren, K., Grant, B. R., and Grant, P. R. 1999. A phylogeny of Darwin's finches based on microsatellite DNA length variation. *Proceedings of the Royal Society B* 266: 321–329.

Pettis, J. S., vanEngelsdorp, D., Johnson, J., and Dively, G. 2012. Pesticide exposure in honey bees results in increased levels of the gut pathogen *Nosema*. *Naturwissenschaften* 99: 153–158.

Pfennig, D. 1990. The adaptive significance of an environmentally-cued developmental switch in an anuran tadpole. *Oecologia* 85: 101–107.

Pfennig, D. W. 1992a. Proximate and functional causes of polyphenism in an anuran tadpole. *Functional Ecology* 6: 167–174.

Pfennig, D. W. 1992b. Polyphenism in spadefoot toad tadpoles as a logically adjusted evolutionarily stable strategy. *Evolution* 46: 1408–1420.

Phillips, B. L., Brown, G. P., Webb, J. K., and Shine, R. 2006. Invasion and the evolution of speed in toads. *Nature* 439: 803.

Pianka, E. R. 1970. On *r*- and *K*-selection. *American Naturalist* 104: 592–597.

Pimm, S. L., and 8 others. 2014. The biodiversity of species and their rates of extinction, distribution, and protection. *Science* 344: 987.

Plowright, W., and McCulloch, G. 1967. Investigations on the incidence of rinderpest virus induction in game animals of N. Tanganyika and S. Kenya 1960/63. *Journal of Hygiene* 65: 343–358.

Post, D. M. 2002. Using stable isotopes to estimate trophic position: models, methods and assumptions. *Ecology* 83: 703–718.

Poulter, B., and 12 others. 2014. Contribution of semi-arid ecosystems to interannual variability of the global carbon cycle. *Nature* 509: 600–603.

Poveda K., Steffan-Dewenter, I., Scheuand, S., and Tscharntke, T. 2007. Plant-mediated interactions between below- and aboveground processes: decomposition, herbivory, parasitism and pollination. In T. Ohgushi, T. Craig, and P. Price (eds), *Biological Communities: Plant Mediation in Indirect Interaction Webs*. Cambridge: Cambridge University Press, pp. 147–163.

Power, M. E., and 9 others. 1996. Challenges in the quest for keystones. *Bioscience* 46: 609–620.

Pöyry, J., Paukkunen, J., Heliölä, J., and Kuussaari, M. 2009. Relative contributions of local and regional factors to species richness and total density of butterflies and moths in semi-natural grasslands. *Oecologia* 160: 577–587.

Pride, R. E. 2005. Optimal group size and seasonal stress in ring-tailed lemurs (*Lemur catta*). *Behavioral Ecology* 16: 550–560.

Pringle, R. M. 2005. The origins of the Nile perch in Lake Victoria. *Bioscience* 55: 780–787.

Pusey, A. E. 1980. Inbreeding avoidance in chimpanzees. *Animal Behaviour* 28: 543–552.

Pusey, A. E. 1983. Mother–offspring relationships in chimpanzees after weaning. *Animal Behaviour* 31: 363–377.

Pusey, A. E., Williams, J., and Goodall, J. 1997. The influence of dominance rank on the reproductive success of female chimpanzees. *Science* 277: 828–831.

Pusey, A. E., Pintea, L., Wilson, M. L., Kamenya, S., and Goodall, J. 2007. The contribution of long-term research at Gombe National Park to chimpanzee conservation. *Conservation Biology* 21: 623–634.

Pusey, A., and 5 others. 2008. Severe aggression among female *Pan troglodytes schweinfurthii* at Gombe National Park, Tanzania. *International Journal of Primatology* 29: 949–973.

Radville, L., Gonda-King, L., Gómez, S., Kaplan, I., and Preisser, E. L. 2014. Are exotic herbivores better competitors? A meta-analysis. *Ecology* 95: 30–36.

Ramalho, C. E., Laliberté, E., Poot, P., and Hobbs, R. J. 2014. Complex effects of fragmentation on remnant woodland plant communities of a rapidly urbanizing biodiversity hotspot. *Ecology* 95: 2466–2478.

Redman, R. S., Sheehan, K. B., Stout, R. G., Rodriguez, R. J., and Henson, J. M. 2002. Thermotolerance generated by plant/fungal symbiosis. *Science* 298: 1581.

Reeve, H. K., Westneat, D. F., Noon, W. A., Sherman, P. W., and Aquadro, C. F. 1990. DNA "fingerprinting" reveals high levels of inbreeding in colonies of the eusocial naked mole rat. *Proceedings of the National Academy of Sciences* 87: 2496–2500.

Regehr, E. V., Hunter, C. M., Caswell, H., Armstrup, S. C., and Stirling, I. 2010. Survival and breeding of polar bears in the southern Beaufort Sea in relation to sea ice. *Journal of Animal Ecology* 79: 117–127.

Reimchen, T. E. 1992. Injuries on stickleback from attacks by a toothed predator (*Oncorhynchus*) and implications for the evolution of lateral plates. *Evolution* 46: 1224–1230.

Reimchen, T. E. 2000. Predator handling failures of lateral plate morphs in *Gasterosteus aculeatus*: functional implications for the ancestral plate condition. *Behaviour* 137: 1081–1096.

Richter-Boix, A., Orizaola, G., and Laurila, A. 2014. Transgenerational phenotypic plasticity links breeding phenology with offspring life-history. *Ecology* 95: 2715–2722.

Ries, L., Taron, D. J. Rendon-Salinas, E., and Oberhauser, K. S. 2015. Connecting eastern monarch population dynamics across their migratory cycle. In K. S. Oberhauser, K. R. Nail, and S. M. Altizer (eds), *Monarchs in a Changing World: Biology and Conservation of an Iconic Butterfly*. Ithaca, NY: Cornell University Press, pp. 268–282.

Ripple, W. J., and Beschta, R. L. 2004. Wolves and the ecology of fear: can predation risk structure ecosystems? *Bioscience* 54: 755–766.

Ripple, W. J., and Beschta, R. L. 2006. Linking wolves to willows via risk-sensitive foraging by ungulates in the northern Yellowstone ecosystem. *Forest Ecology and Management* 230: 96–106.

Ripple, W. J., and 13 others. 2014. Status and ecological effects of the world's largest carnivores. *Science* 343: 1241484.

Rodriguez, R. J., and Roossinck, M. 2012. Viruses, fungi and plants: cross-kingdom communication and mutualism. In G. Witzany (ed.), *Biocommunication of Fungi*. Dordrecht, the Netherlands: Springer, pp. 219–227.

Romme, W. H., and Turner, M. G. 2004. Ten years after the 1988 Yellowstone fires: is restoration needed? In L. L. Wallace (ed.), *After the Fires: The Ecology of Change in Yellowstone National Park*. New Haven, CT: Yale University Press, pp. 318–361.

Rossiter, M. C., Schultz, J. C., and Baldwin, I. T. 1988. Relationships among defoliation, red oak phenolics and gypsy moth growth and reproduction. *Ecology* 69: 267–277.

Rubidge, E. M., and 5 others. 2012. Climate-induced range contraction drives genetic erosion in an alpine mammal. *Nature Climate Change* 2: 285–288.

Rudicell, R. S. and 19 others. 2010. Impact of simian immunodeficiency virus infection on chimpanzee population dynamics. *PLoS Pathogens* 6: e1001116.

Russell, R. W., May, M. L., Soltesz, K. L., and Fitzpatrick, J. W. 1998. Massive swarm migrations of dragonflies (Odonata) in Eastern North America. *American Midland Naturalist* 140: 325–342.

Saab, I. N., Sharp, R. E., Pritchard, J., and Voetberg, G. S. 1990. Increased endogenous abscisic acid maintains primary root growth and inhibits shoot growth of maize seedlings at low water potentials. *Plant Physiology* 93: 1329–1336.

Sabo, J. L., Finlay, J. C., Kennedy, T., and Post, D. M. 2010. The role of discharge variation in scaling of drainage area and food chain length in rivers. *Science* 330: 965–967.

Sadovy, Y., and Cheung, W. L. 2003. Near extinction of a highly fecund fish: the one that nearly got away. *Fish and Fisheries* 4: 86–99.

Sage, R. F. 2004. The evolution of C_4 photosynthesis. *New Phytologist* 161: 341–370.

Sage, R. F., and Monson, R. K. 1999. C_4 *Plant Biology*. San Diego, CA: Academic Press.

Sakata, Y., Yamasaki, M., Isagi, Y., and Ohgushi, T. 2014. An exotic herbivorous insect drives the evolution of resistance in the exotic perennial herb. *Ecology* 95: 2569–2578.

Sanabria, E. A., Quiroga, L. B., and Martino, A. L. 2012. Seasonal changes in the thermal tolerance of the toad *Rhinella arenarum* (Bufonidae) in the Monte Desert of Argentina. *Journal of Thermal Biology* 37: 400–412.

Sanderson, E. W., and 5 others. 2002. The human footprint and the last of the wild. *Bioscience* 52: 891–904.

Saugier, B., Roy, R., and Mooney, H. A. 2001. Estimations of global terrestrial productivity: converging toward a single number? In J. Roy, B. Saugier, and H. A. Mooney (eds), *Terrestrial Global Productivity*. San Diego, CA: Academic Press, pp. 543–558.

Schaffer, W. M., and Rosenzweig, M. L. 1977. Selection for optimal life histories II: multiple equilibria and the evolution of alternative reproduction strategies. *Ecology* 58: 60–72.

Schaffer, W. M., and Schaffer, M. V. 1977. The adaptive significance of variations in reproductive habit in the Agavaceae. In B. Stonehouse and C. M. Perrins (eds), *Evolutionary Ecology*. London: Macmillan Press, pp. 261–276.

Schaller, G. B. 1972. *The Serengeti Lion*. Chicago, IL: University of Chicago Press.

Scheffer, M., and Carpenter, S. R. 2003. Catastrophic regime shifts in ecosystems: linking theory to observation. *Trends in Ecology and Evolution* 18: 648–656.

Schindler, D. E., Carpenter, S. R., Cole, J. J., and Kitchell, J. F. 1997. Influence of food web structure on carbon exchange between lakes and the atmosphere. *Science* 277: 248–251.

Schindler, D. W., and Fee, E. J. 1974. Experimental lakes area: whole lake experiments in eutrophication. *Journal of the Fisheries Research Board of Canada* 31: 937–953.

Schlesinger, W. H., and Bernhardt, E. S. 2013. *Biogeochemistry: An Analysis of Global Change*. 3rd edn. Oxford, UK: Elsevier.

Schluter, D. 1993. Adaptive radiation in sticklebacks: size, shape, and habitat use efficiency. *Ecology* 73: 699–709.

Schmidt-Nielsen, K. 1997. *Animal Physiology: Adaptation and Environment*. 5th edn. Cambridge: Cambridge University Press.

Schmitz, O. J. 2004. Perturbation and abrupt shift in trophic control of biodiversity and productivity. *Ecology Letters* 7: 403–409.

Schnitzer, S. A., and Carson, W. P. 2001. Treefall gaps and the maintenance of species diversity in a tropical forest. *Ecology* 82: 913–919.

Schoennagel, T., Turner, M. G., and Romme, W. H. 2003. The influence of fire interval and serotiny on postfire lodgepole pine density in Yellowstone National Park. *Ecology* 84: 2967–2978.

Schultz, J. C., and Appel, H. M. 2004. Cross-kingdom cross-talk: hormones shared by plants and their insect herbivores. *Ecology* 85: 70–77.

Schuur, E. A. G. 2001. The effect of water on decomposition dynamics of mesic to wet Hawaiian montane forests. *Ecosystems* 4: 259–273.

Schuur, E. A. G. 2003. Productivity and global climate revisited: the sensitivity of tropical forest growth to precipitation. *Ecology* 84: 1165–1170.

Schuur, E. A. G., and Matson, P. A. 2001. Net primary productivity and nutrient cycling across a mesic to wet precipitation gradient in Hawaiian montane forest. *Oecologia* 128: 431–442.

Seddon, P. J., Griffiths, C. J., Soorae, P. S., and Armstrong, D. P. 2014. Reversing defaunation: restoring species in a changing world. *Science* 345: 406–412.

Seitzinger, S., and 7 others. 2006. Denitrification across landscapes and waterscapes: a synthesis. *Ecological Applications* 16: 2064–2090.

Selye, H. 1936. Thymus and adrenals in the response of the organism to injuries and intoxications. *British Journal of Experimental Pathology* 17: 234–248.

Seymour, R. S. 2004. Dynamics and precision of thermoregulatory responses of eastern skunk cabbage *Symplocarpus foetidus*. *Plant, Cell and Environment* 27: 1014–1022.

Sheldon, K. S., and Tewksbury, J. J. 2014. The impact of seasonality in temperature on thermal tolerance and elevational range size. *Ecology* 95: 2134–2143.

Sheriff, M. J., Krebs, C. J., and Boonstra, R. 2009. The sensitive hare: sublethal effects of predator stress on reproduction in snowshoe hares. *Journal of Animal Ecology* 78(6): 1249–1258.

Sherman, P. W. 1980. The limits of ground squirrel nepotism. In G. W. Barlow and J. Silverberg (eds), *Sociobiology: Beyond Nature/Nurture?* Washington DC: Westview Press, pp. 505–544.

Sibley, D. A., Bevier, L. R., Patten, M. A., and Elphick, C. S. 2006. Comment on "Ivory-billed woodpecker (*Campephilus principalis*) persists in Continental North America." *Science* 311: 1555.

Siemann, E., and Rogers, W. E. 2001. Genetic differences in growth of an invasive tree species. *Ecology Letters* 4: 514–518.

Siemann, E., and Rogers, W. E. 2003. Reduced resistance of invasive varieties of the alien tree *Sapium sebiferum* to a generalist herbivore. *Oecologia* 135: 451–457.

Silverstein, S. 1964. *The Giving Tree*. New York: HarperCollins Publishers.

Simberloff, D. S. 1976. Experimental zoogeography of islands: effects of island size. *Ecology* 57: 629–648.

Simberloff, D. S., and Wilson, E. O. 1969. Experimental zoogeography of islands: the colonization of empty islands. *Ecology* 50: 278–296.

Simberloff, D. S., and Wilson, E. O. 1970. Experimental zoogeography of islands: a two-year record of colonization. *Ecology* 51: 934–937.

Sinclair, A. R. E. 1973. Population increases of buffalo and wildebeest in the Serengeti. *East African Wildlife Journal* 11: 93–107.

Sinclair, A. R. E. 1975. The resource limitation of trophic levels in tropical grassland ecosystems. *Journal of Animal Ecology* 44: 497–520.

Sinclair, A. R. E. 1977a. *The African Buffalo: A Study of Resource Limitation of Populations*. Chicago, IL: University of Chicago Press.

Sinclair, A. R. E. 1977b. The eruption of the ruminants. In A. R. E. Sinclair and M. Norton-Griffiths (eds), *Serengeti: Dynamics of an Ecosystem*. Chicago, IL: University of Chicago Press, pp. 82–103.

Sinclair, A. R. E., Dublin, H., and Borner, M. 1985. Population regulation of Serengeti wildebeest: a test of the food hypothesis. *Oecologia* 65: 266–268.

Sinclair, A. R. E., Chitty, D., Stefan, C. I., and Krebs, C. J. 2003a. Mammal population cycles: evidence for intrinsic differences during snowshoe hare cycles. *Canadian Journal of Zoology* 81: 216–220.

Sinclair, A. R. E., Mduma, S., and Brashares, J. S. 2003b. Patterns of predation in a diverse predator–prey system. *Nature* 425: 288–290.

Sinclair, A. R. E., and 5 others. 2007. Long-term ecosystem dynamics in the Serengeti: lessons for conservation. *Conservation Biology* 21: 580–590.

Sinclair, A. R. E., and 5 others. 2008. Historical and future changes to the Serengeti ecosystem. In A. R. E. Sinclair, C. Packer, A. A. R. Mduma, and J. M. Fryxell (eds), *Serengeti III: Human Impacts on Ecosystem Dynamics*. Chicago, IL: University of Chicago Press, pp. 7–46.

Sinclair, A. R. E., and 17 others. 2013. Asynchronous food-web pathways could buffer the response of Serengeti predators to El Niño Southern Oscillation. *Ecology* 94: 1123–1130.

Slobodkin, L. B., Smith, F. E., and Hairston, N. G. 1967. Regulation in terrestrial ecosystems, and the implied balance of nature. *American Naturalist* 101: 109–124.

Smith, D. W, Peterson, R. O., and Houston, D. B. 2003. Yellowstone after wolves. *Bioscience* 53: 330–340.

Society for Conservation Biology. 2010. 2011–2015 SCB Strategic Plan: Enhancing the impact of conservation science. http://www.conbio.org/images/content_about_scb/2011SCBStrategicPlan_Branded_edited.pdf.

Solomon, S. Garcia, R. R., Rowland, F. S., and Wuebbles, D. J. 1986. On the depletion of Antarctic ozone. *Nature* 321: 755–758.

Soto, K. H, and Trites, A. W. 2011. South American sea lions in Peru have a lek-like mating system. *Marine Mammal Science* 27: 306–333.

Sousa, W. P. 1979. Disturbance in marine intertidal boulder fields: the nonequilibrium maintenance of species diversity. *Ecology* 60: 1225–1239.

Spottiswoode, C. N., and Stevens, M. 2012. Host–parasite arms races and rapid changes in bird egg appearance. *American Naturalist* 179: 633–648.

Springer, A. M., and 7 others. 2003. Sequential megafaunal collapse in the north Pacific Ocean: an ongoing legacy of industrial whaling? *Proceedings of the National Academy of Sciences* 100: 12223–12228.

Stachowicz, J. J., and Whitlatch, R. B. 2005. Multiple mutualists provide complementary benefits to their seaweed host. *Ecology* 86: 2418–2427.

Standen, E. M., Du, T. Y., and Larsson, H. C. 2014. Developmental plasticity and the origin of tetrapods. *Nature* 513: 54–58.

Stanford, C. B. 1995. The influence of chimpanzee predation on group size and anti-predator behavior in red colobus monkeys. *Animal Behaviour* 49: 577–587.

Stanford, C. B. 1998. *Chimpanzee and Red Colobus: The Ecology of Predator and Prey*. Cambridge, MA: Harvard University Press.

Stanley, D. W. 1999. *Eicosanoids in Invertebrate Signal Transduction Systems*. Princeton, NJ: Princeton University Press.

Stastny, M., and Agrawal, A. A. 2014. Love thy neighbour? Reciprocal impacts between plant community structure and insect herbivory in co-occurring Asteraceae. *Ecology* 95: 2904–2914.

Steenbergh, W. F., and Lowe, C. H. 1977. *Ecology of the Saguaro: II*. National Park Service Monograph Series Number 8. Washington DC: US Government Printing Office.

Steiner, C. F., Long, Z. T., Krumins, J. A., and Morin, P. J. 2006. Population and community resilience in multitrophic communities. *Ecology* 87: 996–1007.

Steinhaus, E. A. 1958. Crowding as a possible stress factor in insect disease. *Ecology* 39: 503–514.

Stephan C. I., and Krebs, C. J. 2001. Reproductive changes in a cyclic population of snowshoe hares. *Canadian Journal of Zoology* 79: 2101–2108.

Sterner, R. W., and Elser, J. J. 2002. *Ecological Stoichiometry: The Biology of Elements from Molecules to the Biosphere*. Princeton, NJ: Princeton University Press.

Stiling, P. S. 2002. *Ecology: Theories and Applications*. 4th edn. Upper Saddle River, NJ: Prentice Hall.

Stokstad, E. 2005. What's wrong with the Endangered Species Act? *Science* 309: 2150–2152.

Stokstad, E. 2006. Native mussel quickly evolves fear of invasive crab. *Science* 313: 245.

Stouffer, P. C., Strong, C., and Naka, L. N. 2009. Twenty years of understorey bird extinctions from Amazonian rain forest fragments: consistent trends and landscape-mediated dynamics. *Diversity and Distributions* 15: 88–97.

Stratford, J. A., and Stouffer, P. C. 1999. Local extinctions of terrestrial insectivorous birds in a fragmented landscape near Manaus, Brazil. *Conservation Biology* 13: 1416–1423.

Strous, M., and 36 others. 2006. Deciphering the evolution and metabolism of an anammox bacterium from a community genome. *Nature* 440: 790–794.

Stuart, S. N., and 6 others. 2004. Status and trends of amphibian declines and extinctions worldwide. *Science* 306: 1783–1786.

Stuart-Smith, R. D., and 17 others. 2013. Integrating abundance and functional traits reveals new global hotspots of fish diversity. *Nature* 501: 539–542.

Suess, H. E. 1955. Radiocarbon concentration in modern wood. *Science* 122: 415–417.

Sugimoto, K., and 14 others. 2014. Intake and transformation to a glycoside of (Z)-3-hexenol from infested neighbors reveals a mode of plant odor reception and defense. *Proceedings of the National Academy of Sciences*, 111: 7144–7149.

Sumich, J. L., and Morrissey, J. F. 2004. *Introduction to the Biology of Marine Life*. 8th edn. Sudbury, MA: Jones and Bartlett Publishers.

Sutcliffe, W. H. J. 1970. Relationship between growth rate and ribonucleic acid concentration in some invertebrates. *Journal of the Fisheries Research Board of Canada* 27: 606–609.

Swanson, F. J., and Major, J. J. 2005. Physical events, environments, and geological ecological interactions at Mount St. Helens: March 1980–2004. In V. H. Dale, F. J. Swanson, and C. M. Crisafulli (eds), *Ecological Responses to the 1980 Eruption of Mount St. Helens*. New York: Springer Science, pp. 27–44.

Takahashi, S., Bauwe, H., and Badger, M. 2007. Impairment of the photorespiratory pathway accelerates photoinhibition of photosystem II by suppression of repair but not acceleration of damage processes in Arabidopsis. *Plant Physiology* 144: 487–494.

Takashima, T. Hikosaka, K., and Hirose, T. 2004. Photosynthesis or persistence: nitrogen allocation in leaves of evergreen and deciduous *Quercus* species. *Plant, Cell and Environment* 27: 1047–1054.

Talbot, L. M., and Talbot, M. H. 1963. The wildebeest in western Masai-land. Wildlife Monograph No. 12. The Wildlife Society. Hobokken, NJ: Wiley & Sons.

Tanner, J. T. 1942. *The Ivory-billed Woodpecker*. Mineola, NY: Dover Publications.

Teal, J., and Teal, M. 1969. *Life and Death of the Salt Marsh*. Boston, MA: Little, Brown and Company.

Thamdrup, B. 2012. New pathways and processes in the global nitrogen cycle. *Annual Review of Ecology, Evolution, and Systematics* 43: 407–428.

Thomas, F., and Poulin, R. 1998. Manipulation of a mollusc by a trophically transmitted parasite: convergent evolution or phylogenetic inheritance? *Parasitology* 116: 431–436.

Tilman, D. 1987. The importance of the mechanisms of interspecific competition. *American Naturalist* 129: 769–774.

Tilman, D., Reich, P. B., and Knops, J. M. H. 2006. Biodiversity and ecosystem stability in a decade-long grassland experiment. *Nature* 441: 629–632.

Tracey, R. L., and Walsberg, G. E. 2002. Kangaroo rats revisited: re-evaluating a classic case of desert survival. *Oecologia* 133: 449–457.

Treu, R., and 8 others. 2014. Decline of ectomycorrhizal fungi following a mountain pine beetle epidemic. *Ecology* 95: 1096–1103.

Trites, A. W. and 29 others. 2007. Bottom-up forcing and the decline of Steller sea lions (*Eumetopias jubatus*) in Alaska: assessing the ocean climate hypothesis. *Fisheries Oceanography* 16: 46–67.

Tunnicliffe, V. 1992. Hydrothermal-vent communities of the deep sea. *American Scientist* 80: 336–349.

Turlings, T. C. J., Tumlinson, J. H., and Lewis, W. J. 1990. Exploitation of herbivore-induced plant odors by host-seeking parasitic wasps. *Science* 250: 1251–1253.

Turner, M. G., Gardner, R. H., and O'Neill, R. V. 2001. *Landscape Ecology in Theory and Practice*. New York, NY: Springer Press.

Turner, M. G., Romme, W. H., and Gardner, R. H. 1999. Prefire heterogeneity, fire severity, and early postfire plant reestablishment in subalpine forests of Yellowstone National Park, Wyoming. *International Journal of Wildland Fire* 9: 21–36.

Turner, M. G., Romme, W. H., and Tinker, D. B. 2003. Surprises and lessons from the 1988 Yellowstone fires. *Frontiers in Ecology and the Environment* 1: 351–358.

Tweddle. J. C., Turner, R. M., and Dickie, J. B. 2002. Seed Information Database (release 3.0, July 2002), http://www.rbgkew.org.uk/data/sid.

Tweddle, J. C., Dickie, J. B., Baskin, C. C., and Baskin, J. M. 2003. Ecological aspects of seed desiccation sensitivity. *Journal of Ecology* 91: 294–304.

United Nations Department of Economic and Social Affairs, Population Division. 2013. *World Population Prospects: The 2012 Revision*. Geneva: United Nations Publications.

Urquhart, F. A., and Urquhart, N. R. 1978. Autumnal migration routes of the eastern population of the monarch butterfly (*Danaus p. plexippus* L.; Danaidae; Lepidoptera) in North America to the overwintering site in the Neovolcanic Plateau of Mexico. *Canadian Journal of Zoology* 56: 1759–1764.

USDA 2013. Report on the national stakeholders conference on honey bee health. Washington DC: National Honey Bee Health Stakeholders Conference Steering Committee.

US Fish and Wildlife Service, and 12 others. 2013. Northern Rocky Mountain Wolf Recovery Program 2012 interagency annual report. M. D. Jimenez and S. A Becker (eds). Helena, MT: USFWS, Ecological Services.

Valencia, J., de La Cruz, C., Carranza, J., and Mateos, C. 2006. Parents increase their parental effort when aided by helpers in a cooperatively breeding bird. *Animal Behaviour* 71: 1021–1028.

Valentine, D. L., and 7 others. 2012. Dynamic autoinoculation and the microbial ecology of a deep water hydrocarbon irruption. *Proceedings of the National Academy of Sciences* 109: 20286–20291.

Vamosi, S. M., and Schluter, D. 2002. Impacts of trout predation on fitness of sympatric sticklebacks and their hybrids. *Proceedings of the Royal Society B* 269: 923–930.

Vamosi, J. C., and 35 others. 2006. Pollination decays in biodiversity hotspots. *Proceedings of the National Academy of Sciences* 103: 956–961.

Van der Heijden, M. G. A. 2010. Mycorrhizal fungi reduce nutrient loss from model grasslands. *Ecology* 91: 1163–1171.

van Groenigen, K. J., Osenberg, C. W., and Hungate, B. A. 2011. Increased soil emissions of potent greenhouse gases under increased atmospheric CO_2. *Nature* 475: 214–216.

Van Mooy, B. A. S., Rocap, G., Fredricks, H., Evans, C. T., and Devol, A. H. 2006. Sulfolipids dramatically decrease phosphorus demand by picocyanobacteria in oligotrophic marine environments. *Proceedings of the National Academy of Sciences* 103: 8607–8612.

van Nouhuys, S., and Hanski, I. 2002. Colonization rates and distances of a host butterfly and two specific parasitoids in a fragmented landscape. *Journal of Animal Ecology* 71: 639–650.

Van Valen, L. 1973. A new evolutionary law. *Evolutionary theory* 1: 1–30.

vanEngelsdorp, D., and 12 others. 2009. Colony collapse disorder: a descriptive study. *PLoS ONE* 4(8): e6481.

vanEngelsdorp, D., Hayes Jr., J., Underwood, R. M., and Pettis, J. 2008. A survey of honeybee colony losses in the US, fall 2007 to spring 2008. *PLoS ONE* 3: e4071.

Vermeij, G. J., and Grosberg, R. K. 2010. The great divergence: when did diversity on land exceed that in the sea? *Integrative and Comparative Biology* 50: 675–682.

Voesenek, L. A. C. J., Runders, J. H. G. M., Peeters, A. J. M., van de Steeg, H. M., and de Kroon, H. 2004. Plant hormones regulate fast shoot elongation under water: from genes to communities. *Ecology* 85: 16–27.

Volterra, V. 1926. Fluctuations in the abundance of a species considered mathematically. *Nature* 118: 558–560.

Waage, J. K. 1979. Dual function of the damselfly penis: sperm removal and transfer. *Science* 203: 916–918.

Wallace, A. R. (1962 [1876]). *The Geographical Distribution of Animals*. London: MacMillan Press. Reprinted by Hafner Pub., New York.

Warren-Rhodes, K. A., and 9 others. 2006. Hypolithic cyanobacteria, dry limit of photosynthesis and microbial ecology in the hyperarid Atacama Desert. *Microbial Ecology* 52: 389–398.

Wassenaar, L. J., and Hobson, K. A. 1998. Natal origins of migratory monarch butterflies at wintering colonies in Mexico: new isotopic evidence. *Proceedings of the National Academy of Sciences* 95: 15436–15439.

Weather Underground 1998. Weather underground. http://www.wunderground.com Atlanta, GA: The Weather Channel, LLC.

Wegener, A. (1966 [1929]). *The Origin of Continents and Oceans*. New York: Dover Reprint.

Weiner, J. 1994. *The Beak of the Finch. A Story of Evolution in Our Own Time*. New York: Alfred A. Knopf, Inc.

Weldon, A. J., and Haddad, N. M. 2005. The effects of patch shape on indigo buntings: evidence for an ecological trap. *Ecology* 86: 1422–1431.

Westbury, D. B. 2004. *Rhinanthus minor* L. *Journal of Ecology* 92: 906–927.

White, H. K., and 14 others. 2012. Impact of the *Deepwater Horizon* oil spill on a deep-water coral community in the Gulf of Mexico. *Proceedings of the National Academy of Sciences* 109: 20303–20308.

White, K. S., Barten, N. L., Crouse, S., and Crouse, J. 2014. Benefits of migration in relation to nutritional condition and predation risk in a partially migratory moose population. *Ecology* 95: 225–237.

Whitehead, A., and 11 others. 2012 Genomic and physiological footprint of the Deepwater Horizon oil spill on resident marsh fishes. *Proceedings of the National Academy of Sciences* 109: 20298–20302.

Whitehorn, P. R., O'Connor, S., Wackers, F. L., and Goulson, D. 2012. Neonicotinoid pesticide reduces bumble bee colony growth and queen production. *Science* 336: 351–352.

Whitlock, M. C., and Schluter, D. 2009. *The Analysis of Biological Data*. Greenwood Village, CO: Roberts and Company Publishers.

Whitman, W. B., Coleman, D. C., and Wiebe, W. J. 1998. Prokaryotes: the unseen majority. *Proceedings of the National Academy of Science* 95: 6578–6583.

Whittaker, R. H. 1956. Vegetation of the Great Smoky Mountains. *Ecological Monographs* 26: 1–80.

Whittaker, R. H. 1960. Vegetation of the Siskiyou mountains, Oregon and California. *Ecological Monographs* 30: 279–338.

Whittaker, R. H. 1972. Evolution and measurement of species diversity. *Taxon* 21: 213–251.

Whittaker, R. H. 1975. *Communities and Ecosystems* 2nd edn. New York: MacMillan Publishing.

Wiencke, C., Roleda, M. Y., Gruber, A., Clayton, M. N., and Bischof, K. 2006. Susceptibility of zoospores to UV radiation determines upper depth distribution limit of Arctic kelps: evidence through field experiments. *Journal of Ecology* 94: 455–463.

Wikelski, M., and 5 others. 2006. Simple rules guide dragonfly migration. *Biology Letters* 2: 325–329.

Wild, C., Jantzen, C., Struck, U., Hoegh-Guldberg, O., and Huettel, M. 2008. Biogeochemical responses following coral mass spawning on the Great Barrier Reef: pelagic-benthic coupling. *Coral Reefs* 27: 123–132.

Williams, J. M., and 5 others. 2008. Causes of death in the Kasekela chimpanzees of Gombe National Park, Tanzania. *American Journal of Primatology* 70: 766–777.

Williams, K. S., and Simon, C. 1995. The ecology, behavior, and evolution of periodical cicadas. *Annual Review of Entomology* 40: 269–295.

Williams, S., and Kay, R. 2001. A comparative test of adaptive explanation for hypsodonty in ungulates and rodents. *Journal of Mammalian Evolution* 8: 207–229.

Williamson, M., Gaston, K., and Lonsdale, W. M. 2001. The species–area relationship does not have an asymptote! *Journal of Biogeography* 28: 827–830.

Wilson, D. E., and Ruff, S. 1999. *The Smithsonian Book of North American Mammals*. Washington DC: Smithsonian Institution Press.

Wilson, E. O. 1961. The nature of the taxon cycle in the Melanesian ant fauna. *American Naturalist* 95: 169–193.

Wilson, E. O. 1994. *Naturalist*. Washington DC: Island Press.

Wilson, E. O., and Simberloff, D. S. 1969. Experimental zoogeography of islands: defaunation and monitoring techniques. *Ecology* 50: 267–278.

Wilson, M. L., and 29 others. 2014. Lethal aggression in *Pan* is better explained by adaptive strategies than human impacts. *Nature* 513: 414–417.

Winiwarter, W., Erisman, J. W., Galloway, J. N., Klimont, Z., and Sutton, M. A. 2013. Estimating environmentally relevant fixed nitrogen demand in the 21st century. *Climatic Change* 120: 889–901.

Winnie Jr., J. A. 2012. Predation risk, elk, and aspen: tests of a behaviorally mediated trophic cascade in the Greater Yellowstone ecosystem. *Ecology* 93: 2600–2614.

Winnie Jr., J. 2014. Predation risk, elk, and aspen: reply. *Ecology* 95: 2671–2674.

Witte, F., and 5 others. 2007. Differential decline and recovery of haplochromine trophic groups in the Mwanza Gulf of Lake Victoria. *Aquatic Ecosystem Health and Management* 10: 416–433.

Woodwell, G. M., Wurster. C. W., and Isaacson, P. A. 1967. DDT residues in an East Coast estuary: a case of biological concentration of a persistent insecticide. *Science* 156: 821–824.

Woolfenden, G. E., and Fitzpatrick, J. W. 1984. *The Florida Scrub Jay: Demography of a Cooperative-breeding Bird*. Princeton, NJ: Princeton University Press.

World Health Organization (2012). Global tuberculosis report 2012. Geneva: WHO. http://www.who.int/tb/publications/global_report/gtbr12_main.pdf.

Worm, B., Sandow, M., Oschlies, A., Lotze, H. K., and Myers, R. A. 2005. Global patterns of predator diversity in the open oceans. *Science* 309: 1365–1369.

Wrangham, R. W. 1975. *The behavioural ecology of chimpanzees in Gombe National Park, Tanzania*. PhD thesis, Cambridge University.

Wroblewski, E. E., and 5 others. 2009. Male dominance rank and reproductive success in chimpanzees, *Pan troglodytes schweinfurthii*. *Animal Behaviour* 77: 873–885.

Xie, J., Li, Y., Zhai, C., Li, C., and Lan, Z. 2009. CO_2 absorption by alkaline soils and its implication to the global carbon cycle. *Environmental Geology* 56: 953–961.

Yamori, W., Noguchi K., and Terashima, I. 2005. Temperature acclimation of photosynthesis in spinach leaves: analyses of photosynthetic components and temperature dependencies of photosynthetic partial reactions. *Plant, Cell and Environment* 28: 536–547.

Yoon, I., and 5 others. 2004. Webs on the web (wow): 3D visualization of ecological networks on the www for collaborative research and education. In *Electronic Imaging 2004*. Belligham, WA: International Society for Optics and Photonics, pp. 124–132.

Young, H. S., and 5 others. 2013. The roles of productivity and ecosystem size in determining food chain length in tropical terrestrial ecosystems. *Ecology* 94: 692–701.

Young, T. P. 1981. A general model of comparative fecundity for semelparous and iteroparous life histories. *American Naturalist* 118: 27–36.

Young, T. P., and Augspurger, C. K. 1991. Ecology and evolution of long-lived semelparous plants. *Trends in Ecology and Evolution* 6: 285–289.

Zavala, J. A., Patankar, A. G., Gase, K., and Baldwin, I. T. 2004. Constitutive and inducible trypsin proteinase inhibitor production incurs large fitness costs in *Nicotiana attenuata*. *Proceedings of the National Academy of Sciences* 101: 1607–1612.

Zavala, J. A. Giri, A. P., Jongsma, M. A., and Baldwin, I. T. 2008. Digestive duet: midgut digestive proteinases of *Manduca sexta* ingesting *Nicotiana attenuata* with manipulated trypsin proteinase inhibitor expression. *PLoS One* 3(4): 1–10.

Zhao, M., and Running, S. W. 2010. Drought-induced reduction in global terrestrial net primary production from 2000 through 2009. *Science* 329: 940–943.

Zhong, Z., and 5 others. 2014. Positive interactions between large herbivores and grasshoppers, and their consequences for grassland plant diversity. *Ecology* 95: 1055–1064.

Zhu, X., Long, S. P., and Ort, D. R. 2008. What is the maximum efficiency with which photosynthesis can convert solar energy into biomass? *Current Opinion in Biotechnology* 19: 153–159.

Zwolak, R., Pearson, D. E., Ortega, Y. K., and Crone, E. E. 2010. Fire and mice: seed predation moderates fire's influence on conifer recruitment. *Ecology* 91: 1124–1131.

FIGURE AND QUOTATION CREDITS

Chapter 1

Opening image: Migrating wildebeest. T. R. Shankar Raman, Creative Commons License

1.1a	Karl Kössler
1.1b	imageBROKER/Alamy
1.2	Redrawn by Simon Tegg from Sinclair, A. R. E. 1973. Population increases of buffalo and wildebeest in the Serengeti. *East African Wildlife Journal* 11: 93–107.
1.3	Redrawn from Sinclair, A. R. E. 1975. The resource limitation of trophic levels in tropical grassland ecosystems. *Journal of Animal Ecology* 44: 497–520.
1.4, 1.5, 1.6a, 1.7, 1.9	Redrawn from Sinclair, A. R. E. 1977. The eruption of the ruminants. In A. R. E. Sinclair and M. Norton-Griffiths (eds), *Serengeti: Dynamics of an Ecosystem*. Chicago, IL: University of Chicago Press, pp. 82–103.
1.6b	Redrawn from Sinclair, A. R. E., Dublin, H., Borner, M. 1985. Population regulation of Serengeti wildebeest: a test of the food hypothesis. *Oecologia* 65: 266–268.
1.7	Drawn by Simon Tegg
1.8a	Lockenes/Shutterstock
1.8b	Eric Isselee/Shutterstock
1.8c	NASA
1.9	Drawn by David Cox
1.10–12	Redrawn from Sinclair, A. R. E., and 5 others. 2007. Long-term ecosystem dynamics in the Serengeti: lessons for conservation. *Conservation Biology* 21: 580–590.
1.13	Redrawn from Sinclair, A. R. E and 17 others. 2013. Asynchronous food-web pathways could buffer the response of Serengeti predators to El Niño Southern Oscillation. *Ecology* 94: 1123–1130.
	Dealing with Data Figures D1.1.2 and D1.1.3 Redrawn from Sinclair, A. R. E., and 5 others. 2007. Long-term ecosystem dynamics in the Serengeti: lessons for conservation. *Conservation Biology* 21: 580–590.
1.14	Drawn by Simon Tegg

Chapter 2

Opening image: Soft_light/Shutterstock

2.1a	John A. Anderson/Shutterstock
2.1b	Krzysztof Odziomek/Shutterstock
2.2a and b	Redrawn by Simon Tegg from Kleypas, J. A., and 5 others. 2006. *Impacts of ocean acidification on coral reefs and other marine calcifiers: a guide for future research*. Report of a workshop held 18–20 April 2005, St. Petersburg, FL, sponsored by the National Science Foundation, the National Oceanic and Atmospheric Administration, and the US Geological Survey.
2.3	Redrawn by David Cox from World Resources Institute and US Environmental Protection Agency
2.4, 2.5, 2.7a, 2.7b, 2.8, 2.10 and 2.11b	Drawn by Simon Tegg
2.6	courtesy of NOAA, redrawn by Simon Tegg
2.9	Drawn by David Cox
2.11a	Karin Jaehne/Shutterstock
2.12a	Chris Rubino/Shutterstock
2.13a	KPG Payless2/Shutterstock
2.14ai	Dr Morley Read/Shutterstock
2.14	Daniel Janzen
2.15	Wilson, E. O. 1992. The diversity of life. Belknap Press, Cambridge
2.18a	Steve Jurvetson
2.19a	vitmark/Shutterstock
2.20	Redrawn by David Cox from Keeley, J. E., and Fotheringham, C. J. 2001. Historic fire regime in southern California shrublands. *Conservation Biology* 15: 1536–1548.

2.21a   LianeM/Shutterstock
2.22 and   Stefan Kröpelin
2.24
2.23   Drawn by Simon Tegg
2.25   Courtesy of US Department of Agriculture, redrawn by David Cox
2.26   Drawn by Simon Tegg
2.27a   LFRabanedo/Shutterstock
2.27b   Jan Hamrsky/Nature Picture Library
2.28a   Belozorova Elena/Shutterstock
2.28b   Oleg Nekhaev/Shutterstock
2.29a   Redrawn by Simon Tegg from NRC Research Press Schindler, D. W., and Fee, E. J. 1974. Experimental
            lakes area: whole lake experiments in eutrophication. *Journal of the Fisheries Research Board of Canada*
            31: 937–953.
2.29b   David Schindler
2.30a   indykb/Shutterstock
2.30b   Redrawn from Tunnicliffe, V. 1992. Hydrothermal-vent communities of the deep sea. *American
            Scientist* 80: 336–349.
2.30c   Woods Hole Oceanographic Institution, VISUALS UNLIMITED /SCIENCE PHOTO LIBRARY
2.30d   © 2015 Ocean Networks Canada
2.31   229114111/Shutterstock
2.32a, b,   Redrawn by Simon Tegg from Wild, C., Jantzen, C., Struck, U., Hoegh-Guldberg, O., and Huettel,
and c   M. 2008. Biogeochemical responses following coral mass spawning on the Great Barrier Reef: pelagic–
            benthic coupling. *Coral Reefs* 27: 123–132.
2.33   Ethan Daniels/Shutterstock
2.34a   Arto Hakola/Shutterstock
2.34b   Redrawn from Ellison, A. M., Farnsworth, E. J., and Twilley, R. R. 1996. Facultative mutualism
            between red mangroves and root-fouling sponges in Belizean mangal. *Ecology* 77: 2431–2444.
2.35   US Fish and Wildlife Service
2.36   NASA image created by Jesse Allen, Earth Observatory, using data obtained from the University of
            Maryland's Global Land Cover Facility.
2.37   Ellis, E. C., and Ramankutty, N.. 2008. Putting people in the map: anthropogenic biomes of the world.
            Frontiers in Ecology 6: 439–447.

Chapter 3

Opening image: All Canada Photos / Alamy.

3.1   Windsor Aguirre
3.2   Drawn by Simon Tegg
3.3   Redrawn by David Cox from World Health Organisation Global Tuberculosis report 2012
3.5a   Drawn by Simon Tegg
3.5b and c Redrawn by Simon Tegg from Boag, P. T., and Grant, P. R. 1984. The classical case of character release:
            Darwin's finches (Geospiza) on Isla Daphne Major, Galápagos. *Biological Journal of the Linnean Society*
            22: 243–287.
3.6a   David W. Eickhoff
3.6b   Redrawn from Boag, P. T., and Grant, P. R. 1984. The classical case of character release: Darwin's
            finches (*Geospiza*) on Isla Daphne Major, Galápagos. *Biological Journal of the Linnean Society* 22:
            243–287.
3.7   Redrawn from Grant, B. R., and Grant, P. R. 2003. What Darwin's finches can tell us about the
            evolutionary origin and regulation of biodiversity. *Bioscience* 53: 965-975.
3.8a and b Redrawn from Grant, B. R., and Grant, P. R. 2008. *How and Why Species Multiply: The Radiation of
            Darwin's Finches*. Princeton, NJ: Princeton University Press.
3.9   Redrawn from Grant, P. R., and Grant, B. R. 2002. Unpredictable evolution in a 30-year study of
            Darwin's finches. *Science* 296: 707–711.
3.10 and   Redrawn by Simon Tegg from Gagneux, S., and 5 others. 2006. The competitive cost of antibiotic
3.11   resistance in *Mycobacterium tuberculosis*. Science 312: 1944–1946.
3.12   Redrawn by Simon Tegg from Reimchen, T. E. 1992. Injuries on stickleback from attacks by a toothed
            predator (*Oncorhynchus*) and implications for the evolution of lateral plates. *Evolution* 46: 1224–1230.

3.13   Redrawn by Simon Tegg from Barrett, R. D. H., Rogers, S. M., and Schluter, D. 2009. Environment specific pleiotropy facilitates divergence at the ectodysplasin locus in threespine stickleback. *Evolution* 63: 2831–2837.

3.14   Peter Waters/Shutterstock

3.15a   Alexander Wild

3.15b   Moore, J. C., and Pannell, J. R. 2011. Sexual selection in plants. *Current Biology* 21(5): R176–182.

3.15c   Sarah Jessup/Shutterstock

3.15d   lapas77/Shutterstock

3.15e   underworld/Shutterstock

3.15f   Roy Mangersnes/Nature Picture Library

3.16   Drawn by Simon Tegg

3.17a   raulbaenacasado/Shutterstock

3.17b   Drawn by David Cox

3.18   Redrawn by Simon Tegg from Berthold, P., Helbig, A. J., Mohr, G., and Querner, U. 1992. Rapid microevolution of migratory behaviour in a wild bird species. *Nature* 360: 668–670.

3.19   Redrawn from Bearhop, S., and 8 others. 2005. Assortative mating as a mechanism for rapid evolution of a migratory divide. *Science* 310: 502–504.

3.20ai   Stubblefield Photography/Shutterstock

3.20aii   Stubblefield Photography/Shutterstock

3.20aiii   Stubblefield Photography/Shutterstock

3.20aiv   Stubblefield Photography/Shutterstock

3.20b, 3.21   Drawn by Simon Tegg
and 3.23a

3.23b   Redrawn by Simon Tegg from Schluter, D. 1993. Adaptive radiation in sticklebacks: size, shape, and habitat use efficiency. *Ecology* 73: 699–709.

3.24a and b   Redrawn from Phillips, B. L., Brown, G. P., Webb, J. K., and Shine, R. 2006. Invasion and the evolution of speed in toads. *Nature* 439: 803.

Chapter 4

Opening image: Lebendkulturen.de/Shutterstock

4.1   Redrawn by Simon Tegg from Martin, J. H., and Fitzwater, S. E. 1988. Iron deficiency limits phytoplankton growth in the north-east Pacific subarctic. *Nature* 331: 341–343.

4.2, 4.3,   Drawn by Simon Tegg
4.4, 4.5
and 4.6b

4.6ai   t50/Shutterstock

4.6aii   Marc Venema/Shutterstock

4.6aiii   Janelle Lugge/Shutterstock

4.6aiv   unchalee_foto/Shutterstock

4.7   Drawn by David Cox

4.8a   Pierre BRYE/Alamy

4.8b and   Redrawn by Simon Tegg from Takahashi, S., Bauwe, H., and Badger, M. 2007. Impairment of the
4.9   photorespiratory pathway accelerates photoinhibition of photosystem II by suppression of repair but not acceleration of damage processes in *Arabidopsis*. *Plant Physiology* 144: 487–494.

4.10   Redrawn by Simon Tegg from Aerts, R., and Chapin, F. S. 2000. The mineral nutrition of wild plants revisited; a reevaluation of processes and patterns. *Advances in Ecological Research* 30: 1–67.

4.11a–d   Redrawn from Takashima, T. Hikosaka, K., and Hirose, T. 2004. Photosynthesis or persistence: nitrogen allocation in leaves of evergreen and deciduous *Quercus* species. *Plant, Cell and Environment* 27: 1047–1054.

4.12   insima/Shutterstock; kerstiny/Shutterstock; iaRada/Shutterstock; Daria Yakovleva/Shutterstock; Chalintra.B/Shutterstock; MarijaPiliponyte/Shutterstock

4.13   Redrawn from Williams, S., and Kay, R. 2001. A comparative test of adaptive explanation for hypsodonty in ungulates and rodents. *Journal of Mammalian Evolution* 8: 207–229.

4.14a   MERLIN TUTTLE/SCIENCE PHOTO LIBRARY

4.14b   Doug Meek/Shutterstock

4.15	Redrawn from González-Bergonzoni, I., and 5 others. 2012. Meta-analysis shows a consistent and strong latitudinal pattern in fish omnivory across ecosystems. *Ecosystems* 15: 492–503.
4.16a and b	Redrawn by Simon Tegg from Ewbanks, M. D., Styrsky, J. D., and Denno, R. F. 2003. The evolution of omnivory in heteropteran insects. *Ecology* 84: 2549–2555.
4.18	Redrawn from Main, T., Dobberfuhl, D. R., and Elser, J.J. 1997. N:P stoichiometry and ontogeny of crustacean zooplankton: a test of the growth rate hypothesis. *Limnology and Oceanography* 42: 1474–1478.
4.19a	Renata Suman/Shutterstock
4.19b	Shutterschock/Shutterstock
4.20	Redrawn from Schultz, J. C., and Appel, H. M. 2004. Cross-kingdom cross-talk: hormones shared by plants and their insect herbivores. *Ecology* 85: 70–77.
4.21	Redrawn by Simon Tegg from Martin, J. H. and 43 others. 1994. Testing the iron hypothesis in ecosystems of the equatorial Pacific Ocean. *Nature* 371: 123–129.
4.22	Redrawn by Simon Tegg from Coale, K. H. and 18 others. 1996. A massive phytoplankton bloom induced by an ecosystem-scale iron fertilization experiment in the equatorial Pacific Ocean. *Nature* 383: 495–501.
4.23	Redrawn by Simon Tegg from Zhu, X., Long, S. P., and Ort, D. R. 2008. What is the maximum efficiency with which photosynthesis can convert solar energy into biomass? *Current Opinion in Biotechnology* 19: 153–159.

Chapter 5

Opening image: An avalanche falling onto the highway in Nape, Norway. USGS

5.1	Nicholas Dawes, WSL Institute for Snow and Avalanche Research SLF
5.2	Redrawn by Simon Tegg from Barbeito, I., Dawes, M. A., Rixen, C., Senn, J., and Bebi, P. 2012. Factors driving mortality and growth at treeline: a 30-year experiment of 92 000 conifers. *Ecology* 93: 389-401.
5.4	Redrawn from Baldwin, J. 1971. Adaptation of enzymes to temperature: acetylcholinesterases in the central nervous system of fishes. *Comparative Biochemistry and Physiology* 40: 181–187.
5.6a and 5.6b	Drawn by Simon Tegg
5.7a	Ingo Arndt/Nature Picture Library
5.7b	USGS
5.8	Redrawn by Simon Tegg from Yamori, W., Noguchi K., and Terashima, I. 2005. Temperature acclimation of photosynthesis in spinach leaves: analyses of photosynthetic components and temperature dependencies of photosynthetic partial reactions. *Plant, Cell and Environment* 28: 536–547.
5.9a	Scott Prokop/Shutterstock
5.9b	Scott Prokop/Shutterstock
5.10	Sébastien Ibanez
5.11a	Sakaori (Creative Commons Licence)
5.11b and c	Redrawn from Seymour, R. S. 2004. Dynamics and precision of thermoregulatory responses of eastern skunk cabbage *Symplocarpus foetidus*. *Plant, Cell and Environment* 27: 1014–1022.
5.12a and b	Redrawn by Simon Tegg from Angilletta Jr, M. 2009. *Thermal Adaptation: A Theoretical and Empirical Synthesis*. New York: Oxford University Press. Original from Clark, A., and Gaston, K. J. 2006. Climate, energy and diversity. *Proceedings of the Royal Society B* 273: 2257–2266.
5.13	Redrawn by Simon Tegg from Schmidt-Nielsen, K. 1997. *Animal Physiology: Adaptation and Environment*. 5th edition. Cambridge University Press
5.14, 5.15 and 5.16	Drawn by Simon Tegg
5.18 a,b and c	Redrawn from Saab, I. N., Sharp, R. E., Pritchard, J., and Voetberg, G. S. 1990. Increased endogenous abscisic acid maintains primary root growth and inhibits shoot growth of maize seedlings at low water potentials. *Plant Physiology* 93: 1329–1336.
5.19	Redrawn from Voesenek, L. A. C. J., Runders, J. H. G. M., Peeters, A. J. M., van de Steeg, H. M., and de Kroon, H. 2004. Plant hormones regulate fast shoot elongation under water: from genes to communities. *Ecology* 85: 16–27.
5.20a	Bernadette Heath/Shutterstock
5.20b	(inset) Sara Demeter/Shutterstock

5.21a and b Redrawn from Schmidt-Nielsen, K. 1997. *Animal Physiology: Adaptation and Environment*. 5th edn. Cambridge: Cambridge University Press.
5.22a Howard Noel/Alamy
5.22b Drawn by Simon Tegg
5.23a Redrawn from Landrum, J. V. 2001. Wide-band tracheids in leaves in Aizoaceae: the systematic occurrence of a novel cell type and its implications for the monophyly of the subfamily Rushioideae. *Plant Systematics and Evolution* 227: 49–61.
5.23b GFC Collection/Alamy
5.23c Parolin, P. 2006. Ombrohydrochory: rain-operated seed dispersal in plants: with special regard to jet action dispersal in Aizoaceae. *Flora* 201: 511–518.
5.24a, b, and c Redrawn by Simon Tegg from Froeschke, G., Harf, R., Sommer, S., and Matthee, S. 2010. Effects of precipitation on parasite burden along a natural climatic gradient in southern Africa: implications for possible shifts in infestation patterns due to global changes. *Oikos* 119: 1029–1039.
5.25 Redrawn by Simon Tegg from Perry, A. L., Low, P. J., Ellis, J. R., and Reynolds, J. D. 2005. Climate change and distribution shifts in marine fishes. *Science* 308: 1912–1915.

Chapter 6

Opening image: anandoart/Shutterstock
6.1 Stripsa/Shutterstock; Matthew Cole/Shutterstock
6.2a rorue/Shutterstock
6.2b Redrawn from Cooper, W. E. Jr., Perez-Mellado, V., and Hawlena, D. 2006. Magnitude of food reward affects escape behavior and acceptable risk in Balearic lizards, *Podarcis lilfordi*. *Behavioral Ecology* 17: 554–559.
6.3 Redrawn from Michaelidis, C. I., Demary, K. C., and Lewis, S. M. 2006. Male courtship signals and female signal assessment in *Photinus greeni* fireflies. *Behavioral Ecology* 17: 329–335.
6.4 Tomatito/Shutterstock
6.6a and b Redrawn by Simon Tegg from Davies, N. B., and Lundberg, A. 1984. Food distribution and a variable mating system in the dunnock, *Prunella modularis*. *Journal of Animal Ecology* 53: 895–912.
6.7a Roy Pedersen/Shutterstock
6.7b Redrawn by Simon Tegg from Pride, R. E. 2005. Optimal group size and seasonal stress in ring-tailed lemurs (*Lemur catta*). *Behavioral Ecology* 16: 550–560.
6.8 Redrawn by Simon Tegg from Pride, R. E. 2005. Optimal group size and seasonal stress in ring-tailed lemurs (*Lemur catta*). *Behavioral Ecology* 16: 550–560.
6.9 Redrawn from Pride, R. E. 2005. Optimal group size and seasonal stress in ring-tailed lemurs (*Lemur catta*). *Behavioral Ecology* 16: 550–560.
6.10 Drawn by Simon Tegg
6.11 Kacelnik, A. 1984. Central place foraging in starlings (*Sturnus vulgaris*). I. Patch residence time. *Journal of Animal Ecology* 53: 283–299.
6.12a claffra/Shutterstock
6.12b Redrawn from Parker, G. A., and Stuart, R. A. 1976. Animal behavior as a strategy optimizer: evolution of resource assessment strategies and optimal emigration thresholds. *American Naturalist* 110: 1055–1076.
6.13a Redrawn by Simon Tegg from Parker, G. A., and Stuart, R. A. 1976. Animal behavior as a strategy optimizer: evolution of resource assessment strategies and optimal emigration thresholds. *American Naturalist* 110: 1055–1076.
6.13b Drawn by Simon Tegg
6.15 Redrawn from Griffith, S. C., Owens, I. P. F., and Thuman, K. A. 2002. Extra pair paternity in birds: a review of interspecific variation and adaptive function. *Molecular Ecology* 11: 2195–2212.
6.16 Redrawn from Byrne, P. G., and Keogh, J. S. 2009. Extreme sequential polyandry insures against nest failure in a frog. *Proceedings of the Royal Society B* 276: 115–120.
6.17a Doug Lemke/Shutterstock
6.17b Waage, J. K. 1979. Dual function of the damselfly penis: sperm removal and transfer. *Science* 203: 916–918.
6.18a Joseph Marino
6.18b Alaskastock
6.18c Donald E Hurlsbert, Smithsonian Institute

6.19   Redrawn by Simon Tegg Davies, N. B. 1992. *Dunnock Behavior and Social Evolution*. Oxford: Oxford University Press.

6.20   Mumme, R. L. 1992. Do helpers increase reproductive success? An experimental analysis in the Florida scrub jay. *Behavioral Ecology and Sociobiology* 31: 319–328.

6.21   Drawn by Simon Tegg

6.22   Stanton Braude

6.23a and b, 6.24   Redrawn by Simon Tegg from Hauser, M. D., Chen, M. K., Chern, F., and Chuang, E. 2003. Give unto others: genetically unrelated cotton-top tamarin monkeys preferentially give food to those who altruistically give food back. *Proceedings of the Royal Society B* 270: 2363–2370.

6.24   Drawn by Simon Tegg

6.25   Redrawn by Simon Tegg from Valencia, J., de La Cruz, C., Carranza, J., and Mateos, C. 2006. Parents increase their parental effort when aided by helpers in a cooperatively breeding bird. *Animal Behaviour* 71: 1021–1028.

Chapter 7

Photo of Bernd Heinrich: Lynn Jennings

Opening image: Heinrich's wasp, *Hepiopelmus variegatorius*. Bernd Heinrich.

7.1   M. I. WALKER/SCIENCE PHOTO LIBRARY

7.2   Svetlana Foote/Shutterstock

7.4, 7.5, 7.6a-c, 7.7a-b, 7.8, 7.9a-b and 7.10   Drawn by Simon Tegg

7.2b-7.10, 7.13, 7.17   Bernd Heinrich

7.11 and 7.12   Redrawn by Simon Tegg from Heinrich, B. 1979. Foraging strategies of caterpillars: leaf damage and possible predator avoidance strategies. *Oecologia* 42: 325–337.

7.14   Redrawn from Heinrich, B. (1988). Winter foraging at carcasses by three sympatric corvids, with emphasis on recruitment by the raven, *Corvus corax*. *Behavioral Ecology and Sociobiology* 23(3): 141–156.

7.15   Drawn by Simon Tegg

7.16   Redrawn by Simon Tegg from Heinrich, B. and Bugnyar, T. 2005. Testing understanding in ravens: string-pulling to reach food. *Ethology* 111: 962–976.

Chapter 8

Opening image: A 1-year-old North Atlantic right whale breeches in the Bay of Fundy. WDC/Regina Asmutis-Silva

8.1   Center for Coastal Studies, taken under NOAA permit #932-1905

8.2a   Florida Fish and Wildlife Conservation Commission/NOAA

8.2b   Seweryn Olkowicz

8.3a   Redrawn by Simon Tegg from Kleiber, M. 1932. Body size and metabolism. *Hilgardia* 6: 315–353.

8.5 – 8.7   Redrawn by Simon Tegg from Charnov, E. L., and Ernest, S. K. M. 2006. The offspring-size/clutch-size trade-off in mammals. *American Naturalist* 167: 578–582.

8.8   Sekar B/Shutterstock

8.9a   Dr Morley Read/Shutterstock

8.9b   Drawn by Simon Tegg

8.10a and b   Redrawn by Simon Tegg from Karban, R. 1982. Increased reproductive success at high densities and predator satiation for periodical cicadas. *Ecology* 63: 321–328.

8.12ai   Mariusz S. Jurgielewicz/Shutterstock

8.12aii   Tom Van Devender

8.12b and c   Drawn by Simon Tegg, based on data from Schaffer, W. M., and Schaffer, M. V. 1977. The adaptive significance of variations in reproductive habit in the Agavaceae. In B. Stonehouse and C. M. Perrins (eds), *Evolutionary Ecology*. London: Macmillan Press, pp. 261–276.

8.13a   Peter Lesica

8.13b   Redrawn from Lesica, P., and Young, T. 2005. A demographic model explains life history variation in *Arabis fecunda*. *Functional Ecology* 19: 471–477.

8.14	Wild Horizon/Getty
8.15	Redrawn from Pfennig, D. W. 1992b. Polyphenism in spadefoot toad tadpoles as a logically adjusted evolutionarily stable strategy. *Evolution* 46: 1408–1420.
8.16, 8.17a and b	Redrawn from Charmantier, A., and 5 others. 2008. Adaptive phenotypic plasticity in response to climate change in a wild bird population. *Science* 320: 800–803.
8.18 a,b,c	Redrawn from Richter-Boix, A., Orizaola, G., Laurila, A. 2014. Transgenerational phenotypic plasticity links breeding phenology with offspring life-history. *Ecology* 95: 2715–2722.
8.19	Drawn by Simon Tegg
	Dealing with Data Figure D8.1.1 Based on data from Kleiber, M. 1932. Body size and metabolism. *Hilgardia* 6: 315–353.

Chapter 9

Opening image: The packrat *Neotoma stephensi* posing in front of a small portion of its midden. Rick & Nora Bowers / Alamy.

9.1a	Inge Johnsson / Alamy
9.1b	Frans Lanting Studio / Alamy
9.1c	Christian Musat/Shutterstock
9.2 and 9.3	Redrawn by Simon Tegg from Gray, S. T., Betancourt, J. L., Jackson, S. T., and Eddy, R. G. 2006. Role of multidecadal climate variability in a range extension of pinyon pine. *Ecology* 87: 1124–1130.
9.4	Redrawn by Simon Tegg from Campagna, C., and Le Boeuf, B. J. 1988. Reproductive behaviour of Southern sea lions. *Behaviour* 104: 233–261.
9.5a, b and c	Redrawn by Simon Tegg from Soto, K. H., and Trites, A. W. 2011. South American sea lions in Peru have a lek-like mating system. *Marine Mammal Science* 27: 306–333.
9.6	Wikelski, M., and 5 others. 2006. Simple rules guide dragonfly migration. *Biology Letters* 2: 325–329.
9.7a, 9.8a	Redrawn by David Cox from Oberhauser, K. S., and Solensky, M. J. 2004. *The Monarch Butterfly: Biology and Conservation*. Ithaca, NY: Cornell University Press.
9.7b	Jean-Edouard Rozey/Shutterstock
9.8b	Redrawn by David Cox from http://www.learner.org/jnorth/tm/monarch/MigrationMaps.html
9.8c	Redrawn by David Cox from https://www.learner.org/jnorth/maps/monarch_spring2012.html http://www.learner.org/jnorth/tm/monarch/MigrationMaps.html
9.9	Redrawn by David Cox from Laliberté, A. S., and Ripple, W. J. 2004. Range contractions of North American carnivores and ungulates. *Bioscience* 54: 123–138.
9.10	Phil McDonald/Shutterstock
9.11	Ethan Daniels/Shutterstock
9.12, 9.13	Redrawn by Simon Tegg from Wiencke, C., Roleda, M. Y., Gruber, A., Clayton, M. N., and Bischof, K. 2006. Susceptibility of zoospores to UV radiation determines upper depth distribution limit of Arctic kelps: evidence through field experiments. *Journal of Ecology* 94: 455–463.
9.14	Redrawn from Steenbergh, W. F., Lowe, C. H. 1977. Ecology of the saguaro: II. National Park Service Monograph Series; number 8. page 113
9.15a and b	Redrawn by Simon Tegg from Drezner, T. D., and Garrity, C. M. 2003. Saguaro distribution under nurse plants in Arizona's Sonoran Desert: directional and microclimate influences. *The Professional Geographer* 55: 505–512.
9.16a	fritz16/Shutterstock
9.16b	Redrawn by Simon Tegg from Kinnear, J. E., Onus, M. L., and Sumner, N. R. 1998. Fox control and rock-wallaby population dynamics II. An update. *Wildlife Research* 25: 81–88.
9.17	Drawn by Simon Tegg
9.18 and 9.19	Redrawn by Simon Tegg from Halliwell, D. B., Whittier, T. R., and Ringler, N. H. 2001. Distributions of lake fishes of the northeast USA: III. Salmonidae and associated coldwater species. *Northeastern Naturalist* 8: 189–206.
9.21	Redrawn from Mercado-Silva, N., Olden, J. D., Maxted, J. T., Hrabik, T. R., and Zanden, M. J. V. 2006. Forecasting the spread of invasive rainbow smelt in the Laurentian Great Lakes region of North America. *Conservation Biology* 20: 1740–1749.
9.22	Redrawn by Simon Tegg from Wassenaar, L. J., and Hobson, K. A. 1998. Natal origins of migratory monarch butterflies at wintering colonies in Mexico: new isotopic evidence. *Proceedings of the National Academy of Sciences* 95: 15436–15439.
	Dealing with data Figure D9.1.1: Sanderson, E. W., and 5 others. 2002. The human footprint and the last of the wild. *Bioscience* 52: 891–890.

Chapter 10

Opening image: The seven moai of Ahu Akivi on Rapa Nui. steve100/Shutterstock.

10.1a	chris kolaczan/Shutterstock
10.1b	Jung Hsuan/Shutterstock
10.1c	Mars 2002, Creative Commons License
10.2	Drawn by Simon Tegg
10.3	Alexey Stiop/Shutterstock
10.4a	Drawn by Simon Tegg
10.4b	Steve Byland/Shutterstock
10.5	Bull's-Eye Arts/Shutterstock
10.6a	Redrawn by Simon Tegg from Gardner, T. A., Coté, I. M., Gill, J. A., Grant, A., and Watkinson, A. R. 2003. Long-term region-wide declines in Caribbean corals. *Science* 301: 958–960.
10.6b	John A. Anderson/Shutterstock
10.7a	Drawn by Simon Tegg
10.7b	risha green/Shutterstock
10.8a	Courtesy of United States Fish and Wildlife Service, redrawn by David Cox
10.8b	Debbie Steinhausser/Shutterstock
10.9a-b, 10.10, 10.11	Drawn by Simon Tegg
10.12	Redrawn by Simon Tegg from Southern Scientific Press. Davidson, J. 1938. On the growth of the sheep population in Tasmania. *Transactions of the Royal Society of South Australia* 62: 342–346.
10.13	Redrawn by Simon Tegg from Matthes, U., and Larson, D. W., 2006. Microsite and climatic controls of tree population dynamics: an 18-year study on cliffs. *Journal of Ecology* 94: 402–414.
10.14	Redrawn from Steinhaus, E. A. 1958. Crowding as a possible stress factor in insect disease. *Ecology* 39: 503–514.
10.15	Redrawn from Arcese, P., and Smith, J. N. M. 1988, Effects of population density and supplemental food on reproduction in song sparrows. *Journal of Animal Ecology* 57: 119–136.
10.18	Redrawn by Simon Tegg from Gause, G. F. 1934.*The Struggle for Existence*. Baltimore, MD: Williams and Wilkins.
10.19a and b	Redrawn from DeWalt, S. J. 2006. Population dynamics and potential for biological control of an exotic invasive shrub in Hawaiian rainforests. *Biological Invasions* 8: 1145–1158.
10.19c	Elena Mirage/Shutterstock
10.20	US Census Bureau, International Database, June 2012
10.21, 10.22	(redrawn by Simon Tegg) Based on data from United Nations Department of Economic and Social Affairs, Population Division. 2013. *World Population Prospects: The 2012 Revision*. Geneva: United Nations Publications.
10.23	Redrawn from Hunt, T. L., and Lipo, C. P. 2006. Late colonization of Easter Island. *Science* 311: 1603–1606.

Chapter 11

Chapter 11 Introduction Excerpt from The Grail Bird: The Rediscovery of the Ivory-billed Woodpecker by Tim Gallagher, Copyright © 2006 by Tim Gallagher. Reprinted by permission of Houghton Mifflin Harcourt Publishing Company. All rights reserved.
Opening image: ivory-billed woodpecker flying through forested wetland. Larry Chandler.

11.1a	Richard A McMillin/Shutterstock
11.1b	Redrawn from Newman, D., and Pilson, D. 1997. Increased probability of extinction due to decreased genetic effective population size: experimental populations of *Clarkia pulchella. Evolution* 51: 354–362.
11.2	Robert O'Malley
11.3a	Johner Images/Alamy
11.3b	Premaphotos/Alamy
11.3c	Redrawn by Simon Tegg from Gonzalez, A., Lawton, J. H., Gilbert, F. S. Blackburn, T. M., and Evans-Freke, I. 1998. Metapopulation dynamics, abundance, and distribution in a microecosystem. *Science* 281: 2045–2047.
11.4a, b	Drawn using data from using data from Gonzalez, A., Lawton, J. H., Gilbert, F. S. Blackburn, T. M., and Evans-Freke, I. 1998. Metapopulation dynamics, abundance, and distribution in a microecosystem. *Science* 281: 2045–2047.

11.5a	Jim Marden
11.5b	Anne Sorbes
11.5c	Drawn using data from Nieminen, M., Siljander, M., and Hanski, I. 2004. Structure and dynamics of *Melitaea cinxia* metapopulations. In P. R. Ehrlich and I. Hanski (eds), *On the Wings of Checkerspots: A Model System for Population Biology*. New York: Oxford University Press, pp. 63–91.
11.6	Redrawn from Hanski, I. 2011. Eco-evolutionary spatial dynamics in the Glanville fritillary butterfly. *Proceedings of the National Academy of Sciences* 108: 14397–14404.
11.7a and b	Redrawn using data from Nieminen, M., Siljander, M., and Hanski, I. 2004. Structure and dynamics of *Melitaea cinxia* metapopulations. In P. R. Ehrlich and I. Hanski (eds), *On the Wings of Checkerspots: A Model System for Population Biology*. New York: Oxford University Press, pp. 63–91.
11.8	Redrawn from Kuussari, M., Saccheri, I., Camara, M., and Hanski, I. 1998. Allee effect and population dynamics in the Glanville fritillary butterfly. *Oikos* 82: 384–392.
11.9c	Redrawn from van Nouhuys, S., and Hanski, I. 2002. Colonization rates and distance of a host butterfly and two specific parasitoids in a fragmented landscape. *Journal of Animal Ecology* 71: 639–650.
11.10	Redrawn by Simon Tegg from Millennium Ecosystem Assessment. 2005. *Ecosystems and Human Well-being: Biodiversity Synthesis*. Washington DC: World Resources Institute.
11.11	Redrawn by Simon Tegg from Johnson, S. J., Neal, P. R., Peter, C. I., and Edwards, T. J. 2004. Fruiting failure and limited recruitment in remnant populations of the hawkmoth-pollinated tree *Oxyanthus pyriformis* subsp. *pyriformis* (Rubiaceae). *Biological Conservation* 120: 31–39.
11.12a	BRUCE COLEMAN INC./Alamy
11.12b	North Wind Picture Archives/Alamy
11.13	Cheng Tai-sing
11.14a	Tom Gilks/Nature Picture Library
11.14c	Redrawn by Goudswaard, K., Witte, F., and Katunzi, E. F. B. 2008. The invasion of an introduced predator, Nile perch (*Lates niloticus*, L.) in Lake Victoria (East Africa): chronology and causes. *Environmental Biology and Fisheries* 81: 127–139.
11.15b	Rob Hainer/Shutterstock
11.16a and b	Redrawn by Simon Tegg from Mitchell, C. E., and Power, A. G. 2003. Release of invasive plants from fungal and viral pathogens. *Nature* 421: 625–627.
11.17a	DanielCD (Creative Commons Licence)
11.17b and c	Redrawn by Simon Tegg from Siemann, E., and Rogers, W. E. 2001. Genetic differences in growth of an invasive tree species. *Ecology Letters* 4: 514–518.
11.18a and b	Redrawn from Hinz, H. L., and 5 others. 2012. Biogeographical comparison of the invasive *Lepidium draba* in its native, expanded and introduced ranges. *Biological Invasions* 14: 1999–2016.
11.19	Redrawn from Fitzpatrick, J. W., and 17 others. 2005. Ivory-billed woodpecker (*Campephilus principalis*) persists in continental North America. *Science* 308: 1460–1462.

Chapter 12

Photo of Anne Pusey: Anne Pusey

Photo of Jane Goodall: © BRUCE COLEMAN INC. / Alamy

Opening image: The Africa Image Library/Alamy

12.1b	Redrawn by Simon Tegg from Rudicell, R. S., and 19 others. (2010). Impact of simian immunodeficiency virus infection on chimpanzee population dynamics. *PLoS Pathogens* 6(9): e1001116.
12.2a and b	Anthony Collins
12.3, 12.4, 12.7, and 12.8, 12.10b and 12.11a	Jane Goodall Institute
12.4	Drawn by Simon Tegg
12.5, 12.6ai, 12.6aii, and 12.6b	Robert O'Malley
12.9	Seba Koya, courtesy of Cleve Hicks, The Wasmoeth Wildlife Foundation

12.10a	Redrawn by Simon Tegg with permission from Jane Goodall Institute.
12.11b, 12.12a and b	Redrawn from Pusey, A. E. 1983. Mother-offspring relationships in chimpanzees after weaning. *Animal Behaviour* 31: 363–377.
12.13a and b	Redrawn from Wroblewski, E. E., and 5 others. 2009. Male dominance rank and reproductive success in chimpanzees, *Pan troglodytes schweinfurthii*. *Animal Behaviour* 77: 873–885.
12.14	Redrawn from Pusey, A. E., Williams, J., and Goodall, J. 1997. The influence of dominance rank on the reproductive success of female chimpanzees. *Science* 277: 828–831.
12.15	Redrawn from Murray, C. M., Eberly, L. E., and Pusey, A. E. 2006. Foraging strategies as a function of season and rank among wild chimpanzees (*Pan troglodytes*). *Behavioral Ecology* 17: 1020–1028.
12.16a and b	Redrawn by Simon Tegg from Murray, C. M., Gilby, I. C., Mane, S. V., and Pusey, A. E. 2008. Adult male chimpanzees inherit maternal ranging patterns. *Current Biology* 18: 20–24.
12.17	Redrawn by Simon Tegg from Rudicell, R. S. and 19 others. 2010. Impact of simian immunodeficiency virus infection on chimpanzee population dynamics. *PLoS Pathogens* 6: e1001116.
12.18a and b	Redrawn by Simon Tegg from Keele, B. F. and 24 others. 2009. Increased mortality and AIDS-like immunopathology in wild chimpanzees infected with SIVcpz. *Nature* 460: 515–519.
12.19	Redrawn by Simon Tegg from Rudicell, R. S. and 19 others. 2010. Impact of simian immunodeficiency virus infection on chimpanzee population dynamics. *PLoS Pathogens* 6: e1001116.
12.20	Holland Jones, J., Wilson, M. L., Murray, C., and Pusey, A. 2010. Phenotypic quality influences fertility in Gombe chimpanzees. *Journal of Animal Ecology* 79: 1262–1269.
12.21	Drawn by Simon Tegg

Chapter 13

Opening image: Aerial view of Atafu: smallest of the three Tokelau atolls. NASA.

13.1a	Redrawn from McNatty, A., Abbott, K. L., and Lester, P. J. 2009. Invasive ants compete with and modify the trophic ecology of hermit crabs on tropical islands. *Oecologia* 160: 187–194.
13.1ai (inset)	John Tann
13.1b and 13.2	Redrawn from McNatty, A., Abbott, K. L., and Lester, P. J. 2009. Invasive ants compete with and modify the trophic ecology of hermit crabs on tropical islands. *Oecologia* 160: 187–194.
13.1bi	Peter Reijners/Shutterstock
13.3	Drawn by Simon Tegg
13.4	Redrawn from Connell, J. 1961. The influence of interspecific competition and other factors on the distribution of the barnacle *Chthamalus stellatus*. *Ecology* 42: 710–723.
13.5, 13.6a and b	Redrawn from Kandori, I., Hirao, T., Matsunaga, S., and Kurosaki, T. 2009. An invasive dandelion reduces the reproduction of a native congener through competition for pollination. *Oecologia* 159: 559–569.
13.7ai	Josep del Hoyo
13.7aii	Marcel Holyoak
13.7b	Drawn by Simon Tegg
13.8a-d	Redrawn from Hautier, Y., Niklaus, P. A., and Hector, A. 2009. Competition for light causes plant biodiversity loss after eutrophication. *Science* 324: 636–638.
13.9, 13.10, 13.11, 13.12	Drawn by Simon Tegg
13.13a and b	Redrawn by Simon Tegg from Alcaraz, C., Bisazza, A., and Garcia-Berthou, E. 2008. Salinity mediates the competitive interactions between invasive mosquitofish and an endangered fish. *Oecologia* 155: 205–213.
13.13ai (inset)	Chris Appleby, USGS
13.13aii (inset)	Jörg Freyhof
13.14ai	Paul Reeves Photography/Shutterstock
13.14aii	davemhuntphotography/Shutterstock
13.14aiii	Graham Taylor/Shutterstock
13.14b	Redrawn from Harrington, L. A., and 7 others. 2009. The impact of native competitors on an alien invasive: temporal niche shifts to avoid interspecific aggression? *Ecology* 90: 1207–1216.
13.15a	Malcolm Tattersall, Creative Commons License
13.15b, 13.16	Redrawn from Gilbert, B., Srivastava, D. S., and Kirby, K. R. 2008. Niche partitioning at multiple scales facilitates coexistence among mosquito larvae. *Oikos* 117: 944–950.

13.17	Drawn by Simon Tegg
13.18a	Redrawn by David Cox from Johnson, H. E., and 5 others. 2013. Evaluating apparent competition in limiting the recovery of an endangered ungulate. *Oecologia* 171: 295–307.
13.19a and b	Redrawn from LeBrun, E. G., Plowes, R. M., and Gilbert, L. E. 2009. Indirect competition facilitates widespread displacement of one naturalized parasitoid of imported fire ants by another. *Ecology* 90: 1184–1194.

Chapter 14

Opening image: Cockles as ecosystem engineers, Kim Nørgaard Mouritsen.

14.1, 14.2	Drawn by Simon Tegg
14.3	Redrawn from McGraw, J. B., Furedi, M. A. 2005. Deer browsing and population viability of a forest understory plant. *Science* 307: 920–922.
14.4a	Harvey Wood/Alamy
14.4b, c, and d	Redrawn from Bardgett, R. D., and 6 others. 2006. Parasitic plants indirectly regulate below-ground properties in grassland ecosystems. *Nature* 439: 969–972.
14.5a	Christian Ziegler/Getty
14.5b and c	Redrawn from Kalka, M. B., Smith, A. R., and Kalko, E. K. V. 2008. Bats limit arthropods and herbivory in a tropical forest. *Science* 320: 71.
14.6	Redrawn from Sinclair, A. R. E., Mduma, S., and Brashares, J. S. 2003b. Patterns of predation in a diverse predator-prey system. *Nature* 425: 288–290.
14.7a	Traveller Martin/Shutterstock
14.7b	Martin Shields/Alamy
14.7c	dvande/Shutterstock
14.7d	Nick Greaves/Alamy
14.8ai and b	Redrawn by Simon Tegg from Zavala, J. A., Giri, A. P., Jongsma, M. A., and Baldwin, I. T. 2008. Digestive duet: midgut digestive proteinases of *Manduca sexta* ingesting *Nicotiana attenuata* with manipulated trypsin proteinase inhibitor expression. *PLoS One* 3(4): 1–10.
14.8aii	David G Riley
14.8b	Redrawn from Zavala, J. A., Giri, A. P., Jongsma, M. A., and Baldwin, I. T. 2008. Digestive duet: midgut digestive proteinases of *Manduca sexta* ingesting *Nicotiana attenuata* with manipulated trypsin proteinase inhibitor expression. *PLoS One* 3(4): 1–10.
14.9	Redrawn by Simon Tegg from Kessler, A., and Baldwin, I. T. 2001. Defensive function of herbivore-induced plant volatile emissions in nature. *Science* 291: 2141–2144.
14.10	Redrawn by Simon Tegg from Freeman, A. S., and Byers, J. E. 2006. Divergent induced responses to an invasive predator in marine mussel populations. *Science* 313: 831–833.
14.10i	Naturepix/Alamy
14.10ii	Jens Metschurat/Shutterstock
14.10iii	Silvia Waajen
14.11a and b	Redrawn with data from Bro-Jørgensen, J. 2013. Evolution of sprint speed in African savannah herbivores in relation to predation *Evolution* 67: 3371–3376.
14.12a-b and 14.13a-b	Drawn by Simon Tegg
14.14a	TOM & PAT LEESON/SCIENCE PHOTO LIBRARY
14.15	Redrawn by Simon Tegg from Sheriff, M. J., Krebs, C. J., and Boonstra, R. 2009. The sensitive hare: sublethal effects of predator stress on reproduction in snowshoe hares. *Journal of Animal Ecology* 78(6): 1249–1258.
14.16	Drawn by Simon Tegg
14.17b	Redrawn from Holling, C. S. 1959b. The components of predation as revealed by a study of small-mammal predation of the European pine sawfly. *The Canadian Entomologist* 91: 293-320.
14.18	Redrawn from Anderson, R. M., and May, R. M. 1979. Population biology of infectious diseases: Part I. *Nature* 280: 361–367.
14.19	Drawn by Simon Tegg
14.20a	Redrawn by Simon Tegg from Mouritsen, K. N., and Poulin, R. 2003. Parasite-induced trophic facilitation exploited by a non-host predator: a manipulator's nightmare. *International Journal for Parasitology* 33: 1043–1050.
14.20b	Redrawn from Tompkins, D. M., Mouritsen, K. N., and Poulin, R. 2004. Parasite-induced surfacing in the cockle *Austrovenus stuchburyi*: adaptation or not? *Journal of Evolutionary Biology* 17(2): 247–256.

| 14.20c and 14.21 | Redrawn from Mouritsen, K. N. 2002. The parasite-induced surfacing behaviour in the cockle *Austrovenus stutchburyi*: a test of an alternative hypothesis and identification of potential mechanisms. *Parasitology* 124: 521–528. |

Chapter 15

Opening image: Commercial beehives set up in almond orchard.

15.1a	Science Photo Library/Alamy
15.1bi	sosha/Shutterstock
15.1bii	Lehrer/Shutterstock
15.2a	Colin Paterson-Jones
15.2b	Redrawn from Vamosi, J. C., and 5 others. 2006. Pollination decays in biodiversity hotspots. *Proceedings of the National Academy of Sciences* 103: 956–961.
15.3 and 15.4	Redrawn by Simon Tegg from Nuñez, M. A., Horton, T. R., and Simberloff, D. 2009. Lack of belowground mutualisms hinders Pinacae invasions. *Ecology* 90: 2352–2359.
15.5	Redrawn from McClanahan, T. T., Ateweberhan, M., Muhando, C. A., Maina, J., and Mohammed, M. S. 2007. Effects of climate and seawater temperature variation on coral bleaching and mortality. *Ecological Monographs* 77: 503–525.
15.6b	Redrawn by Simon Tegg from Maynard, J. A., Anthony, K. R. N., Marshall, P. A., and Masiri, I. 2008. Major bleaching events can lead to increased thermal tolerance in corals. *Marine Biology* 155: 173–182.
15.7	Rusty Rodriguez
15.8	Redrawn by Simon Tegg from Marquez, L. M., Redman, R. S., Rodriguez, R. J., and Roossinck, M. J. 2007. A virus in a fungus in a plant: three-way symbiosis required for thermal tolerance. *Science* 315: 513–515.
15.9	Redrawn by Simon Tegg from Briggs, J. S., Vander Wall, S. B., Jenkins, S. H. 2009. Forest rodents provide directed dispersal of Jeffrey pine seeds. *Ecology* 90: 675–687.
15.10a–c	Daniel Janzen
15.11a	Megan Frederickson
15.11b	Redrawn by Simon Tegg from Frederickson, M. E., Greene, M. J., and Gordon, D. M. 2005. "Devil's gardens" bedeviled by ants. *Nature* 437: 495–496.
15.12a	Wally Eberhart/Visuals Unlimited/Corbis
15.12b, c and d	Kiers, E. T., Rousseau, R. A., West, S. A., and Denison, R. F. 2003. Host sanctions and the legume–rhizobium mutualism. *Nature* 425: 78–81.
15.13	Redrawn by Simon Tegg from Callaway, R. M., and 12 others. 2002. Positive interactions among alpine plants increase with stress. *Nature* 417: 844–848.
15.14 and 15.15	Redrawn from Canestrari, D., and 5 others. 2014. From parasitism to mutualism: unexpected interactions between a cuckoo and its host. *Science* 343: 1350–1352.
15.16 and 15.17	Redrawn by Simon Tegg from Knight, T. M, McCoy, M. W., Chase, J. M., McCoy, K. A., and Holt, R. D. 2005. Trophic cascades across ecosystems. *Nature* 437: 880–883.
15.18ai	Michael McCoy
15.18aii	Redrawn from Knight, T. M, McCoy, M. W., Chase, J. M., McCoy, K. A., and Holt, R. D. 2005. Trophic cascades across ecosystems. *Nature* 437: 880–888.
15.19	Anthony Dunn/Alamy
15.20i	StevenRussellSmithPhotos
15.20ii	Rick & Nora Bowers/Alamy
15.20iii	Randimal/Shutterstock
15.20iv	Menno Schaefer/Shutterstock

Chapter 16

Opening image: Sea otter in its favorite feeding position. Christopher Boswell/Shutterstock.

16.1ai	Frans Lanting Studio /Alamy
16.1aii	Ethan Daniels/Shutterstock
16.1b–c	Drawn by Simon Tegg
16.2	Redrawn by Simon Tegg from Estes, J. A., Smith, N. S., and Palmisano, J. F. 1978. Sea otter predation and community organization in the western Aleutian Islands, Alaska. *Ecology* 59: 822–833.

16.3	Redrawn from Estes, J. A., Tinker, M., and Bodkin, J. L. 2010. Using ecological function to develop recovery criteria for depleted species: sea otters and kelp forests in the Aleutian archipelago. *Conservation Biology* 24: 852–860.
16.4a	Jan Mastnik/Shutterstock
16.4b	CreativeNature R.Zwerver/Shutterstock
16.5, 16.6, 16.7a	Drawn by Simon Tegg
16.7b	Redrawn by Simon Tegg from Menge, B. A., Berlow, E. L., Blanchette, C. A., Navarrete, S. A., and Yamada, S. B. 1994. The keystone species concept: variation in interaction strength in a rocky intertidal habitat. *Ecological Monographs* 64: 249–286.
16.9	Redrawn by Simon Tegg from Power, M. E., and 9 others. 1996. Challenges in the quest for keystones. *Bioscience* 46: 609–620.
16.10a and b	Redrawn by Simon Tegg from Lafferty, K. D., Hechinger, R. F., Shaw, J. C., Whitney, K. L., and Kuris, A. M. 2006b. Food webs and parasites in a salt marsh ecosystem. In S. Collinge S. and C. Ray (eds), *Disease Ecology: Community Structure and Pathogen Dynamics*. New York: Oxford University Press, pp. 119–134.
16.11	Drawn by Simon Tegg
16.12	Redrawn by Simon Tegg from Lafferty, K. D., Dobson, A. P., and Kuris, A. M. 2006a. Parasites dominate food web links. *Proceedings of the National Academy of Science* 103: 11211–11216.
16.13	Redrawn by Simon Tegg from Link, J. 2002. Does food web theory work for marine ecosystems? *Marine Ecology Progress Series* 230: 1–9.
16.14	Redrawn from Link, J. 2002. Does food web theory work for marine ecosystems? *Marine Ecology Progress Series* 230: 1–9.
16.15a and b	Redrawn from Baez, S., Collins, S., Lightfoot, D., and Koontz, T. L. 2006. Bottom-up regulation of plant community structure in an aridland ecosystem. *Ecology* 87: 2746–2754.
16.16	Redrawn by Simon Tegg from Ludwig, F., De Kroon, H., and Prins, H. H. T. 2008. Impacts of savanna trees on forage quality for a large African herbivore. *Oecologia* 155: 487–496.
16.17	Redrawn by Simon Tegg from Ripple, W. J., and Beschta, R. L. 2004. Wolves and the ecology of fear: can predation risk structure ecosystems? *Bioscience* 54: 755–766.
16.18	Redrawn by Simon Tegg from Peterson, R. O., Vucetich, J. A., Bump, J. M., and Smith, D. W. 2014. Trophic cascades in a multicausal world: Isle Royale and Yellowstone. *Annual Review of Ecology, Evolution, and Systematics* 45: 325–345.
16.19a and b	Redrawn from Laundré, J. W., Hernández, L., and Altendorf, K. B. 2001. Wolves, elk, and bison: reestablishing the "landscape of fear" in Yellowstone National Park USA. *Canadian Journal of Zoology* 79: 1401–1409.
16.20a, b, and c	Redrawn from Creel, S., Christianson, D., Liley, S., and Winnie Jr., J. A. 2007. Predation risk affects reproductive physiology and demography of elk. *Science* 315: 960.
16.21a-c	William Ripple
16.22a	Redrawn by Simon Tegg from Kauffman, M. J., and 5 others. 2007. Landscape heterogeneity shapes predation in a newly restored predator–prey system. *Ecology Letters* 10: 690–700.
16.22b	Redrawn from Kauffman, M. J., Brodie, J. F., and Jules, E. S. 2010. Are wolves saving Yellowstone's aspen? A landscape-level test of a behaviorally mediated trophic cascade. *Ecology* 91: 2742–2755
16.23ai	William Ripple
16.23aii and bii	Redrawn by Simon Tegg from Beschta, R. L., and Ripple, W. J. 2009. Large predators and trophic cascades in terrestrial ecosystems of the western United States. *Biological Conservation* 142: 2401–2414.
16.23bi	Robert Beschta
16.23bii	Redrawn from Beschta, R. L., and Ripple, W. J. 2009. Large predators and trophic cascades in terrestrial ecosystems of the western United States. *Biological Conservation* 142: 2401–2414.
16.24	Redrawn from Springer, A. M., and 7 others. 2003. Sequential megafaunal collapse in the north Pacific Ocean: an ongoing legacy of industrial whaling? *Proceedings of the National Academy of Sciences* 100: 12223–12228.

Chapter 17

Opening image:	Green Bear/Shutterstock
17.1a and 17.3	Drawn by Simon Tegg
17.1b	Ruud Morijn Photographer/Shutterstock

17.2a and b   Redrawn by Simon Tegg from Stuart-Smith, R. D and 17 others. 2013. Integrating abundance and functional traits reveals new global hotspots of fish diversity. *Nature* 501: 539–542.

17.4   Redrawn by David Cox from Barthlott, W., Mutke, J., Rafiqpoor, D., Kier, G., and Kreft, H. 2005. Global centers of vascular plant diversity. *Nova Acta Leopoldina* 92: 61–83.

17.5a, c, d   Redrawn by Simon Tegg from Novotny, V., and 7 others. 2007. Low beta diversity of herbivorous insects in tropical forests. *Nature* 448: 692–695.

17.5b   Redrawn from Novotny, V. et al. 2007. Low beta diversity of herbivorous insects in tropical forests. Nature 448: 692–695

17.6a and b   Redrawn from Pöyry, J., Paukkunen, J., Heliölä, J., and Kuussaari, M. 2009. Relative contributions of local and regional factors to species richness and total density of butterflies and moths in semi-natural grasslands. *Oecologia* 160: 577–587.

17.7a and b,   Redrawn from Ogada, D. L., Gadd, M. E., Ostfeld, R. S., Young, T. P., and Keesing, F. 2008. Impacts of
17.8   large herbivorous mammals on bird diversity and abundance in an African savanna. *Oecologia* 156: 387–397.

17.9a   Gucio_55/Shutterstock

17.9b,   Redrawn from Novotny, V., and 6 others. 2006. Why are there so many species of herbivorous insects
17.10   in tropical rain forests? *Science* 313: 1115–1118.

17.12a   Andrey_Kuzmin/Shutterstock

17.12b   Ehrman Photographic/Shutterstock

17.13a   Redrawn from Harpole, W. S., and Tilman, D. 2007. Grassland species loss resulting from reduced
and b   niche dimension. *Nature* 446: 791–793.

17.14a   Jack Jelly/Shutterstock

17.14b   Drawn by Simon Tegg

17.16a   Jacob Miller

17.16b,   Redrawn from Tilman, D., Reich, P. B., and Knops, J. M. H. 2006. Biodiversity and ecosystem stability
17.18a   in a decade-long grassland experiment. *Nature* 441: 629–632.

17.17 and   Redrawn from Steiner, C. F., Long, Z. T., Krumins, J. A., and Morin, P. J. 2006. Population and
17.18b   community resilience in multitrophic communities. *Ecology* 87: 996–1007.

17.19a   Redrawn by Simon Tegg from Brodribb, T. J., and Feild, T. S. 2010. Leaf hydraulic evolution led a
and b, 17.20   surge in leaf photosynthetic capacity during early angiosperm diversification. *Ecology Letters* 13: 175–183.

Chapter 18

Opening image: Janzen (top) and a pet porcupine, *Coendou mexicanus* (bottom), working at a laboratory at ACG. Courtesy of Daniel Janzen.

18.1, 18.2a   Daniel Janzen
and b,
18.9a,
18.11-14:

18.3   Drawn by David Cox

18.4   Redrawn by Simon Tegg from Janzen, D. H. 1970. Herbivores and the number of tree species in tropical forests. *American Naturalist* 104: 501–528.

18.5 and 6   Redrawn from Packer, A. P., and Clay, K. 2000. Soil pathogens and spatial patterns of seedling mortality in a temperate tree. *Nature* 404: 278–281.

18.7a   Karen Carr

18.7b   Redrawn by David Cox from Martin, P. S. 1973. The discovery of America. *Science* 179: 969–974.

18.8a   H. J. Larsen, Bugwood.org

18.8b   Dejan Milinkovic/Shutterstock

18.9b   Erni/Shutterstock

18.10   Pablo Vasquez

18.15   Redrawn from Hebert, P. D. N., Penton, E. H., Burns, J. M., Janzen, D. H., and Hallwachs, W. 2004. Ten species in one: DNA barcoding reveals cryptic species in the neotropical skipper butterfly *Astraptes fulgerator*. *Proceedings of the National Academy of Sciences* 101: 14812–14817.

Chapter 19

Opening image: Ammit Jack/Shutterstock

19.1a   Nila Newsom/Shutterstock

19.1b	Mivr/Shutterstock
19.1c	David Pearson/Alamy
19.2	Zhao, M., and Running, S. W. 2010. Drought-induced reduction in global terrestrial net primary production from 2000 through 2009. *Science* 329: 940–943.
19.3	Images by Robert Simmon, NASA GSFC Earth Observatory, based on data provided by Watson Gregg, NASA GSFC.
19.4a and b	Schuur, E. A. G., and Matson, P. A. 2001. Net primary productivity and nutrient cycling across a mesic to wet precipitation gradient in Hawaiian montane forest. *Oecologia* 128: 431–442.
19.5a and b	Redrawn by Simon Tegg from del Grosso, S., and 7 others. 2008. Global potential net primary production predicted from vegetation class, precipitation and temperature. *Ecology* 89: 2117–2226
19.6	Redrawn with data from LeBauer, D. S., and Treseder, K. K. 2008. Nitrogen limitation of net primary productivity in terrestrial ecosystems is globally distributed. *Ecology* 89: 371–379.
19.7	Redrawn by Simon Tegg from Elser, J. J., and 9 others. 2007. Global analysis of nitrogen and phosphorus limitation of primary producers in freshwater, marine and terrestrial ecosystems. *Ecology Letters* 10: 1135–1142.
19.8i	Black Sheep Media/Shutterstock
19.8ii	Sergey Uryadnikov/Shutterstock
19.8iii	imageBROKER/Alamy
19.8iv	Redrawn by Simon Tegg from Bjork, K. E., Averbeck G. A., and Stromberg B. E. 2000. Parasites and parasite stages of free-ranging wild lions (*Panthera leo*) of Northern Tanzania. *Journal of Zoo and Wildlife Medicine* 31: 56–61.
19.9i	Andrea Izzotti/Shutterstock
19.9ii	RomanenkoAlexey/Shutterstock
19.9iii	tea maeklong/Shutterstock
19.9iv	Secher/Shutterstock
19.10	Drawn by Simon Tegg
19.11a and b	Redrawn from Duarte, C. M., Agustí, S., Gasol, J. M., Vaqué, D., and Vasquez-Dominguez, E. 2000. Effect of nutrient supply on the biomass structure of planktonic communities: an experimental test on a Mediterranean coastal community. *Marine Ecology Progress Series* 206: 87–95.
19.12, 19.13, 19.14, 19.16	Drawn by Simon Tegg
19.15a and b	Redrawn from Schindler, D. E., Carpenter, S. R., Cole, J. J., and Kitchell, J. F. 1997. Influence of food web structure on carbon exchange between lakes and the atmosphere. *Science* 277: 248–251.
19.17a, b, and c	Redrawn from Sabo, J. L., Finlay, J. C., Kennedy, T., and Post, D. M. 2010. The role of discharge variation in scaling of drainage area and food chain length in rivers. *Science* 330: 965–967.
19.18a	US Fish and Wildlife Service Pacific Region
19.18b	Redrawn by Simon Tegg from Young, H. S., and 5 others. 2013. The roles of productivity and ecosystem size in determining food chain length in tropical terrestrial ecosystems. *Ecology* 94: 692–701.
19.19	Redrawn from by Simon Tegg Elser, J. J., and 9 others. 2007. Global analysis of nitrogen and phosphorus limitation of primary producers in freshwater, marine and terrestrial ecosystems. *Ecology Letters* 10: 1135–1142.

Chapter 20

Opening image: The Black Sea. Dark green in the photograph indicates a phytoplankton bloom near the western shore. The arrow points to the very narrow Bosphorus Strait, which ultimately connects to the Aegean and Mediterranean Seas. NASA image courtesy Jeff Schmaltz, MODIS Land Rapid Response Team at NASA GSFC.

20.1, 20.2	Drawn by Simon Tegg
20.3a	CHAINFOTO24/Shutterstock
20.3b	Redrawn by David Cox from Paulmier, A., and Ruiz-Pino, D. 2009. Oxygen minimum zones (OMZs) in the modern ocean. *Progress in Oceanography* 80: 113–128.
20.4	Redrawn by Simon Tegg from Thamdrup, B. 2012. New pathways and processes in the global nitrogen cycle. *Annual Review of Ecology, Evolution, and Systematics* 43: 407–428.
20.5	Drawn by Simon Tegg
20.6a and b	Redrawn by Simon Tegg from Morford, S. L., Houlton, B. Z., and Dahlgren, R. A. 2011. Increased forest ecosystem carbon and nitrogen storage from nitrogen-rich bedrock. *Nature* 477: 78–81.
20.7	Drawn by Simon Tegg

20.8a   CLAIRE TING/SCIENCE PHOTO LIBRARY
20.8b   Redrawn by Simon Tegg from Van Mooy, B. A. S., Rocap, G., Fredricks, H., Evans, C. T., and
               Devol, A. H. 2006. Sulfolipids dramatically decrease phosphorus demand by picocyanobacteria
               in oligotrophic marine environments. *Proceedings of the National Academy of Sciences* 103:
               8607–8612.
20.9a-b   Drawn by Simon Tegg
20.10   Redrawn by Simon Tegg from Parton, W., and 10 others. 2007. Global-scale similarities in nitrogen
               release patterns during long-term decomposition. *Science* 315: 361–364.
20.12a   Mulholland, P. J. and 30 others. 2008. Stream denitrification across biomes and its response to
and b   anthropogenic nitrate loading. *Nature* 452: 202–205.
20.13a   Hubbard Brook Foundation
20.13b   Redrawn from Likens, G. E., Bormann, F. H., Pierce, R. S., and Reiners, W. A. 1978. Recovery of a
and 20.14   deforested ecosystem. *Science* 199: 492–496.
20.15a   Drawn by Simon Tegg
20.15b   Redrawn by Simon Tegg from Van der Heijden, M. G. A. 2010. Mycorrhizal fungi reduce nutrient loss
               from model grasslands. *Ecology* 91: 1163–1171.
20.16a, b,   Redrawn by Simon Tegg from Mee, L. 2006. Reviving dead zones. *Scientific American* 295: 78–85.
and c

Chapter 21

Opening image: Mount St. Helens erupts on May 18, 1980. USGS.
21.1a   US National Park Service
21.1b   US National Park Service
21.2aiii   Susan Hazlett
21.2b   Drawn by Simon Tegg
21.3a and b   Redrawn from Fastie, C. L. 1995. Causes and ecosystem consequences of multiple pathways of primary
               succession at Glacier Bay, Alaska. *Ecology* 76: 1899–1916.
21.4a,   US Forest Service
b and e
21.4c   Al/Levno/US Forest Service
21.4d   Jerry Franklin/US Forest Service
21.4f   Fred Swanson/US Forest Service
21.4g   Drawn by Simon Tegg (based on US Forest Service data)
21.5   Drawn by Simon Tegg
21.6a and b   Redrawn by Simon Tegg from Turner, M. G., Romme, W. H., and Gardner, R. H. 1999. Prefire
               heterogeneity, fire severity, and early postfire plant reestablishment in subalpine forests of
               Yellowstone National Park, Wyoming. *International Journal of Wildland Fire* 9: 21–36.
21.7a   Frank Singer
21.7b   Redrawn from Schoennagel, T., Turner, M. G., and Romme, W. H. 2003. The influence of fire interval
               and serotiny on postfire lodgepole pine density in Yellowstone National Park. *Ecology* 84: 2967–2978.
21.8   Redrawn from Chapin, F. S., Walker, L. R., Fastie, C. L., and Sharman, L. C. 1994. Mechanisms of
               primary succession following deglaciation at Glacier Bay, Alaska. *Ecological Monographs* 64: 149–175.
21.9, 21.10   Drawn by Simon Tegg
21.11   Redrawn from Edwards, J. S., and Sugg, P. M. 2005. Arthropods as pioneers in the regeneration of life
               on the pyroclastic-flow deposits of Mount St. Helens. In V. H. Dale, F. J. Swanson, and C. M. Crisafulli
               (eds). *Ecological Responses to the 1980 Eruption of Mount St. Helens.* New York: Springer Science,
               pp. 127-138.
21.12a   Dan Schreiber/Shutterstock
21.12b   Redrawn from Dahm, D. N., Larson, D. W., Petersen, R. R., and Wissmar, R. C. 2005. Response and
               recovery of lakes. In V. H. Dale, F. J. Swanson, and C. M. Crisafulli (eds). *Ecological Responses to the
               1980 Eruption of Mount St. Helens.* New York: Springer Science, pp. 255–274.
21.13a   Redrawn from Gilmour, J. P., Smith, L. D., Heyward, A. J., Baird, A. H., and Pratchett, M. S. 2013.
and b   Recovery of an isolated coral reef system following severe disturbance. *Science* 340: 69–71.
21.14   Redrawn from Bihn, J. H., Verhaagh, M., Brandle, M., and Brandl, R. 2008. Do secondary forests act as
               refuges for old growth forest animals? Recovery of ant diversity in the Atlantic forest of Brazil.
               Biological Conservation 141: 733–743.

21.15a and b	Redrawn from Dunn, R. R. 2004. Recovery of faunal communities during tropical forest regeneration. *Conservation Biology* 18: 302–309.
21.16	Redrawn from De Deyn, G. B., and 7 others. 2003. Soil invertebrate fauna enhances grassland succession and diversity. *Nature* 422: 711–713.
21.17a	Design Pics Inc/Alamy
21.17b and c	Redrawn from Zwolak, R., Pearson, D. E., Ortega, Y. K., and Crone, E. E. 2010. Fire and mice: seed predation moderates fire's influence on conifer recruitment. *Ecology* 91: 1124–1131.
21.18a–d	Redrawn from Hughes, T. P., and 9 others. 2007. Phase shifts, herbivory, and the resilience of coral reefs to climate change. *Current Biology* 17: 1–6.
21.19a-b, 21.20, 21.21b	Drawn by Simon Tegg
21.21a	Beatriz Moisset

Chapter 22

Chapter 22 Introduction quotation from *Naturalist* by Edward O. Wilson. Copyright © 1994 Island Press. Reproduced by permission of Island Press, Washington, DC.

Opening image: Aerial view of a small portion of the Fiji archipelago in eastern Melanesia. LOOK Die Bildagentur der Fotografen GmbH/Alamy.

22.1	Uses data from Wilson, E. O. 1961. The nature of the taxon cycle in the Melanesian ant fauna. *American Naturalist* 95: 169–193.
22.2a-b	Drawn by Simon Tegg
22.3	Wilson, E. O., and Simberloff, D. S. 1969. Experimental zoogeography of islands: defaunation and monitoring techniques. *Ecology* 50: 267–268.
22.4	Redrawn by Simon Tegg from Simberloff, D. S., and Wilson, E. O. 1970. Experimental zoogeography of islands: a two-year record of colonization. *Ecology* 51: 934–937.
22.5b	Redrawn from Williamson, M., Gaston, K., and Lonsdale, W. M. 2001. The species–area relationship does not have an asymptote! *Journal of Biogeography* 28: 827–830.
22.6	Redrawn from Bell, T., and 6 others. 2005. Larger islands house more bacterial taxa. *Science* 308: 1884.
22.7	Redrawn from Peay, K. G., Garbelotto, M., and Bruns, T. D. 2010. Evidence of dispersal limitation in soil microorganisms: isolation reduces species richness on mycorrhizal tree islands. *Ecology* 91: 3631–3640.
22.8, 22.9b and c	Redrawn from Drakare, S., Lennon, J. J., and Hillebrand, H. 2006. The imprint of the geographical, evolutionary and ecological context on species–area relationships. *Ecology Letters* 9: 215–227.
22.9ai	blickwinkel/Alamy
22.9aii	Joseph T. Lapp
22.10ai and aii	© The Trustees of the Natural History Museum, London
22.10aiii and 22.10b	Drawn by David Cox
22.11ai	EBFoto/Shutterstock
22.11aii and 22.11bii	Drawn by David Cox
22.11bi	Corbin17/Alamy
22.12	Drawn by Simon Tegg
22.13ai, 22.13bi, 22.13c and 22.13d	Drawn by David Cox
22.14a	José Luis Bartheld (Creative Commons Licence)
22.14b	Drawn by Simon Tegg
22.15	Redrawn by Simon Tegg from Nilsson, M. A., and 6 others. 2010. Tracking marsupial evolution using archaic genomic retroposon insertions. *PLoS Biology* 8: e1000436.
22.16a and b	Redrawn by Simon Tegg from Bonacum, J., O'Grady, P. M., Kambysellis, M., and DeSalle, R. 2005. Phylogeny and age of diversification of the planitibia species group of Hawaiian *Drosophila*. *Molecular Phylogenetics and Evolution* 37: 73–82.

22.17	Redrawn by Simon Tegg from Collinge, S. K. 1998. Spatial arrangement of habitat patches and corridors: clues from ecological field experiments. *Landscape and Urban Planning* 42(2): 157–168.
22.18, 22.19, 22.21a	Drawn by Simon Tegg
22.20a, b, and c	Redrawn from Garcia, D., and Chacoff, N. P. 2007. Scale-dependent effects of habitat fragmentation on hawthorn pollination, frugivory, and seed predation. *Conservation Biology* 21: 400–411.
22.21a	Redrawn from Laurance, W. F., and 6 others. 2006. Rain forest fragmentation and the proliferation of successional trees. *Ecology* 87: 469–482.
22.21b	Rob Bierregaard
22.22	Redrawn from Stratford, J. A., and Stouffer, P. C. 1999. Local extinctions of terrestrial insectivorous birds in a fragmented landscape near Manaus, Brazil. *Conservation Biology* 13: 1416–1423.
22.23a, b, and c	Redrawn from Stouffer, P. C., Strong, C., and Naka, L. N. 2009. Twenty years of understorey bird extinctions from Amazonian rain forest fragments: consistent trends and landscape-mediated dynamics. *Diversity and Distributions* 15: 88–97.
22.24a–c	Redrawn from Laurance, W. F., Delmonica, P., Laurance, S. G., Vasconcelos, H. L., and Lovejoy, T. E. 2000. Rainforest fragmentation kills big trees. *Nature* 404: 836.
22.25aii, 22.25b	Redrawn from Weldon, A. J., and Haddad, N. M. 2005. The effects of patch shape on indigo buntings: evidence for an ecological trap. *Ecology* 86: 1422–1431.
22.26	Redrawn from Becker, C. G., Fonseca, C. R., Haddad, C. F. B., Batista, R. F., and Prado, P. I. 2007. Habitat split and the global decline of amphibians. *Science* 318: 1775–1777.
22.27	Jeff Burrell/WCS

Chapter 23

Opening image: Climate change deniers out in force. Matt Mignanelli

23.1, 23.12a and b, 23.13a and b, 23.14aii, 23.14b	Redrawn from IPCC. 2013. Climate change 2013: the physical science basis. Contribution of Working Group I to the fifth assessment report of the intergovernmental panel on climate change. T. F. Stocker, and 9 others (eds). Cambridge: Cambridge University Press.
23.2	Redrawn from Bacastow, R. B., Keeling, C. D., Woodwell, G. M., and Pecan, E. V. 1973. Atmospheric carbon dioxide and radiocarbon in the natural carbon cycle. II. Changes from AD 1700 to 2070 as deduced from a geochemical model (No. CONF-720510–). University of California, San Diego, La Jolla; Brookhaven National Lab., Upton, NY.
23.3a	JG Photography/Alamy
23.3b	Redrawn from Keeling, C. D. 1998. Rewards and penalties of monitoring the Earth. *Annual Review of Energy and the Environment* 23: 25–82.
23.3c	NOAA
23.4a	NOAA, redrawn by David Cox
23.4b	Redrawn by David Cox from IPCC. 2007. Climate change 2007: the physical science basis. Contribution of Working Group I to the fourth assessment report of the Intergovernmental Panel on Climate Change. S. Solomon and 7 others (eds). Cambridge: Cambridge University Press.
23.5	Redrawn by David Cox from Pan, Y., and 17 others. 2011. A large and persistent carbon sink in the world's forests. *Science* 333: 988–993.
23.6	Redrawn by Simon Tegg from Xie, J., Li, Y., Zhai, C., Li, C., and Lan, Z. 2009. CO_2 absorption by alkaline soils and its implication to the global carbon cycle. *Environmental Geology* 56: 953–961.
23.7	Drawn by Simon Tegg
23.8	Redrawn by Simon Tegg from Hanel R. A. and 4 others. 1971. Spectrum taken with the infrared interferometer spectrometer flown on the satellite Nimbus 4 in 1971. *Applied Optics* 10: 1376–1382.
23.9	David F. Karnosky
23.10a, and b	Redrawn by Simon Tegg using data from Ainsworth, E. A., and Long, S. P. 2005. What have we learned from 15 years of free-air CO_2 enrichment (FACE)? A meta-analytic review of the responses of photosynthesis, canopy properties and plant production to rising CO_2. *New Phytologist* 165: 351–372.
23.11	Shin Okamoto/Shutterstock
23.14ai	David Greedy/Getty
23.13a-b	Drawn by Simon Tegg

23.15	Redrawn by Simon Tegg from Barnett, T. P., and 6 others. 2005. Penetration of human-induced warming into the world's oceans. *Science* 309: 284–287.
23.16ai	ALFRED PASIEKA/SCIENCE PHOTO LIBRARY
23.16aii, 23.16bi	Redrawn by Simon Tegg from Jansen, E. J., and 15 others. 2007. Palaeoclimate. In Climate change 2007: the physical science basis. Contribution of Working Group I to the fourth assessment report of the Intergovernmental Panel on Climate Change. S. Solomon and 7 others (eds). Cambridge: Cambridge University Press.United Kingdom and New York, NY, USA
23.17a and b	Cantin, N. E., Cohen, A. L., Karnauskas, K. B., Tarrant, A. M., and McCorkle, D. C. 2010. Ocean warming slows coral growth in the central Red Sea. *Science* 329: 322–325.
23.18a	BMJ/Shutterstock
23.18b	Redrawn by Simon Tegg from Hunter, C. M., and 5 others. Climate change threatens polar bear populations: a stochastic demographic analysis. *Ecology* 91: 2883–2997.
23.19a and b	Redrawn from Chen, I., Hill, J. K., Ohlemuller, R., Roy, D. B., and Thomas, C. D. 2011. Rapid range shifts of species associated with high levels of global warming. *Science* 333: 1024–1026.
23.20ai	Risa Sargent
23.20aii	TOM MCHUGH/SCIENCE PHOTO LIBRARY
23.20b	Redrawn by Simon Tegg from Rubidge, E. M., and 5 others. 2012. Climate-induced range contraction drives genetic erosion in an alpine mammal. *Nature Climate Change* 2: 285–288.
23.21	Redrawn by Simon Tegg from Hudson, J. M. G., and Henry, G. H. R. 2009. Increased plant biomass in a High Arctic heath community from 1981 to 2008. *Ecology* 90: 2657–2663.
23.22a	All Canada Photos/Alamy
23.22b and c	Redrawn from Kerby, J., and Post, E. 2013. Capital and income breeding traits differentiate trophic match–mismatch dynamics in large herbivores. *Philosophical Transactions of the Royal Society B: Biological Sciences* 368: 20120484.
23.23	NASA

Chapter 24

Opening image: The *Deepwater Horizon* platform after the explosion. US Coastguard.
Photo of Jane Lubchenco courtesy of Bruce Menge

24.1	Redrawn from McNutt, M. K., and 7 others. 2012. Review of flow rate estimates of the *Deepwater Horizon* oil spill. *Proceedings of the National Academy of Sciences* 109(50): 20260–20267.
24.2	Courtesy of NOAA, redrawn by Simon Tegg
24.3	Redrawn by Simon Tegg from Menge, J. L., and Menge, B. A. 1974. Role of resource allocation, aggression and spatial heterogeneity in coexistence of two competing intertidal starfish. *Ecological Monographs* 44: 189–209.
24.3i	Amar and Isabelle Guillen – Guillen Photo LLC/Alamy
24.3ii	Mark Conlin/Alamy
24.4	Redrawn by Simon Tegg using data from Menge, J. L., and Menge, B. A. 1974. Role of resource allocation, aggression and spatial heterogeneity in coexistence of two competing intertidal starfish. *Ecological Monographs* 44: 189–209.
24.5a	Konstantin Novikov/Shutterstock
24.5b	Hhelene/Shutterstock
24.5c	Premaphotos/Alamy
24.5d	Andrey Nekrasov/Alamy
24.6a	Redrawn by David Cox from Menge, B. A. 1976. Organization of the New England rocky intertidal community: role of predation, competition, and environmental heterogeneity. *Ecological Monographs* 46(4): 355–393.
24.6b, 24.7, 24.8a and b	Redrawn by Simon Tegg from Lubchenco, J., and Menge, B. 1978. Community development and persistence in a low rocky intertidal zone. *Ecological Monographs* 59: 67–94.
24.9a	Drawn by Simon Tegg
24.9b	Robert Zottoli, https://razottoli.wordpress.com.
24.10	Redrawn by Simon Tegg from Menge, B. A., and Lubchenco, J. 1981. Community organization in temperate and tropical rocky intertidal habitats: prey refuges in relation to consumer pressure gradients. *Ecological Monographs* 51: 429–450.
24.11	Jane Lubchenco 2012. Reflections on the Sustainable Biosphere Initiative. *Bulletin of the Ecological Society of America* 93:260–267.

24.12   Drawn by Simon Tegg
24.13   Redrawn from Matson, S. E. 2012. West coast groundfish IFQ fishery catch summary for 2011: first look. Supplemental NMFS Report March 2012.
24.13a   Redrawn by Simon Tegg from Hazen, T. C. and 31 others. 2010. Deep-sea oil plume enriches
and b   indigenous oil-degrading bacteria. *Science* 330: 204–208.
24.14   Redrawn by Simon Tegg from Lubchenco, J., and Menge, B. 1978. Community development and persistence in a low rocky intertidal zone. *Ecological Monographs* 59: 67–94.

INDEX

Abies religiosa. See Oyamel fir
abiotic effects
    on mussel abundance, 410
    wave action on community structure, 614
abundance, 4
    abiotic effects on mussels, 410
    buffalo, 4–10
    effect of consumers on tropical algae, 621
    effect of predators on tropical sessile
        invertebrates, 621
    estimating, 246–8
    large scale studies of sea otters, sea urchins,
        and kelp, 405
    North Atlantic right whale, 194
    of Caribbean coral, 253
    of dragonflies in relation to fish presence,
        395
    of elk at Yellowstone, 419
    of humpback whale, 250
    of mussels, 614
    of red algae, 614
    of sea otters off the Aleutian Islands, 402
    parasites, 140
    percent cover, 253
    populations, 246–70
    small scale study of sea otters and sea
        urchins, 404
    southern right whale, 214
    wildebeest, 4–10
Acacia tortilis
    and soil nutrient levels, 417
Acacia trees
    and thorn length, 358
acclimation. *See also* acclimatization
    thermal in spinach, 124
acclimatization, 118, *See also* acclimation
    in the toad *Rhinella arenarum*, 118
    Leptasterias seastars to *Pisaster*, 613
accumulation curve analysis
    estimating population size, 251
Acer saccharum. See sugar maple
Achillea millefolium. See yarrow
acidification
    at Hubbard Brook, 515
    of oceans, 23
    of terrestrial biomes, 24
    of the ocean and carbonate depletion, 596
    of the ocean by CO_2, 581
Actitis macularia. See spotted sandpiper
Adam, D., 579
Adams, P. A., 179
adaptation, 63
    and fitness, 148

antibiotic resistance in bacteria, 63
anti-predator behavior in caterpillars, 185
    as physiological response, 119
    beak depth in Darwin's finches, 64
    dermal plates in sticklebacks, 70
    for mating in barnacles, 330
    for mating in sedentary organisms, 330
    foraging in the Balearic lizard, 148
    inducing rot by microorganisms, 465
    of coral to thermal stress, 384
    ripe apple production of benzoic acid, 466
    self-sacrifice in the redback spider, 72
    semelparity in periodical cicadas, 204
    serotiny and fire, 530
    thermal in marine animals, 127
    thermal in terrestrial animals, 122
    thermal in terrestrial plants, 124
    thorns as constitutive defense, 358
    to variable water availability in plants, 130
    tropical plants and seed dispersal by extinct
        mammals, 464
adaptive management
    of Black Sea ecosystem, 519
adaptive radiation
    and high vein density in flowering plants,
        452
    Hawaiian drosophilids, 565
    of flowering plants, 452
    on the Galapagos, 75
adjusted litter size
    in relation to parental investment, 202
Aerts, R., 98
afforestation, 117
Agave parviflora, 207
age structure
    and population growth, 266
    aspen at Yellowstone, 422
    in Bulgaria, 267
    in Poland, 267
    in Uganda, 267
    rainbow trout in Spirit Lake, 535
aggression
    in chimpanzees, 316
    in sea stars, 613
agouti
    dispersers of guapinol seeds, 467
    seed predator of *Scheelea rostrata*, 465
ahu, 246
Ailanthus altissima. See tree of heaven
Ainsworth, E. A., 589
Aland
    and Glanville fritillary butterflies, 284
Alaria esculenta, 231

albedo
    and clouds, 589
Aldo Leopold Leadership Program, 624
Aleutian Islands, 402
alfalfa caterpillar
    and disease, 261
    mortality, 261
Allee effect
    and extinctions, 290
    and per capita growth rate, 286
    in Glanville fritillary butterflies, 286
    in Lotka-Volterra predation model, 368
Allee, W.C., 286
Allele, 60
Allen, A., 275
Allen, W., 468
Allison, F. R., 351
alpha diversity
    along Bullhead Trail, 435
    species richness within a habitat, 434
alpine chipmunk
    elevational range shift, 599
    fragmentation and loss of genetic diversity,
        600
alternative stable states, 541–4
    hysteresis, 543
    of a meadow ecosystem, 543
    theory, 541
Altevogt, B. M., 307
Altizer, S. M., 412
Ambrosia deltoidea. See triangle-leaf bursage
American chestnut
    destruction and community structural
        integrity, 459
    destruction by chestnut blight fungus,
        459
American ginseng
    population regulation by deer, 352
    population viability analysis, 353
American mink
    and niche shift, 339
ammonification
    and nitrogen cycle, 503
anadromous
    rainbow smelt, 237
Anakena
    stratigraphic column in Rapa Nui, 269
analysis of variance (ANOVA), 149
    F-value, 150
    Tukey-Kramer test, 151
anammox
    chemical reaction, 504
    importance in global nitrogen cycle, 505